IB DIPLOMA P

MATHEMATICS
STANDARD LEVEL

COURSE COMPANION

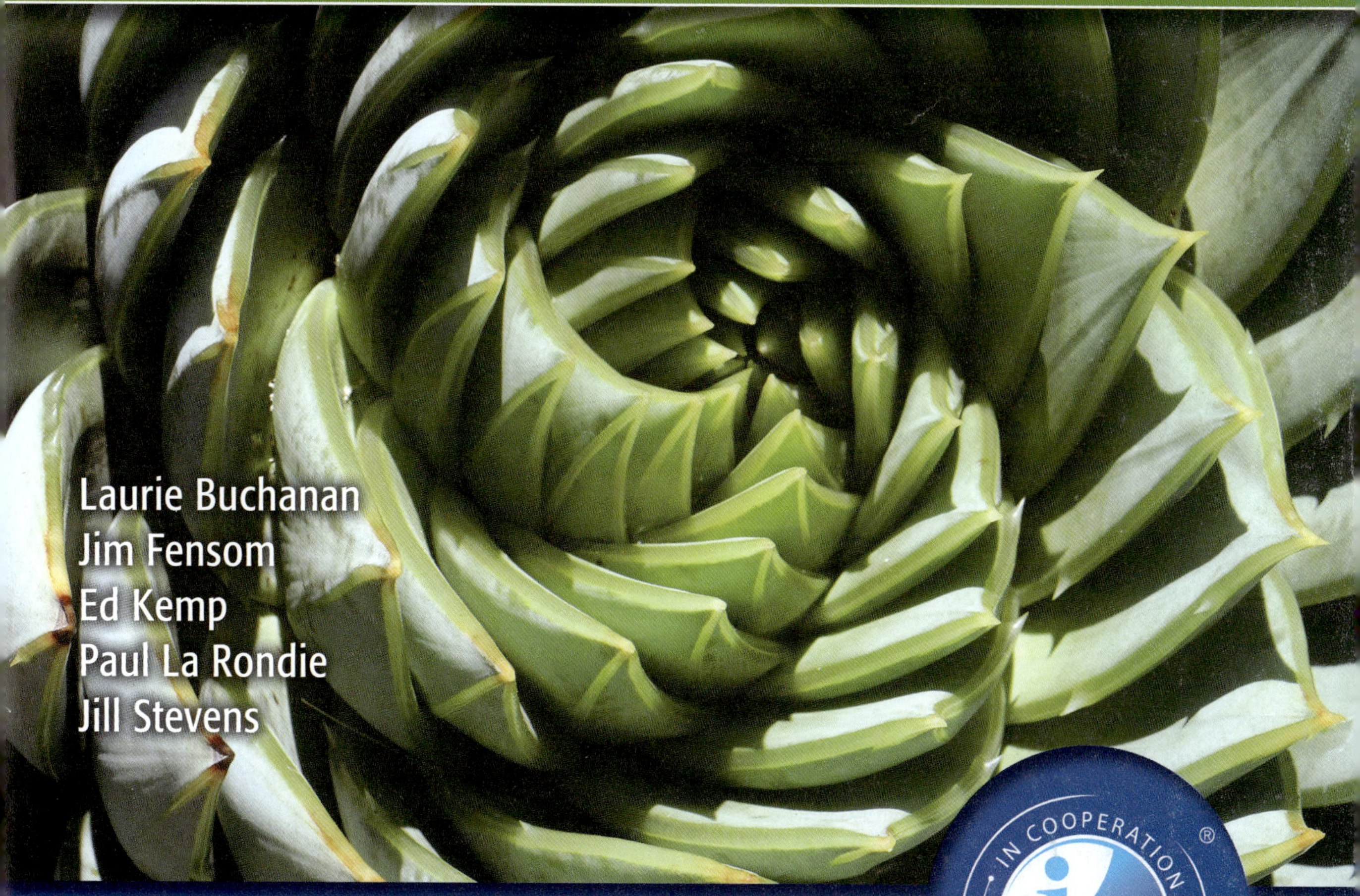

Laurie Buchanan
Jim Fensom
Ed Kemp
Paul La Rondie
Jill Stevens

OXFORD
UNIVERSITY PRESS

Great Clarendon Street, Oxford OX2 6DP

Oxford University Press is a department of the University of Oxford. It furthers the University's objective of excellence in research, scholarship, and education by publishing worldwide in

Oxford   New York

Auckland   Cape Town   Dar es Salaam   Hong Kong   Karachi
Kuala Lumpur   Madrid   Melbourne   Mexico City   Nairobi
New Delhi   Shanghai   Taipei   Toronto

With offices in

Argentina   Austria   Brazil   Chile   Czech Republic   France
Greece   Guatemala   Hungary   Italy   Japan   Poland   Portugal
Singapore   South Korea   Switzerland   Thailand   Turkey
Ukraine   Vietnam

© Oxford University Press 2012

The moral rights of the author have been asserted

Database right Oxford University Press (maker)

First published 2012

All rights reserved. No part of this publication may be reproduced, stored in a retrieval system, or transmitted, in any form or by any means, without the prior permission in writing of Oxford University Press, or as expressly permitted by law, or under terms agreed with the appropriate reprographics rights organization. Enquiries concerning reproduction outside the scope of the above should be sent to the Rights Department, Oxford University Press, at the address above

You must not circulate this book in any other binding or cover and you must impose this same condition on any acquirer

British Library Cataloguing in Publication Data

Data available

ISBN: 978-0-19-912935-5

10 9 8 7 6 5 4 3 2 1

Printed by Vivar Printing Sdn Bhd, Malaysia

Acknowledgments

The publishers would like to thank the following for permission to reproduce photographs:

P3: Nasa; P4: Konstantin Chagin/Shutterstock; P4: Janine Wiedel Photolibrary/Alamy; P13: Hulton Archive/Stringer/Getty Images; P17: Trip/Art Director; P17: Lunar And Planetary Instotute; P31: Nlshop/Shutterstock; P31: Itsmejust/Shutterstock; P33: Robert Crow/Dreamstime.com; P33: Lane Erickson/Dreamstime.com; P41: Sean Nel/Shutterstock; P54: Blasbike/Dreamstime.com; P56: Brad Remy/Dreamstime.com; P61: David Gee/Alamy; P61: Paulpaladin/Dreamstime.com; P61: Konstantin Androsov/Dreamstime.com; P61: Lembit Ansperi/Dreamstime.com; P61: Julián Rovagnati/Dreamstime.com; P60: Ilya Postnikov/Dreamstime.com; P61: Victor Habbick Visions/Science Photo Library; P63: Martin Fischer/Dreamstime.com; P64: Mrshining/Dreamstime.com; P64: Science Photo Library; P73: Francesco Abrignani/Shutterstock; P75: Ints Vikmanis/Shutterstock; P84: 3dimentii/Shutterstock; P85: Nicemonkey/Shutterstock; P92: Supri Suharjoto/Shutterstock; P98: Peter E Noyce/Alamy; P99: James Steidl/Dreamstime.com; P99: Motorolka/Dreamstime.com; P98: Stephen Gray/Shutterstock; P101: Viorel Dudau/Dreamstime.com; P111: Pcheruvi/Dreamstime.com; P112: Classic Image/Alamy; P132: Adisa/Shutterstock; P133: Robyn Mackenzie/Shutterstock; P134: Nigel Spiers/Dreamstime.com; P138: Irochka/Dreamstime.com; P139: Nasa Archive/Alamy; P141: Iofoto/Shutterstock; P142: Science Source/Science Photo Library; P145: Shutterstock/Patrik Dietrich; P149: Shutterstock/Plampy; P152: Shutterstock/John Orsbun; P152: Shutterstock/Wavebreakmedia Ltd; P153: Shutterstock/Filipe B. Varela; P155: Shutterstock/Luckyphoto; P155: Shutterstock/Upthebanner; P158: The Art Gallery Collection/Alamy; P158: Mireille Vautier/Alamy; P159: Yayayoyo/Shutterstock; P159: Travis Manley/Dreamstime.com; P159: Loulouphotos/Shutterstock; P161: James Harbal/Dreamstime.com; P162: Robyn Mackenzie/Dreamstime.com; P164: Science Photo Library; P176: Nito/Shutterstock; P183: Gsplanet/Shutterstock; P182: Gingergirl/Dreamstime.com; P: Christopher King/Dreamstime.com; P192: Art Directors & Trip/Alamy; P192: Alex Garaev/Shutterstock; P193: Adam Eastland Italy/Alamy; P192: Lebrecht Music And Arts Photo Library/Alamy; P195: Yellowj/Shutterstock; P201: Science Source/Science Photo Library; P205: Glasscuter/Dreamstime.com; P205: Jules2000/Shutterstock; P214: Sheila Terry/Science Photo Library; P217: Sheila Terry/Science Photo Library; P221: Glowimages/Getty Images; P222: Ccat82/Shutterstock; P253: Adrian Zenz/Dreamstime.com; P255: Mitchell Gunn/Dreamstime.com; P256: Will & Deni Mcintyre/Corbis; P256: Gravicapa/Fotolia.Com; P256: Brett Critchley/Dreamstime.com; P257: Ivan Hafizov/Dreamstime.com; P264: Mlehmann78/Dreamstime.com; P270: Pamela Tekiel/Dreamstime.com; P271: William Perry/Dreamstime.com; P274: Sergio Azenha/Alamy; P275: Brenda Carson/Dreamstime.com; P280: Tina Norris/Rex Features; P283: Maniec/Dreamstime.com; P282: Asdf_1/Dreamstime.com; P289: Robodread/Dreamstime.com; P288: Niday Picture Library/Alamy; P291: Sean Gladwell/Fotolia; P313: Science Photo Library; P318: Andrew Brookes, National Physical Laboratory/Science Photo Library; P318: Ted Foxx/Alamy; P330: Mediacolor's/Alamy; P330: Fromoldbooks.Org/Alamy; P333: Alex James Bramwell/Shutterstock; P334: Paulmerrett/Dreamstime.com; P349: Science Photo Library; P361: Maxx-Studio/Shutterstock; P363: Lamb/Alamy; P363: Bcampbell65/Shutterstock; P372: Maridav/Shutterstock ; P373: Chris Harvey/Shutterstock; P384: Michel Stevelmans/Shutterstock ; P384: Brandon Bourdages/Shutterstock; P385: Darren Baker/Dreamstime.com; P: Viktor Pravdica/Dreamstime.com; P403: Tomadesign/Shutterstock; P403: Tomadesign/Shutterstock; P403: Tomadesign/Shutterstock; P403: Konstantin Mironov/Shutterstock; P405: Darryl Brooks/Shutterstock; P406: Rafa Irusta/Shutterstock; P436: Dmitry_K/Shutterstock; P444: Pagadesign/Istockphoto; P447: Mr.Xutakupu/Shutterstock; P455: National Portrait Gallery London; P483: Rorem/Shutterstock; P487: Cynthia Burkhardt/Shutterstock; P487: Lori Martin/Shutterstock; P488: Phb.Cz (Richard Semik)/Shutterstock; P493: Mary Evans Picture Library/Alamy; P492: Alan Haynes/Alamy; P497: Mythic Ink/Getty Images; P497: Noah Berger/Associated Press; P506: Mythic Ink/Getty Images; P517: Science Photo Library; P521: Reuters Pictures; P526: Doodledance/Shutterstock; P535: Stanth/Shutterstock; P536: Anke Van Wyk/Dreamstime.com; P528: Cla78/Shutterstock; P529: Vladimir Yessikov/Shutterstock; P547: Monkey Business Images/Dreamstime.com; P552: Monkey Business Images/Dreamstime.com; P555: Vladimir Voronin/Dreamstime.com; P557: Phase4photography/Dreamstime.com; P558: Sculpies/Dreamstime.com; P555: Mario Savoia/Shutterstock; P555: R. Gino Santa Maria/Shutterstock; P554: James Weston/Shutterstock; P567: Beboy/Shutterstock; P567: Buslik/Shutterstock; P567: Dadek/Shutterstock; P566: Scott Camazine/Science Photo Library; P566: Nasa/Science Photo Library; P567: Mikkel Juul Jensen/Science Photo Library.

Cover Image: Joshua McCullough / Photo Library

Every effort has been made to contact copyright holders of material reproduced in this book. If notified, the publishers will be pleased to rectify any errors or omissions at the earliest opportunity.

Course Companion definition

The IB Diploma Programme Course Companions are resource materials designed to provide students with support through their two-year course of study. These books will help students gain an understanding of what is expected from the study of an IB Diploma Programme subject.

The Course Companions reflect the philosophy and approach of the IB Diploma Programme and present content in a way that illustrates the purpose and aims of the IB. They encourage a deep understanding of each subject by making connections to wider issues and providing opportunities for critical thinking.

The books mirror the IB philosophy of viewing the curriculum in terms of a whole-course approach; the use of a wide range of resources; international-mindedness; the IB learner profile and the IB Diploma Programme core requirements; theory of knowledge, the extended essay, and creativity, action, service (CAS).

Each book can be used in conjunction with other materials and indeed, students of the IB are required and encouraged to draw conclusions from a variety of resources. Suggestions for additional and further reading are given in each book and suggestions for how to extend research are provided.

In addition, the Course Companions provide advice and guidance on the specific course assessment requirements and also on academic honesty protocol.

IB mission statement

The International Baccalaureate aims to develop inquiring, knowledgable and caring young people who help to create a better and more peaceful world through intercultural understanding and respect.

To this end the IB works with schools, governments and international organizations to develop challenging programmes of international education and rigorous assessment.

These programmes encourage students across the world to become active, compassionate, and lifelong learners who understand that other people, with their differences, can also be right.

The IB Learner Profile

The aim of all IB programmes is to develop internationally minded people who, recognizing their common humanity and shared guardianship of the planet, help to create a better and more peaceful world. IB learners strive to be:

Inquirers They develop their natural curiosity. They acquire the skills necessary to conduct inquiry and research and show independence in learning. They actively enjoy learning and this love of learning will be sustained throughout their lives.

Knowledgable They explore concepts, ideas, and issues that have local and global significance. In so doing, they acquire in-depth knowledge and develop understanding across a broad and balanced range of disciplines.

Thinkers They exercise initiative in applying thinking skills critically and creatively to recognize and approach complex problems, and make reasoned, ethical decisions.

Communicators They understand and express ideas and information confidently and creatively in more than one language and in a variety of modes of communication. They work effectively and willingly in collaboration with others.

Principled They act with integrity and honesty, with a strong sense of fairness, justice, and respect for the dignity of the individual, groups, and communities. They take responsibility for their own actions and the consequences that accompany them.

Open-minded They understand and appreciate their own cultures and personal histories, and are open to the perspectives, values, and traditions of other individuals and communities. They are accustomed to seeking and evaluating a range of points of view, and are willing to grow from the experience.

Caring They show empathy, compassion, and respect towards the needs and feelings of others. They have a personal commitment to service, and act to make a positive difference to the lives of others and to the environment.

Risk-takers They approach unfamiliar situations and uncertainty with courage and forethought, and have the independence of spirit to explore new roles, ideas, and strategies. They are brave and articulate in defending their beliefs.

Balanced They understand the importance of intellectual, physical, and emotional balance to achieve personal well-being for themselves and others.

Reflective They give thoughtful consideration to their own learning and experience. They are able to assess and understand their strengths and limitations in order to support their learning and personal development

A note on academic honesty

It is of vital importance to acknowledge and appropriately credit the owners of information when that information is used in your work. After all, owners of ideas (intellectual property) have property rights. To have an authentic piece of work, it must be based on your individual and original ideas with the work of others fully acknowledged. Therefore, all assignments, written or oral, completed for assessment must use your own language and expression. Where sources are used or referred to, whether in the form of direct quotation or paraphrase, such sources must be appropriately acknowledged.

How do I acknowledge the work of others?

The way that you acknowledge that you have used the ideas of other people is through the use of footnotes and bibliographies.

Footnotes (placed at the bottom of a page) or endnotes (placed at the end of a document) are to be provided when you quote or paraphrase from another document, or closely summarize the information provided in another document. You do not need to provide a footnote for information that is part of a "body of knowledge". That is, definitions do not need to be footnoted as they are part of the assumed knowledge.

Bibliographies should include a formal list of the resources that you used in your work. "Formal" means that you should use one of the several accepted forms of presentation. This usually involves separating the resources that you use into different categories (e.g. books, magazines, newspaper articles, Internet-based resources, CDs and works of art) and providing full information as to how a reader or viewer of your work can find the same information. A bibliography is compulsory in the extended essay.

What constitutes malpractice?

Malpractice is behavior that results in, or may result in, you or any student gaining an unfair advantage in one or more assessment component. Malpractice includes plagiarism and collusion.

Plagiarism is defined as the representation of the ideas or work of another person as your own. The following are some of the ways to avoid plagiarism:

- Words and ideas of another person used to support one's arguments must be acknowledged.
- Passages that are quoted verbatim must be enclosed within quotation marks and acknowledged.
- CD-ROMs, email messages, web sites on the Internet, and any other electronic media must be treated in the same way as books and journals.
- The sources of all photographs, maps, illustrations, computer programs, data, graphs, audio-visual, and similar material must be acknowledged if they are not your own work.
- Works of art, whether music, film, dance, theatre arts, or visual arts, and where the creative use of a part of a work takes place, must be acknowledged.

Collusion is defined as supporting malpractice by another student. This includes:

- allowing your work to be copied or submitted for assessment by another student
- duplicating work for different assessment components and/or diploma requirements.

Other forms of malpractice include any action that gives you an unfair advantage or affects the results of another student. Examples include, taking unauthorized material into an examination room, misconduct during an examination, and falsifying a CAS record.

About the book

The new syllabus for Mathematics Standard Level is thoroughly covered in this book. Each chapter is divided into lesson size sections with the following features:

- **Investigations**
- **Exploration suggestions**
- **Examiner's tip**
- **Theory of Knowledge**
- **Did you know?**
- **Historical exploration**

Mathematics is a most powerful, valuable instrument that has both beauty in its own study and usefulness in other disciplines. The Sumerians developed mathematics as a recognized area of teaching and learning about 5,000 years ago and it has not stopped developing since then.

The Course Companion will guide you through the latest curriculum with full coverage of all topics and the new internal assessment. The emphasis is placed on the development and improved understanding of mathematical concepts and their real life application as well as proficiency in problem solving and critical thinking. The Course Companion denotes questions that would be suitable for examination practice and those where a GDC may be used. Questions are designed to increase in difficulty, strengthen analytical skills and build confidence through understanding. Internationalism, ethics and applications are clearly integrated into every section and there is a TOK application page that concludes each chapter.

It is possible for the teacher and student to work through in sequence but there is also the flexibility to follow a different order. Where appropriate the solutions to examples using the TI-Nspire calculator are shown. Similar solutions using the TI-84 Plus and Casio FX-9860GII are included on the accompanying interactive CD which includes a complete ebook of the text, prior learning, GDC support, an interactive glossary, sample examination papers, internal assessment support, and ideas for the exploration.

Mathematics education is a growing, ever changing entity. The contextual, technology integrated approach enables students to become adaptable, life-long learners.

Note: US spelling has been used, with IB style for mathematical terms.

About the authors

Laurie Buchanan has been teaching mathematics in Denver, Colorado for over 20 years. She is a team leader and a principal examiner for mathematics SL Paper One and an assistant examiner for Paper Two. She is also a workshop leader and has worked as part of the curriculum review team.

Jim Fensom has been teaching IB mathematics courses for nearly 35 years. He is currently Mathematics Coordinator at Nexus International School in Singapore. He is an assistant examiner for Mathematics HL.

Edward Kemp has been teaching IB Diploma Programme mathematics for 20 years. He is currently the head of mathematics at Ruamrudee International School in Thailand. He is an assistant examiner for IB mathematics, served on the IB curriculum review board and is an online workshop developer for IB.

Paul La Rondie has been teaching IB Diploma Programme mathematics at Sevenoaks School for 10 years. He has been an assistant examiner and team leader for both papers in Mathematics SL and an IA moderator. He has served on the IB curriculum review board and is an online workshop developer for IB.

Jill Stevens has been teaching IB Diploma Programme mathematics at Trinity High School, Euless, Texas for nine years. She is an assistant examiner for Mathematics SL, a workshop leader and has served the IB in curriculum review. Jill has been a reader and table leader for the College Board AP Calculus exam.

Contents

Chapter 1 Functions — 2
1.1 Introducing functions — 4
1.2 The domain and range of a relation on a Cartesian plane — 8
1.3 Function notation — 13
1.4 Composite functions — 14
1.5 Inverse functions — 16
1.6 Transforming functions — 21

Chapter 2 Quadratic functions and equations — 32
2.1 Solving quadratic equations — 34
2.2 The quadratic formula — 38
2.3 Roots of quadratic equations — 41
2.4 Graphs of quadratic functions — 43
2.5 Applications of quadratics — 53

Chapter 3 Probability — 62
3.1 Definitions — 64
3.2 Venn diagrams — 68
3.3 Sample space diagrams and the product rule — 77
3.4 Conditional probability — 85
3.5 Probability tree diagrams — 89

Chapter 4 Exponential and logarithmic functions — 100
4.1 Exponents — 103
4.2 Solving exponential equations — 107
4.3 Exponential functions — 109
4.4 Properties of logarithms — 115
4.5 Logarithmic functions — 118
4.6 Laws of logarithms — 122
4.7 Exponential and logarithmic equations — 127
4.8 Applications of exponential and logarithmic functions — 131

Chapter 5 Rational functions — 140
5.1 Reciprocals — 142
5.2 The reciprocal function — 143
5.3 Rational functions — 147

Chapter 6 Patterns, sequences and series — 160
6.1 Patterns and sequences — 162
6.2 Arithmetic sequences — 164
6.3 Geometric sequences — 167
6.4 Sigma (Σ) notation and series — 170
6.5 Arithmetic series — 172
6.6 Geometric series — 175
6.7 Convergent series and sums to infinity — 178
6.8 Applications of geometric and arithmetic patterns — 181
6.9 Pascal's triangle and the binomial expansion — 184

Chapter 7 Limits and derivatives — 194
7.1 Limits and convergence — 196
7.2 The tangent line and derivative of x^n — 200
7.3 More rules for derivatives — 208
7.4 The chain rule and higher order derivatives — 215
7.5 Rates of change and motion in a line — 221
7.6 The derivative and graphing — 230
7.7 More on extrema and optimization problems — 240

Chapter 8 Descriptive statistics — 254
8.1 Univariate analysis — 256
8.2 Presenting data — 257
8.3 Measures of central tendency — 260
8.4 Measures of dispersion — 267
8.5 Cumulative frequency — 271
8.6 Variance and standard deviation — 276

Chapter 9 integration — 290
9.1 Antiderivatives and the indefinite integral — 291
9.2 More on indefinite integrals — 297
9.3 Area and definite integrals — 302
9.4 Fundamental Theorem of Calculus — 309
9.5 Area between two curves — 313
9.6 Volume of revolution — 318
9.7 Definite integrals with linear motion and other problems — 321

Chapter 10 Bivariate analysis — 332
10.1 Scatter diagrams — 334
10.2 The line of best fit — 339
10.3 Least squares regression — 345
10.4 Measuring correlation — 349

Chapter 11 Trigonometry — 362
11.1 Right-angled triangle trigonometry — 363
11.2 Applications of right-angled triangle trigonometry — 369
11.3 Using the coordinate axes in trigonometry — 373
11.4 The sine rule — 380
11.5 The cosine rule — 386
11.6 Area of a triangle — 389
11.7 Radians, arcs and sectors — 391

Chapter 12 Vectors — 404
12.1 Vectors: basic concepts — 407
12.2 Addition and subtraction of vectors — 420
12.3 Scalar product — 426
12.4 Vector equation of a line — 430
12.5 Application of vectors — 437

Chapter 13 Circular functions — 446
13.1 Using the unit circle — 448
13.2 Solving equations using the unit circle — 454
13.3 Trigonometric identities — 456
13.4 Graphing circular functions — 462
13.5 Translations and stretches of trigonometric functions — 469
13.6 Combined transformations with sine and cosine functions — 478
13.7 Modeling with sine and cosine functions — 483

Chapter 14 Calculus with trigonometric functions — 494
14.1 Derivatives of trigonometric functions — 496
14.2 More practice with derivatives — 500
14.3 Integral of sine and cosine — 505
14.4 Revisiting linear motion — 510

Chapter 15 Probability distributions — 518
15.1 Random variables — 520
15.2 The binomial distribution — 527
15.3 The normal distribution — 538

Chapter 16 The Exploration — 556
16.1 About the exploration — 556
16.2 Internal assessment criteria — 557
16.3 How the exploration is marked — 562
16.4 Academic Honesty — 562
16.5 Record keeping — 563
16.6 Choosing a topic — 564
16.7 Getting started — 568

Chapter 17 Using a graphic display calculator — 570
1 Functions — 572
2 Differential calculus — 598
3 Integral calculus — 606
4 Vectors — 608
5 Statistics and probability — 612

Chapter 18 Prior learning — 632
1 Number — 633
2 Algebra — 657
3 Geometry — 673
4 Statistics — 699

Chapter 19 Practice papers — 708
Practice paper 1 — 708
Practice paper 2 — 712

Answers — 716

Index — 784

What's on the CD?

The material on your CD-ROM includes the entire student book as an eBook, as well as a wealth of other resources specifically written to support your learning. On these two pages you can see what you will find and how it will help you to succeed in your Mathematics Standard Level course.

The whole print text is presented as a user-friendly eBook for use in class and at home.

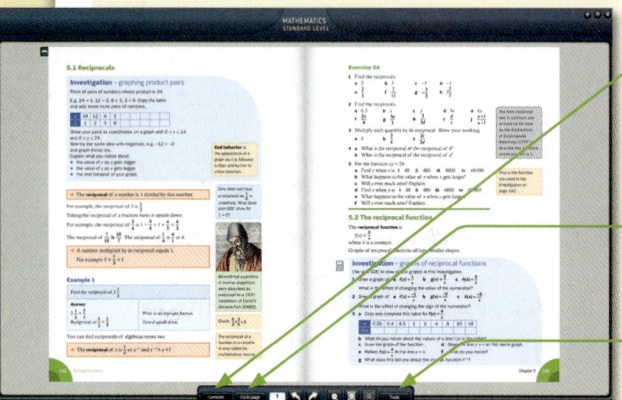

Extra content can be found in the Contents menu or attached to specific pages.

 This icon appears in the book wherever there is extra content.

Navigation is straightforward either through the Contents Menu, or through the Search and Go to page tools.

A range of tools enables you to zoom in and out and to annotate pages with your own notes.

The glossary provides comprehensive coverage of the language of the subject and explains tricky terminology. It is fully editable making it a powerful revision tool.

Extension material is included for each chapter containing a variety of extra exercises and activities. Full worked solutions to this material are also provided.

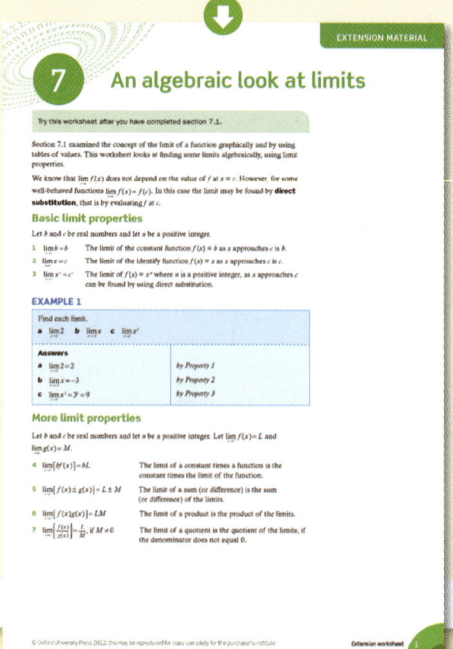

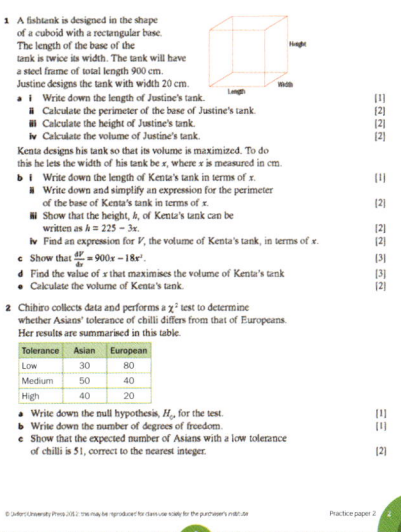

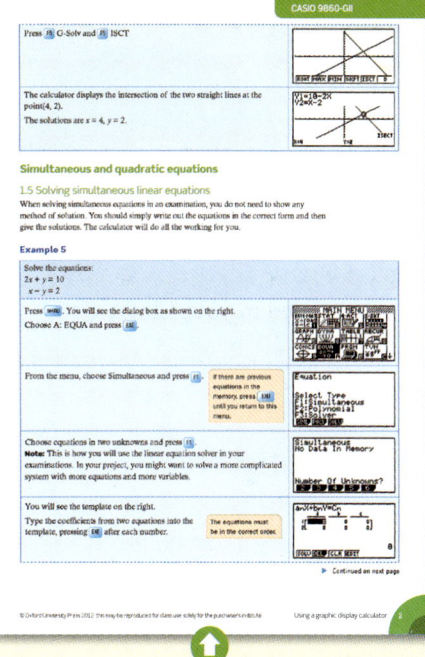

Practice exam papers will help you to fully prepare for your examinations. Worked solutions can be found on the website *www.oxfordsecondary.co.uk/ibmathsl*

Alternative GDC instructions for all material in the book is given for the TI-84 Plus and Casio-9860-GII calculators, so you can be sure you will be supported no matter what calculator you use.

Powerpoint presentations cover detailed worked solutions for the practice papers in the book, showing common errors and providing hints and tips.

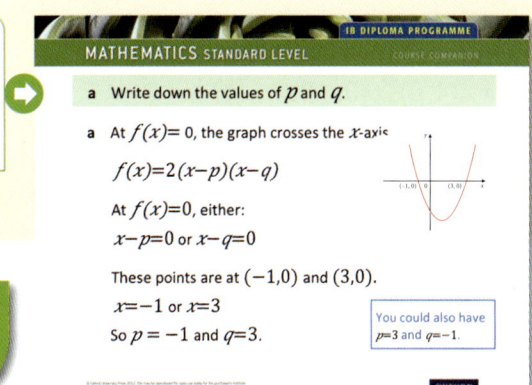

What's on the website?

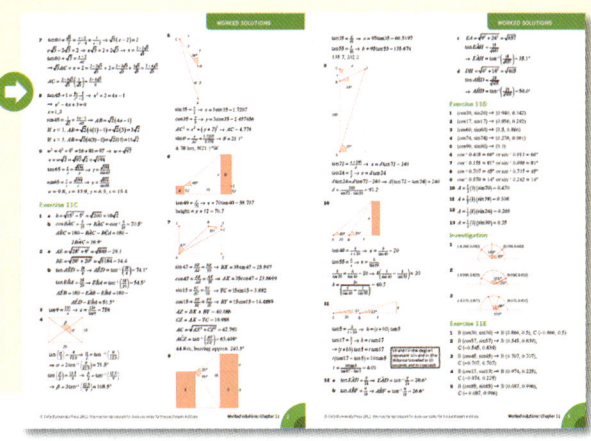

Visit *www.oxfordsecondary.co.uk/ibmathsl* for free access to the full worked solutions to each and every question in the Course Companion.

www.oxfordsecondary.co.uk/ibmathsl also offers you a range of GDC activities for the TI-Nspire to help support your understanding.

1 Functions

CHAPTER OBJECTIVES:
2.1 Functions: domain, range, composite, identity and inverse functions
2.2 Graphs of functions, by hand and using GDC, their maxima and minima, asymptotes, the graph of $f^{-1}(x)$
2.3 Transformations of graphs, translations, reflections, stretches and composite transformations

Before you start

You should know how to:

1 Plot coordinates.
 e.g. Plot the points $A(4, 0)$, $B(0, -3)$, $C(-1, 1)$ and $D(2, 1)$ on a coordinate plane.

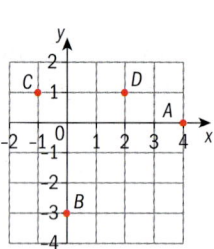

2 Substitute values into an expression.
 e.g. Given $x = 2$, $y = 3$ and $z = -5$, find the value of **a** $4x + 2y$ **b** $y^2 - 3z$
 a $4x + 2y = 4(2) + 2(3) = 8 + 6 = 14$
 b $y^2 - 3z = (3)^2 - 3(-5) = 9 + 15 = 24$

3 Solve linear equations.
 e.g. Solve $6 - 4x = 0$
 $6 - 4x = 0 \Rightarrow 6 = 4x$
 $1.5 = x \Rightarrow x = 1.5$

4 Use your GDC to graph a function.
 e.g. Graph $f(x) = 2x - 1$, $-3 \le x \le 3$

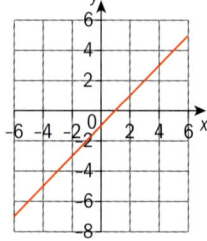

5 Expand linear binomials.
 e.g. Expand $(x + 3)(x - 2)$
 $= x^2 + x - 6$

Skills check

1 **a** Plot these points on a coordinate plane.
 $A(1, 3)$, $B(5, -3)$, $C(4, 4)$, $D(-3, 2)$, $E(2, -3)$, $F(0, 3)$.
 b Write down the coordinates of points A to H.

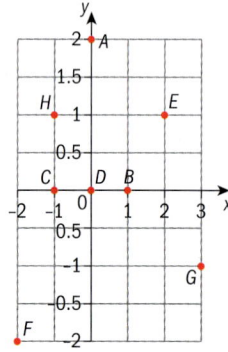

2 Given that $x = 4$, $y = 6$ and $z = -10$, find
 a $4x + 3y$ **b** $z^2 - 3y$ **c** $y - z$ **d** $\dfrac{2x + 5}{yz}$

3 Solve
 a $3x - 6 = 6$ **b** $5x + 7 = -3$ **c** $\dfrac{x}{2} + 6 = 11$

4 Graph these functions on your GDC within the given domain. Then sketch the functions on paper.
 a $y = 2x - 3$, $-4 \le x \le 7$
 b $y = 10 - 2x$, $-2 \le x \le 5$
 c $y = x^2 - 3$, $-3 \le x \le 3$.

5 Expand
 a $(x + 4)(x + 5)$ **b** $(x - 1)(x - 3)$
 c $(x + 5)(x - 4)$

▲ International Space Station

The International Space Station (ISS) has been orbiting the Earth over 15 times a day for more than ten years, yet how many of us have actually seen it? Spotting the ISS with the naked eye is not as difficult as it might seem – provided you know in which direction to look. Although the ISS travels at a speed of $7.7 \, \text{km s}^{-1}$, it is in one of the lowest orbits possible, at approximately 390 km above our heads. Thanks to its large solar wings it is one of the brightest 'stars', which makes it fairly easy to distinguish as it moves across the night sky.

The relation $t = \dfrac{d}{22\,744}$ gives the speed of the ISS, where t is the time measured in hours and d is the distance traveled in kilometres.

This is a mathematical relationship called a **function** and is just one example of how a mathematical function can be used to describe a situation.

In this chapter you will explore functions and how they can be applied to a wide variety of mathematical situations.

One of the first mathematicians to study the concept of function was French philosopher Nicole Oresme (1323–1382). He worked with independent and dependent variable quantities.

1.1 Introducing functions

Investigation – handshakes

In some countries it is customary at business meetings to shake hands with everybody in the meeting. If there are 2 people there is 1 handshake, if there are 3 people there are 3 handshakes and so on.

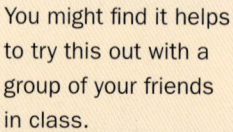

a How many handshakes are there for 4 people?
b Copy and complete this table.

Number of people	Number of handshakes
2	
3	
4	
5	
6	
7	
8	
9	
10	

> You might find it helps to try this out with a group of your friends in class.

> Do not join the points in this case as we are dealing only with whole (discrete) numbers.

c Plot the points on a Cartesian coordinate plane with the number of people on the *x*-axis and the number of handshakes on the *y*-axis.
d Write a formula for the number of handshakes, H, in terms of the number of people, n.

Relations and functions

Distance (m)	Time (s)
100	15
200	34
300	60
400	88

The table shows the amount of time it takes for a student to run certain distances.

Another way of showing this information is as **ordered pairs**: (100, 15), (200, 34), (300, 60) and (400, 88). Each ordered pair has two pieces of data in a specific order. They are separated by a comma and enclosed within brackets in the form (x, y).

→ A **relation** is a set of ordered pairs.

There is nothing special at all about the numbers that are in a relation. In other words, any group of numbers is a relation provided that these numbers come in pairs.

→ The **domain** is the set of all the first numbers (*x*-values) of the ordered pairs.

The domain of the ordered pairs above is {100, 200, 300, 400}. ← *The curly brackets, { }, mean 'the set of'.*

→ The **range** is the set of the second numbers (*y*-values) in each pair.

The range of the ordered pairs above is {15, 34, 60, 88}.

Example 1

Find the domain and range of these relations.
a {(1, 4), (2, 7), (3, 10), (4, 13)}
b {(−2, 4), (−1, 1), (0, 0), (1, 1), (2, 4)}

Answers

a The domain is {1, 2, 3, 4}   *First elements in the ordered pairs*

   The range is {4, 7, 10, 13}   *Second elements in the ordered pairs*

b The domain is {−2, −1, 0, 1, 2}   *Do not repeat values even though*
   The range is {0, 1, 4}   *there are two 4s and two 1s in the*
                                         ordered pairs.

→ A **function** is a mathematical relation such that each element of the domain of the function is associated with exactly one element of the range of the function. In order for a relation to be a function no two ordered pairs may have the same first element.

Example 2

Which of these sets of ordered pairs are functions?
a {(1, 4), (2, 6), (3, 8), (3, 9), (4, 10)}
b {(1, 3), (2, 5), (3, 7), (4, 9), (5, 11)}
c {(−2, 1), (−1, 1), (0, 2), (1, 4), (2, 6)}

Answers
a Not a function because the number 3 occurs twice in the domain.
b A function; all of the first elements are different.
c A function; all of the first elements are different.   *Note that it doesn't matter that some of the y-values are the same.*

Chapter 1

Exercise 1A

1 Which of these sets of ordered pairs are functions?
   a $\{(5,5), (4,4), (3,3), (2,2), (1,1)\}$
   b $\{(-3,4), (-1,6), (0,5), (2,-1), (3,-1)\}$
   c $\{(4,1), (4,2), (4,3), (4,4), (4,5)\}$
   d $\{(-1,1), (0,3), (1,6), (1,7), (2,8)\}$
   e $\{(-4,4), (-4,5), (-3,6), (-3,7), (-2,8)\}$
   f $\{(1,2), (2,2), (3,2), (4,2), (5,2)\}$

2 For each diagram, identify the domain and range and say whether the relation is a function.

   a 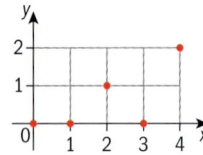   b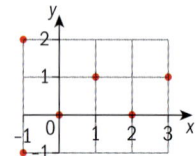

   > Write down the coordinates as ordered pairs.

3 Look back at the table on page 4 that shows the amount of time it takes for a student to run certain distances. Is the relationship between a distance traveled and time taken a function?

The vertical line test

You can represent relations and functions on a Cartesian plane. You can use the vertical line test to determine whether a particular relation is a function or not, by drawing vertical lines across the graph.

→ A relation is a function if any vertical line drawn will not intersect the graph more than once. This is called the **vertical line test**.

> Cartesian coordinates and the Cartesian plane are named after Frenchman René Descartes (1596–1650).

Example 3

Which of these relations are functions?

a 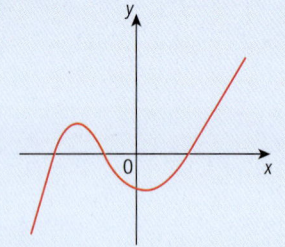   b 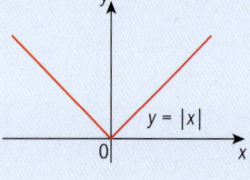   c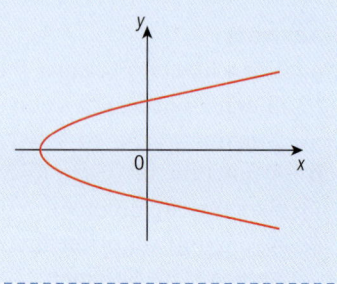

▶ Continued on next page

6 Functions

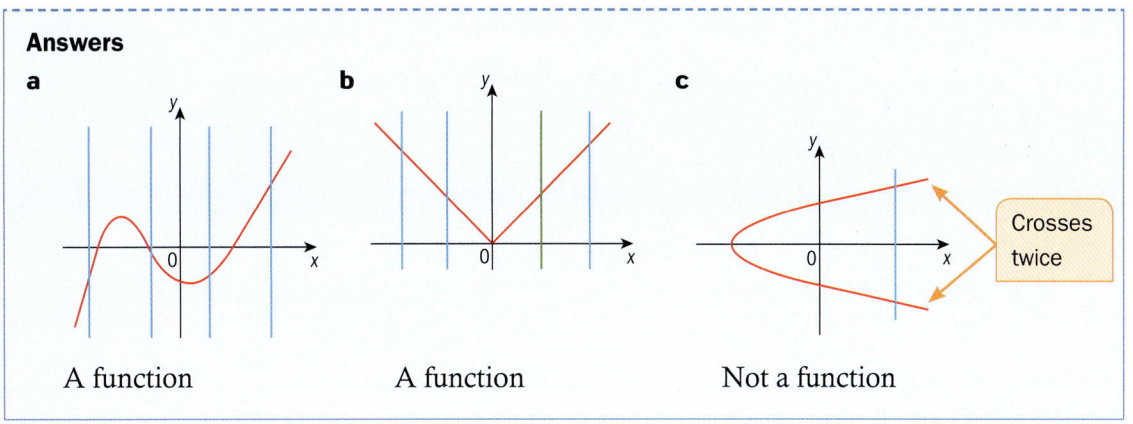

Exercise 1B

1 Which of these relations are functions?

a   b   c

> Draw, or imagine, vertical lines on the graph.

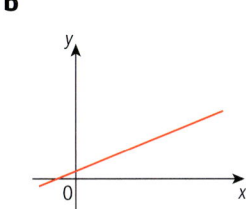

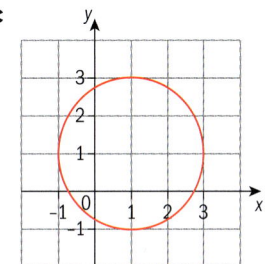

d   e   f

> If the function has a 'solid dot' •, this indicates that the value is included in the function.
> If the function has a 'hollow dot' ○, this indicates that the value is not included in the function.

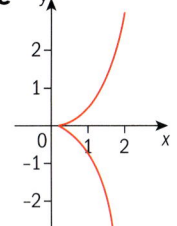

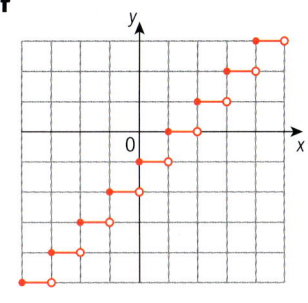

g   h   i

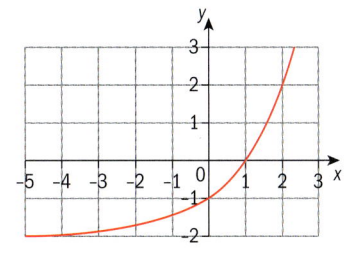

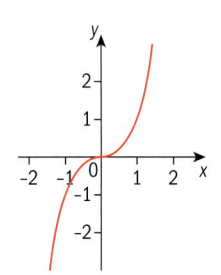

2. Use your GDC to sketch these straight line graphs.
   a $y = x$ b $y = x + 2$ c $y = 2x - 3$ d $y = 4$
   e Are they all functions? Explain your answer.
   f Will all straight lines be functions? Why?

> Indicate where the line crosses the *x*- and/or *y*-axis on your sketch.

3. Sketch the region $y < 3x - 2$. Is this a function? Why?

> When using your GDC, aim to have the ends of your graph near the corners of the view window.

4. Use an algebraic method to show that $x^2 + y^2 = 4$ is not a function.

> Try substituting positive and negative values of *x*.

1.2 The domain and range of a relation on a Cartesian plane

You can often write the domain and range of a relation using interval notation. This is another method of writing down a set of numbers. For example, for the set of numbers that are all less than 3, you can write the inequality $x < 3$, where x is a number in the set. In interval notation, this set of numbers is written $(-\infty, 3)$
Interval notation uses only five symbols.

Brackets	( )
Square brackets	[ ]
Infinity	∞
Negative infinity	$-\infty$
Union	$\cup$

> ▲ A function maps the domain (horizontal, *x*-values) onto the range (vertical, *y*-values)

> To use interval notation:
> → Use the round brackets (,) if the value is not included in the graph as in $(-\infty, 3)$ or when the graph is undefined at that point (a hole or **asymptote**, or a jump).
> Use the square brackets [,] if the value is part of the graph.

> How many numbers are there in the sequence 0, 1, 2, 3, 4,... if we go on forever?
> How many numbers are there in the sequence 0, 0.5, 1, 1.5, 2, 2.5, 3, 3.5, 4,... if we go on forever?

Whenever there is a break in the values, write the interval up to the point. Then write another interval for the values after that point. Put a union sign between each interval to 'join' them together. For example $(-\infty, 3) \cup (4, \infty)$

If a graph goes on forever to the left, the domain (*x*-values) starts with $(-\infty$. If it goes on forever to the right then the domain ends with $\infty)$. If a graph travels downward forever, the range (*y*-values) starts with $(-\infty$. And if a graph goes up forever, then the range ends with $\infty)$.

> Why do we call infinity undefined?

Usually we use interval notation to describe a set of values along the *x*- or *y*-axis. However, you can use it to describe any group of numbers. For example, in interval notation $x \geq 6$ is $[6, \infty)$.

8 Functions

Asymptotes

Asymptotes are visible on your GDC for some functions. An asymptote is a line that a graph approaches, but does not intersect.

For example, in the graph of $y = \dfrac{1}{x}$, the line approaches the x-axis ($y = 0$), but never touches it. As we go to infinity the line will not actually reach $y = 0$, but will always get closer and closer. The x-axis or $y = 0$ is callled the horizontal asymptote.

The y-axis or $x = 0$ is the vertical asymptote for the same reasons. There will be a more in-depth treatment of asymptotes in the chapter on rational functions.

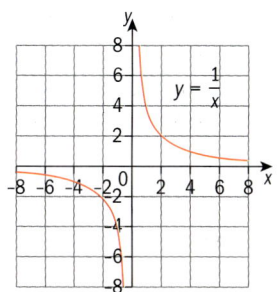

> Finding asymptotes by looking at the graph is called locating asymptotes by inspection.

Example 4

Identify the horizontal and vertical asymptotes for these functions if they exist.

a $y = 2^x$ **b** $y = \dfrac{2x}{x+1}$ **c** $y = \dfrac{x+2}{(x+1)(x-2)}$

Answers

a

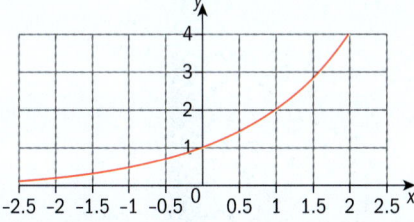

Horizontal asymptote $y = 0$

As we go along the x-axis to the left the curve gets closer but never actually meets the x-axis.

b

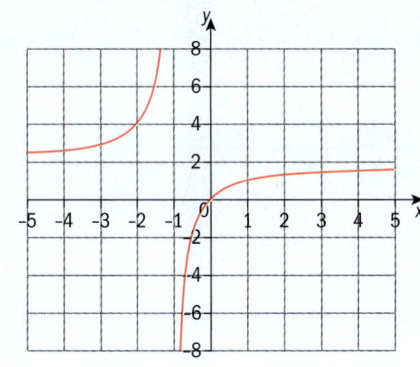

Horizontal asymptote $y = 2$
Vertical asympote $x = -1$

c

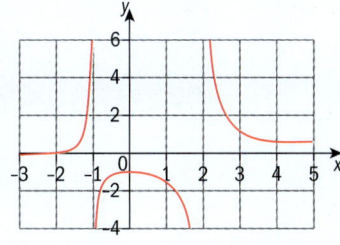

Horizontal asymptote $y = 0$
Vertical asympote $x = -1$ and $x = 2$

Exercise 1C

Identify the horizontal and vertical asympotes for these functions, if they exist.

1. $y = 3^x$
2. $y = \dfrac{3}{x}$
3. $y = \dfrac{4}{x+1}$
4. $y = \dfrac{2x}{x+2}$
5. $y = \dfrac{2x+1}{x-1}$
6. $y = \dfrac{6}{x^2-9}$

Set builder notation

In set builder notation we use curly brackets { } and variables to express the domain and range. We can compile sets of inequalities using inequality and other symbols.

the set of	{ }
less than	<
less than or equal to	≤
greater than	>
greater than or equal to	≥
is a member of the set of real numbers	∈ ℝ

You may wish to explore the 'internationalism' of symbols in the language of mathematics.

→ Set notation:

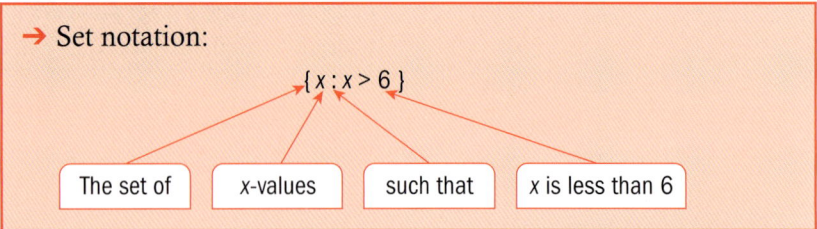

The set of — x-values — such that — x is less than 6

Interval notation is often considered more efficient than set builder notation

Interval notation	Description	Set builder notation
(−2, +∞)	x is greater than −2	{x : x > −2}
(−∞, 4]	x is less than or equal to 4	{x : x ≤ 4}
[−3, 3)	x lies between −3 and 3 including −3 but not 3	{x : −3 ≤ x < 3}
(−∞, 5) ∪ [6, +∞)	x is less than 5 or greater than or equal to 6	{x : x < 5, x ≥ 6}
(−∞, +∞)	x may be any real number	x ∈ ℝ

Around the world there are many different words for the same symbol. Brackets are also called parentheses. Radicals are also called surds. How does this affect understanding? Can you find some more examples?

Some people use 'backwards square brackets' to show greater than or less than. For example:] 2, ∞ [is equivalent to x > 2, and] ∞, −4[is equivalent to x < −4.

Example 5

Find the domain and range of this function.

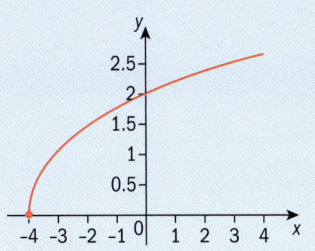

> You may wish to explore the influence of technology on notation and vice versa.

Answer

The domain of the function is $\{x : x \geq -4\}$ or $[-4, +\infty)$.
The range of the function is $\{y : y \geq 0\}$ or $[0, +\infty)$.

x only takes values greater than or equal to −4.
The function only takes y-values greater than or equal to 0.

Example 6

Find the domain and range of each function.

a
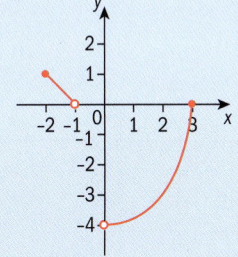

b
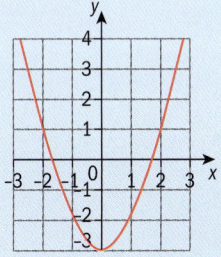

> What values are included in the domain $0 \leq x \leq 1$?
> How many values are there?

Answers

a The domain is $\{x : -2 \leq x < -1$ and $0 < x \leq 3\}$
or $[-2, -1) \cup (0, 3]$.
The range is $\{y : -4 < y \leq 1\}$ or $(-4, 1]$.

b The domain of the function is $x \in \mathbb{R}$ or $(-\infty, +\infty)$.
The range of the function is $\{y : y \geq -3\}$ or $[-3, +\infty)$.

x can take any real value.

> Do we all use the same notation in mathematics? We are using an empty dot to indicate that $x = -1$ is *not* included. Different countries have different notations to represent the same thing. Furthermore, different teachers from the same country use different notations!

Exercise 1D

1. Look back to page 4 at the graph and the formula for the numbers of handshakes for various numbers of people. Is this a function? If so, what is the domain and range?

2. Find the domain and range for each of these relations.

   a 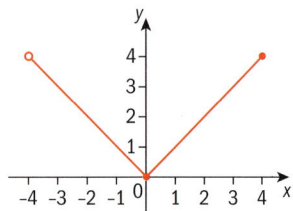 **b** 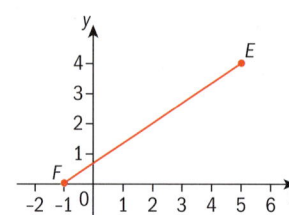 **c**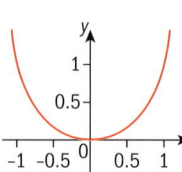

   d 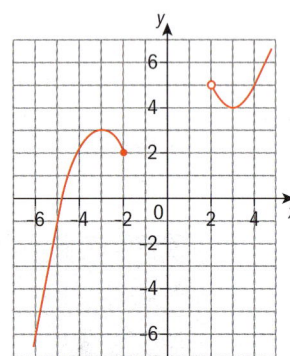 **e** 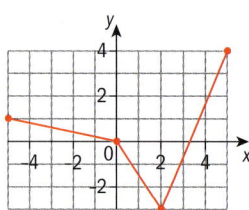 **f**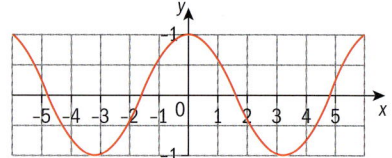

   g 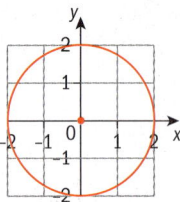 **h** 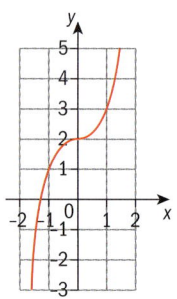 **i**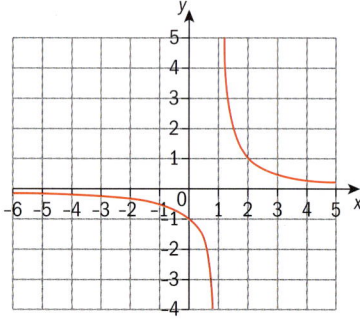

EXAM-STYLE QUESTION

3. Use your GDC to sketch these graphs. Write down the domain and range of each.
   - **a** $y = 2x - 3$
   - **b** $y = x^2$
   - **c** $y = x^2 + 5x + 6$
   - **d** $y = x^3 - 4$
   - **e** $y = \sqrt{x}$
   - **f** $y = \sqrt{4 - x}$
   - **g** $y = \dfrac{1}{x}$
   - **h** $y = e^x$
   - **i** $y = \dfrac{1}{x + 2}$
   - **j** $y = \dfrac{x + 4}{x - 2}$
   - **k** $y = \dfrac{x^2 - 9}{x + 3}$
   - **l** $y = \dfrac{2}{x^2 + 1}$

> Your GDC will find the *x*- and *y*-intercepts. To do this algebraically, use the fact that a function crosses the *x*-axis when $y = 0$ and crosses the *y*-axis when $x = 0$. For example, the function $y = 2x - 4$ crosses the *x*-axis where $2x - 4 = 0$, $x = 2$. It crosses the *y*-axis where $y = 2(0) - 4 = -4$.

> **3k** gives a most unusual answer. Look carefully for a hole where $x = -3$.

1.3 Function notation

Functions are often described by equations. For example, the equation $y = 2x + 1$ describes y as a function of x. By giving the function the symbol 'f' we write this equation in function notation as $f(x) = 2x + 1$ and so $y = f(x)$.

> → $f(x)$ is read as 'f of x' and means the value of function f at x.

$f(x)$ can also be written like this: $f : x \to 2x + 1$.
An ordered pair (x, y) can be written as $(x, f(x))$.
Finding $f(x)$ for a particular value of x means evaluating the function f at that value.

$f : (x) \to 2x + 1$ means that f is a function that maps x to $2x + 1$.

Example 7

a Evaluate the function $f(x) = 2x + 1$ at $x = 3$.
b If $f(x) = x^2 + 4x - 3$, find i $f(2)$ ii $f(0)$ iii $f(-3)$ iv $f(x + 1)$

Answers

a $f(3) = 2(3) + 1 = 7$ *For x, substitute 3.*
b i $f(2) = (2)^2 + 4(2) - 3 = 4 + 8 - 3 = 9$
  ii $f(0) = (0)^2 + 4(0) - 3 = 0 + 0 - 3 = -3$
  iii $f(-3) = (-3)^2 + 4(-3) - 3$
        $= 9 - 12 - 3 = -6$
  iv $f(x + 1) = (x + 1)^2 + 4(x + 1) - 3$
              $= x^2 + 2x + 1 + 4x + 4 - 3$
              $= x^2 + 6x + 2$

The German mathematician and philosopher Gottfried Leibniz first used the mathematical term 'function' in 1673.

Exercise 1E

1 Find i $f(7)$ ii $f(-3)$ iii $f(\frac{1}{2})$ iv $f(0)$ v $f(a)$
   for these functions.

   a $f(x) = x - 2$ b $f(x) = 3x$ c $f(x) = \frac{1}{4}x$
   d $f(x) = 2x + 5$ e $f(x) = x^2 + 2$

2 If $f(x) = x^2 - 4$, find
   a $f(-a)$ b $f(a + 5)$ c $f(a - 1)$
   d $f(a^2 - 2)$ e $f(5 - a)$

EXAM-STYLE QUESTION
3 If $g(x) = 4x - 5$ and $h(x) = 7 - 2x$
   a find x when $g(x) = 3$
   b find x when $h(x) = -15$
   c find x when $g(x) = h(x)$.

4 a If $h(x) = \dfrac{1}{x - 6}$ find $h(-3)$.
   b Is there a value where $h(x)$ does not exist? Explain.

Notice that we do not always use the letter f for a function. Here we have used g and h. When considering velocity in terms of time we often use $v(t)$.

5 The volume of a cube with edges of length x is given by the function $f(x) = x^3$.
   a Find $f(5)$.
   b Explain what $f(5)$ represents.

6 $g(x) = \dfrac{3x+1}{x-2}$

   a Evaluate
      i $g(6)$ **ii** $g(-2)$ **iii** $g(0)$ **iv** $g\left(-\dfrac{1}{3}\right)$
   b Evaluate
      i $g(1)$ **ii** $g(1.5)$ **iii** $g(1.9)$ **iv** $g(1.99)$
      v $g(1.999)$ **vi** $g(1.9999)$
   c What do you notice about your answers to **b**?
   d Is there a value of x for which $g(x)$ does not exist?
   e Graph the function on your GDC and look what happens when $x = 2$. Explain.

> You can use mathematical functions to represent things from your own life. For example, suppose the number of pizzas your family eats depends on the number of football games you watch. If you eat 3 pizzas during every football game, the function would be 'number of pizzas' (p) = 3 times 'number of football games' (g) or $p = 3g$. Can you think of another real-life function? It could perhaps be about the amount of money you spend or the number of minutes you spend talking on the phone.

EXAM-STYLE QUESTION

7 The velocity of a particle is given by $v(t) = t^2 - 9 \text{ m s}^{-1}$.
   a Find the initial velocity.
   b Find the velocity after 4 seconds.
   c Find the velocity after 10 seconds.
   d At what time does the particle come to rest?

> The initial velocity means the velocity at the start, when $t = 0$.

> The particle comes to rest when $v = 0$.

8 Given $f(x) = \dfrac{f(x+h) - f(x)}{h}$ find
   a $f(2+h)$ **b** $f(3+h)$

Extension material on CD:
Worksheet 1 - Polynomials

1.4 Composite functions

A **composite function** is a combination of two functions. You apply one function to the result of another.

> → The composition of the function f with the function g is written as $f(g(x))$, which is read as 'f of g of x', or $(f \circ g)(x)$, which is read as 'f composed with g of x'.

When you evaluate a function $f(x)$, you substitute a number or another variable for x.
For example, if $f(x) = 2x + 3$ then $f(5) = 2(5) + 3 = 13$
You can find $f(x^2 + 1)$ by substituting $x^2 + 1$ for x to get
$f(x^2 + 1) = 2(x^2 + 1) + 3 = 2x^2 + 5$

> → A **composite function** applies one function to the result of another and is defined by $(f \circ g)(x) = f(g(x))$.

Example 8

If $f(x) = 5 - 3x$ and $g(x) = x^2 + 4$, find $(f \circ g)(x)$.

Answer

$(f \circ g)(x) = 5 - 3(x^2 + 4)$	*Substitute $x^2 + 4$ into $f(x)$.*
$ = 5 - 3x^2 - 12$	
$ = -3x^2 - 7$	

g(x) goes in here

You may need to evaluate a composite function for a particular value of x.

Example 9

$f(x) = 5 - 3x$ and $g(x) = x^2 + 4$. Find $(f \circ g)(3)$.

Answer

Method 1

$(f \circ g)(x) = 5 - 3(x^2 + 4)$	*Work out the composite function.*
$ = -3x^2 - 7$	
$(f \circ g)(3) = -3(3)^2 - 7$	*Then substitute 3 for x.*
$ = -27 - 7$	
$ = -34$	

Method 2

$g(3) = (3)^2 + 4 = 13$	*Substitute 3 into $g(x)$.*
$f(13) = 5 - 3(13) = -34$	*Substitute that value into $f(x)$.*

Both methods give the same result — you can use the one you prefer.

Example 10

Given $f(x) = 2x + 1$ and $g(x) = x^2 - 2$, find

a $(f \circ g)(x)$ **b** $(f \circ g)(4)$

Answers

a $(f \circ g)(x) = 2(x^2 - 2) + 1$	*Substitute $x^2 - 2$ into $f(x)$.*
$ = 2x^2 - 3$	
b $(f \circ g)(4) = 2(4)^2 - 3 = 29$	*Substitute 4 for x.*

Or use Method 2:
$g(4) = (4)^2 - 2 = 14$
and then
$f(14) = 2(14) + 1 = 29$

Exercise 1F

1 Given $f(x) = 3x$, $g(x) = x + 1$ and $h(x) = x^2 + 2$, find

- **a** $(f \circ g)(3)$
- **b** $(f \circ g)(0)$
- **c** $(f \circ g)(-6)$
- **d** $(f \circ g)(x)$
- **e** $(g \circ f)(4)$
- **f** $(g \circ f)(5)$
- **g** $(g \circ f)(-6)$
- **h** $(g \circ f)(x)$
- **i** $(f \circ h)(2)$
- **j** $(h \circ f)(2)$
- **k** $(f \circ h)(x)$
- **l** $(h \circ f)(x)$
- **m** $(g \circ h)(3)$
- **n** $(h \circ g)(3)$
- **o** $(g \circ h)(x)$
- **p** $(h \circ g)(x)$

$(f \circ h)(2) \neq (h \circ f)(2)$

2 Given $f(x) = x^2 - 1$ and $g(x) = 3 - x$, find

   a $(g \circ f)(1)$ **b** $(g \circ f)(2)$ **c** $(g \circ f)(4)$ **d** $(f \circ g)(3)$

   e $(g \circ f)(3)$ **f** $(f \circ g)(-4)$ **g** $(f \circ g)(x + 1)$ **h** $(f \circ g)(x + 2)$

EXAM-STYLE QUESTIONS

3 Given the functions $f(x) = x^2$ and $g(x) = x + 2$ find

   a $(f \circ g)(x)$ **b** $(f \circ g)(3)$

4 Given the functions $f(x) = 5x$ and $g(x) = x^2 + 1$ find

   a $(f \circ g)(x)$ **b** $(g \circ f)(x)$

5 $g(x) = x^2 + 3$ and $h(x) = x - 4$

   a Find $(g \circ h)(x)$.
   b Find $(h \circ g)(x)$.
   c Hence solve the equation $(g \circ h)(x) = (h \circ g)(x)$.

> 'Hence' means 'Use the preceding work to obtain the required result'.

6 If $r(x) = x - 4$ and $s(x) = x^2$, find $(r \circ s)(x)$ and state its domain and range.

1.5 Inverse functions

> → The **inverse** of a function $f(x)$ is $f^{-1}(x)$. It reverses the action of that function.

If $f(x) = 3x - 4$ and $g(x) = \dfrac{x+4}{3}$, then

$f(10) = 3(10) - 4 = 26$ and $g(26) = \dfrac{26+4}{3} = 10$, so we are back to where we started.

> $(f \circ g)(10) = 10$

So $g(x)$ is the inverse of $f(x)$.
Not all functions have an inverse.

If g is the inverse function of f, then g will reverse the action of f for all values in the domain of f and f will also be the inverse of g. When f and g are inverse functions, we write $g(x) = f^{-1}(x)$.

> Note that f^{-1} means the inverse of f; the '−1' is not an exponent (power).

> → Functions $f(x)$ and $g(x)$ are inverses of one another if:
> $(f \circ g)(x) = x$ for all of the x-values in the domain of g
> $(g \circ f)(x) = x$ for all of the x-values in the domain of f.

The horizontal line test

> → You can use the **horizontal line test** to identify inverse functions.
> If a horizontal line crosses the graph of a function more than once, there is no inverse function.

Example 11

Which of these functions have inverse functions?

a

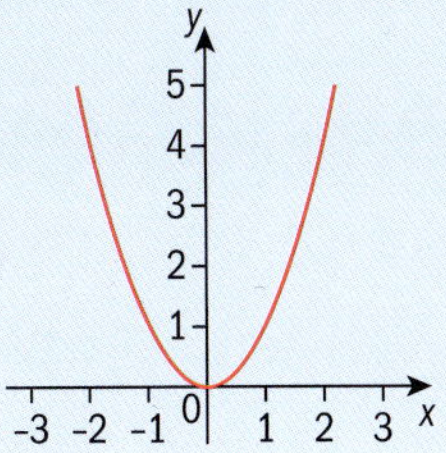

b

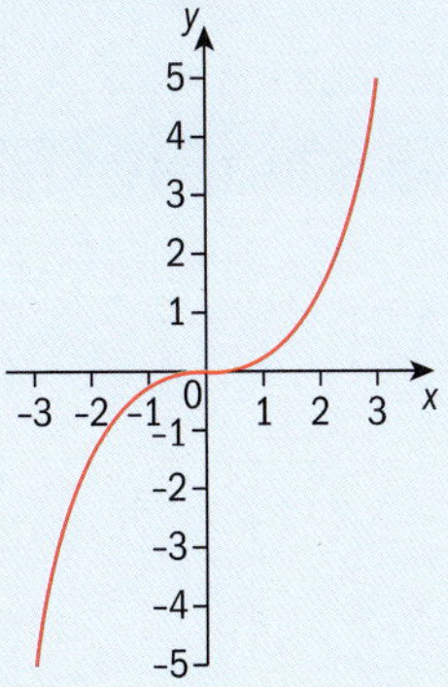

c

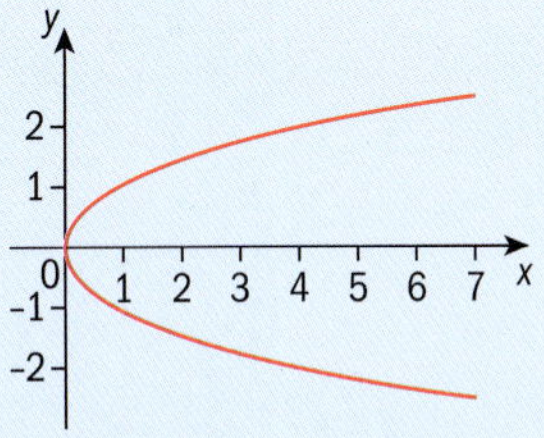

d

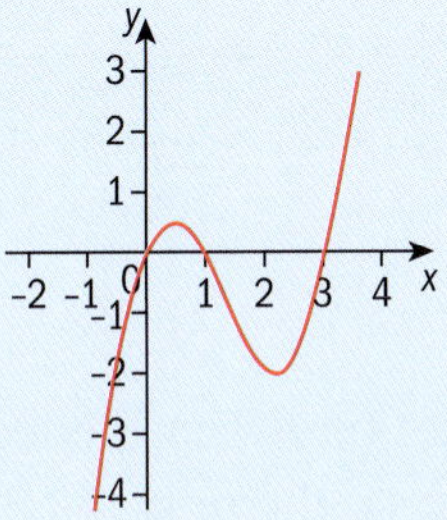

Answers

a

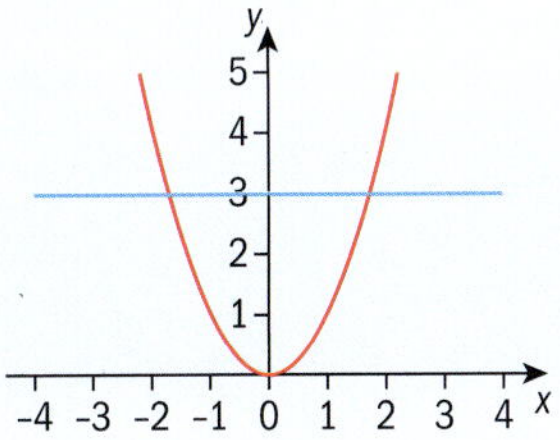

No inverse function

b

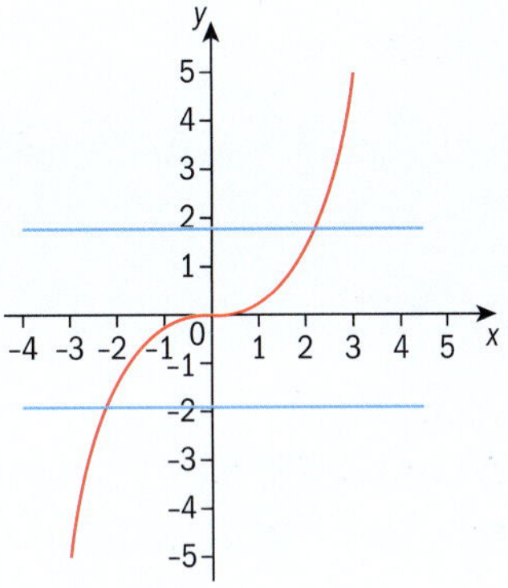

Inverse function

c

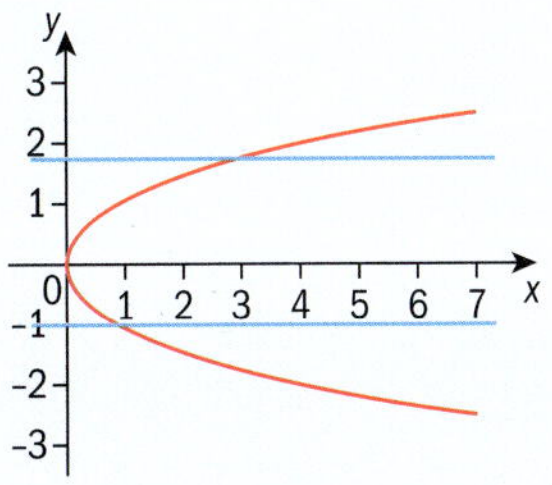

Inverse function

d

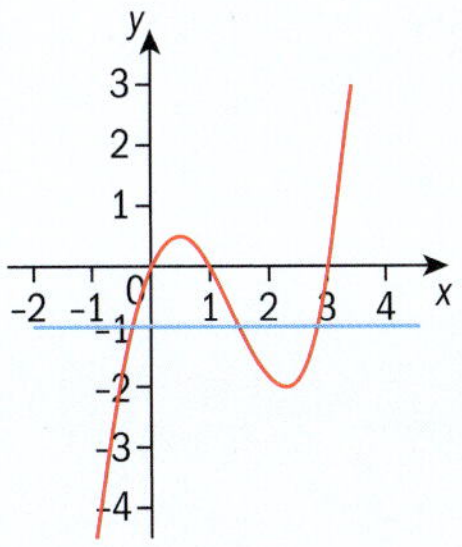

No inverse function

Did you know that Abul Wafa Buzjani, a Persian mathematician from the 10th century, used functions? There is a crater on the moon named after him.

The graphs of inverse functions

→ The graph of the inverse of a function is a reflection of that function in the line $y = x$.

Here are some examples of functions and their inverse functions.

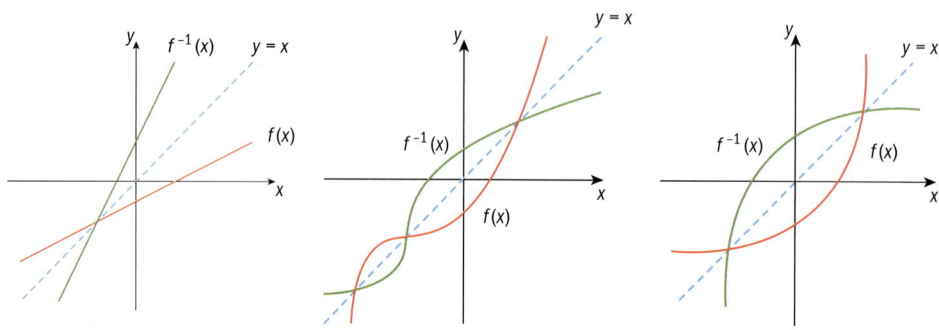

If (x, y) lies on the line $f(x)$, then (y, x) lies on $f^{-1}(x)$. Reflecting the function in the line $y = x$ 'swaps' x and y, so the point $(1, 3)$ reflected in the line $y = x$ becomes point $(3, 1)$.

Exercise 1G

1 Use the horizontal line test to determine which of these functions have inverse functions.

a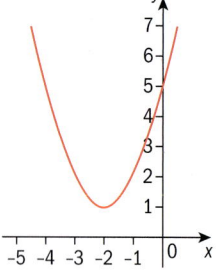

b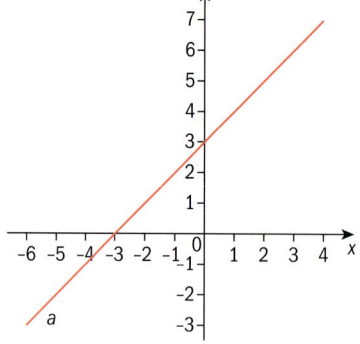

c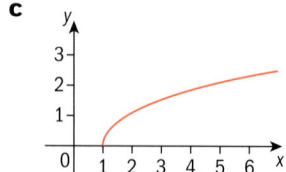

d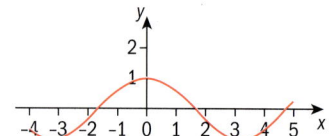

> Pioneering work by Indian scientist Panini in the 6th century BCE included functions.

2 Copy the graphs of these functions. For each, draw the line $y = x$ and the graph of the inverse function.

a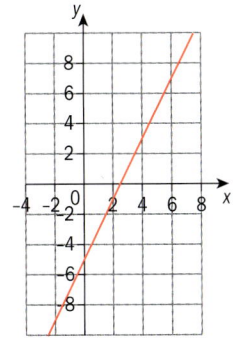

b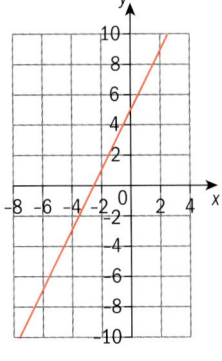

c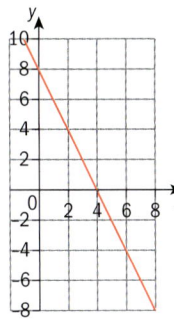

d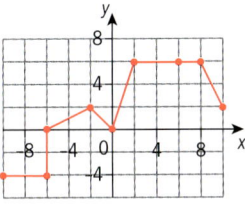

e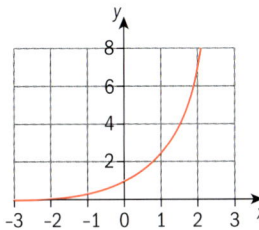

f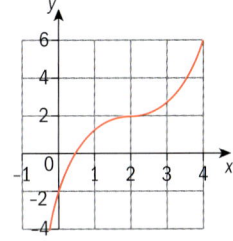

Finding inverse functions algebraically

Look at how the function $f(x) = 3x - 2$ is made up. We start with x on the left.

$$x \longrightarrow \boxed{\times 3} \longrightarrow \boxed{-2} \longrightarrow 3x - 2$$

To form the inverse function we reverse the process, using inverse operations.

$$\frac{x+2}{3} \longleftarrow \boxed{\div 3} \longleftarrow \boxed{+2} \longleftarrow x$$

> The inverse of $+2$ is -2
> The inverse of $\times 3$ is $\div 3$

So $f^{-1}(x) = \dfrac{x+2}{3}$

The next example shows you how to do this without diagrams.

Example 12

If $f(x) = 3x - 2$, find the inverse function $f^{-1}(x)$.

Answer

$y = 3x - 2$	*Replace $f(x)$ with y.*
$x = 3y - 2$	*Replace every x with y and every y with x.*
$x + 2 = 3y$	*Make y the subject.*
$y = \dfrac{x+2}{3}$	
$f^{-1}(x) = \dfrac{x+2}{3}$	*Replace y with $f^{-1}(x)$.*

As you saw in the graphs of functions and their inverses, the inverse function of a given function f is the reflection of the graph $y = f(x)$ in the line $y = x$, which 'swaps' x and y. So in Example 12 we swapped x and y, and then made y the subject.

> → To find the inverse function algebraically, replace $f(x)$ with y and solve for y.

Example 13

If $f(x) = 4 - 3x$, find $f^{-1}(x)$.

Answer

$y = 4 - 3x$	*Replace $f(x)$ with y.*
$x = 4 - 3y$	*Replace every x with y and every y with x.*
$x - 4 = -3y$	*Make y the subject.*
$\dfrac{x-4}{-3} = y$	
$y = \dfrac{4-x}{3}$	
$f^{-1}(x) = \dfrac{4-x}{3}$	*Replace y with $f^{-1}(x)$.*

To check that the inverse function in Example 13 is correct, combine the functions

$$(f \circ f^{-1})(x) = 4 - 3\left(\dfrac{4-x}{3}\right) = 4 - (4-x) = x$$

So $(f \circ f^{-1})(x) = x$ and f and f^{-1} are inverses of each other.

> → The function $I(x) = x$ is called the identity function.
> It leaves x unchanged.
> So $f \circ f^{-1} = I$

Exercise 1H

EXAM-STYLE QUESTION

1 If $f(x) = \dfrac{x+4}{2}$ and $g(x) = 2x - 4$, find

  a **i** $g(1)$ and $(f \circ g)(1)$ **ii** $f(-3)$ and $(g \circ f)(-3)$

  iii $(f \circ g)(x)$ **iv** $(g \circ f)(x)$

  b What does this tell you about functions f and g?

2 Find the inverse for each of these functions.

  a $f(x) = 3x - 1$ **b** $g(x) = x^3 - 2$ **c** $h(x) = \dfrac{1}{4}x + 5$

  d $f(x) = \sqrt[3]{x} - 3$ **e** $g(x) = \dfrac{1}{x} - 2$ **f** $h(x) = 2x^3 + 3$

  g $f(x) = \dfrac{x}{3+x}, x \neq -3$ **h** $g(x) = \dfrac{2x}{5-x}, x \neq 5$

3 What is $f^{-1}(x)$ if

  a $f(x) = 1 - x$ **b** $f(x) = x$ **c** $f(x) = \dfrac{1}{x}, x \neq 0$

> Self-inverse functions are such that a function and its inverse are the same. Look for self-inverse functions in question 3.

4 Evaluate $f^{-1}(5)$ where
  a $f(x) = 6 - x$ **b** $f(x) = \dfrac{10}{x+7}$ **c** $f(x) = \dfrac{2}{4x-3}$

> Note that the image of point $(a, -b)$ after a reflection in the line $y = x$ is the point $(b, -a)$.

5 If $f(x) = \dfrac{x+1}{x-2}$, find $f^{-1}(x)$.

EXAM-STYLE QUESTION

6 a Draw the graph of $f(x) = 2^x$ by making a table of values and plotting several points.
  b Draw the line $y = x$ on the same graph.
  c Draw the graph of f^{-1} by reflecting the graph of f in the line $y = x$.
  d State the domain and range of f and f^{-1}.

7 The function $f(x) = x^2$ has no inverse function. However, the square root function $g(x) = \sqrt{x}$ does have an inverse function. Find this inverse.
By comparing the range and domain explain why the inverse of $g(x) = \sqrt{x}$ is not the same as $f(x) = x^2$.

8 Prove that the graphs of a linear function and its inverse can never be perpendicular.

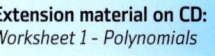

Extension material on CD: *Worksheet 1 - Polynomials*

1.6 Transforming functions

Investigation – functions

You should use your GDC to sketch all the graphs in this investigation.

1 Sketch $y = x$, $y = x + 1$, $y = x - 4$, $y = x + 4$ on the same axes.
Compare and contrast your functions.
What effect do the constant (number) terms have on the graphs of $y = x + b$?

> You will also find this standard equation of a line written as $y = mx + b$ or $y = mx + c$

2 Sketch $y = x + 3$, $y = 2x + 3$, $y = 3x + 3$, $y = -2x + 3$, $y = 0.5x + 3$ on the same axes.
Compare and contrast your functions.
What effect does changing the x-coefficient have?

> The coefficient of x is the number that multiplies the x-value.

3 Sketch $y = |x|$, $y = |x + 2|$, $y = |x - 3|$ on the same axes.
Compare and contrast your functions.
What effect does changing the values of h have on the graphs of $y = |x + h|$?

> |x| means the modulus of x. See chapter 18 for more explanation.

4 Sketch $y = x^2$, $y = -x^2$, $y = 2x^2$, $y = 0.5x^2$ on the same axes.
Compare and contrast your functions.
What effect does the negative sign have on the graph?
What effect does changing the value of a have on the graphs of $y = ax^2$?

Chapter 1

In the investigation you should have found that your graphs in parts 1, 2 and 3 were all the same shape but the position of the graphs changed. The graphs in part 4 should have been reflected or changed by stretching.

These are examples of 'transformations' of graphs. We will now look at these transformations in detail.

Translations

Shift upward or downward

→ $f(x) + k$ translates $f(x)$ vertically a distance of k units upward.

→ $f(x) - k$ translates $f(x)$ vertically a distance of k units downward.

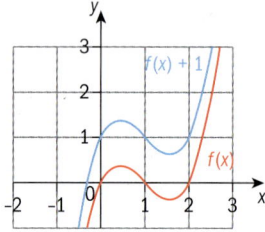
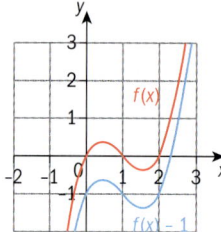

Shift to the right or left

→ $f(x + k)$ translates $f(x)$ horizontally k units to the **left**, when $k > 0$.

→ $f(x - k)$ translates $f(x)$ horizontally k units to the **right**, when $k > 0$.

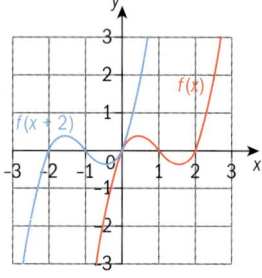
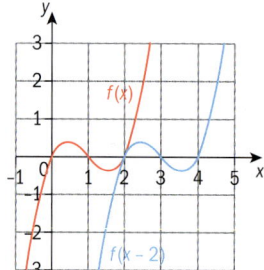

Translations can be represented by vectors in the form $\begin{pmatrix} a \\ b \end{pmatrix}$ where a is the horizontal component and b is the vertical component. $\begin{pmatrix} 3 \\ 0 \end{pmatrix}$ is a horizontal shift of 3 units right. $\begin{pmatrix} 0 \\ -2 \end{pmatrix}$ is a vertical shift of 2 units down.

Translation by the vector $\begin{pmatrix} 3 \\ -2 \end{pmatrix}$ denotes a horizontal shift of 3 units to the right, and a vertical shift of 2 units down.

> Try transforming some functions with different values of k on your GDC.

Reflections

Reflection in the *x*-axis

→ –*f*(*x*) reflects *f*(*x*) in the *x*-axis.

Reflection in the *y*-axis

→ *f*(–*x*) reflects *f*(*x*) in the *y*-axis.

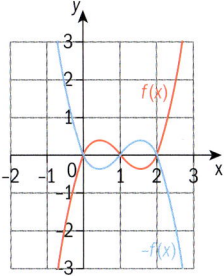
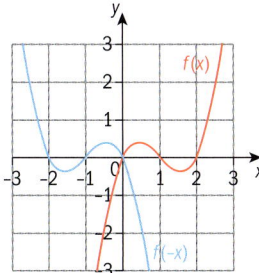

Stretches

Horizontal stretch (or compress)

→ *f*(*qx*) stretches or compresses *f*(*x*) horizontally with scale factor $\frac{1}{q}$.

Vertical stretch (or compress)

→ *pf*(*x*) stretches *f*(*x*) vertically with scale factor *p*.

A stretch with a scale factor *p* where 0 < *p* < 1 will actually compress the graph.

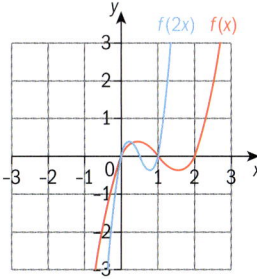
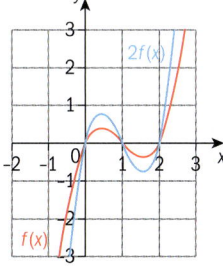

The transformation is a **horizontal stretch of scale factor** $\frac{1}{q}$.
When *q* > 1 the graph is compressed towards the *y*-axis.
When 0 < *q* < 1 the graph is stretched away from the *y*-axis.

The transformation is a **vertical stretch of scale factor** *p*.
When *p* > 1 the graph stretches away from the *x*-axis.
When 0 < *p* < 1 the graph is compressed towards the *x*-axis.

Students often make mistakes with stretches. It is important to remember the different effects of, for example, 2*f*(*x*) and *f*(2*x*).

Example 14

1 Given the graph of the function *f*(*x*) shown here, sketch the graphs of:
  a *f*(*x* + 1)   **b** *f*(*x*) – 2   **c** *f*(–*x*)   **d** –*f*(*x*)   **e** 2*f*(*x*)

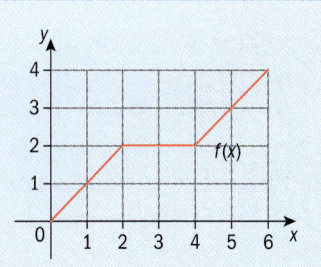

▶ Continued on next page

Answers

a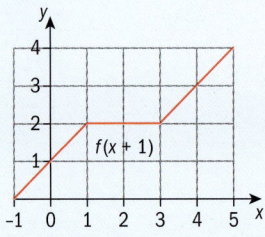
Translated one unit to the left

b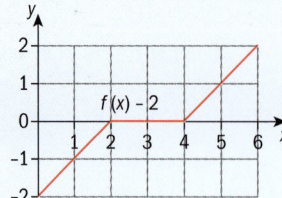
Translated two units down

c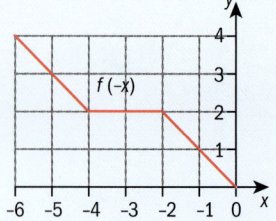
Reflected in the *y*-axis

d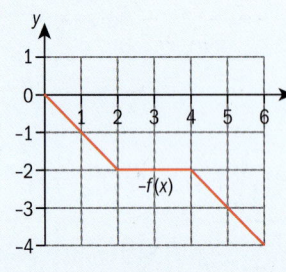
Reflected in the *x*-axis

e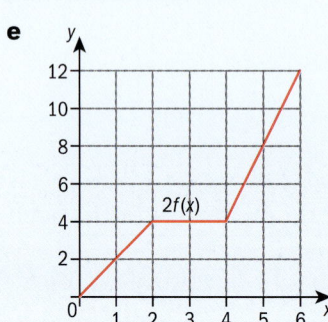
Vertical stretch of scale factor 2

Supply and demand curves in business and economics are reflections.

Radioactive decay curves are reflections.

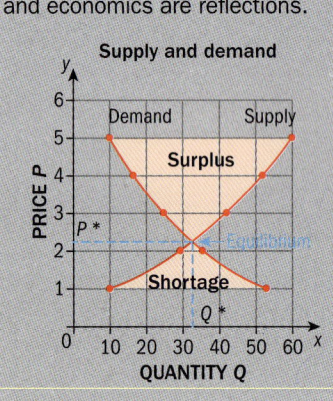
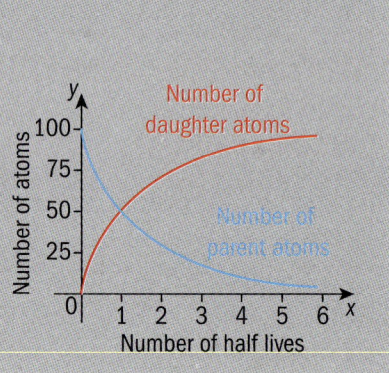

Exercise 1I

EXAM-STYLE QUESTION

1 Copy the graph. Draw these functions on the same axes.
  a $f(x) + 4$ b $f(x) - 2$ c $-f(x)$
  d $f(x + 3)$ e $f(x - 4)$ f $2f(x)$
  g $f(2x)$

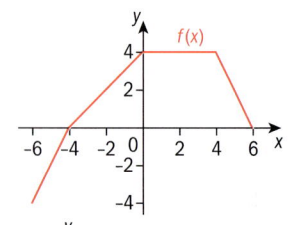

2 Functions *g*, *h* and *q* are transformations of $f(x)$. Write each transformation in terms of $f(x)$.

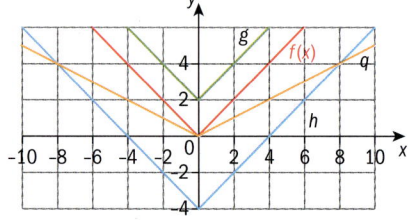

3 Functions q, s and t are transformations of $f(x)$. Write each transformation in terms of $f(x)$.

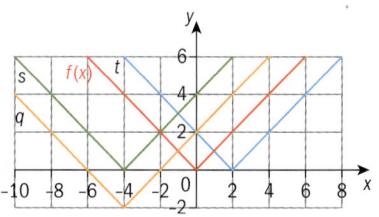

EXAM-STYLE QUESTION

4 Copy the graph of $f(x)$. Sketch the graph of each of these functions, and state the domain and range for each.
  a $2f(x-5)$
  b $-f(2x)+3$

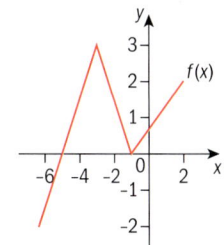

5 The graph of $f(x)$ is shown. A is the point $(1, 1)$. Make separate copies of the graph and draw the function after each transformation.
On each graph, label the new position of A as A_1.
  a $f(x+1)$
  b $f(x)+1$
  c $f(-x)$
  d $2f(x)$
  e $f(x-2)+3$

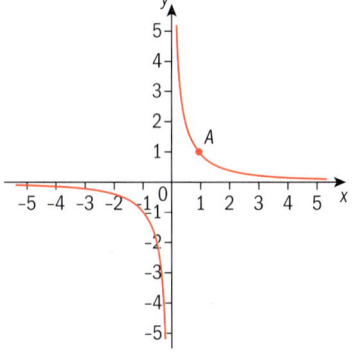

6 In each case, describe the transformation that would change the graph of $f(x)$ into the graph of $g(x)$.
  a $f(x) = x^3$, $g(x) = -(x^3)$
  b $f(x) = x^2$, $g(x) = (x-3)^2$
  c $f(x) = x$, $g(x) = -2x+5$

EXAM-STYLE QUESTION

7 Let $f(x) = 2x + 1$.
  a Draw the graph of $f(x)$ for $0 \leq x \leq 2$.
  b Let $g(x) = f(x+3) - 2$. On the same graph draw $g(x)$ for $-3 \leq x \leq -1$.

> If a domain is given in the question, you must only draw the function for that domain.

Review exercise

1 a If $g(a) = 4a - 5$, find $g(a-2)$.
  b If $h(x) = \dfrac{1+x}{1-x}$, find $h(1-x)$.

2 a Evaluate $f(x-3)$ when $f(x) = 2x^2 - 3x + 1$.
  b For $f(x) = 2x + 7$ and $g(x) = 1 - x^2$, find the composite function defined by $(f \circ g)(x)$.

3 Find the inverses of these functions.
   a $f(x) = \dfrac{3x+17}{2}$
   b $g(x) = 2x^3 + 3$

4 Find the inverse of $f(x) = -\dfrac{1}{5}x - 1$. Then graph the function and its inverse.

5 Find the inverse functions for
   a $f(x) = 3x + 5$ b $f(x) = \sqrt[3]{x+2}$

6 Copy each graph and draw the inverse of each function.
   a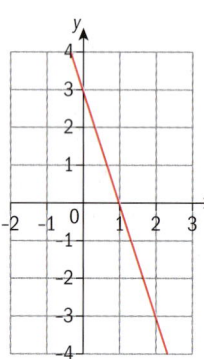
   b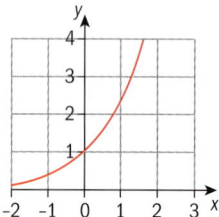

7 Find the domain and range for each of these graphs.
   a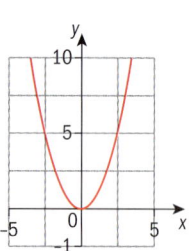
   b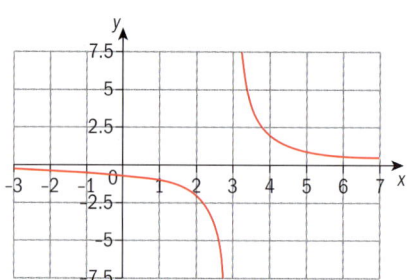

EXAM-STYLE QUESTION

8 For each function, write a single equation to represent the given combination of transformations.
   a $f(x) = x$, reflected in the y-axis, stretched vertically by a factor of 2, horizontally by a factor of $\dfrac{1}{3}$ and translated 3 units left and 2 units up.
   b $f(x) = x^2$, reflected in the x-axis, stretched vertically by a factor of $\dfrac{1}{4}$, horizontally by a factor of 3, translated 5 units right and 1 unit down.

9 a Explain how to draw the inverse of a function from its graph.
   b Graph the inverse of $f(x) = 2x + 3$.

EXAM-STYLE QUESTION

10 Let $f(x) = 2x^3 + 3$ and $g(x) = 3x - 2$.
    a Find $g(0)$.   b Find $(f \circ g)(0)$.   c Find $f^{-1}(x)$.

EXAM-STYLE QUESTIONS

11 The graph shows the function $f(x)$, for $-2 \leq x \leq 4$.

   a Let $h(x) = f(-x)$. Sketch the graph of $h(x)$.

   b Let $g(x) = \frac{1}{2} f(x - 1)$. The point $A(3, 2)$ on the graph of f is transformed to the point P on the graph of g. Find the coordinates of P.

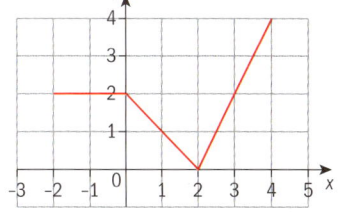

12 The functions f and g are defined as $f(x) = 3x$ and $g(x) = x + 2$.

   a Find an expression for $(f \circ g)(x)$.

   b Show that $f^{-1}(12) + g^{-1}(12) = 14$.

13 Let $g(x) = 2x - 1$, $h(x) = \frac{3x}{x-2}, x \neq 2$

   a Find an expression for $(h \circ g)(x)$. Simplify your answer.

   b Solve the equation $(h \circ g)(x) = 0$.

> The instruction 'Show that...' means 'Obtain the required result (possibly using information given) without the formality of proof'.
> For 'Show that' questions you do not usually need to use a calculator.
> A good method is to cover up the right-hand side of the equation and then work out the left-hand side until your answer is the same as the right-hand side.

Review exercise

1 Use your GDC to sketch the function and state the domain and range of $f(x) = \sqrt{x+2}$.

2 Sketch the function $y = (x + 1)(x - 3)$ and state its domain and range.

3 Sketch the function $y = \frac{1}{x+2}$ and state its domain and range.

EXAM-STYLE QUESTIONS

4 The function $f(x)$ is defined as $f(x) = 2 + \frac{1}{x+1}, x \neq -1$.

   a Sketch the curve $f(x)$ for $-3 \leq x \leq 2$.

   b Use your GDC to help you write down the value of the x-intercept and the y-intercept.

5 a Sketch the graph of $f(x) = \frac{1}{x^2}$

   b For what value of x is $f(x)$ undefined?

   c State the domain and range of $f(x)$.

6 Given the function $f(x) = \frac{2x-5}{x+2}$

   a write down the equations of the asymptotes

   b sketch the function

   c write down the coordinates of the intercepts with both axes.

7 Let $f(x) = 2 - x^2$ and $g(x) = x^2 - 2$.

   a Sketch both functions on one graph with $-3 \leq x \leq 3$.

   b Solve $f(x) = g(x)$.

Chapter 1 27

EXAM-STYLE QUESTIONS

8 Let $f(x) = x^3 - 3$.
   a Find the inverse function $f^{-1}(x)$.
   b Sketch both $f(x)$ and $f^{-1}(x)$ on the same axes.
   c Solve $f(x) = f^{-1}(x)$.

9 $f(x) = e^{2x-1} + \dfrac{2}{x+1}$, $x \neq 1$.

   Sketch the curve of $f(x)$ for $-5 \leq x \leq 2$, including any asymptotes.

10 Consider the functions f and g where $f(x) = 3x - 2$ and $g(x) = x - 3$.
   a Find the inverse function, f^{-1}.
   b Given that $g^{-1}(x) = x + 3$, find $(g^{-1} \circ f)(x)$.
   c Show that $(f^{-1} \circ g)(x) = \dfrac{x-1}{3}$.
   d Solve $(f^{-1} \circ g)(x) = (g^{-1} \circ f)(x)$

   Let $h(x) = \dfrac{f(x)}{g(x)}$, $x \neq 2$.

   d **Sketch** the graph of h for $-6 \leq x \leq 10$ and $-4 \leq y \leq 10$, including any asymptotes.
   e Write down the **equations** of the asymptotes.

> When IB exams have words in **bold** script, it means that you must do exactly what is required. For example the equation could be given as $x = 3$ but not just as 3.

CHAPTER 1 SUMMARY

Introducing functions

- A **relation** is a set of ordered pairs.
- The **domain** is the set of all the first numbers (x-values) of the ordered pairs.
- The **range** is the set of the second numbers (y-values) in each pair.
- A **function** is a relation where every x-value is related to a unique y-value.
- A relation is a function if any vertical line drawn will not intersect the graph more than once. This is called the **vertical line test**.

The domain and range of a relation on a Cartesian plane

Interval notation:

Use round brackets (,) if the value is not included in the graph or when the graph is undefined at that point (a hole or **asymptote**, or a jump).
Use square brackets [,] if the value is included in the graph.
- Set notation:

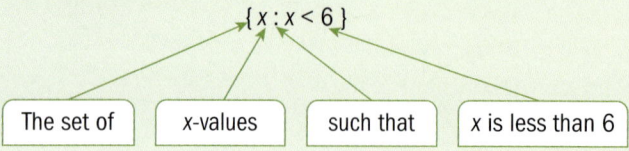

Continued on next page

Function notation
- $f(x)$ is read as 'f of x' and means 'the value of function f at x'.

Composite functions
- The composition of the function f with the function g is written as $f(g(x))$, which is read as 'f of g of x', or $(f \circ g)(x)$, which is read as 'f composed with g of x'.
- A **composite function** applies one function to the result of another and is defined by $(f \circ g)(x) = f(g(x))$.

Inverse functions
- The **inverse** of a function $f(x)$ is $f^{-1}(x)$. It reverses the action of the function.
- Functions $f(x)$ and $g(x)$ are inverses of one another if:
  $(f \circ g)(x) = x$ for all of the x-values in the domain of g and
  $(g \circ f)(x) = x$ for all of the x-values in the domain of f.
- You can use the **horizontal line test** to identify inverse functions. If a horizontal line crosses a function more than once, there is no inverse function.

The graphs of inverse functions
- The graph of the inverse of a function is a reflection of that function in the line $y = x$.
- To find the inverse function algebraically, replace $f(x)$ with y and solve for y.
- The function $I(x) = x$ is called the identity function. It leaves x unchanged. So $f \circ f^{-1} = I$.

Transformations of functions
- $f(x) + k$ translates $f(x)$ vertically a distance of k units upward.
- $f(x) - k$ translates $f(x)$ vertically a distance of k units downward.
- $f(x + k)$ translates $f(x)$ horizontally k units to the left, where $k > 0$.
- $f(x - k)$ translates $f(x)$ horizontally k units to the right, where $k > 0$.
- $-f(x)$ reflects $f(x)$ in the x-axis.
- $f(-x)$ reflects $f(x)$ in the y-axis.
- $f(qx)$ stretches $f(x)$ horizontally with scale factor $\frac{1}{q}$.
- $pf(x)$ stretches $f(x)$ vertically with scale factor p.

Theory of knowledge

Mathematical representation

Mathematics can be represented visually in models, pictures, number lines, and graphs of functions and relationships.

When someone shows you a visual representation, they have decided on the scale to use and what information to show on it, before you see it.

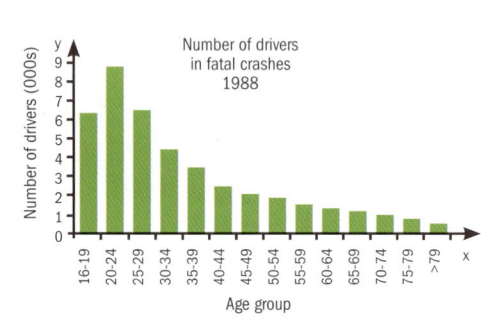 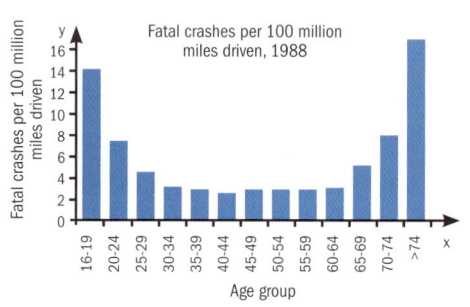

▲ This graph suggests that:
- 16-year-olds are safer drivers than people in their 20s
- 80-year-olds are very safe drivers.
■ Do you think these statements are true?

▲ This graph relates number of crashes to distance driven.
■ What does it tell you about 16-year-old and 75-year-old drivers?

The **Monthly Labor Review** published this data, relating earnings to education

Earnings by Educational Attainment 1996	
Education level	Median annual earnings
Professional	$71,868
Ph.D.	$60,827
Master's	$46,269
Bachelor's	$36,155
High school	$23,317

"Getting a bachelor's degree will increase your earnings by almost $13,000 a year"

■ Is this a true statement?

Accuracy

- How useful are graphs for conveying information?
- How accurate can a graph be?
- What are the benefits and pitfalls of interpolation and extrapolation from data?

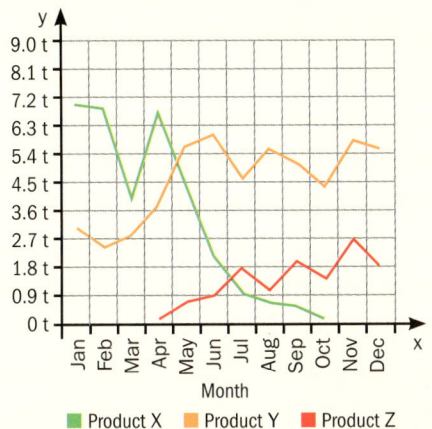

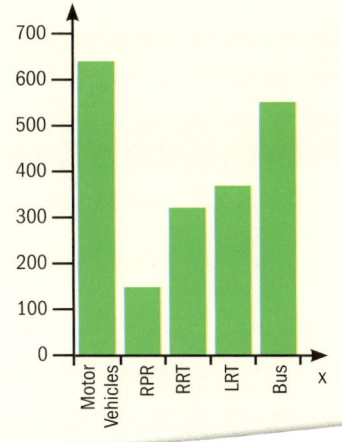

- How accurate are these visual representations:
  - X-rays
  - Snapshots
  - Paintings?

Grids

- What is a grid?
- How are grids used in computing, town planning, biology, the military?
- What are:
  - data grids
  - cluster grids
  - campus grids
  - mapping grids?

A computational grid is "a hardware and software infrastructure that provides dependable, consistent, pervasive and inexpensive access to high-end computational capabilities"

Foster and Kesselman, 1998

- Are there any computers that are not on a grid?
- Is a computer system a grid?

Theory of knowledge

Chapter 1 31

2 Quadratic functions and equations

CHAPTER OBJECTIVES:

2.4 the quadratic function $f(x) = ax^2 + bx + c = 0$: its graph, its vertex, x- and y-intercepts, axis of symmetry
The form $x \mapsto a(x - p)(x - q)$, x-intercepts $(p,0)$ and $(q,0)$
The form $x \mapsto a(x - h)^2 + k$, vertex (h,k)
2.7 solving quadratic equations of the form $ax^2 + bx + c = 0$
2.7 the quadratic formula
2.7 the discriminant and the nature of roots
2.8 applications of graphing skills and solving equations to real-life situations

Before you start

You should know how to:

1 Solve simple equations for a given variable.
e.g. Solve for b:
$3b - 2 = 0$
$3b = 2, b = \dfrac{2}{3}$
e.g. Solve the equation $n^2 + 3 = 5$.
$n^2 + 3 = 5$
$n^2 = 2, n = \pm\sqrt{2}$

2 Factorize mathematical expressions.
e.g. Factorize $p^2 - 5p$:
$p(p - 5)$
e.g. Factorize the expression
$ax - 3x + 2a - 6$:
$x(a - 3) + 2(a - 3)$
$(x + 2)(a - 3)$
e.g. Factorize the expression $x^2 - 3x - 10$:
$(x + 2)(x - 5)$
e.g. Factorize the expression $4a^2 - 25$:
$(2a + 5)(2a - 5)$

Skills check

1 Solve each equation.
 a $3a - 5 = a + 7$
 b $4x^2 + 1 = 21$
 c $3(n - 4) = 5(n + 2)$

2 Factorize each expression.
 a $2k^2 - 10k$
 b $14a^3 + 21a^2 - 49a$
 c $2x^2 + 4xy + 3x + 6y$
 d $5a^2 - 10a - ab + 2b$
 e $n^2 + 4n + 3$
 f $2x^2 - x - 3$
 g $m^2 - 36$
 h $25x^2 - 81y^2$

This World War II Memorial was opened in 2004 in Washington DC. The fountains at the Memorial spray water in beautiful curved trajectories.

This picture shows water streaming from a simple drinking fountain in a similar trajectory. The shapes of the curved paths of these streams of water are called parabolas, and they can be modeled by mathematical functions of the form $f(x) = ax^2 + bx + c$. Functions like these are called **quadratic functions**.

Other situations which can be modeled by quadratic functions include the area of a figure and measuring the height of a dropped object over time.

In this chapter, you will study how to graph quadratic functions that are given in standard form, $f(x) = ax^2 + bx + c$; turning point form, $y = a(x - h)^2 + k$; and factorized form, $f(x) = a(x - p)(x - q)$. Each of these forms are useful in their own way. If you wanted to know the maximum height of a spray of water from a fountain, you might use the turning-point form. If you wanted to find the dimensions of a rectangle with a particular area, the factorized form would be helpful.

2.1 Solving quadratic equations

An equation that can be written in the form $ax^2 + bx + c = 0$, where $a \neq 0$, is called a **quadratic equation**. These are all examples of quadratic equations:

$x^2 - 4x + 7 = 0$
$5x^2 = 3x - 2$
$2x(3x - 7) = 0$
$(x - 7)(2 - 5x) = 14x$

In this section, you will begin solving quadratic equations.

> Some of these equations are not written in the form $ax^2 + bx + c = 0$, but they can be rearranged into this form.

> In a quadratic trinomial $ax^2 + bx + c$, ax^2 is called the quadratic term, bx is the linear term, and c is the constant term.

Solving by factorization

Before you solve quadratic equations by factorizing, it is important to understand an important property:

→ If $xy = 0$, then $x = 0$ or $y = 0$.
This property can be expanded to:
If $(x - a)(x - b) = 0$, then $x - a = 0$ or $x - b = 0$.

> This property is sometimes called the **zero product property**.

Example 1

Solve these equations by factorization.
a $x^2 - 5x - 14 = 0$ **b** $3x^2 + 2x - 5 = 0$ **c** $4x^2 + 4x + 1 = 0$

Answers

a $x^2 - 5x - 14 = 0$
$(x - 7)(x + 2) = 0$
$x - 7 = 0$ or $x + 2 = 0$
$x = 7$ $\qquad$ $x = -2$
$x = -2$ or 7

Factorize the expression on the left-hand side of the equation.

Set each factor equal to zero, using the zero product property.

b $3x^2 + 2x - 5 = 0$
$(3x + 5)(x - 1) = 0$
$3x + 5 = 0$ or $x - 1 = 0$
$x = -\dfrac{5}{3}$ $\qquad$ $x = 1$
$x = -\dfrac{5}{3}, 1$

Factorize the expression on the left-hand side of the equation. Set each factor equal to zero.
You can also find the solutions with your GDC. (See Chapter 17 Section 1.7.)

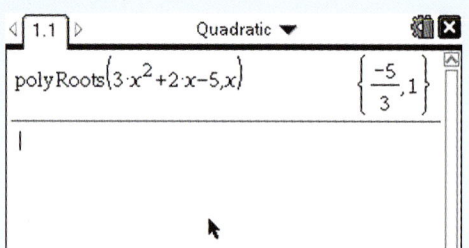

c $4x^2 + 4x + 1 = 0$
$(2x + 1)(2x + 1) = 0$
$(2x + 1)^2 = 0$
$2x + 1 = 0$ $\quad$ $x = -\dfrac{1}{2}$

*When we get the same factor twice, it is a 'perfect square' and there will be only one solution. We sometimes say that this equation has two **equal** roots.*

GDC help on CD: *Alternative demonstrations for the TI-84 Plus and Casio FX-9860GII GDCs are on the CD.*

Exercise 2A

In this exercise, solve all the equations 'by hand' and then check your answers with your GDC.

1. Solve by factorization.
   a $x^2 - 3x + 2 = 0$
   b $a^2 + a - 56 = 0$
   c $m^2 - 11m + 30 = 0$
   d $x^2 - 25 = 0$
   e $x^2 + 2x - 48 = 0$
   f $b^2 + 6b + 9 = 0$

2. Solve by factorization.
   a $6x^2 + 5x - 4 = 0$
   b $5c^2 + 6c - 8 = 0$
   c $2h^2 - 3h - 5 = 0$
   d $4x^2 - 16x - 9 = 0$
   e $3t^2 + 14t + 8 = 0$
   f $6x^2 + x - 12 = 0$

If a quadratic equation is not written in the form $ax^2 + bx + c = 0$, you will have to rearrange the terms before you can factorize as shown in Example 2.

Example 2

Solve these equations by factorization.
a $8x^2 - 5 = 10x - 2$
b $x(x + 10) = 4(x - 2)$

Answers

a $8x^2 - 5 = 10x - 2$
   $8x^2 - 10x - 3 = 0$
   $(4x + 1)(2x - 3) = 0$
   $4x + 1 = 0$ or $2x - 3 = 0$
   $x = -\frac{1}{4}$ $x = \frac{3}{2}$
   $x = -\frac{1}{4}$ or $\frac{3}{2}$

   Collect like terms on one side of the equation.
   Factorize and solve for x.

b $x(x + 10) = 4(x - 2)$
   $x^2 + 10x = 4x - 8$
   $x^2 + 6x + 8 = 0$
   $(x + 4)(x + 2) = 0$
   $x + 4 = 0$ or $x + 2 = 0$
   $x = -4$ $x = -2$
   $x = -4, -2$

   Expand the brackets and collect like terms.
   Factorize and solve for x.

> Ancient Babylonians and Egyptians studied quadratic equations like these thousands of years ago to find, for example, solutions to problems concerning the area of a rectangle.

Exercise 2B

1. Solve by factorization.
   a $x^2 + 2x - 7 = 13 + x$
   b $2n^2 + 11n = 3n - n^2 - 4$
   c $3z(z + 4) = -(z^2 + 9)$
   d $2(a - 5)(a + 5) = 21a$
   e $x + 5 = \dfrac{36}{x}$
   f $2x - 1 = \dfrac{x+1}{2x}$

2. A number and its square differ by 12. Find the number.

> Use 'x' to represent the number, and write an equation to solve for x.

EXAM-STYLE QUESTION

3. The two perpendicular sides of a right-angled triangle have lengths $x + 2$ and $5x - 3$.
   The hypotenuse has length $4x + 1$. Find x.

Investigation – perfect square trinomials

Solve these equations by factorization.

1. $x^2 + 10x + 25 = 0$
2. $x^2 + 6x + 9 = 0$
3. $x^2 + 14x + 49 = 0$
4. $x^2 - 8x + 16 = 0$
5. $x^2 - 18x + 81 = 0$
6. $x^2 - 20x + 100 = 0$

What do you notice? Describe any patterns you see in the original quadratic equations.

A trinomial is a polynomial with three terms.
Why do you think these are called 'perfect square trinomials'?

Solving by completing the square

Some quadratic equations cannot be solved by factorization, but there are other methods you can use to solve a quadratic equation without using your GDC.

Consider the equation $x^2 + 14x + 49 = 0$ from the investigation above. The left side of this equation is a perfect square, because it has two identical factors: $x^2 + 14x + 49 = (x + 7)(x + 7) = (x + 7)^2$.

To solve the equation $x^2 + 14x + 49 = 0$, you could factorize, which would give the equation $(x + 7)^2 = 0$, and lead to the answer $x = -7$.

What if you were asked to solve the equation $x^2 + 14x + 49 = 5$? If you collect all the terms on one side of the equation, you get $x^2 + 14x + 44 = 0$, which does not factorize easily. You could still get an exact answer, however, as shown in Example 3.

Example 3

Solve each equation without using the GDC.
a $x^2 + 14x + 49 = 5$
b $x^2 - 6x + 9 = 6$

Answers

a $x^2 + 14x + 49 = 5$
$(x + 7)^2 = 5$

$x + 7 = \pm\sqrt{5}$

$x = -7 \pm \sqrt{5}$

Factorize the perfect square trinomial on the left-hand side of the equation.
Take the square root of both sides of the equation.
x has two solutions: $-7 + \sqrt{5}$ and $-7 - \sqrt{5}$.

b $x^2 - 6x + 9 = 6$
$(x - 3)^2 = 6$
$x - 3 = \pm\sqrt{6}$
$x = 3 \pm \sqrt{6}$

*Again, we see that the left side of the equation is a perfect square trinomial, so we can use the same method as in part **a**.*
x has two solutions: $3 + \sqrt{6}$ and $3 - \sqrt{6}$.

> Leaving your answers in radical (surd) form gives the exact solutions.

In Example 3, the equations involved perfect square trinomials. You can use perfect square trinomials to solve any quadratic equation, using a method called **completing the square**.

> → To complete the square, take half the coefficient of x, square it, and add the result to both sides of the equation. This step creates a perfect square trinomial on the left side of the equation.

Example 4

Solve each equation by completing the square.
a $x^2 + 10x = 6$ **b** $x^2 - 12x = 3$ **c** $x^2 - 3x - 1 = 0$

Answers

a $x^2 + 10x = 6$
$x^2 + 10x + 25 = 6 + 25$
$(x + 5)^2 = 31$
$x + 5 = \pm\sqrt{31}$
$x = -5 \pm \sqrt{31}$

The coefficient of x is 10. Halve this (5) and square it (25).
Complete the square by adding 25 to both sides.
Solve for x.

b $x^2 - 12x = 3$
$x^2 - 12x + 36 = 3 + 36$
$(x - 6)^2 = 39$
$x - 6 = \pm\sqrt{39}$
$x = 6 \pm \sqrt{39}$

The coefficient of x is 12.
$12 \div 2 = 6, 6^2 = 36$
Complete the square.
Solve for x.

c $x^2 - 3x - 1 = 0$
$x^2 - 3x = 1$
$x^2 - 3x + \dfrac{9}{4} = 1 + \dfrac{9}{4}$
$\left(x - \dfrac{3}{2}\right)^2 = \dfrac{13}{4}$
$x - \dfrac{3}{2} = \dfrac{\pm\sqrt{13}}{2}$
$x = \dfrac{3 \pm \sqrt{13}}{2}$

Add 1 to both sides of the equation.
Half of 3 is $\dfrac{3}{2}$, and $\left(\dfrac{3}{2}\right)^2$ is $\dfrac{9}{4}$.
Add $\dfrac{9}{4}$ to both sides of the equation.
Solve for x.

> Over one thousand years ago, Arab and Hindu mathematicians were developing methods similar to completing the square to solve quadratic equations. They were finding solutions to mathematical problems such as 'What must be the square which, when increased by 10 of its own roots, amounts to 39?' This is written as $x^2 + 10x = 39$.

Exercise 2C

Solve by completing the square.
1 $x^2 + 8x = 3$ **2** $x^2 - 5x = 3$
3 $x^2 - 6x + 1 = 0$ **4** $x^2 + 7x - 4 = 0$
5 $x^2 - 2x - 6 = 0$ **6** $x^2 + x - 3 = 0$

> → In order to complete the square, the coefficient of the x^2 term must be 1. If the x^2 term has a coefficient other than 1, before completing the square, you can factor out the coefficient, or divide through by the coefficient.

Example 5

Solve each equation by completing the square.
  a $2x^2 + 8x = 6$ **b** $3x^2 - 15x = 2$

Answers

a $2x^2 + 8x = 6$
   $x^2 + 4x = 3$
   $x^2 + 4x + 4 = 3 + 4$
   $(x + 2)^2 = 7$
   $x + 2 = \pm\sqrt{7}$
   $x = -2 \pm \sqrt{7}$

Divide both sides of the equation by the coefficient of x^2, which is 2. Use completing the square to solve for x.

b $4x^2 - 20x = 5$
   $4(x^2 - 5x) = 5$
   $x^2 - 5x = \dfrac{5}{4}$
   $x^2 - 5x + \dfrac{25}{4} = \dfrac{5}{4} + \dfrac{25}{4}$
   $\left(x - \dfrac{5}{2}\right)^2 = \dfrac{30}{4} = \dfrac{15}{2}$
   $x - \dfrac{5}{2} = \pm\sqrt{\dfrac{15}{2}}$
   $x = \dfrac{5}{2} \pm \sqrt{\dfrac{15}{2}}$

Divide through by the coefficient of x^2, which is 4.

Half of 5 is $\dfrac{5}{2}$, and $\left(\dfrac{5}{2}\right)^2$ is $\dfrac{25}{4}$.

This answer could also be written as $x = \dfrac{5 \pm \sqrt{30}}{2}$.

> Abu Kamil Shuja (c.850 – c.930), also known as al-Hasib al-Misri, meaning 'the calculator from Egypt', was one of the first to introduce symbols for indices, such as $x^m \, x^n = x^{m+n}$, in algebra.

Exercise 2D

Solve by completing the square.

1 $2x^2 + 12x = 6$ **2** $3x^2 - 6x = 3$

3 $5x^2 - 10x + 2 = 0$ **4** $4x^2 + 6x - 5 = 0$

5 $2x^2 - x - 6 = 0$ **6** $10x^2 + 4x - 5 = 0$

2.2 The quadratic formula

You know that a quadratic equation can be written in the form $ax^2 + bx + c = 0$. Suppose you wanted to solve this general quadratic equation using the completing the square method.

You would have:
  $ax^2 + bx + c = 0$
  $ax^2 + bx = -c$
  $x^2 + \dfrac{b}{a}x = -\dfrac{c}{a}$
  $x^2 + \dfrac{b}{a}x + \left(\dfrac{b}{2a}\right)^2 = -\dfrac{c}{a} + \left(\dfrac{b}{2a}\right)^2$

> Subtract c from both sides of the equation.

> Divide both sides of the equation by a.

> Half of $\dfrac{b}{a}$ is $\dfrac{b}{2a}$. Squaring this gives $\dfrac{b^2}{4a^2}$.

Quadratic functions and equations

$$\left(x+\frac{b}{2a}\right)^2 = -\frac{c}{a}+\frac{b^2}{4a^2}$$

$$\left(x+\frac{b}{2a}\right)^2 = \frac{b^2-4ac}{4a^2}$$

$$x+\frac{b}{2a} = \pm\sqrt{\frac{b^2-4ac}{4a^2}} = \frac{\pm\sqrt{b^2-4ac}}{2a}$$

$$x = \frac{-b\pm\sqrt{b^2-4ac}}{2a}$$

This gives us an extremely useful formula which can be used to solve any quadratic equation.

→ **The quadratic formula**
For any equation in the form $ax^2 + bx + c = 0$,
$$x = \frac{-b\pm\sqrt{b^2-4ac}}{2a}$$

This formula is given in the Formula booklet for the IB exam, so you do not have to memorize it.

Example 6

Solve each equation using the quadratic formula.
a $x^2 + 4x - 6 = 0$ **b** $2x^2 - 3x = 7$ **c** $3x^2 = 7x + 6$

Answers

a $x^2 + 4x - 6 = 0$

$$x = \frac{-4\pm\sqrt{4^2-4(1)(-6)}}{2(1)}$$

$$x = \frac{-4\pm\sqrt{40}}{2}$$

$$x = \frac{-4\pm 2\sqrt{10}}{2} = -2\pm\sqrt{10}$$

Use the quadratic formula with $a = 1$, $b = 4$, and $c = -6$.

This answer is correct, but it can still be simplified.

b $2x^2 - 3x = 7$
$2x^2 - 3x - 7 = 0$

$$x = \frac{3\pm\sqrt{(-3)^2-4(2)(-7)}}{2(2)}$$

$$x = \frac{3\pm\sqrt{65}}{4}$$

First write the equation in standard form $ax^2 + bx + c = 0$.

Use the quadratic formula with $a = 2$, $b = -3$, and $c = -7$.

c $3x^2 = 7x + 6$
$3x^2 - 7x - 6 = 0$

$$x = \frac{7\pm\sqrt{(-7)^2-4(3)(-6)}}{2(3)}$$

$$x = \frac{7\pm\sqrt{121}}{6} = \frac{7\pm 11}{6}$$

$$x = -\frac{2}{3}, 3$$

First write the equation in standard form $ax^2 + bx + c = 0$.

Use the quadratic formula with $a = 3$, $b = -7$, and $c = -6$.

Exercise 2E

Solve each equation using the quadratic formula.

1. $4x^2 + 9x - 7 = 0$
2. $3x^2 + 2x - 8 = 0$
3. $5x^2 + 6x + 1 = 0$
4. $x^2 - 6x = -4$
5. $x^2 = x - 3$
6. $3x^2 + 10x = 5$
7. $2x^2 - 3x = 1$
8. $2x^2 = 9x + 4$
9. $\dfrac{6}{x} - 2x = 9$
10. $\dfrac{x+3}{5x-2} = \dfrac{x}{x+1}$

Example 7

The sum of the squares of two consecutive integers is 613.
Find the two integers.

Answer

$x^2 + (x+1)^2 = 613$
$x^2 + x^2 + 2x + 1 = 613$
$2x^2 + 2x - 612 = 0$
$x^2 + x - 306 = 0$

$x = \dfrac{-1 \pm \sqrt{(1)^2 - 4(1)(-306)}}{2(1)}$

$x = \dfrac{-1 \pm \sqrt{1225}}{2} = \dfrac{-1 \pm 35}{2}$

$x = -18$ or 17

The two integers are -18 and -17, or 17 and 18.

First, you need to write an equation.

Let x be the smaller integer, and $x + 1$ be the next consecutive integer. Expand the brackets and collect like terms.
Divide by 2.
This quadratic equation could also be solved using factorization or completing the square.

Since there are two values for x, there will also be two values for $x + 1$.
There are two possible pairs of consecutive integers.

Exercise 2F

1. Two numbers have a sum of 50 and a product of 576.
   Find the numbers.

2. A rectangle has a perimeter of 70 m and an area of 264 m².
   Find the length and width of the rectangle.

3. Find the value of x in the diagram.

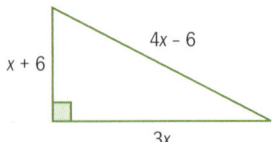

Quadratic functions and equations

EXAM-STYLE QUESTIONS

4 A rectangle has a length of 23 cm and a width of 16 cm. If the length is decreased by x cm, and the width is increased by x cm, the area of the new rectangle is 378 cm². Find the dimensions of the new rectangle.

5 The formula $h = 2 + 14t - 4.9t^2$ gives the height, h metres, of a ball t seconds after it is thrown. For how long is the ball in the air?

2.3 Roots of quadratic equations

Extension material on CD:
Worksheet 2 - Two more challenging quadratics

Investigation – roots of quadratic equations

1 Solve these equations using the quadratic formula.
   a $x^2 - 8x + 16 = 0$ **b** $4x^2 - 12x + 9 = 0$ **c** $25x^2 + 10x + 1 = 0$
2 Solve these equations using the quadratic formula.
   a $x^2 + 5x - 14 = 0$ **b** $3x^2 - 8x + 2 = 0$ **c** $5x^2 - 3x - 4 = 0$
3 Solve these equations using the quadratic formula.
   a $x^2 + 3x + 6 = 0$ **b** $2x^2 - 4x + 5 = 0$ **c** $4x^2 + 2x + 1 = 0$
4 What patterns did you notice in the solutions of the equations in questions **1**, **2** and **3**? Why do you think this happened?

Now let's take another look at the quadratic formula, used for solving equations in the form $ax^2 + bx + c = 0$, where a, b, and c are all constants.

$$x = \frac{-b \pm \sqrt{b^2 - 4ac}}{2a}$$

This formula will give us all the roots of a quadratic equation. One part of the quadratic formula, the **discriminant**, will give us information about the roots of an equation, without actually giving us the solution. The discriminant is the part of the quadratic formula under the radical (square root) sign, $b^2 - 4ac$. We often use the symbol '$\triangle$' to represent the discriminant.

→ For a quadratic equation $ax^2 + bx + c = 0$,
  • if $b^2 - 4ac > 0$, the equation will have two different real roots
  • if $b^2 - 4ac = 0$, the equation will have two equal real roots
  • if $b^2 - 4ac < 0$, the equation will have no real roots.

You can think of an equation with two equal roots as having only one solution.

Example 8

Use the discriminant to determine the nature of the roots of each equation
a $9x^2 + 6x + 1 = 0$
b $3x - 5 = \dfrac{4}{x}$

Answers

a $9x^2 + 6x + 1 = 0$

$\triangle = 6^2 - 4(9)(1) = 36 - 36 = 0$
The equation will have two equal roots.

This is a quadratic equation with $a = 9$, $b = 6$ and $c = 1$. Calculate the discriminant. Discriminant = 0 means two equal roots.

b $3x - 5 = \dfrac{4}{x}$
$3x^2 - 5x = 4$
$3x^2 - 5x - 4 = 0$
$\triangle = (-5)^2 - 4(3)(-4)$
$= 25 + 48 = 73$
The equation will have two different real roots.

First, get the equation into standard form. Multiply by x on both sides, then add 4.
Remember, $\triangle = b^2 - 4ac$.

$\triangle > 0$ means two different real roots.

Example 9

Find the value(s) of k for which the equation $2x^2 - kx + 3 = 0$ will have two different real roots.

Answer
Solution:
$b^2 - 4ac > 0$
$(-k)^2 - 4(2)(3) > 0$
$k^2 - 24 > 0$
$k^2 > 24$
$|k| > \sqrt{24}$
$|k| > 2\sqrt{6}$
$k > 2\sqrt{6}$ or $k < -2\sqrt{6}$

For the equation to have two different real roots, you must have $\triangle > 0$.

You can use the absolute value when taking the square root in an inequality.

> For more on absolute value, see Chapter 18, Section 2.7.

Exercise 2G

1 Find the value of the discriminant, and state the nature of the roots for each equation.
 a $x^2 + 5x - 3 = 0$
 b $2x^2 + 4x + 1 = 0$
 c $4x^2 - x + 5 = 0$
 d $x^2 + 8x + 16 = 0$
 e $x^2 - 3x + 8 = 0$
 f $12x^2 - 20x + 25 = 0$

EXAM-STYLE QUESTION

2 Find the values of p such that the equation has two different real roots.
  a $x^2 + 4x + p = 0$
  b $px^2 + 5x + 2 = 0$
  c $x^2 + px + 8 = 0$
  d $x^2 + 3px + 1 = 0$

3 Find the values of k such that the equation has two equal real roots.
  a $x^2 + 10x + k = 0$
  b $2x^2 - 3x + k = 0$
  c $3x^2 - 2kx + 5 = 0$
  d $x^2 - 4kx - 3k = 0$

4 Find the values of m such that the equation has no real roots.
  a $x^2 - 6x + m = 0$
  b $x^2 + 5mx + 25 = 0$
  c $3mx^2 - 8x + 1 = 0$
  d $x^2 + 6x + m - 3 = 0$

EXAM-STYLE QUESTION

5 Find the values of q for which the quadratic equation $qx^2 - 4qx + 5 - q = 0$ will have no real roots.

Investigation – graphs of quadratic functions

Each of these functions is given in the form $y = ax^2 + bx + c$.
For each function,
  i find the value of $b^2 - 4ac$
  ii graph the function on your GDC.
  a $y = x^2 - 3x - 5$
  b $y = 3x^2 - 6x + 4$
  c $y = x^2 + 2x + 7$
  d $y = 4x^2 + 3x + 5$
  e $y = x^2 - 6x + 9$
  f $y = 2x^2 - 4x + 2$
  g $y = -x^2 + 5x + 2$
  h $y = x^2 + 7x + 3$

What do these examples suggest to you about the relationship between the value of the discriminant and the graph of a quadratic function?

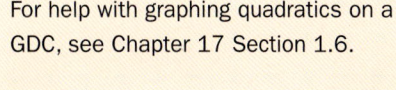

For help with graphing quadratics on a GDC, see Chapter 17 Section 1.6.

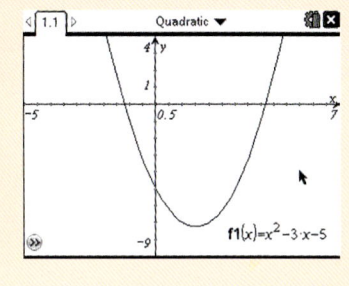

2.4 Graphs of quadratic functions

A function of the form $y = ax^2 + bx + c$, or $f(x) = ax^2 + bx + c$, where $a \neq 0$, is called a quadratic function. In this section, we will look at the graphs of quadratic functions.

The simplest quadratic function is $y = x^2$. Its graph is shown.

This graph has a minimum at the point (0, 0), and it is symmetrical about the y-axis.

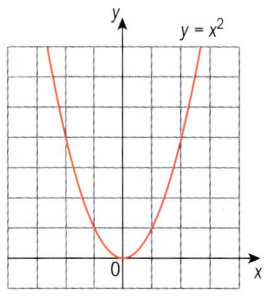

If you look at the graphs of other quadratic functions, you should notice some similarities.

$y = x^2 + 2x - 1$

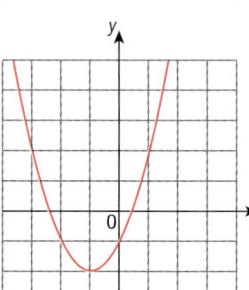

$y = 3x^2 - 4x + 2$

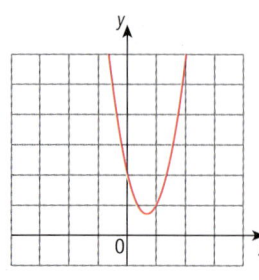

$y = -2x^2 + 2x + 3$

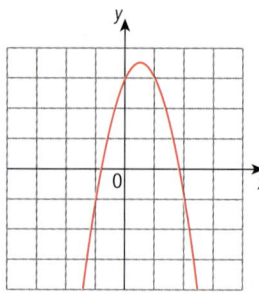

Each of these graphs has a curved shape known as a **parabola**. Each graph also has a minimum or a maximum point called a **vertex**. If the coefficient of x^2 is positive, the parabola will open upwards, with the vertex as the minimum point on the graph. If the coefficient of x^2 is negative, the parabola will open downwards, and the vertex will be a maximum point.

If you imagine a vertical line running through the vertex of a parabola, you will notice the graph is symmetrical on the left and right sides of this vertical line. This imaginary vertical line is called the **axis of symmetry**. This axis of symmetry is shown in red on this graph.

We will now look at different forms of quadratic functions. Consider the graphs of these quadratic functions in the form $y = ax^2 + bx + c$:

$y = x^2 + x - 3$

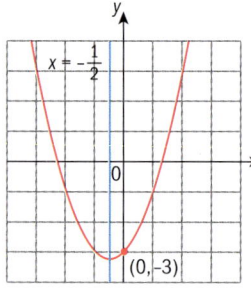

$y = -0.5x^2 - 2x + 4$

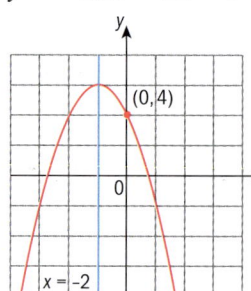

$y = x^2 - 3x + 1$

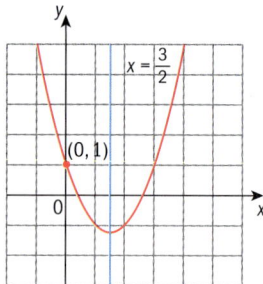

→ For quadratic functions in standard form $y = ax^2 + bx + c$, the graph crosses the y-axis at $(0, c)$.
The equation of the axis of symmetry is $x = \dfrac{-b}{2a}$.

→ When the basic quadratic function $y = x^2$ undergoes transformations, the resulting functions can be written as $y = a(x - h)^2 + k$.

> You might want to look back at the section about transformations of graphs in Chapter 1 of this book.

Look at the graphs of these quadratic functions in the form $y = a(x - h)^2 + k$:

$y = (x - 2)^2 - 1$ $y = 2(x + 1)^2 - 4$ $y = -(x - 3)^2 + 2$

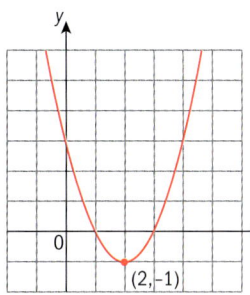   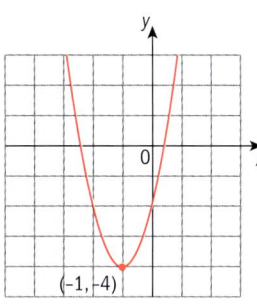   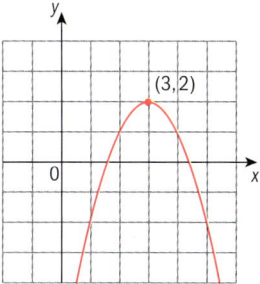

→ For quadratic functions in the form $y = a(x - h)^2 + k$, the graph has its vertex at (h, k).

> This form of a quadratic function is sometimes called 'turning-point form'.

Example 10

a Write the function $y = x^2 - 6x + 4$ in the form $y = (x - h)^2 + k$.
b Sketch the graph of the function, labeling the vertex and the y-intercept.

Answers

a $y = x^2 - 6x + 4$

$y = (x^2 - 6x + 9) + 4 - 9$
$y = (x - 3)^2 - 5$

By looking at the equation in standard form, you know the y-intercept will be (0, 4). Use 'completing the square' to rewrite the equation. By adding 9, then subtracting 9, the value of the right-hand side of the equation has not changed.

b

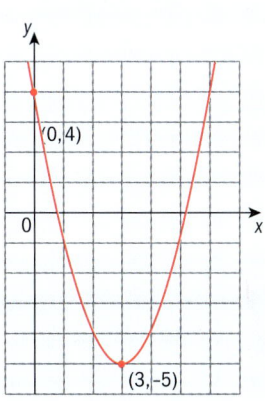

Note: The equation of the axis of symmetry is $x = 3$.

Chapter 2

Example 11

a Write the function $f(x) = 2x^2 + 8x + 11$ in the form $f(x) = a(x-h)^2 + k$.
b Sketch the graph of the function, labeling the vertex and the y-intercept.

Answers

a $f(x) = 2x^2 + 8x + 11$
$f(x) = 2(x^2 + 4x + 4) + 11 - 8$
$f(x) = 2(x + 2)^2 + 3$

b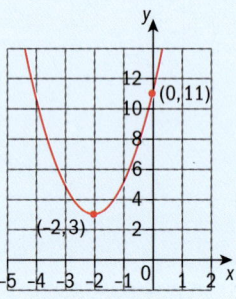

Note: The equation of the axis of symmetry is $x = -2$.

The y-intercept of the graph is (0, 11).

Be careful when completing the square if the x^2 term has a coefficient! Factor out the coefficient from the first two terms.
By adding 2×4, then subtracting 8, the value of the right-hand side of the equation has not changed.

> The name 'parabola' was introduced by Apollonius of Perga (Greek, c. 262 BCE–190 BCE) in his work on conic sections.

Exercise 2H

1 For each function, write the equation of the axis of symmetry and give the y-intercept of the graph of each function.
   a $f(x) = x^2 + 8x + 5$
   b $f(x) = x^2 - 6x - 3$
   c $f(x) = 5x^2 + 10x + 6$
   d $f(x) = -3x^2 + 10x + 9$

> You can find the vertex and the y-intercept points using your GDC. See Chapter 17 Section 1.8.

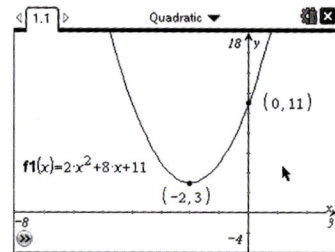

2 For each function, write the coordinates of the vertex and give the coordinates of the y-intercept of the graph.
   a $y = (x - 7)^2 - 2$
   b $y = (x + 5)^2 + 1$
   c $y = 4(x - 1)^2 + 6$
   d $y = 3(x + 2)^2 - 7$

> It may be helpful to substitute $x = 0$, or write the function in standard form to find the y-intercept.

3 Write each function in the form $f(x) = a(x - h)^2 + k$. Then sketch the graph of the function, labeling the vertex and the y-intercept.
   a $f(x) = x^2 + 10x - 6$
   b $f(x) = x^2 - 5x + 2$
   c $f(x) = 3x^2 - 6x + 7$
   d $f(x) = -2x^2 + 8x - 3$

We will now consider quadratic functions in the form $y = a(x - p)(x - q)$.
For obvious reasons, we sometimes call this the 'factorized form'.

Look at the graphs of these quadratic functions in the form
$y = a(x - p)(x - q)$:

$y = (x + 3)(x - 1)$ $\qquad$ $y = -3(x + 1)(x - 4)$ $\qquad$ $y = (x + 2)(x - 5)$

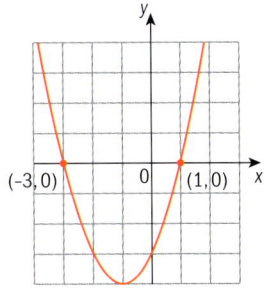 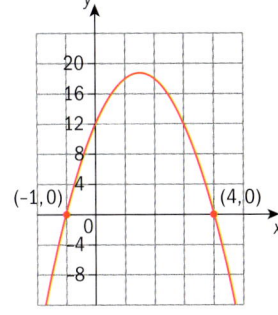 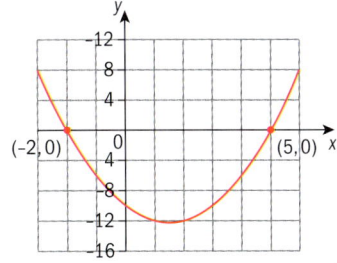

> → For quadratic functions in the form $y = a(x - p)(x - q)$, the graph crosses the x-axis at $(p, 0)$ and at $(q, 0)$.
> For quadratic functions in the form $y = a(x - p)(x - q)$, the axis of symmetry will have the equation $x = \dfrac{p+q}{2}$.

Note: The x-intercepts of the graph of a quadratic function $y = f(x)$ tell us the roots of the quadratic equation in the form $f(x) = 0$.
For example, in the first graph above, the function $y = (x + 3)(x - 1)$ crosses the x-axis at $(-3, 0)$ and at $(1, 0)$. The equation $(x + 3)(x - 1) = 0$ has roots $x = -3$ and $x = 1$.

Example 12

Write the function $f(x) = x^2 + 3x - 10$ in the form $f(x) = (x - p)(x - q)$.
Then sketch the graph of the function, labeling the x- and y-intercepts.

Answer
$f(x) = x^2 + 3x - 10$
$f(x) = (x + 5)(x - 2)$

The graph will cross the y-axis at $(0, -10)$.
Factorize the right-hand side of the equation.

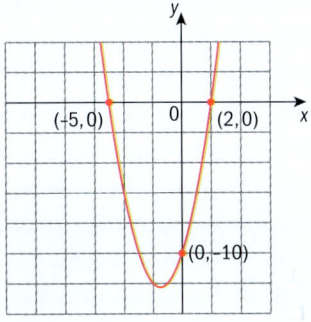

Note: The equation of the axis of symmetry is $x = \dfrac{(-5)+2}{2} = -\dfrac{3}{2}$.

use $x = \dfrac{p+q}{2}$

Chapter 2

Example 13

Write the function $y = 2x^2 - x - 3$ in the form $y = a(x - p)(x - q)$. Then sketch the graph of the function, labeling the x- and y-intercepts.

Answer

$y = 2x^2 - x - 3$

$y = (2x - 3)(x + 1)$

$y = 2(x - 1.5)(x + 1)$

The y-intercept of the graph will be $(0, -3)$.
Factorize the right-hand side of the equation.
Factor out the coefficient of x in the first factor.

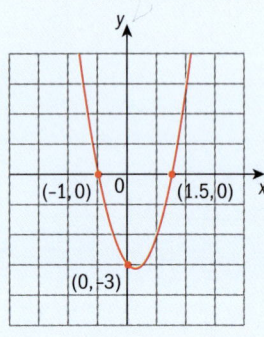

Note: The equation of the axis of symmetry is $x = \dfrac{1}{4}$.

Exercise 2I

1. Write the coordinates of the x- and y-intercepts of the graph of each function.
   a $f(x) = (x + 3)(x - 7)$
   b $f(x) = 2(x - 4)(x - 5)$
   c $f(x) = -3(x + 2)(x + 1)$
   d $f(x) = 5(x + 6)(x - 2)$

   It may be helpful to substitute x = 0, or write the function in standard form to find the y-intercept.

2. Write each function in the form $y = a(x - p)(x - q)$. Then sketch the graph of the function, labeling the x- and y-intercepts.
   a $y = x^2 - 7x - 8$
   b $y = x^2 - 8x + 15$
   c $y = -2x^2 + 3x + 5$
   d $y = 5x^2 + 6x - 8$

3. Write each function in the form $y = a(x - h)^2 + k$ and in the form $y = a(x - p)(x - q)$. Then make a neat sketch of the graph of the function, labeling the vertex and the x- and y-intercepts.
   a $y = x^2 + 6x - 16$
   b $y = -x^2 - 4x + 21$
   c $y = -0.5x^2 + 3.5x - 3$
   d $y = 4x^2 - 18x + 8$

EXAM-STYLE QUESTION

4. Let $f(x) = 2x^2 - 12x$. Part of the graph of f is shown.
   a The graph crosses the x-axis at A and B. Find the x-coordinate of
      i A
      ii B.
   b Write down the equation of the axis of symmetry.
   c The vertex of the graph is at C. Find the coordinates of C.

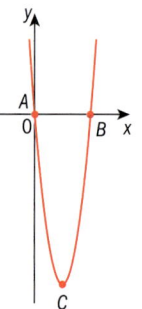

EXAM-STYLE QUESTION

5 Let $f(x) = x^2 + 3$, and let $g(x) = x - 2$.
  a Find $(f \circ g)(x)$.
  b Write down the coordinates of the vertex of the graph of $(f \circ g)$.
  The graph of the function h is formed by translating the graph of $(f \circ g)$ by 5 units in the positive x-direction, and by 2 units in the negative y-direction.
  c Write the equation of the function $h(x)$ in the form $h(x) = ax^2 + bx + c$.
  d Hence, write down the y-intercept of the graph of h.

Finding the equation of a quadratic function from a graph

You can tell a lot about the graph of a function by looking at the equation of the function in different forms.

> - When the equation is written in standard form $f(x) = ax^2 + bx + c$, you know the y-intercept of the graph is $(0, c)$, and the equation of the axis of symmetry is $x = \frac{-b}{2a}$.
> - When the equation is in the form $f(x) = a(x - h)^2 + k$, also known as turning-point form, the vertex will be (h, k).
> - When the equation is written in factorized form $f(x) = a(x - p)(x - q)$, the graph will cross the x-axis at $(p, 0)$ and at $(q, 0)$.

Now you will look at how you can find the equation of a quadratic function from the information given in its graph.

If you know the x-intercepts, begin with the equation in factorized form.

If you are given the vertex, you can start with the equation in turning-point form.

Example 14

Using the information provided in the graph, write the equation of the quadratic function. Write your final answer in standard form $y = ax^2 + bx + c$.

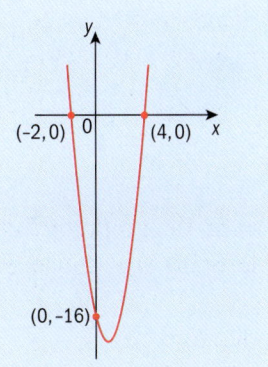

▶ Continued on next page

Answer

$y = a(x + 2)(x - 4)$

$-16 = a(0 + 2)(0 - 4)$
$-8a = -16$
$\quad a = 2$

$y = 2(x + 2)(x - 4)$
$y = 2x^2 - 4x - 16$

Since the x-intercepts are given, start with the equation in factorized form. You know that $y = -16$ when $x = 0$. Substitute these values into your equation to solve for a.

You can check this answer by graphing the equation on your GDC, and comparing the x- and y-intercepts to those in the given graph.

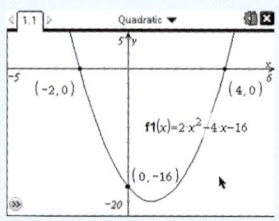

GDC help on CD: *Alternative screenshots for the TI-84 Plus and Casio FX-9860GII GDCs are on the CD.*

Example 15

Write the equation of the quadratic function shown in the graph.
Write your final answer in standard form $y = ax^2 + bx + c$.

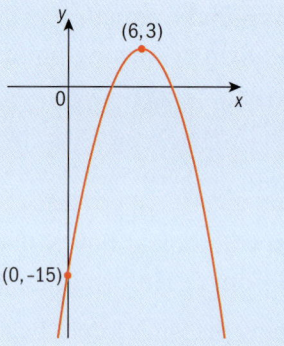

Answer

$y = a(x - 6)^2 + 3$

$-15 = a(0 - 6)^2 + 3$
$36a + 3 = -15$

$36a = -18$

$a = -\dfrac{1}{2}$

$y = -\dfrac{1}{2}(x - 6)^2 + 3$

$y = -\dfrac{1}{2}x^2 + 6x - 15$

Since the vertex is given, start with the equation in turning-point form.

You know that $y = -15$ when $x = 0$. Substitute these values into your equation to solve for a.
You can check this answer by graphing the equation on your GDC, and checking the vertex and the y-intercept.

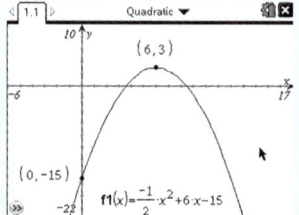

GDC help on CD: *Alternative screenshots for the TI-84 Plus and Casio FX-9860GII GDCs are on the CD.*

Finally, let's look at what happens if you don't know the vertex or the axial intercepts of the graph. This next example also leads to three equations in three variables to solve using a GDC.

Example 16

Write the equation of the quadratic function shown in the graph.

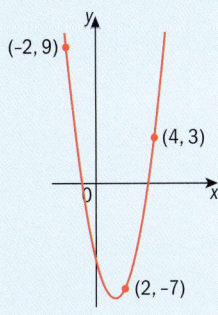

Answer

For the point $(-2, 9)$,
$9 = a(-2)^2 + b(-2) + c$
$9 = 4a - 2b + c$

For the point $(2, -7)$,
$-7 = a(2)^2 + b(2) + c$
$-7 = 4a + 2b + c$

For the point $(4, 3)$,
$3 = a(4)^2 + b(4) + c$
$3 = 16a + 4b + c$

In this case, you are given the coordinates of three points on the graph of the function.

Substitute the x- and y-coordinates of these three points into the standard-form quadratic equation $y = ax^2 + bx + c$.

You now have three equations with three variables. You can use your GDC to solve for a, b and c.

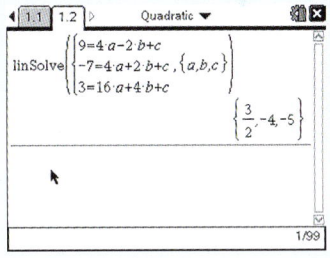

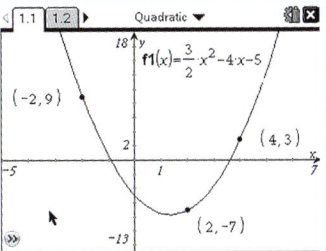

GDC help on CD: Help with solving simultaneous equations on the TI-84 Plus and Casio FX-9860GII GDCs is given on the CD.

To find these points on the graph, see Chapter 17 Section 1.5.
If you graph this function on your GDC, you will see that it passes through all three points, as described.

Using the GDC, $a = 1.5$, $b = -4$, and $c = -5$.
$y = 1.5x^2 - 4x - 5$

Chapter 2

Exercise 2J

Use the information provided in the graphs to write the equation of each function in standard form $y = ax^2 + bx + c$.

1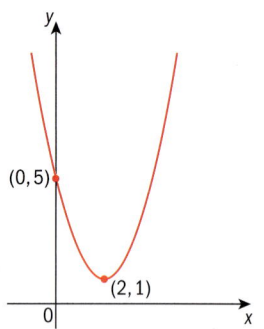

2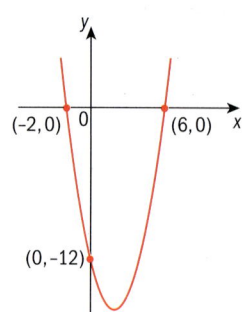

3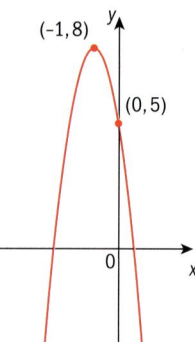

4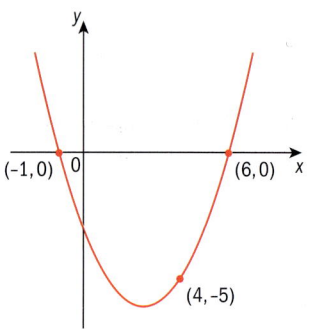

5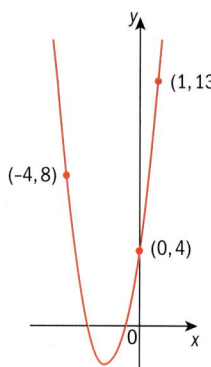

6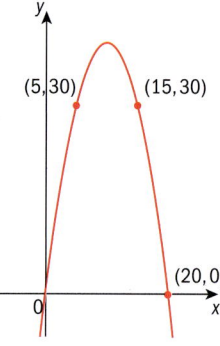

7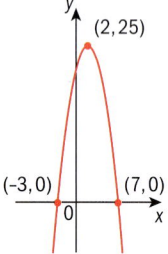

8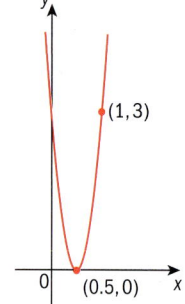

2.5 Applications of quadratics

At the beginning of this chapter, you saw that the shape formed by water in a fountain can be modeled by a quadratic function. Quadratic functions and their graphs can be used to model many different situations.

When solving problems using quadratics, you can use the methods you learned throughout this chapter. You will also be expected to use your GDC to help you answer many questions.

Example 17

A farmer wishes to enclose a rectangular garden with 100 metres of fencing.
a If the garden is x metres wide, find the length and the area of the garden in terms of x.
b Find the width of a garden with an area of 525 m².
c Find the maximum area the garden can have.

Answers

a

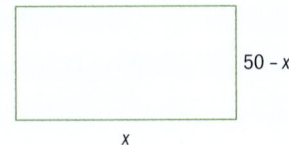

length = $50 - x$
area = $x(50 - x)$

If the farmer has 100 m of fencing, the perimeter of the rectangle must be 100. The sum of the length and width will therefore be 50 m.

Area = width × length

b $x(50 - x) = 525$
$50x - x^2 = 525$
$x^2 - 50x + 525 = 0$
$(x - 15)(x - 35) = 0$

Set the area equal to 525.
Write as a quadratic equation in standard form, and solve for x.

You could also solve this equation by completing the square, or by using the quadratic formula, or by using your GDC.

$x = 15$ m *or* 35 m

If the width is 15, the length is 35.
If the width is 35, the length is 15.

c

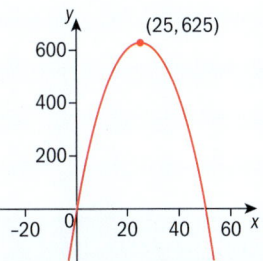

The maximum area is 625 m².

The easiest way to find the maximum area is to graph the function $y = x(50 - x)$, where y is the area and x is the width. You can do this on your GDC. See Chapter 17 Section 1.6.
The vertex (25, 625) is the highest point on the graph, and tells you the maximum area occurs when the width of the garden is 25 metres.

Example 18

The height of a ball t seconds after it is thrown is modeled by the function $h = 24t - 4.9t^2 + 1$, where h is the height of the ball in metres.
a Find the maximum height reached by the ball.
b For what length of time will the ball be higher than 20 metres?

Answers

a

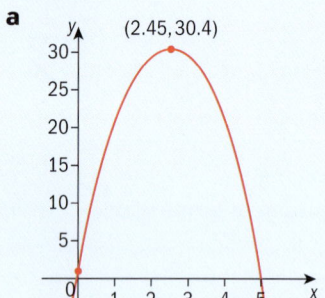

The maximum height is 30.4 metres.

Graph the function $y = 24x - 4.9x^2 + 1$, where y is the height of the ball and x is the time in seconds.
The vertex is approximately (2.45, 30.4). This tells you the maximum height occurs when the ball has been in the air for 2.45 seconds. You can find the vertex using your GDC. See Chapter 17 Section 1.8.

b $20 = 24t - 4.9t^2 + 1$
$4.9t^2 - 24t + 19 = 0$

$t \approx 0.9930$ seconds and 3.905 seconds
$3.905 - 0.9930 = 2.912$
The ball will be higher than 20 metres for about 2.91 seconds.

Let $h = 20$.
Write as a quadratic equation in standard form, and solve for t.
You can solve this using your GDC. See Chapter 17, Section 1.7.
The ball is at a height of 20 metres twice, once on the way up, and once on the way down.

What other kinds of real-life situations might be modeled by quadratic functions?

Example 19

It takes Luisa 3 hours to ride her bicycle up a hill and back down. Her average speed riding down the hill is 35 km h^{-1} faster than her average speed riding up the hill. If the distance from the bottom to the top of the hill is 40 km, find Luisa's average uphill and downhill speeds.

Answer

Let x represent Luisa's uphill riding speed.
$$\frac{40}{x} + \frac{40}{x+35} = 3$$

Remember time $= \frac{distance}{speed}$, and when you add the uphill and downhill times, the total is 3 hours.

▶ Continued on next page

Quadratic functions and equations

$40 + \dfrac{40x}{x+35} = 3x$

$40x + 1400 + 40x = 3x^2 + 105x$

$3x^2 + 25x - 1400 = 0$

$x \approx 17.8$ km h^{-1}

Luisa averages 17.8 km h^{-1} riding uphill, and 52.8 km h^{-1} riding downhill.

You can multiply through by x, and then by (x + 35), to get rid of the denominators.
Write as a quadratic equation in standard form and solve for x using your GDC. See Chapter 17, Section 1.7.

Exercise 2K

1. The height of a ball t seconds after it is thrown is modeled by the function $h = 15t - 4.9t^2 + 3$, where h is the height of the ball in metres.
   a. Find the maximum height reached by the ball.
   b. For what length of time will the ball be higher than 12 metres?

2. The area, A cm², of a rectangular picture is given by the formula $A = 32x - x^2$, where x is the width of the picture in centimetres. Find the dimensions of the picture if the area is 252 cm².

3. A piece of wire 40 cm long is cut into two pieces. The two pieces are formed into two squares.
   a. If the side length of one of the squares is x cm, what is the side length of the other square?
   b. Show that the combined area of the two squares is given by $A = 2x^2 - 20x + 100$.
   c. What is the minimum combined area of the two squares?

4. A rectangular portrait measures 50 cm by 70 cm. It is surrounded by a rectangular frame of uniform width. If the area of the frame is the same as the area of the portrait, what is the approximate width of the frame?

5. The length of a rectangle is five less than three times its width. Find the dimensions of the rectangle if its area is 782 m².

6. The sum of the squares of three consecutive positive odd integers is 251. Find the integers.

7 A 'golden rectangle' has the property that if it is divided into a square and a smaller rectangle, the smaller rectangle will be similar in proportion to the original rectangle. In the golden rectangle *ABCD* below, *PQ* forms a square *APQD* and a rectangle *PBCQ*, as shown.

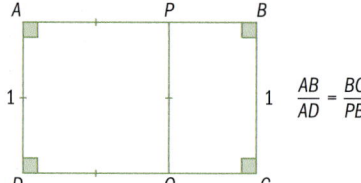

$\dfrac{AB}{AD} = \dfrac{BC}{PB}$

> The length-to-width ratio of a golden rectangle is known as the **golden ratio**. You may want to investigate other situations in which this interesting ratio appears.

Given that *AD* = 1, find *AB*.

8 A homebuilder wants to build a rectangular deck on the back of a house. One side of the deck will share a wall with the house, and the other three sides will have a wooden railing. If the builder has enough wood for 15 metres of railing, what is the area of the largest deck he could build?

9 Jaswinder takes a trip to visit his sister, who lives 500 km away. He travels 360 km by bus, and 140 km by train. The train averages 10 km h^{-1} faster than the bus. If the entire journey takes 8 hours, find the average speeds of the bus and the train.

10 Working alone, John takes two more hours to clean the house than Jane does. If they work together, John and Jane can clean the house in 2 hours 24 minutes. How long does it take John to clean the house if he is working alone?

Review exercise

1 Solve each equation.
   a $(x + 2)^2 = 16$
   b $x^2 - 16x + 64 = 0$
   c $3x^2 + 4x - 7 = 0$
   d $x^2 - 7x + 12 = 0$
   e $x^2 + 2x - 12 = 0$
   f $3x^2 - 7x + 3 = 0$

EXAM-STYLE QUESTION
2 Let $f(x) = x^2 + 3x - 4$. Part of the graph of f is shown.
   a Write down the *y*-intercept of the graph of *f*.
   b Find the *x*-intercepts of the graph.
   c Write down the equation of the axis of symmetry.
   d Write down the *x*-coordinate of the vertex of the graph.

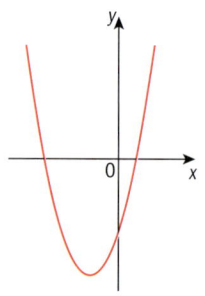

EXAM-STYLE QUESTIONS

3 Let $f(x) = a(x-p)(x-q)$. Part of the graph of f is shown.
   The graph passes through the points $(-5, 0)$, $(1, 0)$ and $(0, 10)$.
   a Write down the value of p and of q.
   b Find the value of a.

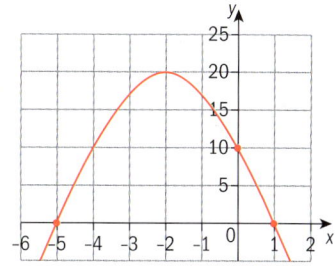

4 Let $f(x) = a(x + 3)^2 - 6$
   a Write down the coordinates of the vertex of the graph of f.
   b Given that $f(1) = 2$, find the value of a.
   c Hence find the value of $f(3)$.

> Quadratic functions are closely related to other functions called 'conic sections' (see page 60). How are these functions used in the real world?

5 The equation $x^2 + 2kx + 3 = 0$ has two equal real roots. Find the possible values of k.

6 Let $f(x) = 2x^2 + 12x + 5$.
   a Write the function f, giving your answer in the form $f(x) = a(x-h)^2 + k$.
   b The graph of g is formed by translating the graph of f by 4 units in the positive x-direction and 8 units in the positive y-direction. Find the coordinates of the vertex of the graph of g.

7 Write the equation of the quadratic function shown in the graph.
   Give your answer in the form $y = ax^2 + bx + c$.

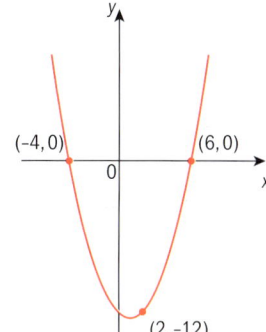

Review exercise

1 Solve each equation, giving your answers to 3 significant figures.
   a $3x^2 - 5x - 7 = 0$ b $2x^2 + 8x = 3$
   c $\dfrac{x}{x+3} = 2x - 1$ d $\dfrac{1}{x} + \dfrac{1}{x+2} = 5$

EXAM-STYLE QUESTION

2 The height, h metres above the water, of a stone thrown off a bridge is modeled by the function $h(t) = 15t + 20 - 4.9t^2$, where t is the time in seconds after the stone is thrown.
   a What is the initial height from which the stone is thrown?
   b What is the maximum height reached by the stone?
   c For what length of time is the height of the stone greater than 20 m?
   d How long does it take for the stone to hit the water below the bridge?

3 The length of a rectangle is 5 cm more than 3 times its width. The area of the rectangle is 1428 cm². Find the length and width of the rectangle.

EXAM-STYLE QUESTION

4 The function f is given by $f(x) = ax^2 + bx + c$. Part of the graph of f is shown.
The graph of f passes through the points $P(-10, 12)$, $Q(-5, -3)$ and $R(5, 27)$.
Find the values of a, b and c.

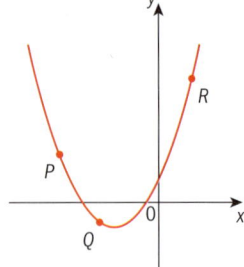

5 Thomas drives his car 120 km to work. If he could increase his average speed by 20 km h^{-1}, he would make it to work 30 minutes faster. What is his average driving speed?

CHAPTER 2 SUMMARY

Solving quadratic equations

- If $xy = 0$, then $x = 0$ or $y = 0$.
  This property is sometimes called the **zero product property**.
- This property can be expanded to:
  If $(x - a)(x - b) = 0$, then $x - a = 0$ or $x - b = 0$.
- To solve an equation by completing the square, take half the coefficient of x, square it, and add the result to both sides of the equation. This step creates a perfect square trinomial on the left side of the equation.
- In order to complete the square, the coefficient of the x^2 term must be 1. If the x^2 term has a coefficient other than 1, you can factor out the coefficient, or divide through by the coefficient.

The quadratic formula

- For any equation in the form $ax^2 + bx + c = 0$,
  $$x = \frac{-b \pm \sqrt{b^2 - 4ac}}{2a}$$

> This formula is given in the IB Formula booklet so you do not need to memorize it!

Roots of quadratic equations

- For a quadratic equation $ax^2 + bx + c = 0$,
  - if $b^2 - 4ac > 0$, the equation will have two different real roots
  - if $b^2 - 4ac = 0$, the equation will have two equal real roots
  - if $b^2 - 4ac < 0$, the equation will have no real roots.

> You can think of an equation with two equal roots as having only one solution.

Continued on next page

Graphs of quadratic equations

- For quadratic functions in standard form $y = ax^2 + bx + c = 0$, the graph will cross the y-axis at $(0, c)$.
- The equation of the axis of symmetry is $x = \dfrac{-b}{2a}$.
- When the basic quadratic function $y = x^2$ undergoes transformations, the resulting functions can be written as $y = a(x - h)^2 + k$.
- For quadratic functions in the form $y = a(x - h)^2 + k$, the graph will have its vertex at (h, k).
- For quadratic functions in the form $y = a(x - p)(x - q)$, the graph crosses the x-axis at $(p, 0)$ and at $(q, 0)$.
  For quadratic functions in the form $y = a(x - p)(x - q)$, the axis of symmetry will have the equation $x = \dfrac{p+q}{2}$.
- When the equation is in the form $f(x) = a(x - h)^2 + k$, also known as turning-point form, the vertex will be (h, k).
- When the equation is written in factorized form $f(x) = a(x - p)(x - q)$, the graph will cross the x-axis at $(p, 0)$ and at $(q, 0)$.

Theory of knowledge

Conic sections: mathematical shapes in the real world

The graph of a quadratic function is in the shape of a parabola. We also see parabolas in the real world – the path of a baseball flying through the air, or the shape of water streaming from a fountain.

Parabolas are just one of the four mathematical shapes known as **conic sections**. These conic sections are formed by the intersection of a cone (or two cones) and a plane. The other conic sections are the circle, the ellipse and the hyperbola.

▶ A parabola is the shape that results when a plane intersects a cone parallel to one of its slanted edges.

The ancient Greeks studied conic sections, and Apollonius of Perga (c.262 – 190 BCE) first named them.

Hypatia (born between 350 and 370, died 415 CE) was a mathematician and astronomer and head of the Platonist School in Alexandria, Egypt, at a time when few women received an education. She developed Apollonius' work on conic sections.

They were further studied by the Persian mathematician and poet Omar Khayyám (c.1048–1131).

Mathematical equations can be used to describe each of these shapes.

Parabola: $y = ax^2 + bx + c$

Circle: $(x - h)^2 + (y - k)^2 = r^2$

Ellipse: $\dfrac{(x - h)^2}{a^2} + \dfrac{(y - k)^2}{b^2} = 1$

Hyperbola: $\dfrac{(x - h)^2}{a^2} - \dfrac{(y - k)^2}{b^2} = 1$

Many would consider a circle to be the most 'perfect' of these conic sections. It is certainly the most well-known, and we see circles around us every day.

- Why is a circle 'perfect'?

- Did you know that the orbits of planets are elliptical shapes?

  This wasn't shown until the early 17th century. Apollonius had hypothesized that planets had such orbits when he was studying and naming conic sections.

- How do you think this knowledge developed over time?

  Nowadays, we see ellipses, hyperbolas and parabolas in suspension bridges, the paths of spacecraft and other bodies in space, and the shape of satellite dishes.

  Who knew that slicing a cone could result in such useful mathematical equations, and would give us shapes that can help us understand the universe?

- Look around you – what other figures and shapes do you see that might be modeled by mathematical equations?

- Why do you think people try to use mathematics to describe shapes and patterns in the world around us?

- How can using mathematics help us understand our world and our universe?

3 Probability

CHAPTER OBJECTIVES:

5.5a Concepts of trial, outcome, equally likely outcomes, sample space (*U*) and event. The probability of an event *A* is $P(A) = \frac{n(A)}{n(U)}$. The complementary events *A* and *A'* (not *A*). The use of Venn diagrams, tree diagrams and tables of outcomes.

5.6 Combined events, the formula for $P(A \cup B)$. Mutually exclusive events: $P(A \cap B) = 0$. Conditional probability; the definition $P(A|B) = \frac{P(A \cap B)}{P(B)}$. Independent events; the definition $P(A|B) = P(A) = P(A|B')$. Probabilities with and without replacement.

Before you start

You should know how to:

1 Add, subtract, multiply and divide fractions
$$\frac{2}{3} + \frac{1}{5} = \frac{10}{15} + \frac{3}{15} = \frac{13}{15}$$
$$1 - \frac{2}{9} = \frac{9}{9} - \frac{2}{9} = \frac{7}{9}$$
$$\frac{3}{4} \times \frac{3}{5} = \frac{3 \times 3}{4 \times 5} = \frac{9}{20}$$
$$\frac{4}{7} \div \frac{3}{7} = \frac{4}{7} \times \frac{7}{3} = \frac{4}{3} = 1\frac{1}{3}$$

2 Add, subtract and multiply decimals

$$\begin{array}{r} 0.2 \\ +0.7 \\ \hline 0.9 \end{array} \qquad \begin{array}{r} 0.35 \\ +0.4 \\ \hline 0.75 \end{array} \qquad \begin{array}{r} 1.\overset{0}{\cancel{0}}\overset{9}{\cancel{0}}\overset{1}{\cancel{0}} \\ -0.62 \\ \hline 0.38 \end{array}$$

0.2×0.34
Since $2 \times 34 = 68$
then $0.2 \times 0.34 = 0.068$

3 Calculate percentages
52% of 60 = 0.52 × 60 = 31.2

Skills check

1 Work these out without a calculator.
   a $1 - \frac{3}{7}$ b $\frac{2}{5} + \frac{5}{7}$ c $\frac{1}{5} \times \frac{2}{3}$

   d $1 - \left(\frac{1}{3} \times \frac{5}{9}\right)$ e $\dfrac{\frac{3}{20}}{\frac{7}{20}}$

2 Work these out.
   a $1 - 0.375$ b $0.65 + 0.05$
   c 0.7×0.6 d 0.25×0.64
   e 50% of 30   f 22% of 0.22
   g 12% of 10% of 0.8

3 Check your answers to questions **1** and **2** using a calculator.

- What is the chance of it raining tomorrow?
- What is the likelihood that I will pass my test?
- What is the probability of us winning the football match this afternoon?
- Am I certain to get to school on time if I catch the bus rather than the train?

We consider questions like these all the time. We use the words 'chance', 'likelihood', 'probability' and 'certain' in everyday speech. These words also describe mathematical probability. This important branch of mathematics helps us to understand risk, and everything from sporting averages to the weather report and the chance of being struck by lightning.

This chapter looks at the language of probability, how to quantify probability (give it a numerical value) and the basic tools you need to solve problems involving probability.

▲ According to the US government's National Weather Service, the probability of being struck by lightning in a given year is $\frac{1}{750\,000}$. The probability of being struck by lightning in an 80-year lifetime is $\frac{1}{6250}$. These probabilities have been estimated from data on size of population and number of people struck by lightning in the past 30 years.

Investigation – rolling dice

During the mid-1600s, mathematicians Blaise Pascal, Pierre de Fermat and Antoine Gombaud puzzled over this simple gambling problem:

Which is more likely, rolling a 'six' on four throws of one dice, or rolling a 'double six' on 24 throws with two dice?

Which option do you **think** is more likely? Why?

3.1 Definitions

→ An **event** is an outcome from an experiment.
An **experiment** is the process by which we obtain an outcome.
A **random experiment** is one where there is uncertainty over which event may occur.

Some examples of random experiments are:

- rolling a dice three times
- tossing a coin once
- picking two cards from a pack of 52 playing cards
- recording the number of cars that pass the school gate in a 5-minute period.

We can express the chance of an event occurring using a number between 0 and 1. On this scale, 0 represents an impossible event and 1 represents an event that is certain to happen. This is called the **probability** that the event will happen.

impossible	even chance	certain
0	$\frac{1}{2}$	1

We write $P(A)$ to represent the probability of an event A occurring. Hence $0 \leq P(A) \leq 1$.

There are three ways of finding the value of the probability of an event:

- theoretical probability
- experimental probability
- subjective probability.

Theoretical probability

A fair dice has six numbered sides, all of which are equally likely to occur. The list of equally likely possible outcomes is 1, 2, 3, 4, 5, 6.

The first book written on probability, *The Book of Chance and Games*, was written by Jerome Cardan (1501–75). Cardan was an Italian astrologer, philosopher, physician, mathematician and gambler. His book contained techniques on how to cheat and how to catch others at cheating.

A probability cannot be greater than 1.

You can write probability as a decimal, fraction or percentage.

On a fair (unbiased) dice the probability of each outcome is the same. On a biased dice, some outcomes are more likely than others.

We call a list of possible outcomes the **sample space**, U. The notation $n(U) = 6$ shows that there are six members of the sample space.

Let event A be defined as 'the number 6'. In this sample space there is one 6. $n(A) = 1$ shows that there is one 6 in the sample space. The probability of getting a 6 when you roll the dice is one out of six, or $\frac{1}{6}$. In probability notation,

$$P(A) = \frac{n(A)}{n(U)} = \frac{1}{6}.$$

→ The theoretical probability of an event A is $P(A) = \frac{n(A)}{n(U)}$ where $n(A)$ is the number of ways that event A can occur and $n(U)$ is the total number of possible outcomes.

→ If the probability of an event is P, in n trials you would expect the event to occur $n \times P$ times.

Example 1

A fair 20-sided dice with faces numbered 1 to 20 is rolled. The event A is defined as 'the number obtained is a multiple of 4'.
a Determine $P(A)$.
The dice is rolled 100 times.
b How many times would you expect a multiple of 4?

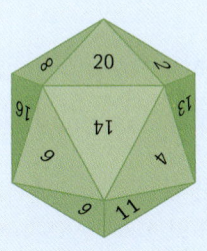

Answers

a $n(A) = 5$ and $n(U) = 20$ — *Find $n(A)$*
$P(A) = \frac{n(A)}{n(U)} = \frac{5}{20} = \frac{1}{4}$ — *There are 20 possible outcomes. 5 of these are multiples of 4 (4, 8, 12, 16 and 20).*

b $\frac{1}{4} \times 100 = 25$ — *Probability × number of trials*

A 20-sided polyhedron is called an 'icosahedron'.

Processes that are too complicated to allow exact analysis may be solved using probability methods that employ the 'law of large numbers'. These methods, developed in the 1930s and 40s, are known as **Monte Carlo methods** after the famous casino. They are used in a wide variety of situations, from estimating the strength of a hand in the card game 'Bridge' to modeling the statistics of a nuclear chain reaction. You may wish to explore the applications of the Monte Carlo method further.

Experimental (empirical) probability

Sometimes outcomes are not equally likely but you can use an experiment to estimate probabilities.

For example, to find the probability that a particular component that is being produced by a factory is faulty we would test some components. If the first component we test is faulty we could conclude that all of the components are faulty. However, this may not be the case. If the second component is not faulty then we could conclude that the probability of a component being faulty is $\frac{1}{2}$ since half of all components so far are faulty.

Continuing this process a number of times and calculating the ratio

$$\frac{\text{the number of faulty components}}{\text{the number of components tested}}$$

gives the **relative frequency** of the component being faulty.

As the number of components tested increases, the relative frequency gets closer and closer to the probability that a component is faulty.

> → You can use relative frequency as an estimate of probability. The larger the number of trials, the closer the relative frequency is to the probability.

The US National Weather Service used this method to find the probability of being struck by lightning, using

$$\frac{\text{No. of people struck}}{\text{No. of people in population}}$$

Example 2

The colors of cars passing the school gate one morning are given in the table:

Color	Frequency
Red	45
Black	16
Yellow	2
Green	14
Blue	17
Gray	23
Other	21
Total	138

a Estimate the probability that the next car to pass the school gates will be red.

b The next morning 350 cars pass the school gates. Estimate the number of red cars that morning.

Answers

a The relative frequency of red cars is $\frac{45}{138}$.

So the probability of a red car is $\frac{45}{138}$.

b When 350 cars pass the school gate, the number of red cars will be approximately $\frac{45}{138} \times 350 = 114$.

These numbers are estimates, because we are using the relative frequency as an estimate of the probability.

This probability is given as a fraction. In IB exams you need to give exact answers or decimals to 3 sf for probabilities.

Subjective probability

You can't always repeat an experiment a large number of times. In these cases you can estimate the probability of an event based on subjective judgment, experience, information and belief.

For example Liverpool are due to play soccer against Arsenal in the English Premiership. What is the probability that Liverpool will win? You could look at past matches between the two teams as well as the last few games each team has played and how the teams have performed in the particular weather conditions the match will be played in, but eventually you will need to make a 'guess'.

Exercise 3A

1. An octahedral (eight-sided) dice is thrown. The faces are numbered 1 to 8. What is the probability that the number thrown is:
   a an even number
   b a multiple of 3
   c a multiple of 4
   d not a multiple of 4
   e less than 4?

   > In probability questions, all dice and coins are 'fair', unless you are told otherwise.

2. A used car dealer has 150 used cars on his lot. The dealer knows that 30 of the cars are defective. One of the 150 cars is selected at random. What is the probability that it is defective?

   > 'At random' means that any car has an equal chance of being selected. One of the 30 defective cars is as likely to be chosen as one of the cars that is not defective.

3. The table below shows the relative frequencies of the ages of the students at a high school.

Age (in years)	Relative frequency
13	0.15
14	0.31
15	0.21
16	0.19
17	0.14
Total	1

   a A student is randomly selected from this school. Find the probability that
      i   the student is 15 years old,
      ii the student is 16 years of age or older.

   There are 1200 students at this school.
   b Calculate the number of 15-year-old students.

4. The sides of a six-sided spinner are numbered from 1 to 6. The table shows the results for 100 spins.

Number on spinner	1	2	3	4	5	6
Frequency	27	18	17	15	16	7

   a What is the relative frequency of getting a 1?
   b Do you think the spinner is fair? Give a reason for your answer.
   c The spinner is spun 3000 times. Estimate the number of times the result will be a 4.

5. Each letter of the word CONSECUTIVE is written on a separate card. The 11 cards are placed face downwards. A card is drawn at random.
   What is the probability of picking a card with
   a the letter C   b the letter P   c a vowel?

6 The spinner shown is biased. The probabilities of getting red and getting blue are shown in the table. The probability of getting green is twice that of getting yellow.

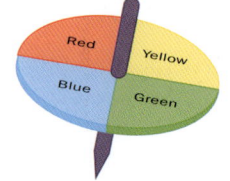

Color	red	yellow	blue	green
Frequency	0.4		0.3	

Find the probability of getting green.

7 A bag contains 40 discs numbered 1 to 40. A disc is selected at random. Find the probability that the number on the disc
   a is an even number,   b has the digit 1 in it.

3.2 Venn diagrams

There are 100 students in a year group.
38 of them do archery.
This information can be shown on a **Venn diagram**.

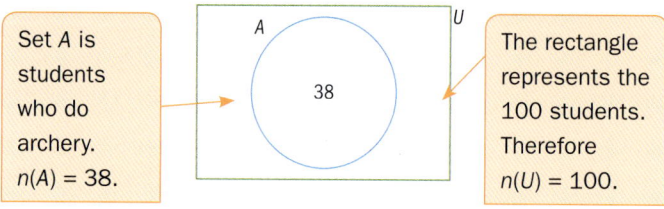

Set A is students who do archery. $n(A) = 38$.

The rectangle represents the 100 students. Therefore $n(U) = 100$.

John Venn was born in Hull, England in 1834. His father and grandfather were priests and John was also encouraged to follow in their footsteps. In 1853 he went to Gonville and Caius College, Cambridge, and graduated in 1857 becoming a fellow of the college. For the next five years he went into the priesthood and returned to Cambridge in 1862 to teach logic and probability theory.
John Venn developed a graphical way to look at sets. This graph became known as a Venn diagram.

A student is chosen at random. The probability that the student does archery is written $P(A)$.

$$P(A) = \frac{38}{100} = \frac{19}{50}$$

Remember $P(A) = \frac{n(A)}{n(U)}$.

Complementary event A'

The area outside set A (but still in the sample space U) represents the students that do not do archery. This is A', the **complement** of set A.

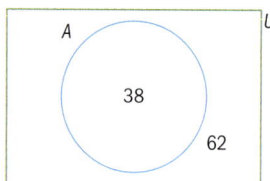

$n(A') = n(U) - n(A)$
From the Venn diagram we see that $n(A') = 100 - 38 = 62$
The probability that a student does not do archery,

$$P(A') = \frac{n(A')}{n(U)} = \frac{62}{100} = \frac{31}{50}$$

Note that

$$P(A') + P(A) = \frac{31}{50} + \frac{19}{50} = 1$$

Every student either does archery or doesn't do archery.

→ As an event, A, either happens or it does not happen.
$P(A) + P(A') = 1$
$P(A') = 1 - P(A)$

Intersection of events

Of the 100 students, 30 students play badminton. Of those, 16 do both archery and badminton.

You can show this on a Venn diagram like this:

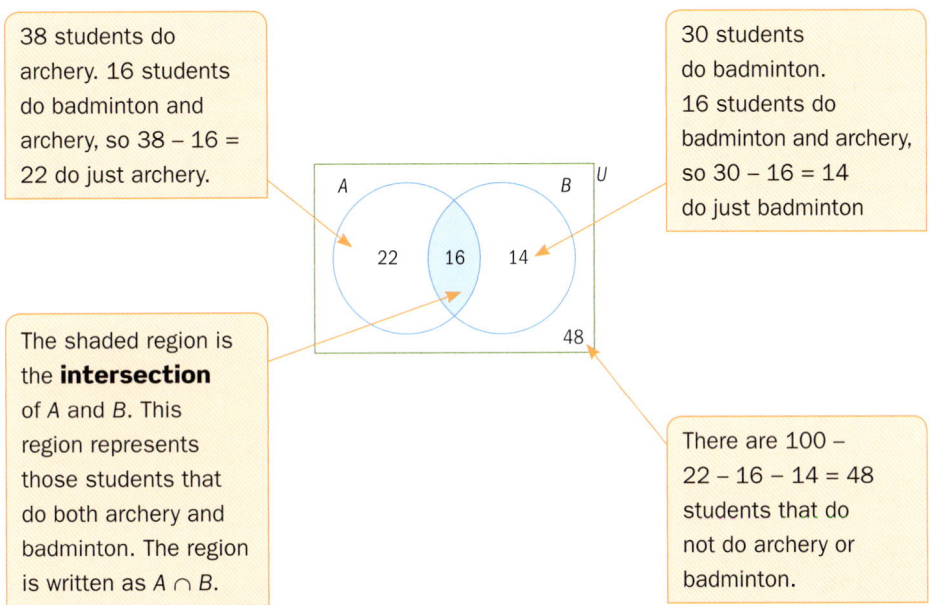

38 students do archery. 16 students do badminton and archery, so 38 − 16 = 22 do just archery.

30 students do badminton. 16 students do badminton and archery, so 30 − 16 = 14 do just badminton

The shaded region is the **intersection** of A and B. This region represents those students that do both archery and badminton. The region is written as $A \cap B$.

There are 100 − 22 − 16 − 14 = 48 students that do not do archery or badminton.

The probability that a student chosen at random does both archery and badminton is written $P(A \cap B)$.

$n(A \cap B) = 16$

$n(A \cap B)$ is the number in the intersection of the sets A and B.

$P(A \cap B) = \dfrac{n(A \cap B)}{n(U)} = \dfrac{16}{100} = \dfrac{4}{25}$

If a student is chosen at random then the probability that a student does not do badminton but does do archery is written $P(A \cap B')$.

22 students out of 100 do archery but not badminton.

$P(A \cap B') = \dfrac{22}{100} = \dfrac{11}{50}$

$A' \cap B'$ represents the students who do not do badminton or archery.

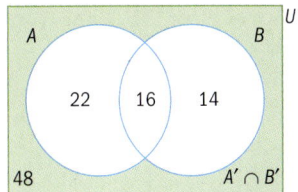

Chapter 3

Union of events

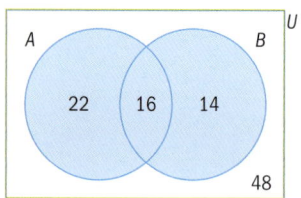

The shaded region is the **union** of A and B, The region represents those students that do either archery or badminton or both. The region is written $A \cup B$.

The probability that a student chosen at random does either archery or badminton is written $P(A \cup B)$.

Notice that 'or' in mathematics includes the possibility of both – we call it the 'inclusive' or.

From the diagram, $n(A \cup B) = 22 + 16 + 14 = 52$ and hence

This is from the definition of probability.

$$P(A \cup B) = \frac{n(A \cup B)}{n(U)} = \frac{52}{100} = \frac{13}{25}$$

$A \cup B'$ represents all those students that either do archery **or** do not do badminton.

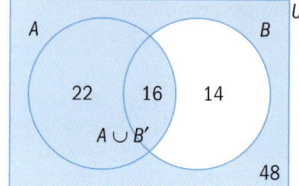

$n(A \cup B') = 22 + 16 + 48 = 86$ and hence

$$P(A \cup B') = \frac{n(A \cup B')}{n(U)} = \frac{86}{100} = \frac{43}{50}$$

Example 3

In a group of 30 students, 17 play computer games, 10 play board games and 9 play neither.
Draw a Venn diagram to show this information.
Use your diagram to find the probability that:
a   a student chosen at random from the group plays board games,
b   a student plays both computer games and board games,
c   a student plays board games but not computer games.

Answers

Let C = plays computer games,
B = plays board games.
Let $x = n(C \cap B)$
$n(C \cap B') = 17 - x$ and
$n(C' \cap B) = 10 - x$

First define your notation.

You don't know how many do computer games AND board games; use x for this value.

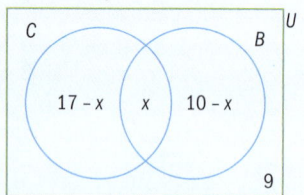

▶ Continued on next page

$(17 - x) + x + (10 - x) + 9 = 30$
$36 - x = 30$
$x = 6$

The four regions of the Venn diagram make the universal set U and so must add up to 30. Solve for x.

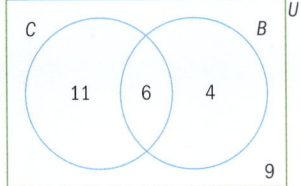

Substitute x = 6 to get the number in each section of the diagram.

Use the Venn diagram and
$$P(A) = \frac{n(A)}{n(U)}$$

a $P(B) = \dfrac{10}{30} = \dfrac{1}{3}$

b $P(C \cap B) = \dfrac{6}{30} = \dfrac{1}{5}$

c $P(C' \cap B) = \dfrac{4}{30} = \dfrac{2}{15}$

Exercise 3B

1 In a group of 35 children, 10 have blonde hair, 14 have brown eyes, and 4 have both blonde hair and brown eyes.
Draw a Venn diagram to represent this situation.
A child is selected at random. Find the probability that the child has blonde hair or brown eyes.

2 In a class of 25 students, 15 of them study French, 13 of them study Malay and 5 of them study neither language.
One of these students is chosen at random from the class. What is the probability that he studies both French and Malay?

3 There are 25 girls in a PE group. 13 have taken aerobics before and 17 have taken gymnastics. One girl has done neither before. How many have done both activities?
One girl is chosen at random. Find the probability that:
 a she has taken both activities,
 b she has taken gymnastics but not aerobics.

EXAM-STYLE QUESTION
4 Of the 32 students in a class, 18 play golf, 16 play the piano and 7 play both. How many play neither?
One student is chosen at random. Find the probability that:
 a he plays golf but not the piano,
 b he plays the piano but not golf.

EXAM-STYLE QUESTION

5 The universal set U is defined as the set of positive integers less than or equal to 15. The subsets A and B are defined as:

A = {integers that are multiples of 3}
B = {integers that are factors of 30}

a List the elements of
  i A
  ii B
b Place the elements of A and B in the appropriate region on a Venn diagram.
c A number is chosen at random from U.
  Find the probability that the number is
  i both a multiple of 3 and a factor of 30,
  ii neither a multiple of 3 nor a factor of 30.

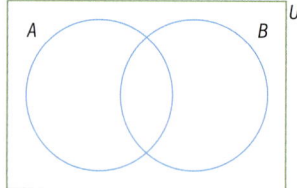

6 In a town 40% of the population read newspaper 'A', 30% read newspaper 'B', 10% read newspaper 'C'.
It is found that 5% read both 'A' and 'B'; 4% read both 'A' and 'C'; and 3% read both 'B' and 'C'. Also, 2% of the people read all three newspapers. Find the probability that a person chosen at random from the town
a reads only 'A',
b reads only 'B',
c reads none of the three newspapers.

> For this question you will need to use three circles in the Venn diagram – one to represent each newspaper.

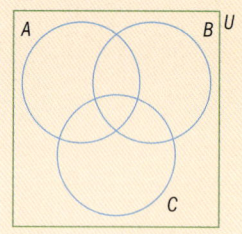

The addition rule

Here is the Venn diagram for the students who do archery and badminton from page 69.

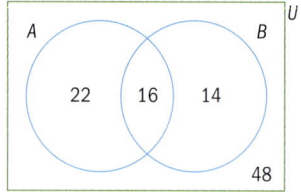

$n(A \cup B) = 38 + 30 - 16$,
or $n(A \cup B) = n(A) + n(B) - n(A \cap B)$
So, $P(A \cup B) = P(A) + P(B) - P(A \cap B)$

> The probability that a student does archery and the probability that a student does badminton **each includes the probability that a student does both archery and badminton**. We only wish to include this probability once so we subtract one of these probabilities.

→ For any two events A and B
  $P(A \cup B) = P(A) + P(B) - P(A \cap B)$

Playing cards

In the next example you need to be familiar with an ordinary pack of 52 playing cards. In a pack there are four suits – clubs, spades, hearts and diamonds. The clubs and spades are black cards, hearts and diamonds are red cards. There are 13 cards in each suit: Ace, 2, 3, 4, 5, 6, 7, 8, 9, 10, Jack, Queen and King. The Jack, Queen and King are called the picture cards. Are playing cards in your country similar to, or the same as these?

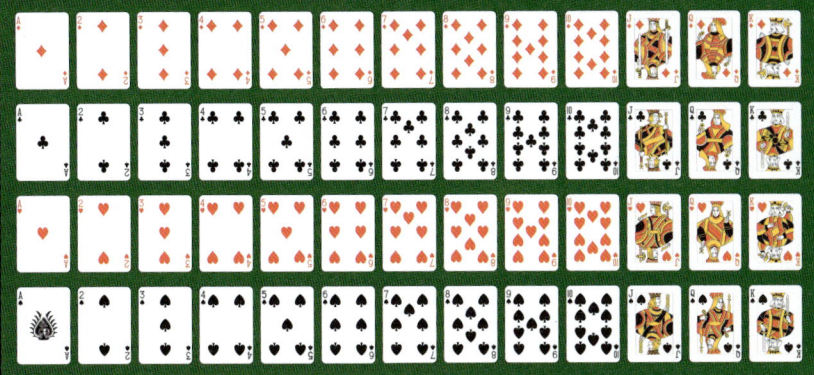

Example 4

A card is drawn at random from an ordinary pack of 52 playing cards.
Find the probability that the card is a heart or a king.

Answer

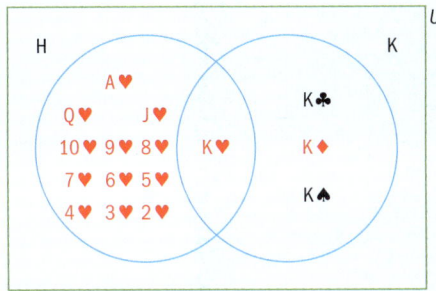

We require $P(H \cup K)$.
Draw a Venn diagram.

$P(H) = \dfrac{13}{52}$

There are 13 hearts in the pack.

$P(K) = \dfrac{4}{52}$

There are 4 kings in the pack.

$P(H \cap K) = \dfrac{1}{52}$

*There is 1 card that is both a king **and** a heart.*

So

$P(H \cup K) = \dfrac{13}{52} + \dfrac{4}{52} - \dfrac{1}{52} = \dfrac{16}{52} = \dfrac{4}{13}$

Using $P(H \cup K) = P(H) + P(K) - P(H \cap K)$

Example 5

If A and B are two events such that $P(A) = \frac{9}{20}$ and $P(B) = \frac{3}{10}$ and $P(A \cup B) = 2P(A \cap B)$ find
a $P(A \cup B)$ **b** $P(A \cup B)'$ **c** $P(A \cap B')$.

Answers

a Let $P(A \cap B) = x$

$2x = \frac{9}{20} + \frac{3}{10} - x$

$3x = \frac{15}{20}$

$x = \frac{3}{4} \div 3$

$x = \frac{1}{4} = P(A \cap B)$

$P(A \cup B) = \frac{1}{2}$

Use
$P(A \cup B) = P(A) + P(B) - P(A \cap B)$

Since $P(A \cup B) = 2P(A \cap B)$.

b If $P(A \cup B) = \frac{1}{2}$ then

$P(A \cup B)' = 1 - \frac{1}{2} = \frac{1}{2}$

Since $P(A') = 1 - P(A)$.

c If $P(A \cap B) = \frac{1}{4}$

$P(A \cap B') = P(A) - P(A \cap B)$

$= \frac{9}{20} - \frac{1}{4} = \frac{1}{5}$

Use result from part **a**.

This is the region on the Venn diagram that is A without its intersection with B.

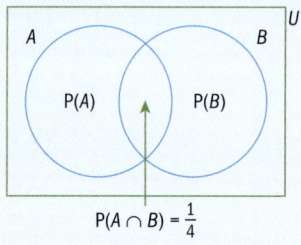

Exercise 3C

1 Two dice are thrown 500 times. For each throw, the sum of the two numbers shown on the dice is written down.

The following frequencies are obtained:

Sum	2	3	4	5	6	7	8	9	10	11	12
Frequencies	6	8	21	34	65	80	63	77	68	36	42

Using these frequencies, calculate the probability of
a the sum being exactly divisible by 5,
b the sum being an even number,
c the sum being exactly divisible by 5 or being an even number.

2 A ten-sided dice, numbered 1 to 10, is rolled. Calculate the probability that:
  a the number scored is a prime number,
  b the number scored is either a prime number or a multiple of 4,
  c the number scored is either a multiple of 4 or a multiple of 3.

3 In a group of 80 tourists 40 have cameras, 50 are female and 22 are females with cameras. Find the probability that a tourist picked from this group at random is either a camera owner or female.

4 A letter is chosen at random from the 26-letter English alphabet. Find the probability that it is
  a in the word **MATHEMATICS**
  b in the word **TRIGONOMETRY**
  c in the word **MATHEMATICS** *and* in the word **TRIGONOMETRY**
  d in the word **MATHEMATICS** *or* in the word **TRIGONOMETRY**.

5 A student goes to the library. The probability that she checks out a work of fiction is 0.40, a work of non-fiction is 0.30, and both fiction and non-fiction is 0.20.
  a What is the probability that the student checks out a work of fiction, non-fiction, or both?
  b What is the probability that the student does not check out a book?

EXAM-STYLE QUESTIONS

6 In a certain road $\frac{1}{3}$ of the houses have no newspapers delivered. If $\frac{1}{4}$ have a national paper delivered and $\frac{3}{5}$ have a local paper delivered, what is the probability that a house chosen at random has both?

7 If X and Y are two events such that $P(X) = \frac{1}{4}$ and $P(Y) = \frac{1}{8}$ and $P(X \cap Y) = \frac{1}{8}$, find
  a $P(X \cup Y)$
  b $P(X \cup Y)'$.

8 If $P(A) = 0.2$ and $P(B) = 0.5$ and $P(A \cap B) = 0.1$, find
  a $P(A \cup B)$
  b $P(A \cup B)'$
  c $P(A' \cup B)$.

Mutually exclusive events

In a student survey it is found that 32 students play chess. Chess and archery clubs are on the same day at the same time so a student cannot do both archery and chess.

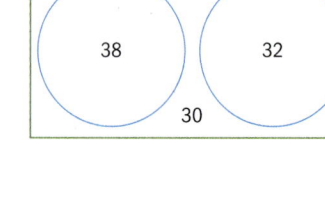

The events A and C are called **mutually exclusive** events. These are events where two outcomes cannot occur at the same time. Here we can see that the circles do not overlap, so $n(A \cap C) = 0$ and hence $P(A \cap C) = 0$.

Now $P(A \cup C) = P(A) + P(C) - 0$.

Generally if two events A and B are mutually exclusive then it follows that $P(A \cap B) = 0$.

Hence we can adapt the addition rule in these cases
$P(A \cup B) = P(A) + P(B)$.

→ In general, if A and B are mutually exclusive, then $P(A \cap B) = 0$ and $P(A \cup B) = P(A) + P(B)$.

Example 6

A box contains board-pens of various colors. A teacher picks out a pen at random. The probability of drawing out a red pen is $\frac{1}{5}$, and the probability of drawing out a green pen is $\frac{3}{7}$.
What is the probability of drawing neither a red nor a green pen?

Answer
Let R = red pen drawn,
G = green pen drawn.
$P(R \cup G) = P(R) + P(G)$
$= \frac{1}{5} + \frac{3}{7} = \frac{22}{35}$

$P(R \cup G)' = 1 - \frac{22}{35} = \frac{13}{35}$

First define your notation.

R and G are mutually exclusive events.
The teacher either draws out a red pen or a green pen, but not both colors.
Since $P(A') = 1 - P(A)$.

Exercise 3D

1 Here are some events relating to throwing two dice:
   A: both dice show a 4
   B: the total is 7 or more
   C: there is at least one 6
   D: the two dice show the same number
   E: both dice are odd

   Which of these pairs of events are mutually exclusive?
   a A and B   **b** A and C   **c** A and D   **d** A and E
   e B and E   **f** C and D   **g** B and C

EXAM-STYLE QUESTION

2 Two events N and M are such that $P(N) = \frac{1}{5}$ and $P(M) = \frac{1}{10}$ and $P(N \cup M) = \frac{3}{10}$.

Are N and M mutually exclusive events?

3 In a group of 89 students, 30 are freshmen (first-year students) and 27 are sophomores (second-year students). Find the probability that a student picked from this group at random is either a freshman or sophomore.

EXAM-STYLE QUESTION

4 In an inter-school quiz, the probability of school A winning the competition is $\frac{1}{3}$, the probability of school B winning is $\frac{1}{4}$ and the probability of school C winning is $\frac{1}{5}$.

Find the probability that

a A or B wins the competition,   **b** A, B or C wins,

c none of these wins the competition.

3.3 Sample space diagrams and the product rule

You can list all the possible outcomes of an experiment if there are not too many.

> A question may tell you to list the possible outcomes.

Example 7

A fair spinner with the numbers 1, 2 and 3 on it as shown is spun three times. List all the possible outcomes from this experiment.
Hence find the probability that the score on the last spin is greater than the scores on the first two spins.

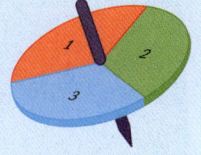

Answer
The 27 outcomes are:

1 1 1	1 2 1	1 3 1
1 1 2	1 2 2	1 3 2
1 1 3	1 2 3	1 3 3
2 1 1	2 2 1	2 3 1
2 1 2	2 2 2	2 3 2
2 1 3	2 2 3	2 3 3
3 1 1	3 2 1	3 3 1
3 1 2	3 2 2	3 3 2
3 1 3	3 2 3	3 3 3

Of these, the five in red have the score on the last spin greater than the scores on the first two spins.

Hence the probability is $\frac{5}{27}$.

When listing all the outcomes, you need to be systematic so that you do not miss any out.

Sample space diagrams

A sample space diagram is another way of showing all the possible outcomes of an event.

> Sample space diagrams are also called probability space diagrams.

Example 8

Draw a sample space diagram to represent the scores when two dice are thrown.
Find the probability of:
a obtaining a score of 6   **b** throwing a double   **c** scoring less than 6.

Answers

		DICE 1				
	1	**2**	**3**	**4**	**5**	**6**
1	(1, 1)	(2, 1)	(3, 1)	(4, 1)	(5, 1)	(6, 1)
2	(1, 2)	(2, 2)	(3, 2)	(4, 2)	(5, 2)	(6, 2)
3	(1, 3)	(2, 3)	(3, 3)	(4, 3)	(5, 3)	(6, 3)
4	(1, 4)	(2, 4)	(3, 4)	(4, 4)	(5, 4)	(6, 4)
5	(1, 5)	(2, 5)	(3, 5)	(4, 5)	(5, 5)	(6, 5)
6	(1, 6)	(2, 6)	(3, 6)	(4, 6)	(5, 6)	(6, 6)

(DICE 2 labels the rows)

(1, 1) gives a score of 2, (4, 6) gives a score of 10.

There are 36 possible outcomes illustrated on this diagram.

a $P(6) = \dfrac{5}{36}$

The five possible ways of getting a score of 6 are highlighted in yellow.

(Highlighted: (5, 1), (4, 2), (3, 3), (2, 4), (1, 5))

b $P(\text{double}) = \dfrac{6}{36} = \dfrac{1}{6}$

The six possible ways of getting a double are highlighted in red.

(Highlighted: (1, 1), (2, 2), (3, 3), (4, 4), (5, 5), (6, 6))

▶ Continued on next page

c P(score < 6) = $\frac{10}{36} = \frac{5}{18}$

The 10 ways of getting a score less than 6 are highlighted in green.

		DICE 1					
		1	**2**	**3**	**4**	**5**	**6**
DICE 2	**1**	(1, 1)	(2, 1)	(3, 1)	(4, 1)	(5, 1)	(6, 1)
	2	(1, 2)	(2, 2)	(3, 2)	(4, 2)	(5, 2)	(6, 2)
	3	(1, 3)	(2, 3)	(3, 3)	(4, 3)	(5, 3)	(6, 3)
	4	(1, 4)	(2, 4)	(3, 4)	(4, 4)	(5, 4)	(6, 4)
	5	(1, 5)	(2, 5)	(3, 5)	(4, 5)	(5, 5)	(6, 5)
	6	(1, 6)	(2, 6)	(3, 6)	(4, 6)	(5, 6)	(6, 6)

Example 9

In an experiment a coin is tossed and a dice is rolled.
Draw the sample space diagram for this experiment.
Find the probability that in a single experiment you obtain a head and a number less than 3 on the dice.

Answer

	1	2	3	4	5	6
H	(1, H)	(2, H)	(3, H)	(4, H)	(5, H)	(6, H)
T	(1, T)	(2, T)	(3, T)	(4, T)	(5, T)	(6, T)

P(head and number less than 3) = $\frac{2}{12} = \frac{1}{6}$

The outcomes that give a head and a number less than 3 are highlighted.

Exercise 3E

EXAM-STYLE QUESTION

1 Three unbiased coins are tossed one at a time and the results are noted. One possible outcome is that all the coins are heads. This is written HHH. Another is that the first two coins are heads and the last one is a tail. This is written HHT.
List the complete sample space for this random experiment.
Find the probability that:
  a the number of heads is greater than the number of tails,
  b at least two heads are tossed consecutively,
  c heads and tails are tossed alternately.

> An unbiased coin is one that is just as likely to land heads up as tails up.

2 Draw the sample space diagram for the random experiment 'Two tetrahedral dice, one blue and the other red, are each numbered 1 to 4. They are rolled and the result noted'.
Find the probability that:
  a the number on the red dice is greater than the number on the blue dice,
  b the difference between the numbers on the dice is one,
  c the red dice shows an odd number and the blue dice shows an even number,
  d the sum of the numbers on the dice is prime.

EXAM-STYLE QUESTION

3 A box contains three cards bearing the numbers 1, 2, 3. A second box contains four cards with the numbers 2, 3, 4, 5. A card is chosen at random from each box.
Draw the sample space diagram for the random experiment.
Find the probability that:
  a the cards have the same number,
  b the larger of the two numbers drawn is 3,
  c the sum of the two numbers on the cards is less than 7,
  d the product of the numbers on the cards is at least 8,
  e at least one even number is chosen.

4 Six cards, numbered 0, 1, 2, 3, 4 and 5, are placed in a bag. One is drawn at random, its number noted, and then it is replaced in the bag. Then a second card is chosen.
Draw the sample space diagram for the random experiment.
Find the probability that:
  a the cards have the same number,
  b the larger of the two numbers drawn is prime,
  c the sum of the two numbers on the cards is less than 7,
  d the product of the numbers on the cards is at least 8,
  e at least one even number is chosen.

5 Tilman plays a game with a dice called 'Come and Go'. He rolls the dice. If the score is 1 he moves up one metre. If it is 2 he moves right one metre. If it is 3 he moves down one metre. If it is 4 he moves left one metre. If it is 5 or 6 he stays where he is.

Tilman rolls the dice twice. He makes two steps.

What is the probability that he is
  a at the same point where he started,
  b exactly 2 metres away from his starting point,
  c more than 1 but less than 2 metres away from his starting point?

> **Genetic fingerprinting**
> Genetic fingerprinting was developed in 1984 by Professor Alec Jeffreys at the University of Leicester. Each one of us has a unique genetic make-up which is contained in the DNA and which is inherited from our parents. The DNA can be extracted from cells and body fluids and analyzed to produce the characteristics (seen below) – our 'genetic fingerprint'. When matching 'fingerprints' it is usual to compare these bands. Some of these comparisons have been used as evidence to convict criminals but the field is under scrutiny due to its reliance on probability. Usually between 10 and 20 bands are examined and compared. Experimental evidence has suggested that the probability of one band matching by coincidence is $\frac{1}{4}$ (although this figure is subject to debate). The probability of two bands matching will therefore be $\frac{1}{16}$.

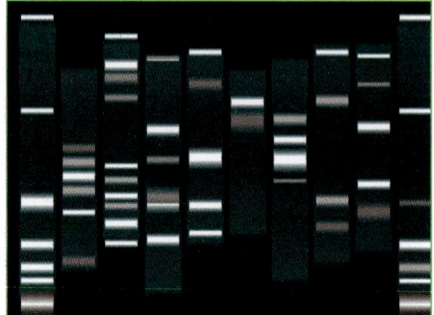

Product rule for independent events

When a dice is rolled and a coin is tossed, as in Example 9 on the previous page, the events are **independent**. This is because the outcome of tossing the coin does not influence the outcome of rolling the dice and vice versa.

Probability

→ Two events A and B are independent if the occurrence of one does not the affect the chance that the other occurs.

Here is the sample space for a dice and a coin.

	1	2	3	4	5	6
H	(1, H)	(2, H)	(3, H)	(4, H)	(5, H)	(6, H)
T	(1, T)	(2, T)	(3, T)	(4, T)	(5, T)	(6, T)

Let H stand for the event 'coin lands heads'.
From the diagram:
$$P(H) = \frac{6}{12} = \frac{1}{2}$$

Let L stand for the event 'dice score less than 3'.
$$P(L) = \frac{4}{12} = \frac{1}{3}$$
$$P(H \cap L) = \frac{2}{12} = \frac{1}{6}$$

But we can also note that
$$P(H \cap L) = P(H) \times P(L)$$
$$= \frac{1}{2} \times \frac{1}{3} = \frac{1}{6}$$

There are two outcomes where the coin is heads and the dice score is less than 3.

→ When two events A and B are independent
$$P(A \cap B) = P(A) \times P(B)$$

This is the **product rule for independent events**.
This is also called the multiplication rule.

Sample space diagrams can help you visualize the number of possible outcomes, but you don't always need to draw one.

Example 10

One bag contains 3 red and 2 white balls, another bag contains 1 red and 4 white balls. A ball is selected at random from each bag, Find the probability that
a both the balls are red,   **b** the balls are different colors,
c at least one ball is white.

Answers

a From the first bag $P(R_1) = \frac{3}{5}$

From the second bag $P(R_2) = \frac{1}{5}$

Therefore $P(R_1 \cap R_2)$

$= \frac{3}{5} \times \frac{1}{5} = \frac{3}{25}$

The events 'picking a red from the first bag' (R_1) and 'picking a red from the second bag' (R_2) are independent events. In R_1 there are 3 red balls in 5. In R_2 there is 1 red ball in 5.
The events R_1 and R_2 are independent, so $P(R_1 \cap R_2) = P(R_1) \times P(R_2)$.

▶ Continued on next page

b From the first bag $P(R_1) = \dfrac{3}{5}$

From the second bag $P(W_2) = \dfrac{4}{5}$
Therefore $P(R_1 \cap W_2)$
$= \dfrac{3}{5} \times \dfrac{4}{5} = \dfrac{12}{25}$

From the first bag $P(W_1) = \dfrac{2}{5}$

From the second bag $P(R_2) = \dfrac{1}{5}$
Therefore $P(W_1 \cap R_2)$
$= \dfrac{2}{5} \times \dfrac{1}{5} = \dfrac{2}{25}$

P(different colors) =
$P(R_1 \cap W_2) + P(W_1 \cap R_2)$
$\dfrac{12}{25} + \dfrac{2}{25} = \dfrac{14}{25}$

If the balls are different colors, this means either the first one is red and the second one white or the first one is white and the second one red.

These are mutually exclusive events.

c P(at least one white)
= 1 − probability that both are red
= 1 − $P(R_1 \cap R_2)$
= 1 − $\dfrac{3}{25} = \dfrac{22}{25}$

*For 'at least one of the balls is white' we **could** calculate the probability that both are white, the probability that the first is white and the second red and the probability that the first is red and the second is white.*
OR
If at least one is white then it means that both cannot be red.
This is a common method of solving problems that involve the words '… at least…'. Calculate 1 − the complement of the event.

Exercise 3F

1. My wardrobe contains five shirts with one blue, one brown, one red, one white and one black. I reach into the wardrobe and choose a shirt without looking. I replace this shirt and then choose another. What is the probability that I will choose the red shirt both times?

2. A card is chosen at random from a deck of 52 cards. It is then replaced and a second card is chosen. What is the probability of choosing a king and a ten?

3. A large school conducts a survey of the food provided by the school cafeteria. It is found that $\dfrac{4}{5}$ of the students like pasta.
Three students are chosen at random. What is the probability that all three students like pasta?

> For questions 2 and 8, you may need to remind yourself about playing cards – see page 73.

EXAM-STYLE QUESTIONS

4 Adam is playing in a cricket match and a game of hockey at the weekend.
The probability that his team will win the cricket match is 0.75, and the probability of winning the hockey match is 0.85.
Assume that the results in the matches are independent. What is the probability that his team will win in both matches?

5 Three events A, B and C are such that A and B are mutually exclusive and $P(A) = 0.2$, $P(C) = 0.3$, $P(A \cup B) = 0.4$ and $P(B \cup C) = 0.34$.
   a Calculate $P(B)$ and $P(B \cap C)$.
   b Determine whether B and C are independent.

6 I toss a coin and roll a six-sided dice. Find the probability that I get a head on the coin, and don't get a 6 on the dice.

7 An air-to-air missile has probability $\frac{8}{9}$ of hitting a target. If five missiles are launched, what is the probability that the target is not destroyed?

8 Four cards are chosen from a standard deck of 52 playing cards with replacement. What is the probability of choosing 4 hearts in a row?

EXAM-STYLE QUESTION

9 Given that $P(E') = P(F) = 0.6$ and $P(E \cap F) = 0.24$
   a write down $P(E)$,
   b explain why E and F are independent,
   c explain why E and F are not mutually exclusive,
   d find $P(E \cup F')$.

10 Three bags each contain 4 red and 8 blue marbles. One marble is randomly chosen from each bag.

What is the probability that the first marble will be red, the second marble blue and the third marble red?

11 A six-sided dice is numbered: 1, 2, 2, 5, 6, 6. It is thrown three times. What is the probability that the scores add up to 6?

EXAM-STYLE QUESTION

12 A and B are independent events such that $P(A) = 0.9$ and $P(B) = 0.3$. Find:
   a $P(A \cap B)$ **b** $P(A \cap B')$ **c** $P(A \cup B)'$.

EXAM-STYLE QUESTION

13 Independent events G and H are such that $P(G \cap H') = 0.12$ and $P(G' \cap H) = 0.42$.
Draw a Venn diagram to represent the events G and H.
Let $P(G \cap H) = x$.
Find two possible values of x.

14 I throw four dice. Find the probability that
  a all four dice show a 6,
  b all four dice show the same number.

15 Which is more likely: rolling a 'six' on four throws of one dice, or rolling a 'double six' on 24 throws with two dice?

> This is the question you considered in the dice Investigation on page 64.

16 A program produces (independently) three random digits from 0 to 9. For example
  247   or   309   or   088   or   936
  a Find the probability that none of the three digits is a 5.
  b Find the probability that at least one digit is a 5.

Investigation – the Monty Hall dilemma

The following is a famous probability puzzle based on the American television game show 'Let's Make a Deal'.
The name comes from the show's original host, Monty Hall.

Suppose you're on the game show and you're given the choice of three doors. Behind one door is the main prize (a car) and behind the other two doors there are dud, unwanted prizes.
The car and the unwanted prizes are placed randomly behind the doors before the show.

The rules of the game are: After you have chosen a door, the door remains closed for the time being. Monty Hall, who knows what is behind the doors, then opens one of the two remaining doors and always reveals an unwanted prize. After he has opened one of the doors and shown a dud, Monty Hall asks the participants whether they want to stay with their first choice or to switch to the last remaining door.

What should you do?

  a Stick with your first choice.
  b Switch to the other remaining closed door.
  c It does not matter. Chances are even.

> We will revisit this problem at the end of this chapter.

3.4 Conditional probability

Here is the Venn diagram showing students who do archery and badminton.

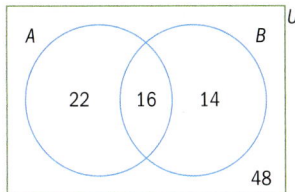

If we know that a particular student does badminton, how does this affect the probability that they also do archery?

Altogether 30 students do badminton; of these 16 also do archery.

We write the probability that a student does archery given that they do badminton as $P(A|B)$.

Note that
$$P(A|B) = \frac{n(A \cap B)}{n(B)} = \frac{16}{30} = \frac{8}{15}$$

This is known as **conditional probability** since the outcome of A is **dependent on** the outcome of B.

It also follows that $P(A|B) = \frac{P(A \cap B)}{P(B)} = \frac{\frac{16}{100}}{\frac{30}{100}}$

$$= \frac{16}{30} = \frac{8}{15}$$

> In general for two events A and B the probability of A occurring given that B has occurred can be found using
> $$P(A|B) = \frac{P(A \cap B)}{P(B)}.$$

Rearranging the formula gives

$P(A \cap B) = P(A|B) \times P(B)$.

> If A and B are independent events,
> $P(A|B) = P(A)$, $P(B|A) = P(B)$, $P(A|B') = P(A)$
> and $P(B|A') = P(B)$.

Recall that for independent events $P(A \cap B) = P(A) \times P(B)$. By definition for independent events, A and B, the probability of A occurring given that B has occurred will equal the probability of A occurring, since A is not affected by the occurrence of B.

Example 11

Of the 53 staff at a school, 36 drink tea, 18 drink coffee, and 10 drink neither tea nor coffee.
a How many staff drink both tea and coffee?
One member of staff is chosen at random. Find the probability that:
b he drinks tea but not coffee,
c if he is a tea drinker he drinks coffee as well,
d if he is a tea drinker he does not drink coffee.

Answers

a

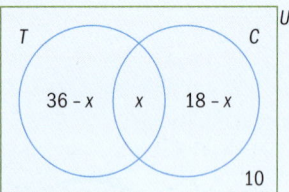

Let $n(T \cap C) = x$
so
$36 - x + x + 18 - x + 10 = 53$
$64 - x = 53$
$x = 11$

Therefore $P(T \cap C) = \dfrac{11}{53}$

Draw a Venn diagram to show the information

$n(T \cap C)$ is the number who drink both tea and coffee.

53 is the total number of staff on the Venn diagram.
Solve for x.

Since $x = 11$ and total = 53.

b $P(T \cap C') = \dfrac{25}{53}$

$36 - 11 = 25$

c $P(C \mid T) = \dfrac{P(C \cap T)}{P(T)} = \dfrac{\frac{11}{53}}{\frac{36}{53}}$

$= \dfrac{11}{53} \times \dfrac{53}{36} = \dfrac{11}{36}$

d $P(C' \mid T) = \dfrac{P(C' \cap T)}{P(T)} = \dfrac{\frac{25}{53}}{\frac{36}{53}}$

$P(C' \cap T) = P(T \cap C')$

$= \dfrac{25}{53} \times \dfrac{53}{36} = \dfrac{25}{36}$

Exercise 3G

EXAM-STYLE QUESTIONS

1 There are 27 students in a class. 15 take Art and 20 take Theater. Four do neither subject. How many students do both subjects? One person is chosen at random. Find the probability that
  a he or she takes Theater but not Art,
  b he or she takes at least one of the two subjects,
  c he or she takes Theater, given that he or she takes Art.

2 For events A and B it is known that: $P(A' \cap B') = 0.35$; $P(A) = 0.25$; $P(B) = 0.6$. Find
  a $P(A \cap B)$ **b** $P(A|B)$ **c** $P(B'|A')$.

3 48% of all teenagers own a skateboard and 39% of all teenagers own a skateboard and roller blades. What is the probability that a teenager owns roller blades given that the teenager owns a skateboard?

4 A number is chosen at random from this list of eight numbers:
   1   2   4   7   11   16   22   29
Find:
  a P(it is even | it is not a multiple of 4)
  b P(it is less than 15 | it is greater than 5)
  c P(it is less than 5 | it is less than 15)
  d P(it lies between 10 and 20 | it lies between 5 and 25).

5 In my town 95% of all households have a desktop computer. 61% of all households have a desktop computer and a laptop computer. What is the probability that a household has a laptop computer given that it has a desktop computer?

6 The probability that a student takes Design Technology and Spanish is 0.1. The probability that a student takes Design Technology is 0.6. What is the probability that a student takes Spanish given that the student is taking Technology?

7 U and V are mutually exclusive events. $P(U) = 0.26$; $P(V) = 0.37$. Find
  a $P(U \text{ and } V)$ **b** $P(U|V)$ **c** $P(U \text{ or } V)$.

8 A teacher gave her class an IB Paper 1 and an IB Paper 2. 35% of the class passed both tests and 52% of the class passed the first test. What percentage of those who passed the first test also passed the second test?

9 A jar contains black and white marbles. Two marbles are chosen without replacement. The probability of selecting a black marble and then a white marble is 0.34, and the probability of selecting a black marble on the first draw is 0.47. What is the probability of selecting a white marble on the second draw, given that the first marble drawn was black?

EXAM-STYLE QUESTION

10 The table below shows the number of left- and right-handed table-tennis players in a sample of 50 males and females.

	Left-handed	Right-handed	Total
Male	5	32	37
Female	2	11	13
Total	7	43	50

A table-tennis player was selected at random from the group. Find the probability that the player is:
  a male and left-handed,   **b** right-handed,
  c right-handed, given that the player selected is female.

11 J and K are independent events. Given that $P(J \mid K) = 0.3$ and $P(K) = 0.5$, find $P(J)$.

12 Your neighbour has two children. You learn that he has a son, Sam. What is the probability that Sam's sibling is a brother?

> This is not as obvious as it might seem!

Investigation – the Monty Hall problem revisited!

Take a typical situation in the game. Suppose the contestant has chosen Door 3 and Monty Hall reveals that there is an unwanted prize behind Door 2. What is the conditional probability that the car is behind Door 1?
Let A stand for the condition that there is a car behind Door 1 and the contestant has chosen Door 3.
Let B stand for the condition that Monty Hall has revealed that there is a dud behind Door 2 given that the contestant has chosen Door 3.

The probability of A and B ($P(A \cap B)$) is just $\frac{1}{3} \times \frac{1}{3} = \frac{1}{9}$ because if the car is behind Door 1 and the contestant has chosen Door 3 Monty Hall has to show what is behind Door 2.

> Analysis of the monty Hall problem using conditional probability

The problem is the computation of the probability of being shown a dud behind Door 2 given that the choice was Door 3. This situation can arise in two ways:

1 when the car is behind Door 1
2 when the car is behind Door 3.

The first way has a probability of $\frac{1}{9}$, as shown above.
In the second way, the host could reveal either what is behind Door 1 or Door 2. If he is equally likely to choose either of these doors then the probability of showing what is behind Door 2 is $\frac{1}{2} \times \frac{1}{9} = \frac{1}{18}$.
Therefore the probability of there being revealed an unwanted prize behind Door 2 when the contestant has chosen Door 3 is
$$\frac{1}{9} + \frac{1}{2} \times \frac{1}{9} = \frac{3}{18}$$
This is $P(B)$, the probability of B.

We want the conditional probability, $P(A \mid B)$. This is given by
$$P(A \mid B) = \frac{P(A \cap B)}{P(B)} = \frac{\frac{1}{9}}{\frac{3}{18}} = \frac{2}{3}.$$

This means that the conditional probability that the car is behind Door 3 given that the contestant has chosen Door 3 and has been shown that there is an unwanted prize behind Door 2 is only $\frac{1}{3}$.
Therefore it is worthwhile to switch!

3.5 Probability tree diagrams

Tree diagrams are useful for problems where more than one event occurs. It is sometimes easier to use these than to list all the possible outcomes. It is important to read the question carefully and distinguish between the different types of problems.

'With replacement' and repeated events

Example 12

> The probability that Samuel, a keen member of the school Archery Club, hits the bullseye is 0.8. Samuel takes two shots. Assume that success with each shot is independent from the previous shot.
>
> Represent this information on a tree diagram.
>
> Find the probability that Samuel
> **a** hits two bullseyes
> **b** hits only one bullseye
> **c** hits at least one bullseye.

Answers

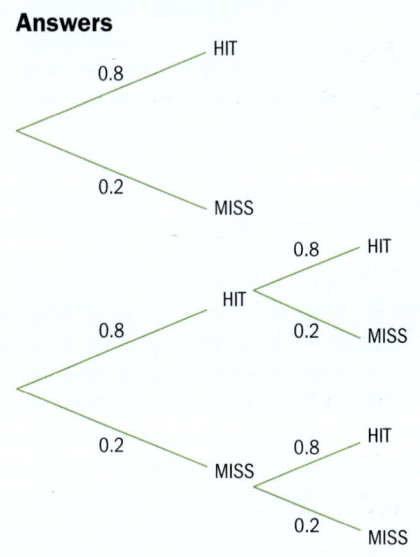

a We require P(H and H). So P(H and H) = 0.8 × 0.8 = 0.64	*Since a hit with the first shot is independent of getting a hit with the second shot we can multiply the probabilities together (the product rule). Multiply along the top two branches.*
b P(H and M) + P(M and H) = (0.8 × 0.2) + (0.2 × 0.8) = 0.32	*Only one hit could be either a hit on the first or a hit on the second and missing the other one.* *These two events, (H and M) and (M and H) are mutually exclusive: they can't both happen at the same time. Multiply along each branch (as again events are independent) and then add between them (as 2 outcomes are mutually exclusive).*
c P(at least one bullseye) = 1 − (0.2 × 0.2) = 1 − 0.04 = 0.96	*Here we need 1 − P(miss the bullseye both times)* *So we have 1 − P(M and M)*

The first section of the tree diagram represents Samuel's first shot. He will either hit the bullseye or miss it. The probability that he misses is 1 − 0.8 = 0.2.
The outcome is on the end of the branch, the probability is beside the branch.

The second shot will also either hit or miss the bullseye. There are therefore four possible outcomes of this 'experiment':
a hit followed by a hit (H and H),
a hit followed by a miss (H and M),
a miss followed by a hit (M and H),
a miss followed by a miss (M and M).

Exercise 3H

1 Lizzie is attempting two exam questions. The probability that she gets any exam question correct is $\frac{2}{3}$.
  a Copy and complete the diagram.
  b What is the probability that she will get only one of them correct?
  c What is the probability she will get at least one correct?

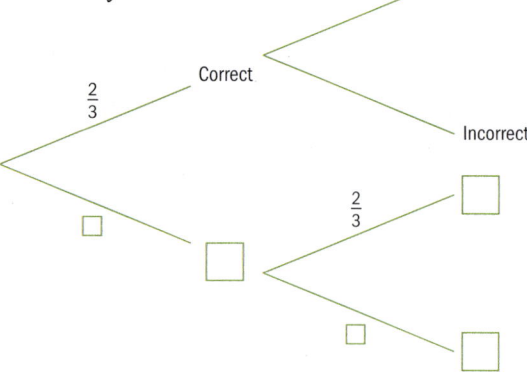

2 When Laura and Michelle play in the hockey team the probability that Laura scores is $\frac{1}{3}$ and that Michelle scores is $\frac{1}{2}$.

Draw a tree diagram to illustrate this information and use it to find the probability that neither will score in the next game.

EXAM-STYLE QUESTION

3 There are equal numbers of boys and girls in a school and it is known that $\frac{1}{10}$ of the boys and $\frac{1}{10}$ of the girls walk in every day.
Also $\frac{1}{3}$ of the boys and $\frac{1}{2}$ of the girls get a lift.
The rest come by coach.
Determine
  a the proportion of the school population that are girls who come by coach,
  b the proportion of the school population that come by coach.

> In question 3 there will be two branches in the first section and three branches from each of these in the second section.

4 Determine the probability of getting two heads in three tosses of a biased coin for which $P(head) = \frac{2}{3}$.

5 A 10-sided dice has the numbers 1–10 written on it. It is rolled twice. Find the probability that:
  a exactly one prime number is rolled,
  b at least one prime number is rolled.

EXAM-STYLE QUESTION

6 The probability of a day being windy is 0.6. If it's windy the probability of rain is 0.4. If it's not windy the probability of rain is 0.2.
  a Copy and complete the tree diagram.
  b What is the probability of a given day being rainy?
  c What is the probability of two successive days **not** being rainy?

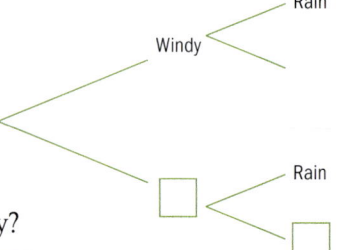

'Without replacement' and conditional probability

Example 13

A bag contains 5 green and 6 red balls. If two balls are taken out successively, without replacement, what is the probability that
a at least one green is chosen,
b red is picked on the first pick given that at least one green is chosen?

> This means that the probability of the second draw is dependent on the results of the first draw, since a ball has been removed after the first draw.

Answers

Since a red ball has been taken initially there will be 5 red balls (and 5 green balls) left

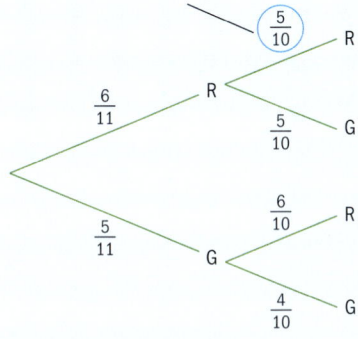

Draw a tree diagram. The probabilities on the second section of branches depend on what has happened in the first section.

a P(at least one green)
   = 1 − P(both red)
   $= 1 - \left(\dfrac{6}{11} \times \dfrac{5}{10}\right) = 1 - \dfrac{3}{11} = \dfrac{8}{11}$

It is quicker to calculate the probability in this way than it is to work out the probability that the ball is a green on the first pick, or a green on the second pick or a green on both picks.

b P(red followed by green)

$= \dfrac{P(\text{red on 1st and at least one green})}{P(\text{at least one green})}$

$= \dfrac{\dfrac{6}{11} \times \dfrac{\cancel{5}^{1}}{\cancel{10}^{2}}}{\dfrac{8}{11}} = \dfrac{\dfrac{3}{11}}{\dfrac{8}{11}} = \dfrac{3}{8}$

When red is picked first, the probability of the second being green is $\dfrac{5}{10}$, so multiply these probabilities.

Some tree diagrams do not have the same 'classic' shape as the we have seen so far.

Example 14

Toby is a rising star of the school Tennis Club. He has found that when he gets his first serve in the probability that he wins that point is 0.75. When he uses his second serve there is a 0.45 chance of him winning the point. He is successful at getting his first serve in on 3 out of 5 occasions and his second serve in on 3 out of 4 occasions.

a Find the probability that the next time it is Toby's turn to serve he wins the point.

b Given that Toby wins the point, what is the probability that he got his first serve in?

Answers

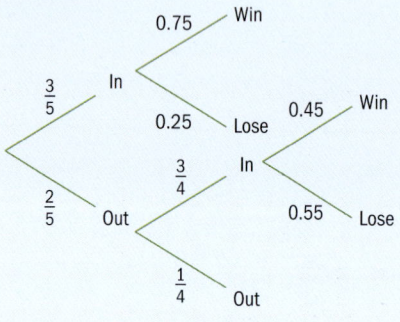

On this tree diagram, it is not necessary to continue the branches once the point has been won.

a P (win) = (get first serve in and win) + (miss first serve, get second serve in and win)

$\left(\dfrac{3}{5} \times 0.75\right) + \left(\dfrac{2}{5} \times \dfrac{3}{4} \times 0.45\right)$

= 0.45 + 0.135
= 0.585

Multiply along the branches.

b P(1st serve in | win point)

$= \dfrac{P(\text{1st serve in and win point})}{P(\text{win point})}$

*Both of these values have been found in part **a**.*

$= \dfrac{\left(\dfrac{3}{5} \times 0.75\right)}{0.585} = 0.769 \text{ (3 sf)}$

This answer has been given to 3 sf as the exact answer (fractional) is not obvious.

Exercise 3I

1 Three cards are drawn at random from a pack of playing cards. Each card is not replaced. Find the probability of obtaining
  a three picture cards
  b two picture cards.

See page 73 for the ordinary pack of 52 playing cards.

: **EXAM-STYLE QUESTION**
: **2** A pencil case contains 5 faulty and 7 working pens. A boy and
:   then a girl each need to take a pen.
:   **a** What is the probability that two faulty pens are chosen?
:   **b** What is the probability that at least one faulty pen is chosen?
:   **c** If exactly one faulty pen is chosen, what is the probability that
:   the girl chose it?

> Even if the question does not ask for it you may find it useful to use a tree diagram to answer some of these questions.

3 In a bag are 4 red balls, 3 green balls and 2 yellow balls. A ball is chosen at random, and not replaced. A second ball is then chosen.
  a Find P(the balls are both green).
  b Find P(the balls are the same color).
  c Find P(neither ball is red).
  d Find P(at least one ball is yellow).

4 Four balls are drawn at random, one after the other and without replacement, from a bag containing the following balls:
5 red, 4 blue, 3 orange, 2 purple.
Find the probability that you obtain one ball of each color.

5 A club has 10 members, of which 6 are girls and 4 are boys. One of the members is chosen at random to be President of the club.
  a Find the probability that the chosen President is a boy.
  b Two people are chosen at random to represent the club in a competition. Find the probability that one boy and one girl are chosen.

6 Billy answers on average 5 problems correctly out of 7. Natasha's average is 5 questions out of 9. They both attempt the same problem.
  a What is the probability that at least one of the students answers the question correctly?
  b If the question is answered correctly, what is the probability that Billy got the correct answer?
  c If the question is answered correctly, what is the probability that Natasha got the correct answer?
  d If there was at least one correct answer, what is the probability that there were two?

Extension material on CD:
Worksheet 3 - Conditional probability

Review exercise

1. A two-digit number between 10 and 99 inclusive is written down at random. What is the probability that it:
   a. is divisible by 5,
   b. is divisible by 3,
   c. is greater than 50,
   d. is a square number?

2. In a class of 30 students, 18 have a dog, 20 have a cat and 3 have neither. A student is selected at random. What is the probability that this student has both a cat and a dog?

EXAM-STYLE QUESTIONS

3. For events C and D it is known that:
   $P(C) = 0.7 \quad P(C' \cap D') = 0.25 \quad P(D) = 0.2$.
   a. Find $P(C \cap D')$.
   b. Explain why C and D are not independent events.

4. The two events A and B are such that $P(A) = 0.6$, $P(B) = 0.2$ and $P(A|B) = 0.1$.

   Calculate the probabilities that:
   a. both of the events occur,
   b. at least one of the events occur,
   c. exactly one of the events occur,
   d. B occurs given that A has occurred.

5. A group of 100 students are asked which of three types of TV programme, drama, comedy and reality, they watch regularly. They provide the following information:

   15 watch all three types;
   18 watch drama and comedy;
   22 watch comedy and reality TV;
   35 watch drama and reality TV;
   10 watch of none these programmes regularly.

   There are three times as many students who watch drama only than comedy only and two times as many who watch comedy only than reality TV only.

   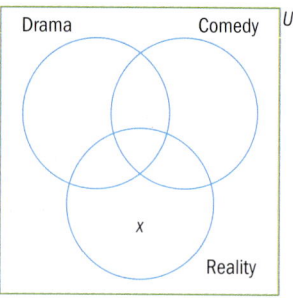

   a. If x is taken as the number of students who watch reality TV only, write an expression for the number of students who watch drama only.
   b. Using all the above information copy and complete the Venn diagram.
   c. Calculate the value of x.

Review exercise

1. Let P(*C*) = 0.4, P(*D*) = 0.5, P(*C* | *D*) = 0.6.
   a Find P(*C* and *D*).
   b Are *C* and *D* mutually exclusive? Give a reason for your answer.
   c Are *C* and *D* independent events? Give a reason for your answer.
   d Find P(*C* or *D*).
   e Find P(*D* | *C*).

2. Jack does $\frac{3}{5}$ of the jobs around the house and Jill does the rest. If 35% of Jack's jobs are finished properly and 55% of Jill's jobs are finished properly, find the probability that a job done around the house will be done:
   a properly,
   b by Jill if it was not done properly.

EXAM-STYLE QUESTIONS

3. Max travels to school each day by bicycle, by bus or by car. The probability that he travels by bus on any day is 0.6. The probability that he travels by bicycle on any day is 0.3.
   a Draw a tree diagram which shows the possible outcomes for Max's journeys on Monday and Tuesday. Label the tree clearly, writing in the probabilities of each outcome.
   b What is the probability that he travels
      i by bicycle on Monday and Tuesday,
      ii by bicycle on Monday and by bus on Tuesday,
      iii by the same method of travel on Monday and Tuesday?
   c Max traveled to school by bicycle on Monday and Tuesday. What is the probability that he does not travel to school by bicycle on Wednesday and Thursday and Friday?
   d What is the probability that in any three days Max travels twice by car and once by bus or twice by bicycle and once by car?

4. A bag contains 6 red apples and 10 green apples. Without looking into the bag, Maddy randomly selects one apple.
   a What is the probability that it is red?

   The apple is red and Maddy eats it. Next the bag is passed to Janet. Without looking into the bag, she randomly selects one apple.
   b What is the probability that it is green?

   The apple is green and Janet replaces it in the bag. Next the bag is passed to Tarish. Without looking into the bag, he randomly selects two apples.
   c What is the probability that they are both red?

EXAM-STYLE QUESTION

5 On a walk I count 70 rabbits, 42 are female, 34 are not eating carrots and 23 are female and not eating carrots.
Draw a Venn diagram and hence find the number that are both female and eating carrots.
  a What is the probability that a rabbit is male and not eating carrots?
  b What is the probability that a rabbit is female given that it is eating carrots?
  c Is being female independent of eating carrots? Justify your answer.

CHAPTER 3 SUMMARY
Definitions

- An **event** is an outcome from an experiment.
  An **experiment** is the process by which we obtain an outcome.
  A **random experiment** is one where there is uncertainty over which event may occur.
- The theoretical probability of an event A is $P(A) = \dfrac{n(A)}{n(U)}$
  where $n(A)$ is the number of ways that event A can occur and $n(U)$ is the total number of possible outcomes.
- If the probability of an event is P, in n trials you would expect the event to occur $n \times P$ times.
- You can use relative frequency as an estimate of probability. The larger the number of trials, the closer the relative frequency is to the probability.

Venn diagrams

- As an event, A, either happens or it does not happen
  $P(A) + P(A') = 1$
  $P(A') = 1 - P(A)$

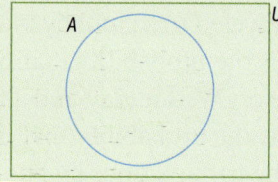

- For any two events A and B
  $(A \cup B) = P(A) + P(B) - P(A \cap B)$.

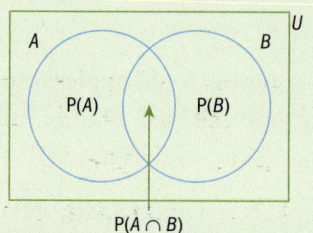

- In general, if A and B are mutually exclusive, then
  $P(A \cap B) = 0$ and $P(A \cup B) = P(A) + P(B)$.

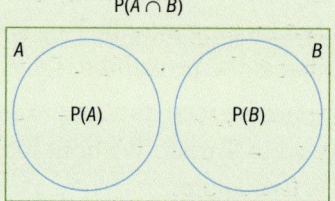

Continued on next page

Sample space diagrams and the product rule

- Two events A and B are independent if the occurrence of one does not the affect the chance that the other occurs.
- When two events A and B are independent
$$P(A \cap B) = P(A) \times P(B)$$
This is the **product rule for independent events**.
This is also called the multiplication rule.

Conditional probability

- For two events A and B the probability of A occurring given that B has occurred can be found using
$$P(A|B) = \frac{P(A \cap B)}{P(B)}$$
- If A and B are independent events, $P(A|B) = P(A)$, $P(B|A) = P(B)$, $P(A|B') = P(A)$, $P(B|A') = P(B)$
- In general for two events A and B the probability of A occurring given that B has occurred can be found using
$$P(A|B) = \frac{P(A \cap B)}{P(B)}.$$

Theory of knowledge

Probability – uses and abuses

Math textbook probability problems often involve picking colored balls from a bag – how useful is that in real life? But probability has some surprising uses – such as finding answers to sensitive questions.

People also misuse or misunderstand probability, by relying on their intuition rather than calculating the odds.

- Why do people buy lottery tickets when the chances of winning are so small?
- What is the probability of winning your national lottery?

Asking sensitive questions

How do you get the truth from a survey that asks a sensitive question?

A headteacher wants to know how many of the 600 students at her school have cheated in examinations. She is not interested in whether a specific individual has cheated, she just wants an overall estimate for the whole school.

If she sends a questionnaire to each student asking:

> Have you ever cheated in school exams?
> ☐ Yes   ☐ No

she is unlikely to get honest answers.

The randomized response method

This relies on each student knowing that the headteacher doesn't know whether they are asking a sensitive question or a perfectly harmless one.

Each student in the survey flips a coin twice, showing no-one the result.

They then follow the instructions on this card.

1. If you got a head with the first flip, answer the question: 'Have you ever cheated in an exam?' honestly.

2. If you got a tail with the first flip, answer the question: 'Did the second flip give a tail?' honestly.

The tree diagram helps to estimate p, the fraction of students who have cheated in an exam.

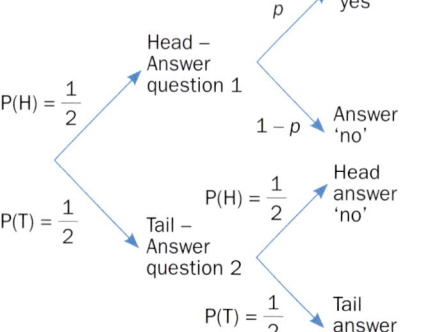

$$P(\text{Yes to Q1}) = \frac{1}{2} \times p = \frac{p}{2}$$

$$P(\text{Yes to Q2}) = \frac{1}{2} \times \frac{1}{2} = \frac{1}{4}$$

Probability of answer 'yes'
= P(Yes to Q1) + P(Yes to Q2)
$$= \frac{p}{2} + \frac{1}{4}$$

Suppose 220 students answer 'yes' out of 600 interviewed.

$$\frac{p}{2} + \frac{1}{4} = \frac{220}{600}$$

$$\frac{p}{2} = \frac{220}{600} - \frac{1}{4}$$

$$\frac{p}{2} = \frac{7}{60}$$

$$p = \frac{7}{30}$$

The estimated number of students who have cheated in an exam is $600 \times \frac{7}{30} = 140$

Provided everybody tells the truth when answering their question, this method estimates the number of students who have cheated in an exam.

- Would students answer the question honestly?
- Are there any problems with this method of finding the truth?

Probability and intuition – the birthday problem

- In a class of 23 people, what is the probability that two people share the same birthday?
  What do you think? 1%? Maybe 5%, or even as much as 10%?

Let's do the math and work it out.

23 students means there are 253 possible pairs of students:

$$\frac{23 \times 22}{2} = 253$$

There are 23 choices for the first person in a pair and then 22 choices for the second.

The pair (Tim, Jane) is exactly the same as the pair (Jane, Tim), so halve the total.

The probability of two people having different birthdays is

$$\frac{364}{365} = 0.997260$$

Ignoring leap years – there are 364 out of 365 birthdays that are 'different'.

So for 253 possible pairs, the probability that the two people in each pair have different birthdays is

$$\left(\frac{364}{365}\right)^{253} = 0.4995$$

So the probability that, for the 253 pairs, two people in a pair have the same birthday is

$1 - 0.4995 = 0.5005$, or 50.05% – just over half!

- Do you rely on intuition to help you make decisions?
- Are there other areas of mathematics where your intuition has let you down?
- What about in other areas of knowledge?

Chapter 3 99

4 Exponential and logarithmic functions

CHAPTER OBJECTIVES:

1.2 Elementary treatment of exponents and logarithms
Laws of exponents; laws of logarithms; change of base

2.6 Exponential functions and their graphs
$x \mapsto a^x, a > 0, x \mapsto e^x$
Logarithmic functions and their graphs
$x \mapsto \log_a x, x > 0, x \mapsto \ln x, x > 0$
Relationship between these functions
$a^x = e^{x \ln a}, \log_a a^x = x; a^{\log_a x} = x, x > 0$

2.7 Solving equations of the form $a^x = b, a^x = b^y$

2.8 Applications of graphing skills and solving equations to real-life situations

Before you start

You should know how to:

1 Evaluate simple positive exponents
e.g. Evaluate 3^4
$3^4 = 3 \times 3 \times 3 \times 3 = 81$
e.g. Evaluate $\left(\dfrac{2}{5}\right)^3$
$\left(\dfrac{2}{5}\right)^3 = \dfrac{2^3}{5^3} = \dfrac{2 \times 2 \times 2}{5 \times 5 \times 5} = \dfrac{8}{125}$

2 Convert numbers to exponential form
e.g. Find n given $2^n = 128$
$128 = 2^7$ so $n = 7$

3 Transform graphs
e.g. Given the graph of $y = x^2$ sketch the graph of $y = x^2 + 3$

Skills check

1 Evaluate

a $\left(\dfrac{3}{4}\right)^4$ **b** $\left(\dfrac{1}{2}\right)^7$

c 0.001^3

2 State the value of n in these equations.
a $7^n = 343$ **b** $3^n = 243$
c $5^n = 625$

3 Transform the graph of $y = x^2$ to give the graph of $y = (x - 2)^2$

Facebook, the social media giant, celebrated its sixth birthday in February 2010 with more than 450 million users, up from about 100 million in August 2008 and a huge increase from December 2004 when there were one million members.

This graph shows how the number of Facebook users has increased over time.

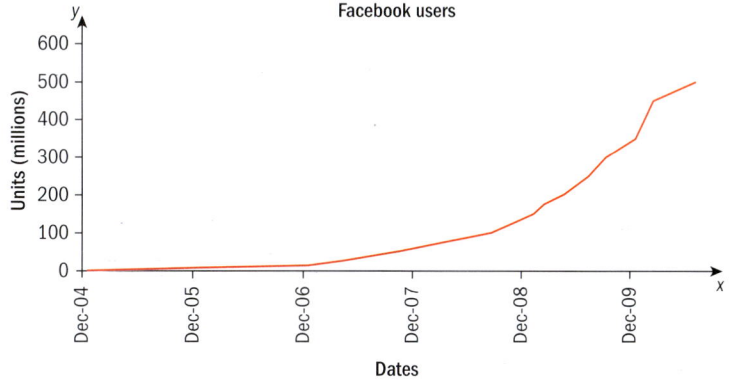

(Source: http://www.facebook.com/press/info.php?timeline)

Growth like this (certainly until February 2010) is **exponential growth**. As you move along the curve its gradient increases with the growth rate. The growth rate at any time is roughly proportional to the number of users at that time.

A good model for the data on Facebook users is
$$n = 1.32 \times 1.1^x$$
where n is the number of users in millions, and x is the number of months after December 2004.

You could use the equation $n = 1.32 \times 1.1^x$ to estimate the number of users at any specified date or to find the date at which a particular number of users was reached.

You will come across many other examples of exponential growth and its opposite, **exponential decay** (where the gradient decreases as you move along the curve).

> You could also use the model to make predictions about the future growth of Facebook. This is called 'extrapolation'. What are the problems with using a model of this type to estimate future growth?
> What other factors do you need to consider?

Investigation – folding paper

Malcolm Gladwell posed this problem in his book *The Tipping Point*. Imagine taking a large piece of paper, and folding it over again and again, until you have folded it 50 times.
How high do you think the final stack would be?

1. Fold a sheet of paper (any size) in half as many times as possible.
2. For each fold, complete this table to show the number of folds, the number of layers and the thickness of the paper formed. You can assume that a sheet of paper is about 0.1 mm thick, that is 1×10^{-7} km.

   The first few entries have been done

Number of folds	Number of layers	Thickness (km)	As thick as a
0	1	1×10^{-7}	Piece of paper
1	2	2×10^{-7}	
2	4	4×10^{-7}	Credit card
3	8		
4	16		
5			
6			
7			
8			
9			

3. How many folds would you need to make the paper
   a. as thick as the height of a table?
   b. just taller than the height of a man?
4. How thick would the paper be after 50 folds?

You can probably get to about six or seven folds before you can't fold the paper any more. At seven folds the paper is already about as thick as this textbook. In fact after only 13 folds the paper is roughly the height of a table and after 15 folds it will be much taller than any man. After 17 folds it is roughly 13 m thick, roughly the height of a two-storey house!

> Does it depend on how big the paper is to start with? Try it.

After 50 folds the paper would be approximately 113 million km thick. This is about the distance from the Earth to the Sun.

Paper folding is an example of exponential growth. The 'number of layers' of paper form a **sequence**. The terms in the sequence are a function of the number of folds, n, where $f(n) = 2^n$.

$f(n)$ is an **exponential growth function**.

In this chapter you will learn more about exponential functions and their inverses, which are called **logarithmic functions**.

4.1 Exponents

Exponents are a shorthand way of representing the repeated multiplication of a number by itself.
The expression 3^5, for example, represents $3 \times 3 \times 3 \times 3 \times 3$.
The 3 in this expression is the **base** number and the 5 is the **exponent**.
Other names for exponent are **power** and **index**.

You can also use a variable as the base, for example,
$x^4 = x \times x \times x \times x$

> It is quicker to write x^4 than $x \times x \times x \times x$

Laws of exponents
Multiplication

Simplify $x^5 \times x^3$

$x^5 \times x^3 = (x \times x \times x \times x \times x) \times (x \times x \times x)$ ← Remove brackets.
$\qquad = x \times x \times x \times x \times x \times x \times x \times x$
$\qquad = x^8$

So $x^5 \times x^3 = x^{(5+3)} = x^8$

→ $a^m \times a^n = a^{m+n}$

> Notice that in $x^5 \times x^3$ the two base variables are the same. You cannot simplify $x^5 \times y^3$, for example, using this law.
> $x^5 \times y^3 = x^5 y^3$

Division

Simplify $x^5 \div x^3$

$$x^5 \div x^3 = \frac{x \times x \times x \times x \times x}{x \times x \times x} = \frac{\cancel{x} \times \cancel{x} \times \cancel{x} \times x \times x}{\cancel{x} \times \cancel{x} \times \cancel{x}} = x \times x = x^2$$

So $x^5 \div x^3 = x^{(5-3)}$
$= x^2$

> Cancel common factors.

→ $a^m \div a^n = a^{m-n}$

> Notice that you can't simplify $x^5 \div y^3$ because the bases are not the same.

Raising to a power

Simplify $(x^5)^3$

$(x^5)^3 = (x \times x \times x \times x \times x) \times (x \times x \times x \times x \times x) \times (x \times x \times x \times x \times x)$
$= x \times x \times x \times x \times x \times x \times x \times x \times x \times x \times x \times x \times x \times x \times x$
$= x^{15}$

So $(x^5)^3 = x^{5 \times 3} = x^{15}$

→ $(a^m)^n = a^{mn}$

Example 1

Expand $(2xy^2)^3$

Answer

$(2xy^2)^3 = (2xy^2) \times (2xy^2) \times (2xy^2)$	*You don't need to show this line of working.*
$= 2^3 \times x^3 \times (y^2)^3 = 8x^3y^6$	*Apply the power of 3 to every term in the bracket.*

> Don't forget you have to raise the numbers in the bracket to the power as well as the x- and y-terms.

Exercise 4A

1 Simplify
   a $x^3 \times x^2$ b $3p^2 \times 2p^4q^2$ c $\frac{1}{2}(xy^2) \times \frac{2}{3}(x^2y)$ d $(x^3y^2)(xy^4)$

2 Simplify
   a $x^5 \div x^2$ b $2a^7 \div 2a^3$ c $2a^7 \div (2a)^3$ d $\frac{4x^3y^5}{2xy^2}$

3 Simplify
   a $(x^3)^4$ b $(3t^2)^3$ c $3(x^3y^2)^2$ d $(-y^2)^3$

> Remember to multiply the constants (numbers) together as well as the variables.

The power zero

Simplify $x^2 \div x^2$

$\frac{x^2}{x^2} = x^{2-2} = x^0$

But $\frac{x^2}{x^2} = 1$

Therefore $x^0 = 1$

> $a^0 = 1$
> Any base raised to the power of zero is equal to 1.

'Anything to the power zero is equal to 1.'
'Zero to any power is 0.'
So what about 0^0? How should we decide what this is equal to? Who should decide?

Fractional exponents

Simplify $x^{\frac{1}{2}} \times x^{\frac{1}{2}}$

Using **Law 1** $x^{\frac{1}{2}} \times x^{\frac{1}{2}} = x^{\frac{1}{2}+\frac{1}{2}} = x^1$

But $\sqrt{x} \times \sqrt{x} = (\sqrt{x})^2 = x$

so $\sqrt{x} = x^{\frac{1}{2}}$

Similarly $x^{\frac{1}{3}} \times x^{\frac{1}{3}} \times x^{\frac{1}{3}} = x$ and $\sqrt[3]{x} \times \sqrt[3]{x} \times \sqrt[3]{x} = (\sqrt[3]{x})^3 = x$

and so $\sqrt[3]{x} = x^{\frac{1}{3}}$

> $\sqrt[n]{a} = a^{\frac{1}{n}}$

You can always assume that a is positive when considering even roots of a.

Roots

Simplify $\sqrt[3]{x^6}$

Since $x^6 = x^2 \times x^2 \times x^2$

$\sqrt[3]{x^6} = \sqrt[3]{x^2 \times x^2 \times x^2}$

$= x^2$

$= x^{\frac{6}{3}}$

> $\sqrt[n]{a^m} = \left(\sqrt[n]{a}\right)^m = \left(a^m\right)^{\frac{1}{n}} = a^{\frac{m}{n}}$

Example 2

Without using a calculator, evaluate:

a $36^{\frac{1}{2}}$ **b** $\left(\frac{1}{27}\right)^{\frac{4}{3}}$

'Evaluate' means 'work out the value of'.

Answers

a $36^{\frac{1}{2}} = \sqrt{36} = 6$ Since $\sqrt[n]{a} = a^{\frac{1}{n}}$

b $\left(\frac{1}{27}\right)^{\frac{4}{3}} = \left(\left(\frac{1}{27}\right)^{\frac{1}{3}}\right)^4$ Since $\left(a^m\right)^n = a^{mn}$

$= \left(\frac{1}{\sqrt[3]{27}}\right)^4$

$= \left(\frac{1}{3}\right)^4$

$= \frac{1}{81}$

Negative exponents

Simplify $x^3 \div x^5$

$$x^3 \div x^5 = \frac{\cancel{x} \times \cancel{x} \times \cancel{x}}{\cancel{x} \times \cancel{x} \times \cancel{x} \times x \times x}$$

$$= \frac{1}{x \times x}$$

$$= \frac{1}{x^2}$$

Also $x^3 \div x^5 = x^{3-5} = x^{-2}$

And therefore $x^{-2} = \frac{1}{x^2}$

→ $a^{-n} = \frac{1}{a^n}$

You must learn the laws for exponents as they are not in the Formula booklet.

Example 3

Without using a calculator evaluate

a 6^{-2} **b** $\left(\dfrac{3}{4}\right)^{-2}$

Answers

a $6^{-2} = \dfrac{1}{6^2} = \dfrac{1}{36}$

Use $a^{-n} = \dfrac{1}{a^n}$

b $\left(\dfrac{3}{4}\right)^{-2} = \dfrac{1}{\left(\dfrac{3}{4}\right)^2} = \dfrac{1}{\left(\dfrac{9}{16}\right)}$

$= \dfrac{16}{9}$

Exercise 4B

1 Evaluate

  a $9^{\frac{1}{2}}$ **b** $125^{\frac{1}{3}}$ **c** $64^{\frac{2}{3}}$

  d $8^{\frac{2}{3}}$ **e** $\left(\dfrac{8}{27}\right)^{\frac{2}{3}}$

2 Evaluate

  a 2^{-3} **b** $32^{-\frac{2}{5}}$ **c** $81^{-\frac{1}{4}}$

  d $(2^3)^{-\frac{4}{3}}$ **e** $\left(\dfrac{64}{125}\right)^{-\frac{2}{3}}$

Example 4

Simplify these expressions.

a $5d^0$ **b** $6x^{-3} \div (2x^2)^3$ **c** $\sqrt[3]{27a^6}$ **d** $\left(\dfrac{9v^2}{16w^4}\right)^{-\frac{1}{2}}$

Here 'simplify' means write these expressions using only positive exponents.

Answers

a $5d^0 = 5 \times 1 = 5$ *Use $a^0 = 1$.*

b $6x^{-3} \div (2x^2)^3 = 6x^{-3} \div 8x^6$ *Use $(a^m)^n = a^{mn}$.*

$\qquad = \dfrac{6}{8}x^{-9} = \dfrac{3}{4x^9}$ *Use $a^m \div a^n = a^{m-n}$.*

c $\sqrt[3]{27a^6} = (27a^6)^{\frac{1}{3}} = 27^{\frac{1}{3}}(a^6)^{\frac{1}{3}}$ *Use $\sqrt[n]{a^m} = (a^m)^{\frac{1}{n}}$.*

$\qquad = 3a^2$

d $\left(\dfrac{9v^2}{16w^4}\right)^{-\frac{1}{2}} = \left(\dfrac{16w^4}{9v^2}\right)^{\frac{1}{2}}$ *Use $a^{-n} = \dfrac{1}{a^n}$.*

$\qquad = \dfrac{(16w^4)^{\frac{1}{2}}}{(9v^2)^{\frac{1}{2}}} = \dfrac{4w^2}{3v}$

Exercise 4C

1 Simplify these exponential expressions.

a $(64a^6)^{\frac{1}{2}}$ **b** $\sqrt[4]{16x^{-8}}$ **c** $\dfrac{q\sqrt{q}}{q^{-1.5}}$ **d** $\left(\dfrac{27c^3}{d^3}\right)^{-\frac{1}{3}}$ **e** $\dfrac{(8p)^{\frac{2}{3}}}{(4p)^2}$

In this exercise, make sure your answers have positive exponents.

2 Simplify these expressions.

a $\dfrac{a^{\frac{3}{2}}}{b^3} \div \dfrac{a^{-1}}{b^2}$ **b** $\sqrt{\dfrac{x^{-2}y^2}{25x^4}}$ **c** $\dfrac{6x^2y^{-2}}{\sqrt[3]{8x^{-3}}}$

4.2 Solving exponential equations

Exponential equations are equations involving 'unknowns' as exponents, for example, $5^x = 25$.
You can write an exponential equation in the form $a^x = b^y$.

Example 5

Solve $3^{x-1} = 3^{5x}$

Answer

$3^{x-1} = 3^{5x}$

$x - 1 = 5x$ *Both sides of the equation are powers of 3 so the two exponents are equal.*

$-1 = 4x$

$x = -\dfrac{1}{4}$

Example 6

Solve $3^{3x+1} = 81$	
Answer	
$3^{3x+1} = 81$	
$3^{3x+1} = 3^4$	*Write 81 as a power of 3.*
$3x+1 = 4$	*Equate exponents.*
$3x = 3$	
$x = 1$	

For this example and many of the following questions you need to learn these powers.

$2^0 = 1$ $\quad$ $3^0 = 1$
$2^1 = 2$ $\quad$ $3^1 = 3$
$2^2 = 4$ $\quad$ $3^2 = 9$
$2^3 = 8$ $\quad$ $3^3 = 27$
$2^4 = 16$ $\quad$ $3^4 = 81$
$2^5 = 32$ $\quad$ $3^5 = 243$
$2^6 = 64$
$2^7 = 128$

$5^0 = 1$ $\quad$ $7^0 = 1$
$5^1 = 5$ $\quad$ $7^1 = 7$
$5^2 = 25$ $\quad$ $7^2 = 49$
$5^3 = 125$ $\quad$ $7^3 = 343$
$5^4 = 625$

Exercise 4D

1 Solve these equations for x.
  a $2^x = 32$
  b $3^{1-2x} = 243$
  c $3^{x^2 - 2x} = 27$
  d $5^{2x-1} - 25 = 0$
  e $7^{1-x} = \dfrac{1}{49}$

2 Solve these equations for x.
  a $3^{x-3} = 3^{2-x}$
  b $5^{3x} = 25^{x-2}$
  c $9(3^{3x+1}) = \dfrac{1}{9^x}$
  d $2^{2-3x} = 4^{x-1}$

EXAM-STYLE QUESTION

3 Solve $8(2^{x+1}) = 2\sqrt{2^x}$

Example 7

Solve $3x^{-\frac{3}{5}} = 24$	
Answer	
$3x^{-\frac{3}{5}} = 24$	*Divide both sides by 3.*
$x^{-\frac{3}{5}} = 8$	*Multiply the exponent by its reciprocal since $-\dfrac{a}{b} \times -\dfrac{b}{a} = 1$*
$\left(x^{\frac{-3}{5}}\right)^{\frac{-5}{3}} = 8^{-\frac{5}{3}}$	
$x = (2^3)^{-\frac{5}{3}}$	*Replace 8 with 2^3.*
$x = 2^{-5}$	
$x = \dfrac{1}{32}$	

Exponential and logarithmic functions

Exercise 4E

1 Solve these equations for x.
- **a** $2x^4 = 162$
- **b** $x^5 - 32 = 0$
- **c** $x^{-2} = 16$
- **d** $8x^{-3} = (8x)^3$
- **e** $27x^{-2} = 81x$
- **f** $27x^{-3} = 64$

2 Solve these equations for x.
- **a** $x^{\frac{1}{3}} = 2$
- **b** $5x^{\frac{1}{2}} = 125$
- **c** $x^{-\frac{1}{4}} = 4$
- **d** $x^{\frac{2}{3}} = 16$
- **e** $x^{-\frac{3}{5}} = \frac{1}{8}$
- **f** $3x^{-\frac{1}{4}} = 6$

3 Solve these equations for x.
- **a** $x^{-\frac{3}{2}} = 125$
- **b** $6x^{-\frac{2}{3}} = 216$
- **c** $3x^{\frac{2}{3}} = 192$
- **d** $9x^{-\frac{2}{3}} = 16$

4.3 Exponential functions

Graphs and properties of exponential functions

> → An **exponential function** is a function of the form
>
> $f(x) = a^x$
>
> where a is a positive real number (that is, $a > 0$) and $a \neq 1$.

We could also write $f : x \to a^x$

Investigation – graphs of exponential functions 1

Using a GDC, sketch the graphs of these exponential functions.

- **a** $y = 3^x$
- **b** $y = 5^x$
- **c** $y = 10^x$

Look at your three graphs.
What can you deduce about the exponential function, $f(x) = a^x$, when $a > 1$?

Think about the domain, range, intercepts on the axes, asymptotes, shape and behavior of each graph as x tends to infinity.

Whatever positive value a has in the equation $f(x) = a^x$, the graph will always have the same shape.
$f(x) = a^x$ is an **exponential growth function**.

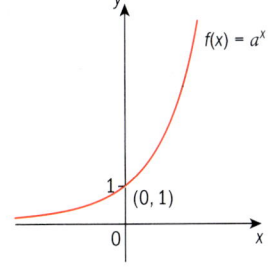

Chapter 4

→ The **domain** of $f(x) = a^x$ is the set of all real numbers.
The **range** is the set of all positive real numbers.
The curve does not intercept the x-axis.
The graph approaches closer and closer to the x-axis as the value of x decreases.
The y-intercept is 1.
The graph of f passes through the points $\left(-1, \dfrac{1}{a}\right)$, $(0, 1)$ and $(1, a)$.
The graph increases continually.

Now look at the graphs of exponential functions when the base a is between 0 and 1.

Investigation - graphs of exponential functions 2

Using a GDC sketch the graphs of these exponential functions.

a $y = 3^{-x}$
b $y = 5^{-x}$
c $y = 10^{-x}$

What can you deduce about the exponential function, $f(x) = a^{-x}$, when $a > 1$, from these three graphs?

$y = 3^{-x}$ is the equivalent of $y = \dfrac{1}{3^x}$ or $y = \left(\dfrac{1}{3}\right)^x$ so the base is between 0 and 1.

Whatever positive value a has, the graph of $f(x) = a^{-x}$ will always have this shape.

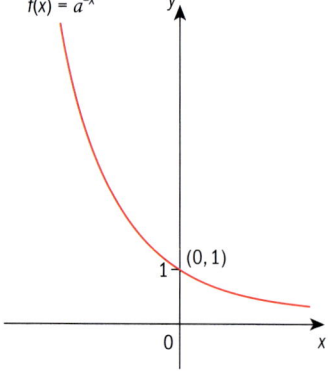

$f(x) = a^{-x}$ is an **exponential decay function**.

The natural exponential function

The base e is one that you will come across often in exponential functions.

Investigation – compound interest

When you invest money it earns interest.

We use this formula $A = C\left(1 + \dfrac{r}{n}\right)^{nt}$ to calculate the interest,

where A is the final amount (capital + interest), C is the capital, r is the interest rate expressed as a decimal, n is the number of compoundings in a year, and t is the total number of years. What happens if you start compounding more and more frequently?

1 £1 is invested at an interest rate of 100% for 1 year.

   a How much will you have if this is compounded yearly?

   $P = 1, r = 100\% = \dfrac{100}{100} = 1, n = 1, t = 1$

   $A = C\left(1 + \dfrac{1}{1}\right)^1 = 2$ (since $r = 1$ and $n = 1$)

   b How much will you have if this is compounded quarterly?
   $C = 1, r = 100\% = 1, n = 4, t = 1$

   $A = \left(1 + \dfrac{1}{4}\right)^4 = 2.44140625$

2 Copy and complete the table.

Compounding	Calculation	Final amount (write all figures on calculator)
Yearly	$\left(1 + \dfrac{1}{1}\right)^1$	2
Half-Yearly	$\left(1 + \dfrac{1}{2}\right)^2$	2.25
Quarterly	$\left(1 + \dfrac{1}{4}\right)^4$	2.44140625
Monthly		
Weekly		
Daily		
Hourly		
Every minute		
Every second		

The final amount increases as the interval between compoundings decreases but each separate increase is smaller and the final amount converges on a value. This value is called 'e'.

The value of e is approximately 2.71828 and it is an exceptionally important number in mathematics which has applications in many subject areas.

e is an **irrational** number.

Jacob Bernoulli (1654–1705) was one of the great mathematicians of the Swiss Bernoulli family. When he was looking at the problem of compound interest, he tried to find the limit of $\left(1+\dfrac{1}{n}\right)^n$ as n tends to infinity. He used the binomial theorem to show that the limit had to lie between 2 and 3. This is considered to be the first approximation found for e.

> **Mathematics sometimes throws out some surprising and beautiful results**.
>
> Here is one such result.
> To 20 decimal places e = 2.718 281 828 459 045 235 36...
> There is no obvious pattern to this chain of numbers.
> However look at this series, which gives a value of e:
>
> $$e = 1 + \frac{1}{1} + \frac{1}{2\times 1} + \frac{1}{3\times 2\times 1} + \frac{1}{4\times 3\times 2\times 1} + \frac{1}{5\times 4\times 3\times 2\times 1} + ...$$
>
> You might wonder about the connection between this series and the value of e. [See the Theory of Knowledge page at the end of this chapter for thoughts and discussion on beauty in mathematics.]

→ The graph of the exponential function $f(x) = e^x$ is a graph of exponential growth and the graph of $f(x) = e^{-x}$ is a graph of exponential decay.

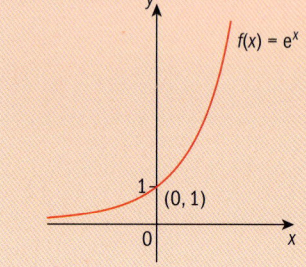 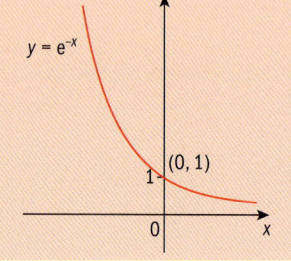

An irrational number cannot be expressed exactly as a fraction or a decimal.

Transformations of exponential functions

Now you know the general shape of the graph of an exponential function, you can use the rules for transformations of graphs from Chapter 1 to help you sketch graphs of other exponential functions.

→ $f(x) \pm k$ translates $f(x)$ through k units vertically up or down

$f(x \pm k)$ translates $f(x)$ through k units horizontally to the left or right

$-f(x)$ reflects $f(x)$ in the x-axis

$f(-x)$ reflects $f(x)$ in the y-axis

$pf(x)$ stretches $f(x)$ vertically with scale factor p

$f(qx)$ stretches $f(x)$ horizontally with scale factor $\dfrac{1}{q}$

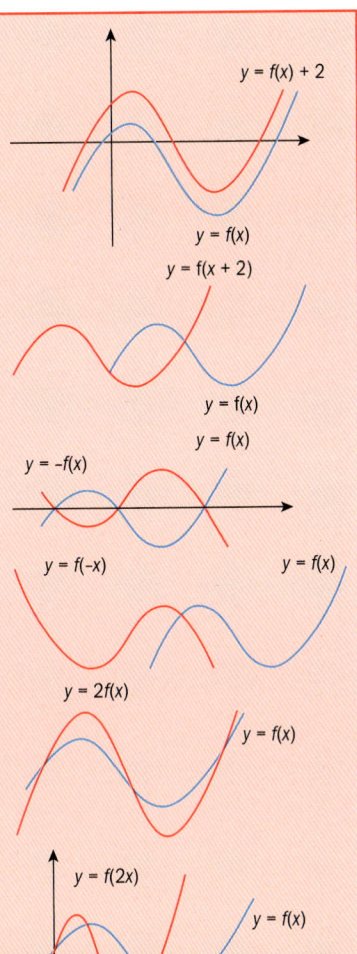

Example 8

The diagram shows the sketch of $f(x) = 2^x$
On the same axes sketch the graph of $g(x) = 2^{x-2}$

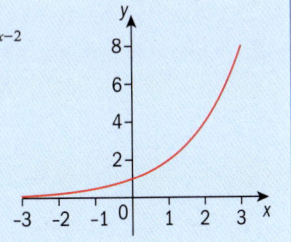

Answer

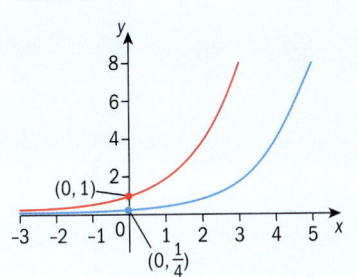

You find $g(x)$ by translating $f(x)$ through 2 units to the right.

The graph of $g(x)$ will pass through the point $\left(0, \dfrac{1}{4}\right)$.

Both graphs get closer and closer to the x-axis as the value of x decreases.

Exercise 4F

1 Given the graph of $f(x)$, and without using a calculator, sketch the graph of $g(x)$ on the same set of axes showing clearly any intercepts on the axes and any asymptotes.

a $f(x) = 2^x$ $g(x) = 2^x + 3$

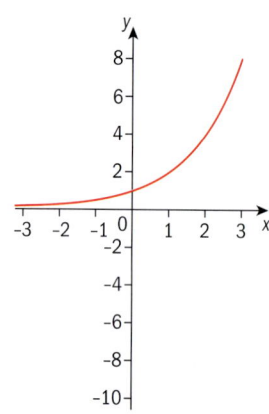

b $f(x) = 3^x$ $g(x) = 3^{-x}$

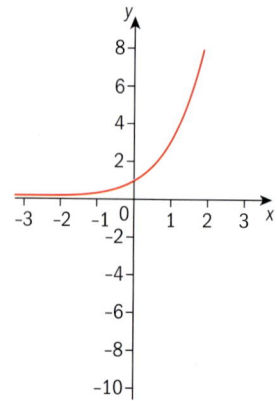

c $f(x) = \left(\dfrac{1}{2}\right)^x$ $g(x) = -\left(\dfrac{1}{2}\right)^{-x}$

d $f(x) = e^x$ $g(x) = e^{x+1}$

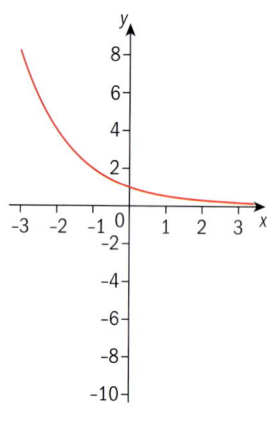

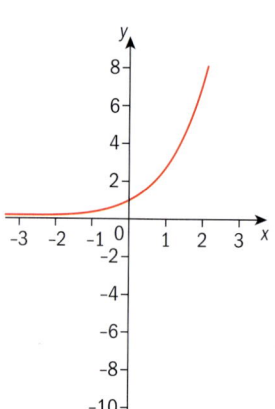

e $f(x) = \left(\dfrac{1}{3}\right)^x$ $g(x) = 2\left(\dfrac{1}{3}\right)^x$ f $f(x) = \left(\dfrac{1}{e}\right)^x$ $g(x) = \left(\dfrac{1}{e}\right)^{2x}$

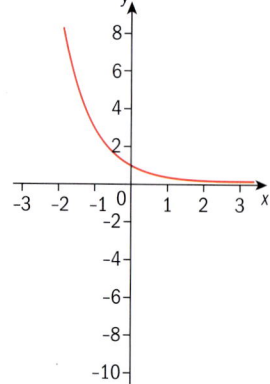

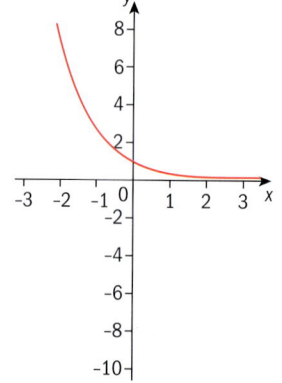

2 State the domain and range of each $g(x)$ function in question 1.

4.4 Properties of logarithms

Look at this equation: $2^3 = 8$

2 is the base and 3 is the exponent or **logarithm**.
So we say that the **logarithm** of 8 to the base 2 is 3 and write this as $\log_2 8 = 3$
In general, given that $a > 0$:

> → If $b = a^x$ then $\log_a b = x$
>
> or, if b is a to the power x, then x is the logarithm of b, to base a.

Being able to change between these two forms allows you to simplify log statements.

Example 9

Evaluate $\log_5 125$

Answer

$x = \log_5 125$	Write '$x =$' the log statement.
$5^x = 125$	Change equation to exponent form.
$5^x = 5^3$	Equate exponents.
$x = 3$	

Example 10

Evaluate $\log_{64} 4$

Answer

$x = \log_{64} 4$	
$64^x = 4$	Change equation to exponent form.
$(4^3)^x = 4^1$	Write 64 as 4^3
$3x = 1$	Equate the exponents
$x = \dfrac{1}{3}$	and solve for x.

Exercise 4G

1 Evaluate these expressions.

    a $\log_7 49$ **b** $\log_5 \sqrt{5}$ **c** $\log_2 64$ **d** $\log_9 1$

2 Evaluate these expressions.

    a $\log_3 \dfrac{1}{81}$ **b** $\log_5 125^{\frac{1}{2}}$ **c** $\log_{32} 8$ **d** $\log_3 3^4$

Example 11

Evaluate $\log_4 4$	
Answer	
$x = \log_4 4$ $4^x = 4$ $x = 1$	*Write 'x =' log statement.* *Change equation to exponent form.* *Equate exponents ($4 = 4^1$).*

In general, the log to base a of any number $a = 1$.

→ $\log_a a = 1$

Example 12

Evaluate $\log_5 1$	
Answer	
$x = \log_5 1$ $5^x = 1$ $x = 0$	*Write equation in exponent form.*

Any number raised to the power 0 is equal to 1 so the log of 1 in any base is 0.

→ $\log_a 1 = 0$

Exercise 4H

1 Evaluate
   a $\log_6 6$
   b $\log_{10} 10$
   c $\log_n n$
   d $\log_8 1$
   e $\log_2 1$
   f $\log_b 1$

Some log expressions are **undefined** – this means that you can't find solutions for them.

1 What happens when you try to evaluate the expression $\log_3(-27)$?
   First write the log equation.
   $x = \log_3(-27)$

   Then rewrite the equation in exponent form.
   $3^x = -27$

   This equation has no solution.
   You can only find logarithms of **positive** numbers.

→ $\log_a b$ is undefined for any base a if b is negative.

2 What is the value of $\log_3 0$?

First write an equation.

$x = \log_3 0$

Rewrite in exponent form.

$3^x = 0$

This equation has no solution.

→ $\log_a 0$ is undefined.

Example 13 illustrates another property of logarithms.

Example 13

Evaluate $\log_2 2^5$

Answer

$x = \log_2 2^5$	*Write log equation.*
$2^x = 2^5$	*Rewrite in exponent form.*
$x = 5$	*Solve.*

→ $\log_a(a^n) = n$

Summary of properties of logarithms

Given that $a > 0$
- If $x = a^b$ then $\log_a x = b$
- $\log_a a = 1$
- $\log_a 1 = 0$
- $\log_a b$ is undefined if b is negative
- $\log_a 0$ is undefined
- $\log_a(a^n) = n$

Example 14

Find the value of x if $\log_2 x = 5$

Answer

$\log_2 x = 5$	
$2^5 = x$	*Rewrite in exponent form.*
$x = 32$	*Solve.*

Exercise 4I

1 Write these equations in log form.

   a $x = 2^9$ **b** $x = 3^5$ **c** $x = 10^4$ **d** $x = a^b$

2 Write these equations in exponent form.

   a $x = \log_2 8$ **b** $x = \log_3 27$ **c** $x = \log_{10} 1000$ **d** $x = \log_a b$

3 Solve these equations.
   a $\log_4 x = 3$
   b $\log_3 x = 4$
   c $\log_x 64 = 2$
   d $\log_x 6 = \dfrac{1}{2}$
   e $\log_2 x = -5$

4.5 Logarithmic functions

Investigation – inverse functions

What kind of function would undo an exponential function such as $f : x \mapsto 2^x$?

a Copy and complete this table of values for the function $y = 2^x$.

x	−3	−2	1	0	1	2	3
y	$\dfrac{1}{8}$						

$f : x \mapsto 2^x$ means that f is a function under which x is mapped to 2^x.

The **inverse** function of $y = 2^x$ will take all the x- and y-values and switch them.

b Copy and complete this table of values for the inverse function of $y = 2^x$.

x	$\dfrac{1}{8}$						
y	−3						

c Using these tables of values sketch a graph of both $y = 2^x$ and its inverse function on the same set of axes.
d What do you notice?

Now let's find the equation of the graph of the inverse function.

→ To find an **inverse** of a function algebraically, switch x and y and then rearrange to make y the subject.

To get the inverse function, f^{-1}, of $f : x \mapsto 2^x$:

Write $y = 2^x$
$x = 2^y$ Switch x and y
$\log_2 x = y \log_2 2$ Take logs to the base 2 of both sides
so $y = \log_2 x$ Since $\log_2 2 = 1$

So $f^{-1} : x \mapsto \log_2 x$

$f : x \mapsto 2^x$ is another way of writing $y = 2^x$

y is the exponent that the base 2 is raised by in order to get x

Log is short for logarithm.

→ Generally if $f : x \mapsto a^x$ then $f^{-1} : x \mapsto \log_a x$
$y = \log_a x$ is the inverse of $y = a^x$.

Exponential and logarithmic functions

The graph of $y = \log_a x$ is a reflection of $y = a^x$ in the line $y = x$.

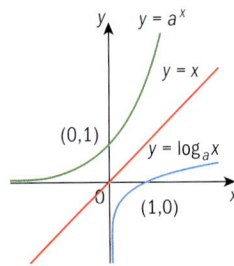

→ A logarithmic function, $f(x) = \log_a x$, has these properties:
the domain is the set of all positive real numbers
the range is the set of all real numbers
the curve does not intercept the y-axis
the y-axis is a vertical asymptote
the x-intercept is 1
the graph is continually increasing.

John Napier (1550–1617) is credited with much of the early work on logarithms. Would you say that he *invented* logarithms or *discovered* them?

Transformations of logarithmic functions

Again once you know the general shape of the graph of a logarithmic function you can use what you learnt in Chapter 1 to consider the graphs of other logarithmic functions.

Exercise 4J

1 Given the function $f(x) = \log_a x$ describe the transformation required in each case to obtain the graph of $g(x)$
  a $g(x) = \log_a(x) - 2$
  b $g(x) = \log_a(x - 2)$
  c $g(x) = 2\log_a x$

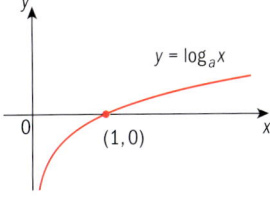

EXAM-STYLE QUESTION

2 Sketch the graph of $y = -2\log(x - 1)$ without using a calculator. Include on your graph the intercepts with the two axes (if they exist).

When no base is given the logarithms are base 10.

3 Sketch the graph of $y = \log_2(x + 1) + 2$ clearly labeling any asymptotes on the graph.

4 The sketch shows the graph of $y = \log_a x$. Find the value of a.

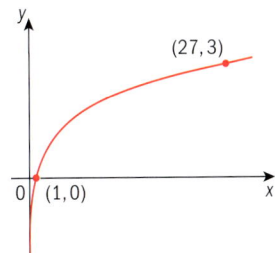

5 Given that $f(x) = \log_3 x$ find $f^{-1}(2)$

Logarithms to base 10

$y = \log_{10} x$ is the inverse of $y = 10^x$. This is an important logarithm as it is one of the only ones that you can use the calculator to find. Base 10 logs are called common logs and you can omit the base and just write $\log x$ for $\log_{10} x$.
There is a 'log' key on the calculator.

Example 15

Use a calculator to evaluate $\log 2$ to 3 dp.

Answer

$\log 2 = 0.301$ to 3 dp.

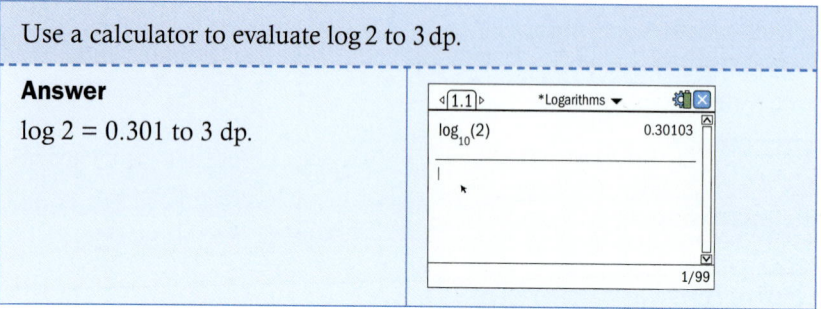

GDC help on CD: Alternative demonstrations for the TI-84 Plus and Casio FX-9860GII GDCs are on the CD.

Natural logarithms

The **natural logarithm**, $\log_e x$ (log to the base e), is the other important logarithm.
You write $\ln x$ for $\log_e x$. There is an 'ln' key on the calculator

Example 16

Use a calculator to evaluate $\dfrac{\ln 4}{\ln 2}$

Answer

$\dfrac{\ln 4}{\ln 2} = 2$

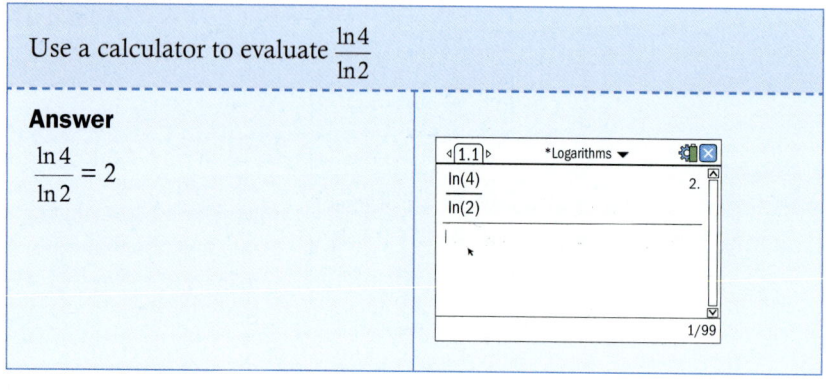

Make sure you close the brackets after the 4 otherwise the calculator will calculate $\ln\left(\dfrac{4}{\ln 2}\right)$

GDC help on CD: Alternative demonstrations for the TI-84 Plus and Casio FX-9860GII GDCs are on the CD.

Exercise 4K

1 Use a calculator to evaluate these expressions correct to 3 significant figures.

   a $\log 3$ **b** $4\log 2$ **c** $\ln \sqrt{5}$

   d $\dfrac{\log 4}{\log 5}$ **e** $\dfrac{\ln 4}{\ln 5}$ **f** $\log \dfrac{4}{5}$

   g $(\log 3)^2$ **h** $\log 3^2$

→ $y = \ln x$ is the inverse of the exponential function $y = e^x$

This relation gives us three important results:

→ $\log_a(a^x) = x$ and $a^{\log_a x} = x$
$\ln(e^x) = x$ and $e^{\ln x} = x$
$\log(10^x) = x$ and $(10^{\log x}) = x$

Example 17

Solve these equations, giving your answers to 3 significant figures.
a $e^x = 2.3$ **b** $\ln x = -1.5$ **c** $10^x = 0.75$ **d** $\log x = 3$

Answers

a $e^x = 2.3$
$\ln(e^x) = \ln 2.3$ *Write in natural log form.*
$x = 0.833 (3\,sf)$

b $\ln x = -1.5$ *Use $\ln(e^x) = x$ and evaluate.*
$e^{\ln x} = e^{-1.5}$ *Use $(e^{\ln x}) = x$ and evaluate.*
$x = 0.223 (3\,sf)$

c $10^x = 0.75$ *Use $\log(10^x) = x$ and evaluate.*
$\log(10^x) = \log 0.75$
$x = -0.125 (3\,sf)$

d $\log x = 3$ *Use $10^{\log x} = x$ and evaluate.*
$10^{\log x} = 10^3$
$x = 1000$

Example 18

Given that $f(x) = \frac{1}{3}e^{2x}$, find $f^{-1}(x)$.

Answer

$f(x) = \frac{1}{3}e^{2x}$

$y = \frac{1}{3}e^{2x}$

$x = \frac{1}{3}e^{2y}$ *Interchange x and y.*

▶ Continued on next page

$3x = e^{2y}$
$\ln(3x) = \ln e^{2y}$
$\ln(3x) = 2y$ Use $\ln(e^x) = x$.
$\frac{1}{2}\ln(3x) = y$ Solve for y.
So $f^{-1}(x) = \frac{1}{2}\ln(3x)$, $x > 0$

Exercise 4L

1 Solve these equations giving answers to 3 sf where necessary.
   a $e^x = 1.53$ **b** $e^x = 0.003$ **c** $e^x = 1$
   d $e^x = \frac{1}{2}$ **e** $5e^x = 0.15$

2 Solve these equations giving answers to 3 sf where necessary.
   a $10^x = 2.33$ **b** $10^x = 0.6$ **c** $10^x = 1$ **d** $10^x = \frac{1}{2}$

3 Find x if
   a $\log x = 2$ **b** $\log x = -1$ **c** $\log x = 0$ **d** $\log x = -5.1$

4 Without using a calculator evaluate these expressions.
   a $5^{\log_5 12}$ **b** $5^{\log_5 4}$ **c** $e^{\ln\sqrt{3}}$ **d** $e^{\ln 4}$

5 Without using a calculator evaluate these expressions.
   a $\ln e^5$ **b** $\log 100$ **c** $\ln 1$ **d** $\ln e$ **e** $\ln \frac{1}{e^3}$

EXAM-STYLE QUESTIONS

6 Given that $f(x) = e^{2x-1}$ find $f^{-1}(x)$ and state its domain.

7 Given that $f(x) = e^{0.25x}$, $-2 \leq x \leq 4$, state the domain and range of f^{-1}.

8 Given that $f(x) = \ln 3x$, $x > 0$, find $f^{-1}(x)$.

9 Given that $f(x) = \ln(x-1)$, $x > 1$, and $g(x) = 2e^x$ find $(g \circ f)(x)$.

4.6 Laws of logarithms

We can deduce the laws of logarithms from the exponential equations, $x = a^p$ and $y = a^q$.

        $x = a^p$ and $y = a^q$
then $p = \log_a x$ and $q = \log_a y$
and $xy = a^p \times a^q = a^{p+q}$
so $\log_a xy = p + q$
and hence $\log_a xy = \log_a x + \log_a y$

This equation is true for logarithms in any base so

→ $\log x + \log y = \log xy$

> Notice that
> $\log xy \neq \log x \times \log y$
> and that $\log \dfrac{x}{y} \neq \dfrac{\log x}{\log y}$

$$\dfrac{x}{y} = a^p \div a^q = a^{p-q}$$

so $\log_a \dfrac{x}{y} = p - q$

and hence $\log_a \dfrac{x}{y} = \log_a x - \log_a y$

→ $\log x - \log y = \log \dfrac{x}{y}$

$$x^n = (a^p)^n = a^{pn}$$

so $\log_a x^n = pn$

and hence $\log_a x^n = n \log_a x$

→ $n \log x = \log x^n$

We can also derive this key result from the third law.

→ $\log_a \dfrac{1}{x} = \log_a x^{-1} = -1 \times \log_a x = -\log_a x$

All these laws are true for logarithms in any base and so the bases can be omitted. You must learn these laws as they are not in the Formula booklet.

Example 19

Express $\log_2 5 + \dfrac{1}{2}\log_2 36 - \log_2 10$ as a single logarithm.

Answer

$\log_2 5 + \dfrac{1}{2}\log_2 36 - \log_2 10$

$= \log_2 5 + \log_2 36^{\frac{1}{2}} - \log_2 10$ $n\log_a x = \log_a x^n$

$= \log_2 5 + \log_2 6 - \log_2 10$

$= \log_2 30 - \log_2 10$ $\log x + \log y = \log xy$

$= \log_2 3$ $\log x - \log y = \dfrac{x}{y}$

Chapter 4

Exercise 4M

1 Express as single logarithms:

a $\log 5 + \log 6$
b $\log 24 - \log 2$
c $2\log 8 - 4\log 2$
d $\frac{1}{2}\log 49$
e $3\log x - 2\log y$
f $\log x - \log y - \log z$
g $\log x + 2\log y - 3\log xy$

2 Express as single logarithms:

a $\log_2 6 + 2\log_2 3 - \log_2 4$
b $\log_3 40 - \log_3 15 + 2\log_3 \left(\frac{3}{5}\right)$
c $\log_a 4 + 2\log_a 3 - 2\log_a 6$
d $2\ln 3 - \ln 18$
e $3\ln 2 - 2$
f $4\log_2 x + \frac{1}{3}\log_2 y - 5\log_2 z$

3 Find the value of each expression (each answer is an integer).

a $\log_6 2 + \log_6 18$
b $\log_2 24 - \log_2 3$
c $\log_8 2 + \log_8 32$
d $2\log_6 3 + \log_6 24$
e $\frac{1}{2}\log 36 - \log 15 + 2\log 5$

Example 20

> Given that $a = \log_5 x$, $b = \log_5 y$ and $c = \log_5 z$,
> write $\log_5\left(\dfrac{\sqrt{x}}{y^2 z^3}\right)$ in terms of a, b and c.
>
> **Answer**
>
> $\log_5\left(\dfrac{\sqrt{x}}{y^2 z^3}\right) = \log_5 \sqrt{x} - \log_5 y^2 z^3$
>
> $\qquad = \log_5 x^{\frac{1}{2}} - (\log_5 y^2 + \log_5 z^3)$
>
> $\qquad = \frac{1}{2}\log_5 x - 2\log_5 y - 3\log_5 z$
>
> $\qquad = \frac{1}{2}a - 2b - 3c$

Exercise 4N

EXAM-STYLE QUESTION

1 Given that $p = \log_2 a$ and $q = \log_2 b$, find an expression in terms of p and/or q for:

a $\log_2 ab$
b $\log_2 a^3$
c $\log_2 \dfrac{b}{a}$
d $\log_2 \sqrt{b}$
e $\log_2 \dfrac{b^2}{\sqrt{a}}$

2 Let $x = \log P$, $y = \log Q$ and $z = \log R$.

Express $\log\left(\dfrac{P^2}{QR^2}\right)^3$ in terms of x, y and z.

3 Write these expressions in the form $a + b\log x$ where a and b are integers.

   a $\log 10x$ **b** $\log \dfrac{100}{x^2}$ **c** $\log \sqrt{10x}$ **d** $\log \dfrac{1}{10\sqrt{x}}$

EXAM-STYLE QUESTIONS

4 Given that $y = \log_3 \dfrac{27^a}{81}$ write y in the form $y = pa + q$

where p and q are integers to be found.

5 Write $\log_3 \dfrac{1}{27x^2}$ in the form $a + b\log_3 x$ where a and b are integers.

6 Show that $e^{x\ln 2} = 2^x$

Notice that question 6 in Exercise 4N demonstrates the general result
$$a^x = e^{x\ln a}$$

Change of base

Sometimes you need to change the base of a logarithm and there is a formula that enables you to do this.

Suppose $y = \log_b a$ and you want to change the log to base c.
If $y = \log_b a$ then $a = b^y$.
Start with $a = b^y$.
Take logs to base c of both sides:
$$\log_c a = \log_c b^y$$
$$\log_c a = y\log_c b$$
$$y = \dfrac{\log_c a}{\log_c b}$$

But $y = \log_b a$ so

> **Change of base formula:**
> $$\log_b a = \dfrac{\log_c a}{\log_c b}$$

This formula is useful as most calculators only give logs to base 10 or e.

You can use this formula to evaluate a logarithm or to change a logarithm to any base.

Example 21

Use the change of base formula to evaluate $\log_4 9$ to 3 significant figures.

Answer

$\log_4 9 = \dfrac{\log 9}{\log 4}$ Change the log to base 10.

$\qquad = 1.58\ (3\text{ sf})$ Use calculator to evaluate answer.

> For base 10 logs, the 10 is omitted.

Example 22

$\log_x 3 = a$ and $\log_x 6 = b$.
Find $\log_3 6$ in terms of a and b.

Answer

$\log_3 6 = \dfrac{\log_x 6}{\log_x 3}$ Use the change of base formula.

$\qquad = \dfrac{b}{a}$

Exercise 40

1 Use the change of base formula to evaluate these expressions to 3 significant figures.

  a $\log_2 7$ b $\log_5\left(\dfrac{1}{7}\right)$ c $\log_3(0.7)$

  d $\log_7 e$ e $\log_3 7^7$

2 Given that $\log_3 x = y$, express $\log_9 x$ in terms of y.

EXAM-STYLE QUESTION

3 If $\log_a 2 = x$ and $\log_a 6 = y$, find in terms of x and y:
  a $\log_2 6$ b $\log_6 2$ c $\log_2 36$
  d $\log_a 24$ e $\log_6 12$ f $\log_2 3$

4 Use your GDC to sketch these graphs.
  a $y = \log_4 x$ b $y = 2\log_5 x$

5 Given that $\log_4 a = b$ express y in terms of b.
  a $y = \log_4 a^2$ b $y = \log_{16} a$
  c $y = \log_{\frac{1}{4}} a^2$ d $y = \log_{\frac{1}{16}} \sqrt{a}$

4.7 Exponential and logarithmic equations

Solving exponential equations

You can use logarithms to solve exponential equations.
In Section 4.2 you solved exponential equations where the base numbers were the same or could be made the same. In this section you will learn how to solve equations where the base numbers are different.

Example 23

Solve $5^x = 9$

Answer

$5^x = 9$ — Take logs of both sides.
$\log 5^x = \log 9$ — Now bring down the exponent.
$x \log 5 = \log 9$ — Rearrange the equation.
$x = \dfrac{\log 9}{\log 5}$
$x = 1.3652\ldots$
$x = 1.37$ (3 sf) — Check whether the question requires an exact answer

> Choose base 10 or natural logs so that you can use your GDC.

Example 24

Solve $6^x = 3^{x+1}$ giving your answer in the form $\dfrac{\ln a}{\ln b}$ where a and b are integers.

Answer

$6^x = 3^{x+1}$
$\ln 6^x = \ln 3^{x+1}$ — Take natural logs of both sides.
$x \ln 6 = (x+1) \ln 3$ — Bring down the exponents.
$x \ln 6 = x \ln 3 + \ln 3$ — Multiply out brackets.
$x \ln 6 - x \ln 3 = \ln 3$ — Collect x-terms together.
$x(\ln 6 - \ln 3) = \ln 3$ — Factorize and divide.
$x = \dfrac{\ln 3}{(\ln 6 - \ln 3)}$
$x = \dfrac{\ln 3}{\ln 2}$ — $\ln a - \ln b = \ln \dfrac{a}{b}$

Example 25

Solve $e^{3x} = 5^{1-x}$, giving an exact answer.

Answer

$e^{3x} = 5^{1-x}$	Use natural logs since $\ln e^x = x$
$\ln e^{3x} = \ln 5^{1-x}$	
$3x = (1-x)\ln 5$	Bring down the exponents.
$3x = \ln 5 - x\ln 5$	
$3x + x\ln 5 = \ln 5$	Multiply out brackets.
$x(3 + \ln 5) = \ln 5$	Collect x-terms together.
$x = \dfrac{\ln 5}{(3 + \ln 5)}$	Factorize and divide.

Leave your answer in log form since an exact answer is required.

Exercise 4P

1 Solve these equations to find the value of x to 3 significant figures.

   a $2^x = 5$ **b** $3^x = 50$ **c** $5^{-x} = 17$ **d** $7^{x+1} = 16$

   e $\left(\dfrac{1}{3}\right)^x = \dfrac{7}{9}$ **f** $2^{2x-1} = 3.2 \times 10^{-3}$ **g** $e^x = 6$ **h** $e^{\frac{x}{5}} = 0.11$

EXAM-STYLE QUESTION

2 Solve these equations to find the value of x to 3 significant figures.

   a $2^{x+2} = 5^{x-3}$ **b** $3^{2-x} = 4^{2x-5}$ **c** $3^{\frac{x}{3}} = 5^{x+3}$ **d** $7^x = (0.5)^{x-1}$

   e $e^{3x-1} = 3^x$ **f** $4e^{3x-2} = 244$ **g** $35e^{-0.001x} = 95$

Example 26

Solve $3 \times 6^{x-1} = 2 \times 3^{x+2}$, giving your answer in the form $x = \dfrac{\ln a}{\ln b}$, where $a, b \in \mathbb{Z}$.

Answer

$\ln(3 \times 6^{x-1}) = \ln(2 \times 3^{x+2})$	Take natural logs of both sides.
$\ln 3 + \ln(6^{x-1}) = \ln 2 + \ln(3^{x+2})$	
$\ln 3 + (x-1)\ln 6 = \ln 2 + (x+2)\ln 3$	Collect x-terms and factorize.
$\ln 3 + x\ln 6 - \ln 6 = \ln 2 + x\ln 3 + 2\ln 3$	
$x\ln 6 - x\ln 3 = \ln 2 + 2\ln 3 + \ln 6 - \ln 3$	
$x(\ln 6 - \ln 3) = \ln 2 + \ln 9 + \ln 6 - \ln 3$	
$x = \dfrac{\ln\left(\dfrac{108}{3}\right)}{\ln\left(\dfrac{6}{3}\right)} = \dfrac{\ln 36}{\ln 2}$	You can't simplify this any further — $\dfrac{\ln a}{\ln b} \neq \ln \dfrac{a}{b}$

Exponential and logarithmic functions

Exercise 4Q

EXAM-STYLE QUESTIONS

1 Solve these equations to find the value of x to 3 significant figures.

    a $7 \times 3^x = 25$ **b** $4 \times 3^x = 5^{2x-1}$ **c** $3 \times 2^x = 4 \times 5^x$

    d $5 \times 2^{x-1} = 3 \times 7^{2x}$ **e** $3^x 4^{x-1} = 2 \times 7^{x+2}$

2 Solve these equations to find the value of x in the form $x = \dfrac{\ln a}{\ln b}$, where $a, b \in \mathbb{Q}$

    a $2^{x+2} = 5^{x-3}$ **b** $5 \times 3^x = 8 \times 7^x$

    c $5 \times 3^{x+1} = 2 \times 6^{3-2x}$ **d** $(6^x)(2^{x-1}) = 2(4^{x+2})$

3 Solve for x

    a $e^{2x} - e^x = 0$ **b** $4^x - 3(2^x) = 0$

Solving logarithmic equations

Some logarithmic equations can be solved by ensuring that both sides of the equation contain logarithms written to the same base. Then you can equate the **arguments**.

> The argument is the expression inside the brackets.

Example 27

Solve $\log_a(x^2) = \log_a(3x+4)$

Answer

$\log_a(x^2) = \log_a(3x+4)$
$x^2 = 3x + 4$ *Equate the arguments.*
$x^2 - 3x - 4 = 0$ *Solve the quadratic.*
$(x-4)(x+1) = 0$
$x = 4$ or $x = -1$

You **must** check that both solutions are possible.
Remember you cannot find the logarithm of a negative number.
Substituting $x = 4$ and $x = -1$ into both sides of the original equation gives the log of a positive number so here both solutions are possible.

Example 28

Solve $\ln(12 - x) = \ln x + \ln(x - 5)$

Answer

$\ln(12 - x) = \ln x + \ln(x - 5)$
$\ln(12 - x) = \ln x(x - 5)$
$\ln(12 - x) = \ln(x^2 - 5x)$
$12 - x = x^2 - 5x$ *Equate arguments.*
$x^2 - 4x - 12 = 0$ *Solve the quadratic.*
$(x - 6)(x + 2) = 0$
$x = 6$ or $x = -2$

▶ Continued on next page

When $x = 6$	Check solutions.
$\ln x$ and $\ln(x-5)$ are positive. When $x = -2$ $\ln x$ and $\ln(x-5)$ are negative so $x = 6$ is the only solution.	

Exercise 4R

EXAM-STYLE QUESTION

1 Solve these equations for x.

   a $\log_2(x) = \log_2(6x - 1)$ **b** $\ln(x+1) = \ln(3-x)$

   c $\log_5(2-x) = \log_5(6x-1)$ **d** $\log_2(2x+3) + \log_2(x-1) = \log_2(x+1)$

   e $\log_3 x - \log_3(x-1) = \log_3(x+1)$

Sometimes it is easier to solve a log equation using exponents.

Example 29

Solve $\log_5(x - 2) = 3$

Answer

$\log_5(2x - 1) = 3$	
$\quad\quad 5^3 = 2x - 1$	Since $\log_a x = b \Rightarrow x = a^b$
$\quad 125 = 2x - 1$	
$\quad\quad 2x = 126$	
$\quad\quad\quad x = 63$	

Example 30

Solve $\log_2 x + \log_2(x - 2) = 3$

Answer

$\log_2 x + \log_2(x-2) = 3$	
$\quad\log_2[x(x-2)] = 3$	Using the first law on page 123.
$\quad\log_2(x^2 - 2x) = 3$	
$\quad\quad x^2 - 2x = 2^3$	Since $\log_a x = b \Rightarrow x = a^b$
$\quad\quad x^2 - 2x = 8$	
$\quad x^2 - 2x - 8 = 0$	
$\quad (x+2)(x-4) = 0$	
$\quad\quad x = -2$ or $x = 4$	x must be positive.
$\quad x = 4$ is the only solution	

Exercise 4S

1 Solve these equations for x.
  a $\log_9(x-2) = 2$ **b** $\log_3(2x-1) = 3$ **c** $\log_{\frac{1}{2}}(3-x) = 5$

2 Solve these equations for x.
  a $\log_6(x-5) + \log_6 x = 2$ **b** $\log_2(4x-8) - \log_2(x-5) = 4$
  c $\log_7(2x-3) - \log_7(4x-5) = 0$

EXAM-STYLE QUESTIONS

3 Given that $\log_2 x + \log_2(2x+7) = \log_2 A$
  find an expression for A in terms of x.
  Hence or otherwise solve $\log_2 x + \log_2(2x+7) = 2$

4 Solve $\log_4 x + \log_x 4 = 2$

> You will need to change the base here first.

5 Solve $\log_2 x^2 + \log_4 \sqrt{x} = 9$

4.8 Applications of exponential and logarithmic functions

Extension material on CD:
Worksheet 4 - Reduction to linear form

Exponential growth and decay

Models of exponential growth and decay use exponential functions.

Here are just a few applications of exponential growth and decay models.

Biology
- Growth of micro-organisms in a culture
- Human population
- Spread of a virus

Physics
- Nuclear chain reactions
- Heat transfer

Economics
- Pyramid schemes

Computer technology
- Processing power of computers
- Internet traffic growth

> You may wish to pick one of these as the basis of your Mathematical Exploration.

Two areas of mathematics that appear to be completely disconnected might be exponentials and probability.
But consider this problem…
A group of people go to lunch and afterwards pick up their hats at random. What is the probability that no one gets their own hat?
It can be shown that this probability is $\dfrac{1}{e}$.
(You might like to explore this once you have studied probability further.)
Can you think of any other areas of knowledge that are surprisingly connected?

Exponential growth

Example 31

The population, $A(t)$, in thousands, of a city is modeled by the function $A(t) = 30e^{(0.02)t}$ where t is the number of years after 2010. Use this model to answer these questions:

a What was the population of the city in 2010?
b By what percentage is the population of the city increasing each year?
c What will the population of the city be in 2020?
d When will the city's population be 60 000?

Answers

a $A(0) = 30e^0$
$= 30$
The population in 2010 was 30 000.

*t is the number of years **after** 2010, so for 2010, $t = 0$*

b $A(1) = 30e^{(0.02)}$
$\dfrac{30e^{(0.02)}}{30} = e^{(0.02)}$
$= 1.0202...$
The population is increasing at 2.02% each year.

Write an equation for the population one year after 2010.
Calculate the multiplying factor.

c $A(10) = 30e^{(0.02) \times 10}$
$= 36.642...$
In 2020 the population will be 36 642

In 2020, $t = 10$

d $60 = 30e^{(0.02)t}$
$2 = e^{(0.02)t}$
$\ln 2 = \ln e^{(0.02)t}$
$\ln 2 = 0.02t$
$t = \dfrac{\ln 2}{0.02}$
$t = 34.657...$

The population will be 60 000 after 34.65 years, that is, during 2044.

When population is 60 000, $A(t) = 60$
Take logarithmics of each side.
Bring down the exponent.
Solve for t.

Exponential decay

Example 32

A casserole is removed from the oven and cools according to the model with equation $T(t) = 85e^{-0.1t}$, where t is the time in minutes and T is the temperature in °C.
a What is the temperature of the casserole when it is removed from the oven?
b If the temperature of the room is 25 °C, how long will it take for the casserole to reach room temperature?

Answers

a $T(0) = 85e^0$
     $= 85$
   The temperature of the casserole is 85 °C

When the casserole is removed from the oven, $t = 0$

b $85e^{-0.1t} = 25$
   $e^{-0.1t} = \dfrac{25}{85} = \dfrac{5}{17}$
   $\ln e^{-0.1t} = \ln \dfrac{5}{17}$
   $-0.1t = \ln \dfrac{5}{17}$
   $\quad\quad = -1.22377...$
   $t = 12.2$ (3 sf)

$T = 25$ if the temperature of the room is 25 °C.
Take logarithms of both sides.

Solve for t.

The casserole will reach room temperature after 12.2 min.

Exercise 4T

1 The sum of €450 is invested at 3.2% interest, compounded annually.
   a Write down a formula for the value of the investment after n years.
   b After how many years will the value first exceed €600?

2 In the early stages of a measles epidemic there were 100 infected people and each day the number rose by 10%.
   a How many people were infected
      i   after 2 days   ii after a week?
   b How long would it take for 250 people to be infected?

3. Forest fires spread exponentially. Every hour that the fire is left to burn unchecked 15% of the remaining area is burnt.
   If 10 hectares are burnt and the fire becomes out of control how long will it take until 10 000 hectares are burning?

4. Joseph did a parachute jump for charity. After jumping out of the aircraft his velocity at time t seconds after his parachute opened was v m s^{-1} where
   $$v = 9 + 29e^{-0.063t}$$
   a. Sketch the graph of v against t.
   b. What was Joseph's speed at the instant the parachute opened?
   c. What was his lowest possible speed if he fell from a very great height?
   d. If he actually landed after 45 seconds what was his speed on landing?
   e. How long did it take him to reach half the speed he had when the parachute opened?

5. Two variables x and n are connected by the formula $x = a \times n^b$
   When $n = 2$, $x = 32$ and when $n = 3$, $x = 108$. Find the values of a and b.

The American geologist Charles Richter defined the magnitude of an earthquake to be

$$M = \log \frac{I}{S}$$

where M is the magnitude (as a decimal), I is the intensity of the earthquake (measured by the amplitude of a seismograph reading in mm taken 100 km from the epicenter of the earthquake) and S is the intensity of a 'standard' earthquake. The intensity of a standard earthquake (S) is 0.001 millimetres.
Explore the Richter Scale further.

Severity	Richter Scale
Mild	0–4.3
Moderate	4.3–4.8
Intermediate	4.8–6.2
Severe	6.2–7.3
Catastrophic	7.3+

Review exercise

1. Evaluate $\log_5 287$

2. Solve these equations.
   a. $3^{2x+3} = 90$
   b. $5^{x-1} = 3^{3x}$
   c. $2 \times 3^{2x} = 5^x$

3 Solve these equations.
   a $\log x + \log(3x - 13) = 1$
   b $\log_5(x + 6) - \log_5(x + 2) = \log_5 x$
   c $\ln(4x - 7) = 2$
   d Solve $\log_2(x^2) = (\log_2 x)^2$
   e Solve $\log_{10} x = 4\log_x 10$

EXAM-STYLE QUESTIONS

4 The functions f and g are defined as

   $f(x) = e^{2x}$ for all real x
   $g(x) = \dfrac{3}{2}\ln x$ for $x > 0$

   a State the ranges of $f(x)$ and $g(x)$.
   b Explain why both functions have inverses.
   Find expressions for the inverse functions $f^{-1}(x)$ and $g^{-1}(x)$.
   c Find an expression for $(f \circ g)(x)$ and $(g \circ f)(x)$
   d Solve the equation $(f \circ g)(x) = (g \circ f)(x)$

5 The number, n, of insects in a colony, is given by $n = 4000e^{0.08t}$
   where t is the number of days after observation commences.
   a Find the population of the colony after 50 days.
   b How long does it take the population to double from when the observations commenced?

Review exercise

1 Solve $25^{4x-3} = \left(\dfrac{1}{125}\right)^{x+2}$

2 Find the exact value x satisfying the equation $(5^{x+1})(7^x) = 3^{2x+1}$
   Give your answer in the form $\dfrac{\log a}{\log b}$ where $a, b \in \mathbb{Z}$

3 Find the exact value of $2\log_3 27 + \log_3\left(\dfrac{1}{3}\right) - \log_3 \sqrt{3}$

EXAM-STYLE QUESTION

4 Write $4\log_3 x + \dfrac{1}{3}\log_3 y - 5\log_3 z$ as a single logarithm.

5 Solve
   a $\log_3(4x - 1) = 3$ **b** $\log_{x+1}(x - 1) = 2$
   c $\log_3(2\log x) = 4$ **d** $\log_2(x - 2) + \log_{\frac{1}{2}}(x - 1) = 3$

EXAM-STYLE QUESTION

6 If $m = \log_x 4$ and $n = \log_x 8$, find expressions in terms of m and n for
   a $\log_4 8$ **b** $\log_x 2$ **c** $\log_x 16$ **d** $\log_8 32$

7 The function f is defined for all real values of x by $f(x) = e^{3(x-1)} + 2$
Describe a series of transformations whereby the graph of $y = f(x)$ can be obtained from the graph of $y = e^x$

EXAM-STYLE QUESTIONS

8 Find the inverse function $f^{-1}(x)$ if
   a $f(x) = 3e^{2x}$ **b** $f(x) = 10^{3x}$ **c** $f(x) = \log_2(4x)$

9 Solve these simultaneous equations for a and b, given that a and b are positive real numbers.
$\log_a 64 + \log_a b = 8 \quad \log_{ba} = \dfrac{1}{2}$

CHAPTER 4 SUMMARY
Exponents

Laws of exponents
- $a^m \times a^n = a^{m+n}$
- $a^m \div a^n = a^{m-n}$
- $(a^m)^n = a^{mn}$
- $a^0 = 1$
- $\sqrt[n]{a} = a^{\frac{1}{n}}$
- $\sqrt[n]{(a^m)} = (\sqrt[n]{a})^m = (a^m)^{\frac{1}{n}} = \left(a^{\frac{1}{n}}\right)^m = a^{\frac{m}{n}}$
- $a^{-n} = \dfrac{1}{a^n}$

Exponential functions
- An **exponential function** is a function of the form $f(x) = a^x$ where a is a positive real number (that is, $a > 0$) and $a \neq 1$.
- The **domain** of the exponential function is the set of all real numbers.
- The **range** is the set of all positive real numbers.
- The graph of the exponential function $f(x) = e^x$ is a graph of exponential growth and the graph of $f(x) = e^{-x}$ is a graph of exponential decay.

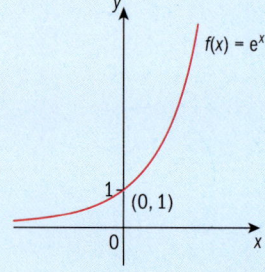
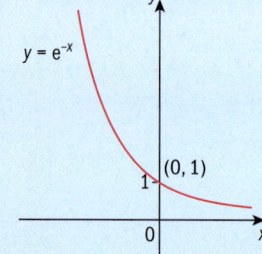

Continued on next page

Logarithms

Properties of logarithms
- If $b = a^x$ then $\log_a b = x$
- $\log_a a = 1$
- $\log_a 1 = 0$
- $\log_a b$ is undefined for any base a if b is negative
- $\log_a 0$ is undefined
- $\log_a(a^n) = n$

Logarithmic functions
- To find an **inverse** of a function algebraically, switch x and y and then rearrange to make y the subject.
- Generally if $f : x \mapsto a^x$ then $f^{-1} : x \mapsto \log_a x$
  $y = \log_a x$ is the inverse of $y = a^x$.
- $y = \ln x$ is the inverse of the exponential function $y = e^x$

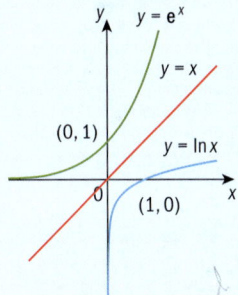

- $\log_a(a^x) = x$ and $a^{\log_a x} = x$
  $\ln(e^x) = x$ and $e^{\ln x} = x$
  $\log(10^x) = x$ and $(10^{\log x}) = x$

Laws of logarithms
- $\log x + \log y = \log xy$
- $\log x - \log y = \log \dfrac{x}{y}$
- $\log x^n = n \log x$
- $\log \dfrac{1}{x} = -\log x$

Change of base formula
- $\log_b a = \dfrac{\log_c a}{\log_c b}$

Theory of knowledge

The beauty of mathematics

> "The greatest mathematics has the simplicity and inevitableness of supreme poetry and music, standing on the borderland of all that is wonderful in science, and all that is beautiful in art."
>
> Herbert Westren Turnbull (1885–1961)
> *The Great Mathematicians*, 1929

Beautiful and simple solutions

Have you ever solved a problem in mathematics and been pleased with your solution?

Was it just because it was correct, or was it because your solution was efficient, stylish, even beautiful?

Look at these two solutions to the problem:

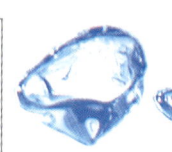

Expand and simplify $(x + y + z)(x - y - z)$

Solution 1	Solution 2
$(x + y + z)(x - y - z)$	$(x + y + z)(x - y - z)$
$= x^2 - xy - xz + xy - y^2 - yz + xz - yz - z^2$	$= (x + (y + z))(x - (y + z))$
$= x^2 - 2yz - y^2 - z^2$	$= x^2 - (y + z)^2$
$= x^2 - (y^2 + 2yz + z^2)$	
$= x^2 - (y + z)^2$	

- Which solution is better?

They both give us the same right answer and yet somehow the second solution seems better. It's more elegant and insightful than the first one.

> "Pure mathematics is, in its way, the poetry of logical ideas."
>
> Albert Einstein (1879–1955)

Simple, beautiful equations that model the world

> "The essence of mathematics is not to make simple things complicated, but to make complicated things simple."
>
> Stan Gudder, Professor of mathematics, University of Denver

Here are some famous equations

Einstein's equation: $E = mc^2$

Newton's second law: $F = ma$

Boyle's law: $V = \dfrac{k}{P}$

Schrödinger's equation: $H\psi = E\psi$

Newton's law of universal gravitation: $F = G\dfrac{m_1 m_2}{r^2}$

Isn't it startling that the universe can be described using mathematical equations such as these?

These equations have helped to put man on the moon and bring him back, develop wireless internet and understand the workings of the human body.

- These are just five equations – which is your favorite?
- Is it possible that mathematics and science will one day discover the ultimate theory of everything:
    - A theory that fully explains and links together all known physical phenomena?
    - A theory that has predictive power for the outcome of any experiment that could be carried out?

Now wouldn't that be wonderful?

◀ Boyle's Law explains why bubbles increase in size as they rise to the surface.

5 Rational functions

CHAPTER OBJECTIVES:

2.5 The reciprocal function $x \mapsto \dfrac{1}{x}$, $x \neq 0$, its graph and self-inverse nature

The rational function $x \mapsto \dfrac{ax+b}{cx+d}$ and its graph

Vertical and horizontal asymptotes

Applying rational functions to real-life situations

Before you start

You should know how to:

1 Expand polynomials.
   e.g. Multiply the polynomials
   $-2(3x - 1)$ and $3x(x^2 + 1)$:
   $-2(3x - 1) = -6x + 2$
   $3x(x^2 + 1) = 3x^3 + 3x$

2 Graph horizontal and vertical lines.
   e.g. Graph the lines $y = x$, $y = -x$, $x = 2$, $x = -1$, $y = 3$ and $y = -2$ on the same graph.

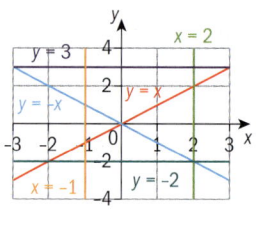

3 Recognize and describe a translation.
   e.g. Find the translations that map $y = x^2$ onto A and B.
   A is a horizontal shift of 2 units to the right. Function A is $y = (x - 2)^2$.
   B is a vertical shift of 3 units up. Function B is $y = x^2 + 3$.

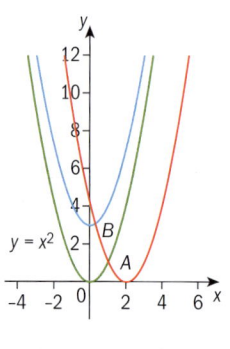

Skills check

1 Expand the polynomials.
   a $-4(2x - 5)$ **b** $6(2x - 3)$
   c $-x(x^2 + 7)$ **d** $x^2(x + 3)^2$
   e $x(x - 3)(x + 8)$

2 Draw these lines on one graph.
   $x = 0$, $y = 0$,
   $x = 3$, $x = -2$,
   $y = -3$, $y = 4$

3 Describe the transformations that map $y = x^3$ onto functions A and B and write down the equations of A and B.

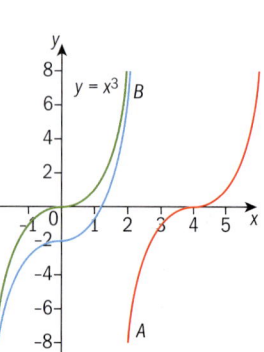

If you have an MP3 player, do you know how many songs, albums, sounds and so on can you fit on it? The answer depends on the quality of the recording setting and the length of the song. However, a rough idea is that a 4GB MP3 player will hold 136 hours or 8160 minutes of music. That's approximately

    2000 songs of 4 minutes
  or 1000 songs of 8 minutes
  or 4000 songs of 2 minutes.

This leads us to the function $s = \dfrac{8000}{m}$ where s is the number of songs and m is the number of minutes that a song lasts.

This function is an example of the reciprocal function, $f(x) = \dfrac{k}{x}$. In this chapter, you will use a GDC to explore the graphs of reciprocal functions and other rational functions that can be expressed in the form $f(x) = \dfrac{ax+b}{cx+d}$. You will examine horizontal and vertical asymptotes for the graphs of these functions and the domain and ranges of the functions.

Chapter 5

5.1 Reciprocals

Investigation – graphing product pairs

Think of pairs of numbers whose product is 24.

E.g. 24 × 1, 12 × 2, 8 × 3, 3 × 8. Copy the table and add some more pairs of numbers.

x	24	12	8	3				
y	1	2	3	8				

Show your pairs as coordinates on a graph with $0 \leq x \leq 24$ and $0 \leq y \leq 24$.
Now try the same idea with negatives, e.g. -12×-2 and graph these too.
Explain what you notice about
- the value of x as y gets bigger
- the value of y as x gets bigger
- the end behavior of your graph.

End behavior is the appearance of a graph as it is followed further and further in either direction.

→ The **reciprocal** of a number is 1 divided by that number.

For example, the reciprocal of 2 is $\frac{1}{2}$

Taking the reciprocal of a fraction turns it upside down.

For example, the reciprocal of $\frac{3}{4}$ is $1 \div \frac{3}{4} = 1 \times \frac{4}{3} = \frac{4}{3}$

The reciprocal of $\frac{7}{10}$ is $\frac{10}{7}$. The reciprocal of $\frac{1}{4}$ is $\frac{4}{1}$ or 4.

Zero does not have a reciprocal as $\frac{1}{0}$ is undefined. What does your GDC show for $1 \div 0$?

→ A number multiplied by its reciprocal equals 1.
For example $3 \times \frac{1}{3} = 1$

Geometrical quantities in inverse proportion were described as *reciprocali* in a 1570 translation of Euclid's *Elements* from 300 BCE.

Example 1

Find the reciprocal of $2\frac{1}{2}$

Answer
$2\frac{1}{2} = \frac{5}{2}$ *Write as an improper fraction.*
Reciprocal of $\frac{5}{2} = \frac{2}{5}$ *Turn it upside down.*

Check: $\frac{5}{2} \times \frac{2}{5} = 1$

You can find reciprocals of algebraic terms too.

→ The **reciprocal** of x is $\frac{1}{x}$ or x^{-1} and $x^{-1} \times x = 1$

The reciprocal of a number or a variable is also called its multiplicative inverse.

Exercise 5A

1. Find the reciprocals.
   a 2
   b 3
   c −3
   d −1
   e $\dfrac{2}{3}$
   f $\dfrac{7}{11}$
   g $-\dfrac{3}{2}$
   h $3\dfrac{1}{2}$

2. Find the reciprocals.
   a 6.5
   b x
   c y
   d $3x$
   e $4y$
   f $\dfrac{2x}{9}$
   g $\dfrac{3a}{5}$
   h $\dfrac{2}{3d}$
   i $\dfrac{d}{t}$
   j $\dfrac{x+1}{x-1}$

> The term *reciprocal* was in common use at least as far back as the third edition of *Encyclopaedia Britannica* (1797) to describe two numbers whose product is 1.

3. Multiply each quantity by its reciprocal. Show your working.
   a 6
   b $\dfrac{3}{4}$
   c $\dfrac{2c}{3d}$

4. a What is the reciprocal of the reciprocal of 4?
   b What is the reciprocal of the reciprocal of x?

5. For the function $xy = 24$
   a Find y when x is i 48 ii 480 iii 4800 iv 48 000
   b What happens to the value of y when x gets larger?
   c Will y ever reach zero? Explain.
   d Find x when y is i 48 ii 480 iii 4800 iv 48 000
   e What happens to the value of x when y gets larger?
   f Will x ever reach zero? Explain.

> This is the function you used in the Investigation on page 142.

5.2 The reciprocal function

The **reciprocal function** is
$$f(x) = \dfrac{k}{x}$$
where k is a constant.

Graphs of reciprocal functions all have similar shapes.

Investigation – graphs of reciprocal functions

Use your GDC to draw all the graphs in this investigation.

1. Draw a graph of   a $f(x) = \dfrac{1}{x}$ b $g(x) = \dfrac{2}{x}$ c $h(x) = \dfrac{3}{x}$

   What is the effect of changing the value of the numerator?

2. Draw a graph of   a $f(x) = \dfrac{-1}{x}$ b $g(x) = \dfrac{-2}{x}$ c $h(x) = \dfrac{-3}{x}$

   What is the effect of changing the sign of the numerator?

3. a Copy and complete this table for $f(x) = \dfrac{4}{x}$

x	0.25	0.4	0.5	1	2	4	8	10	16
f(x)									

   b What do you notice about the values of x and $f(x)$ in the table?
   c Draw the graph of the function.
   d Draw the line $y = x$ on the same graph.
   e Reflect $f(x) = \dfrac{4}{x}$ in the line $y = x$.
   f What do you notice?
   g What does this tell you about the inverse function f^{-1}?

Asymptotes

The graphs of the functions $f(x)$, $g(x)$ and $h(x)$ in the Investigation on page 143 all consist of two curves. The curves get closer and closer to the axes but never actually touch or cross them.

The axes are asymptotes to the graph.

> → If a curve gets continually closer to a straight line but never meets it, the straight line is called an **asymptote**.

The word *asymptote* is derived from the Greek *asymptotos*, which means 'not falling together'.

$y = b$ is an asymptote to the function $y = f(x)$

As $x \to \infty$, $f(x) \to b$

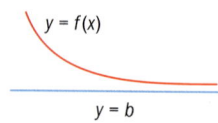

The symbol → means 'approaches'.

> → The graph of any reciprocal function of the form $y = \dfrac{k}{x}$ has a vertical asymptote $x = 0$ and a horizontal asymptote $y = 0$

The horizontal line $y = b$ is a horizontal asymptote of the graph of $y = f(x)$.

> → The graph of a reciprocal function is called a **hyperbola**.
> - The x-axis is the horizontal asymptote.
> - The y-axis is the vertical asymptote.
> - Both the domain and range are all the real numbers except zero.
> - The two separate parts of the graph are reflections of each other in $y = -x$
> - $y = -x$ and $y = x$ are lines of symmetry for this function.

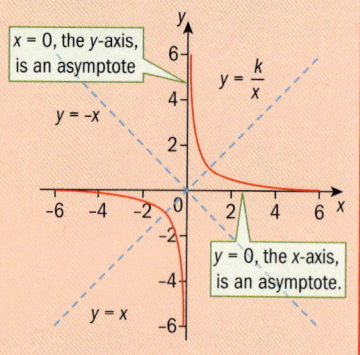

The reciprocal function has many applications in computer science algorithms, particularly those related to number theory. You may wish to investigate these further.

In Chapter 1 you saw that to draw the inverse function of $f(x)$, you reflect its graph in the line $y = x$. If you reflect $f(x) = \dfrac{1}{x}$ in the line $y = x$ you get the same graph as for $f(x)$.

> → The reciprocal function is a **self-inverse function**.

The **reciprocal function**, $f(x) = \dfrac{1}{x}$, is one of the simplest examples of a function that is self-inverse.

The equation of the function in the Investigation on page 142 is $xy = 24$. It can be written as $y = \dfrac{24}{x}$ and is a reciprocal function. It has a graph similar to the one shown above.

The design of the Yas Hotel in Abu Dhabi (designed by Asymptote Architecture) is based on mathematical models. It also has a Formula 1 racetrack running through the center of the hotel!

 Example 2

For each function:
- write down the equations of the vertical and horizontal asymptotes
- sketch the graph
- state the domain and range.

a $y = \dfrac{9}{x}$ **b** $y = \dfrac{9}{x} + 2$

Answers

a Asymptotes are $x = 0$ and $y = 0$

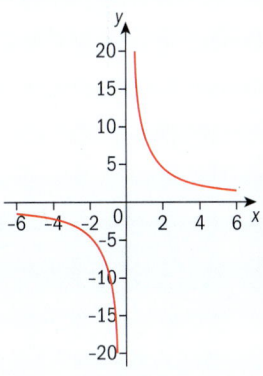

Domain $x \in \mathbb{R}, x \neq 0$,
range $y \in \mathbb{R}, y \neq 0$

b Asymptotes are $x = 0$ and $y = 2$

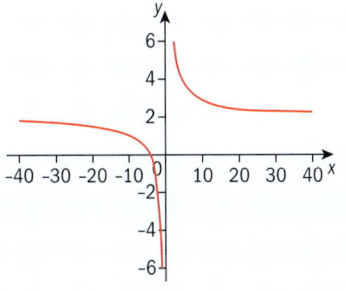

The graph of $f(x) + 2$ is the same as the graph of $f(x)$ but shifted 2 units in the y-direction.

Domain $x \in \mathbb{R}, x \neq 0$,
range $y \in \mathbb{R}, y \neq 2$

Exercise 5B

1 Draw these on separate graphs.

   a $y = \dfrac{5}{x}$ **b** $y = \dfrac{6}{x}$ **c** $xy = 8$

2 On the same graph show $y = \dfrac{12}{x}$ and $y = \dfrac{-12}{x}$

> You need to be able to do questions **3b** and **4b** and **c** both analytically (using algebra and by sketching and using transformations) and using your GDC.

3 a Sketch the graph of $f(x) = \dfrac{1}{x}$ and write down its asymptotes.

   b Sketch the graph of $f(x) = \dfrac{1}{x} + 2$ and write down its asymptotes.

4 Identify the horizontal and vertical asymptotes of these functions and then state their domain and range.

   a $y = \dfrac{20}{x}$ **b** $y = \dfrac{3}{x} + 2$ **c** $y = \dfrac{4}{x} - 2$

> It may help to draw the graphs.

5 The Corryvreckan, the third largest whirlpool in the world, is between the islands of Jura and Scarba off the coast of Scotland. Flood tides and inflow from the west and the roar of the resulting maelstrom can be heard 16 km away.

The speed of the surrounding water increases as you approach the center and is modeled by $s = \dfrac{250}{d}$ where s is the speed of the water in m s^{-1} and d is the distance from the center in metres.

   a Use your GDC to sketch the function with $0 \leq d \leq 50$ and $0 \leq s \leq 200$.

   b At what distance is the speed 10 m s^{-1}?

   c What is the speed of the water 100 m from the center?

▲ Archimedes is believed to have said "*Give me a place to stand, and a lever long enough and I shall move the earth.*"

6 The force (F) required to raise an object of mass 1500 kg is modeled by $F = \dfrac{1500}{l}$ where l is the length of the lever in metres and the force is measured in newtons.

   a Sketch the graph with $0 \leq l \leq 6$ and $0 \leq F \leq 5000$

   b How much force would you need to apply if you had a 2 m lever?

   c How long would the lever need to be if you could manage a force of   **i** 1000 N   **ii** 2000 N   **iii** 3000 N?

> N is the symbol for the unit of force, the newton.

5.3 Rational functions

Have you noticed the way the sound of a siren changes as a fire engine or police car passes you? The observed frequency is higher than the emitted frequency during the approach, it is identical at the instant of passing by, and it is lower during the time it moves away. This is called the Doppler effect. The equation for the observed frequency of sound when the source is traveling toward you is:

$$f_1 = \frac{330 f}{330 - v}$$

where

- 330 is the speed of sound in $m\,s^{-1}$
- f_1 is the observed frequency in Hz
- f is the emitted frequency
- v is the velocity of the source toward you

f_1 is a rational function.

> Sound frequency is measured in hertz (Hz), the number of waves per second.

→ A **rational function** is a function of the form $f(x) = \dfrac{g(x)}{h(x)}$ where g and h are polynomials.

> $h(x)$ cannot be zero since a value divided by zero is undefined.

In this course $g(x)$ and $h(x)$ will be restricted to linear functions of the form $px + q$ so we can investigate rational functions $f(x)$ where

$$f(x) = \frac{ax + b}{cx + d}$$

Example 3

A vehicle is coming towards you at $96\,km\,h^{-1}$ (60 miles per hour) and sounds its horn with a frequency of 8000 Hz. What is the frequency of the sound you hear if the speed of sound is $330\,m\,s^{-1}$?

Answer

$96\,km\,h^{-1} = 96\,000\,m\,h^{-1}$

$96\,000\,m\,h^{-1} = \dfrac{96\,000}{3600} = 26.7\,m\,s^{-1}$

Observed frequency $= \dfrac{330 f}{330 - v}$

$= \dfrac{330 \times 8000}{330 - 26.7}$

$= 8700\,Hz\,(3\,sf)$

Convert kilometres per hour to metres per second.
Since 1 hour = 3600 seconds

> The units of speed must all be the same in the equation. You can round numbers to get an approximate answer.

Investigation – graphing rational functions 1

a Use your GDC to show sketches of $y = \frac{1}{x}$, $y = \frac{1}{x-2}$, $y = \frac{1}{x+3}$ and $y = \frac{2}{x+3}$.

b Copy and complete the table.

Rational function	Vertical asymptote	Horizontal asymptote	Domain	Range
$y = \frac{1}{x}$				
$y = \frac{1}{x-2}$				
$y = \frac{1}{x+3}$				
$y = \frac{2}{x+3}$				

c What effect does changing the denominator have on the vertical asymptote?
d What do you notice about the horizontal asymptotes?
e What do you notice about the domain and the value of the vertical asymptote?
f What do you notice about the range and the value of the horizontal asymptote?

Rational functions of the form $y = \frac{k}{x-b}$

A rational function $y = \frac{k}{x-b}$, where k and b are constants, will have a vertical asymptote when the denominator equals zero, that is, when $x = b$.

The horizontal asymptote will be the x-axis.

$\frac{1}{0}$ is undefined. We will consider this in more detail in the Theory of Knowledge section at the end of the chapter.

Example 4

a Identify the horizontal and vertical asymptotes of $y = \frac{1}{x-3}$.
b State the domain and range.
c Sketch the function with the help of your GDC.

You may wish to explore the concept of infinity.

Answers

a The x-axis ($y = 0$) is the horizontal asymptote.
$x = 3$ is the vertical asymptote.

Since the numerator will never be 0, the graph of this function never touches the x-axis.
The denominator is zero when $x = 3$.

▶ Continued on next page

148 Rational functions

b Domain $x \in \mathbb{R}, x \neq 3$
Range $y \in \mathbb{R}, y \neq 0$

c

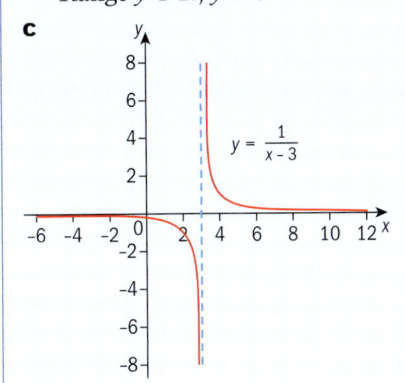

Exercise 5C

1 Identify the horizontal and vertical asymptotes of these functions and state their domain and range.

   a $y = \dfrac{1}{x+1}$ **b** $y = \dfrac{1}{x-4}$ **c** $y = \dfrac{-2}{x-5}$ **d** $y = \dfrac{4}{x+1}$

   e $y = \dfrac{4}{x+1} + 2$ **f** $y = \dfrac{4}{x+1} - 2$ **g** $y = \dfrac{4}{x-3} + 2$ **h** $y = \dfrac{-2}{x+3} - 2$

   > You should use algebra (this is called 'using an analytic method') to do question 1, although you may want to check your answers with a GDC.

2 Sketch each function with the help of your GDC and state the domain and range.

   a $y = \dfrac{4}{x}$ **b** $y = \dfrac{3}{x-3} + 1$ **c** $y = \dfrac{-4}{x+5} - 8$

   d $y = \dfrac{1}{x-7} + 3$ **e** $y = \dfrac{6}{x+2} - 6$ **f** $y = \dfrac{5}{x} + 4$

   g $y = \dfrac{1}{4x+12} - 2$ **h** $y = \dfrac{3}{2x}$ **i** $y = \dfrac{4}{3x-6} + 5$

   > Use your GDC with the correct viewing window.

3 When lightning strikes, the light reaches your eyes virtually instantaneously. But the sound of the thunder travels at approximately 331 m s⁻¹. However, sound waves are affected by the temperature of the surrounding air. The time sound takes to travel one kilometre is modeled by $t = \dfrac{1000}{0.6c + 331}$ where t is the time in seconds and c is the temperature in degrees Celsius.
   a Sketch the graph of t for temperatures from −20 °C to 40 °C.
   b If you are one kilometre away and it is 3 seconds before you hear the thunder, what is the temperature of the surrounding air?

4 **a** On the same set of axes, sketch $y = x + 2$ and $y = \dfrac{1}{x+2}$

   Compare the two graphs and make connections between the linear function and its reciprocal function.

   b Now do the same for $y = x + 1$ and $y = \dfrac{1}{x+1}$

Rational functions of the form $y = \dfrac{ax+b}{cx+d}$

> → Every rational function of the form $y = \dfrac{ax+b}{cx+d}$ has a graph called a hyperbola.

The graph of any rational function $y = \dfrac{ax+b}{cx+d}$ has a vertical and a horizontal asymptote.

Investigation – graphing rational functions 2

a Use your GDC to show sketches of

$y = \dfrac{x}{x+3}$, $y = \dfrac{x+1}{x+3}$, $y = \dfrac{2x}{x+3}$ and $y = \dfrac{2x-1}{x+3}$

b Copy and complete the table.

Rational function	Vertical asymptote	Horizontal asymptote	Domain	Range
$y = \dfrac{x}{x+3}$				
$y = \dfrac{x+1}{x+3}$				
$y = \dfrac{2x}{x+3}$				
$y = \dfrac{2x-1}{x+3}$				

c What do you notice about the horizontal asymptotes?
d What do you notice about the domain and the value of the vertical asymptote?

> → The vertical asymptote occurs at the x-value that makes the denominator zero.
> → The horizontal asymptote is the line $y = \dfrac{a}{c}$

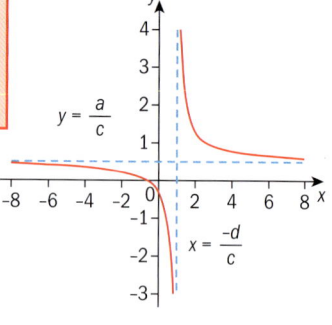

To find the horizontal asymptote rearrange the equation to make x the subject.

$$y = \dfrac{ax+b}{cx+d}$$
$$y(cx+d) = ax+b$$
$$cyx - ax = b - dy$$
$$x = \dfrac{b-dy}{cy-a}$$

The horizontal asymptote occurs when the denominator is zero, that is, when

$$cy = a \text{ or } y = \dfrac{a}{c}$$

Rational functions

Example 5

For the function $y = \dfrac{x+1}{2x-4}$

a sketch the graph
b find the vertical and horizontal asymptotes
c state the domain and range.

Answers

a

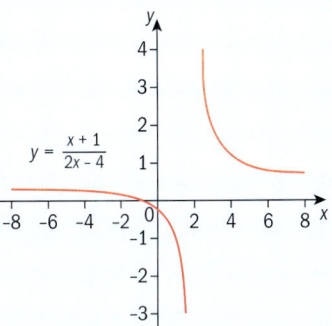

b Vertical asymptote $x = 2$

Horizontal asymptote $y = \dfrac{1}{2}$

c Domain $x \in \mathbb{R}, x \neq 2$

Range $y \in \mathbb{R}, y \neq \dfrac{1}{2}$

When $2x - 4 = 0$, $x = 2$

$a = 1, c = 2, y = \dfrac{a}{c}$

Exercise 5D

1 Identify the horizontal and vertical asymptotes of these functions and then state the domain and range.

a $y = \dfrac{x+2}{x-3}$ **b** $y = \dfrac{2x+2}{3x-1}$ **c** $y = \dfrac{-3x+2}{-4x-5}$ **d** $y = \dfrac{34x-2}{16x+4}$

2 Match the function with the graph.

a $y = \dfrac{5}{x}$ **b** $y = \dfrac{x+2}{x-2}$ **c** $y = \dfrac{x-1}{x-3}$ **d** $y = \dfrac{1}{x-4}$

i

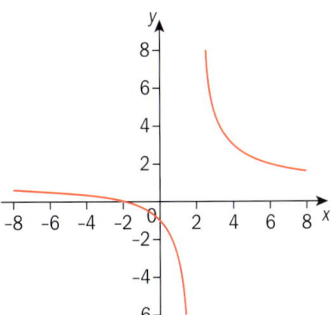

ii

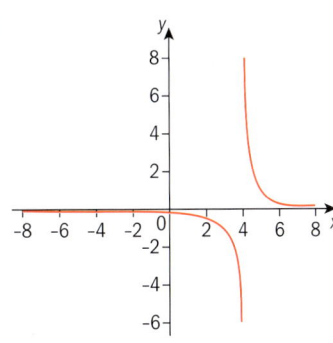

iii 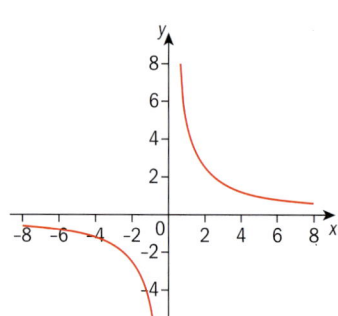 iv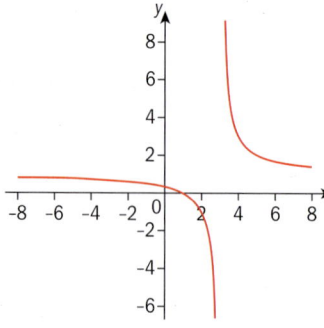

3 Sketch each function using your GDC and state the domain and range.

a $y = \dfrac{x+2}{x+3}$ b $y = \dfrac{x}{4x+3}$ c $y = \dfrac{x-7}{3x-8}$

d $y = \dfrac{9x+1}{3x-2}$ e $y = \dfrac{-3x+10}{4x-12}$ f $y = \dfrac{5x+2}{4x}$

g $y = \dfrac{3x}{2x-4}$ h $y = \dfrac{7x}{-x-15}$ i $y = \dfrac{14x-4}{2x-1}$

> Check your answer by using your GDC to graph the function.

4 Write a rational function that has a vertical asymptote at $x = -4$ and a horizontal asymptote at $y = 3$

5 Chris and Lee design T-shirts for surfers and set up a T-shirt printing business in their garage. It will cost $450 to set up the equipment and they estimate that it will cost $5.50 to print each T-shirt.
   a Write a linear function $C(x)$ giving the total cost of producing x T-shirts. Remember to take the set-up cost into account.
   b Write a rational function $A(x)$ giving the **average cost** per T-shirt of producing x of them.
   c What is the domain of $A(x)$ in the context of the problem? Explain.
   d Write down the vertical asymptote of $A(x)$.
   e Find the horizontal asymptote for $A(x)$. What meaning does this value have in the context of the problem?

> Sketch the function.

EXAM-STYLE QUESTION

6 Young's rule is a way of calculating doses of medicine for children over the age of two, based on the adult dose.
'Take the age of the child in years and divide by their age plus 12. Multiply this number by the adult dose.'
This is modeled by the function $c = \dfrac{at}{t+12}$ where c is the child's dose, a is the adult dose in mg and t is the age of the child in years.

Rational functions

a Make a table of values for ages 2 to 12 with an adult dose of 100 mg.
b Use your values from **a** to draw a graph of the function.
c Use the graph to estimate the dose for a $7\frac{1}{2}$-year old.
d Write down the equation of the horizontal asymptote.
e What does the value of the horizontal asymptote mean for Young's rule?

7 The average cost of electricity per year for a refrigerator is $92.
  a A new refrigerator costs $550. Determine the total annual cost for a refrigerator that lasts for 15 years. Assume costs include purchase and electricity.
  b Develop a function that gives the annual cost of a refrigerator as a function of the number of years you own the refrigerator.
  c Sketch a graph of that function. What is an appropriate window? Label the scale.
  d Since this is a rational function, determine its asymptotes.
  e Explain the meaning of the horizontal asymptote in terms of the refrigerator.
  f A company offers a refrigerator that costs $1200, but says that it will last at least twenty years. Is this refrigerator worth the difference in cost?

Review exercise

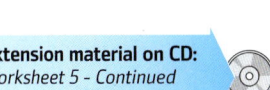
Extension material on CD:
Worksheet 5 - Continued fractions and aysmptotes

EXAM-STYLE QUESTION

1 Match the function with the graph.

  i $f(x) = \dfrac{2}{x+2}$ **ii** $f(x) = \dfrac{1}{x-3}$ **iii** $f(x) = \dfrac{4x+1}{x}$
  iv $f(x) = \dfrac{1-x}{x}$ **v** $f(x) = \dfrac{x-2}{x-4}$ **vi** $f(x) = \dfrac{x+2}{x+4}$

a
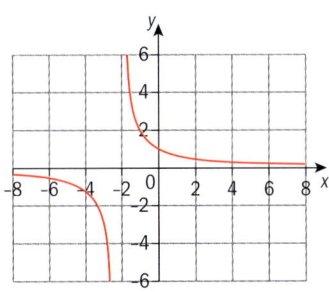

b
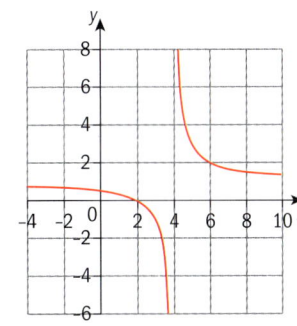

EXAM-STYLE QUESTIONS

c
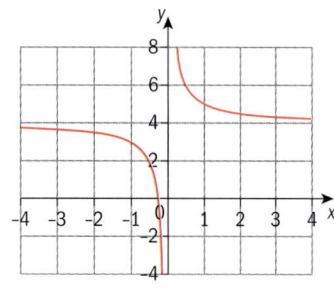

d
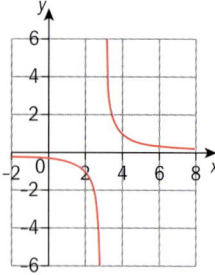

e
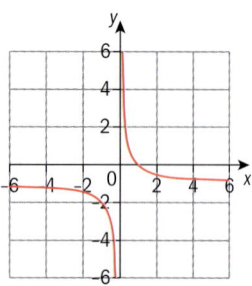

f
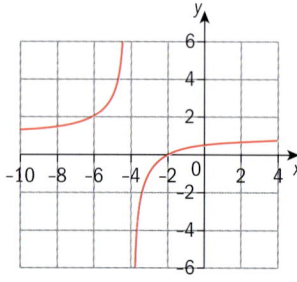

2 Given   **a** $f(x) = \dfrac{5}{x}$ **b** $f(x) = \dfrac{1}{x+1}$ **c** $f(x) = \dfrac{x+3}{3-x}$

  i Sketch the function.
  ii Determine the vertical and horizontal asymptotes of the function.
  iii Find the domain and range of the function.

3 For each of these functions, write down the asymptotes, domain and range.

a
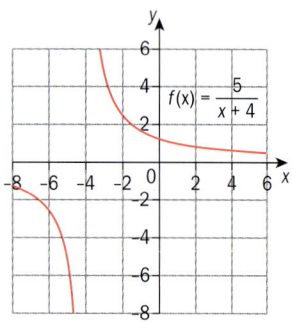

b
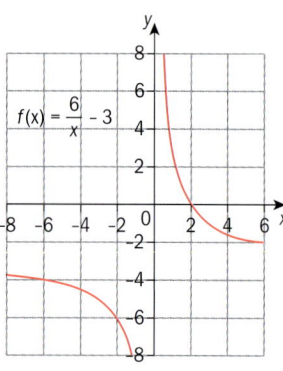

c
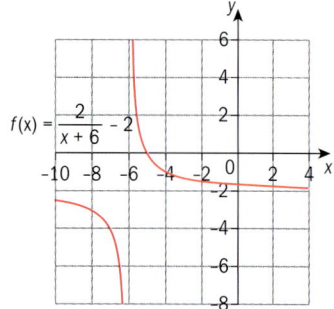

d
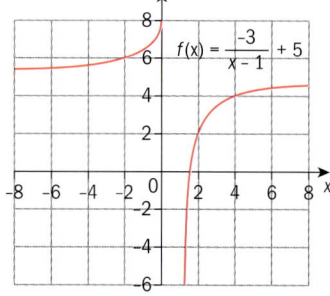

154 Rational functions

4 A group of students want to give their teacher a voucher for a weekend at a health spa. The voucher costs $300.
   a If c represents the cost for each student and s represents the number of students, write an equation to show the cost in terms of the number of students.
   b Draw a graph of the function.
   c Explain any limitations on the range and domain of this function.

5 The function f is given by
$$f(x) = \frac{2x-1}{x+2}, x \in \mathbb{R}, x \neq -2$$
   a i Find the horizontal asymptote of the graph of $y = f(x)$
     ii Find the vertical asymptote of the graph.
     iii Write down the coordinates of the point P at which the asymptotes intersect.
   b Find the points of intersection of the graph with the axes.
   c Hence sketch the graph of $y = f(x)$, showing the asymptotes by dotted lines.

Review exercise

EXAM-STYLE QUESTION

1 Sketch each function with the help of your GDC. State the domain and range.
   a $f(x) = \frac{6}{x} - 5$
   b $f(x) = \frac{2}{x} + 3$
   c $f(x) = \frac{-2}{x-5}$
   d $f(x) = \frac{3}{x-7} - 8$
   e $f(x) = \frac{8}{x+3}$
   f $f(x) = \frac{-6}{x+4} - 2$

2 An airline flies from London to New York, which is a distance of 5600 km.
   a Show that this information can be written as $s = \frac{5600}{t}$ where s is the average speed of the plane in km h^{-1} and t is the time in hours.
   b Sketch a graph of this function with $0 \leq s \leq 1200$ and $0 \leq t \leq 20$.
   c If the flight takes 10 hours, what is the average speed of the plane?

EXAM-STYLE QUESTIONS

3 People with sensitive skin must be careful about the amount of time spent in direct sunlight. The relation
$$m = \frac{22.2s + 1428}{s}$$
where m is the time in minutes and s is the sun scale value, gives the maximum amount of time that a person with sensitive skin can spend in direct sunlight without skin damage.
  a Sketch this relation when $0 \leq s \leq 120$ and $0 \leq m \leq 300$
  b Find the number of minutes that skin can be exposed when
   i $s = 10$ **ii** $s = 40$ **iii** $s = 100$
  c What is the horizontal asymptote?
  d Explain what this represents for a person with sensitive skin.

4 The city mayor is giving out face masks during a flu outbreak in Bangkok. The cost (c) in Thai baht for giving masks to m percent of the population is given by
$$c = \frac{750\,000m}{100 - m}$$
  a Choose a suitable scale and use your GDC to help sketch the function.
  b Find the cost of supplying
   i 20% **ii** 50% **iii** 90%
   of the population.
  c Would it be possible to supply all of the population using this model? Explain your answer.

5 The function $f(x)$ is defined as
$$f(x) = 2 + \frac{1}{2x - 5}, \quad x \neq \frac{5}{2}$$
  a Sketch the curve of f for $-3 \leq x \leq 5$, showing the asymptotes.
  b Using your sketch, write down
   i the equation of each asymptote
   ii the value of the x-intercept
   iii the value of the y-intercept.

CHAPTER 5 SUMMARY
Reciprocals

- The **reciprocal** of a number is 1 divided by that number.
- A number multiplied by its reciprocal equals 1.
  For example $3 \times \frac{1}{3} = 1$
- The **reciprocal** of x is $\frac{1}{x}$ or x^{-1} and $x^{-1} \times x = 1$

The reciprocal function

- If a curve gets continually closer to a straight line but never meets it, the straight line is called an **asymptote**.
- The graph of any reciprocal function of the form $y = \frac{k}{x}$ has a vertical asymptote $x = 0$ and a horizontal asymptote $y = 0$
- The graph of a reciprocal function is called a **hyperbola**.
  - The x-axis is the horizontal asymptote.
  - The y-axis is the vertical asymptote.
  - Both the domain and range are all the real numbers except zero.
  - The two separate parts of the graph are reflections of each other in $y = -x$
  - $y = x$ and $y = -x$ are lines of symmetry for this function.
- The reciprocal function is a **self-inverse function**.

Rational functions

- A **rational function** is a function of the form $f(x) = \frac{g(x)}{h(x)}$ where g and h are polynomials.
- Every rational function of the form $y = \frac{ax+b}{cx+d}$ has a graph called a hyperbola.
- The vertical asymptote occurs at the x-value that makes the denominator zero.
- The horizontal asymptote is the line $y = \frac{a}{c}$

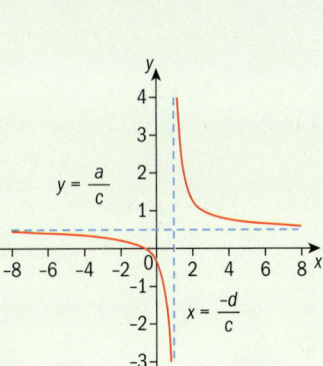

Theory of knowledge

Number systems

Egyptian fractions

The ancient Egyptians only used fractions with a numerator of 1, for example: $\frac{1}{2}, \frac{1}{3}, \frac{1}{4}$ etc. This meant that instead of $\frac{3}{4}$ they wrote $\frac{1}{2} + \frac{1}{4}$. Their fractions were all in the form $\frac{1}{n}$ and are called **unit fractions**. Numbers such as $\frac{2}{7}$ were represented as sums of unit fractions (e.g. $\frac{2}{7} = \frac{1}{4} + \frac{1}{28}$). Also, the same fraction could not be used twice (so $\frac{2}{7} = \frac{1}{7} + \frac{1}{7}$ was not allowed). For example, $\frac{5}{8}$ would be $\frac{1}{2} + \frac{1}{8}$.

- Write these as unit fractions.

  $\frac{5}{6}$ $\frac{5}{8}$ $\frac{2}{5}$ $\frac{6}{7}$

In algebra: $\frac{3}{4x} = \frac{1}{2x} + \frac{1}{4x}$

- Write each algebraic expression as an Egyptian fraction.

  $\frac{4}{3x}$ $\frac{5}{4x}$ $\frac{7}{4x}$ $\frac{23}{24x}$

Where do you think this could be useful?

What are the limitations of these fractions?

Is it possible to write every fraction as an Egyptian fraction? How do you know?

In an Inca quipu, the knots in the strings represent numbers

The Rhind Mathematical Papyrus dated 1650 BCE contains a table of Egyptian fractions copied from another papyrus 200 years older!

Is there a difference between zero and nothing?

More than 2000 years ago, Babylonian and Hindu cultures had systems for representing an absence of a number. In the ninth century CE, the Islamic mathematician and philosopher Muhammad al-Khwarizmi remarked that if, in a calculation, no number appears in the place of tens, a little circle should be used 'to keep the rows'. The Arabs called this circle *sifr* (empty). The name *sifr* eventually became our word *zero*.

- Does this mean that zero was nothing?

- Who first used zero?
- What was used before that?
- Make a list of all of the subsets of {0, 1, 2, 3}.
- Notice that one subset is {0} and another is { }.
- Now try this. Solve the equation $9 + x = 3^2$ and the equation $3x = 0$.
- We have 1 CE and 1 BCE. What about a year zero?
- The ancient Greeks were not sure what to do with zero and they questioned how nothing could be something. Zeno's paradoxes (a good topic to research) depend in some part on the tentative use of zero.
- How did the Mayan and Inca cultures understand zero?
- Where is zero in the decimal system? Is it positive or negative?
- What happens if you divide zero by anything?
- What happens if you divide anything by zero?
- What happens if you divide zero by zero?

▶ The Mayans used a shell symbol to represent zero.

Chapter 5 159

6 Patterns, sequences and series

CHAPTER OBJECTIVES:

1.1 Arithmetic sequences and series; sum of finite arithmetic series; geometric sequences and series; sum of finite and infinite geometric series. Sigma notation.
Applications

1.3 The binomial theorem: expansion of $(a+b)^n$, $n \in \mathbb{N}$;

Calculation of binomial coefficients using Pascal's triangle and $\binom{n}{r}$

Before you start

You should know how to:

1 Solve linear and quadratic equations and change the subject of a formula.
   e.g. Solve the equation $n(n-4) = 12$
   $n^2 - 4n = 12$
   $n^2 - 4n - 12 = 0$
   $(n-6)(n+2) = 0$
   $n = -2, n = 6$
   e.g. Make b the subject of this formula.
   $ac = b - 3$
   $b = ac + 3$

2 Substitute known values into formulae.
   e.g. Using the formula $A = 3p^4 - 10q$, find the value of A if $p = 2$ and $q = 1.5$
   $A = 3p^4 - 10q$
   $A = 3(2)^4 - 10(1.5)$
   $A = 3(16) - 15$
   $A = 48 - 15$
   $A = 33$

Skills check

1 Solve each equation.
   a $3x - 5 = 5x + 7$
   b $p(2-p) = -15$
   c $2n + 9 = 41$

2 Solve for k.
   a $6m + 8k = 30$
   b $2pk - 5 = 3$

3 If $T = 2x(x + 3y)$, then find the value of T when
   a $x = 3$ and $y = 5$
   b $x = 4.7$ and $y = -2$

4 Using the formula $m = 2^x - y^3$, find the value of m if
   a $x = 5$ and $y = 3$
   b $x = 3$ and $y = -2$
   c $x = -5$ and $y = \frac{1}{2}$

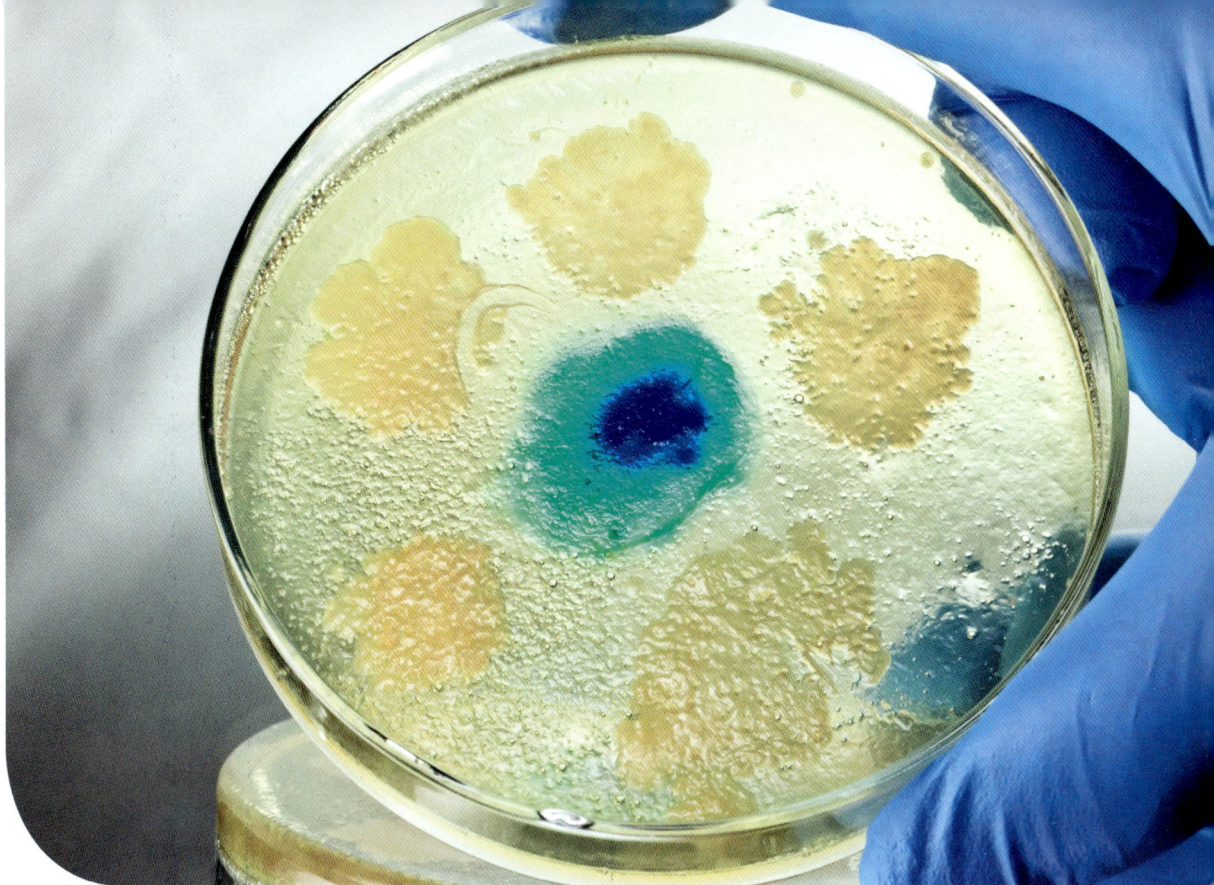

▲ Bacteria growing in a petri dish

The bacteria in this petri dish are growing and reproducing; in this case the total mass doubles every two hours. At 8 a.m. the mass is measured as 3 grams, so the total mass at 10:00 will be 6 grams, the mass at 12:00 will be 12 grams, and so on.

The mass of the bacteria in the dish forms a numerical pattern. This pattern could be used to predict the mass of bacteria in the dish after 8 hours, 12 hours or 24 hours.

In this chapter you will study mathematical patterns. Patterns can help us make predictions about the near and distant future. For example, we can use patterns to:

- predict the population of a country in 20 years
- work out how long it will take to pay off a bank loan
- predict how long a natural resource will last
- calculate the total distance that a bouncing ball will travel
- calculate how long it will take for an investment to double in value.

6.1 Patterns and sequences

Investigation – saving money

Joel decides to start saving money.
He saves $20 the first week, $25 the second week, $30 the third week, and so on.

a Copy and complete the table below to show how much Joel saves each week, and how much he has saved in total, for the first 8 weeks.

Week number	Weekly savings	Total savings
1	20	20
2	25	45
3	30	75
4		
5		
6		
7		
8		

b How much will Joel save in the 10th week? In the 17th week?
c How much money will Joel save in total in the first year?
d How long will it take for him to save a total of at least $1000?
e Try to write a formula for the amount of money Joel saves **each** week. Let M represent the amount of money he saves each week, and let n represent the week number.
f Try to write a formula for the **total** amount of money Joel has saved. Let T represent his total savings, and let n represent the number of weeks.

In the investigation above, the amounts of money Joel saves each week form a **sequence**. The total amounts of money he has saved as time passes form a different sequence.

→ A **number sequence** is a pattern of numbers arranged in a particular order according to a rule.

Here are some number sequences.

8, 11, 14, 17, …
800, 400, 200, 100, …
1, 4, 9, 16, 25, …
5, 10, 15, 20, 25, …

> Each individual number, or element, of a sequence is called a **term**.

In the sequence 8, 11, 14, 17, …, the first term is 8, the second term is 11, the third term is 14, and so on.

You can also use the notation u_n to denote the nth term of a sequence, where n is a positive integer.

So for 8, 11, 14, 17, … you could say
$u_1 = 8$, $u_2 = 11$, $u_3 = 14$, and so on.

You can continue the pattern if you notice that the value of each term is three greater than the value of the previous term:

8, 11, 14, 17, <u>20</u>, <u>23</u>, <u>26</u>

For this sequence, you could write: $u_1 = 8$ and $u_{n+1} = u_n + 3$
This is called a **recursive** formula, in which the value of a term depends on the value of the previous term.

In the sequence 800, 400, 200, 100, …, the value of each term is one-half the value of the previous term.

In this case, $u_1 = 800$ and $u_{n+1} = \frac{1}{2}u_n$

Sometimes, we use letters other than u to represent the terms of a sequence.
For example, we might use a_n, t_n, or x_n to represent the nth term of a sequence.

Example 1

Write a recursive formula for the nth term of each sequence.
a 9, 15, 21, 27, …
b 2, 6, 18, 54, …

Answers
a $u_1 = 9$ and $u_{n+1} = u_n + 6$ | *To get from one term to the next, you add 6.*
b $u_1 = 2$ and $u_{n+1} = 3u_n$ | *To get from one term to the next, you multiply by 3.*

Sometimes it is more useful to write a **general formula for the nth term** of a sequence. With a general formula, you can find the value of a term without having to know the value of the previous term.

In the sequence 1, 4, 9, 16, 25, … , each term is a perfect square. The first term is 1^2, the second is 2^2, and so on. A general formula for the nth term of this sequence is $u_n = n^2$.

In the sequence 5, 10, 15, 20, 25, … , each term is a multiple of 5. The first term is 5×1, the second is 5×2, and so on. A general formula for the nth term of this sequence is $u_n = 5n$.

Sometimes this is called the 'general rule for the nth term'.

Remember that n, the term number, will always be a whole number. We could not have a '$\frac{3}{4}$th' term, or a '7.5th' term.

Example 2

Write a general formula for the nth term of each sequence.
a 4, 8, 12, 16, …
b $\frac{1}{3}, \frac{1}{6}, \frac{1}{9}, \frac{1}{12}, …$

Answers
a $u_n = 4n$ — Each term is a multiple of 4.
b $u_n = \frac{1}{3n}$ — The denominators are the multiples of 3.

Exercise 6A

1 Write down the next three terms in each sequence.
   a 3, 7, 11, 15, …
   b 1, 2, 4, 8, …
   c 3, 4, 6, 9, 13, …
   d 5, −10, 20, −40, …
   e $\frac{1}{2}, \frac{3}{5}, \frac{5}{8}, \frac{7}{11}, …$
   f 6.0, 6.01, 6.012, 6.0123, …

2 Write down the first four terms in each sequence.
   a $u_1 = 10$ and $u_{n+1} = 3(u_n)$
   b $u_1 = 3$ and $u_{n+1} = 2u_n + 1$
   c $u_1 = \frac{3}{4}$ and $u_{n+1} = \frac{2}{3}(u_n)$
   d $u_1 = x$ and $u_{n+1} = (u_n)^2$

3 Write a recursive formula for each sequence.
   a 2, 4, 6, 8, …
   b 1, 3, 9, 27, …
   c 64, 32, 16, 8, …
   d 7, 12, 17, 22, …

4 Write down the first four terms in each sequence.
   a $u_n = 3^n$
   b $u_n = -6n + 3$
   c $u_n = 2^{n-1}$
   d $u_n = n^n$

> To find the first term substitute $n = 1$; to find the second term use $n = 2$ and so on.

5 Write a general formula for the nth term of each sequence.
   a 2, 4, 6, 8, …
   b 1, 3, 9, 27, …
   c 64, 32, 16, 8, …
   d 7, 12, 17, 22, …
   e $\frac{1}{2}, \frac{2}{3}, \frac{3}{4}, \frac{4}{5}, …$
   f $x, 2x, 3x, 4x, …$

6 The sequence 1, 1, 2, 3, 5, 8, 13, … is known as the Fibonacci sequence.
   a Find the 15th term of the Fibonacci sequence.
   b Write a recursive formula for the Fibonacci sequence.

▲ Fibonacci, also known as Leonardo of Pisa (Italian c. 1175 – c. 1250)

6.2 Arithmetic sequences

In the sequence 8, 11, 14, 17, …, the value of each term is three greater than the value of the previous term. This sequence is an example of an **arithmetic sequence**, or arithmetic progression.

→ In an arithmetic sequence, the terms increase or decrease by a constant value. This value is called the **common difference**, or **d**. The common difference can be a positive or a negative value.

Examples of arithmetic progressions appear on the Ahmes Papyrus, which dates from about 1650 BCE.

For example:
8, 11, 14, 17, …   In this sequence, $u_1 = 8$ and $d = 3$
35, 30, 25, 20, …   In this sequence, $u_1 = 35$ and $d = -5$
4, 4.1, 4.2, 4.3, …   In this sequence, $u_1 = 4$ and $d = 0.1$
$c, 2c, 3c, 4c, \ldots$ In this sequence, $u_1 = c$ and $d = c$

For any arithmetic sequence, $u_{n+1} = u_n + d$
We can find any term of the sequence by adding the common difference, d, to the previous term.
In an arithmetic sequence:
$u_1 = $ the first term
$u_2 = u_1 + d$
$u_3 = u_2 + d = (u_1 + d) + d = u_1 + 2d$
$u_4 = u_3 + d = (u_1 + 2d) + d = u_1 + 3d$
$u_5 = u_4 + d = (u_1 + 3d) + d = u_1 + 4d$
…
…
$u_n = u_1 + (n-1)d$

→ You can find the nth term of an arithmetic sequence using the formula: $u_n = u_1 + (n-1)d$

Example 3

a Find the 12th term of the arithmetic sequence 13, 19, 25, …
b Find an expression for the nth term.

Answers

a $u_1 = 13$ and $d = 6$
$u_{12} = 13 + (12 - 1)6$
$= 13 + 66$
$u_{12} = 79$

Find these values by looking at the sequence.
For the 12th term, substitute $n = 12$ into the formula
$u_n = u_1 + (n-1)d$

b $u_n = 13 + (n-1)6$
$= 13 + 6n - 6$
$u_n = 6n + 7$

For the nth term, substitute the values of u_1 and d into the formula
$u_n = u_1 + (n-1)d$

Example 4

Find the number of terms in the arithmetic sequence 84, 81, 78, …, 12.

Answer

$u_1 = 84$ and $d = -3$
$u_n = 84 + (n-1)(-3) = 12$

$84 - 3n + 3 = 87 - 3n = 12$
$3n = 75$
$n = 25$

There are 25 terms in the sequence.

Find these values by looking at the sequence.
Substitute the values of u_1 and d into the formula $u_n = u_1 + (n-1)d$
Solve for n.

> If a sequence continues indefinitely and there is no final term, it is an *infinite* sequence.
> If a sequence ends, or has a 'last term' it is a *finite* sequence.

Exercise 6B

1 For each sequence:
  i Find the 15th term.
  ii Find an expression for the nth term.
  - **a** 3, 6, 9, …
  - **b** 25, 40, 55, …
  - **c** 36, 41, 46, …
  - **d** 100, 87, 74, …
  - **e** 5.6, 6.2, 6.8, …
  - **f** $x, x + a, x + 2a, …$

2 Find the number of terms in each sequence.
  - **a** 5, 10, 15, …, 255
  - **b** 4.8, 5.0, 5.2, …, 38.4
  - **c** $\frac{1}{2}, \frac{7}{8}, \frac{5}{4}, …, 14$
  - **d** 250, 221, 192, …, −156
  - **e** $2m, 5m, 8m, …, 80m$
  - **f** $x, 3x + 3, 5x + 6, …, 19x + 27$

Example 5

In an arithmetic sequence, $u_9 = 48$ and $u_{12} = 75$. Find the first term and the common difference.

Answer

$u_9 + 3d = u_{12}$
$48 + 3d = 75$
$3d = 27$
$d = 9$

$u_9 = u_1 + (9-1)9 = 48$
$u_1 + 72 = 48$
$u_1 = -24$

The first term of the sequence is −24, and the common difference is 9.

To get from the 9th term, u_9, to the 12th term, u_{12}, you would need to add the common difference 3 times.

To find the first term, use the formula.

Exercise 6C

1. An arithmetic sequence has first term 19 and 15th term 31.6. Find the common difference.

EXAM-STYLE QUESTION

2. In an arithmetic sequence, $u_{10} = 37$ and $u_{21} = 4$.
   Find the common difference and the first term.

3. Find the value of x in the arithmetic sequence 3, x, 8, ...

4. Find the value of m in the arithmetic sequence m, 13, $3m - 6$, ...

6.3 Geometric sequences

In the sequence 2, 6, 18, 54, ..., each term is three times the previous term. This sequence is an example of a **geometric sequence**, or geometric progression.

> → In a **geometric sequence**, each term can be obtained by multiplying the previous term by a constant value. This value is called the **common ratio**, or r.

The common ratio, r, can be positive or negative.
For example:

1, 5, 25, 125, ...	$u_1 = 1$ and $r = 5$
3, −6, 12, −24, ...	$u_1 = 3$ and $r = -2$
81, 27, 9, 3, ...	$u_1 = 81$ and $r = \frac{1}{3}$
$k, k^2, k^3, k^4, ...$	$u_1 = k$ and $r = k$

For any geometric sequence, $u_{n+1} = (u_n)r$. You can find any term of the sequence by multiplying the previous term by the common ratio, r.
For any geometric sequence:

$u_1 =$ the first term
$u_2 = u_1 \times r$
$u_3 = u_2 \times r = (u_1 \times r) \times r = u_1 \times r^2$
$u_4 = u_3 \times r = (u_1 \times r^2) \times r = u_1 \times r^3$
$u_5 = u_4 \times r = (u_1 \times r^3) \times r = u_1 \times r^4$
...
...
$u_n = u_1 \times r^{n-1}$

> → You can find the nth term of a geometric sequence using the formula: $u_n = u_1(r^{n-1})$

Example 6

Find the 9th term of the sequence 1, 4, 16, 64, ...

Answer

$u_1 = 1$ and $r = 4$

$u_9 = 1(4^{9-1}) = 1(4^8)$
$= 1(65\,536)$
$u_9 = 65\,536$

Find these values by looking at the sequence.
For the 9th term, substitute $n = 9$ into the formula $u_n = u_1(r^{n-1})$

Example 7

Find the 12th term of the sequence 7, −14, 28, −56, ...

Answer

$u_1 = 7$ and $r = -2$

$u_{12} = 7((-2)^{12-1}) = 7((-2)^{11})$
$= 7(-2048)$
$u_{12} = -14\,336$

Find these values by looking at the sequence.
For the 12th term, substitute $n = 12$ into the formula $u_n = u_1(r^{n-1})$

Exercise 6D

1 For each sequence, find the common ratio and the 7th term.
  a 16, 8, 4, ...
  b −4, 12, −36, ...
  c 1, 10, 100, ...
  d 25, 10, 4, ...
  e 2, $6x$, $18x^2$, ...
  f a^7b, a^6b^2, a^5b^3, ...

Example 8

In a geometric sequence, $u_1 = 864$ and $u_4 = 256$
Find the common ratio.

Answer

$u_4 = u_1(r^{4-1}) = u_1(r^3)$
$256 = 864(r^3)$
$r^3 = \dfrac{256}{864} = \dfrac{8}{27}$
$r = \sqrt[3]{\dfrac{8}{27}}$
$r = \dfrac{2}{3}$

Substitute $n = 4$, $u_1 = 864$, and $u_4 = 256$ into the formula $u_n = u_1(r^{n-1})$

Solve for r.

Example 9

For the geometric sequence 5, 15, 45, ... find the least value of n such that the nth term is greater than 50 000.

Answer

$u_1 = 5$ and $r = 3$
$u_n = 5 \times 3^{n-1}$

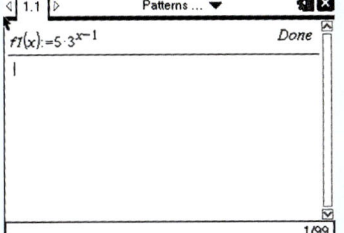

Find u_1 and r by looking at the sequence.
Substitute $u_1 = 5$ and $r = 3$ into the formula $u_n = u_1(r^{n-1})$
You can use the GDC to help find the value of n. First, enter the formula for u_n into a function. Let the variable x represent n, as shown.

GDC help on CD: *Alternative demonstrations for the TI-84 Plus and Casio FX-9860GII GDCs are on the CD.*

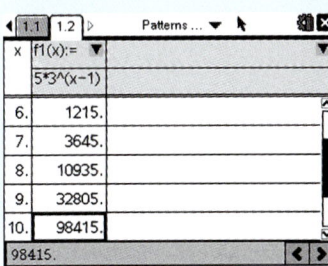

Now look at the TABLE to see the values of the first n terms.

The 9th term is 32 805, and the 10th term is 98 415.

$n = 10$, since $u_{10} > 50\,000$

Exercise 6E

1 A geometric sequence has 2nd term 50 and 5th term 3.2. Find the first term and the common ratio.

2 A geometric sequence has 3rd term –18 and 6th term 144. Find the first term and the common ratio.

3 For each geometric sequence, find the least value of n such that the nth term is greater than 1000.
    a 16, 24, 36, ...   **b** 1, 2.4, 5.76, ...
    c 112, –168, 252, ...   **d** 50, 55, 60.5, ...

4 A geometric sequence has first term 9 and third term 144. Show that there are two possible values for the common ratio, and find the two possible values for the second term.

5 Find the value of p in the geometric sequence $18, p, 40.5, \ldots$

> **EXAM-STYLE QUESTION**
> **6** Find the positive value of x in the geometric sequence
> $7x - 2, 4x + 4, 3x, \ldots$

6.4 Sigma (Σ) notation and series

This section looks at ways to add the terms of a sequence.
Adding the terms of a sequence gives a **series**.

$u_1, u_2, u_3, u_4, \ldots, u_n$ is a *sequence*.
$u_1 + u_2 + u_3 + u_4 + \cdots + u_n$ is a *series*.

The Greek letter Σ, called 'sigma', is often used to represent sums of values.

→ $\sum_{i=1}^{n} u_i$ means the sum of the first n terms of a sequence.

You read this 'the sum of all the terms u_i from $i = 1$ to $i = n$'.

When you represent a sum of values in this form, you are using **sigma notation**.

The arithmetic sequence $8, 14, 20, \ldots$ has first term 8 and common difference 6. A general rule for the nth term of this sequence is $u_n = 6n + 2$

The sum of the first five terms of this sequence is $\sum_{n=1}^{5}(6n + 2)$

This means 'the sum of all the terms $6n + 2$ from $n = 1$ to $n = 5$'.

To calculate this sum, substitute all the integer values of n from 1 to 5 into the expression $6n + 2$, and add them:

$$\sum_{n=1}^{5}(6n + 2) = [6(1) + 2] + [6(2) + 2] + [6(3) + 2] + [6(4) + 2]$$
$$+ [6(5) + 2]$$
$$= 8 + 14 + 20 + 26 + 32 = 100$$

Example 10

a Write the expression $\sum_{x=1}^{4}(x^2 - 3)$ as a sum of terms.
b Calculate the sum of these terms.

Answers

a $\sum_{x=1}^{4}(x^2 - 3)$

$= (1^2 - 3) + (2^2 - 3)$
$ + (3^2 - 3) + (4^2 - 3)$
$= -2 + 1 + 6 + 13$

Substitute consecutive integers beginning with $x = 1$ and ending with $x = 4$

b $-2 + 1 + 6 + 13 = 18$

Example 11

Evaluate the expression $\sum_{a=3}^{8}(2^a)$.

Answer

$\sum_{a=3}^{8}(2^a) = 2^3 + 2^4 + 2^5 + 2^6 + 2^7 + 2^8$
$= 8 + 16 + 32 + 64 + 128 + 256$
$= 504$

Substitute consecutive integers beginning with a = 3 and ending with a = 8

'Evaluate' tells you to find the value so the final answer will be a number.

Example 12

Write the series 3 + 15 + 75 + 375 + 1875 + 9375 using sigma notation.

Answer

$u_n = 3(5^{n-1})$

$\sum_{n=1}^{6}\left(3(5^{n-1})\right)$

The terms are a geometric progression, with first term 3 and common ratio 5.
This series is the first six terms of the geometric progression.

Exercise 6F

1 Write an expression for each series using sigma notation.
   a 1 + 2 + 3 + 4 + 5 + 6 + 7 + 8
   b 9 + 16 + 25 + 36 + 49
   c 27 + 25 + 23 + 21 + 19 + 17
   d 240 + 120 + 60 + 30 + 15 + 7.5
   e $5x + 6x + 7x + 8x + 9x + 10x$
   f 4 + 7 + 10 + 13 + ⋯ + 55
   g 1 + 3 + 9 + 27 + ⋯ + 59 049
   h $a + 2a^2 + 3a^3 + 4a^4 + 5a^5$

2 Write each series as a sum of terms.
   a $\sum_{n=1}^{8}(3n+1)$ b $\sum_{a=1}^{5}(4^a)$ c $\sum_{r=3}^{7}(5(2^r))$ d $\sum_{n=5}^{11}(x^n)$

3 Evaluate.
   a $\sum_{n=1}^{9}(8n-5)$ b $\sum_{r=1}^{5}(3^r)$ c $\sum_{m=1}^{7}(m^2)$ d $\sum_{x=4}^{10}(7x-4)$

Remember, the word *evaluate* tells you to find the value, so you need to give numerical answers.

6.5 Arithmetic series

The sum of the terms of a sequence is called a series. The sum of the terms of an arithmetic sequence is called an **arithmetic series**.

For example, 5, 12, 19, 26, 33, 40 is an arithmetic sequence, so 5 + 12 + 19 + 26 + 33 + 40 is an arithmetic series.

When a series has only a few elements, adding the individual terms is not a difficult task. However, if a series has 50 terms or 100 terms it would be very time-consuming to add all these terms. It will be helpful to find a rule, or formula, for evaluating arithmetic series.

S_n denotes the sum of the first n terms of a series. For a series with n terms,

$$S_n = u_1 + u_2 + u_3 + u_4 + u_5 + \cdots + u_n$$

For an arithmetic series this would be:

$$S_n = u_1 + (u_1 + d) + (u_1 + 2d) + (u_1 + 3d) + (u_1 + 4d) + \cdots + (u_1 + (n-1)d)$$

If we reverse the order of the terms in the equation, the value of the sum would be the same, and it would look like this:

$$S_n = u_n + (u_n - d) + (u_n - 2d) + (u_n - 3d) + (u_n - 4d) + \cdots + u_1$$

Adding these two equations for S_n vertically, term by term,

$$2S_n = (u_1 + u_n) + (u_1 + u_n) + (u_1 + u_n) + (u_1 + u_n) + (u_1 + u_n) + \cdots + (u_1 + u_n)$$

This is $(u_1 + u_n)$ added n times, so:

$$2S_n = n(u_1 + u_n)$$

Dividing both sides by 2 gives:

$$S_n = \frac{n}{2}(u_1 + u_n)$$

Substitute $u_1 + (n-1)d$ for u_n, then

$$S_n = \frac{n}{2}(u_1 + u_1 + (n-1)d) = \frac{n}{2}(2u_1 + (n-1)d)$$

> → You can find the sum of the first n terms of an arithmetic series using the formula:
>
> $$S_n = \frac{n}{2}(u_1 + u_n) \quad \text{or} \quad S_n = \frac{n}{2}(2u_1 + (n-1)d)$$

Carl Friedrich Gauss (1777–1885) is often said to be the greatest mathematician of the 19th century. Find out how Gauss worked out the sum of the first 100 integers.

Remember that n must be a positive integer.

Start with the final term u_n, then the next-to-last term is $u_n - d$, and so on.

Patterns, sequences and series

Example 13

Calculate the sum of the first 15 terms of the series
29 + 21 + 13 + …

Answer

$u_1 = 29$ and $d = -8$

$S_{15} = \dfrac{15}{2}(2(29)+(15-1)(-8))$

$= 7.5(58 - 112)$

$= -405$

For the sum of 15 terms, substitute $n = 15$ into the formula

$S_n = \dfrac{n}{2}(2u_1 + (n-1)d)$

Example 14

a Find the number of terms in the series
14 + 15.5 + 17 + 18.5 + ⋯ + 50

b Find the sum of the terms.

Answers

a $u_1 = 14$ and $d = 1.5$

$u_n = 50$
$u_n = 14 + (n - 1)(1.5) = 12.5 + 1.5n$
$12.5 + 1.5n = 50$
$1.5n = 37.5$
$n = 25$

b $S_{25} = \dfrac{25}{2}(14+50)$

$= 12.5(64)$

$= 800$

Find these values by looking at the sequence.
To find n, substitute the values you know into the formula
$u_n = u_1 + (n - 1)d$
Solve for n.

Substitute the first term, the last term and the value of n into the formula
$S_n = \dfrac{n}{2}(u_1 + u_n)$

Exercise 6G

1 Find the sum of the first 12 terms of the arithmetic series
3 + 6 + 9 + ⋯

2 Find the sum of the first 18 terms of the arithmetic series
2.6 + 3 + 3.4 + ⋯

3 Find the sum of the first 27 terms of the arithmetic series
100 + 94 + 88 + ⋯

4 Find the sum of the first 16 terms of the series
$(2 - 5x) + (3 - 4x) + (4 - 3x) + \cdots$

: EXAM-STYLE QUESTION
5 Consider the series $120 + 116 + 112 + \cdots + 28$.
  a Find the number of terms in the series
  b Find the sum of the terms.

6 Find the sum of the series $15 + 22 + 29 + \cdots + 176$

Example 15

a Write an expression for S_n, the sum of the first n terms, of the series $64 + 60 + 56 + \cdots$
b Hence, find the value of n for which $S_n = 0$

Answers

a $u_1 = 64$ and $d = -4$

$S_n = \dfrac{n}{2}(2(64) + (n-1)(-4))$

$= \dfrac{n}{2}(128 - 4n + 4)$

$= \dfrac{n}{2}(132 - 4n)$

$S_n = 66n - 2n^2$

Substitute the values for u_1 and d into the formula
$S_n = \dfrac{n}{2}(2u_1 + (n-1)d)$

b $66n - 2n^2 = 0$
$2n(33 - n) = 0$
$n = 0$ or $n = 33$

$n = 33$

Set $S_n = 0$, and solve for n.
(You can also solve this equation using your GDC.) When we solve by factoring, the equation usually has two solutions.
Since the number of terms must be a positive integer, we disregard $n = 0$

> The word *hence* in the question tells you to use your previous answer in this part.

Exercise 6H

1 An arithmetic series has $u_1 = 4$ and $S_{30} = 1425$
  Find the value of the common difference.

: EXAM-STYLE QUESTION
2 a Write an expression for S_n, for the series $1 + 7 + 13 + \cdots$
  b Hence, find the value of n for which $S_n = 833$

3 a Write an expression for S_n, for an arithmetic series with $u_1 = -30$ and $d = 3.5$
  b Hence, find the value of n for which $S_n = 105$

4 In January 2012, a new coffee shop sells 500 drinks. In February, they sell 600 drinks, then 700 in March, and so on in an arithmetic progression.
  a How many drinks will they expect to sell in December 2012?
  b Calculate the total number of drinks they expect to sell in 2012.

5 In an arithmetic sequence, the 2nd term is four times the 5th term, and the sum of the first ten terms is –20. Find the first term and the common difference.

6 In an arithmetic series, the sum of the first 12 terms is equal to ten times the sum of the first 3 terms. If the first term is 5, find the common difference and the value of S_{20}.

6.6 Geometric series

Just as an arithmetic series is the sum of the terms of an arithmetic sequence, a **geometric series** is the sum of the terms of a geometric sequence.

Adding the terms of a geometric sequence gives the following equation:

$$S_n = u_1 + u_1 r + u_1 r^2 + u_1 r^3 + \cdots + u_1 r^{n-2} + u_1 r^{n-1}$$

Multiply both sides of this equation by r.

$$rS_n = u_1 r + u_1 r^2 + u_1 r^3 + u_1 r^4 + \cdots + u_1 r^{n-1} + u_1 r^n$$

Subtract the first equation from the second.

$$rS_n - S_n = -u_1 + u_1 r^n = u_1 r^n - u_1$$

$$S_n(r-1) = u_1(r^n - 1)$$

Factorize both sides of the equation.

$$S_n = \frac{u_1(r^n - 1)}{r - 1}$$

→ You can find the sum of the first n terms of a geometric series using the formula:

$$S_n = \frac{u_1(r^n - 1)}{r - 1} \quad \text{or} \quad S_n = \frac{u_1(1 - r^n)}{1 - r}, \text{ where } r \neq 1$$

You may find it more convenient to use the first formula when $r > 1$, as it avoids using a negative denominator.

Example 16

Calculate the sum of the first 12 terms of the series $1 + 3 + 9 + \cdots$

Answer

$u_1 = 1$ and $r = 3$

$S_{12} = \dfrac{1(3^{12} - 1)}{3 - 1}$

$= \dfrac{531\,440}{2}$

$= 265\,720$

Substitute the values of u_1, r and n into the formula

$S_n = \dfrac{u_1(r^n - 1)}{r - 1}$

Example 17

a Find the number of terms in the series
$8192 + 6144 + 4608 + \cdots + 1458$.
b Calculate the sum of the terms.

> Geometric series are often seen in the study of fractals, such as the Koch snowflake.

Answers

a $u_1 = 8192$ and $r = \dfrac{6144}{8192} = \dfrac{3}{4}$ — *Find r by dividing u_2 by u_1.*

$1458 = 8192\left(\dfrac{3}{4}\right)^{n-1}$ — *Substitute the values you know into the formula $u_n = u_1(r^{n-1})$*

$\dfrac{1458}{8192} = \dfrac{729}{4096} = \left(\dfrac{3}{4}\right)^{n-1}$

$\dfrac{729}{4096} = \dfrac{3^6}{4^6} = \left(\dfrac{3}{4}\right)^6$ — *$3^6 = 729$ and $4^6 = 4096$. You could also solve this equation using logarithms (see Example 19).*

$n - 1 = 6$
$n = 7$

b $S_7 = \dfrac{8192\left(1 - \left(\dfrac{3}{4}\right)^7\right)}{1 - \dfrac{3}{4}}$ — *Substitute the values of u_1, r and n into the formula*

$S_n = \dfrac{u_1(r^n - 1)}{r - 1}$

$= \dfrac{8192\left(\dfrac{14\,197}{16\,384}\right)}{\dfrac{1}{4}}$

*You can also calculate sums using the **seq** (and **sum**) functions on your GDC.*

$= 28\,394$

▲ Koch snowflake

Exercise 6I

1 Calculate the value of S_{12} for each geometric series.
   a $0.5 + 1.5 + 4.5 + \cdots$
   b $0.3 + 0.6 + 1.2 + \cdots$
   c $64 - 32 + 16 - 8 + \cdots$
   d $(x+1) + (2x+2) + (4x+4) + \cdots$

2 Calculate the value of S_{20} for each series.
   a $0.25 + 0.75 + 2.25 + \cdots$
   b $\dfrac{16}{9} + \dfrac{8}{3} + 4 + \cdots$
   c $3 - 6 + 12 - 24 + \cdots$
   d $\log a + \log(a^2) + \log(a^4) + \log(a^8) + \cdots$

EXAM-STYLE QUESTION

3 For each geometric series:
   i find the number of terms
   ii calculate the sum
   a $1024 + 1536 + 2304 + \cdots + 26\,244$
   b $2.7 + 10.8 + 43.2 + \cdots + 2764.8$
   c $\dfrac{125}{128} + \dfrac{25}{64} + \dfrac{5}{32} + \cdots + \dfrac{1}{625}$
   d $590.49 + 196.83 + 65.61 + \cdots + 0.01$

> So far we have looked at arithmetic and geometric sequences and series. Are there other types of mathematical sequences and series? How are they used?

Example 18

For the geometric series $3 + 3\sqrt{2} + 6 + 6\sqrt{2} + \ldots$, determine the least value of n for which $S_n > 500$

Answer

$u_1 = 3$ and $r = \sqrt{2}$

$$S_n = \frac{3(\sqrt{2}^n - 1)}{\sqrt{2} - 1} > 500$$

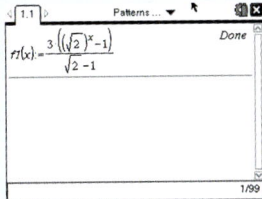

$n = 13$, since $S_{13} > 500$

Substitute the known values into the S_n formula. Enter the S_n equation into the GDC.

Remember:
On the GDC, the X represents 'n', the number of terms, and $f1(x)$ represents S_n.

Look at the TABLE to see the sums of the first n terms.

The sum of the first 12 terms is approximately 456.29, and the sum of the first 13 terms is approximately 648.29

GDC help on CD: Alternative demonstrations for the TI-84 Plus and Casio FX-9860GII GDCs are on the CD.

An old Indian fable tells us that a prince was so taken with the new game of chess that he asked its inventor to choose his reward. The man said he would like one grain of rice on the first square of the chess board, two grains on the second, four on the third etc., doubling the number each time. This seemed so little to ask that the prince agreed straight away. Servants started to bring the rice – and to the prince's great surprise the grain soon overflowed the chess board to fill the palace.
How many grains of rice did the prince have to give the man?

When the sum of a geometric series includes an exponent n, you can use logarithms.

Example 19

A geometric progression has first term of 0.4 and common ratio 2. Find the value of n such that $S_n = 26\,214$

Answer

$$S_n = \frac{0.4(2^n - 1)}{2 - 1} = 26\,214$$

$0.4(2^n - 1) = 26\,214$

$2^n - 1 = 65\,535$

$2^n = 65\,536$

$n = \log_2(65\,536)$

$n = \dfrac{\log 65\,536}{\log 2}$

$n = 16$

Express this using logarithms. Use the change-of-base rule and your GDC to find this value.

Chapter 6 | 177

Exercise 6J

1 For each series, determine the least value of n for which $S_n > 400$
   a $25.6 + 38.4 + 57.6 + \cdots$
   b $14 - 42 + 126 - 378 + \cdots$
   c $\dfrac{2}{3} + \dfrac{8}{9} + \dfrac{32}{27} + \cdots$
   d $0.02 + 0.2 + 2 + \cdots$

2 A geometric series has third term 1.2 and eighth term 291.6
   Find the common ratio and the value of S_{10}.

3 In a geometric series, $S_4 = 20$ and $S_7 = 546.5$
   Find the common ratio, if $r > 1$

EXAM-STYLE QUESTION

4 a Find the common ratio for the geometric series $\dfrac{1}{12} + \dfrac{1}{8} + \dfrac{3}{16} + \cdots$
   b Hence, find the least value of n such that $S_n > 800$

> 'Hence' tells you to use your previous answer in this part.

5 In a geometric series, the sum of the first three terms is 304, and the sum of the first 6 terms is 1330. Find the sum of the first seven terms.

6 In a geometric series, the sum of the first four terms is ten times the sum of the first two terms. If $r > 1$, find the common ratio.

Extension material on CD:
Worksheet 6 - Finance

6.7 Convergent series and sums to infinity

Investigation – converging series

Here are three geometric series.
a $2 + 1 + 0.5 + \cdots$ b $75 + 30 + 12 + \cdots$
c $240 - 60 + 15 - 3.75 + \cdots$

1 For each of these series:
   i Find the common ratio, r.
   ii Use your GDC to calculate the values of S_{10}, S_{15}, S_{20}.
      Write the full values you see on your GDC screen.

2 Do you notice any patterns? Why do you think this is happening?

3 Now use your GDC to calculate the value of S_{50} for each series.
   Do you think your calculator is correct? Explain why or why not.

For each of the series in the investigation you should have noticed that the values of S_{10}, S_{15} and S_{20} are very close. This is because when a geometric series has a common ratio of $|r| < 1$, the difference between each term decreases (becomes closer to zero) as n increases. This means that, as you add more terms, the value of the sum changes very little. The sum is actually approaching some constant value as n gets very large. We call geometric series such as these **convergent series**.

> **Paradox**
> Suppose you are walking down a 30-metre hallway. Every ten seconds, you walk half the distance to the end of the hallway. How long will it take you to reach the end of the hallway? Will you ever get there?

In the series $2 + 1 + 0.5 + 0.25 + \cdots$, you might suspect that the sum is approaching 4 as n gets very large.

If you try to find S_{50} on the GDC, you get:

$$S_{50} = \frac{2(1-0.5^{50})}{0.5} = 4(1-0.5^{50}) = 4$$

Is the sum actually 4? NO! The GDC rounds the last digit of long decimals like 3.999 999 999 99 to fit on its screen, so all you see is the rounded value of 4.

Convergent series

The sum of the terms of a geometric series is $S_n = \dfrac{u_1(1-r^n)}{1-r}$

As n gets very large, you can say that n 'approaches infinity', or $n \to \infty$.

If $|r| < 1$, then as $n \to \infty$, $r^n \to 0$, so $S_n \to \dfrac{u_1(1-0)}{1-r} = \dfrac{u_1}{1-r}$

We can write this as:

$$\lim_{n \to \infty}\left(\frac{u_1(1-r^n)}{1-r}\right) = \frac{u_1}{1-r}, \quad \text{or } S_\infty = \frac{u_1}{1-r}$$

This means that as n gets very large (it approaches infinity), the value of the series is approaching $\dfrac{u_1}{1-r}$. The series is **converging** to the value $\dfrac{u_1}{1-r}$. We write this as S_∞, and call it 'the sum to infinity'.

> **IMPORTANT!**
> This is only true for **geometric** series, and only when $|r| < 1$. (Remember, if $|r| < 1$, then $-1 < r < 1$.)

> We say 'The limit of $\dfrac{u_1(1-r^n)}{1-r}$ as n approaches infinity is equal to $\dfrac{u_1}{1-r}$'.

→ For a geometric series with $|r| < 1$, $S_\infty = \dfrac{u_1}{1-r}$

Example 20

For the series $18 + 6 + 2 + \cdots$, find S_{10}, S_{15} and S_∞.

Answer

$u_1 = 18$ and $r = \dfrac{1}{3}$

$$S_{10} = \frac{18\left(1-\left(\frac{1}{3}\right)^{10}\right)}{1-\frac{1}{3}}$$

$\approx 26.999\,542\,75$

$$S_{15} = \frac{18\left(1-\left(\frac{1}{3}\right)^{15}\right)}{1-\frac{1}{3}}$$

$\approx 26.999\,998\,12$

$$S_\infty = \frac{18}{\left(1-\frac{1}{3}\right)} = 27$$

Substitute $u_1 = 18$ and $r = \dfrac{1}{3}$ into the formulae $S_n = \dfrac{u_1(1-r^n)}{1-r}$ and $S_\infty = \dfrac{u_1}{1-r}$

Write down all the digits from the GDC display.

Chapter 6

Example 21

The sum of the first three terms of a geometric series is 148, and the sum to infinity is 256.
Find the first term and the common ratio of the series.

Answer

$S_3 = \dfrac{u_1(1-r^3)}{1-r} = 148$ — This is the equation for S_3.

$S_\infty = \dfrac{u_1}{1-r} = 256$ — Multiply both sides of this equation by $(1-r^3)$

$\dfrac{u_1(1-r^3)}{1-r} = 256(1-r^3)$ — The left side of this equation is now identical to the left side of the S_3 equation.

$256(1-r^3) = 148$ — Set the right sides of these equations equal to each other. Solve for r.

$1 - r^3 = \dfrac{148}{256} = \dfrac{37}{64}$

$r^3 = 1 - \dfrac{37}{64} = \dfrac{27}{64}$

$r = \dfrac{3}{4}$

$\dfrac{u_1}{\left(1-\dfrac{3}{4}\right)} = 256$ — Substitute $r = \dfrac{3}{4}$ into the equation $S_\infty = \dfrac{u_1}{1-r} = 256$

$\dfrac{u_1}{\left(\dfrac{1}{4}\right)} = 256$

$4u_1 = 256$

$u_1 = 64$

Exercise 6K

1 Explain how you know if a geometric series will be a convergent series.

2 Find S_4, S_7 and S_∞ for each of these series.

    a $144 + 48 + 16 + \cdots$ **b** $500 + 400 + 320 + \cdots$

    c $80 + 8 + 0.8 + \cdots$ **d** $\dfrac{9}{2} + 3 + 2 + \cdots$

3 A geometric series has $S_\infty = \dfrac{27}{2}$ and $S_3 = 13$. Find S_5.

EXAM-STYLE QUESTION

4 For a geometric progression with $u_3 = 24$ and $u_6 = 3$, find S_∞.

> What real-life situations might be modeled by covergent series?

5 In a geometric progression, $u_2 = 12$ and $S_\infty = 64$. Find u_1.

> **EXAM-STYLE QUESTION**
>
> **6** A geometric series has a common ratio of 0.4 and a sum to infinity of 250. Find the first term.

7 The sum of the first five terms of a geometric series is 3798, and the sum to infinity is 4374. Find the sum of the first seven terms.

6.8 Applications of geometric and arithmetic patterns

We see examples of geometric patterns in many real-life situations, such as compound interest and population growth.

If a person deposits $1000 in a savings account which pays interest at a rate of 4% annually, and makes no other withdrawals or deposits, how much will be in the account after ten years?

When the interest is compounded annually (once per year), the amount in the account at the end of each year will be 104% of the amount at the start of the year. (Multiply the previous amount by 1.04.) The amount in the account after 10 years would be $1000(1.04)^{10} \approx \$1480.24$.

You can think of the amount of money in the account at the end of each year as a geometric sequence with $u_1 = 1000$ and $r = 1.04$:

$u_1 = \$1000$
$u_2 = \$1000(1.04) = \$1040$
$u_3 = \$1040(1.04) = \$1081.60$
$u_4 = \$1081.60(1.04) \approx \$1124.86$
    and so on.

Now consider what happens when interest is compounded more than once each year.

Let
   A = the amount of money in the account
   r = the interest rate (a percentage, written as a decimal)
   n = the number of times per year that interest is compounded
   t = the number of years
   p = the principal (initial amount of money)

you can find the amount of money in the account using the formula:

$$A = p\left(1 + \frac{r}{n}\right)^{nt}$$

> In questions like this, the term 'per annum' is sometimes used rather than 'annually' or 'once a year'.

Example 22

A person deposits $1000 in an account which pays interest at 4% APR, compounded quarterly. Assuming the person makes no additional withdrawals or deposits, how much will be in the account after ten years?

> APR means 'annual percentage rate'. 4% APR is the same as 4% per annum.

Answer

$$A = 1000\left(1 + \frac{0.04}{4}\right)^{4(10)}$$
$$= 1000(1.01)^{40}$$
$$\approx \$1488.86$$

Substitute the known values into the formula $A = p\left(1 + \frac{r}{n}\right)^{nt}$

This equation works because the annual interest rate (4%) is divided up into 4 parts, one for each quarter, and so the quarterly interest rate is 1%. If interest is compounded four times every year (quarterly) for a period of 10 years, this quarterly interest rate will be applied 40 times.

> What other types of mathematics are useful in finance?

Population growth

Example 23

The population of a small town increases by 2% per year. If the population at the start of 1980 was 12 500, what is the predicted population at the start of 2020?

Answer

$12\,500(1.02)^{40} \approx 27\,600.496$

The population of the town will be approximately 27 600

At the start of each year, the population will be 102% of the previous year's starting population. From 1980 to 2020, 40 years will have passed.

In questions like Example 23, you need to think of n as the number of years, rather than as a term number.

Exercise 6L

1. In an arithmetic sequence, $u_6 = 3u_4$
   Find u_1 if $u_8 = 50$

2. A plastic cup is 12 cm high.
   When five cups are stacked together, the stack is 15 cm high.
   a. How high would a stack of 20 cups stand?
   b. How many cups would it take to make a stack at least 1 metre high?

3 George deposits $2500 in an account which pays interest at 6% APR. Assuming he makes no additional withdrawals or deposits, how much will be in the account after 8 years if the
   a interest is compounded annually
   b interest is compounded quarterly
   c interest is compounded monthly?

4 An arithmetic sequence is defined by $u_n = 12n - 7$ and a geometric sequence is defined by $v_n = 0.3(1.2)^{n-1}$
Find the least number of terms such that $v_n > u_n$

5 In a geometric sequence, the first term is 6 and the common ratio is 1.5. In an arithmetic sequence, the first term is 75 and the common difference is 100. After how many terms will the sum of the terms in the geometric sequence be greater than the sum of the terms in the arithmetic sequence?

> This question uses v_n, rather than u_n, to represent the nth term of the geometric sequence.

6 At the beginning of 2012, a lake contains 200 fish. The number of fish in the lake is expected to increase at a rate of 5% per year. What will be the number of fish in the lake at the beginning of 2015?

7 The population of a city is 275 000 people. The population is increasing at a rate of 3.1% per year. Assuming the population continues to increase at this rate, how long will it take for the population of the city to reach 500 000 people?

8 A series has the formula $S_n = 3n^2 - 2n$
   a Find the values of S_1, S_2 and S_3.
   b Find the values of u_1, u_2 and u_3.
   c Write an expression for u_n.

9 A series has the formula $S_n = 2^{n+2} - 4$
   a Find the values of S_1, S_2 and S_3.
   b Find the values of u_1, u_2 and u_3.
   c Write an expression for u_n.

10 Two species of spiders inhabit a remote island. The population of species A is 12 000 and is increasing at a rate of 1.25% per month. The population of species B is 50 000 and is decreasing at a rate of 175 spiders each month. When will the population of species A be greater than the population of species B? 11. Mohira invests $3000 in an account which pays 3% annual interest, compounded yearly.

11 Mohira invests $3000 in an account which pays 3% annual interest, compounded yearly. Ryan invests $3000 in an account which also pays 3% annual interest, but is compounded monthly. Assuming neither person has made any additional withdrawals or deposits, how much more money does Ryan have in his account than Mohira has in hers after ten years?

6.9 Pascal's triangle and the binomial expansion

Now we will look at a famous mathematical pattern known as Pascal's triangle. Here are rows 1 to 7 of Pascal's triangle.

```
                    1   1
                  1   2   1
                1   3   3   1
              1   4   6   4   1
            1   5  10  10   5   1
          1   6  15  20  15   6   1
        1   7  21  35  35  21   7   1
```

> Pascal's triangle is named after Blaise Pascal (French, 1623–62).

Any number in Pascal's triangle is the sum of the two numbers immediately above it.

You generate the numbers in the triangle by starting at the top and adding pairs of numbers to get the next row. But what if we wanted to find the numbers in the 15th row? Or the 27th row? It would be very time-consuming to make a triangle that big!

> Can you predict what the numbers in the 8th row will be?

Here are the numbers in the 4th row of the triangle, 1, 4, 6, 4, 1. These numbers can also be found using **combinations**, or the $_nC_r$ function on the GDC.

$$_4C_0 = 1 \quad _4C_1 = 4 \quad _4C_2 = 6 \quad _4C_3 = 4 \quad _4C_4 = 1$$

$\binom{n}{r}$, or C_r^n, represents the number of ways n items can be taken r at a time. For example, suppose a bag contains 5 balls labeled A, B, C, D and E. If you reach in and take 2 balls from the bag, there are $\binom{5}{2} = 10$ different combinations of balls you could select.

These combinations are AB, AC, AD, AE, BC, BD, BE, CD, CE or DE.

You can find the values of expressions like $\binom{5}{2}$ without using a calculator.

> $_nC_r$ is commonly written as $\binom{n}{r}$, or sometimes as C_r^n.

> Make sure you know how to use the nCr function on your GDC.

→ The number of combinations of n items taken r at a time is found by:

$$\binom{n}{r} = \frac{n!}{r!(n-r)!}, \text{ where } n! = n \times (n-1) \times (n-2) \times \ldots \times 1$$

> ! is the **factorial** sign. The expression $n!$ is called 'n factorial'.

Patterns, sequences and series

Example 24

Find the value of $\binom{7}{5}$ using the formula, and check with your GDC.

Answer

$\binom{7}{5} = \dfrac{7!}{5!(7-5)!}$

$= \dfrac{7 \times 6 \times 5 \times 4 \times 3 \times 2 \times 1}{(5 \times 4 \times 3 \times 2 \times 1)(2 \times 1)}$

$= \dfrac{7 \times 6}{2 \times 1} = \dfrac{42}{2}$

$= 21$

$\binom{7}{5} = 21$

Using the calculator:

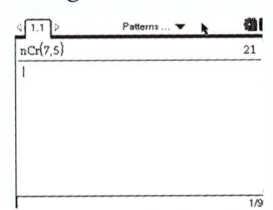

Substitute $n = 7$ and $r = 5$ into the formula.

Cancel out like factors from the numerator and the denominator.

Remember, we can also find the value using Pascal's triangle.

On the TI Nspire, nCr is on the Probability, Combinations menu.

You may see dots used instead of multiplication signs.
For example:
$3 \cdot 2 \cdot 1$ for
$3 \times 2 \times 1$

GDC help on CD: Alternative demonstrations for the TI-84 Plus and Casio FX-9860GII GDCs are on the CD.

Exercise 6M

Find each value using the formula, and check with your GDC.

1. $\binom{5}{3}$
2. $\binom{8}{2}$
3. $_7C_3$
4. $_9C_6$
5. $\binom{6}{4}$
6. $\binom{10}{3}$

Investigation – patterns in polynomials

Expand each of the following expressions (write each expression as a polynomial).
Time how long it takes you to do each expansion.

1. $(a + b)^1$
2. $(a + b)^2$
3. $(a + b)^3$
4. $(a + b)^4$
5. $(a + b)^5$
6. $(a + b)^6$

Look at your answers and note any patterns you see.
Do you notice any similarities to Pascal's triangle?
Based on these patterns, predict what the expansion of $(a + b)^7$ might be.

Chapter 6

Binomial expansion

We will look at what happens when we expand an expression like $(a + b)^n$, where n is a positive integer.

In the Investigation on page 185, you expanded these expressions.
$(a + b)^1 = a + b$
$(a + b)^2 = a^2 + 2ab + b^2$
$(a + b)^3 = a^3 + 3a^2b + 3ab^2 + b^3$
$(a + b)^4 = a^4 + 4a^3b + 6a^2b^2 + 4ab^3 + b^4$
$(a + b)^5 = a^5 + 5a^4b + 10a^3b^2 + 10a^2b^3 + 5ab^4 + b^5$

If you look closely at each expansion, you will see some patterns:

1 The number of terms is one greater than the value of n.

2 The powers of a begin with a^n, and the powers of a decrease by 1 until you reach a^0 ($a^0 = 1$) in the last term.

3 The powers of b begin with b^0 ($b^0 = 1$), and the powers of b increase by 1 until you reach b^n in the last term.

4 The coefficients are all numbers from Pascal's triangle!

The coefficients of $(a + b)^n$ are numbers from the nth row of Pascal's triangle. You can find these using the triangle, or the formula for combinations, or the nCr function on the GDC.

5 The exponents in each term add to the power of the binomial.

For example, in the expansion $(a + b)^3 = a^3 + 3a^2b + 3ab^2 + b^3$, the exponents in each term add to 3.

> For example, when $n = 4$, the expansion has 5 terms.

> $(a + b)^4 = a^4 + 4a^3b + 6a^2b^2 + 4ab^3 + b^4$
> The coefficients 1, 4, 6, 4, 1 are the 4th row of Pascal's triangle.

> In $(a + b)^5$ the coefficients 1, 5, 10, 10, 5, 1 are the 5th row of Pascal's triangle.

You can use these patterns to expand the expression $(a + b)^6$.
This expansion will have 7 terms.
The powers of a will decrease, the powers of b will increase.
The coefficients will be the 6th row of Pascal's triangle (1, 6, 15, 20, 15, 6, 1).
Therefore, $(a + b)^6 = a^6 + 6a^5b + 15a^4b^2 + 20a^3b^3 + 15a^2b^4 + 6ab^5 + b^6$.

These patterns and observations can help you to understand the general binomial theorem for expanding powers of binomials.

> $n \in \mathbb{Z}^+$ means that n is a positive integer.

→ The binomial theorem states that for any power of a binomial, where $n \in \mathbb{N}$,

$$(a+b)^n = \binom{n}{0}a^nb^0 + \binom{n}{1}a^{n-1}b^1 + \binom{n}{2}a^{n-2}b^2 + \cdots + \binom{n}{n}a^0b^n$$

→ You can also write the binomial expansion using sigma notation:

$$(a+b)^n = \sum_{r=0}^{n}\left(\binom{n}{r}(a)^{n-r}(b)^r\right)$$

> Combinations are used in many areas of mathematics beyond the binomial theorem, including probability. You can even use combinations to calculate your chance of winning the lottery!

Patterns, sequences and series

Example 25

Use the binomial theorem to expand $(x + 3)^5$. Write your answer in its simplest form.

Answer

$(x+3)^5 = \binom{5}{0}x^5 3^0 + \binom{5}{1}x^4 3^1 + \binom{5}{2}x^3 3^2 + \binom{5}{3}x^2 3^3 + \binom{5}{4}x^1 3^4 + \binom{5}{5}x^0 3^5$

$= (1)(x^5)(1) + (5)(x^4)(3) + (10)(x^3)(9) + (10)(x^2)(27) + (5)(x^1)(81) + (1)(1)(243)$

$= x^5 + 15x^4 + 90x^3 + 270x^2 + 405x + 243$

Substitute into the binomial theorem. You should be able to find these values both with and without your GDC.

Example 26

Use the binomial theorem to expand $(2x - 5y)^3$. Write your answer in its simplest form.

Answer

$(2x-5y)^3 = \binom{3}{0}(2x)^3(-5y)^0 + \binom{3}{1}(2x)^2(-5y)^1 + \binom{3}{2}(2x)^1(-5y)^2$

$+ \binom{3}{3}(2x)^0(-5y)^3$

$= (1)(8x^3)(1) + (3)(4x^2)(-5y) + (3)(2x)(25y^2) + (1)(1)(-125y^3)$

$= 8x^3 - 60x^2y + 150xy^2 - 125y^3$

Be careful when you see an expression like $(2x)^3$. The exponent needs to be applied to both the variable and the coefficient!
$(2x)^3 = 2^3 x^3 = 8x^3$

Exercise 6N

Use the binomial theorem to expand each expression.

1 $(y + 3)^5$ 2 $(2b - 1)^4$ 3 $(3a + 2)^6$ 4 $\left(x^2 + \dfrac{2}{x}\right)^3$

5 $(x + y)^8$ 6 $(3a - 2b)^4$ 7 $\left(3c + \dfrac{2}{d}\right)^5$ 8 $\left(4x^2 + \dfrac{1}{2y}\right)^3$

Sometimes, you will not need the entire expansion of a power of a binomial. You may simply be looking for one particular term.

Example 27

Find the x^3 term in the expansion of $(4x - 1)^9$

Answer

$\binom{9}{6}(4x)^3(-1)^6$

$= (84)(64x^3)(1)$

$= 5376x^3$

In order to get x^3, raise the first term of the binomial, $4x$, to the third power. So the second term of the binomial, -1, will be raised to the sixth power. You could use $\binom{9}{3}$ instead of $\binom{9}{6}$ as these values are equal.

Example 28

In the expansion of $(2x + 1)^n$, the coefficient of the x^3 term is 80. Find the value of n.

Answer

$\binom{n}{3}(2x)^3 1^{n-3} = 80x^3$

$\left(\dfrac{n!}{(3)!(n-3)!}\right)(8x^3)(1) = 80x^3$

$\left(\dfrac{n!}{(3)!(n-3)!}\right)(8) = 80$

$\dfrac{n \times (n-1) \times (n-2) \times (n-3) \times (n-4) \times \ldots}{(3 \times 2 \times 1) \times [(n-3) \times (n-4) \times \ldots]}(8) = 80$

$\dfrac{n \times (n-1) \times (n-2) \times (n-3) \times (n-4) \times \ldots}{(3 \times 2 \times 1) \times [(n-3) \times (n-4) \times \ldots]} = 10$

$\dfrac{n \times (n-1) \times (n-2)}{6} = 10$

$n \times (n-1) \times (n-2) = 60$

$n^3 - 3n^2 + 2n - 60 = 0$

$n = 5$

You could have used $\binom{n}{n-3}$ instead of $\binom{n}{3}$, as these values are equal.

Use the formula $\binom{n}{r} = \dfrac{n!}{r!(n-r)!}$

As you only have to find the coefficient, you can leave off the x^3.

Divide both sides by 8

Simplify by canceling out like factors from the numerator and the denominator.

You can solve polynomial equations such as this using a GDC.

Exercise 60

1 Find the x^5 term in the expansion of $(x - 4)^7$

EXAM-STYLE QUESTIONS

2 Find the y^4 term in the expansion of $(4y - 1)^5$

3 Find the a^2b^4 term in the expansion of $(2a - 3b)^6$

4 Find the constant term in the expansion of $(x - 2)^9$

5 In the expansion of $(px + 1)^6$, the coefficient of the x^3 term is 160. Find the value of p.

> The 'constant term' is just the numerical term with no variables.

6 In the expansion of $(3x + q)^7$, the coefficient of the x^5 term is 81 648. Find the value of q.

EXAM-STYLE QUESTION

7 Find the constant term in the expansion of $\left(4x + \dfrac{1}{x}\right)^8$

8 Find the constant term in the expansion of $\left(2x^2 - \dfrac{3}{x}\right)^6$

EXAM-STYLE QUESTION

9 In the expansion of $(x + 1)^n$, the coefficient of the x^3 term is two times the coefficient of the x^2 term. Find the value of n.

10 In the expansion of $(x + 2)^n$, the coefficient of the x^3 term is two times the coefficient of the x^4 term. Find the value of n.

Review exercise

EXAM-STYLE QUESTIONS

1 Consider the arithmetic sequence 3, 7, 11, 15, …
  a Write down the common difference.
  b Find u_{71} **c** Find the value of n such that $u_n = 99$

2 The first three terms of an infinite geometric sequence are 64, 16 and 4.
  a Write down the value of r.   **b** Find u_4.
  c Find the sum to infinity of this sequence.

3 In an arithmetic sequence, $u_6 = 25$ and $u_{12} = 49$
  a Find the common difference.
  b Find the first term of the sequence.

4 Consider the arithmetic sequence 22, x, 38, …
  a Find the value of x.   **b** Find u_{31}.

5 Evaluate the expression $\sum_{a=1}^{4}(3^a)$

6 Consider the geometric series $800 + 200 + 50 + \cdots$
  a Find the common ratio.   **b** Find the sum to infinity.

7 Find all possible values of x such that this sequence is geometric: x, 12, $9x$, …

EXAM-STYLE QUESTION

8 Find the x^3 term in the expansion of $(2x + 3)^5$

9 A grocery store has a display of soup cans stacked in a pyramid. The top row has three cans, and each row has two more cans than the row above it.
  a If there are 35 cans in the bottom row, how many rows are in the display?
  b How many cans are in the display in total?

Review exercise

1 In an arithmetic series, the first term is 4 and the sum of the first 25 terms is 1000.
  a Find the common difference.   **b** Find the value of the 17th term.

EXAM-STYLE QUESTION

2 Consider the arithmetic sequence 3, 4.5, 6, 7.5, …
  a Find u_{63}.   **b** Find the value of n such that $S_n = 840$

3 In an arithmetic series, the tenth term is 25 and the sum of the first 10 terms is 160.
   a Find the first term and the common difference.
   b Find the sum of the first 24 terms.

EXAM-STYLE QUESTIONS

4 In a geometric sequence, the first term is 3 and the sixth term is 96.
   a Find the common ratio.
   b Find the least value of n such that $u_n > 3000$

5 In an arithmetic sequence, the first term is 28 and the common difference is 50. In a geometric sequence, the first term is 1 and the common ratio is 1.5
Find the least value of n such that the nth term of the geometric sequence is greater than the nth term of the arithmetic sequence.

6 In a geometric series, the 3rd term is 45 and the sum of the first 7 terms is 2735.
Find the first term and the common ratio, r, if $r \in \mathbb{Z}$

EXAM-STYLE QUESTION

7 Find the term in x^4 in the expansion of $\left(\dfrac{x}{2} - 3\right)^7$

8 In the expansion of $(ax + 2)^8$, the x^5 term has a coefficient of $\dfrac{7}{16}$.
Find the value of a.

9 At the beginning of 2010, the population of a country was 3.4 million.
   a If the population grows at a rate of 1.6% annually, estimate the country's population at the beginning of 2040.
   b If population growth continues at this rate, in what year would the population of the country be expected to exceed 7 million?

CHAPTER 6 SUMMARY

Patterns and sequences

- A **number sequence** is a pattern of numbers arranged in a particular order according to a rule.
- Each individual number, or element, of a sequence is called a **term**.

Arithmetic sequences

- In an arithmetic sequence, the terms increase or decrease by a constant value. This value is called the **common difference**, or d. The common difference can be a positive or a negative value.
- You can find the nth term of an arithmetic sequence using the formula:
$u_n = u_1 + (n-1)d$

Continued on next page

Geometric sequences

- In a **geometric sequence**, each term can be obtained by multiplying the previous term by a constant value. This value is called the **common ratio**, or r.
- You can find the nth term of a geometric sequence using the formula: $u_n = u_1(r^{n-1})$

Sigma (Σ) notation and series

- $\sum_{i=1}^{n} u_i$ means the sum of the first n terms of a sequence.

  You read this 'the sum of all the terms u_i from $i = 1$ to $i = n$'.

Arithmetic series

- You can find the sum of the first n terms of an arithmetic series using the formula:
  $$S_n = \frac{n}{2}(u_1 + u_n) \text{ or } S_n = \frac{n}{2}(2u_1 + (n-1)d)$$

Geometric series

- You can find the sum of the first n terms of a geometric series using the formula:
  $$S_n = \frac{u_1(r^n - 1)}{r - 1} \text{ or } S_n = \frac{u_1(1 - r^n)}{1 - r}, \text{ where } r \neq 1.$$

Convergent series and sums to infinity

- For a geometric series with $|r| < 1$, $S_\infty = \frac{u_1}{1 - r}$

Pascal's triangle and the binomial expansion

- The number of combinations of n items taken r at a time is found by:
  $$\binom{n}{r} = \frac{n!}{r!(n-r)!}, \text{ where } n! = n \times (n-1) \times (n-2) \times \ldots \times 1$$

- The binomial theorem states that for any power of a binomial, where $n \in \mathbb{N}$,
  $$(a+b)^n = \binom{n}{0}a^n b^0 + \binom{n}{1}a^{n-1}b^1 + \binom{n}{2}a^{n-2}b^2 + \ldots + \binom{n}{n}a^0 b^n$$

- You can also write the binomial expansion using sigma notation:
  $$(a+b)^n = \sum_{r=0}^{n} \left(\binom{n}{r}(a)^{n-r}(b)^r \right)$$

Theory of knowledge

Whose idea was it anyway?

Pascal's triangle is named after the Frenchman Blaise Pascal, who wrote about it in 1654 in his *Treatise on the Arithmetic Triangle*.

However, the properties of this pattern were known and studied by mathematicians in India, China and other parts of the world for centuries before Pascal's time.

In China, Pascal's triangle is called 'Yang Hui's triangle' after a 13th century mathematician, but it was known long before this.

In the 11th century, the Persian mathematician and poet Omar Khayyám wrote of the pattern seen in Pascal's triangle.

- What is Tartaglia's triangle?
- How is Pascal's triangle used?

This is not the first case of a relatively long-standing mathematical idea being attributed to a particular person. This has often happened when a well-known mathematician has published an important work, introducing a mathematical idea to the public.

Throughout the years, mathematicians have been given credit for mathematical discoveries or inventions.

- Do you think that many of these ideas have been attributed to the wrong person?

▼ Omar Khayyám (1048–c.1131)

▲ Blaise Pascal (1623–62)

Pascal's triangle

```
            1
           1 1
          1 2 1
         1 3 3 1
        1 4 6 4 1
       1 5 10 10 5 1
      1 6 15 20 15 6 1
```

Fibonacci: patterns in nature

The Italian mathematician Fibonacci, Leonardo of Pisa, introduced the Fibonacci sequence in his book *Liber Abaci*, published in 1202.

In it he set this problem:

If you begin with a single pair of rabbits, and each month each pair produces a new pair which becomes productive from the second month on, how many pairs of rabbits will be produced in a year?

Fibonacci was not the only mathematician to work with this pattern.

The diagram shows how the sequence grows.

		Number of pairs
1st month:	1 pair of original two rabbits	1
2nd month:	still 1 pair as they are not yet productive	1
3rd month:	2 pairs – original pair and the new pair they produce	2
4th month:	3 pairs – original pair, pair they produced in 3rd month, pair they produced in 4th month	3
		5

The number of pairs gives the Fibonacci sequence

1, 1, 2, 3, 5, 8, 13, 21, 34, 55, 89, 144, 233, …

where each term is the sum of the two preceding terms.

The numbers of the Fibonacci sequence are also frequently seen in nature. The number of spirals in pinecones or in the heads of flowers are quite often numbers from the Fibonacci sequence.

- Is it simply an accident that this well-known mathematical sequence appears in nature?
- Could it be that there is a relationship between mathematics and nature?
- What is the golden section? Where does it appear in nature?
- How are Pascal's triangle and the Fibonacci sequence related? **Hint:** look at the sums of the diagonals in the triangle.

▶ Fibonacci (c.1170–c.1250)

7 Limits and derivatives

CHAPTER OBJECTIVES:

6.1 Informal ideas of limit and convergence; limit notation; Definition of derivative from first principles as $f'(x) = \lim\limits_{h \to 0}\left(\dfrac{f(x+h) - f(x)}{h}\right)$

Derivative interpreted as gradient function and as rate of change; Tangents and normals, and their equations

6.2 Derivative of x^n ($n \in \mathbb{R}$); Differentiation of a sum and a real multiple of these functions; Derivatives of e^x and $\ln x$; The chain rule for composite functions; the product and quotient rules; The second derivative; use of both forms of notation, $\dfrac{d^2 y}{dx^2}$ and $f''(x)$

6.3 Local maximum and minimum points; Points of inflexion with zero and non-zero gradients; Graphical behavior of functions, including the relationship between the graphs of f, f' and f''; Optimization and applications

6.6 Kinematic problems involving displacement s, velocity v and acceleration a

Before you start

You should know how to:

1 Factorize an expression.
  e.g. $2x^3 + 4x^2 + 2x = 2x(x^2 + 2x + 1)$

2 Expand binomials.
  e.g. Expand $(2x - 1)^4$.

  $(2x - 1)^4$
  $= 1(2x)^4(-1)^0 + 4(2x)^3(-1)^1$
  $+ 6(2x)^2(-1)^2 + 4(2x)^1(-1)^3$
  $+ 1(2x)^0(-1)^4$
  $= 16x^4 - 32x^3 + 24x^2 - 8x + 1$

  $\begin{array}{ccccccc} & & & 1 & & & \\ & & 1 & & 1 & & \\ & 1 & & 2 & & 1 & \\ 1 & & 3 & & 3 & & 1 \\ 1 & 4 & & 6 & & 4 & 1 \end{array}$

3 Use rational exponents to rewrite expressions in the form cx^n.
  e.g. $\dfrac{2}{x^5} = 2x^{-5}$; $\sqrt{x} = x^{\frac{1}{2}}$

Skills check

1 Factorize:
  a $9x^4 - 15x^3 + 3x$ **b** $4x^2 - 9$
  c $x^2 - 5x + 6$ **d** $2x^2 - 9x - 5$

2 Expand each binomial.
  a $(x + 2)^3$ **b** $(3x - 1)^4$ **c** $(2x + 3y)^3$

3 Use rational exponents to rewrite each expression in the form cx^n.
  a $\dfrac{1}{x^6}$ **b** $\dfrac{4}{x^3}$ **c** $5\sqrt{x}$
  d $\sqrt[7]{x^5}$ **e** $\dfrac{7}{\sqrt{x^3}}$

If you pluck a string on a guitar and let it vibrate, the sound gets quieter as time passes. This can be modeled by the function $f(t) = \frac{\sin t}{t}$, where t represents time. As t becomes larger and larger, $\frac{\sin t}{t}$ becomes closer to zero – this is the limit value of the function. We write this as $\lim_{t \to \infty} \frac{\sin t}{t} = 0$. Limits are a fundamental concept of calculus. You will learn more about the sine function, whose graph is a sine wave, in a later chapter.

Calculus is the branch of mathematics that takes algebra and geometry together with the limit process and looks at two types of problems. Differential calculus uses limits to find the rate at which a variable quantity is changing. Integral calculus uses limits to solve problems that involve repeated change. In this chapter we will learn to evaluate basic limits and then focus on differential calculus.

7.1 Limits and convergence

In this section you will investigate the concepts of limits and convergence and use limit notation. The concept of limits is the basis of calculus.

Investigation – creating a sequence

Work with a partner. You will need one rectangular piece of paper, a pair of scissors and a copy of this table.

Round number	Portion of the paper you have at the end of the round	
	Fraction	Decimal (3 sf)
1		
2		
3		
4		
5		
6		

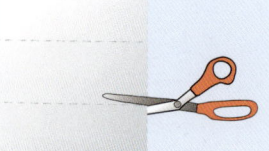

Round 1: Cut the rectangular piece of paper into three pieces of roughly equal size. Take one piece each, leaving one piece on the table. Record the portion of the original rectangle you now have as both a fraction and a decimal (to three significant figures).

Round 2: Cut the 'spare' piece of paper into three pieces of equal size. Each of you add one piece of this to your portion of the original rectangle. Record the total portion of the original rectangle you now have, in the same way as before.

Repeat the same process for four more rounds.

1 As you complete more and more rounds of this activity, what can you say about the portion of the original rectangle you have?
2 If you repeat this process forever, what can you say about the portion of the original rectangle you have?

> When cutting the paper into three equal size pieces you do not have to be exact, just approximate.

> As you complete more and more rounds of the activity you could say the number of rounds approaches infinity. Can you give an example in real life that grows or develops like this?

Limits of sequences

The data you collected in the investigation forms a **sequence**, where u_1 is the portion of the rectangle you have after round 1, u_2 the portion you have after round 2, and so on.

Sequences like this are called **convergent** because as the term number in the sequence increases, the terms in the sequence approach a fixed value known as the **limit**, L, of the sequence. We can write this as: $\lim_{n \to \infty} u_n = L$
Sequences that are not convergent are **divergent**.

What is the limit of the sequence formed in the investigation?

> The notation $\lim_{n \to \infty} u_n = L$ is read as 'the limit as n approaches infinity of u_n equals L'.

> The ancient Greeks used the idea of limits to find areas, using the 'method of exhaustion'. You may wish to research this.

Example 1

Determine whether each sequence is convergent or divergent.
If a sequence is convergent, give the limit of the sequence.
a 0.3, 0.33, 0.333, 0.3333, …
b 2, 4, 8, 16, …
c $\dfrac{1}{5}, \dfrac{6}{25}, \dfrac{31}{125}, \dfrac{156}{625}, \dfrac{781}{3125}, \ldots$
d 1, −1, 1, −1, …

Answers

a Convergent; $\lim\limits_{n \to \infty} u_n = \dfrac{1}{3}$

The pattern indicates that the sequence is approaching 0.3333…, or $0.\dot{3}$, which is the decimal form of $\dfrac{1}{3}$.

> Other notations for recurring decimals include $0.\overline{3}$

b Divergent

Each term in the sequence is larger than the previous term, so they are not approaching a limit.

c Convergent; $\lim\limits_{n \to \infty} u_n = \dfrac{1}{4}$

*To compare fractions with different denominators, use a GDC to convert them to decimals: 0.2, 0.24, 0.248, 0.2496, 0.24992, …
The values are approaching 0.25 or $\dfrac{1}{4}$.*

d Divergent

The terms in the sequence are oscillating between two values and are not approaching a fixed value.

Exercise 7A

Determine whether each sequence is convergent or divergent.
If a sequence is convergent, give the limit of the sequence.

1 1, 3, 5, 7, …

2 3.49, 3.499, 3.499, 3.4999, …

3 $\dfrac{1}{10}, -\dfrac{1}{100}, \dfrac{1}{1000}, -\dfrac{1}{10\,000}, \ldots$

4 $\dfrac{20}{27}, \dfrac{121}{162}, \dfrac{182}{243}, \dfrac{1093}{1458}, \dfrac{1640}{2187}, \ldots$

5 3, 4, 3, 4, 3, 4, …

Limits of functions

$\lim\limits_{x \to c} f(x) = L$ means that as the value of x becomes sufficiently close to c (from either side), the function, $f(x)$, becomes close to a fixed value L. If $f(x)$ does not become close to a fixed value L, we say that the limit does not exist.

> You can use a GDC to help find the limit of a function.
> **Graphically**: You can graph the function and examine the values of $f(x)$ when x is near c.
> **Numerically**: You can make a table of values and examine the values of $f(x)$ when x is near c.

Example 2

Use a GDC to examine each function graphically and numerically.
Find the limit or state that it does not exist.

a $\lim_{x \to 2} x^2$ **b** $\lim_{x \to 1} \dfrac{x^2-1}{x-1}$ **c** $\lim_{x \to 0} f(x)$; where $f(x) = \begin{cases} 1 & \text{for } x \geq 0 \\ -1 & \text{for } x < 0 \end{cases}$

Answers

a $\lim_{x \to 2} x^2$

Plot the graph of f(x) = x² using a GDC, and look at the values of f(x) as x approaches 2 from the right and from the left.

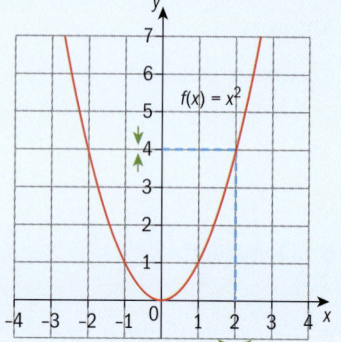

Graphically, f(x) approaches 4 as x approaches 2:
Numerically, as x becomes close to 2 from either side, f(x) becomes close to 4.

$\to 2 \leftarrow$

x	1.8	1.9	1.99	1.999	2.001	2.01	2.1	2.2
f1(x)	3.24	3.61	3.960	3.996	4.004	4.040	4.41	4.84

$\to 4 \leftarrow$

To build the table above using a GDC, enter f1(x) = x². Then set the independent variables to 'Ask'. Enter the values for x.

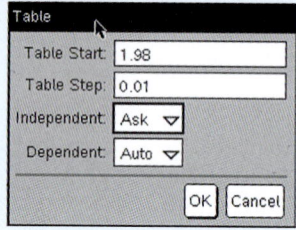

GDC help on CD: *Alternative demonstrations for the TI-84 Plus and Casio FX-9860GII GDCs are on the CD.*

So, $\lim_{x \to 2} x^2 = 4$

The graph and table are shown on the same screen.

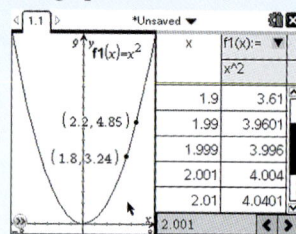

For f(x) = x² we can substitute and find that $\lim_{x \to 2} x^2 = 2^2 = 4$

▶ Continued on next page

b $\lim_{x \to 1} \dfrac{x^2-1}{x-1}$

$f(x)$ approaches 2 as x approaches 1:

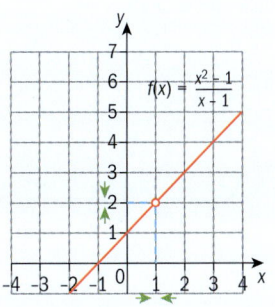

Since division by zero is not defined, $f(x) = \dfrac{x^2-1}{x-1}$ is undefined when $x - 1 = 0$ or $x = 1$. Therefore there is a **discontinuity** in the graph when $x = 1$. Notice that $f(x) = \dfrac{x^2-1}{x-1} = \dfrac{(x+1)(x-1)}{x-1} = x+1$, when $x \neq 1$

Even though $f(x) = \dfrac{x^2-1}{x-1}$ is undefined when $x = 1$, the limit exists since as x becomes close to 1 from either side, $f(x)$ becomes close to 2.

$\to 1 \leftarrow$

x	0.8	0.9	0.99	0.999	1.001	1.01	1.1	1.2
f(x)	1.8	1.9	1.99	1.999	2.001	2.01	2.1	2.2

$\to 2 \leftarrow$

So, $\lim_{x \to 1} \dfrac{x^2-1}{x-1} = 2$

Note that $\lim_{x \to 1} \dfrac{x^2-1}{x-1} = \lim_{x \to 1} \dfrac{(x+1)(x-1)}{x-1}$
$= \lim_{x \to 1}(x+1) = 1 + 1 = 2$

c $\lim_{x \to 0} f(x)$ where

$f(x) = \begin{cases} 1 & \text{for } x \geq 0 \\ -1 & \text{for } x < 0 \end{cases}$

$f(x)$ does not approach the same value as x approaches 0 from the left and right:

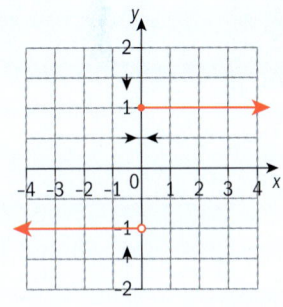

$\to 0 \leftarrow$

x	-0.2	-0.1	-0.01	-0.001	0.001	0.01	0.1	0.2
f(x)	-1	-1	-1	-1	1	1	1	1

So, $\lim_{x \to 0} f(x)$ does not exist.

Note that $f(0) = 1$, but $\lim_{x \to 0} f(x)$ does not exist.

This is because $f(x)$ is close to 1 for values of x to the right of $x = 0$ and $f(x)$ is close to -1 for values of x to the left of $x = 0$.

Exercise 7B

Use a GDC to examine each function graphically and numerically. Find the limit or state that it does not exist.

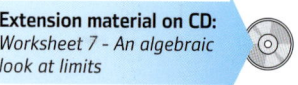

Extension material on CD:
Worksheet 7 - An algebraic look at limits

1 $\lim\limits_{x \to 3}(x^2 + 1)$

2 $\lim\limits_{x \to 0} \dfrac{x^3 - 4x^2 + x}{x}$

3 $\lim\limits_{x \to 2} \dfrac{x^2 - 3x + 2}{x - 2}$

4 $\lim\limits_{x \to 4} \dfrac{1}{x - 4}$

5 $\lim\limits_{x \to 1} f(x)$; where $f(x) = \begin{cases} x + 3 & \text{for } x \geq 1 \\ -x + 5 & \text{for } x < 1 \end{cases}$

6 $\lim\limits_{x \to 2} f(x)$; where $f(x) = \begin{cases} x^2 + 3 & \text{for } x \geq 2 \\ x & \text{for } x < 2 \end{cases}$

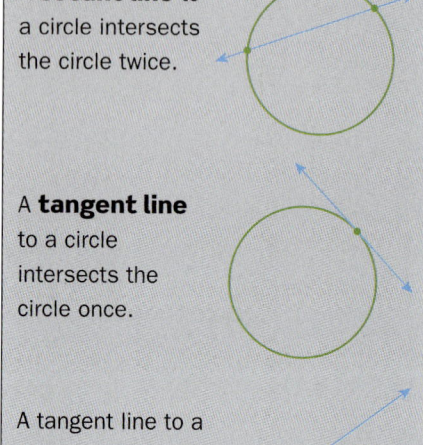

A **secant line** to a circle intersects the circle twice.

A **tangent line** to a circle intersects the circle once.

A **tangent line** to a curve may intersect the curve more than once.

7.2 The tangent line and derivative of x^n

In this section we will work with secant, tangent and normal lines. We will define the derivative of a function and learn some rules for finding derivatives of certain functions.

Investigation – secant and tangent lines

Here is the graph of $f(x) = x^2 + 1$

1 Copy the graph to paper and draw lines AP, BP, CP, DP, EP and FP. These lines are called **secant lines** to the graph of $f(x) = x^2 + 1$.

2 Copy and complete the table.

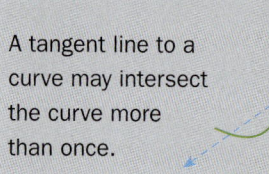

Point	Coordinates	Line	Gradient or slope
P		—	—
A		AP	
B		BP	
C		CP	
D		DP	
E		EP	
F		FP	

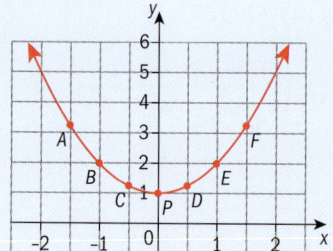

Recall that the gradient of a line through the points (x_1, y_1) and (x_2, y_2) is $\dfrac{y_2 - y_1}{x_2 - x_1}$

3 As points on the curve get closer and closer to point P, what value does the gradient of the secant lines seem to be approaching?

4 Draw the line at point P that has the gradient you found in question 3. This line is called the **tangent line** to the graph of $f(x) = x^2 + 1$ at P.

Lines have a constant gradient, but other curves do not. The gradient of a curve at a given point is the gradient of the tangent line to the curve at that point. This is the concept that Sir Isaac Newton worked with when he wanted to find the instantaneous velocity of a moving object whose velocity was always changing.

Gradient of a secant line

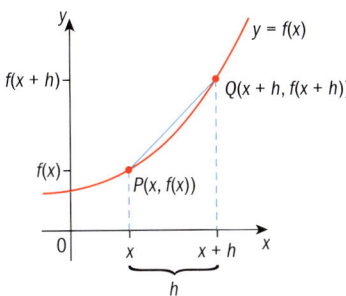

▲ Sir Isaac Newton 1642–1727, English mathematician, is one of the mathematicians credited with developing calculus.

The gradient of the secant line PQ is written as

$$\frac{f(x+h)-f(x)}{(x+h)-x} = \frac{f(x+h)-f(x)}{h}$$

The expression $\frac{f(x+h)-f(x)}{h}$ is known as the **difference quotient**.

Example 3

Write an expression for the gradient of a secant line for $f(x) = x^2 + 1$. Simplify your expression.

Answer

$$\begin{aligned}\frac{f(x+h)-f(x)}{h} &= \frac{\left[(x+h)^2+1\right]-\left(x^2+1\right)}{h} \\ &= \frac{\left(x^2+2xh+h^2+1\right)-\left(x^2+1\right)}{h} \\ &= \frac{2xh+h^2}{h} \\ &= \frac{h(2x+h)}{h} \\ &= 2x+h\end{aligned}$$

Replace the x in $x^2 + 1$ with $x + h$ to write an expression for $f(x + h)$

Expand $(x + h)^2$

Combine like terms.

Factorize.

Simplify.

Exercise 7C

Write an expression for the gradient of a secant line for each function. Simplify your expression.

1 $f(x) = 3x + 4$

2 $f(x) = 2x^2 - 1$

3 $f(x) = x^2 + 2x + 3$

Gradient of a tangent line and the derivative

Suppose that point Q slides down the curve and approaches point P. The secant lines PQ will get closer to the tangent line at point P. As Q gets closer to P, h gets closer to 0. We can take the limit as h approaches 0 of the gradient of the secant line to get the gradient of the tangent line:

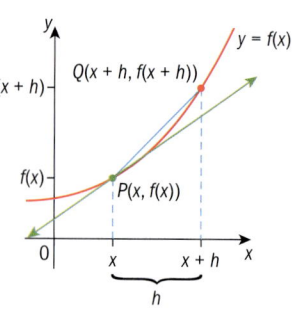

$$\lim_{h \to 0} \frac{f(x+h) - f(x)}{h}$$

> $\lim_{h \to 0} \frac{f(x+h) - f(x)}{h}$ is not a constant. It is a function that gives the gradient of f at x.

→ The function defined by $\lim_{h \to 0} \frac{f(x+h) - f(x)}{h}$ is known as the **derivative** of f. The derivative is defined by
$$f'(x) = \lim_{h \to 0} \frac{f(x+h) - f(x)}{h} \quad \text{or} \quad \frac{dy}{dx} = \lim_{h \to 0} \frac{f(x+h) - f(x)}{h}$$

> $f'(x)$ is read as: the derivative of f, or: f prime of x. $\frac{dy}{dx}$ is read as the 'derivative of y with respect to x' or 'd y d x'.
> Recall that slope is $\frac{\text{change in } y}{\text{change in } x}$. This is expressed as $\frac{\Delta y}{\Delta x}$.
> $\frac{dy}{dx} = \lim_{\Delta x \to 0} \frac{\Delta y}{\Delta x}$.

Example 4

Find the derivative of $f(x) = x^2 + 1$ and hence find the gradient of the tangent line when $x = 3$

Answer

$f'(x) = \lim_{h \to 0} \frac{\left[(x+h)^2 + 1\right] - \left(x^2 + 1\right)}{h}$ *Simplify the quotient as shown in Example 3.*

$= \lim_{h \to 0} (2x + h) = 2x + 0$ *Evaluate the limit by substituting 0 for h.*

$f'(x) = 2x$

$f'(3) = 2(3) = 6$

So the gradient of the tangent line when $x = 3$ is 6.

> The derivative, $f'(x) = 2x$, is a function that gives the gradient of the curve $f(x) = x^2 + 1$ at any point x.

Exercise 7D

Use the definition of derivative to find the derivative of f and hence find the gradient of the tangent line at the given value of x.

1 $f(x) = 2x - 3$; $x = 2$

2 $f(x) = 3x^2 + 2x$; $x = -3$

3 $f(x) = x^2 - x + 2$; $x = 1$

Some rules for derivatives

Investigation – the derivative of $f(x) = x^n$

1. Use the definition of derivative to find the derivatives of $f(x) = x^2$, $f(x) = x^3$ and $f(x) = x^4$
2. Make a conjecture about the derivative of $f(x) = x^n$
   Express your conjecture in words and as a function.
3. Use your conjecture to predict the derivative of $f(x) = x^5$
   Use the definition of derivative to see if your prediction was correct.

> Recall the definition of derivative is
> $f'(x) = \lim_{h \to 0} \dfrac{f(x+h) - f(x)}{h}$.

We investigated only positive integer values for n, but the following is true for any real number n.

→ **Power rule**
If $f(x) = x^n$, then $f'(x) = nx^{n-1}$, where $n \in \mathbb{R}$.

Example 5

Use the power rule to find the derivative of each function.

a $f(x) = x^{12}$ **b** $f(x) = \dfrac{1}{x^3}$ **c** $f(x) = \sqrt{x}$

Answers

a $f(x) = x^{12}$ *Use the power rule.*
$f'(x) = 12x^{12-1} = 12x^{11}$

b $f(x) = \dfrac{1}{x^3} = x^{-3}$ *Rewrite using rational exponents.*
Use the power rule.
$f'(x) = -3x^{-3-1} = -3x^{-4} = -\dfrac{3}{x^4}$ *Simplify.*

c $f(x) = \sqrt{x} = x^{\frac{1}{2}}$ *Rewrite using rational exponents.*
$f'(x) = \dfrac{1}{2}x^{\frac{1}{2}-1} = \dfrac{1}{2}x^{-\frac{1}{2}}$ *Use the power rule.*
$= \dfrac{1}{2x^{\frac{1}{2}}}$ or $\dfrac{1}{2\sqrt{x}}$ *Simplify.*

Exercise 7E

Find the derivative of each function.

1 $f(x) = x^5$ **2** $f(x) = x^8$ **3** $f(x) = \dfrac{1}{x^4}$

4 $f(x) = \sqrt[3]{x}$ **5** $f(x) = \dfrac{1}{\sqrt{x}}$ **6** $f(x) = \sqrt[5]{x^3}$

Using the power rule and the following two rules we can find the derivative of many functions. The process of finding the derivative of a function is called **differentiation**.

→ **Constant rule**
If $f(x) = c$, where c is any real number, then $f'(x) = 0$

Constant rule
The derivative of any constant is zero. The graph of the constant function $f(x) = c$ is a horizontal line which has a gradient of zero.

→ **Constant multiple rule**
If $y = cf(x)$, where c is any real number, then $y' = cf'(x)$

Constant multiple rule
The derivative of a constant times a function is the constant times the derivative of the function.

→ **Sum or difference rule**
If $f(x) = u(x) \pm v(x)$ then $f'(x) = u'(x) \pm v'(x)$

Sum or difference rule
The derivative of a function that is the sum or difference of two or more terms is the sum or difference of the derivatives of the terms.

Example 6

Differentiate each function.

a $f(x) = 4x^3 + 2x^2 - 3$
b $f(x) = 3\sqrt[5]{x} + 8$
c $f(x) = (x-2)(x+4)$
d $f(x) = \dfrac{4x^3 + 2x^2 - 3}{x}$

Answers

a $f(x) = 4x^3 + 2x^2 - 3$
$f'(x) = 4(3x^{3-1}) + 2(2x^{2-1}) - 0$
$= 12x^2 + 4x$

Find the derivative of each term. Note the derivative of the constant term is 0.

b $f(x) = 3\sqrt[5]{x} + 8 = 3x^{\frac{1}{5}} + 8$

Rewrite using rational exponents.

$f'(x) = 3 \cdot \dfrac{1}{5} x^{\frac{1}{5}-1} + 0 = \dfrac{3}{5} x^{-\frac{4}{5}}$

Find the derivative of each term. Note the derivative of the constant term is 0.

$= \dfrac{3}{5x^{\frac{4}{5}}}$ or $\dfrac{3}{5\sqrt[5]{x^4}}$

Simplify.

c $f(x) = (x-2)(x+4) = x^2 + 2x - 8$
$f'(x) = 2x^{2-1} + 2 \cdot 1x^{1-1} - 0 = 2x + 2$

First expand so that the function is the sum or difference of terms in the form ax^n.

▶ Continued on next page

d $f(x) = \dfrac{4x^3 + 2x^2 - 3}{x} = \dfrac{4x^3}{x} + \dfrac{2x^2}{x} - \dfrac{3}{x}$

$ = 4x^2 + 2x - 3x^{-1}$

$f'(x) = 4 \cdot 2x^{2-1} + 2 \cdot x^{1-1} - 3 \cdot (-1) \cdot x^{-1-1}$

$ = 8x + 2 + 3x^{-2} = 8x + 2 + \dfrac{3}{x^2}$

or $\dfrac{8x^3 + 2x^2 + 3}{x^2}$

Rewrite so that the function is the sum or difference of terms in the form ax^n.

Exercise 7F

Differentiate each function.

1 $f(x) = \dfrac{2}{x^8}$ **2** $f(x) = 5$ **3** $f(x) = x^3 - \dfrac{3}{x^2}$

4 $f(x) = \pi x^5$ **5** $f(x) = (x-4)^2$ **6** $f(x) = \sqrt{x} - 4\sqrt[3]{x}$

7 $f(x) = \dfrac{3}{4x^2}$ **8** $f(x) = \dfrac{3}{(4x)^2}$ **9** $f(x) = 12 - x^4$

10 $f(x) = \sqrt{x}\left(\sqrt[3]{x} + \sqrt[4]{x}\right)$ **11** $f(x) = 3x^4 - 2x^2 + 5$ **12** $f(x) = 2x^2 + 3x + 7$

13 $f(x) = x^{\frac{2}{3}} + 2x^{\frac{1}{3}} + 1$ **14** $f(x) = 2x(x^2 - 3x)$ **15** $f(x) = (x^2 + 3x)(x-1)$

Equations of tangent and normal lines

The **normal line** at a point on a curve is the line perpendicular to the tangent line at that point.

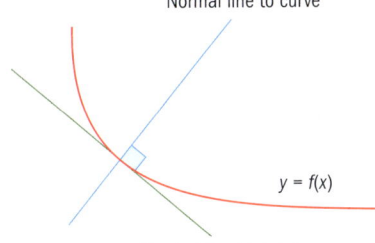

▲ Sparks created by a grinding wheel are **tangent** to the wheel.

Example 7

Write an equation for each line.
a The tangent line to the curve $f(x) = x^2 + 1$ at the point $(1, 2)$
b The normal line to the curve $f(x) = 2\sqrt{x}$ when $x = 9$
c The tangent and normal lines to the curve $f(x) = x + \dfrac{27}{2x^2}$ when $x = 3$
d The tangent to $f(x) = x^3 - 3x^2 - 13x + 15$ that is parallel to the tangent at $(4, -21)$

▶ *Continued on next page*

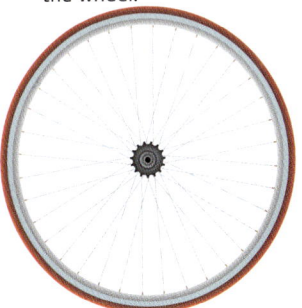

▲ Spokes on a bicycle wheel are **normal** to the rim.

Answers

a $f(x) = x^2 + 1$
   $f'(x) = 2x$
   $m_{tangent} = f'(1)$
   $\phantom{m_{tangent}} = 2(1)$
   $\phantom{m_{tangent}} = 2$
   $\therefore y - 2 = 2(x - 1)$

To find the gradient of the tangent line, find the derivative of f and evaluate when $x = 1$

Use the point $(1, 2)$ and $m = 2$ to write the equation of the tangent line.

> The symbol $\therefore$ is used to mean 'therefore'.

> The equation of a line through the point (x_1, y_1) with gradient m is $y - y_1 = m(x - x_1)$. (See Chapter 18, Section 3.11.)

b $f(x) = 2\sqrt{x}$
   $ = 2x^{\frac{1}{2}}$
   $f'(x) = x^{-\frac{1}{2}}$ or $\dfrac{1}{\sqrt{x}}$
   $m_{tangent} = f'(9)$
   $\phantom{m_{tangent}} = \dfrac{1}{\sqrt{9}}$
   $\phantom{m_{tangent}} = \dfrac{1}{3}$
   $m_{normal} = -3$
   $f(9) = 2\sqrt{9} = 6$

   $\therefore y - 6 = -3(x - 9)$

Rewrite the function using rational exponents.

To find the gradient of the tangent line, find the derivative of f and evaluate when $x = 9$.

Since the normal line is perpendicular to the tangent line, find the gradient by taking the opposite reciprocal of the gradient of the tangent line.
Find a point on the normal line by evaluating f when $x = 9$

Use the point $(9, 6)$ and $m = -3$ to write the equation of the tangent line.

> If a line has gradient m, the gradient of the perpendicular line is $-\dfrac{1}{m}$. (See Chapter 18, Section 3.11.)

c $f(x) = x + \dfrac{27}{2x^2}$
   $ = x + \dfrac{27}{2}x^{-2}$
   $f'(x) = 1 - \dfrac{27}{x^3}$
   $m_{tangent} = f'(3)$
   $\phantom{m_{tangent}} = 1 - \dfrac{27}{3^3}$
   $\phantom{m_{tangent}} = 0$
   $f(3) = 3 + \dfrac{27}{2(3^2)}$
   $ = \dfrac{9}{2}$

   $\therefore$ Normal line is $x = 3$ and tangent line is $y = \dfrac{9}{2}$

Rewrite the function using rational exponents.

To find the gradient of the tangent line, find the derivative of f and evaluate when $x = 3$

Since the gradient is 0, the tangent line is horizontal, so the normal line must be vertical.

Find a point on the lines by evaluating f when $x = 3$

▶ Continued on next page

d $f(x) = x^3 - 3x^2 - 13x + 15$

$f'(x) = 3x^2 - 6x - 13$

$f'(4) = 3(4)^2 - 6(4) - 13$
$ = 11$ 　　*Find the gradient of the tangent line when $x = 4$*

$3x^2 - 6x - 13 = 11$
$3x^2 - 6x - 24 = 0$
$3(x^2 - 2x - 8) = 0$
$3(x-4)(x+2) = 0$
$x = 4, -2$

Set the derivative equal to 11 to find x-coordinates of points with parallel tangent lines.

Notice that one of the values, $x = 4$, is the x-coordinate of the given point of tangency $(4, -21)$.

The x-coordinate of the point of tangency for the parallel line is $x = -2$

$f(-2) = (-2)^3 - 3(-2)^2 - 13(-2) + 15$
$ = 21$

$\therefore y - 21 = 11(x+2)$

Evaluate f at $x = -2$ to find the y-coordinate of the point of tangency.

Use the point $(-2, 21)$ and $m = 11$ to write the equation of the tangent line.

> Recall that parallel lines have the same gradient.

Exercise 7G

1 Find the equations of the tangent and normal lines to the graph of $f(x) = x^2 - 4x$ at the point $(3, -3)$. Graph the function and the lines by hand.

2 Find the equation for the tangent line to the curve at the given point.

　a $f(x) = x^2 + 2x + 1$ at $(-3, 4)$ 　　**b** $f(x) = 2\sqrt{x} + 4$ at $x = 1$

　c $f(x) = \dfrac{x^2 + 6}{x}$ at $(3, 5)$ 　　**d** $f(x) = \sqrt[4]{x} + \dfrac{8}{\sqrt{x}}$ at $x = 1$

3 Find the equation for the normal line to the curve at the given point.

　a $f(x) = 2x^2 - x - 3$ at $(2, 3)$ 　　**b** $f(x) = \dfrac{4}{x} - \dfrac{1}{x^2}$ at $x = -1$

　c $f(x) = (2x+1)^2$ at $(2, 25)$ 　　**d** $f(x) = 2\sqrt[3]{x} - \dfrac{4}{x^2}$ at $x = 1$

EXAM-STYLE QUESTIONS

4 Find the equations for all the vertical normal lines to the graph of $f(x) = x^3 - 3x$

5 The gradient of the tangent line to the graph of $f(x) = 2x^2 + kx - 3$ at $x = -1$ is 1. Find the value of k.

7.3 More rules for derivatives

You can use a GDC to evaluate a derivative of a function at a given value. We know that the derivative of $f(x) = \frac{1}{4}x^3 - 3x$ is $f'(x) = \frac{3}{4}x^2 - 3$ and so $f'(4) = \frac{3}{4}(4)^2 - 3 = 9$.

Click to display the templates.

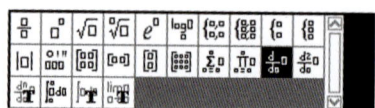

Choose the first-derivative template and enter the function, variable and the value of x.

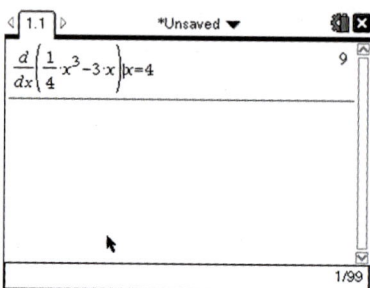

GDC help on CD: *Alternative demonstrations for the TI-84 Plus and Casio FX-9860GII GDCs are on the CD.*

Since the calculator is using a secant line to approximate the value of the derivative, it will not always be exact.

You can graph the function and find its derivative by pressing menu : Analyze Graph | 5: $\frac{dy}{dx}$ and choosing the point on the graph.

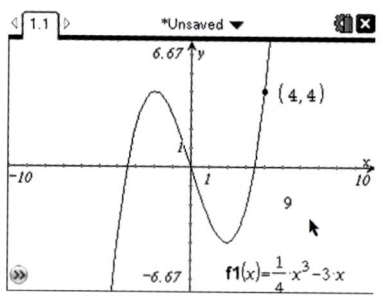

To find the derivative at a specific value of x, use the context menu of the point to show its coordinates, and then edit the x-coordinate.

You can look at the graphs and a table of values for the function and its derivative. To graph f and f', use the first-derivative template to write the function.

This time there will be no space to enter a *value* of x. You can save time by entering $f1(x)$ instead of re-typing the equation.

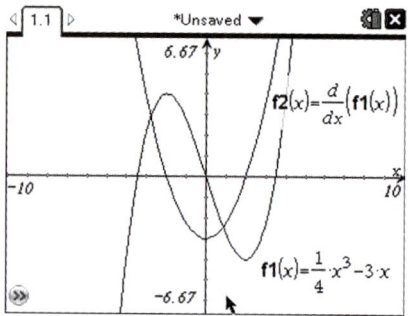

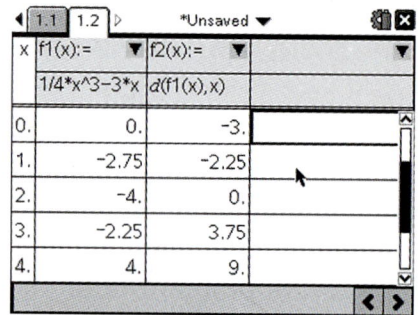

Limits and derivatives

Investigation – the derivatives of e^x and $\ln x$

1. Use a GDC to graph $f(x) = e^x$ and the derivative of $f(x) = e^x$
   Examine the graphs and the table of values for the functions to make a conjecture about the derivative of $f(x) = e^x$
2. Use a GDC to graph $f(x) = \ln x$ and the derivative of $f(x) = \ln x$
   Examine the graphs and the table of values for the functions to make a conjecture about the derivative of $f(x) = \ln x$

→ **Derivative of e^x**
If $f(x) = e^x$, then $f'(x) = e^x$

→ **Derivative of $\ln x$**
If $f(x) = \ln x$, then $f'(x) = \dfrac{1}{x}$

Recall that $y = e^x$ and $y = \ln x$ are inverses.
$e^{\ln x} = x$
$\ln e^x = x$

Example 8

Find the derivative of each function.
a $f(x) = 3e^x$ b $f(x) = x^2 + \ln x$ c $f(x) = \ln e^{3x}$

Answers

a $f(x) = 3e^x$
   $f'(x) = 3 \cdot e^x = 3e^x$ *Use the constant multiple rule and the fact that the derivative of e^x is e^x*

b $f(x) = x^2 + \ln x$
   $f'(x) = 2x + \dfrac{1}{x}$ or $\dfrac{2x^2 + 1}{x}$ *Find the derivative of each term. Use that fact that the derivative of $\ln x$ is $\dfrac{1}{x}$*

c $f(x) = \ln e^{3x} = 3x$
   $f'(x) = 3$ *Use the fact that the functions are inverses to simplify first. Then find the derivative.*

The letter e is used as the base of the exponential function $f(x) = e^x$, in honor of the Swiss mathematician Leonhard Euler (1707–83).

Exercise 7H

Find the derivative of each function.

1 $f(x) = 4\ln x$

2 $f(x) = e^x + \sqrt{x}$

3 $f(x) = \ln e^{3x^4} + \ln x$

4 $f(x) = e^{\ln 4x^2} + 3x + 1$

5 $f(x) = 2e^x + \ln x$

6 $f(x) = 5e^x + 4\ln e^x$

Write an equation for each line in questions 7–10.

7 The line tangent to the curve $f(x) = 4e^x - 7$ at $x = \ln 3$

8 The normal line to the curve $f(x) = \ln\left(e^{x^2}\right)$ at the point $(-3, 9)$

9 The line tangent to the curve $f(x) = \ln x$ at $x = e$

10 The line normal to the curve $f(x) = 2x^2 + e^{\ln x} - 3$ at $x = 2$

How are exponential functions used in determining the concentration of a drug in a patient's body?

Find the exact value of the derivative at the given value of x in questions **11** and **12** and then use a GDC to find an approximate value to check your work.

11 Find $f'(3)$ if $f(x) = 2e^x - 5$

12 Find $f'(8)$ if $f(x) = \sqrt[3]{x} + \ln x$

Investigation – the derivative of the product of two functions

For steps 1–4 let $u(x) = x^4$, $v(x) = x^7$ and $f(x) = u(x) \cdot v(x)$

1 The function f can be written as $f(x) = x^n$. Find n.
2 Find $f'(x)$.
3 Find $u'(x)$ and $v'(x)$.
4 Find $u'(x) \cdot v'(x)$
5 Is $f'(x)$ the same as $u'(x) \cdot v'(x)$?
6 Using the three derivatives found in steps 2 and 3, fill in the blanks below to make a true mathematical statement.
$f'(x) = x^4 \cdot$ _____ $+ x^7 \cdot$ _____ $=$ _____
7 Complete the conjecture.
If $f(x) = u(x) \cdot v(x)$ then $f'(x) =$ ____ $\cdot$ ____ $+$ ____ $\cdot$ ____
8 Use the function $f(x) = (3x + 1)(x^2 - 1)$ to reject or help confirm your conjecture from step 7.

The derivative of the sum of two functions is the sum of the derivatives of the two functions.
If $f(x) = u(x) + v(x)$ then $f'(x) = u'(x) + v'(x)$.
Is a similar rule true for the product of two functions?

The conjecture in the investigation is known as the product rule. Many proofs are straightforward, but the proof of this rule uses a creative step. You can research the proof and find an example of a clever step needed to complete the proof.

For functions like $f(x) = x^4 \cdot x^7$ and $f(x) = (3x + 1)(x^2 - 1)$ you can rewrite the function and use the power rule to take the derivative. But for other functions such as $f(x) = (3x + 1)(\ln x)$ you would need a rule like the one developed in the conjecture to find the derivative. The following rules are used to find the derivative of the product or quotient of two functions.

> **The product rule**
> If $f(x) = u(x) \cdot v(x)$ then $f'(x) = u(x) \cdot v'(x) + v(x) \cdot u'(x)$

> **The quotient rule**
> If $f(x) = \dfrac{u(x)}{v(x)}$ then $f'(x) = \dfrac{v(x) \cdot u'(x) - u(x) \cdot v'(x)}{[v(x)]^2}$

Product rule
The derivative of the product of two factors is the first factor times the derivative of the second factor plus the second factor times the derivative of the first factor.

Quotient rule
The derivative of the quotient of two factors is the denominator times the derivative of the numerator minus the numerator times the derivative of the denominator, all by divided by the denominator squared.

Example 9

Find the derivative of each function.
  a $f(x) = (3x + 1)(\ln x)$
  b $f(x) = (x^4 + 3x^3 + 6)(2x - 1)$
  c $f(x) = \dfrac{5x+3}{x^2+1}$
  d $f(x) = \dfrac{x+2}{2e^x - 3}$

Answers

a $f(x) = \underbrace{(3x+1)}_{\text{First factor}} \underbrace{(\ln x)}_{\text{Second factor}}$

$f'(x) = \underbrace{(3x+1)}_{\text{First factor}} \cdot \underbrace{\left(\dfrac{1}{x}\right)}_{\text{Derivative of second}} + \underbrace{(\ln x)}_{\text{Second factor}} \cdot \underbrace{(3)}_{\text{Derivative of first}}$

$= 3 + \dfrac{1}{x} + 3\ln x$ or $\dfrac{3x + 1 + 3x \ln x}{x}$

$f(x) = u(x) \cdot v(x)$, where $u(x) = 3x + 1$ is the first factor and $v(x) = \ln x$ is the second factor.
Apply the product rule.
$f'(x) = u(x) \cdot v'(x) + v(x) \cdot u'(x)$

b $f(x) = \underbrace{(x^4 + 3x^3 + 6)}_{\text{First factor}} \underbrace{(2x-1)}_{\text{Second factor}}$

$f'(x) = \underbrace{(x^4 + 3x^3 + 6)}_{\text{First factor}} \cdot \underbrace{(2)}_{\text{Derivative of second}}$
$+ \underbrace{(2x-1)}_{\text{Second factor}} \cdot \underbrace{(4x^3 + 9x^2)}_{\text{Derivative of first}}$

$= (2x^4 + 6x^3 + 12) +$
$(8x^4 - 4x^3 + 18x^3 - 9x^2)$

$= 10x^4 + 20x^3 - 9x^2 + 12$

$f(x) = u(x) \cdot v(x)$, where $u(x) = x^4 + 3x^3 + 6$ is the first factor and $v(x) = 2x - 1$ is the second factor.
Apply the product rule.
$f'(x) = u(x) \cdot v'(x) + v(x) \cdot u'(x)$.

Expand the brackets.

Simplify.

▶ Continued on next page

c $f(x) = \dfrac{5x+3}{x^2+1}$

$f'(x) =$

$= \dfrac{\overbrace{(x^2+1)}^{\text{Denominator}} \cdot \overbrace{(5)}^{\substack{\text{Derivative of}\\ \text{numerator}}} - \overbrace{(5x+3)}^{\text{Numerator}} \cdot \overbrace{(2x)}^{\substack{\text{Derivative of}\\ \text{denominator}}}}{\underbrace{(x^2+1)^2}_{\text{Denominator squared}}}$

$= \dfrac{(5x^2+5)-(10x^2+6x)}{(x^2+1)^2}$

$= \dfrac{-5x^2-6x+5}{(x^2+1)^2}$

$f(x) = \dfrac{u(x)}{v(x)}$, where $u(x) = 5x+3$ is the numerator and $v(x) = x^2+1$ is the denominator.
Apply the quotient rule.
$f'(x) = \dfrac{v(x)\cdot u'(x) - u(x)\cdot v'(x)}{[v(x)]^2}$

Expand the numerator so that you can combine like terms. Do not expand the denominator.

Simplify.

d $f(x) = \dfrac{x+2}{2e^x-3}$

$f'(x) =$

$= \dfrac{\overbrace{(2e^x-3)}^{\text{Denominator}} \cdot \overbrace{(1)}^{\substack{\text{Derivative of}\\ \text{numerator}}} - \overbrace{(x+2)}^{\text{Numerator}} \cdot \overbrace{(2e^x)}^{\substack{\text{Derivative of}\\ \text{denominator}}}}{\underbrace{(2e^x-3)^2}_{\text{Denominator squared}}}$

$= \dfrac{(2e^x-3)-(2xe^x+4e^x)}{(2e^x-3)^2}$

$= \dfrac{-2xe^x-2e^x-3}{(2e^x-3)^2}$

$f(x) = \dfrac{u(x)}{v(x)}$, where $u(x) = x+2$ is the numerator and $v(x) = 2e^x - 3$ is the denominator.

Apply the quotient rule.
$f'(x) = \dfrac{v(x)\cdot u'(x) - u(x)\cdot v'(x)}{[v(x)]^2}$

Expand the numerator so that you can combine like terms. Do not expand the denominator.

Simplify.

Exercise 7I

Find the derivative of each function in questions 1 to 8.

1 $f(x) = \dfrac{x^2}{x-4}$

2 $f(x) = (2x^3 + x^2 + x)(x^2 + 1)$

3 $f(x) = \dfrac{\ln x}{x}$

4 $f(x) = e^x \ln x$

5 $f(x) = \dfrac{x-2}{x+4}$

6 $f(x) = \dfrac{e^x}{e^x + 1}$

7 $f(x) = e^x(5x^3 + 4x)$

8 $f(x) = \dfrac{2-x^2}{x^3+1}$

EXAM-STYLE QUESTIONS

9 The function $f(x) = xe^x$ has a horizontal tangent line at $x = k$. Find k.

10 Write the equations for the tangent lines to the graph of $f(x) = \dfrac{x+1}{x-1}$ that are parallel to the line $x + 2y = 10$

The product and quotient rules are not needed for all products and quotients. It is sometimes more convenient to rewrite the function before differentiating.

Example 10

Find the derivative. If it is more convenient to rewrite the function first, do so.

a $f(x) = \sqrt{x}(4x^2 - 2x)$

b $f(x) = \dfrac{3x+4}{x^2-2}$

c $f(x) = \dfrac{9}{\sqrt[3]{x^4}}$

d $f(x) = \dfrac{3x^2 + 2x + 1}{x^2}$

Answers

a $f(x) = \sqrt{x}(4x^2 - 2x)$

$\qquad = x^{\frac{1}{2}}(4x^2 - 2x)$

$\qquad = 4x^{\frac{5}{2}} - 2x^{\frac{3}{2}}$

Rewrite using rational exponents and expand.

$f'(x) = 4 \cdot \dfrac{5}{2} x^{\frac{5}{2}-1} - 2 \cdot \dfrac{3}{2} x^{\frac{3}{2}-1}$

$\qquad = 10x^{\frac{3}{2}} - 3x^{\frac{1}{2}}$

Use the constant multiple and power rules to find the derivative and simplify.

b $f(x) = \dfrac{3x+4}{x^2-2}$

$f'(x) = \dfrac{(x^2-2) \cdot (3) - (3x+4) \cdot (2x)}{(x^2-2)^2}$

Use the quotient rule.

$\qquad = \dfrac{(3x^2 - 6) - (6x^2 + 8x)}{(x^2-2)^2}$

$\qquad = \dfrac{-3x^2 - 8x - 6}{(x^2-2)^2}$

c $f(x) = \dfrac{9}{\sqrt[3]{x^4}} = 9x^{-\frac{4}{3}}$

Rewrite using rational exponents.

$f'(x) = 9 \cdot -\dfrac{4}{3} x^{-\frac{4}{3}-1}$

$\qquad = -12x^{-\frac{7}{3}} = -\dfrac{12}{x^{\frac{7}{3}}}$

d $f(x) = \dfrac{3x^2 + 2x + 1}{x^2}$

$\qquad = \dfrac{3x^2}{x^2} + \dfrac{2x}{x^2} + \dfrac{1}{x^2}$

Rewrite by separating terms and then use rational exponents.

$\qquad = 3 + 2x^{-1} + x^{-2}$

$f'(x) = 0 - 2x^{-2} - 2x^{-3}$

$\qquad = \dfrac{-2}{x^2} - \dfrac{2}{x^3}$ or $\dfrac{-2x - 2}{x^3}$

We have been using the 'prime' notation, $f'(x)$, to denote derivatives. We can use Leibniz notation, $\dfrac{dy}{dx}$ or $\dfrac{d}{dx}[f(x)]$ and we can also use variables other than x and y. The notation $\dfrac{dy}{dx}$ is read as 'the derivative of y with respect to x', or 'd y by d x', or simply 'd y d x'. The notation $\dfrac{d}{dx}[f(x)]$ is read as 'the derivative of f with respect to x'.

Example 11

a Find $\dfrac{d}{dx}[(\ln x)(7x-2)]$

b If $s(t) = (4t^2 - 1)^2$, find $\dfrac{ds}{dt}$

c If $A = \pi r^2$, find $\left.\dfrac{dA}{dr}\right|_{r=3}$

▲ Gottfried Wilhelm Leibniz (1646–1716), a German mathematician, debated with Isaac Newton over who was the first to develop calculus. It is widely believed that Leibniz and Newton independently developed calculus at about the same time.

Answers

a $\dfrac{d}{dx}[(\ln x)(7x-2)]$

$= (\ln x)(7) + (7x-2)\left(\dfrac{1}{x}\right)$

$= \dfrac{7x \ln x + 7x - 2}{x}$

Use the product rule to find the derivative of $(\ln x)(7x - 2)$ with respect to x.

b $s(t) = (4t^2 - 1)^2$

$= 16t^4 - 8t^2 + 1$

$\dfrac{ds}{dt} = 64t^3 - 16t$

Expand and use the power rule to find the derivative of s with respect to t.

c $A = \pi r^2$

$\dfrac{dA}{dr} = 2\pi r$

$\left.\dfrac{dA}{dr}\right|_{r=3} = 2\pi(3)$

$= 6\pi$

Find the derivative of πr^2 with respect to r.

The bar tells you to evaluate the derivative when $r = 3$

Exercise 7J

Differentiate each function in questions 1 to 12. If it is more convenient to rewrite the function first, do so.

1 $f(x) = \dfrac{2x^3 - 5x}{3}$

2 $f(x) = (x^2 - 5)(x^2 + 5)$

3 $f(x) = 2e^x(x^2)$

4 $f(x) = \dfrac{2e^x}{x^2}$

5 $f(x) = e^{\ln x^3} + \dfrac{4}{\sqrt[5]{x^4}}$

6 $f(x) = \dfrac{x^2}{e^x}$

7 $f(x) = \dfrac{x^2}{x^2+1}$

8 $f(x) = 3x \ln x$

9 $f(x) = \dfrac{x^2 - 2x + 1}{x}$

10 $f(x) = \sqrt{x}(x^2 + 1)$

11 $f(x) = \dfrac{x}{x^2 - 2x + 1}$

12 $f(x) = (x^3 - 3x)(2x^2 + 3x + 5)$

EXAM-STYLE QUESTIONS

13 Write the equation of the line normal to the graph of $f(x) = xe^x - e^x$ at $x = 1$

14 Write the equation of the tangent line to the graph of $f(x) = x^3 \ln x$ at $x = 1$

15 If $c(n) = -4.5n^2 + 3.5n - 2$, find $\dfrac{dc}{dn}$

16 If $A = \dfrac{4}{3}\pi r^3$, find $\dfrac{dA}{dr}$

17 If $v(t) = 2t^2 - t + 1$, find $\dfrac{dv}{dt}\bigg|_{t=2}$

EXAM-STYLE QUESTION

18 $\dfrac{d}{dt}\left[(e^t)(t+3)\right]$ can be written as $e^t(t+k)$. Find k.

7.4 The chain rule and higher order derivatives

The power rule alone will not give the correct derivative for $f(x) = (2-x)^3$. This is because the function is not a power of x, but rather a power of another function $v(x) = 2 - x$. The function f is a composite function, $(u \circ v)(x)$ or $u(v(x))$, where $u(x) = x^3$ and $v(x) = 2 - x$

The symbol ∘ is used to show a composite function. If $u(x) = x^3$ and $v(x) = 2 - x$, then
$f(x) = (u \circ v)(x)$
$= u(v(x))$
$= u(2 - x)$
$= (2 - x)^3$

Investigation – finding the derivative of a composite function

1. Let $f(x) = (2 - x)^3$
   a. Expand $f(x) = (2 - x)^3$ Differentiate each term to find the derivative of f.
   b. You can also find the derivative of $f(x) = (2 - x)^3$ by applying the power rule to $(2 - x)^3$ and multiplying by another factor. Compare the following to your answer in step 1 and find the missing factor: $f'(x) = 3(2 - x)^2$. _____

2. Repeat the process for $f(x) = (2x + 1)^2$
   a. Expand f and find the derivative.
   b. Apply the power rule to $(2x + 1)^2$ to find the missing factor: $f'(x) = 2(2x + 1)$. _____

3. Repeat the process for $f(x) = (3x^2 + 1)^2$
   a. Expand f and find the derivative.
   b. Apply the power rule to $(3x^2 + 1)^2$ to find the missing factor: $f'(x) = 2(3x + 1)$. _____

4. Make a conjecture about finding the derivative of a composite function.

5. Verify that your conjecture works for $f(x) = (x^4 + x^2)^3$

> If $u(x) = x^2$ and $v(x) = 2x + 1$, then
> $f(x) = u(v(x))$
> $\quad = u(2x + 1)$
> $\quad = (2x + 1)^2$

> If $u(x) = x^2$ and $v(x) = 3x^2 + 1$, then
> $f(x) = u(v(x))$
> $\quad = u(3x^2 + 1)$
> $\quad = (3x^2 + 1)^2$

To find the derivative of a composite function we use the chain rule.

→ **The chain rule**
If $f(x) = u(v(x))$ then $f'(x) = u'(v(x)) \cdot v'(x)$

→ The chain rule can also be written as:
If $y = f(u)$, $u = g(x)$ and $y = f(g(x))$, then $\dfrac{dy}{dx} = \dfrac{dy}{du} \cdot \dfrac{du}{dx}$

Chain rule
The derivative of a composite function is the derivative of the outside function with respect to the inside function (inside function remains the same), multiplied by the derivative of the inside function with respect to x.

Example 12

Each function is in the form $f(x) = u(v(x))$
Identify $u(x)$ and $v(x)$, then find the derivative of f.

a. $f(x) = 4(5x^3 + 2)^6$ b. $f(x) = \sqrt{4x^2 + 1}$ c. $f(x) = e^{x^2}$

Answers

a. $f(x) = 4(5x^3 + 2)^6$
   $u(x) = 4x^6$ u is the outside function.
   $v(x) = 5x^3 + 2$ v is the inside function.
   $f'(x) = \underbrace{24(5x^3 + 2)^5}_{\text{Derivative of outside function with respect to inside function}} \cdot \underbrace{(15x^2)}_{\text{Derivative of inside function with respect to }x}$ Apply chain rule.

   $= 360x^2(5x^3 + 2)^5$ Simplify.

▶ Continued on next page

b $f(x) = \sqrt{4x^2+1}$ *Rewrite using rational exponents.*

$\quad\quad = (4x^2+1)^{\frac{1}{2}}$

$u(x) = x^{\frac{1}{2}}$ *u is the outside function.*
$v(x) = 4x^2+1$ *v is the inside function.*

$f'(x) = \underbrace{\frac{1}{2}(4x^2+1)^{-\frac{1}{2}}}_{\text{Derivative of outside function with respect to inside function}} \cdot \underbrace{(8x)}_{\text{Derivative of inside function with respect to } x}$ *Apply the chain rule.*

$\quad\quad = \dfrac{4x}{(4x^2+1)^{\frac{1}{2}}}$ or $\dfrac{4x}{\sqrt{4x^2+1}}$ *Simplify.*

c $f(x) = e^{x^2}$

$\quad\quad = e^{(x^2)}$

$u(x) = e^x$ *u is the outside function.*
$v(x) = x^2$ *v is the inside function.*

$f'(x) = \underbrace{e^{(x^2)}}_{\text{Derivative of outside function with respect to inside function}} \cdot \underbrace{(2x)}_{\text{Derivative of inside function with respect to } x}$ *Apply the chain rule.*

$\quad\quad = 2xe^{x^2}$ *Simplify.*

Exercise 7K

Each function is in the form $f(x) = u(v(x))$.
Identify $u(x)$ and $v(x)$, then find the derivative of f.

1 $f(x) = (3x^4 + 2x)^5$ **2** $f(x) = 4(2x^2 + 3x + 1)^3$

3 $f(x) = \ln(3x^5)$ **4** $f(x) = \sqrt[3]{2x+3}$

5 $f(x) = e^{4x}$ **6** $f(x) = (\ln x)^3$

7 $f(x) = (9x+2)^{\frac{2}{3}}$ **8** $f(x) = \sqrt[4]{2x^2+3}$

9 $f(x) = 5(x^3+3x)^4$ **10** $f(x) = e^{4x^3}$

You can find the derivative of some functions more efficiently by rewriting the function into a form where you can apply the chain rule.

> Maria Agnesi (1718–99), an Italian mathematician, published a text on calculus that included the methods of calculus of both Isaac Newton and Gottfried Leibniz. Maria also studied curves of the form $y = \dfrac{a^3}{x^2+a^2}$ whose graphs came to be known as *witches of Agnesi*. The function $f(x) = \dfrac{1}{x^2+1}$ in Example 13 is an example of such a graph.

Example 13

Use the chain rule to find the derivative of $f(x) = \dfrac{1}{x^2+1}$

Answer

$f(x) = \dfrac{1}{x^2+1}$ *Rewrite using rational exponents.*

$= (x^2+1)^{-1}$

$f'(x) = -1(x^2+1)^{-2} \cdot 2x$ *Apply the chain rule.*

$= -\dfrac{2x}{(x^2+1)^2}$ *Simplify.*

For some functions the chain rule must be combined with the product or quotient rule, or the chain rule may need to be repeated.

Example 14

a $f(x) = x\sqrt{1-x^2}$ **b** $f(x) = e^{2(3x-1)^4}$ **c** $f(x) = \ln\left(\dfrac{x}{x^2+1}\right)$

Answers

a $f(x) = x\sqrt{1-x^2} = x(1-x^2)^{\frac{1}{2}}$ *Rewrite using rational exponents.*

$f'(x) = \underbrace{x}_{\text{First factor}} \cdot \underbrace{\dfrac{1}{2}(1-x^2)^{-\frac{1}{2}}(-2x)}_{\substack{\text{Derivative of second factor}\\\text{using chain rule}}}$ *Apply the product rule, using the chain rule to find the derivative of the second factor.*

$+ \underbrace{(1-x^2)^{\frac{1}{2}}}_{\text{Second factor}} \cdot \underbrace{1}_{\substack{\text{Derivative of}\\\text{first factor}}}$

$= \dfrac{-x^2}{(1-x^2)^{\frac{1}{2}}} + (1-x^2)^{\frac{1}{2}}$ *Simplify.*

$= \dfrac{-x^2}{(1-x^2)^{\frac{1}{2}}} + (1-x^2)^{\frac{1}{2}} \cdot \dfrac{(1-x^2)^{\frac{1}{2}}}{(1-x^2)^{\frac{1}{2}}}$ *Find a common denominator.*

$= \dfrac{-x^2 + (1-x^2)}{(1-x^2)^{\frac{1}{2}}}$

$= \dfrac{1-2x^2}{(1-x^2)^{\frac{1}{2}}}$ or $\dfrac{1-2x^2}{\sqrt{1-x^2}}$ *Simplify.*

b $f(x) = e^{2(3x-1)^4}$

$u(x) = e^x$

$v(x) = 2(3x-1)^4$

$f'(x) = \underbrace{e^{2(3x-1)^4}}_{\substack{\text{Derivative of the}\\\text{outside function}\\\text{with respect to the}\\\text{inside function}}} \cdot \underbrace{8(3x-1)^3(3)}_{\substack{\text{Derivative of the}\\\text{inside function with}\\\text{respect to }x}}$

The outside and inside functions are shown. Note the inside function $v(x) = 2(3x-1)^4$ is the composition of $2x^4$ and $3x - 1$.
Apply the chain rule to f and apply it again when finding the derivative of the inside function.

$= 24(3x-1)^3 e^{2(3x-1)^4}$

▶ Continued on next page

c $f(x) = \ln\left(\dfrac{x}{x^2+1}\right)$

$f'(x) = \underbrace{\dfrac{1}{\frac{x}{x^2+1}}}_{\substack{\text{Derivative of the} \\ \text{outside function} \\ \text{with respect to the} \\ \text{inside function}}} \cdot \underbrace{\dfrac{(x^2+1)\cdot 1 - x\cdot(2x)}{(x^2+1)^2}}_{\substack{\text{Derivative of the} \\ \text{inside function} \\ \text{with respect } x}}$

Apply the chain rule and use the quotient rule to find the derivative of the inside function.

$= \dfrac{x^2+1}{x} \cdot \dfrac{x^2+1-2x^2}{(x^2+1)^2}$

Simplify.

$= \dfrac{1-x^2}{x(x^2+1)}$

Exercise 7L

Find the derivative of each function in questions 1 to 10.

1 $f(x) = x^2(2x-3)^4$

2 $f(x) = x^2 e^{-x}$

3 $f(x) = \dfrac{4}{x^2+3}$

4 $f(x) = \dfrac{x}{\sqrt{2x+1}}$

5 $f(x) = \sqrt{e^{2x} + e^{-2x}}$

6 $f(x) = \ln(1-2x^3)$

7 $f(x) = \ln(\ln x^2)$

8 $f(x) = \dfrac{2}{e^x + e^{-x}}$

9 $f(x) = \dfrac{1}{x^2 - 3x - 2}$

10 $f(x) = x^4 \sqrt{x^2+3}$

EXAM-STYLE QUESTIONS

11 For the the curve $f(x) = e^{x^2 - 2x}$

    a Find $f'(x)$.   **b** Find $f'(2)$.
    c Hence find the equation of the tangent line to f when $x = 2$

12 Find the x-coordinate of the point(s) on the graph of $f(x) = x^3 \ln x$ where the tangent line is horizontal.

13 Let $f(x) = \dfrac{1}{x^3}$, $g(x) = 1 - 2x$ and $h(x) = (f \circ g)(x)$

Find $h(x)$ and show that the gradient of $h(x)$ is always positive.

14

x	f(x)	g(x)	f'(x)	g'(x)
3	1	4	-3	2
4	2	-1	3	4

In the table above, the values of f and g and their derivatives at $x = 3$ and $x = 4$ are given.

    a Find the gradient of $(f \circ g)(x)$ when $x = 3$

    b Find the gradient of $\dfrac{1}{[g(x)]^2}$ when $x = 4$

Higher order derivatives

The derivative $f'(x)$ or $\dfrac{dy}{dx}$ is called the **first derivative** of y with respect to x. We are sometimes interested in the gradient of the first derivative. This is known as the **second derivative** of y with respect to x and can be written as $f''(x)$ or $\dfrac{d^2y}{dx^2}$. The third derivative of y with respect to x is written as $f'''(x)$ or $\dfrac{d^3y}{dx^3}$. The second and third derivatives are examples of **higher order derivatives**.

> The second derivative is the derivative of the first derivative. Writing this as $\dfrac{d}{dx}\left[\dfrac{dy}{dx}\right]$ helps you see where the notation $\dfrac{d^2y}{dx^2}$ comes from.

> The 'prime' notation is not very useful for derivatives of order higher than three. For those derivatives we write $f^{(n)}(x)$. For example, instead of writing $f''''(x)$ we write $f^{(4)}(x)$.

Example 15

a Find the first three derivatives of $f(x) = x^4 + 3x^2 + x$

b If $f'(x) = \sqrt{x^2+4}$, find $f''(x)$.

c If $y = 4e^{2x}$, find $\left.\dfrac{d^3x}{dx^3}\right|_{x=1}$

d If $s(t) = -16t^2 + 16t + 32$, find $\dfrac{d^2s}{dt^2}$.

Answers

a $f(x) = x^4 + 3x^2 + x$

$f'(x) = 4x^3 + 6x + 1$

$f''(x) = 12x^2 + 6$

$f'''(x) = 24x$

The first three derivatives are $f'(x)$, $f''(x)$ and $f'''(x)$.

b $f'(x) = \sqrt{x^2+4}$

$= (x^2+4)^{\frac{1}{2}}$

$f''(x) = \dfrac{1}{2}(x^2+4)^{-\frac{1}{2}}(2x)$

$= \dfrac{x}{\sqrt{x^2+4}}$

Note that the first derivative was given, so you only differentiate once to get the second derivative.

c $y = 4e^{2x}$

$\dfrac{dy}{dx} = 4e^{2x} \cdot 2 = 8e^{2x}$

$\dfrac{d^2y}{dx^2} = 8e^{2x} \cdot 2 = 16e^{2x}$

$\dfrac{d^3y}{dx^3} = 16e^{2x} \cdot 2 = 32e^{2x}$

$\left.\dfrac{d^3y}{dx^3}\right|_{x=1} = 32e^{2(1)} = 32e^2$

Find the first three derivatives using the chain rule.

Then evaluate the third derivative when $x = 1$

d $s(t) = -16t^2 + 16t + 32$

$\dfrac{ds}{dt} = -32t + 16$

$\dfrac{d^2s}{dt^2} = -32$

Find the first and then the second derivative of s with respect to t.

Exercise 7M

1. Find the second derivative of $f(x) = 4x^{\frac{3}{2}}$

2. If $f(x) = 3x^5 + x^4 + 2x + 1$, find $f'''(x)$.

3. If $C(n) = (3+2n)e^{-3n}$, find $\dfrac{d^2C}{dn^2}$

4. If $\dfrac{dy}{dx} = \dfrac{4}{x}$, find $\dfrac{d^3y}{dx^3}$

5. If $\dfrac{d^4y}{dx^4} = \ln(4x^3)$, find $\dfrac{d^6y}{dx^6}$

6. If $R(t) = \dfrac{1}{2}t\ln(t^2)$, find $\dfrac{dR}{dt}\bigg|_{t=-1}$

EXAM-STYLE QUESTIONS

7. What is true about the nth derivative of $y = x^3 + 3x^2 + 2x + 4$, for $n \geq 4$?

8. Find the first four derivatives of $y = e^x + e^{-x}$ then write a generalization for finding $\dfrac{d^n y}{dx^n}$ of this function.

9. Find the first four derivatives of $y = \dfrac{1}{x}$ then write a generalization for finding $\dfrac{d^n y}{dx^n}$ of this function.

10. Find the gradient of the slope of the function $f(x) = 3\sqrt[5]{x^2}$

7.5 Rates of change and motion in a line

The derivative gives the slope of a function. It also gives the rate of change of one variable with respect to another variable. In this section we will study **average and instantaneous rates of change** and **motion in a line**.

Example 16

A diver jumps from a platform at time $t = 0$ seconds. The distance of the diver above water level at time t is given by $s(t) = -4.9t^2 + 4.9t + 10$, where s is in metres.

a Find the **average velocity** of the diver over the given time intervals.
   i [1, 2]   ii [1.5, 1]   iii [1.1, 1]   iv [1.01, 1]

b Find the **instantaneous velocity** of the diver at $t = 1$ second.

▶ Continued on next page

Answers

a Average velocity is the $\dfrac{\text{change in distance}}{\text{change in time}} \dfrac{\text{(metres)}}{\text{(seconds)}}$ ← *The units for velocity are $m\,s^{-1}$.*

   i $\dfrac{s(2)-s(1)}{2-1} = -9.8\,\text{ms}^{-1}$

   ii $\dfrac{s(1.5)-s(1)}{1.5-1} = -7.35\,\text{ms}^{-1}$

   iii $\dfrac{s(1.1)-s(1)}{1.1-1} = -5.39\,\text{ms}^{-1}$

   iv $\dfrac{s(1.01)-s(1)}{1.01-1} = -4.949\,\text{ms}^{-1}$

Find the slopes of the secant lines $\dfrac{s(t_2)-s(t_1)}{t_2-t_1}$ on each interval. Use a GDC to evaluate the slopes.

b Instantaneous velocity
$v(t) = s'(t)$
$s'(t) = -9.8t + 4.9$
$s'(1) = -9.8 + 4.9 = -4.9\,\text{ms}^{-1}$

Find the slope of the tangent line to s at $t = 1$
*Note that the slopes of the secant lines in part **a** approach the slope of the tangent line in part **b**.*

> The average rate of change of s, or average velocity, is the slope of a secant line:
> $\dfrac{s(t+h)-s(t)}{(t+h)-t} = \dfrac{s(t+h)-s(t)}{h}$
> The instantaneous rate of change of s, or the velocity, is the slope of a tangent line:
> $v(t) = \lim\limits_{h\to 0} \dfrac{s(t+h)-s(t)}{h} = s'(t)$

Example 17

During one month, the temperature of the water in a pond is modeled by the function $C(t) = 20 + 9te^{-\frac{t}{3}}$, where t is measured in days and C is measured in degrees Celsius.
a Find the average rate of change in temperature in the first 15 days of the month.
b Find the rate of change in temperature on day 15.

Answers

a Average rate of change
$= \dfrac{C(15)-C(0)}{15-0} \approx 0.0606\,°C/\text{day}$

Find the slopes of the secant line on the interval [0, 15]. The units for $\dfrac{\text{change in temperature}}{\text{change in time}}$ are $°C/\text{day}$.

b Instantaneous rate of change:
$C'(t) = 9t\left(e^{-\frac{t}{3}} \cdot -\dfrac{1}{3}\right) + e^{-\frac{t}{3}} \cdot 9$

$= -3te^{-\frac{t}{3}} + 9e^{-\frac{t}{3}}$

$C'(15) = -3 \cdot 15e^{-5} + 9e^{-5}$

$= -36e^{-5}$

$\approx -0.243\,°C/\text{day}$

On day 15 the temperature is dropping at a rate of 0.243 degrees Celsius per day.

Find the slope of the tangent line to C at $t = 15$.

Exercise 7N

Use a GDC to help evaluate function values.

EXAM-STYLE QUESTION

1. A ball is thrown vertically upwards. Its height in metres above the ground t seconds after it is thrown is modeled by the function $h(t) = -4.9t^2 + 19.6t + 1.4$
   a Find the height of the ball when $t = 0$ seconds and when $t = 2$ seconds.
   b Find the average rate of change of the height of the ball from $t = 0$ seconds to $t = 2$ seconds.
   c Find the instantaneous rate of change of the height of the ball when $t = 1$ second, $t = 2$ seconds and $t = 3$ seconds. Explain what these values tell you about the motion of the ball.

2. The amount of water in a tank after t minutes is modeled by the function $V(t) = 4000\left(1 - \dfrac{t}{60}\right)^2$, where V is measured in litres.
   Answer the following to the nearest whole number.
   a Find the amount of water in the tank when $t = 0$ minutes and when $t = 20$ minutes.
   b Find the average rate of change of the amount of water in the tank from when $t = 0$ minutes to $t = 20$ minutes. Explain the meaning of your answer.
   c Find the instantaneous rate of change of the amount of water in the tank when $t = 20$ minutes. Explain the meaning of your answer.
   d Show that the amount of water in the tank is never increasing from $t = 0$ minutes to $t = 40$ minutes.

3. The number of bacteria in a science experiment on day t is modeled by $P(t) = 100e^{0.25t}$
   a Find the average rate of change of the number of bacteria over the interval 0 to 10 days of the experiment.
   b Find the instantaneous rate of change of the number of bacteria at any time t.
   c Find the instantaneous rate of change of the number of bacteria on day 10. Explain the meaning of your answer.

4. The cost (in dollars) of producing n units of a product is modeled by the function $C(n) = 0.05n^2 + 10n + 5000$
   a Find the average rate of change of C with respect to n when the production level changes from $n = 100$ units to $n = 105$ units and when the production level changes from $n = 100$ units to $n = 101$ units.
   b Find the instantaneous rate of change of C with respect to n for any number of units n.
   c Find the instantaneous rate of change of C with respect to n when $n = 100$ units. Explain the meaning of your answer.

Motion in a line

If an object moves along a straight line, its position from an origin at any time t can be modeled by a **displacement function**, $s(t)$. The function $s(t) = -4.9t^2 + 4.9t + 10$ from Example 16 is an example of a displacement function. The **initial position** of the diver is the position when $t = 0$, or $s(0) = 10$ metres. The origin is at water level, so the diver is initially on a diving platform 10 metres above water level.

> You can use a horizontal or vertical line to model motion in a line.
> For $s(t) > 0$, the object is to the right of the origin or above the origin. For $s(t) < 0$, the object is to the left of the origin or below the origin. The initial position is $s(0)$.

→ The instantaneous rate of change of displacement is the **velocity function**,
$$v(t) = \lim_{h \to 0} \frac{s(t+h) - s(t)}{h} = s'(t)$$

> For $v(t) > 0$, the object is moving to the right or up.
> For $v(t) < 0$, the object is moving down or to the left.
> For $v(t) = 0$, the object is at rest.
> The **initial velocity** is $v(0)$.

Example 18

A particle moves in a straight line with a displacement of s metres t seconds after leaving a fixed point. The displacement function is given by $s(t) = 2t^3 - 21t^2 + 60t + 3$, for $t \geq 0$.
- **a** Find the velocity of the particle at any time t.
- **b** Find the initial position and initial velocity of the particle.
- **c** Find when the particle is at rest.
- **d** Find when the particle is moving left and when the particle is moving right.
- **e** Draw a motion diagram for the particle.

> This is an area of mathematics known as **kinematics**, which is about the motion of objects.

Answers

a $v(t) = s'(t)$
$v(t) = 6t^2 - 42t + 60, \; t \geq 0$

Velocity is the derivative of displacement.

b $s(0) = 2(0)^3 - 21(0)^2 + 60(0) + 3 = 3 \, m$

The initial position is the displacement when $t = 0$.

$v(0) = 6(0)^2 - 42(0) + 60 = 60 \, ms^{-1}$

The initial velocity is the velocity when $t = 0$.

c $6t^2 - 42t + 60 = 0$
$6(t^2 - 7t + 10) = 0$
$6(t-2)(t-5) = 0$
$t = 2, 5$

The particle is at rest when velocity is 0.
Set the velocity function equal to 0 and solve for t.

The particle is at rest at 2 seconds and 5 seconds.

▶ Continued on next page

d signs of v

```
    +        −        +
┌───────┬────────┬──────→
0       2        5       t
```

The particle is moving right for $(0, 2)$ and $(5, \infty)$ seconds because $v(t) > 0$. The particle is moving left for $(2, 5)$ seconds because $v(t) < 0$.

e $s(2) = 2(2)^3 - 21(2)^2 + 60(2) + 3 = 55$ m

$s(5) = 2(5)^3 - 21(5)^2 + 60(5) + 3 = 28$ m

```
   t = 0      t = 5           t = 2
  ●─────→   ←───────────   ───────→
┌─┬─┬──────┬────────────┬──────────→
0 3        28           55          s
```

Make a sign diagram for velocity. Put the values when the particle is at rest on the diagram. Choose a value in each interval and find the sign of v(t).

$(0, 2)$ $t = 1$ $v(1) = 6(1-2)(1-5) = (+)(-)(-) = +$

$(2, 5)$ $t = 3$ $v(3) = 6(3-2)(3-5) = (+)(+)(-) = -$

$(5, \infty)$ $t = 6$ $v(6) = 6(6-2)(6-5) = (+)(+)(+) = +$

Find the displacement or position of the particle when the particle changes direction.
Use these positions and the initial position to plot the motion. Although the motion is actually on the line, we draw it above the line.

Exercise 70

1 A particle moves on a line with displacement function $s(t) = t^3 - 6t^2 + 9t$ centimetres for $t \geq 0$ seconds.
  a Find the initial position and the initial velocity for the particle.
  b Find when the particle is at rest.
  c Draw a motion diagram for the particle.

EXAM-STYLE QUESTION

2 A ball is thrown vertically upwards. The height of the ball in feet, t seconds after it is released, is given by $s(t) = -16t^2 + 40t + 4$ for $t \geq 0$ seconds.
  a Find the initial height of the ball.
  b Show that the height of the ball after 2 seconds is 20 feet.
  c There is a second time when the height of the ball is 20 feet.
   i Write down an equation that t must satisfy when the height of the ball is 20 feet.
   ii Solve the equation algebraically.
  d **i** Find $\dfrac{ds}{dt}$.
   ii Find the initial velocity of the ball.
   iii Find when the velocity of the ball is 0.
   iv Find the maximum height of the ball.

3 A particle moves along a line with displacement function $s(t) = \dfrac{t}{e^t}$, where s is in metres and t is in seconds.
  a Show that $v(t) = \dfrac{1-t}{e^t}$.
  b Hence find when the particle is at rest.

→ The instantaneous rate of change of the velocity is the **acceleration function**, $a(t) = \lim_{h \to 0} \dfrac{v(t+h) - v(t)}{h} = v'(t) = s''(t)$

For $a(t) > 0$ the velocity of the object is increasing.
For $a(t) < 0$ the velocity of the object is decreasing.
For $a(t) = 0$ the velocity is constant.

Example 19

For the displacement function from Example 18,
$s(t) = 2t^3 - 21t^2 + 60t + 3$, with s in metres and $t \geq 0$ seconds,
we found that $v(t) = 6t^2 - 42t + 60$.

a Find the **average acceleration** of the particle from $t = 1$ seconds to $t = 4$ seconds.

b Find the **instantaneous acceleration** of the particle at $t = 3$ seconds. Explain the meaning of your answer.

Answers

a Average acceleration is
$\dfrac{\text{change in velocity}}{\text{change in time}}$ $\dfrac{(\text{m s}^{-1})}{(\text{seconds})}$ ← *The units for acceleration are* m s^{-2}.

$\dfrac{v(4) - v(1)}{4 - 1} = -12 \text{ m s}^{-2}$ *Use a GDC to evaluate.*

b Instantaneous acceleration
$a(t) = v'(t)$
$a(t) = v'(t) = 12t - 42$
$a(3) = -6 \text{ m s}^{-2}$

This means the velocity is decreasing 6 metres per second each second at time 3 seconds.

Note that a negative acceleration does not mean an object in motion is slowing down. It means that the velocity is decreasing.

Speed is the absolute value of velocity. Velocity tells us how fast an object is moving **and** the direction in which it is moving. Speed tells us only how fast it is moving. To determine if an object in motion is speeding up or slowing down you can compare the signs of velocity and acceleration.

For more on 'absolute value' see Chapter 18, Section 2.7.

Investigation – velocity, acceleration and speed

1. Copy and complete the tables. Recall that acceleration is the change of velocity. Speed is the absolute value of velocity.

    a Velocity and acceleration are both positive.
    Let acceleration be $2\,m\,s^{-2}$.

Time (sec)	Velocity ($m\,s^{-1}$)	Speed ($m\,s^{-1}$)
0	10	10
1	12	
2		
3		
4		

    b Velocity is positive and acceleration is negative.
    Let acceleration be $-2\,m\,s^{-2}$.

Time (sec)	Velocity ($m\,s^{-1}$)	Speed ($m\,s^{-1}$)
0	10	10
1	8	
2		
3		
4		

    c Velocity and acceleration are both negative.
    Let acceleration be $-2\,m\,s^{-2}$.

Time (sec)	Velocity ($m\,s^{-1}$)	Speed ($m\,s^{-1}$)
0	−10	10
1	−12	
2		
3		
4		

    d Velocity is negative and acceleration is positive.
    Let acceleration be $2\,m\,s^{-2}$.

Time (sec)	Velocity ($m\,s^{-1}$)	Speed ($m\,s^{-1}$)
0	−10	10
1	−8	
2		
3		
4		

2. State whether the object is speeding up or slowing down.
    a Velocity and acceleration are both positive.
    b Velocity is positive and acceleration is negative.
    c Velocity and acceleration are both negative.
    d Velocity is negative and acceleration is positive.
3. Complete the statements:
    a If velocity and acceleration have the same sign then the object is _____.
    b If velocity and acceleration have opposite signs then the object is _____.

> If the speed of an object is increasing, the object is speeding up.
> If the speed of an object is decreasing, the object is slowing down.

When velocity and acceleration have the same sign, the object in motion is speeding up.
When velocity and acceleration have different signs, the object in motion is slowing down.

Example 20

For the displacement function from Example 18,
$s(t) = 2t^3 - 21t^2 + 60t + 3$, with s in metres and $t \geq 0$ seconds,
we found that $v(t) = 6t^2 - 42t + 60$ and $a(t) = 12t - 42$

a Find the speed of the particle at $t = 3$ seconds and determine whether the particle is speeding up or slowing down when $t = 3$ seconds.

b During $0 \leq t \leq 10$ seconds, find the intervals when the particle is speeding up and when it is slowing down.

Answers

a $v(3) = 6(3)^2 - 42(3) + 60$
$= -12 \text{ m s}^{-1}$
speed $= |-12| = 12 \text{ m s}^{-1}$
$a(3) = 12(3) - 42 = -6 \text{ m s}^{-2}$
The particle is speeding up at $t = 3$ seconds since $v(t) < 0$ and $a(t) < 0$.

To find the speed of the particle at a given time, find the velocity and take the absolute value.

The particle is speeding up at $t = 3$ since velocity and acceleration have the same sign.

b Compare the signs of velocity and acceleration.

Use the sign diagram for velocity from Example 18.
Below it, align a sign diagram for $a(t)$.
Find when $a(t) = 0$
$12t - 42 = 0 \Rightarrow t = 3.5$

signs of v +++++ −−−−−−−− +++++++++++
t 0   2   5   10

signs of a −−−−−−−−−− +++++++++++++
t 0   3.5   10

Place this value on the interval $0 \leq t \leq 10$.
Check a value in each interval:
$(0, 3.5)$ $t = 1$
$a(1) = 12(1) - 42 = -30$ $(-)$

$(3.5, 10)$ $t = 4$
$a(4) = 12(4) - 42 = 6$ $(+)$

The particle is speeding up in the interval $(2, 3.5)$ seconds because $v(t) < 0$ and $a(t) < 0$, and in the interval $(5, 10)$ seconds because $v(t) > 0$ and $a(t) > 0$.

The particle is slowing down in the interval $(0, 2)$ seconds because $v(t) > 0$ and $a(t) < 0$, and in the interval $(3.5, 5)$ seconds because $v(t) < 0$ and $a(t) > 0$.

Exercise 7P

Use a GDC to help evaluate function values.

1. A particle moves along a line with the displacement function $s(t) = 2t^4 - 6t^2$, in centimetres, for $t \geq 0$ seconds.
   a. Write expressions for the velocity and acceleration of the particle at time t.
   b. Find the acceleration at time $t = 2$ seconds and explain the meaning of your answer.
   c. Find when velocity and acceleration equal zero. Then find when the particle is speeding up and slowing down.

2. A particle moves along a line with the displacement function $s(t) = -t^3 + 12t^2 - 36t + 20$, in metres, for $0 \leq t \leq 8$ seconds.
   a. Write an expression for the velocity and acceleration of the particle at time t.
   b. Find the initial position, velocity and acceleration for the particle.
   c. Find when the particle changes direction for $0 \leq t \leq 8$ seconds. Then find intervals on which the particle travels right and left.
   d. Find when acceleration is 0 for $0 \leq t \leq 8$ seconds. Then find intervals on which the particle is speeding up and slowing down.

EXAM-STYLE QUESTIONS

3. A diver jumps from a platform at time $t = 0$ seconds. The distance of the diver above water level at time t is given by $s(t) = -4.9t^2 + 4.9t + 10$, where s is in metres.
   a. Write an expression for the velocity and acceleration of the diver at time t.
   b. Find when the diver hits the water.
   c. Find when velocity equals zero. Hence find the maximum height of the diver.
   d. Show that the diver is slowing down at $t = 0.3$ seconds.

   > Look again at the diver in Example 16.

4. A particle moves along a line with displacement function $s(t) = \frac{1}{4}t^2 - \ln(t + 1)$, $t \geq 0$ where, s is in metres and t is in seconds.
   a. i. Write an expression for the velocity of the particle at time t.
      ii. Hence find when the particle is at rest.
   b. i. Write an expression for the acceleration of the particle at time t.
      ii. Hence show that velocity is never decreasing.

7.6 The derivative and graphing

One of the most powerful uses of the derivative is to analyze the graphs of functions. In this section you will see how to connect f' and f'' to the graph of f.

A function is **increasing** on an interval if an increase in x results in an increase in y. A function is **decreasing** on an interval if an increase in x results in a decrease in y.

> Although the Cartesian plane was named after René Descartes (French mathematician, 1596–1650), he used only positive numbers and an x-axis. Isaac Newton (English mathematician, 1642–1727) is attributed with the first use of negative coordinates. In his book *Enumeratio linearum tertii ordinis,* or *Enumeration of Curves of Third Degree,* Newton used both an x-axis and a y-axis and positive and negative coordinates.

Example 21

Write down the intervals on which the function is increasing or decreasing.

a
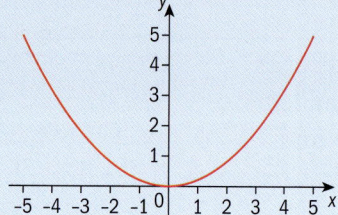

b
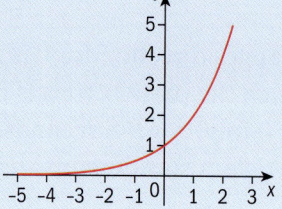

c
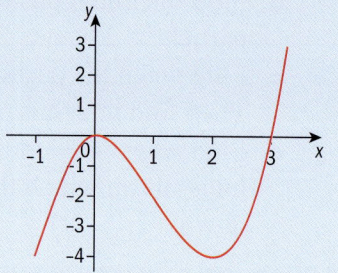

Answers

a Decreasing for $x < 0$
Increasing for $x > 0$

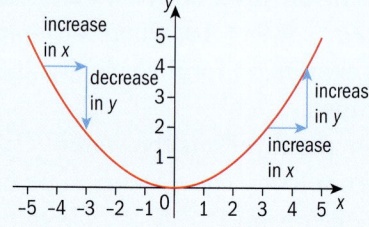

b Increasing for all real numbers

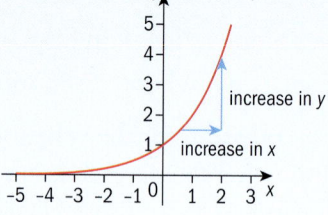

c Increasing for $x < 0$ and $x > 2$
Decreasing for $0 < x < 2$

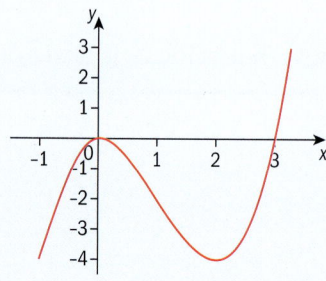

→ When a function is decreasing, the tangent lines to the curve have negative slope. When a function is increasing, the tangent lines to the curve have positive slope. It follows that:
If $f'(x) > 0$ for all x in (a, b) then f is increasing on (a, b).
If $f'(x) < 0$ for all x in (a, b) then f is decreasing on (a, b).

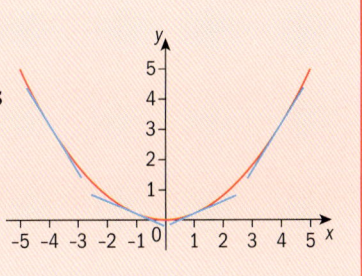

Example 22

Use the derivative of f to find the intervals on which f is increasing or decreasing.

a $f(x) = 2x^3 - 3x^2 - 12x$ **b** $f(x) = \dfrac{x^2 - 4}{x^2 - 1}$ **c** $f(x) = x^3$

A **stationary point** is a point where $f'(x) = 0$

A **critical number** of f is a point where $f'(x) = 0$ or $f'(x)$ is undefined.

Answers

a $f(x) = 2x^3 - 3x^2 - 12x$

$f'(x) = 6x^2 - 6x - 12$ *Find the derivative of f.*

$6x^2 - 6x - 12 = 0$ *Find the critical numbers by setting $f'(x)$ equal to 0 and solving for x.*
$6(x^2 - x - 2) = 0$
$6(x - 2)(x + 1) = 0$
$x = 2, -1$

signs of f': + − +
x: −1   2

Make a sign diagram for $f'(x)$.

f is increasing on $(-\infty, -1)$ and $(2, \infty)$ since $f'(x) > 0$
f is decreasing on $(-1, 2)$ since $f'(x) < 0$

We can use interval notation to describe the intervals.

b $f(x) = \dfrac{x^2 - 4}{x^2 - 1}$

$f'(x) = \dfrac{(x^2 - 1)(2x) - (x^2 - 4)(2x)}{(x^2 - 1)^2}$ *Find the derivative of f.*

$= \dfrac{6x}{(x^2 - 1)^2}$

$f'(x) = 0$: $f'(x)$ undefined when:
$6x = 0$ $(x^2 - 1)^2 = 0$
$x = 0$ $x^2 - 1 = 0$
                    $x = \pm 1$

Find the critical numbers by setting f' equal to 0 and solving for x, and by finding where f' is undefined.

signs of f': − − + +
x: −1 0 1

Make a sign diagram for f'. Notice that f and f' are not defined at $x = \pm 1$. Use open circles on the sign diagram to remember this.

▶ Continued on next page

f is increasing on $(-\infty, -1)$ and $(-1, 0)$ since $f'(x) > 0$. f is decreasing on $(0, 1)$ and $(1, \infty)$ since $f'(x) < 0$.	We cannot say that f is increasing on $(-\infty, 0)$ or decreasing on $(0, \infty)$ since f is not defined at $x = -1$ or $x = 1$.
c $f(x) = x^3$ $f'(x) = 3x^2$ $3x^2 = 0$ $x = 0$ signs of f' + + x 0 f is increasing on $(-\infty, 0)$ and $(0, \infty)$.	Find the derivative of f. Find the critical numbers by setting f' equal to 0 and solving for x. Make a sign diagram for f'. Even though f is defined at $x = 0$, we cannot include 0 in the interval because the gradient is zero at $x = 0$, so $f(x)$ is not increasing at $x = 0$.

Exercise 7Q

Write down the intervals on which f is increasing or decreasing.

1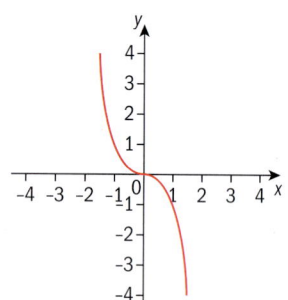

2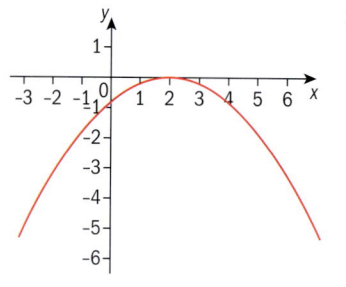

3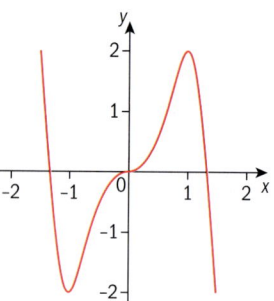

In questions 4–9, use the derivative of f to find all intervals on which f is increasing or decreasing.

4 $f(x) = x^4$

5 $f(x) = x^4 - 2x^2$

6 $f(x) = \dfrac{x+2}{x-3}$

7 $f(x) = \dfrac{1}{\sqrt{x}}$

8 $f(x) = x^3 e^x$

9 $f(x) = \dfrac{x^3}{x^2 - 1}$

Use a GDC to look at a graph of the function to verify your results.

> **EXAM-STYLE QUESTION**
>
> **10** The graph of the *derivative* of f is shown. Write down the intervals on which f is decreasing and increasing.

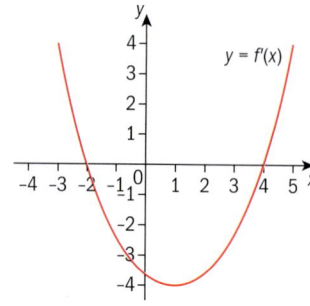

A function has a **relative maximum point** (or local maximum) when the function changes from increasing to decreasing. A function has a **relative minimum point** (or local minimum) when the function changes from decreasing to increasing. Relative minimum and maximum points are called the **relative extrema** of a function.

Note that if $f'(x)$ does not change sign at a critical number $x = c$, then the point $(c, f(c))$ is neither a relative minimum nor a relative maximum.

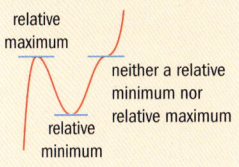

→ The **first derivative test** is used to locate relative extrema of f. If f is defined at a critical number c then:
1 If $f'(x)$ changes from positive to negative at $x = c$, then f has a relative maximum point at $(c, f(c))$.
2 If $f'(x)$ changes from negative to positive at $x = c$, then f has a relative minimum point at $(c, f(c))$.

Example 23

Use the first derivative test to find the relative extrema for the functions in Example 22.

a $f(x) = 2x^3 - 3x^2 - 12x$ **b** $f(x) = \dfrac{x^2 - 4}{x^2 - 1}$ **c** $f(x) = x^3$

Answers

a $f(x) = 2x^3 - 3x^2 - 12x$

$f'(x) = 6x^2 - 6x - 12$
$ = 6(x - 2)(x + 1)$

signs of f': + − + at $x = -1, 2$

Since $f'(x)$ changes from positive to negative at $x = -1$ there is a relative maximum at $x = -1$. Since $f'(x)$ changes from negative to positive at $x = 2$ there is a relative minimum at $x = 2$.

$f(-1) = 2(-1)^3 - 3(-1)^2 - 12(-1)$
$ = 7$

$f(2) = 2(2)^3 - 3(2)^2 - 12(2)$
$ = -20$

So the relative maximum point is $(-1, 7)$ and the relative minimum point is $(2, -20)$.

Use the sign diagram for f' from Example 22. Locate the relative extrema by looking for sign changes for f'.

Evaluate f at $x = -1$ and $x = 2$ to find the maximum and minimum values.

b $f(x) = \dfrac{x^2 - 4}{x^2 - 1}$

$f'(x) = \dfrac{6x}{(x^2 - 1)^2}$

signs of f': − − + + at $x = -1, 0, 1$

Since $f'(x)$ changes from negative to positive at $x = 0$ there is a relative minimum at $x = 0$. $f(0) = \dfrac{0^2 - 4}{0^2 - 1} = 4$

So the relative minimum point is $(0, 4)$.

There would not be relative extrema at $x = -1$ and $x = 1$ even if the sign of $f'(x)$ had changed, since f is undefined at $x = -1$ and $x = 1$.

▶ Continued on next page

c $f(x) = x^3$
$f'(x) = 3x^2$

signs of f'

f has no relative extrema since the derivative does not change sign at $x = 0$.

Note that $f'(x) = 0$ is not a sufficient condition to have a relative extrema at $x = 0$. It must also be true that $f'(x)$ changes sign at $x = 0$.

Exercise 7R

In questions 1 to 8, use the first derivative test to find the relative extrema for each function.

1 $f(x) = 2x^2 - 4x - 3$

2 $f(x) = x^3 - 12x - 5$

3 $f(x) = x^{\frac{5}{3}}$

4 $f(x) = x^4 - 2x^2$

5 $f(x) = x(x+3)^3$

6 $f(x) = x^2 e^{-x}$

7 $f(x) = \dfrac{1}{(x+1)^2}$

8 $f(x) = \dfrac{x^2 - 2x + 1}{x+1}$

→ If $f''(x) > 0$ for all x in (a, b) then f is **concave up** on (a, b).
If $f''(x) < 0$ for all x in (a, b) then f is **concave down** on (a, b).
The points on a graph where the concavity changes are called **inflexion points**. A point on the graph of f is an inflexion point if $f''(x) = 0$ and $f''(x)$ changes sign.

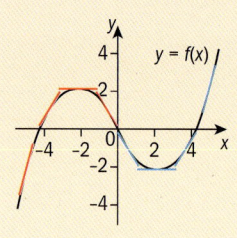

The graph is concave down for $(-\infty, 0)$. The tangent lines shown in red have decreasing gradients. This means that f' is decreasing, so its derivative f'' is negative.

The graph is concave up for $(0, \infty)$. The tangent lines shown in blue have increasing gradients. This means that f' is increasing, so its derivative f'' is positive.

The point $(0, 0)$ is an inflexion point since f changes concavity at $x = 0$.

Example 24

For the functions from Example 22, use the second derivative to find the intervals where the function is concave up and concave down. Find the inflexion points.

a $f(x) = 2x^3 - 3x^2 - 12x$ **b** $f(x) = \dfrac{x^2 - 4}{x^2 - 1}$ **c** $f(x) = x^3$

Answers

a $f(x) = 2x^3 - 3x^2 - 12x$

$f'(x) = 6x^2 - 6x - 12$

$f''(x) = 12x - 6$ *Find the second derivative of f.*

$12x - 6 = 0$ *Find where $f''(x) = 0$.*

$x = \dfrac{1}{2}$

signs of f'' —   +   *Make a sign diagram for f''.*
x $\frac{1}{2}$

f is concave down on $\left(-\infty, \dfrac{1}{2}\right)$ since $f''(x) < 0$ and

f is concave up on $\left(\dfrac{1}{2}, \infty\right)$ since $f''(x) > 0$.

Since $f''(x)$ changes sign at $x = \dfrac{1}{2}$, there is an inflexion point there. $f\left(\dfrac{1}{2}\right) = 2\left(\dfrac{1}{2}\right)^3 - 3\left(\dfrac{1}{2}\right)^2 - 12\left(\dfrac{1}{2}\right) = -\dfrac{13}{2}$ *Evaluate f at $x = \dfrac{1}{2}$ to find the y-coordinate of the inflexion point.*

So the inflexion point is $\left(\dfrac{1}{2}, -\dfrac{13}{2}\right)$

b $f(x) = \dfrac{x^2 - 4}{x^2 - 1}$

$f'(x) = \dfrac{6x}{(x^2 - 1)^2}$

$f''(x) = \dfrac{(x^2 - 1)^2 (6) - (6x)[2(x^2 - 1)(2x)]}{(x^2 - 1)^4} = \dfrac{-6(3x^2 + 1)}{(x^2 - 1)^3}$ *Find the second derivative of f.*

$f''(x) = 0$ $f''(x)$ is undefined when   *To make a sign diagram for f'' you must find where $f''(x) = 0$ **and** where $f''(x)$ is undefined.*

$\dfrac{-6(3x^2 + 1)}{(x^2 - 1)^3} = 0$ $(x^2 - 1)^3 = 0$

$-6(3x^2 + 1) = 0$ $x^2 - 1 = 0$

$x^2 = -\dfrac{1}{3}$ $x = \pm 1$

No real solutions

signs of f'' —   +   —   *Even though $f''(x)$ changes sign at $x = \pm 1$ there are no inflexion points. This is because $f(x)$ is undefined at $x = \pm 1$. In this case the concavity is changing on either side of a vertical asymptote.*
x   −1   1

f is concave down on $(-\infty, -1)$ and $(1, \infty)$ since $f''(x) < 0$, and f is concave up on $(-1, 1)$ since $f''(x) > 0$.

▶ Continued on next page

c $f(x) = x^3$
$f'(x) = 3x^2$
$f''(x) = 6x$
$6x = 0$
$x = 0$

signs of f'' — | +
            x 0

Find the second derivative of f.
Find where $f''(x) = 0$.

Make a sign diagram for f''.

f is concave down on $(-\infty, 0)$ since $f''(x) < 0$, and f is concave up on $(0, \infty)$ since $f''(x) > 0$. Since $f''(x)$ changes signs at $x = 0$, there is an inflexion point there.
$f(0) = (0)^3 = 0$.
So the inflexion point is $(0, 0)$.

Evaluate f at $x = 0$ to find the y-coordinate of the inflexion point.

Exercise 7S

In questions 1 to 6, use the second derivative to find the intervals where the function is concave up and concave down.
Find the inflexion points.

1 $f(x) = 2x^2 - 4x - 3$

2 $f(x) = -x^4 + 4x^3$

3 $f(x) = x^3 - 6x^2 + 12x$

4 $f(x) = x^4$

5 $f(x) = 2xe^x$

6 $f(x) = \dfrac{1}{x^2 + 1}$

EXAM-STYLE QUESTIONS

7 Let $f(x) = \dfrac{24}{x^2 + 12}$

  a Use that fact that $f'(x) = \dfrac{-48x}{(x^2 + 12)^2}$ to show that the second derivative is $f''(x) = \dfrac{144(x^2 - 4)}{(x^2 + 12)^3}$

  b i Find the relative extrema of the graph of f.
  ii Find the inflexion points of the graph of f.

8 The graph of the *second derivative of f* is shown. Write down the intervals on which f is concave up and concave down. Give the x-coordinates of any inflexion points.

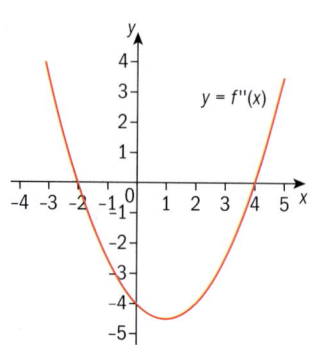

The first and second derivatives of a function tell us much about the graph of the function. We can also use intercepts and asymptotes to help complete the graph.

Example 25

Sketch the graph of each function. Use the information you found in Examples 22–24 and intercepts and asymptotes to help draw the graph.

a $f(x) = 2x^3 - 3x^2 - 12x$ **b** $f(x) = \dfrac{x^2 - 4}{x^2 - 1}$ **c** $f(x) = x^3$

Answers

a $f(x) = 2x^3 - 3x^2 - 12x$
increasing: $(-\infty, -1)$ and $(2, \infty)$
decreasing: $(-1, 2)$
relative maximum: $(-1, 7)$
relative minimum: $(2, -20)$
concave down: $\left(-\infty, \dfrac{1}{2}\right)$
concave up: $\left(\dfrac{1}{2}, \infty\right)$
inflexion point: $\left(\dfrac{1}{2}, -\dfrac{13}{2}\right)$
x-intercepts: $(0, 0), (-1.8, 0), (3.31, 0)$
y-intercept: $(0, 0)$

To find the x-intercepts, set the function equal to 0 and solve:
$2x^3 - 3x^2 - 12x = 0$
$x(2x^2 - 3x - 12) = 0$
$x = 0$ or $x = \dfrac{3 \pm \sqrt{9 - 4(2)(-12)}}{2(2)}$
$x = 0$ or $x \approx -1.81, 3.31$

To find the y-intercept evaluate $f(0)$.

b $f(x) = \dfrac{x^2 - 4}{x^2 - 1}$
increasing: $(-\infty, -1)$ and $(-1, 0)$
decreasing: $(0, 1)$ and $(1, \infty)$
relative minimum: $(0, 4)$
concave down: $(-\infty, -1)$ and $(1, \infty)$
concave up: $(-1, 1)$
inflexion points: none
x-intercepts: $(2, 0), (-2, 0)$
y-intercept: $(0, 4)$

To find the x-intercepts, set the function equal to 0 and solve:
$\dfrac{x^2 - 4}{x^2 - 1} = 0 \Rightarrow x^2 - 4 = 0 \Rightarrow x = \pm 2$

To find the y-intercept evaluate $f(0)$.

▶ Continued on next page

vertical asymptotes: $x = \pm 1$

horizontal asymptote: $y = 1$

To find the vertical asymptotes, find where the denominator equals 0 (check to see that the numerator is not 0 for that same value):
$x^2 - 1 = 0 \Rightarrow x = \pm 1$

We learned that the horizontal asymptote of a function of the form $y = \dfrac{ax + b}{cx + d}$ is found by using the leading coefficients, $y = \dfrac{a}{c}$. This method works for any rational function where the degree of the numerator is equal to the degree of the denominator.
$y = \dfrac{1}{1} \Rightarrow y = 1$

> Limit notation can be used to describe the asymptotes. The horizontal asymptote $y = 1$ is showing us that for large positive values of x, y gets close to 1, and for small negative values of x, y gets close to 1. Using limit notation to say this we write: $\lim\limits_{x \to \infty} f(x) = 1$ and $\lim\limits_{x \to -\infty} f(x) = 1$. For the vertical asymptote $x = 1$, as x gets close to 1 from the left side of 1, y grows large and positive without bound, and as x gets close to 1 from the right side of 1, y grows large and negative without bound. Using limits to say this we write: $\lim\limits_{x \to 1^-} f(x) = \infty$ and $\lim\limits_{x \to 1^+} f(x) = -\infty$.
> Similarly, for $x = -1$ we write: $\lim\limits_{x \to -1^-} f(x) = -\infty$ and $\lim\limits_{x \to -1^+} f(x) = \infty$.

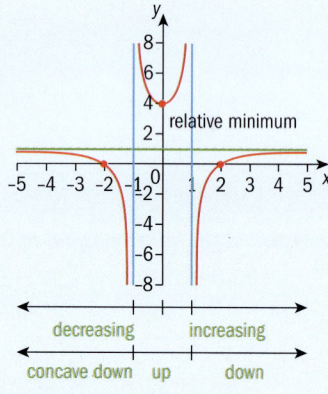

c $f(x) = x^3$
increasing: $(-\infty, \infty)$
no relative extrema
concave down: $(-\infty, 0)$
concave up: $(0, \infty)$
inflexion point: $(0, 0)$
x-intercept: $(0, 0)$
y-intercept: $(0, 0)$

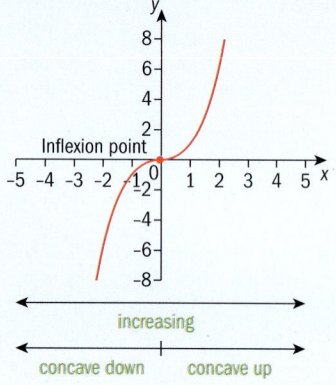

238 Limits and derivatives

Exercise 7T

In questions 1 to 4, sketch the graph of the function. Use the first and second derivative to analyze key features of the graph. Find any intercepts and asymptotes.

1 $f(x) = 3x^2 + 10x - 8$

2 $f(x) = x^3 + x^2 - 5x - 5$

3 $f(x) = \dfrac{x+2}{x-4}$

4 $f(x) = (3-x)^4$

5 $f(x) = \dfrac{e^x - e^{-x}}{2}$

6 $f(x) = \dfrac{x^2 - 1}{x^2 + 1}$

Given the graph of any one of the three functions f, f' or f'', it is possible to sketch a graph of the other two functions.

Example 26

a Given that the graph shown is a graph of f, sketch the graphs of f' and f''.

b Given that the graph shown is a graph of f', sketch the graphs of f and f''.

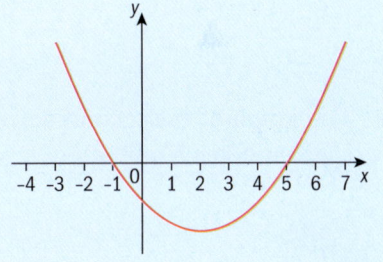

Answers

a

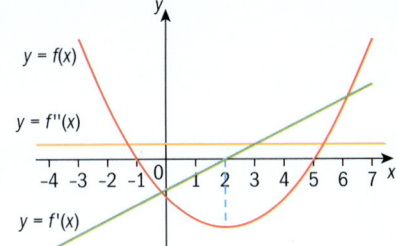

The graph of f changes from decreasing to increasing and has a relative minimum at $x = 2$. This means that $f'(x)$ equals zero at $x = 2$ and changes from negative to positive. The graph of f is always concave up. This means that $f''(x)$ is always positive. Since $f''(x)$ is the derivative of $f'(x)$, a linear function, $f''(x)$ must be a positive constant.

b

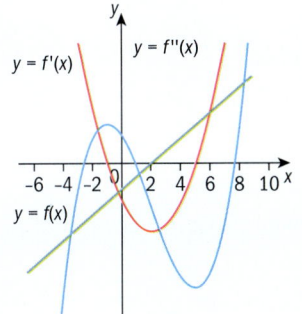

*Since $f'(x)$ equals zero when $x = -1$ and changes from positive to negative, the graph of f has a relative maximum point when $x = -1$.
Since $f'(x)$ equals zero when $x = 5$ and changes from negative to positive, the graph of f has a relative minimum point when $x = 5$.
Since $f'(x)$ has a relative minimum when $x = 2$, the graph of $f''(x)$ equals zero when $x = 2$. Since f is concave down for $x < 2$, $f''(x)$ is negative for $x < 2$. Since f is concave up for $x > 2$, $f''(x)$ is positive for $x > 2$.*

Exercise 7U

EXAM-STYLE QUESTIONS

1 The graph of $y = f(x)$ is given.
   Sketch a graph of $y = f'(x)$ and $y = f''(x)$.

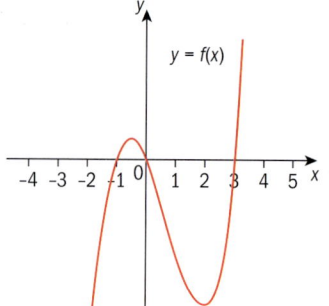

2 The graph of the derivative of f, $y = f'(x)$, is given.
   Sketch a graph of $y = f(x)$ and $y = f''(x)$.

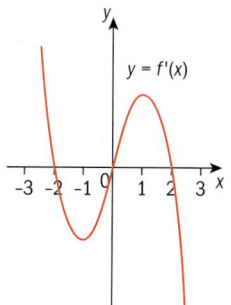

3 The graph of the second derivative of f, $y = f''(x)$, is given. Sketch a graph of $y = f(x)$ and $y = f'(x)$.

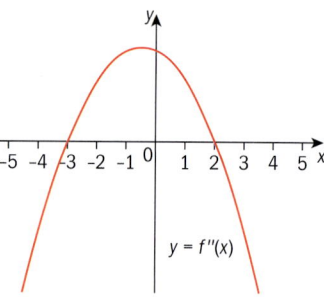

7.7 More on extrema and optimization problems

You have seen how to use the second derivative to determine concavity and inflexion points for a graph of a function. The second derivative of a function can also be used to find relative extrema. This is called the **second derivative test**.

> See Chapter 17, Section 2.6.

Second derivative test

If $f'(c) = 0$ and the second derivative of f exists near c, then

1. If $f''(c) > 0$, then f has a relative minimum at $x = c$.
2. If $f''(c) < 0$, then f has a relative maximum at $x = c$.
3. If $f''(c) = 0$, the second derivative test fails and the first derivative test must be used to locate the relative extrema.

> If $f''(c) > 0$ near c, then f is concave up near c. So f has a relative minimum.
>
> If $f''(c) < 0$ near c, then f is concave down near c. So f has a relative maximum.

Example 27

Find the relative extreme points of each function. Use the second derivative test whenever possible.

a $f(x) = x^3 - 3x^2 - 2$ **b** $f(x) = 3x^5 - 5x^3$

Answers

a $f(x) = x^3 - 3x^2 - 2$

$f'(x) = 3x^2 - 6x$

$f''(x) = 6x - 6$

Find the first and second derivative of f.

$3x^2 - 6x = 0$

$3x(x - 2) = 0$

$x = 0, 2$

Find the values of x where the first derivative equals zero.

$f''(0) = -6 < 0 \Rightarrow$ relative maximum

$f''(2) = 6 > 0 \Rightarrow$ relative minimum

Evaluate the second derivative at each zero of the first derivative.
$f'' < 0$ *implies a relative maximum and* $f'' > 0$ *implies a relative minimum.*

$f(0) = -2 \Rightarrow (0, -2)$ is a relative maximum

$f(2) = -6 \Rightarrow (0, -6)$ is a relative minimum

Evaluate the function where the extrema occur to find the relative minimum and maximum values.

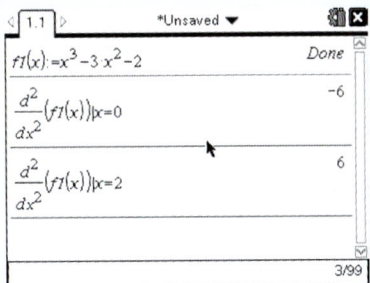

 GDC help on CD: *Alternative demonstrations for the TI-84 Plus and Casio FX-9860GII GDCs are on the CD.*

b $f(x) = 3x^5 - 5x^3$

$f'(x) = 15x^4 - 15x^2 = 15x^2(x+1)(x-1)$

$f''(x) = 60x^3 - 30x$

Find the first and second derivative of f.

$15x^4 - 15x^2 = 0$

$15x^2(x+1)(x-1) = 0$

$x = 0, \pm 1$

Find the values of x where the first derivative equals zero.

$f''(0) = 0 \Rightarrow$ second derivative test fails

$f''(-1) = -30 < 0 \Rightarrow$ relative maximum

$f''(1) = 30 > 0 \Rightarrow$ relative minimum

Evaluate the second derivative at each zero of the first derivative.
$f'' = 0$ *implies the second derivative test fails,* $f'' < 0$ *implies a relative maximum, and* $f'' > 0$ *implies a relative minimum.*

signs of f'

```
          -   0   -
      <---⊕---⊕---⊕--->
  x      -1   0   1
```

▶ Continued on next page

Since there is no sign change in f' at $x = 0$, there is no relative minimum or maximum at that point.

$f(-1) = 2 \Rightarrow (-1, 2)$ is a relative maximum
$f(1) = -2 \Rightarrow (1, -2)$ is a relative minimum

Since the second derivative test failed at $x = 0$, use the first derivative test to see if the sign of f' changes at $x = 0$.

Evaluate the function where the extrema occur to find the relative minimum and maximum values.

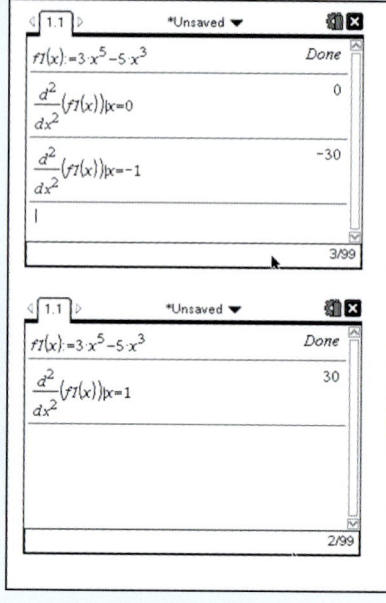

GDC help on CD: *Alternative demonstrations for the TI-84 Plus and Casio FX-9860GII GDCs are on the CD.*

Exercise 7V

Find the relative extreme points of each function. Use the second derivative test whenever possible.

1. $f(x) = 3x^2 - 18x - 48$
2. $f(x) = (x^2 - 1)^2$
3. $f(x) = x^4 - 4x^3$
4. $f(x) = xe^x$
5. $f(x) = (x - 1)^4$
6. $f(x) = \dfrac{1}{x^2 + 1}$

We have been finding relative or local extrema of functions. We can also find the **absolute or global extrema** of a function. Absolute extrema are the greatest and least values of the function over its entire domain. Absolute extrema occur at either the relative extrema or endpoints of a function.

> The relative extrema of a function are the minimum and maximum values of the function on an interval near the critical point. Relative extrema never occur at an endpoint of a function.

Example 28

a Identify each labeled point as an absolute maximum or minimum, a relative maximum or minimum, or neither.

b Find the absolute maximum and minimum for $f(x) = x^2 - 2x$ on $-1 \le x \le 2$.

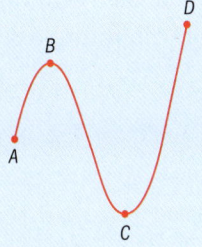

Answers

a *A* is neither.

The points on the graph above the black line have values greater than the value of the function at A and those below the black line have values less than the value of the function at A. So A is neither an absolute maximum nor an absolute minimum. A cannot be a relative extrema since A is at an endpoint of the function.

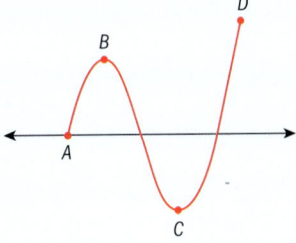

B is a relative maximum.

B cannot be an absolute maximum since there are values of the function greater than the value of the function at B.

C is an absolute minimum and relative minimum.

C is an absolute minimum since the value of the function of C is the least value of the function over its entire domain.

D is an absolute maximum.

The value of the function at D is the greatest value of the function over its entire domain.

b $f(x) = x^2 - 2x$ on $-1 \le x \le 2$

$f'(x) = 2x - 2$

$2x - 2 = 0$

$x = 1$

Find the critical numbers where $f'(x) = 0$.

$f(-1) = (-1)^2 - 2(-1) = 3$

$f(1) = (1)^2 - 2(1) = -1$

$f(2) = (2)^2 - 2(2) = 0$

Evaluate the function at each endpoint and critical number in the interval. The largest value is the maximum and the smallest is the minimum.

The absolute maximum of $f(x) = x^2 - 2x$ on $-1 \le x \le 2$ is 3 and the absolute minimum is -1.

Exercise 7W

Identify each labeled point in questions 1 and 2 as an absolute maximum or minimum, a relative maximum or minimum, or neither.

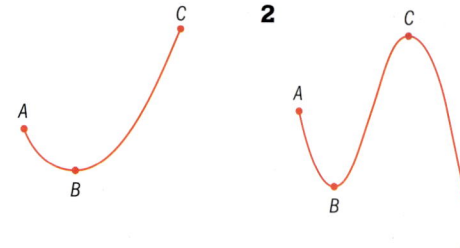

1 **2**

Find the absolute maximum and minimum of the function on the given interval.

3 $f(x) = (x-2)^3$ for $0 \leq x \leq 4$

4 $f(x) = 8x - x^2$ for $-1 \leq x \leq 7$

5 $f(x) = x^3 - \dfrac{3}{2}x^2$ for $-1 \leq x \leq 2$

Many practical problems involve finding maximum or minimum values. For example, we may want to maximize an area or minimize cost. Such problems are called **optimization problems**.

> → For optimization problems:
> 1. Assign variables to given quantities and quantities to be determined. When possible draw a sketch.
> 2. Write an equation to be **optimized** (minimized or maximized) in terms of two variables.
> 3. Find values that are sensible or feasible for the problem situation where the derivative of the equation to be optimized equals zero.
> 4. Verify that you have a minimum or maximum using the first or second derivative test. If the domain is such that $a \leq x \leq b$, remember that the endpoints must be tested since the maximum or minimum on a closed interval can occur when $f'(x) = 0$ or at an endpoint.

Example 29

The product of two positive numbers is 48. Find the two numbers so that the sum of the first number plus three times the second number is a minimum.

Answer

x = the first positive number
y = the second positive number

$S = x + 3y$

$xy = 48 \Rightarrow y = \dfrac{48}{x}$

$S = x + 3\left(\dfrac{48}{x}\right) = x + \dfrac{144}{x}$

$S'(x) = 1 - \dfrac{144}{x^2}$

$1 - \dfrac{144}{x^2} = 0$

$x^2 = 144$

$x = \pm 12$

Since the numbers are positive consider only $x = 12$.

$S''(x) = \dfrac{288}{x^3}$

$S''(12) = \dfrac{288}{12^3} > 0 \Rightarrow$ relative minimum

$y = \dfrac{48}{x} \Rightarrow y = \dfrac{48}{12} = 4$

The numbers are 12 and 4.

Assign variables to the quantities to be determined.

Write an equation for the sum, the quantity to be minimized.

Use the other given information to rewrite the equation for the sum using only two variables.

Find the derivative of the equation to be minimized and then find the critical numbers, where the derivative equals 0.

Use the second derivative test to verify that the critical number 12 gives a minimum. Note that the first derivative test could also be used.

Find the second number.

Example 30

A rectangular plot of farmland is enclosed by 180 m of fencing material on three sides. The fourth side of the plot is bordered by a stone wall. Find the dimensions of the plot that enclose the maximum area. Find the maximum area.

Answer

$A = lw$
$2w + l = 180 \Rightarrow l = 180 - 2w$
$A = (180 - 2w)w = 180w - 2w^2$

$A'(w) = 180 - 4w$
$180 - 4w = 0$
$\quad w = 45$

Make a sketch and assign variables to the quantities to be determined.

Write an equation for the area, the quantity to be maximized.
Use the other given information to rewrite the equation for the area using only two variables.
Find the derivative of the equation to be maximized and then find the critical numbers, where the derivative equals 0.

▶ Continued on next page

$A''(w) = -4$ $A''(45) = -4 < 0 \Rightarrow$ relative maximum $l = 180 - 2w \Rightarrow$ $l = 180 - 2(45) = 90$ $A = 90(45) = 4050$ A 45 m by 90 m plot will have the maximum area of 4050 m².	Use the second derivative test to verify that the critical number 45 gives a maximum. Find the length and the area.

Exercise 7X

1 The sum of two positive numbers is 20. Find the two numbers so that the sum of the first number and the square root of the second number is a maximum.

2 The sum of one positive number and twice a second positive number is 200. Find the two numbers so that their product is a maximum.

3 A rectangular pen is partitioned into two sections and built from 400 feet of fencing as shown in the figure. What dimensions should be used so that the area will be a maximum?

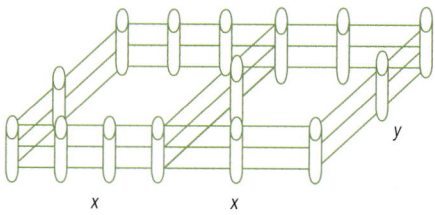

Example 31

Find the dimensions of an open box with a square base and surface area of 192 square centimetres that has a maximum volume.

Answer

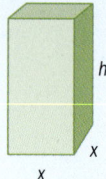

$V = x^2 h$

$x^2 + 4xh = 192$

$h = \dfrac{192 - x^2}{4x}$

$V(x) = x^2 \left(\dfrac{192 - x^2}{4x} \right)$

$= 48x - \dfrac{1}{4}x^3$

Make a sketch and assign variables to the quantities to be determined.

Write an equation for the volume, the quantity to be maximized.
Since the box is open, the surface area is the sum of the area of the bottom of the box, x^2, and the area of the four lateral faces, $4xh$.

Use this to rewrite the equation for the area using only two variables.

▶ Continued on next page

$V'(x) = 48 - \dfrac{3}{4}x^2$ $48 - \dfrac{3}{4}x^2 = 0$ $\dfrac{3}{4}x^2 = 48$ $x^2 = 64$ $x = \pm 8$ The feasible critical number is $x = 8$. $V''(x) = -\dfrac{3}{2}x$ $V''(8) = -\dfrac{3}{2}(8) = -12 < 0$ $\Rightarrow$ relative maximum $h = \dfrac{192 - x^2}{4x} \Rightarrow h = \dfrac{192 - 8^2}{4(8)} = 4$ The dimensions of the box with maximum area are 8 cm by 8 cm by 4 cm.	*Find the derivative of the equation to be maximized and then find the critical numbers, where the derivative equals 0.* *Use the second derivative test to verify that the critical number 8 gives a maximum.* *Find the height of the box.*

Example 32

The cost C of ordering and storing x units of a product is $C(x) = x + \dfrac{10\,000}{x}$. A delivery truck can deliver at most 200 units per order. Find the order size that will minimize the cost.

Answer

$C(x) = x + \dfrac{10\,000}{x}$ where x is the number of units. $C'(x) = 1 - \dfrac{10\,000}{x^2}$ $1 - \dfrac{10\,000}{x^2} = 0$ $\dfrac{10\,000}{x^2} = 1$ $x^2 = 10\,000$ $x = \pm 100$ The feasible critical number is $x = 100$. Since the order must include at least one unit and no more than 200 units, we need to find the absolute minimum for $1 \leq x \leq 200$.	*C is the function to be minimized.* *Find the derivative of the equation to be minimized and then find the critical numbers, where the derivative equals 0.* *Since the function is defined on a closed interval, the endpoints and zeros of the derivative in the interval must be considered for the minimum value.*

▶ Continued on next page

$$C(1) = 1 + \frac{10\,000}{1} = 10\,001$$

$$C(100) = 100 + \frac{10\,000}{100} = 200 \Leftarrow \text{minimum cost}$$

$$C(200) = 200 + \frac{10\,000}{200} = 250$$

The minimum cost occurs when there are 100 units.

Exercise 7Y

1. An open box with a square base and open top has a volume of $32\,000\,\text{cm}^3$. Find the dimensions of the box that minimize the surface area.

2. Suppose that the average cost of producing x units of an item is given by $C(x) = x^3 - 3x^2 - 9x + 30$. If at most 10 items can be produced in a day, how many items should be produced to minimize the cost for the day?

3. A particle moves on a horizontal line so that its position from the origin at time t is given by $s(t) = t^3 - 12t^2 + 36t - 10$ for $0 \le t \le 7$. Find the maximum distance between the particle and the origin.

EXAM-STYLE QUESTIONS

4. A cylinder is inscribed in a cone with radius 6 centimetres and height 10 centimetres.
   a. Find an expression for r, the radius of the cylinder in terms of h, the height of the cylinder.
   b. Find an expression of the volume, V, of the cylinder in terms of h.
   c. Find $\dfrac{dV}{dh}$ and $\dfrac{d^2V}{dh^2}$.
   d. Hence find the radius and height of cylinder with maximum volume.

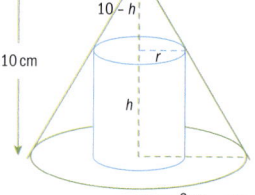

5. Let x be the number of thousands of units of an item produced. The revenue for selling x units is $r(x) = 4\sqrt{x}$ and the cost of producing x units is $c(x) = 2x^2$.
   a. The profit $p(x) = r(x) - c(x)$ Write an expression for $p(x)$.
   b. Find $\dfrac{dp}{dx}$ and $\dfrac{d^2p}{dx^2}$.
   c. Hence find the number of units that should be produced in order to maximize the profit.

Review exercise

1. Differentiate with respect to x.
   a. $4x^3 + 3x^2 - 2x + 6$
   b. $\sqrt[3]{x^4}$
   c. $\dfrac{3}{x^4}$
   d. $(x^2 - 1)(2x^3 - x^2 + x)$
   e. $\dfrac{x-4}{x+7}$
   f. e^{4x}

g $(x^3 + 1)^4$ **h** $\ln(2x+3)$ **i** $\dfrac{\ln x}{x^2}$

j $\dfrac{4x^2 - 2x}{6}$ **k** $(3x^2 + 1)(e^x)$ **l** $\dfrac{2e^x}{e^x - 3}$

m $3\sqrt{2x-5}$ **n** $x^2 e^{2x}$ **o** $\ln\left(\dfrac{1}{x}\right)$

:::EXAM-STYLE QUESTIONS

2 Let $f(x) = 2x^3 - 6x$
  a Expand $(x + h)^3$.
  b Use the formula $f'(x) = \lim\limits_{h \to 0} \dfrac{f(x+h) - f(x)}{h}$ to show that the derivative of $f(x)$ is $6x - 6$.
  c The graph of f is decreasing for $p < x < q$. Find the values of p and q.
  d Write down $f''(x)$.
  e Find the interval on which f is concave up.

3 Find the equation of the normal line to the curve $f(x) = 4xe^{x^2 - 1}$ at the point $(1, 4)$.

4 Find the coordinates on the graph of $f(x) = 2x^3 - 3x + 1$ at which the tangent line is parallel to the line $y = 5x - 2$.

5 A graph of $y = f(x)$ is given.
  a Write down $f(2)$, $f'(2)$ and $f''(2)$ in order from the greatest value to the least value.
  b Justify your answer from part **a**.

6 A curve has the equation $y = x^3(x - 4)$
  a Find **i** $\dfrac{dy}{dx}$ **ii** $\dfrac{d^2y}{dx^2}$
  b For this curve find:
    i the x-intercepts **ii** the coordinates of the relative minimum point
    iii the coordinates of the points of inflexion.
  c Use your answers to part **b** to sketch a graph of the curve, clearly indicating the features you found in part **b**.

7 A particle moves along a horizontal line such that its displacement from the origin O is given by $s(t) = 20t - 100 \ln t$, $t \geq 1$.
  a Find the velocity function for s.
  b Find when the particle is moving to the left.
  c Show that the velocity of the particle is always increasing.

Review exercise

1 Use a GDC to examine each function graphically and numerically. Find the limit or state that it does not exist.

  a $\lim\limits_{x \to 2} \dfrac{1}{x - 2}$ **b** $\lim\limits_{x \to 3} \dfrac{1}{x - 2}$ **c** $\lim\limits_{x \to 4} \dfrac{x^2 - 16}{x - 4}$ **d** $\lim\limits_{x \to 1} \dfrac{x^2 + 3}{x - 1}$

EXAM-STYLE QUESTION

2 A 10 foot post and a 25 foot post stand 30 feet apart and perpendicular to the ground. Wires of lengths y and z run from the top of each pole and are attached by a single stake at a point on the ground between the two poles, as shown in the figure.

a **i** Write an expression for y in terms of x.
 ii Write an expression for z in terms of x.
 iii Hence write an expression for $L(x)$, the total length of wire used for both poles.

b **i** Find $\dfrac{dL}{dx}$.
 ii Hence find the distance x the stake should be placed from the 10 foot pole in order to minimize the amount of wire used.

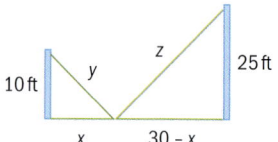

CHAPTER 7 SUMMARY

The tangent line and derivative of x^n

- The function defined by $\lim\limits_{h\to 0} \dfrac{f(x+h)-f(x)}{h}$ is known as the **derivative** of f.

  The derivative is defined by $f'(x) = \lim\limits_{h\to 0} \dfrac{f(x+h)-f(x)}{h}$ or $\dfrac{dy}{dx} = \lim\limits_{h\to 0} \dfrac{f(x+h)-f(x)}{h}$

- **Power rule**
  If $f(x) = x^n$, then $f'(x) = nx^{n-1}$, where $n \in \mathbb{R}$.

- **Constant rule**
  If $f(x) = c$, where c is any real number, then $f'(x) = 0$

- **Constant multiple rule**
  If $y = cf(x)$, where c is any real number, then $y' = cf'(x)$

- **Sum or difference rule**
  If $f(x) = u(x) \pm v(x)$ then $f'(x) = u'(x) \pm v'(x)$

More rules for derivatives

- **Derivative of e^x**
  If $f(x) = e^x$, then $f'(x) = e^x$

- **Derivative of ln x**
  If $f(x) = \ln x$, then $f'(x) = \dfrac{1}{x}$

- **The product rule**
  If $f(x) = u(x) \cdot v(x)$ then $f'(x) = u(x) \cdot v'(x) + v(x) \cdot u'(x)$

- **The quotient rule**
  If $f(x) = \dfrac{u(x)}{v(x)}$ then $f'(x) = \dfrac{v(x) \cdot u'(x) - u(x) \cdot v'(x)}{[v(x)]^2}$

Continued on next page

The chain rule and higher order derivatives

- **The chain rule**
  If $f(x) = u(v(x))$ then $f'(x) = u'(v(x)) \cdot v'(x)$
- The chain rule can also be written as:
  If $y = f(u)$, $u = g(x)$ and $y = f(g(x))$, then $\dfrac{dy}{dx} = \dfrac{dy}{du} \cdot \dfrac{du}{dx}$

Rates of change and motion in a line

- The instantaneous rate of change of displacement is the **velocity function**, $v(t) = \lim\limits_{h \to 0} \dfrac{s(t+h) - s(t)}{h} = s'(t)$
- The instantaneous rate of change of the velocity is the **acceleration function**, $a(t) = \lim\limits_{h \to 0} \dfrac{v(t+h) - v(t)}{h} = v'(t) = s''(t)$

The derivative and graphing

- When a function is decreasing, the tangent lines to the curve have negative slope. When a function is increasing, the tangent lines to the curve have positive slope. It follows that:
  If $f'(x) > 0$ for all x in (a, b) then f is increasing on (a, b).
  If $f'(x) < 0$ for all x in (a, b) then f is decreasing on (a, b).
- The **first derivative test** is used to locate relative extrema of f. If f is defined at a critical number c then:
  1 If $f'(x)$ changes from positive to negative at $x = c$, then f has a relative maximum point at $(c, f(c))$.
  2 If $f'(x)$ changes from negative to positive at $x = c$, then f has a relative minimum point at $(c, f(c))$.
- If $f''(x) > 0$ for all x in (a, b) then f is **concave up** on (a, b). If $f''(x) < 0$ for all x in (a, b) then f is **concave down** on (a, b). The points on a graph where the concavity changes are called **inflexion points**. A point on the graph of f is an inflexion point if $f''(x) = 0$ and $f''(x)$ changes sign.

More on extrema and optimization problems

- For optimization problems:
  1 Assign variables to given quantities and quantities to be determined. When possible draw a sketch.
  2 Write an equation to be **optimized** (minimized or maximized) in terms of two variables.
  3 Find values that are sensible or feasible for the problem situation where the derivative of the equation to be optimized equals zero.
  4 Verify that you have a minimum or maximum using the first or second derivative test. If the domain is such that $a \leq x \leq b$, remember that the endpoints must be tested since the maximum or minimum on a closed interval can occur when $f'(x) = 0$ or at an endpoint.

Theory of knowledge

Truth in mathematics
Inductive reasoning

Inductive reasoning looks at particular cases to make a generalization. Use inductive reasoning to make a conjecture about this problem.

1. Copy the circles and table. Draw all possible chords connecting the points on each circle. Count the number of non-overlapping regions in the interior of each circle

   Record the results in the table.

The first two circles are already completed for you.

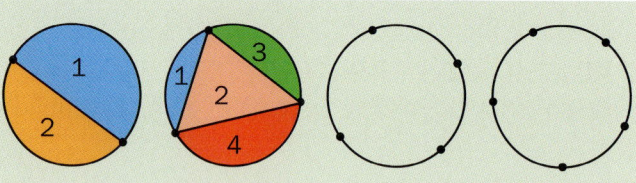

Number of points on the circle	Number of regions formed
2	2
3	4
4	
5	

If you based your conjecture on the most obvious pattern for the number of regions formed, you will find that it was not true for n = 6.

2. Describe, in words, any patterns you observe for the number of regions formed.

3. Make a conjecture about the number of non-overlapping regions that are formed by connecting n points on a circle. Write it as a mathematical expression.

4. Use your conjecture to predict the number of regions formed when all possible chords connecting six points on a circle are drawn.

5. Draw a circle with six points. Draw all the possible chords connecting the points to check your prediction from question 4.

- How many times does a pattern have to repeat before we know it is true?
- Can we ever know it is always true by inspecting the pattern?
- Does this mean we should never use inductive reasoning?

Deductive reasoning

In Section 7.1 we conjectured that the derivative of $f(x) = x^n$ is $f'(x) = nx^{n-1}$. We confirmed that the conjecture was true for $f(x) = x^5$. We can use **deductive reasoning** to provide validity to our conjecture.

In deductive reasoning we reason from the general to the specific. In mathematics we base deductive reasoning on basic axioms, definitions and theorems.

Use the definition of derivative and the binomial theorem to show that if $f(x) = x^n$ then $f'(x) = nx^{n-1}$ for $n \in \mathbb{Z}^+$

$f'(x) = \lim_{h \to 0} \dfrac{f(x+h) - f(x)}{h}$ — Apply the definition of derivative to $f(x) = x^n$ and then use the binomial theorem to expand $(x + h)^n$

$= \lim_{h \to 0} \dfrac{(x+h)^n - x^n}{h}$

$= \lim_{h \to 0} \dfrac{\left[\binom{n}{0}x^n h^0 + \binom{n}{1}x^{n-1}h^1 + \binom{n}{2}x^{n-2}h^2 + \ldots + \binom{n}{n-1}x^1 h^{n-1} + \binom{n}{n}x^0 h^n\right] - x^n}{h}$

$= \lim_{h \to 0} \dfrac{\left[x^n + nx^{n-1}h + \binom{n}{2}x^{n-2}h^2 + \ldots + \binom{n}{n-1}xh^{n-1} + h^n\right] - x^n}{h}$ — Simplify where possible

$= \lim_{h \to 0} \dfrac{nx^{n-1}h + \binom{n}{2}x^{n-2}h^2 + \ldots + \binom{n}{n-1}xh^{n-1} + h^n}{h}$ — Collect like terms

$= \lim_{h \to 0} \dfrac{h\left[nx^{n-1} + \binom{n}{2}x^{n-2}h + \ldots + \binom{n}{n-1}xh^{n-2} + h^{n-1}\right]}{h}$ — Factorize

$= \lim_{h \to 0} \left[nx^{n-1} + \binom{n}{2}x^{n-2}h + \ldots + \binom{n}{n-1}xh^{n-2} + h^{n-1}\right]$ — Simplify

$= nx^{n-1} + \binom{n}{2}(x^{n-2})(0) + \ldots + \binom{n}{n-1}(x)(0)^{n-2} + (0)^{n-1}$ — Evaluate the limit

$f'(x) = nx^{n-1}$

- Can we now say for certain that the conjecture will be true for all $n \in \mathbb{Z}^+$? Why or why not?

A classic 'math joke'

An astronomer, a physicist and a mathematician were traveling through Wales by train, when they saw a black sheep in the middle of a field.

The astronomer said: "All Welsh sheep are black!"

The physicist disagreed: "No! *Some* Welsh sheep are black!"

While the mathematician asserted: "In Wales there is at least one field, containing at least one sheep, *at least one side of which is black!*"

- What kind of reasoning was the mathematician using?

8 Descriptive statistics

CHAPTER OBJECTIVES:

5.1 Population, sample, random sample, discrete and continuous data; presentation of data: frequency distributions (tables); frequency histograms with equal class intervals; box and whisker plots; outliers; grouped data: use of mid-interval values for calculations, interval width, boundaries and modal class.

5.2 Statistical measures. Central tendency: mean, median, mode; quartiles and percentiles. Dispersion: range, interquartile range, variance, standard deviation.

5.3 Cumulative frequency; cumulative frequency graphs.

Before you start

You should know how to:

1 Draw a bar chart.
e.g. Draw a bar chart for the number of children in the families of 30 students in the frequency table below.

Children	f
1	8
2	12
3	5
4	3
5	2

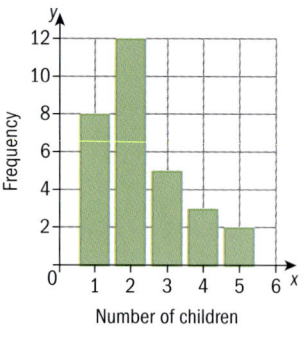

Number of children

2 Find the mean, mode and median.
e.g. Find **a** the mean **b** the mode and **c** the median of 2, 3, 3, 5, 6, 7, 9

a Mean $= \dfrac{2+3+3+5+6+7+9}{7} = \dfrac{35}{7} = 5$

b Mode = 3

c Median = 5

Skills check

1 Draw a bar chart for this frequency table.

Favorite color	f
Red	6
Blue	8
Pink	10
Purple	9
Black	4

2 a Find the mean of 4, 7, 7, 8, 6
b Find the mode of 5, 6, 8, 8, 9
c Find the median of
  i 6, 4, 8, 7, 11, 2, 4
  ii 5, 7, 9, 11, 13, 15
  iii 6, 8, 11, 11, 14, 17

Statistics are visible all around us. Averages (such as the mean, mode and median) and charts (such as bar graphs, line graphs and pie charts) are used everywhere – from business to sports to fashion to media. We use statistics every day without realizing it. You have probably made some statistical statements in your everyday conversation or thinking. Statements such as 'I sleep for about eight hours per night on average' and 'You are more likely to pass the exam if you start preparing earlier' are actually statistical in nature.

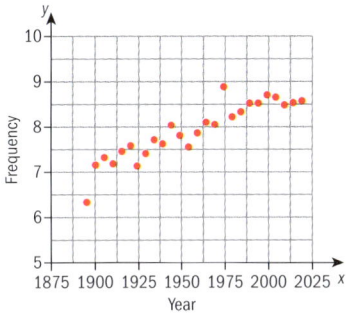

Statistics is concerned with:

- designing experiments and other data collection
- representing and analyzing information to aid understanding
- drawing conclusions from data
- estimating the present or predicting the future.

This chapter explains these techniques and how to apply them to real-life situations.

> Statistics is the science of data. It is a set of tools used to organize and analyze data.

> In this chapter you can do most of the calculations on your GDC, but if you know how to do them by hand too, it will help your understanding. The emphasis is on understanding and interpreting the results you obtain, in context. Statistical tables are not allowed in examinations – you will need to use your GDC.

Chapter 8   255

Investigation – what should we do with our test scores?

32 students took a test scored out of 10. Their results were:
0, 1, 1, 2, 2, 2, 3, 3, 4, 4, 4, 5, 5, 5, 5, 5, 6, 6, 6, 6, 7, 7, 7, 7, 7, 7, 8, 8, 8, 8, 9, 10

What should the teacher do with this data?
How could you organize the data to give a better picture of the scores?
How should you display the scores?
Should you use an average?
What should you do about converting the scores to letter grades?
Can you draw any conclusions from the scores?

8.1 Univariate analysis

Univariate analysis involves a single variable, for example, the height of all of the students in your class. You can draw charts, find averages and much more with this data. Comparing their heights and weights is called **bivariate analysis** and you will see this in Chapter 10.

→ **Data** is the information that you gather and is classified as either **qualitative** data or **quantitative** data.

Qualitative data	Quantitative data
Qualitative data is seen as categories and is sometimes called categorical data. Questions that give qualitative data include: What is your favorite pen color? How do you travel to school? What brand of computer do you own?	Quantitative data describes information that can be counted or measured. Questions that give quantitative data include: How many pens do you own? How long does it take you to get to school? How many computers have you owned?

Is the data from our test scores qualitative or quantitative?

→ Quantitative data can be split up into two categories: **discrete** and **continuous**.

→ A quantitative discrete variable has exact numerical values.

Here we are working with **values of 0, 1, 2, 3,...**, for example, the number of CDs that you have or the number of children in your family.

→ A quantitative continuous variable can be measured and its accuracy depends on the accuracy of the measuring instrument used.

Continuous variables, such as length, weight and time, may have fractions or decimals.

▲ Discrete.
How many pairs of shoes do you own?

▲ Continuous.
What is the speed of the train?

What is the difference between a population and a sample?

When we think of the term **population**, we usually think of people in our town, region, state or country.

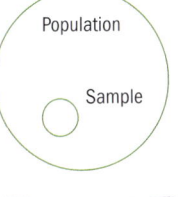

→ In statistics, the term **population** includes all members of a defined group that we are studying for data driven decisions.

→ A part of the population is called a **sample**. It is a subset of the population, a selection of individuals from the population.

Random samples must have two characteristics:
1 Every individual has an equal opportunity of selection.
2 The sample has essentially the same characteristics as the population.

Exercise 8A

1 Classify each of the following as either discrete or continuous data.
   a The number of fish caught by an angler.
   b The length of the fish.
   c The time taken to catch a fish.
   d The number of friends that the angler took with him.

2 Are the test scores at the start of the chapter discrete or continuous data?

8.2 Presenting data

A **frequency table** is an easy way to view your data quickly and look for patterns.
You can also show discrete data in a **bar chart**.

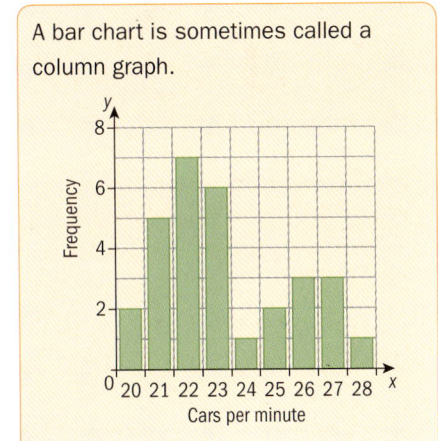

A bar chart is sometimes called a column graph.

Example 1

A student counted how many cars passed his house in one-minute intervals for 30 minutes. His results were:
23, 22, 22, 22, 24, 22, 21, 21, 23, 23, 27, 21, 21, 22, 23, 25, 27, 26, 23, 23, 22, 27, 26, 25, 28, 26, 22, 20, 21, 20.
Display this data in a frequency table.
Draw a bar chart for this data.

▶ Continued on next page

Answer

Number of cars per minute	Tally	Frequency
20	II	2
21	IIII	5
22	IIII II	7
23	IIII I	6
24	I	1
25	II	2
26	III	3
27	III	3
28	I	1

Tally each data item in the correct row. Write the totals in the frequency column.

The number 21 appears 5 times in the data.

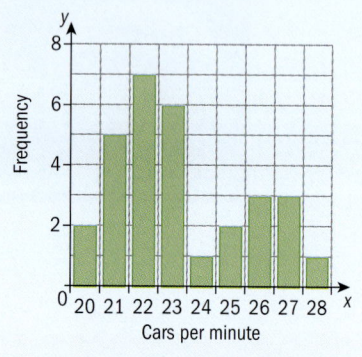

A bar chart is suitable for discrete data and may have gaps between the bars.

Use the vertical scale for the frequency and the horizontal scale for number of cars per minute

→ When you have a lot of data, you can organize it into groups in a **grouped frequency table**.
For continuous data, you can draw a **histogram**. It is similar to a bar chart but it doesn't have gaps between the bars.

Why are there no gaps in continuous data?

Example 2

The ages of 200 members of a tennis club are:
20, 22, 23, 24, 25, 25, 25, 26, 26, 26, 26, 28, 28, 29, 29, 29, 30, 30, 30, 30, 30, 30, 30, 32, 32, 33, 33, 33, 34, 34, 34, 34, 34, 34, 34, 34, 35, 35, 35, 35, 36, 36, 36, 36, 36, 37, 37, 37, 38, 38, 38, 39, 39, 39, 40, 40, 40, 41, 41, 41, 42, 42, 42, 42, 42, 42, 42, 42, 43, 43, 43, 43, 43, 43, 44, 44, 44, 44, 44, 44, 45, 45, 45, 45, 45, 45, 45, 45, 46, 46, 46, 46, 46, 46, 46, 46, 47, 47, 47, 47, 47, 47, 47, 47, 48, 48, 48, 48, 48, 48, 48, 48, 49, 49, 49, 49, 49, 49, 49, 50, 50, 50, 50, 50, 50, 51, 51, 51, 51, 51, 51, 51, 52, 52, 52, 52, 52, 53, 53, 53, 53, 53, 53, 53, 53, 54, 54, 54, 54, 55, 55, 55, 55, 55, 56, 56, 56, 57, 57, 57, 57, 57, 57, 57, 57, 57, 57, 58, 58, 58, 59, 59, 59, 60, 60, 60, 60, 60, 61, 61, 61, 62, 62, 62, 63, 63, 63, 63, 64, 64, 64, 64, 65, 65, 68, 69.
Draw a grouped frequency table and histogram for the data.

Only frequency histograms with equal class intervals will be examined.

Having one line for each age would give us a table 50 lines deep!

▶ Continued on next page

Answer

Age	Tally	Frequency
20 ≤ age < 25	IIII	4
25 ≤ age < 30	IIII IIII II	12
30 ≤ age < 35	IIII IIII IIII IIII	20
35 ≤ age < 40	IIII IIII IIII III	18
40 ≤ age < 45	IIII IIII IIII IIII IIII I	26
45 ≤ age < 50	IIII IIII IIII IIII IIII IIII IIII II	42
50 ≤ age < 55	IIII IIII IIII IIII IIII IIII I	31
55 ≤ age < 60	IIII IIII IIII IIII IIII	24
60 ≤ age < 65	IIII IIII IIII IIII	19
65 ≤ age < 70	IIII	4

Equal class intervals of 5 years

25 is in the class 25 ≤ age < 30

Numbers on the edges of the bars or an x-axis like scale.

No gaps between the bars

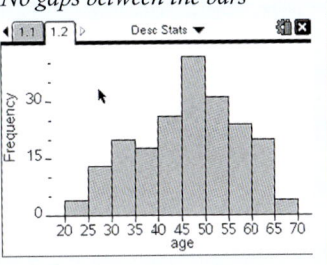

You can use a GDC to draw histograms. See Chapter 17, Section 5.4.

GDC help on CD: *Alternative demonstrations for the TI-84 Plus and Casio FX-9860GII GDCs are on the CD.*

Exercise 8B

EXAM-STYLE QUESTION

1 All of the IB students in a school were asked how many minutes a day they studied mathematics. The results are given in the table.

Time spent studying mathematics (min)	0 ≤ t < 15	15 ≤ t < 30	30 ≤ t < 45	45 ≤ t < 60	60 ≤ t < 75	75 ≤ t < 90
Number of students	21	32	35	41	27	11

a Is this data continuous or discrete?
b Use your GDC to help you draw a fully labeled histogram to represent this data.

Chapter 8

EXAM-STYLE QUESTION

2 The following table shows the age distribution of mathematics teachers who work at Caring High School.
   a Is the data discrete or continuous?
   b How many mathematics teachers work at Caring High School?
   c Use your GDC to help you draw a fully labeled histogram to represent this data.

Age	Number of teachers
$20 \leq x < 30$	5
$30 \leq x < 40$	4
$40 \leq x < 50$	3
$50 \leq x < 60$	2
$60 \leq x < 70$	3

3 The following histogram shows data about frozen chickens in a supermarket. The masses in kg are grouped such that $1 \leq w < 2$, $2 \leq w < 3$ and so on.
   a Is the mass of frozen chickens discrete or continuous data?
   b Draw the grouped frequency table for this histogram.
   c How many frozen chickens are there in the supermarket?

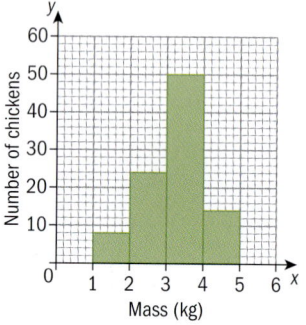

4 The histogram on the right shows how many minutes it takes for students to return home after school.
   a Is the data discrete or continuous?
   b Represent the data in a grouped frequency table.
   c What is the shortest time that a student could take to get home?

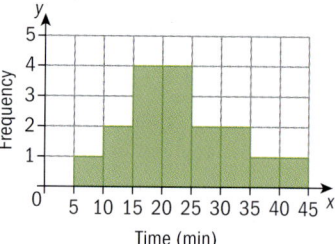

8.3 Measures of central tendency

A measure of central tendency tells us where the middle of a set of data lies. The three most common measures of central tendency are the **mode**, the **mean** and the **median**.

> Another word is 'average'.

The mode

> → The mode is the value that occurs most frequently in a set of data.

The mode in a list of numbers is the number that occurs most often. Remember that **mo**de starts with the same first two letters that **mo**st does.

> There can be more than one mode. If no number occurs more than once in the set, then there is no mode for that set of numbers.

260 Descriptive statistics

Example 3

Find the mode of: 9, 3, 9, 41, 17, 17, 44, 15, 15, 15, 27, 40, 13

Answer

The mode is 15 (15 occurs the most at 3 times).

When presented with a frequency table, the mode is the group with the biggest frequency.

Example 4

Find the mode or modal class for these frequency tables.

a
Goals	Frequency
0	4
1	7
2	3
3	3
4	1

b
Times	Frequency
$0 \leq t < 5$	1
$5 \leq t < 10$	5
$10 \leq t < 15$	6
$15 \leq t < 20$	7
$20 \leq t < 25$	6

Answers

a The mode is 1 goal.

b The modal class is $15 \leq t < 20$.

Common errors:

1 *The mode is 7.*
   No. *The biggest frequency is 7.*
2 *The mode is 3.*
   No. *The most common frequency is 3.*
   The mode from a grouped frequency table is called the modal class.

Exercise 8C

1 Find the modes of these sets of data.
   a 7, 13, 18, 24, 9, 3, 18
   b 8, 11, 9, 14, 9, 15, 18, 6, 9, 10
   c 24, 15, 18, 20, 18, 22, 24, 26, 18, 26, 24
   d −3, 4, 0, −2, 12, 0, 0, 3, 0, 5
   e 2, 7, 4, 2, 1, 9, 3.5, $\frac{1}{2}$, $\frac{3}{4}$, $\frac{1}{2}$, 11

A set of data is **bimodal** if it has two modes.

2 Find the mode of each frequency table.

a
Goals	Frequency
0	4
1	7
2	3
3	3
4	1

b
Height	Frequency
$140 \leq h < 150$	6
$150 \leq h < 160$	6
$160 \leq h < 170$	5
$170 \leq h < 180$	10
$180 \leq h < 190$	8

The mean

The arithmetic **mean** is usually called the mean or **average** and is the most common measure of central tendency.

> → The mean is the sum of the numbers divided by the number of numbers in a set of data.
>
> $$\text{Mean} = \frac{\text{Sum of the data values}}{\text{Number of data values}}$$

The mean gives us a single number that indicates a center of the data set. It is usually not a member of the data set but a representative value. For example, your average mathematics score for the year may be 85.73% even though your teacher always gives scores that are whole numbers. The lower case Greek letter μ is the symbol for the population mean.

Population mean $\mu = \dfrac{\Sigma x}{N}$

where Σx is the sum of the data values and N is the number of data values in the population.

> μ is pronounced 'mu', Σ (which tells us to find the sum here) is pronounced 'sigma' and N is 'nu'.

> There is often a misunderstanding between the population mean and sample mean. The population mean uses Greek letters whereas the sample mean uses $\bar{x}$ and n. Our course uses only the population mean.

Example 5

Find the mean of
a 89, 73, 84, 91, 87, 77, 94
b 2, 3, 3, 4, 6, 7

Answers

a $\mu = \dfrac{\Sigma x}{N} = \dfrac{89 + 73 + 84 + 91 + 87 + 77 + 94}{7}$

$= \dfrac{595}{7} = 85$

b $\mu = \dfrac{\Sigma x}{N} = \dfrac{2 + 3 + 3 + 4 + 6 + 7}{6} = \dfrac{25}{6} = 4.1\dot{6}$

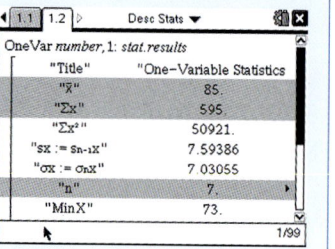

> **GDC help on CD:** Alternative demonstrations for the TI-84 Plus and Casio FX-9860GII GDCs are on the CD.

> You can calculate the mean from a list on your GDC. In one-variable statistics on the GDC, the mean is $\bar{x}$. The GDC also calculates Σx and n.

You can also calculate the mean from a frequency table.

Example 6

Find the mean of each set of data displayed below.

a
Grade (g)	Frequency
0	11
1	10
2	19
3	10

b
Age (a)	Frequency
$10 \leq t < 12$	4
$12 \leq t < 14$	8
$14 \leq t < 16$	5
$16 \leq t < 18$	3

Answers

a
Grade (g)	Frequency	fg
0	11	0
1	10	10
2	19	38
3	10	30
Total	50	78

Add a 3rd column; fg means $f \times g$.

The total of the fg column is the sum of all of the grades.
The total of the f column is the number of grades.

Mean $= \dfrac{\sum fg}{\sum f} = \dfrac{78}{50} = 1.56$

b
Age (a)	f	Mid point (m)	fm
$10 \leq t < 12$	4	11	44
$12 \leq t < 14$	8	13	104
$14 \leq t < 16$	5	15	75
$16 \leq t < 18$	3	17	51
Total	20		274

When the data are grouped, we can calculate the mean by assuming that all of the data values are equally spread around the midpoint.

Mean $= \dfrac{\sum fm}{\sum f} = \dfrac{274}{20} = 13.7$

> This is the formula as it appears in the IB Formula booklet:
> $$\mu = \dfrac{\sum_{i=1}^{n} f_i x_i}{\sum_{i=1}^{n} f_i}$$

> This method leads to small inaccuracies and that is why examination questions often say, 'estimate the mean'. It does not mean 'guess' – it means work out, as in this example or with your GDC.

Example 7

Lana's mathematics test grades are 87, 93, 89 and 85. What score must she get on the fifth test in order to get a mean of 90 for the term?

Answer

$\mu = \dfrac{\sum x}{N}$ *Select the mean formula.*

$90 = \dfrac{87 + 93 + 89 + 85 + x}{5}$ *Substitute the information into the formula.*

$450 = 354 + x$ *Solve for x.*

$x = 96$

Lana must score 96 on her fifth test.   *Answer the question.*

Exercise 8D

1. Find the mean driving speed for 6 different cars on the same road if their speeds are 66 km h^{-1}, 57 km h^{-1}, 71 km h^{-1}, 69 km h^{-1}, 58 km h^{-1} and 54 km h^{-1}.

2. The price of buying music from different sites was seen as $1.79, $1.61, $1.96 and $2.08 per track. What was the mean price?

3. A computer repair service received the following number of calls per day over a period of 30 days.

    6 5 6 9 7 4 2 4 7 8
    3 4 9 8 2 3 5 9 7 8
    9 7 5 6 7 7 4 6 2 4

    a Is the data discrete or continuous?
    b Construct a frequency table and find the mean number of calls per day.

> Ronald Fisher (1890–1962) lived in the UK and Australia and is often called the 'Father of statistics'. He used statistics to analyze practical problems in agriculture, astronomy, biology and social science. Who else could be considered as the father, or inventor of statistics.?

4. The table below shows the number of minutes of sunshine in the first 100 days of the year in Newtown.

Minutes (m)	f
$0 \leq m < 30$	12
$30 \leq m < 60$	16
$60 \leq m < 90$	20
$90 \leq m < 120$	36
$120 \leq m < 150$	16

   a Is the data discrete or continuous?
   b What is the modal class?
   c Find the mean number of minutes of sunshine.

5. Kelly's test scores are 95, 82, 76 and 88. What score must she get on the fifth test in order to achieve an average of 84 on all five tests?

> Calculation of mean using both formula and technology may be seen on exam papers.

6. The mean mass of eleven players in a sports team is 80.3 kg. A new player joins the team and the mean goes up to 81.2 kg. Find the mass of the new player.

EXAM-STYLE QUESTIONS

7. The Onceonly family must drive an average of 250 km per day to complete their vacation on time. On the first five days, they travel 220 km, 300 km, 210 km, 275 km and 240 km. How many km must they travel on the sixth day in order to finish their vacation on time?

8. The mean of Tigger's last 8 rounds of golf is 71 shots. What is the total number of shots that he took in the 8 rounds?

9 After 8 matches, a basketball player had a mean score of 27 points. After 3 more matches her average was 29. How many points did she score in the last 3 games?

10 Billy's mean sales price for 12 computers is $310 and Jean sold 13 with a mean of $320. Their boss tells them to combine their sales at the end of the week. What is the mean after Billy and Jean combine their sales?

The median

→ The median is the number in the middle when the numbers in a set of data are arranged in order of size. If the number of numbers in a data set is even, then the median is the mean of the two middle numbers.

Example 8

Find the median of
2, 13, 7, 5, 19, 23, 39, 23, 42, 23, 14, 12, 55, 23, 29.

Answer

2, 5, 7, 12, 13, 14, 19, 23, 23, 23, 23, 29, 39, 42, 55

The median value of this set of numbers is **23**.

Write the numbers in order.

*There are **15** numbers. Our middle number will be the **8**th number:*

GDC help on CD: Alternative demonstrations for the TI-84 Plus and Casio FX-9860GII GDCs are on the CD.

You can find the median on your GDC.

→ If there are a lot of numbers and it is difficult to find the middle member we can use the formula Median = $\left(\dfrac{n+1}{2}\right)$th member, where n is the number of members in the set.

Common error. This formula does not give the median. It gives the position of the median in the data set.

Exercise 8E

1 Find the median of the following.
   a 2, 3, 4, 5, 6, 7, 2, 3, 4   **b** 2, 5, 5, 2, 7, 3, 8
   c 9, 3, 4, 6, 7, 2, 3, 0
   d 8, 1, 2, 4, 5, 9, 12, 0, 4, 1.5, 8.4
   e 12, 4, 9, 1, 20, 7, 2, 5

The 19th-century German psychologist Gustav Fechner popularized the median into the formal analysis of data, although French mathematician and astronomer Pierre-Simon Laplace had used it previously.

2 Su has been counting the number of tracks on the CDs in her collection. Find the median number of tracks on Su's CDs.

Number of tracks	7	8	9	10	11	12	13
Number of CDs	3	2	2	1	3	5	3

3 Find the mode, mean and median of our test scores at the start of the chapter.

Summary of measures of central tendency

	Advantages	Disadvantages
Mode The mode can be used for qualitative data or when asked to choose the most popular item.	• Extreme values do not affect the mode.	• Does not use all members of the data set. • Not necessarily unique – may be more than one answer. • When no values repeat in the data set, there is no mode. • When there is more than one mode, it is difficult to interpret and/or compare.
Mean The mean describes the middle of a set of data.	• Most popular measure in fields such as business, engineering and computer science. • Uses all members of the data set. • It is unique – there is only one answer. • Useful when comparing sets of data.	• Affected by extreme values. In the data set of salaries €15 000, €20 000, €22 000, €17 000, €75 000 how does the extreme value of €75 000 affect the mean?
Median The median describes the middle of a set of data.	• Extreme values do not affect the median as strongly as they do the mean. • Useful when comparing sets of data. • It is unique – there is only one answer. • As the median is the middle value, 50% of the data is either side of it.	• Not as popular as mean. • Less used in further calculations.

Investigation – measures of central tendency

What will happen to the measures of central tendency if we add the same amount to all data values, or multiply each data value by the same amount? Copy and complete this table. You should use your GDC to calculate the mean, mode and median each time.

	Data	Mean	Mode	Median
Data set	6, 7, 8, 10, 12, 14, 14, 15, 16, 20			
Add 4 to each piece of data in the set.				
Multiply each piece of data in the original data set by 2.				

Now copy and complete the following sentences to explain what happens to the mean, mode and median of the original data set.
a If you add 4 to each data value………………………………………
b If you multiply each data value by 2……………………………

8.4 Measures of dispersion

Measures of central tendency (mean, median, mode) explore the middle of a data set. Measures of dispersion describe the spread of the data around a central value.

→ The **range** is the difference between the largest and smallest values.

When you describe a data set you should give at least one measure of central tendency and one of dispersion.

The range is the simplest measure of dispersion to calculate but it can be affected by extreme values. It doesn't tell you how the remaining data is distributed.

For example, for the test scores at the start of the chapter, the lowest score is 0 and the highest score is 10. Therefore the range is $10 - 0 = 10$.

Quartiles

The median of a set of data separates the data into two halves – half less than the median, half greater. **Quartiles** separate the original set of data into four equal parts. Each of these parts contains one-quarter (25%) of the data.

First quartile	The **first quartile** is the value one-quarter of the way into the data. One quarter of the data lies below the first quartile and three-fourths lies above. It is also called the 25th percentile and often has the symbol Q_1.
Second quartile	The **second quartile** is another name for the median of the entire set of data and is also called the 50th percentile.
Third quartile	The **third quartile** is three-quarters of the way in. Three-fourths of the data lies below the third quartile and one-fourth lies above. It is also called the 75th percentile and has the symbol Q_3.

$Q_1 = \frac{1}{4}(n+1)$th value and $Q_3 = \frac{3}{4}(n+1)$th value where n is the number of data values in the data set.

You can get a sense of a data set's distribution by examining a **five statistical summary**,

1 *minimum*,
2 *maximum*,
3 *median* (or *second quartile*),
4 the *first quartile*,
5 the *third quartile*.

This shows the extent to which the data is located near the median or near the extremes.

The GDC calculates these five values in One-Variable Statistics.

Here is a five statistical summary for a set of test scores.

Minimum	First quartile	Median	Third quartile	Maximum
65	70	80	90	100

You do not know every test score, but median = 80 tells you that half of the scores are below 80 and half are above 80. First quartile = 70 and third quartile = 90 tells you that the middle 50% of the scores are between 70 and 90.

> **GDC help on CD:** Alternative demonstrations for the TI-84 Plus and Casio FX-9860GII GDCs are on the CD.

> You can find the median and quartiles on a GDC. See Chapter 17, Sections 5.7 and 5.8.

→ The difference between the third and first quartiles is called the **interquartile range** (IQR) = $Q_3 - Q_1$.

The IQR is sometimes called 'the middle half'. Here the interquartile range is 20.

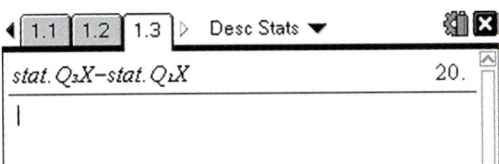

You can use the GDC to calculate the interquartile range. See Chapter 17, Section 5.9.

→ A five statistical summary can be represented graphically as a **box and whisker** plot.

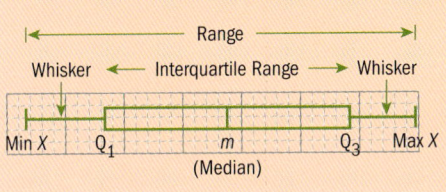

Sometimes a box and whisker plot is just called a box plot.

The diagram should be drawn to scale, for example on graph paper.

The first and third quartiles are at the ends of the box, the median is shown by a vertical line in the box, and the maximum and minimum are at the ends of the whiskers. This box and whisker plot shows the data from page 268.

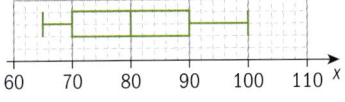
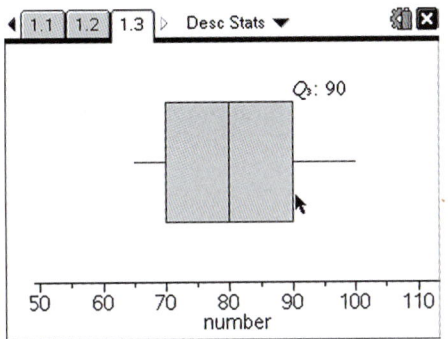

You can draw box and whisker plots on a GDC. See Chapter 17, Sections 5.5 and 5.6.

GDC help on CD: *Alternative screenshots for the TI-84 Plus and Casio FX-9860GII GDCs are on the CD.*

Extreme or distant data values are called **outliers**.

→ An outlier is any value at least 1.5 IQR above Q_3 or below Q_1.

Chapter 8

Example 9

a Find the range, the median, the lower quartile, the upper quartile and the interquartile range of this set of scores.
18, 27, 34, 52, 54, 59, 61, 68, 78, 82, 85, 87, 91, 93, 100
b Show the data in a box and whisker plot.
c Check if 18 is an outlier.

> You may wish to explore some of the misuses of statistics.

Answers

a Range = 100 − 18 = 82

18, 27, 34, 52, 54, 59, 61, 68, 78, 82, 85, 87, 91, 93, 100
Median
$= \left(\frac{n+1}{2}\right)$th $= \left(\frac{15+1}{2}\right)$th value
= 8th value = 68

$Q_1 = \frac{1}{4}(n+1)$th value

$= \frac{1}{4}(15+1)$th = 4th value = 52

$Q_3 = \frac{3}{4}(n+1)$th value

$= \frac{3}{4}(15+1)$th = 12th value = 87

IQR $= Q_3 - Q_1 = 87 - 52 = 35$

Range = largest value − smallest value
Write the data in order.

There are 15 numbers in the data set.
∴ n = 15.

b

```
         Lower        Upper
        quartile    quartile
Minimum           Median        Maximum
   ↓       ↓        ↓       ↓       ↓
   ┌───────┬────────┬───────┐
   │       │        │       │
   └───────┴────────┴───────┘
0  10 20 30 40 50 60 70 80 90 100 110
```

c $Q_1 - 1.5(\text{IQR}) = 52 - 1.5(35) = 52 - 52.5$
$= -0.5$
∴ 18 is not an outlier.

Outliers are more than $1.5 \times \text{IQR}$ below Q_1 or above Q_3.

Exercise 8F

EXAM-STYLE QUESTION

1 The depths of snow at a ski resort are collected every year for 12 years on 31 January. All data is in centimetres.
30, 75, 125, 55, 60, 75, 65, 65, 45, 120, 70, 110.
Find **a** the range, **b** the median, **c** the lower quartile, **d** the upper quartile and **e** the interquartile range of the data set and show the data in a box and whisker plot.

EXAM-STYLE QUESTIONS

2 Here are Albie's test scores for the year.
76 79 76 74 75 71 85 82 82 79 81
Find **a** the range, **b** the median, **c** the lower quartile,
d the upper quartile and **e** the interquartile range of the data set
of scores and show the data in a box and whisker plot.

> You can use a GDC to draw histograms and box and whisker plots.

3 Here are the temperatures in °C at a hill resort in Montana taken every hour for eleven hours.
10, 11, 12, 14, 18, 22, 21, 25, 27, 28, 29.
Find **a** the range, **b** the median, **c** the lower quartile, **d** the upper quartile and **e** the interquartile range of the data set.
Show the data in a box and whisker plot.

4 Use the box plot below to find **a** the range, **b** the median, **c** the lower quartile, **d** the upper quartile and **e** the interquartile range of the data.

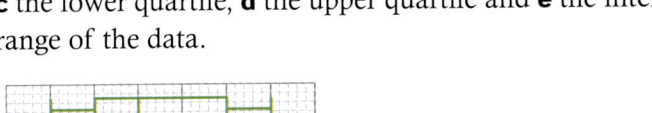

5 Match each box plot with the correct histogram.

a
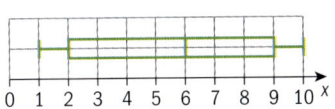

b
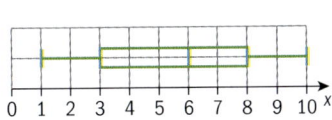

c
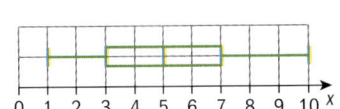

i
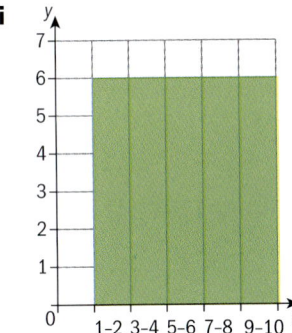

ii
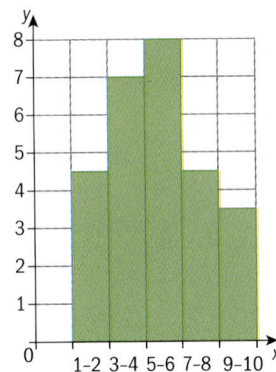

iii
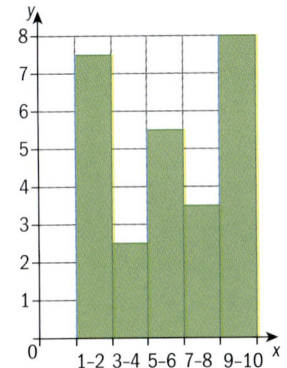

8.5 Cumulative frequency

> → To calculate the cumulative frequency add up the frequencies of the data values as you go along.

> A cumulative frequency diagram is often called a cumulative frequency graph.

A cumulative frequency diagram or **ogive** is most useful when trying to calculate the median, quartiles and percentiles of a large set of grouped or continuous data.

Chapter 8   271

Example 10

50 batteries were tested to see how long they lasted. The results (in hours) are shown in the table. Draw a cumulative frequency diagram and find the median and interquartile range.

Time (h)	f
$0 \leq h < 5$	3
$5 \leq h < 10$	5
$10 \leq h < 15$	8
$15 \leq h < 20$	10
$20 \leq h < 25$	12
$25 \leq h < 30$	7
$30 \leq h < 35$	5

Answer

Time (h)	f	Cumulative frequency
$0 \leq h < 5$	3	3
$5 \leq h < 10$	5	8
$10 \leq h < 15$	8	16
$15 \leq h < 20$	10	26
$20 \leq h < 25$	12	38
$25 \leq h < 30$	7	45
$30 \leq h < 35$	5	50

Add a 'cumulative frequency' column to the table. Work out the cumulative frequency by adding up as you go.

f	Cumulative frequency	
3	3	3 batteries lasted less than 5 hours
5	3 + 5 = 8	8 batteries lasted less than 10 hours
8	3 + 5 + 8 = 16	
10	3 + 5 + 8 + 10 = 26	
12	3 + 5 + 8 + 10 + 12 = 38	38 batteries lasted less than 25 hours
7	3 + 5 + 8 + 10 + 12 + 7 = 45	
5	3 + 5 + 8 + 10 + 12 + 7 + 5 = 50	

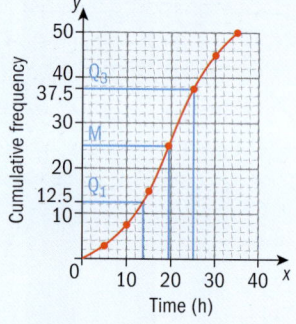

Median = 19 hours

IQR = (25 − 13) hours = 12 hours

Plot the upper limit of the time intervals against the cumulative frequencies. The first two points are (5, 3) and (10, 8)
$n = 50$
Median = $\frac{50}{2}$ = 25th data value
Draw a line across from 25 on the vertical axis to the graph then down to the time axis.
Read off Q_3 and Q_1 from the graph in the same way.
$Q_3 = 25$, $Q_1 = 13$
IQR = $Q_3 − Q_1$

> For large data sets, the median is the $\frac{n}{2}$ th value.

> Values of the median and quartiles from a GDC may be different from values read from a cumulative frequency graph.

Descriptive statistics

Exercise 8G

1 The cumulative frequency plot shows the reach in cm of 100 boxers.
   a Estimate the median reach of a boxer.
   b What is the interquartile range?
   c What does the interquartile range tell you?

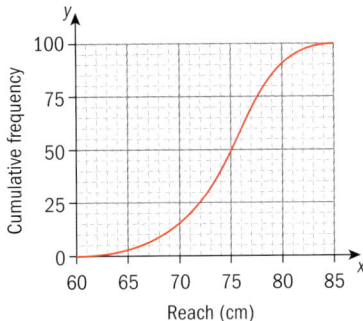

2 The table below shows the length of 40 flash drives in a computer store.
Show this data on a cumulative frequency diagram.

Length (mm)	f	Upper class boundary	Length (*l* mm)	Cumulative frequency
6–10	0	10.5	*l* ≤ 10.5	0
11–15	2	15.5	*l* ≤ 15.5	2
16–20	4	20.5	*l* ≤ 20.5	6
21–25	8	25.5	*l* ≤ 25.5	14
26–30	14	30.5	*l* ≤ 30.5	28
31–35	6	35.5	*l* ≤ 35.5	34
36–40	4	40.5	*l* ≤ 40.5	38
41–45	2	45.5	*l* ≤ 45.5	40

> Sometimes continuous data is given in groups like this. Plot the points at the upper class boundary, usually the midpoint between classes.

3 a The table below shows the cumulative frequency distribution for the times taken by 100 students to eat lunch.

Time (min)	Number of students
2 and under	0
4 and under	6
6 and under	18
8 and under	24
10 and under	40
12 and under	60
14 and under	78
16 and under	92
18 and under	100

Using a scale of 1 cm for 10 students on the vertical axis and 1 cm for 2 minutes on the horizontal axis, plot and draw a cumulative frequency diagram.
Use your graph to estimate
  i the median   **ii** the interquartile range.

b The data in **a** can be represented in the form of the table below. Find the values of p and q.

Time	$2 \leq t < 8$	$8 \leq t < 12$	$12 \leq t < 16$	$16 \leq t < 20$
Frequency	24	36	p	q

EXAM-STYLE QUESTION

4 A class of 30 IB mathematics students has the semester averages shown in the table.

Marks	Frequency
$20 \leq m < 30$	2
$30 \leq m < 40$	3
$40 \leq m < 50$	5
$50 \leq m < 60$	7
$60 \leq m < 70$	6
$70 \leq m < 80$	4
$80 \leq m < 90$	2
$90 \leq m < 100$	1

a Construct a cumulative frequency table.
b Draw a cumulative frequency diagram.
c Use your graph to estimate
  i the median
  ii the upper and lower quartiles
  iii the interquartile range.

EXAM-STYLE QUESTIONS

5 Forty students throw the javelin at the school sports day. The results are shown below.

Distance (m)	0 ≤ d < 20	20 ≤ d < 40	40 ≤ d < 60	60 ≤ d < 80	80 ≤ d < 100
Frequency	4	9	15	10	2

   a Construct a cumulative frequency table.
   b Draw a cumulative frequency diagram.
   c If the top 20% of the students are considered for the final, use your graph to estimate the qualifying distance.
   d Find the interquartile range.
   e Find the median distance thrown.

6 The graph shows the time that students listen to music during school.

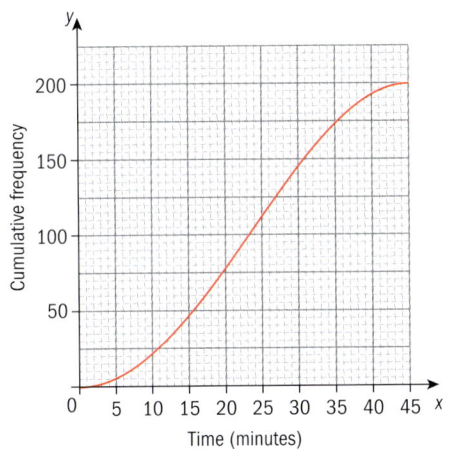

   a Estimate
      i the median time that students listen to music
      ii the interquartile range
      iii the time a student must listen to music to be in the top 10%.
   b The minimum listening time is zero and the maximum listening time is 45 minutes. Draw a box and whisker plot to represent this information.

EXAM-STYLE QUESTION

7 The cumulative frequency diagram below shows the heights of 220 sunflowers.

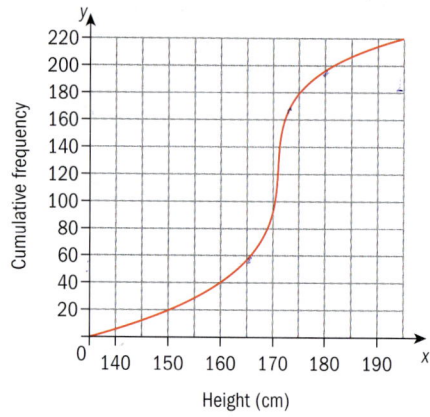

a Find the median height of a sunflower.
b The smallest 25% are sent to home garden shops. How many go to those garden shops? Between what heights are they?
c The tallest 10% go to hotel displays. How many go to the hotels? What is the smallest sunflower that goes to a hotel display?
d The middle half of the sunflowers are sold immediately. How many is this?
e The height of the tallest sunflower is 195 cm and the height of the shortest is 136 cm. Draw a box and whisker plot to represent the heights of the sunflowers.

> You may wish to explore different visual representations of statistics.

> **Extension material on CD:**
> *Worksheet 8 - Measures of central tendency and spread*

8.6 Variance and standard deviation

The range and the interquartile range are good measures of spread but each one is calculated from only two data values.

> → The **variance** combines all the values in a data set to produce a measure of spread. It is the arithmetic mean of the squared differences between each value and the mean value.

Squaring the difference between each value and the mean value has at least three advantages:

1 Squaring makes each term positive so that values above the mean do not cancel values below the mean.
2 Squaring adds more weighting to the larger differences. In many cases this extra weighting is appropriate since points further from the mean may be more significant.
3 The mathematics is relatively manageable when using this measure in subsequent statistical calculations.

Because the differences are squared, the units of variance are not the same as the units of the data.

> You should use a GDC to calculate the population standard deviation and variance.

→ The **standard deviation** is the square root of the variance and has the same units as the data.

→ The formulae for the variance and standard deviation are:

$$\sigma^2 = \text{Population variance} = \frac{\sum_{i=1}^{n}(x-\mu)^2}{n}$$

$$\sigma = \text{Population standard deviation} = \sqrt{\frac{\sum_{i=1}^{n}(x-\mu)^2}{n}}$$

Example 11

Thirty farmers were asked how many farm workers they hire during a typical harvest season. Their responses were:

4, 5, 6, 5, 3, 2, 8, 0, 4, 6, 7, 8, 4, 5, 7, 9, 8, 6, 7, 5, 5, 4, 2, 1, 9, 3, 3, 4, 6, 4

Calculate the mean and standard deviation for this data.

Answer
Solution – 'by hand'

Workers (x)	Frequency (f)	(fx)	(x−μ)	(x−μ)²	f(x−μ)²
0	1	0	−5	25	25
1	1	1	−4	16	16
2	2	4	−3	9	18
3	3	9	−2	4	12
4	6	24	−1	1	6
5	5	25	0	0	0
6	4	24	1	1	4
7	3	21	2	4	12
8	3	24	3	9	27
9	2	18	4	16	32
	30	150			152

The IB syllabus covers 'Calculation of standard deviation/variance using only technology'. This is how you would calculate the standard deviation for a discrete variable 'by hand'.

1. *Calculate the mean.*
2. *Subtract the mean from each observation.*
3. *Square each of the results from step 2.*
4. *Add these squared results together.*
5. *Divide this total by the number of observations. This gives the variance, σ^2.*
6. *Take the positive square root to get the standard deviation, σ.*

To calculate the mean:
$$\mu = \frac{150}{30} = 5$$

$$\mu = \frac{\sum fx}{n}$$

To calculate the standard deviation:
$$\sigma = \sqrt{\frac{152}{30}} = 2.25$$

$$\sigma = \sqrt{\frac{\sum f(x-\mu)^2}{n}}$$

▶ Continued on next page

Solution using GDC

Enter the data in lists called workers and freq. Add a new calculator page to your document.

Press `tab` 6:Statistics | 1:Stat Calculations | 1:One-Var Statistics…

Press `enter`.

This opens a dialogue box.

Leave the number of lists as 1 and press `enter`.

This opens another dialogue box. Choose *number* from the drop down box for X1 List and *freq* from the drop down box for the Frequency list. Press `enter`.

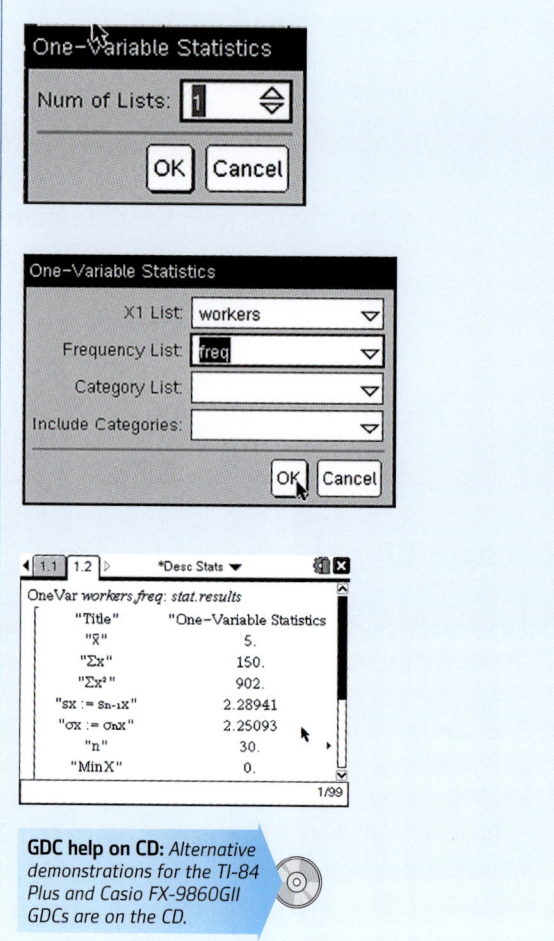

The information shown will not fit onto a single screen. You can scroll up and down to see it all. The standard deviation is the value shown as '$\sigma x: \sigma_n x$' (the population standard deviation). $\sigma = 2.25$ (3 sf)

> You should always use the value σx in this course, never use the value sx.

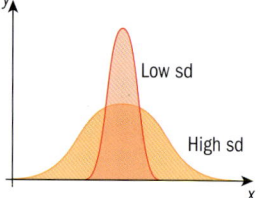

GDC help on CD: *Alternative demonstrations for the TI-84 Plus and Casio FX-9860GII GDCs are on the CD.*

The standard deviation shows how much variation there is from the mean and gives an idea of the shape of the distribution.

- Low standard deviation shows that the data points tend to be very close to the mean.
- High standard deviation indicates that the data is spread out over a large range of values.

Properties of standard deviation

- Standard deviation is only used to measure spread or dispersion around the mean of a data set.
- Standard deviation is never negative.
- Standard deviation is sensitive to outliers. A single outlier can increase the standard deviation and in turn, distort the representation of spread.
- For data with approximately the same mean, the greater the spread, the greater the standard deviation.
- If all values of a data set are the same, the standard deviation is zero because each value is equal to the mean.

> The standard deviation is used widely to describe data in business, science, sports and medicine.

278 Descriptive statistics

Exercise 8H

Use your GDC for this exercise.

1. Find the mean, variance and standard deviation for the following sets of numbers.
   - **a** 7, 9, 12, 25, 37
   - **b** 20, 30, 40, 50, 60

2. Find the variance and standard deviation for the following sets of numbers.
   - **a** 27, 44, 32, 49
   - **b** 19, 28, 30, 44
   - **c** 35, 65, 84, 27, 66

3. The table illustrates the shoe sizes of 73 students in a ballet class. Find the standard deviation of their shoe sizes.

Size	4	5	6	7	8
f	9	14	22	11	17

EXAM-STYLE QUESTIONS

4. The number of children in the families in a class of 29 children is shown below. Find the mean and standard deviation.

Children	1	2	3	4	5	6	7
f	5	12	8	3	0	0	1

5. The table below shows the number of words that the students studying Spanish could remember in a year group. Find the standard deviation.

Words	f
5–9	9
10–14	11
15–19	10
20–24	20
25–29	10
30–34	12
35–39	6
40–44	3
45–49	1
50–54	1
55–59	2
60–64	3
65–69	0
70–74	1
75–79	1

> Pafnuty Lvovich Chebyshev (1821–94) was a Russian mathematician. Chebyshev's theorem shows how the value of a standard deviation can be applied to any data set. Several statistical advances were made in Russia and France in the 19th century. You may wish to research some more.

6 A survey was conducted of the number of bedrooms in 208 randomly chosen houses. The results are shown in the table.

Number of bedrooms	1	2	3	4	5	6
Number of houses	41	60	52	32	15	8

a State whether the data is discrete or continuous.
b Write down the mean number of bedrooms per house.
c Write down the standard deviation of the number of bedrooms per house.
d Find how many houses have a number of bedrooms greater than one standard deviation above the mean.

EXAM-STYLE QUESTIONS

7 A random sample of 167 people who own mobile phones was used to collect data on the amount of time they spent per day using their phones. The results are displayed in the table.

Time spent per day (t minutes)	$0 \leq t < 15$	$15 \leq t < 30$	$30 \leq t < 45$	$45 \leq t < 60$	$60 \leq t < 75$	$75 \leq t < 90$
Number of people	21	32	35	41	27	11

Use your graphic display calculator to calculate approximate values of the mean and standard deviation of the time spent per day on these mobile phones.

8 The figure below shows the lengths in centimetres of fish found in the net of a small trawler.

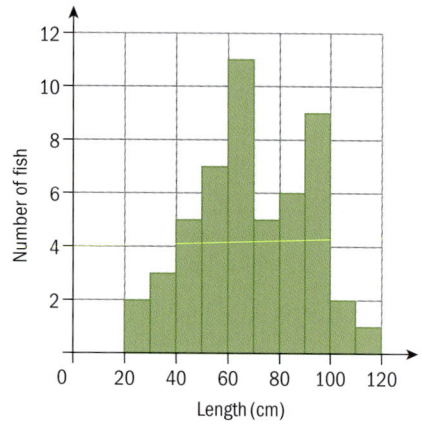

a Find the total number of fish in the net.
b Write down an estimate of the mean length.
c **i** Write down an estimate for the standard deviation of the lengths.
   ii How many fish (if any) have length **greater than** three standard deviations **above** the mean?

> **Extension material on CD:**
> Worksheet 8 – Measures of central tendency and spread

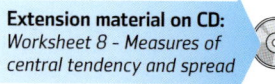

Investigation – the effect of adding or multiplying the data set on a standard deviation

Here is a set of numbers: 4, 2, 0, 9, 3, 5, 5, 1, 4, 6
a Calculate the standard deviation of these numbers.
b Now add 100 to all the numbers in the series: to get 104, 102, 100, 109, 103, 105, 105, 101, 104, 106.
   What happens to the mean?
c Calculate the standard deviation of this new set.
d Explain what you notice and why this happens.
e Now multiply all of the values in the original list by 2 to get 8, 4, 0, 18, 6, 10, 10, 2, 8, 12.
   What happens to the mean?
f Calculate the standard deviation.
g What will happen to the variance? Why?

→ **Effect of constant changes to the original data:**
If you **add/subtract** a constant value k to/from all the numbers in a list, the arithmetic mean increases/decreases by k but the standard deviation **remains the same**.

If you **multiply/divide** all the numbers in the list by a constant value k, both the arithmetic mean and the standard deviation are **multiplied/divided by k**.

You may be expected to use these rules in your exam. See question 3 in the non-GDC review exercise.

Review exercise

1 Find **a** the mode, **b** the median, **c** the mean and **d** the range of 1, 7, 8, 2, 3, 6, 5, 10, 3

2 A class collected the data on the number of pets in their home as shown in the table below.

Pets	2	3	4	5	6	7	8	9	10
f	3	9	10	2	3	1	1	0	1

   a Calculate the mean number of pets.
   b Calculate the median.
   c Write down the mode.

EXAM-STYLE QUESTION
3 The mean age of a group of friends at the end of school is 17.5 years and the standard deviation is 0.4 years. They all meet again at the school reunion after 10 years. What is the new mean and standard deviation of their ages?

Chapter 8

EXAM-STYLE QUESTIONS

4 A farmer grows two different types of sweetcorn and the season's results are shown below.

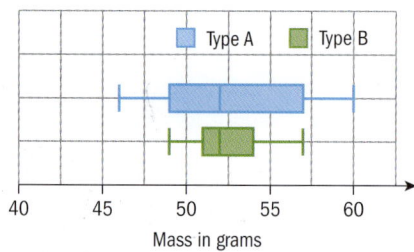

Find **a** the median, **b** the range, **c** the interquartile range for each type.

5 The mean of six numbers is 71. One number is 46, another is 92 and the other four numbers are all the same.

> You could be asked to calculate the mean using the formula or a GDC. You will only be expected to calculate the standard deviation or variance using a GDC.

  a Find the total of all six numbers.
  b Find the value of one of the four missing numbers.
  c If each of the six numbers is decreased by 9 find the mean of the new set of numbers.

6 a Draw a cumulative frequency graph for the data in the table.

Height (cm)	$150 \leq h < 155$	$155 \leq h < 160$	$160 \leq h < 165$	$165 \leq h < 170$	$170 \leq h < 175$
f	4	22	56	32	5

> 'Estimate from your graph' means that you should show the horizontal and vertical lines as working on your graph.

  b Estimate the median from your graph.
  c Estimate the interquartile range from your graph.

7 A dice is rolled 100 times. Each dice has a number between one and six written on it.
The following table shows the frequencies for each number.

Number	1	2	3	4	5	6
Frequency	26	10	20	k	29	11

  a Calculate the value of k.
  b Find   **i** the median   **ii** the interquartile range.

8 The table gives the midday temperature (°F) in the Omani mountains in November. Find the median and IQR.

Temperature	f
$12.5 \leq t < 27.5$	6
$27.5 \leq t < 42.5$	3
$42.5 \leq t < 57.5$	5
$57.5 \leq t < 72.5$	8
$72.5 \leq t < 87.5$	6
$87.5 \leq t < 102.5$	2

Review exercise

1. Calculate the median and interquartile range of
   9, 11, 12, 13, 13, 17, 19, 21, 21, 25, 27, 30, 33, 35

EXAM-STYLE QUESTION

2. June runs a cats' home. The numbers of kittens per litter for last year were

Kittens	4	5	6	7	8	9
f	3	7	11	12	6	3

   a Find the mean number of kittens per litter.
   b Find the standard deviation.

3. The numbers of tennis racquets broken by 410 players in a season were.

Broken racquets	2	3	4	5	6	7	8	9	10
f	3	11	43	90	172	13	64	10	4

   Find   a the mode   b the median   c the mean.

EXAM-STYLE QUESTION

4. The number of hours students study mathematics each night is given in the table.

Hours	0	1	2	3	4	5	6
f	2	5	4	3	4	2	1

   a Find the mean, median, mode, standard deviation and variance.
   b Find the range, lower quartile and the interquartile range.

5. The histogram below shows the heights of the students in a high school in Peru.

   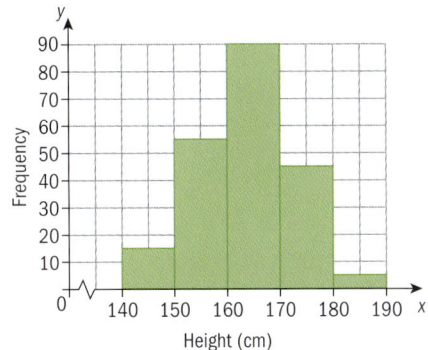

   a Write down the modal class height.
   b Construct a grouped frequency table and calculate an estimate for the mean height of the Peruvian students.

EXAM-STYLE QUESTION

6 A school with 150 students is tested to see how many French words they can remember in one minute. The results are given in the table below.

Number of words	Number of students	Cumulative number of students
15	11	11
16	21	32
17	33	p
18	q	99
19	38	137
20	13	150

a i Write down the value of p.   ii Find the value of q.
b Find the median number of words.
c Find the mean number of words.

CHAPTER 8 SUMMARY
Univariate analysis

- Univariate analysis involves a single variable.
- **Data** is the information that you gather and is classified as either **qualitative** data or **quantitative** data.
- Quantitative data can be split up into two categories: **discrete** and **continuous**.
- A quantitative discrete variable has exact numerical values.
- A quantitative continuous variable can be measured and its accuracy depends on the accuracy of the measuring instrument used.
- Continuous variables, such as length, weight and time, may have fractions or decimals.
- In statistics, the term **population** includes all members of a defined group that we are studying for data driven decisions.
- A part of the population is called a **sample**. It is a subset of the population, a selection of individuals from the population.

Presenting data

- When you have a lot of data, you can organize it into groups in a **grouped frequency table**.
- For continuous data, you can draw a **histogram**. It is similar to a bar chart but it doesn't have gaps between the bars.

Continued on next page

Measures of central tendency

- The **mode** is the value that occurs most frequently in a set of data.
- The **mean** is the sum of the numbers divided by the number of numbers in a set of data.
$$\text{Mean} = \frac{\text{Sum of the data values}}{\text{Number of data values}}$$
- The **median** is the number present in the middle when the numbers in a set of data are arranged in order of size. If the number of numbers in a data set is even, then the median is the mean of the two middle numbers.
- If there are a lot of numbers and it is difficult to find the middle member we can use the formula Median = $\left(\frac{n+1}{2}\right)$th member, where n is the number of members in the set.

	Advantages	Disadvantages
Mode The mode can be used for qualitative data or when asked to choose the most popular item.	• Extreme values do not affect the mode.	• Does not use all members of the data set. • Not necessarily unique – may be more than one answer. • When no values repeat in the data set, there is no mode. • When there is more than one mode, it is difficult to interpret and/or compare.
Mean The mean describes the middle of a set of data.	• Most popular measure in fields such as business, engineering and computer science. • Uses all members of the data set. • It is unique – there is only one answer. • Useful when comparing sets of data.	• Affected by extreme values.
Median The median describes the middle of a set of data.	• Extreme values do not affect the median as strongly as they do the mean. • Useful when comparing sets of data. • It is unique – there is only one answer. • As the median is the middle value, 50% of the data is either side of it.	• Not as popular as mean. • Less used in further calculations.

Continued on next page

Measures of dispersion

- The **range** is the difference between the largest and smallest values.

First quartile	The **first quartile** is the value one-quarter of the way into the data. One quarter of the data lies below the first quartile and three-fourths lies above. It is also called the 25th percentile and often has the symbol Q_1.
Second quartile	The **second quartile** is another name for the median of the entire set of data and is also called the 50th percentile.
Third quartile	The **third quartile** is three-quarters of the way in. Three-fourths of the data lies below the third quartile and one-fourth lies above. It is also called the 75th percentile and has the symbol Q_3.

$Q_1 = \frac{1}{4}(n+1)$th value and $Q_3 = \frac{3}{4}(n+1)$th value where n is the number of data values in the data set.

- The difference between the third and first quartiles is called the **interquartile range** (IQR).
- A five statistical summary can be represented graphically as a **box and whisker** plot.

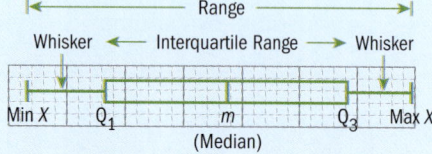

- An outlier is any value at least 1.5 IQR above Q_3 or below Q_1.

Cumulative frequency

- To calculate the cumulative frequency add up the frequencies of the data values as you go along.

Variance and standard deviation

- The **variance** combines all the values in a data set to produce a measure of spread. It is the arithmetic mean of the squared differences between each value and the mean value.

Continued on next page

- The **standard deviation** is the square root of the variance and has the same units as the data.
- The formulae for the variance and standard deviation are:

$$\sigma^2 = \text{Population variance} = \frac{\sum_{i=1}^{n}(x-\mu)^2}{n}$$

$$\sigma = \text{Population standard deviation} = \sqrt{\frac{\sum_{i=1}^{n}(x-\mu)^2}{n}}$$

Effect of constant changes to the original data:

If you **add/subtract** a constant value k to/from all the numbers in a list, the arithmetic mean increases/decreases by k but the standard deviation **remains the same**.

If you **multiply/divide** all the numbers in the list by a constant value k, both the arithmetic mean and the standard deviation are **multiplied/divided by** k.

Theory of knowledge

Facts and misconceptions in statistics

Statistics is a relatively modern branch of mathematics as its main advances have been made in the past 400 years.

- Find out how **Florence Nightingale** used statistics and what it led to.
- What did **Francis Galton** invent?

- Is it easy to mix up μ and $\bar{x}$?
- What is the difference between a sample and a population?
- Do different measures of central tendency (mean, median, mode) express different properties of the data?
- Were measures of central tendency invented or discovered? Where do they come from?
- Could mathematics make alternative, equally true formulae?
- What does this tell us about mathematical truths?

Darrell Huff's book *How to Lie with Statistics* (Norton, 1954) attempted to expose the tricks of the statistical spin-doctors for the 'self-defense' of 'honest men'.

"Statistical thinking will one day be as necessary for efficient citizenship as the ability to read and write."

H. G. Wells (1866–1946)

- What do you think H. G. Wells meant?
- Do you agree with him?

How easy is it to lie with statistics?

- Criticize these graphs:

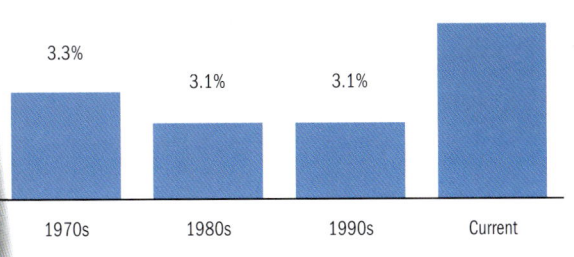

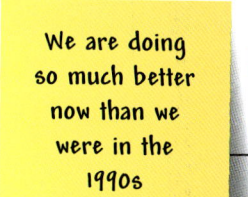

We are doing so much better now than we were in the 1990s

What a huge increase in the number of frogs

"There are three kinds of lies: lies, damned lies, and statistics."

US author Mark Twain attributed this to the 19th century British Prime Minister, Benjamin Disraeli

Take a survey of your friends about their favorite subject.
- Use Microsoft Excel to produce different style charts to show the situation (or draw graphs by hand).
- Try changing the *y*-axis scale or the value the *y*-axis starts at.
- Show 3D charts.
- See what happens to a subject that may have zero votes on a pie chart.

Here are some of the 'tricks' that you could use, and how they mislead:
- Showing an unsuitably large or small amount of data. This hides or highlights the change being reported.
- Using a nonlinear scale. Anyone expecting a linear scale would be misled.
- Not showing the scale at all. Keep them uninformed.
- Making bars of a histogram three-dimensional. It makes the difference between data values look larger.

Statistics can be very helpful in providing an influential interpretation of reality but also can be used to distort our perceptions.
- How can statistics be used or misused to assist and mislead us?
- How can we decide whether to accept the statistical evidence that is presented to us?

9 Integration

CHAPTER OBJECTIVES:

6.4 Indefinite integration as antidifferentiation. Indefinite integral of x^n ($n \in \mathbb{Q}$), $\frac{1}{x}$ and e^x. The composites of any of these with the linear function $ax + b$.

Integration by inspection, or substitution of the form $\int f(g(x))g'(x)\,dx$.

6.5 Antidifferentiation with a boundary condition to determine the constant term. Definite integrals, both analytically and using technology. Areas under curves (between the curve and the x-axis). Areas between curves. Volumes of revolution about the x-axis.

6.6 Kinematic problems involving displacement s, velocity v and acceleration a. Total distance traveled.

Before you start

You should know how to:

1 Write a series given in sigma notation as a sum of terms. e.g.
$$\sum_{i=2}^{4}(2i+1) = [2(2)+1]+[2(3)+1]+[2(4)+1]$$
$$= 5+7+9$$
e.g. $\sum_{j=1}^{4} f(x_j) = f(x_1)+f(x_2)+f(x_3)+f(x_4)$

2 Use geometric formulae to find area.
e.g. Area of trapezium:
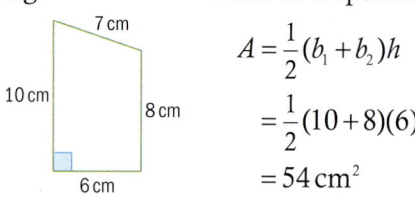
$$A = \frac{1}{2}(b_1+b_2)h$$
$$= \frac{1}{2}(10+8)(6)$$
$$= 54\,cm^2$$

3 Use geometric formulae to find volume.
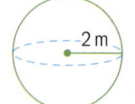
e.g. Volume of sphere:
$$V = \frac{4}{3}\pi r^3 = \frac{4}{3}\pi(2)^3 = \frac{32\pi}{3}\,m^3$$

Skills check

1 Write as a sum of terms.

   a $\sum_{i=1}^{5}(2i^2)$ b $\sum_{k=2}^{6}(3k-2)$

   c $\sum_{i=1}^{5}[(i)^2 g(x_i)]$ d $\sum_{j=1}^{3}[f(x_j)(\Delta x_j)]$

2 Find the area.

   a 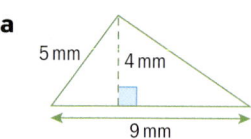   b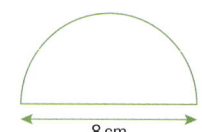

3 Find the volume.

   a 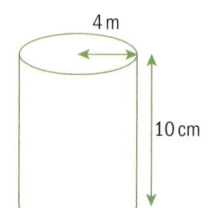   b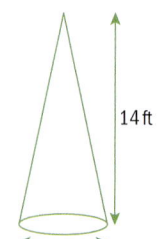

We know we can find the velocity of a moving object by taking the derivative of the displacement function (carrying out differentiation on the displacement function). Now consider the reverse process. Can you find the displacement function for a moving object, if you know the velocity function?

Suppose the velocity function is given by $v(t) = 2t + 1$. We need to find a function, $s(t)$ such that $s'(t) = 2t + 1$. Working backwards, we find one possible displacement function is $s(t) = t^2 + t$, since $\frac{d}{dt}(t^2 + t) = 2t + 1$. Why do we say that $s(t) = t^2 + t$ is **one** possible displacement function?

The function $s(t) = t^2 + t$ is called an **antiderivative** of $v(t) = 2t + 1$. The process of finding an antiderivative is called **integration**. In this chapter you will learn about the process of integration and how integration can be used to solve problems involving motion on a line, area and volume.

9.1 Antiderivatives and the indefinite integral

Suppose the derivative of a function f is given by $2x + 3$. Working backwards, we find that f may be the function $f(x) = x^2 + 3x$, since $\frac{d}{dx}(x^2 + 3x) = 2x + 3$. But there are other functions that have the same derivative, such as $f(x) = x^2 + 3x + 1$ or $f(x) = x^2 + 3x - 6$ since $\frac{d}{dx}(x^2 + 3x + 1) = 2x + 3$ and $\frac{d}{dx}(x^2 + 3x - 6) = 2x + 3$.

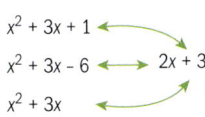

The functions $f(x) = x^2 + 3x$, $f(x) = x^2 + 3x + 1$ and $f(x) = x^2 + 3x - 6$ are all called **antiderivatives** of $2x + 3$.

Any function of the form $f(x) = x^2 + 3x + C$, where C is an arbitrary constant, is an antiderivative of $2x + 3$.

> A function F is called an **antiderivative** of f if $F'(x) = f(x)$.

Investigation – antiderivatives of x^n

1 Copy and complete the table below. The first entry is completed for you.

$f(x)$	Antiderivative of f
x	$\frac{1}{2}x^2 + C$
x^2	
x^3	
x^4	

2 Write a general expression or rule for the antiderivatives of x^n.
3 Show whether your rule gives the correct antiderivatives for x^{-3} and $x^{\frac{1}{2}}$.
4 Are there any values of n where your rule does not apply?

The antiderivatives of x^n are given by $\dfrac{1}{n+1}x^{n+1} + C$,

where C is an arbitrary constant and $n \neq -1$.

> Just as the process of finding a derivative is called **differentiation**, the process of finding an antiderivative is called **antidifferentiation**.

Example 1

Find the antiderivative of each function.

  a x^{10} **b** $\dfrac{1}{x^5}$ **c** $\sqrt[4]{x^3}$

Answers

a $\dfrac{1}{10+1}x^{10+1} + C = \dfrac{1}{11}x^{11} + C$ *Apply the rule $\dfrac{1}{n+1}x^{n+1} + C$, where $n = 10$.*

b $\dfrac{1}{x^5} = x^{-5}$ *Rewrite using rational exponents.*

$\dfrac{1}{-5+1}x^{-5+1} + C = -\dfrac{1}{4}x^{-4} + C$ *Apply the rule $\dfrac{1}{n+1}x^{n+1} + C$, where $n = -5$.*

$= -\dfrac{1}{4x^4} + C$ *Simplify.*

c $\sqrt[4]{x^3} = x^{\frac{3}{4}}$ *Rewrite using rational exponents.*

$\left(\dfrac{1}{\frac{3}{4}+1}\right)x^{\frac{3}{4}+1} + C = \left(\dfrac{1}{\frac{7}{4}}\right)x^{\frac{7}{4}} + C$ *Apply the rule $\dfrac{1}{n+1}x^{n+1} + C$, where $n = \dfrac{3}{4}$.*

$= \dfrac{4}{7}x^{\frac{7}{4}} + C$ *Simplify.*

> **Remember**
> $\sqrt{x} = x^{\frac{1}{2}}$
> $\sqrt[3]{x} = x^{\frac{1}{3}}$
> $\sqrt[4]{x} = x^{\frac{1}{4}}$ etc.

Exercise 9A

Find the antiderivative of each function.

1. x^7
2. x^4
3. x^{-2}
4. $x^{-\frac{1}{2}}$
5. $x^{\frac{1}{3}}$
6. $x^{\frac{2}{5}}$
7. $\dfrac{1}{x^4}$
8. $\dfrac{1}{x^{12}}$
9. $\sqrt[3]{x}$
10. $\sqrt[7]{x^3}$
11. $\dfrac{1}{\sqrt[5]{x}}$
12. $\dfrac{1}{\sqrt[3]{x^2}}$

Antidifferentiation is also known as **indefinite integration** and is denoted with an integral symbol, $\int \, dx$. For example,
$$\int x^3 \, dx = \frac{1}{4}x^4 + C$$
means that the indefinite integral (or antiderivative) of x^3 is $\frac{1}{4}x^4 + C$.

These rules will help you find indefinite integrals.

> **Power rule**
> $$\int x^n \, dx = \frac{1}{n+1} x^{n+1} + C, \; n \neq 1$$
>
> **Constant rule**
> $$\int k \, dx = kx + C$$
>
> **Constant multiple rule**
> $$\int k f(x) \, dx = k \int f(x) \, dx$$
>
> **Sum or difference rule**
> $$\int (f(x) \pm g(x)) \, dx = \int f(x) \, dx \pm \int g(x) \, dx$$

If $F'(x) = f(x)$, we write
$$\int f(x) \, dx = F(x) + C.$$
The expression $\int f(x) \, dx$ is called an **indefinite integral**.

$\int f(x) \, dx$ is read as 'the antiderivative of f with respect to x' or 'the integral of f with respect to x'.

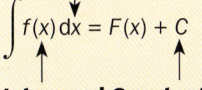

Variable of integration ↓
$$\int f(x) \, dx = F(x) + C$$
↑ ↑
Integrand Constant of integration

Example 2

Find the indefinite integral.

a $\int x^6 \, dx$ 　　b $\int 4 \, dt$ 　　c $\int 3x^5 \, dx$

d $\int (3u^4 + 6u^2 + 2) \, du$ 　　e $\int (x + \sqrt[3]{x}) \, dx$

Answers

a $\int x^6 \, dx = \dfrac{1}{6+1} x^{6+1} + C$

　　　　$= \dfrac{1}{7} x^7 + C$ 　　　　*Apply the power rule with $n = 6$.*

b $\int 4 \, dt = 4t + C$ 　　　　*Apply the constant rule. The dt tells you that the variable of integration is t.*

▶ Continued on next page

c $\int 3x^5 \, dx = 3\int x^5 \, dx$ — Apply the constant multiple rule.

$= 3\left(\dfrac{1}{5+1}x^{5+1} + C_1\right)$ — Apply the power rule with $n = 5$.

$= \dfrac{1}{2}x^6 + 3C_1$ — $3C_1$ is some arbitrary constant C. We usually just show the final arbitrary constant.

$= \dfrac{1}{2}x^6 + C$

d $\int (3u^4 + 6u^2 + 2) \, du$ — Apply the sum rule.

$= \int 3u^4 \, du + \int 6u^2 \, du + \int 2 \, du$

$= 3\int u^4 \, du + 6\int u^2 \, du + \int 2 \, du$ — Apply the constant multiple rule. Apply the power rule and constant rule, with variable of integration u.

$= 3\left(\dfrac{1}{4+1}u^{4+1}\right) + 6\left(\dfrac{1}{2+1}u^{2+1}\right) + 2u + C$

$= \dfrac{3}{5}u^5 + 2u^3 + 2u + C$ — We actually get a constant of integration for each term, but $C_1 + C_2 + C_3$ is some arbitrary constant C.

e $\int (x + \sqrt[3]{x}) \, dx = \int (x^1 + x^{\frac{1}{3}}) \, dx$ — Rewrite using rational exponents.

$= \dfrac{1}{1+1}x^{1+1} + \dfrac{1}{\frac{1}{3}+1}x^{\frac{1}{3}+1} + C$ — Apply the power rule to each term.

$= \dfrac{1}{2}x^2 + \dfrac{3}{4}x^{\frac{4}{3}} + C$

Exercise 9B

Find the indefinite integral in questions 1 to 10.

1 $\int x^3 \, dx$

2 $\int \dfrac{1}{t^2} \, dt$

3 $\int \sqrt[5]{x^4} \, dx$

4 $\int 2 \, du$

5 $\int (3x^2 + 2x + 1) \, dx$

6 $\int \dfrac{4}{x^3} \, dx$

7 $\int (t^2 + \sqrt[4]{t}) \, dt$

8 $\int (\sqrt[3]{x^2} + 1) \, dx$

9 $\int (5x^4 + 12x^3 + 6x - 2) \, dx$

10 $\int dt$

You can check the answers to your indefinite integrals by differentiating your answer and checking to see that it equals the integrand.

$\int dt = \int 1 \times dt = \int t^0 \, dt$

11 Let $f(x) = x^3 + \dfrac{4}{x^2}$.

Find **a** $f'(x)$ **b** $\int f(x) \, dx$

12 Let $g(x) = 30\sqrt[5]{x}$.

Find **a** $g'(x)$ **b** $\int g(x) \, dx$

You saw at the beginning of this section that if the velocity of a moving object is given by $v(t) = 2t + 1$, then the displacement of the particle is $s(t) = t^2 + t + C$, for some arbitrary constant C. You can now write this as $\int (2t + 1)\,dt = t^2 + t + C$, where $t^2 + t + C$ is called the **general solution** for $\int (2t + 1)\,dt$.

Suppose you are also given that, for this particle, the position at time $t = 1$ is 6. You can then find C.

$s(t) = t^2 + t + C$
$s(1) = 1^2 + 1 + C$
$6 = 2 + C$
$C = 4$

Therefore, $s(t) = t^2 + t + 4$. The fact that the position at time $t = 1$ is 6 is called a **boundary condition** and $t^2 + t + 4$ is a **particular solution** of $\int (2t + 1)\,dt$, given the boundary condition.

Example 3

a If $f'(x) = 3x^2 + 2x$ and $f(2) = -3$, find $f(x)$.

b The curve $y = f(x)$ passes through the point $(32, 30)$. The gradient of the curve is given by $f'(x) = \dfrac{1}{\sqrt[5]{x^3}}$.
Find the equation of the curve.

c The rate of growth of a population of fish is given by $\dfrac{dP}{dt} = 150\sqrt{t}$ for $0 \le t \le 5$ years. The initial population was 200 fish. Find the number of fish at $t = 4$ years.

> Sometimes a boundary condition is given as an **initial condition**. This represents a condition when t is zero. For example, if you are told that the initial displacement is 4, this means that displacement is 4 when $t = 0$.

Answers

a $f'(x) = 3x^2 + 2x$
$f(x) = \int (3x^2 + 2x)\,dx$
$f(x) = x^3 + x^2 + C$
$f(2) = 2^3 + 2^2 + C$
$-3 = 8 + 4 + C$
$C = -15$
$\therefore f(x) = x^3 + x^2 - 15$

Apply the power rule to find the general solution for $\int (3x^2 + 2x)\,dx$.
Use the fact that $f(2) = -3$ to find C.

b $f'(x) = \dfrac{1}{\sqrt[5]{x^3}}$
$f(x) = \int \dfrac{1}{\sqrt[5]{x^3}}\,dx$
$= \int x^{-\frac{3}{5}}\,dx$
$f(x) = \dfrac{5}{2}x^{\frac{2}{5}} + C$

Rewrite with rational exponents and apply the power rule to find the general solution for $\int \dfrac{1}{\sqrt[5]{x^3}}\,dx$.

▶ Continued on next page

$$f(32) = \frac{5}{2}(32)^{\frac{2}{5}} + C$$
$$30 = 10 + C$$
$$C = 20$$
$$\therefore f(x) = \frac{5}{2}x^{\frac{2}{5}} + 20$$

Use the fact that the curve passes through the point (32, 30) to find C.

c $\dfrac{dP}{dt} = 150\sqrt{t}$

$$P(t) = \int 150\sqrt{t}\, dt$$
$$= 150 \int t^{\frac{1}{2}}\, dt$$
$$P(t) = 100t^{\frac{3}{2}} + C$$

Rewrite with rational exponents and find the general solution for $\int 150\sqrt{t}\, dt$.

$$P(0) = 100(0)^{\frac{3}{2}} + C$$
$$200 = 0 + C$$
$$C = 200$$
$$P(t) = 100t^{\frac{3}{2}} + 200$$

The initial population was 200 fish means P(0) = 200. Use this to find C.

$$P(4) = 100(4)^{\frac{3}{2}} + 200$$
$$= 1000$$

There are 1000 fish when $t = 4$ years.

Find P when t is 4.

Exercise 9C

EXAM-STYLE QUESTIONS

1. The derivative of the function f is given by $f'(x) = 4x^5 + 8x$.
   The graph of f passes through the point $(0, 8)$.
   Find an expression for $f(x)$.

2. It is given that $\dfrac{dy}{dx} = x^4 + \sqrt[4]{x}$ and that $y = 10$ when $x = 1$.
   Find y in terms of x.

3. The velocity v m s^{-1}, of a moving object at time t seconds is given by $v(t) = 3t^2 - 2t$.
   When $t = 3$, the displacement, s, of the object is 12 metres.
   Find an expression for s in terms of t.

4. The rate at which the volume of a sphere is increasing in cm^3 s^{-1} is given by $\dfrac{dV}{dt} = 2\pi(4t^2 + 4t + 1)$, for $0 \leq t \leq 12$. The initial volume was π cm^3.
   Find the volume of the sphere when $t = 3$.

> **EXAM-STYLE QUESTION**
>
> **5** The velocity v m s^{-1} of a moving object at time t seconds is given by $v(t) = 20 - 5t$.
>
>   **a** Find its acceleration in m s^{-2}.
>
>   **b** The initial displacement s is 5 metres.
>   Find an expression for s in terms of t.

9.2 More on indefinite integrals

The power rule for integration tells us that $\int x^n \, dx = \frac{1}{n+1} x^{n+1} + C$, $n \neq -1$. The rule does not work when $n = -1$ because it would result in dividing by 0. So what is $\int x^{-1} \, dx$?

You have seen that $\frac{d}{dx}(\ln x) = \frac{1}{x} = x^{-1}$ for $x > 0$, so

> $\int \frac{1}{x} \, dx = \ln x + C$, $x > 0$

> Why do we say that $\frac{1}{0}$ is undefined?
>
> Is $\frac{0}{0}$ the same as $\frac{1}{0}$? Why or why not?

Also $\frac{d}{dx}(e^x) = e^x$, so

> $\int e^x \, dx = e^x + C$

Example 4

Find the indefinite integral.

a $\int \frac{4}{x} \, dx$ **b** $\int \frac{e^t}{2} \, dt$

Answers

a $\int \frac{4}{x} \, dx = 4 \int \frac{1}{x} \, dx$ *Apply the constant multiple rule.*

$= 4\ln x + C$, $x > 0$ *Use the fact that $\int \frac{1}{x} \, dx = \ln x + C$, $x > 0$.*

b $\int \frac{e^t}{2} \, dt = \frac{1}{2} \int e^t \, dt$ *Apply the constant multiple rule.*

$= \frac{1}{2} e^t + C$ *Use the fact that $\int e^x \, dx = e^x + C$.*

> **Integration rules**
>
> $\int \frac{1}{x} \, dx = \ln x + C$, $x > 0$
>
> $\int e^x \, dx = e^x + C$

For some integrals such as $\int (x^2 + 1)^2 \, dx$, $\int \frac{3x^2 + 2x + 1}{x} \, dx$ and $\int \ln(e^{2t-1}) \, dt$ you may have to rewrite the integrand by expanding the bracket, separating the terms or simplifying before you can integrate. The next example shows how.

Example 5

Find the indefinite integral.

a $\int (x^2 + 1)^2 \, dx$ **b** $\int \dfrac{3x^2 + 2x + 1}{x} \, dx$ **c** $\int \ln(e^{2t-1}) \, dt$

Answers

a $\int (x^2 + 1)^2 \, dx = \int (x^4 + 2x^2 + 1) \, dx$ *Expand and then integrate each term.*

$= \dfrac{1}{5} x^5 + \dfrac{2}{3} x^3 + x + C$

b $\int \dfrac{3x^2 + 2x + 1}{x} \, dx$

$= \int \left(\dfrac{3x^2}{x} + \dfrac{2x}{x} + \dfrac{1}{x} \right) dx$ *Separate the terms.*

$= \int \left(3x + 2 + \dfrac{1}{x} \right) dx$ *Simplify and then integrate each term.*

$= \dfrac{3}{2} x^2 + 2x + \ln x + C, \; x > 0$

c $\int \ln(e^{2t-1}) \, dx = \int (2t - 1) \, dx$ *Simplify using the fact that e^x and $\ln x$ are inverses.*

$= t^2 - t + C$

Exercise 9D

Find the indefinite integral.

1 $\int \dfrac{2}{x} \, dx$ **2** $\int 3e^x \, dx$

3 $\int \dfrac{1}{4t} \, dt$ **4** $\int e^{\ln x} \, dx$

5 $\int (2x + 3)^2 \, dx$ **6** $\int \dfrac{2x^3 + 6x^2 + 5}{x} \, dx$

7 $\int \ln e^{u^2} \, du$ **8** $\int (x - 1)^3 \, dx$

9 $\int \dfrac{e^x + 1}{2} \, dx$ **10** $\int \dfrac{x^2 + x + 1}{\sqrt{x}} \, dx$

Now we look at indefinite integrals of functions that are compositions with the linear function $ax + b$.

→ $\int (ax + b)^n \, dx = \dfrac{1}{a} \left(\dfrac{1}{n+1} (ax + b)^{n+1} \right) + C$

→ $\int e^{ax+b} \, dx = \dfrac{1}{a} e^{ax+b} + C$

→ $\int \dfrac{1}{ax+b} \, dx = \dfrac{1}{a} \ln(ax+b) + C, \; x > -\dfrac{b}{a}$

> You can verify each rule by differentiating the right-hand side of the equation and showing that you get the integrand.

> Note that $\ln(ax + b)$ is defined when $ax + b > 0$ or $x > -\dfrac{b}{a}$.

Example 6

Find the indefinite integral.

a $\int (3x+1)^4 \, dx$ **b** $\int e^{2x+5} \, dx$ **c** $\int \dfrac{3}{4x-2} \, dx$ **d** $\int \dfrac{1}{(6x+3)^4} \, dx$

Integration rules

$\int (ax+b)^n \, dx = \dfrac{1}{a}\left(\dfrac{1}{n+1}(ax+b)^{n+1}\right) + C$

$\int e^{ax+b} \, dx = \dfrac{1}{a} e^{ax+b} + C$

$\int \dfrac{1}{ax+b} \, dx = \dfrac{1}{a} \ln(ax+b) + C,\ x > -\dfrac{b}{a}$

Answers

a $\int (3x+1)^4 \, dx$

$= \dfrac{1}{3}\left(\dfrac{1}{5}(3x+1)^5\right) + C$

$= \dfrac{1}{15}(3x+1)^5 + C$

Find $\dfrac{1}{a}\left(\dfrac{1}{n+1}(ax+b)^{n+1}\right) + C$ for $a = 3$, $b = 1$ and $n = 4$.

Check by differentiating.

$\dfrac{d}{dx}\left[\dfrac{1}{15}(3x+1)^5\right] = \dfrac{1}{15}(5(3x+1)^4(3))$

$= (3x+1)^4$

b $\int e^{2x+5} \, dx = \dfrac{1}{2} e^{2x+5} + C$

Find $\dfrac{1}{a} e^{ax+b} + C$ for $a = 2$ and $b = 5$.

Check by differentiating.

$\dfrac{d}{dx}\left[\dfrac{1}{2} e^{2x+5}\right] = \dfrac{1}{2}[e^{2x+5}(2)] = e^{2x+5}$

c $\int \dfrac{3}{4x-2} \, dx = 3 \int \dfrac{1}{4x-2} \, dx$

$= 3\left[\dfrac{1}{4}\ln(4x-2)\right] + C,\ x > \dfrac{1}{2}$

$= \dfrac{3}{4}\ln(4x-2) + C,\ x > \dfrac{1}{2}$

Apply the constant multiple rule.
Find $\dfrac{1}{a}\ln(ax+b)$ for $a = 4$ and $b = -2$.
Check by differentiating.

$\dfrac{d}{dx}\left[\dfrac{3}{4}\ln(4x-2)\right] = \dfrac{3}{4}\left(\dfrac{1}{4x-2}(4)\right)$

$= \dfrac{3}{4x-2}$

d $\int \dfrac{1}{(6x+3)^4} \, dx = \int (6x+3)^{-4} \, dx$

$= \dfrac{1}{6}\left(\dfrac{1}{-3}(6x+3)^{-3}\right) + C$

$= -\dfrac{1}{18(6x+3)^3} + C$

$\dfrac{d}{dx}\left[-\dfrac{1}{18(6x+3)^3}\right]$

Rewrite using rational exponents.
Find $\dfrac{1}{a}\left(\dfrac{1}{n+1}(ax+b)^{n+1}\right) + C$ for $a = 6$, $b = 3$ and $n = -4$.
Check by differentiating.

$= \dfrac{d}{dx}\left[-\dfrac{1}{18}(6x+3)^{-3}\right]$

$= -\dfrac{1}{18}(-3(6x+3)^{-4}(6)) = \dfrac{1}{(6x+3)^4}$

Exercise 9E

Find the indefinite integral in questions 1–10.

1. $\int (2x+5)^2 \, dx$
2. $\int (-3x+5)^3 \, dx$
3. $\int e^{\frac{1}{2}x-3} \, dx$
4. $\int \frac{1}{5x+4} \, dx$
5. $\int \frac{3}{7-2x} \, dx$
6. $\int 4e^{2x+1} \, dx$
7. $\int 6(4x-3)^7 \, dx$
8. $\int (7x+2)^{\frac{1}{2}} \, dx$
9. $\int \left(e^{4x} + \frac{4}{3x-5}\right) dx$
10. $\int \frac{2}{3(4x-5)^3} \, dx$

EXAM-STYLE QUESTIONS

11. Given that $f(x) = (4x+5)^3$ find

    a $f'(x)$;   **b** $\int f(x) \, dx$.

12. The velocity v of a particle at time t is given by $v(t) = e^{-3t} + 6t$. The displacement of the particle at time t is s. Given that $s = 4$ metres when $t = 0$ seconds, express s in terms of t.

The substitution method

We use the **substitution method** to evaluate integrals of the form $\int f(g(x)) \, g'(x) \, dx$. The next example shows you how.

Example 7

Find the indefinite integral.

a $\int (3x^2 + 5x)^4 (6x+5) \, dx$

b $\int \sqrt[3]{x^2 - 3x} \, (2x-3) \, dx$

c $\int x e^{4x^2+1} \, dx$

d $\int \frac{12x^3 - 3x^2}{3x^4 - x^3} \, dx$

Answers

a $\int (3x^2 + 5x)^4 (6x+5) \, dx$

This integral is of the form $\int f(g(x)) \, g'(x) \, dx$, where $g(x) = 3x^2 + 5x$ and $g'(x) = 6x + 5$.

$= \int u^4 \frac{du}{dx} \, dx = \int u^4 \, du$

Let $u = 3x^2 + 5x$, then $\frac{du}{dx} = 6x + 5$ and substitute.

$= \frac{1}{5} u^5 + C$

Simplify and integrate.

$= \frac{1}{5} (3x^2 + 5x)^5 + C$

Substitute $3x^2 + 5x$ for u.

▶ Continued on next page

b $\int \sqrt[3]{x^2 - 3x}\,(2x - 3)\,dx$

Check by differentiating.
$$\frac{d}{dx}\left[\frac{1}{5}(3x^2 + 5x)^5\right]$$
$$= \frac{1}{5}(5(3x^2 + 5x)^4(6x + 5))$$
$$= (3x^2 + 5x)^4(6x + 5)$$

This integral is of the form
$$\int f(g(x))\,g'(x)\,dx,$$
where $g(x) = x^2 - 3x$ and $g'(x) = 2x - 3$.

$$= \int u^{\frac{1}{3}}\frac{du}{dx}\,dx$$

Let $u = x^2 - 3x$, then $\frac{du}{dx} = 2x - 3$ and substitute.

$$= \int u^{\frac{1}{3}}\,du$$

Simplify and integrate.

$$= \frac{3}{4}u^{\frac{4}{3}} + C$$

$$= \frac{3}{4}(x^2 - 3x)^{\frac{4}{3}} + C$$

Substitute $x^2 - 3x$ for u.
Check by differentiating.

$$\frac{d}{dx}\left[\frac{3}{4}(x^2 - 3x)^{\frac{4}{3}}\right] = \frac{3}{4}\left(\frac{4}{3}(x^2 - 3x)^{\frac{1}{3}}(2x - 3)\right)$$

$$= (x^2 - 3x)^{\frac{1}{3}}(2x - 3) = \sqrt[3]{x^2 - 3x}\,(2x - 3)$$

c $\int x\,e^{4x^2 + 1}\,dx = \int \left(\frac{1}{8} \times 8x\right)e^{4x^2 + 1}\,dx$

If $g(x) = 4x^2 + 1$ then $g'(x) = 8x$. Rewrite the integrand so that it is in the form $c\int f(g(x))g'(x)\,dx$.

$$= \frac{1}{8}\int e^{4x^2 + 1}(8x)\,dx$$

$$= \frac{1}{8}\int e^u \frac{du}{dx}\,dx$$

Let $u = 4x^2 + 1$, then $\frac{du}{dx} = 8x$ and substitute.

$$= \frac{1}{8}\int e^u\,du$$

Simplify and integrate.

$$= \frac{1}{8}e^u + C$$

$$= \frac{1}{8}e^{4x^2 + 1} + C$$

Substitute $4x^2 + 1$ for u.

d $\int \frac{12x^3 - 3x^2}{3x^4 - x^3}\,dx$

This integral is of the form
$$\int f(g(x))g'(x)\,dx, \text{ where}$$
$g(x) = 3x^4 - x^3$ and $g(x) = 12x^3 - 3x^2$.

$$\int \frac{12x^3 - 3x^2}{3x^4 - x^3}\,dx = \int \frac{\frac{du}{dx}}{u}\,dx$$

$$= \int \frac{1}{u}\,du$$

Let $u = 3x^4 - x^3$, then $\frac{du}{dx} = 12x^3 - 3x^2$ and substitute.

$= \ln u + C,\ u > 0$
$= \ln(3x^4 - x^3) + C,$
  $3x^4 - x^3 > 0$

Simplify and integrate.
Substitute $3x^4 - x^3$ for u.

With practice you may be able find indefinite integrals of the form $\int f(g(x))g'(x)\,dx$ by inspection. That is, you may just be able to think about what you would choose for u, check to see that the derivative of u is the other factor of the integrand and then mentally integrate f with respect to u.

Exercise 9F

Find the indefinite integral in questions 1–10.

1 $\int (2x^2 + 5)^2 (4x)\,dx$

2 $\int \dfrac{3x^2 + 2}{x^3 + 2x}\,dx$

3 $\int (6x + 5)\sqrt{3x^2 + 5x}\,dx$

4 $\int 4x^3 e^{x^4}\,dx$

5 $\int \dfrac{2x + 3}{(x^2 + 3x + 1)^2}\,dx$

6 $\int \dfrac{e^{\sqrt{x}}}{2\sqrt{x}}\,dx$

7 $\int x^2 (2x^3 + 5)^4\,dx$

8 $\int \dfrac{2x + 1}{\sqrt[4]{x^2 + x}}\,dx$

9 $\int (8x^3 - 4x)(x^4 - x^2)^3\,dx$

10 $\int \dfrac{4 - 3x^2}{x^3 - 4x}\,dx$

EXAM-STYLE QUESTIONS

11 Let $f'(x) = \dfrac{8x}{4x^2 + 1}$. Given that $f'(0) = 4$, find $f(x)$.

12 The gradient of a curve is given by $f'(x) = 3x^2 e^{x^3}$. The curve passes through the point $(1, 5e)$. Find an expression for $f(x)$.

9.3 Area and definite integrals

This section is about the definite integral, which is written as $\int_a^b f(x)\,dx$, and its relationship to the area under a curve.

> Indefinite integrals are a family of functions that differ by a constant. Definite integrals are real numbers. In the next section we will learn about the relationship between definite and indefinite integrals and how to evaluate a definite integral without a GDC.

Investigation – area and the definite integral

1 Consider the area bounded by the function $f(x) = x^2 + 1$, $x = 0$, $x = 2$ and the x-axis, which is shaded in green in the graph.
   a **i** Write down the width of each of the four rectangles shown in the graph.
   ii Calculate the height of each of the four rectangles.
   iii Find the sum of the areas of the four rectangles to find a lower bound of the area of the shaded region.

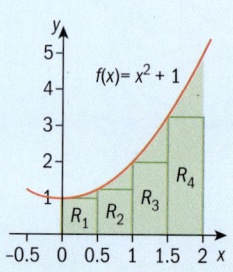

 Continued on next page

b i Write down the width of the four rectangles shown in this graph.
  ii Calculate the height of each of the four rectangles.
  iii Find the sum of the areas of the four rectangles to find an upper bound on the area of the region.
c Use a GDC to evaluate the **definite integral**
$\int_0^2 (x^2+1)\,dx$. Compare your result with your answers in parts **a** and **b**.
What do you think the definite integral might represent?

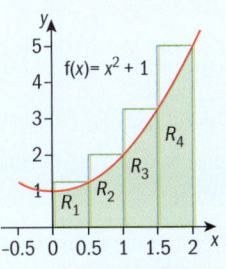

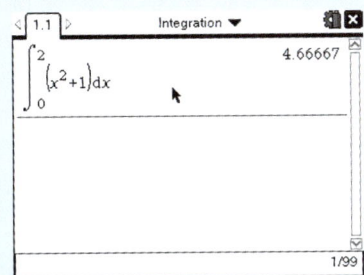

GDC help on CD: *Alternative demonstrations for the TI-84 Plus and Casio FX-9860GII GDCs are on the CD.*

The GDC uses an approximation method to determine the values of definite integrals, so the values from the GDC are not always exact.

We could not use a geometric formula to find the area of the region in question 1; we could only use geometric formulae to approximate the area. Now we will consider some regions whose area can be found geometrically.

2 Find the area of the shaded region under the line $f(x) = 2x + 2$ between $x = -1$ and $x = 2$ using a geometric formula. Then write down a definite integral you think may represent the area. Evaluate the integral on a GDC and compare answers.

3 We refer to the area between a function f and the x-axis as the **area under the curve**. If $f(x)$ is a non-negative function for $a \le x \le b$ then write down the definite integral that gives the area under the curve f from $x = a$ to $x = b$.

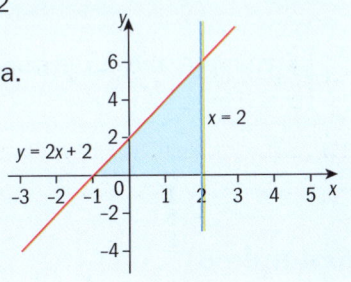

4 Verify that your answer from question 3 works for the following by finding the area using a geometric formula and then writing down a definite integral and evaluating it on a GDC.

In mathematics, a **curve** is a graph on a coordinate plane, so curves include lines.

a $f(x) = -\dfrac{1}{2}x + 3$ from $x = 1$ to $x = 4$

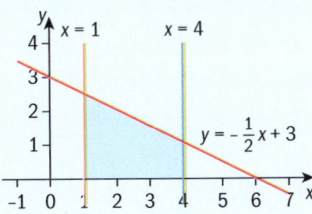

b $f(x) = \sqrt{16 - x^2}$ from $x = -4$ to $x = 4$

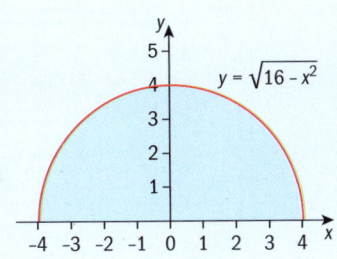

In the investigation you approximated the area under a curve $f(x) = x^2 + 1$ from $x = 0$ to $x = 2$ by summing the areas of four rectangles. Using sigma notation, we can express this as $\sum_{i=1}^{4} f(x_i) \Delta x_i$, where $f(x_i)$ represents the height of each rectangle and Δx_i represents the width of each rectangle.

To get better approximations of the area we can use more rectangles. Using an infinite number of rectangles, $\lim_{n \to \infty} \sum_{i=1}^{n} f(x_i) \Delta x_i$, leads to the exact area.

If a function f is defined for $a \leq x \leq b$ and $\lim_{n \to \infty} \sum_{i=1}^{n} f(x_i) \Delta x_i$ exists, we say that f is **integrable** on $a \leq x \leq b$. We call this limit the **definite integral** and denote it as $\lim_{n \to \infty} \sum_{i=1}^{n} f(x_i) \Delta x_i = \int_{a}^{b} f(x) \, dx$ or $\int_{a}^{b} y \, dx$. The number a is called the **lower limit** of integration and the number b is called the **upper limit** of integration.

> When f is a non-negative function for $a \leq x \leq b$, $\int_{a}^{b} f(x) \, dx$ gives the area under the curve from $x = a$ to $x = b$.

Approximations for area under $f(x) = x^2 + 1$ from $x = 0$ to $x = 2$ for different numbers of rectangle.

# Rectangles	Lower sum	Upper sum
4	3.75	5.75
10	4.28	5.08
50	4.5872	4.7472
100	4.6268	4.7068
500	4.65867	4.67467

Exact area = $\int_{0}^{2} (x^2 + 1) \, dx = \frac{14}{3} \approx 4.66667$

Notice that both the lower and upper sums appear to approach 4.66667.

The symbol $\int$ is an elongated S and is also used to indicate a sum. The definite integral notation was introduced by the German mathematician Gottfried Wilhelm Leibniz towards the end of the 17th century. $\int_{a}^{b} f(x) \, dx$ is read as 'the integral from a to b of $f(x)$ with respect to x'.

Example 8

Write down a definite integral that gives the area of the shaded region and evaluate it using a GDC. Whenever possible find the area using a geometric formula to verify your answer.

a

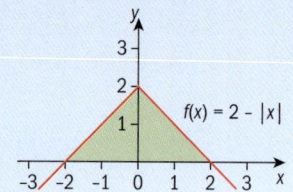

$f(x) = 2 - |x|$

b

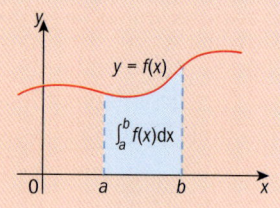

$f(x) = \dfrac{2}{1 + x^2}$

Answers

a $\int_{-2}^{2} (2 - |x|) \, dx = 4$

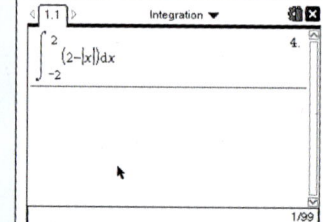

Area $= \dfrac{1}{2}(4 \times 2) = 4$

The function intersects the x-axis at -2 and 2 and forms a triangle. So the limits of integration are -2 and 2. The area formula for a triangle is $A = \dfrac{1}{2}(b \times h)$.

Continued on next page

b $\int_{-1}^{1} \frac{2}{1+x^2} dx \approx 3.14$

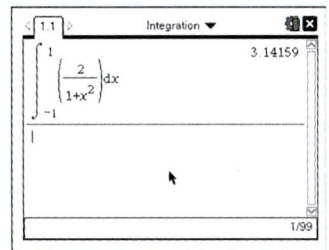

GDC help on CD: Alternative demonstrations for the TI-84 Plus and Casio FX-9860GII GDCs are on the CD.

The region is bounded by the function $f(x) = \frac{2}{1+x^2}$, the x-axis and the vertical lines $x = -1$ and $x = 1$. So the limits of integration are -1 and 1. The area cannot be determined from a geometric formula.

Exercise 9G

Write down a definite integral that gives the area of the shaded region and evaluate it using your GDC. Where possible find the area using a geometric formula to verify your answer.

1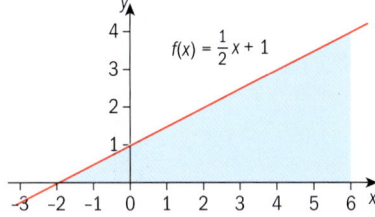

2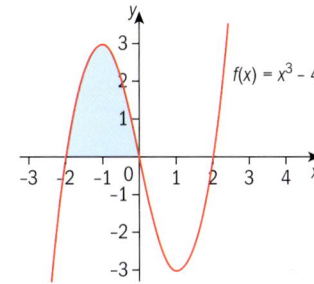

3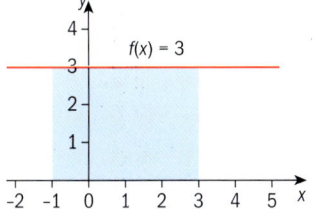

4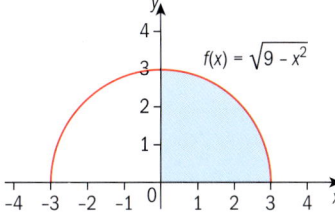

5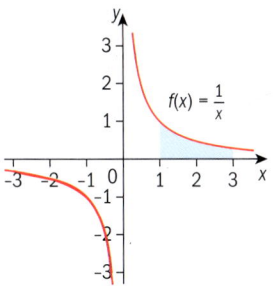

6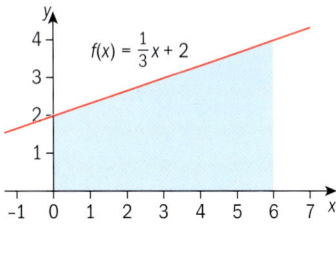

When f is a non-negative function for $a \leq x \leq b$, $\int_a^b f(x)dx$ gives the area under the curve from $x = a$ to $x = b$.

Consider what happens when f is not non-negative.

i $\int_{-3}^{-1} (2x + 2) dx$

The area of the shaded triangle is 4, but

$\int_{-3}^{-1} (2x + 2) dx = -4$ since $f(x) < 0$ when $-3 < x < -1$.

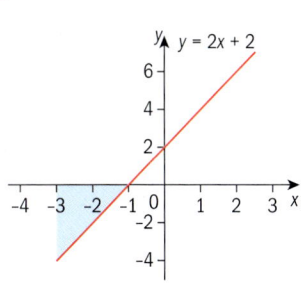

ii $\int_{-1}^{2}(2x+2)\,dx$

$\int_{-1}^{2}(2x+2)\,dx = 9$ is the area of the shaded triangle

since f is a non-negative function for $-1 \le x \le 2$.

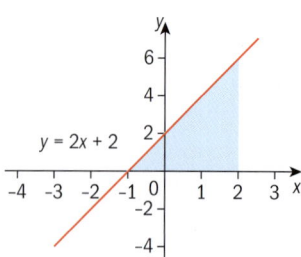

iii $\int_{-3}^{2}(2x+2)\,dx$

$\int_{-3}^{2}(2x+2)\,dx = 5$ because it is equal to

$\int_{-3}^{-1}(2x+2)\,dx + \int_{-1}^{2}(2x+2)\,dx = -4 + 9 = 5$. This is the negative of the area of the region labeled A_1 plus the area of the region labeled A_2.

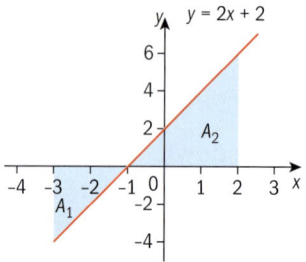

This illustrates one of the properties of definite integrals.

→ $\int_{a}^{b} f(x)\,dx = \int_{a}^{c} f(x)\,dx + \int_{c}^{b} f(x)\,dx$

Example 9

The graph of f consists of line segments as shown in the figure.

Evaluate $\int_{0}^{8} f(x)\,dx$ using geometric formulae.

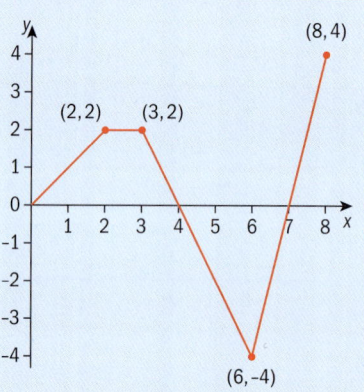

Answer

$\int_{0}^{8} f(x)\,dx = A_1 - A_2 + A_3$

$= \dfrac{1}{2}(4+1)(2) - \dfrac{1}{2}(3)(4) + \dfrac{1}{2}(1)(4)$

$= 5 - 6 + 2$

$= 1$

Find the area of the trapezium A_1, minus the area of the triangle A_2, plus the area of the triangle, A_3.

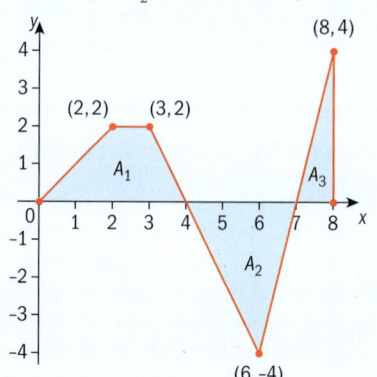

306 Integration

→ **Some properties of definite integrals**

1. $\int_a^b kf(x)\,dx = k\int_a^b f(x)\,dx$

2. $\int_a^b (f(x) \pm g(x))\,dx = \int_a^b f(x)\,dx \pm \int_a^b g(x)\,dx$

3. $\int_a^a f(x)\,dx = 0$

4. $\int_a^b f(x)\,dx = -\int_b^a f(x)\,dx$

5. $\int_a^b f(x)\,dx = \int_a^c f(x)\,dx + \int_c^b f(x)\,dx$

> You do not need to learn the numbers that go with these, just the properties.

Example 10

Given that $\int_0^2 f(x)\,dx = 4$, $\int_2^5 f(x)\,dx = 12$, $\int_0^2 g(x)\,dx = -3$ and $\int_0^4 g(x)\,dx = 6$, evaluate these definite integrals without using your GDC.

a $\int_0^2 (3f(x) - g(x))\,dx$ **b** $\int_2^2 g(x)\,dx + \int_5^2 f(x)\,dx$

c $\int_0^5 f(x)\,dx$ **d** $\int_2^4 g(x)\,dx$ **e** $\int_{-3}^{-1} \frac{1}{2} f(x+3)\,dx$

Answers

a $\int_0^2 (3f(x) - g(x))\,dx$

$= \int_0^2 3f(x)\,dx - \int_0^2 g(x)\,dx$ *Apply property 2.*

$= 3\int_0^2 f(x)\,dx - \int_0^2 g(x)\,dx$ *Apply property 1.*

$= 3(4) - (-3)$ *Substitute and evaluate.*

$= 15$

b $\int_2^2 g(x)\,dx + \int_5^2 f(x)\,dx$

$= 0 - \int_2^5 f(x)\,dx$ *Apply property 3 to the first term and property 4 to the second term.*

$= 0 - 12$ *Substitute and evaluate.*

$= -12$

c $\int_0^5 f(x)\,dx$

$= \int_0^2 f(x)\,dx + \int_2^5 f(x)\,dx$ *Apply property 5.*

$= 4 + 12$ *Substitute and evaluate.*

$= 16$

▶ Continued on next page

d $\int_0^2 g(x)\,dx + \int_2^4 g(x)\,dx$	Apply property 5.
$= \int_0^4 g(x)\,dx$	
So $\int_2^4 g(x)\,dx$	
$= \int_0^4 g(x)\,dx - \int_0^2 g(x)\,dx$	Rearrange terms.
$= 6 - (-3)$	Substitute and evaluate.
$= 9$	
e $\int_{-3}^{-1} \frac{1}{2} f(x+3)\,dx$	Apply property 1. The graph of $f(x+3)$ is a result of translating the graph of $f(x)$ to the left 3 units. The limits of integration, $x = 0$ and $x = 2$ are translated to $x = -3$ and $x = -1$. So the values of these integrals are equal.
$= \frac{1}{2}\int_{-3}^{-1} f(x+3)\,dx$	
$= \frac{1}{2}\int_0^2 f(x)\,dx$	
$= \frac{1}{2}(4)$	
$= 2$	

Exercise 9H

The graph of f consists of line segments as shown. Evaluate the definite integrals in questions 1 and 2 using geometric formulae.

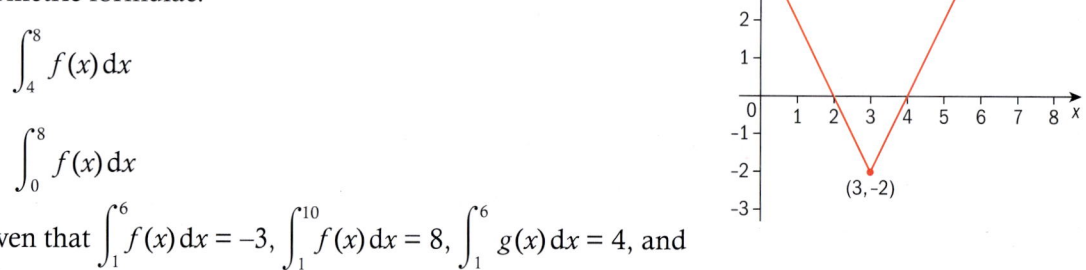

1 $\int_4^8 f(x)\,dx$

2 $\int_0^8 f(x)\,dx$

Given that $\int_1^6 f(x)\,dx = -3$, $\int_1^{10} f(x)\,dx = 8$, $\int_1^6 g(x)\,dx = 4$, and $\int_6^{10} g(x)\,dx = 8$ evaluate the definite integrals in questions 3–10.

3 $\int_1^6 \left(2f(x) + \frac{1}{2}g(x)\right) dx$

4 $\int_{10}^6 g(x)\,dx$

5 $\int_1^{10} g(x)\,dx$

6 $\int_{10}^{10} f(x)\,dx$

7 $\int_6^{10} f(x)\,dx$

8 $\int_5^{10} f(x-4)\,dx$

9 $\int_6^{10} (g(x) + 3)\,dx$

10 $\int_{-1}^4 3g(x+2)\,dx$

EXAM-STYLE QUESTIONS

11 Given that $\int_0^2 h(x)\,dx = -2$ and $\int_2^5 h(x)\,dx = 6$, deduce the value of

   a $\int_0^5 h(x)\,dx$ **b** $\int_2^5 (h(x)+2)\,dx$.

12 Let f be a function such that $\int_0^4 f(x)\,dx = 16$.

   a Deduce the value of $\int_0^4 \frac{1}{4} f(x)\,dx$

   b i If $\int_a^b f(x-3)\,dx = 16$, write down the value of a and of b.

   ii If $\int_0^4 (f(x)+k)\,dx = 28$, write down the value of k.

9.4 Fundamental Theorem of Calculus

The quotient $\frac{\Delta y}{\Delta x}$, the slope of a secant line, gives us an approximation for the slope of a tangent line.
The product $(\Delta y)(\Delta x)$, the area of a rectangle, helps give us an approximation for the area under a curve.
In much the same sense as division and multiplication are inverse operations, Isaac Newton and Gottfried Leibniz independently came to realize that differentiation and definite integrals are inverse operations.

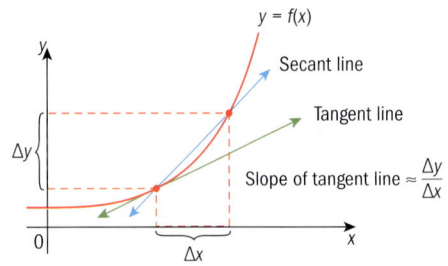

This fact is established in the following theorem.

> **Fundamental Theorem of Calculus**
>
> If f is a continuous function on the interval $a \le x \le b$ and F is an antiderivative of f on $a \le x \le b$, then
>
> $$\int_a^b f(x)\,dx = [F(x)]_a^b = F(b) - F(a).$$

The notation $[F(x)]_a^b$ means $F(b) - F(a)$.

Consider the definite integral $\int_0^2 (x^2+1)\,dx$ that you evaluated using a GDC in the investigation in the last section.
This gave the area under the curve $f(x) = x^2+1$ between $x = 0$ and $x = 2$.
You found $\int_0^2 (x^2+1)\,dx \approx 4.67$.

When applying the Fundamental Theorem of Calculus, although F can be any member of the family of functions of the antiderivatives of f, we choose to use the 'simplest' one, that is, one where the constant of integration is $C = 0$. We can do this because, for any C,

$f(x)\,dx = [F(x) + C]_a^b$
$= [F(b) + C] - [F(a) + C]$
$= F(b) - F(a)$

Using the Fundamental Theorem of Calculus we get:

$$\int_0^2 (x^2 + 1)\,dx = \left[\frac{1}{3}x^3 + x\right]_0^2$$

$$= \left(\frac{1}{3}(2^3) + 2\right) - \left(\frac{1}{3}(0^3) + 0\right)$$

$$= \frac{14}{3}$$

$$\approx 4.67$$

> $\frac{1}{3}x^3 + x$ is the 'simplest' antiderivative of $x^2 + 1$. Evaluate $\frac{1}{3}x^3 + x$ at $x = 2$ and $x = 0$, then find the difference.

Example 11

Evaluate the definite integral without using a GDC.

a $\int_{-2}^{1} (u - 1)\,du$ **b** $\int_{2}^{3} \frac{1}{t}\,dt$ **c** $\int_{1}^{3} 4x^2(x - 1)\,dx$

Answers

a $\int_{-2}^{1} (u - 1)\,du = \left[\frac{1}{2}u^2 - u\right]_{-2}^{1}$

$= \left(\frac{1}{2}(1^2) - 1\right) - \left(\frac{1}{2}(-2)^2 - (-2)\right)$

$= \left(\frac{1}{2} - 1\right) - (2 + 2) = -\frac{9}{2}$

> Find the 'simplest' antiderivative of $u - 1$.
>
> Evaluate $\frac{1}{2}u^2 - u$ at $u = 1$ and $u = -2$, then find the difference.

b $\int_{2}^{3} \frac{1}{t}\,dt = [\ln t]_2^3$

$= \ln 3 - \ln 2 = \ln \frac{3}{2}$

> Recall that $\ln a - \ln b = \ln \frac{a}{b}$.

c $\int_{1}^{3} 4x^2(x - 1)\,dx = 4\int_{1}^{3} (x^3 - x^2)\,dx$

$= 4\left[\frac{1}{4}x^4 - \frac{1}{3}x^3\right]_1^3$

$= 4\left[\left(\frac{1}{4}(3^4) - \frac{1}{3}(3^3)\right) - \left(\frac{1}{4}(1^4) - \frac{1}{3}(1^3)\right)\right]$

$= 4\left[\left(\frac{81}{4} - 9\right) - \left(\frac{1}{4} - \frac{1}{3}\right)\right] = \frac{136}{3}$

> Rewrite the integrand in order to integrate.

Exercise 9I

Evaluate the definite integrals in questions 1–8.

1 $\int_0^1 2x\,dx$

2 $\int_{-1}^{1} (u^2 - 2)\,du$

3 $\int_1^2 \left(\frac{3}{x^2} - 1\right)dx$

4 $\int_0^8 \left(x^{\frac{1}{3}} - x^{\frac{2}{3}}\right)dx$

5 $\displaystyle\int_0^3 4e^x\,dx$

6 $\displaystyle\int_e^{e^2} \frac{1}{x}\,dx$

7 $\displaystyle\int_0^1 (t+3)(t+1)\,dt$

8 $\displaystyle\int_4^9 \frac{2\sqrt{x}+3}{\sqrt{x}}\,dx$

> The force between electric charges depends on the amount of the charge and the distance between the charges. How are definite integrals used to calculate the work done when charges are separated?

EXAM-STYLE QUESTIONS

9 It is given that $\displaystyle\int_0^2 f(x)\,dx = 8$

  a Write down the value of $\displaystyle\int_0^2 3f(x)\,dx$.

  b Find the value of $\displaystyle\int_0^2 (f(x) + x^2)\,dx$

10 Given $\displaystyle\int_2^k \frac{1}{x}\,dx = \ln 6$, find the value of k.

Now we look at definite integrals that involve compositions with the linear function $ax + b$ or the substitution method.

Example 12

Evaluate the definite integral without using a GDC.

a $\displaystyle\int_1^5 \left(e^{2x} + \frac{1}{x^2}\right)dx$ **b** $\displaystyle\int_{-1}^1 (2x-3)^3\,dx$

c $\displaystyle\int_0^3 \sqrt{3x+16}\,dx$ **d** $\displaystyle\int_0^1 (2x^2+1)^3(4x)\,dx$

Answers

a $\displaystyle\int_1^5 \left(e^{2x} + \frac{1}{x^2}\right)dx$

$= \displaystyle\int_1^5 (e^{2x} + x^{-2})\,dx$

$= \left[\dfrac{1}{2}e^{2x} - \dfrac{1}{x}\right]_1^5$

$= \left(\dfrac{1}{2}e^{2(5)} - \dfrac{1}{5}\right) - \left(\dfrac{1}{2}e^{2(1)} - \dfrac{1}{1}\right)$

$= \dfrac{1}{2}e^{10} - \dfrac{1}{2}e^2 + \dfrac{4}{5}$

or $\dfrac{5e^{10} - 5e^2 + 8}{10}$

Recall that $\displaystyle\int e^{ax+b}\,dx = \dfrac{1}{a}e^{ax+b} + C.$

▶ Continued on next page

b $\int_{-1}^{1} (2x-3)^3 \, dx$

$= \left[\frac{1}{2} \left(\frac{1}{4}(2x-3)^4 \right) \right]_{-1}^{1}$

$= \left(\frac{1}{8}(2(1)-3)^4 \right) - \left(\frac{1}{8}(2(-1)-3)^4 \right)$

$= \frac{1}{8} - \frac{625}{8} = -78$

Recall that $\int (ax+b)^n \, dx = \frac{1}{a}\left(\frac{1}{n+1}(ax+b)^{n+1}\right) + C.$

c $\int_0^3 \sqrt{3x+16} \, dx$

$= \int_0^3 (3x+16)^{\frac{1}{2}} \, dx$

$= \left[\frac{1}{3}\left(\frac{2}{3}(3x+16)^{\frac{3}{2}}\right) \right]_0^3$

$= \frac{2}{9}\left((3(3)+16)^{\frac{3}{2}} - (3(0)+16)^{\frac{3}{2}} \right)$

$= \frac{2}{9}\left(25^{\frac{3}{2}} - 16^{\frac{3}{2}} \right) = \frac{122}{9}$

Recall that $\int (ax+b)^n \, dx = \frac{1}{a}\left(\frac{1}{n+1}(ax+b)^{n+1}\right) + C.$

Recall $25^{\frac{3}{2}} = (\sqrt{25})^3 = 125$ and $16^{\frac{3}{2}} = (\sqrt{16})^3 = 64.$

d $\int_0^1 (2x^2+1)^3 (4x) \, dx$

$= \int_{x=0}^{x=1} u^3 \frac{du}{dx} \, dx$

$= \int_{u=1}^{u=3} u^3 \, du = \left[\frac{1}{4}u^4\right]_1^3$

$= \frac{1}{4}[(3)^4 - (1)^4] = 20$

Let $u = 2x^2 + 1$ and $\frac{du}{dx} = 4x$ and substitute.

Change the limits of integration and then you can evaluate the integral in terms of u. When $x = 0$, $u = 2(0^2) + 1 = 1$, and when $x = 1$, $u = 2(1^2) + 1 = 3$

Exercise 9J

Evaluate the definite integrals in questions 1–8.

1 $\int_{-1}^{1} \frac{1}{t+2} \, dt$

2 $\int_3^4 e^{-x+1} \, dx$

3 $\int_{-1}^{2} (-2x+1)^3 \, dx$

4 $\int_{-1}^{1} (e^x + e^{-x}) \, dx$

5 $\int_0^2 \sqrt{6x+4} \, dx$

6 $\int_1^2 (x^2+x)^3 (2x+1) \, dx$

7 $\int_3^4 \frac{8t-6}{2t^2-3t-2} \, dt$

8 $\int_0^1 4xe^{x^2+3} \, dx$

What are some applications of center of mass (centroid)? How can definite integrals be used to find the centroid of a curved area?

EXAM-STYLE QUESTIONS

9 The diagram shows part of the graph of $f(x) = -2x^2(x-2)$.
  a Write down an integral which represents the area of the shaded region.
  b Find the area of the shaded region.

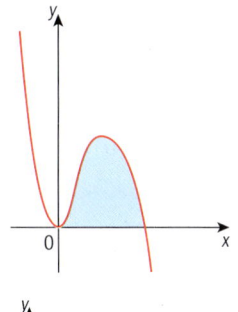

10 The diagram shows part of the graph of $y = \dfrac{1}{x-1}$.
  The area of the shaded region is ln 4 units.
  Find the exact value of k.

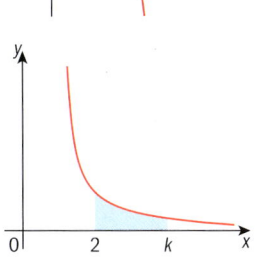

9.5 Area between two curves

We now extend the concept of area *under* a curve to the area *between* two curves.

> The sums of the areas of rectangles used to approximate area are named Riemann sums after German mathematician Georg Riemann. He proved the existence of the limits of such sums.
>
>
>
> ▶ Georg Riemann (1826–66)

Investigation: Area between two curves

Consider the area between the two curves
$f(x) = x^2 + 3x$ and
$g(x) = x - 2$ from $x = -1.5$ to $x = 3.5$.

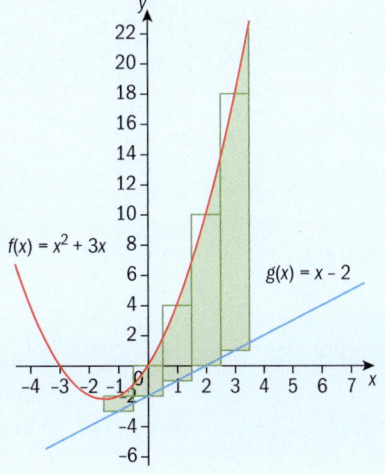

Continued on next page

1 Copy and complete the table to give the dimensions and area of each of the five rectangles shown in the graph.

Interval	Width	Height	Area
$-1.5 \leq x < -0.5$	1	$f(-1) - g(-1) = -2 - (-3) = 1$	$1(1) = 1$
$-0.5 \leq x < 0.5$			
$0.5 \leq x < 1.5$			
$1.5 \leq x < 2.5$			
$2.5 \leq x < 3.5$			

> Notice that, regardless of whether f and g are positive, negative or zero, the height of the rectangle is always given by $f(x)$, the top curve, minus $g(x)$, the bottom curve.

2 Approximate the area between the curves by finding the sum of the areas of the rectangles.

3 Write down a definite integral you think can be used to find the exact area between the two curves $f(x) = x^2 + 3x$ and $g(x) = x - 2$ from $x = -1.5$ to $x = 3.5$.
Evaluate the integral on your GDC.
Compare the answer to your approximation from question 2.

→ If y_1 and y_2 are continuous on $a \leq x \leq b$ and $y_1 \geq y_2$ for all x in $a \leq x \leq b$, then the area between y_1 and y_2 from $x = a$ to $x = b$ is given by $\int_a^b (y_1 - y_2)\,dx$.

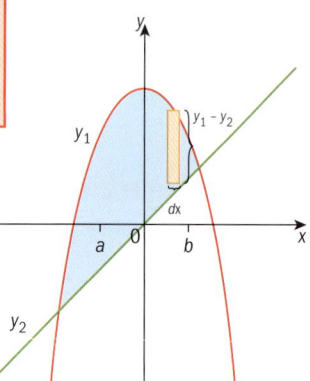

Height of each rectangle = 'top curve' – 'bottom curve' = $y_1 - y_2$
Width of each rectangle = dx
Area of each rectangle = $(y_1 - y_2)\,dx$
Sum of the areas of an infinite number of rectangles from $x = a$ to $x = b$ and exact area between the two curves = $\int_a^b (y_1 - y_2)\,dx$

Example 13

a Graph the region bounded by the curves $y = x^2 - 2$ and $y = -x$.
Write down an expression that gives the area of the region and then find the area.
Solve this problem without using a GDC.

b Sketch the graph of the region bounded by the curves $f(x) = 2e^{-\frac{x}{2}}$ and $g(x) = x^2 - 4x$.
Write down an expression that gives the area of the region.
Find the area, using a GDC.

▶ Continued on next page

Answers

a $x^2 - 2 = -x$
$x^2 + x - 2 = 0$
$(x + 2)(x - 1) = 0$
$x = -2, 1$
Points of intersection: $(-2, 2)$ and $(1, -1)$

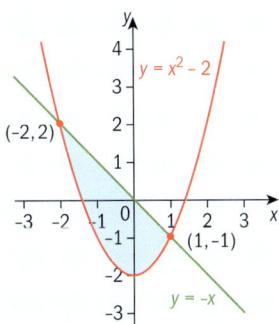

Find the intersection of the two curves by setting the equations equal and solving for x. Substitute the x-values into either equation to get the y-coordinates.

The graph of $y = x^2 - 2$ is the graph of $y = x^2$ translated down 2 units. The graph of $y = -x$ is a line with y-intercept $(0, 0)$ and gradient -1. The graphs intersect at $(-2, 2)$ and $(1, -1)$.

$$\text{Area} = \int_{-2}^{1} ((-x) - (x^2 - 2))\,dx = \int_{-2}^{1} (-x^2 - x + 2)\,dx$$

$$= \left[-\frac{1}{3}x^3 - \frac{1}{2}x^2 + 2x \right]_{-2}^{1}$$

$$= \left(-\frac{1}{3}(1)^3 - \frac{1}{2}(1)^2 + 2(1) \right) - \left(-\frac{1}{3}(-2)^3 - \frac{1}{2}(-2)^2 + 2(-2) \right)$$

$$= \left(-\frac{1}{3} - \frac{1}{2} + 2 \right) - \left(\frac{8}{3} - 2 - 4 \right) = \frac{9}{2}$$

$y = -x$ is greater than or equal to $y = x^2 - 2$ for $-2 \leq x \leq 1$, so the 'height of each rectangle' is represented by $(-x) - (x^2 - 2)$.

b

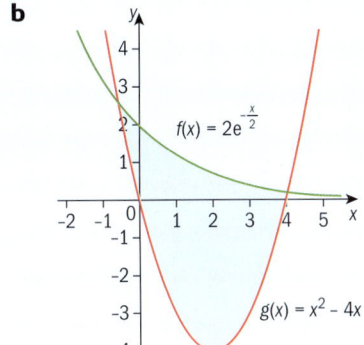

$2e^{-\frac{x}{2}} = x^2 - 4x$
$x \approx -0.5843, 4.064$

$$\text{Area} = \int_{-0.5843}^{4.064} \left((2e^{-\frac{x}{2}}) - (x^2 - 4x) \right) dx \approx 14.7$$

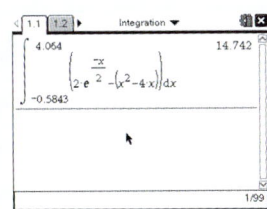
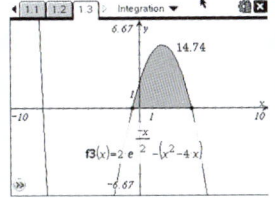

Use a GDC to help sketch the graphs and to find the x-coordinates of the points of intersection. Write down at least 4 significant digits since these values will be used to compute the area.

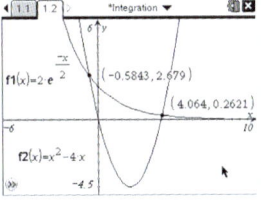

$f(x) = 2e^{-\frac{x}{2}}$ is greater than or equal to $g(x) = x^2 - 4x$ for $-0.5843 \leq x \leq 4.064$, so the 'height of each rectangle' is represented by $\left(2e^{-\frac{x}{2}}\right) - (x^2 - 4x)$.

GDC help on CD: Alternative demonstrations for the TI-84 Plus and Casio FX-9860GII GDCs are on the CD.

Exercise 9K

In questions 1–4, graph the region bounded by the given curves. Write down an expression that gives the area of the region. Find the area using a GDC.

1. $y = -\frac{1}{2}x^2 + 2$ and $y = \frac{1}{2}x^2 - 2$

2. $f(x) = x^2$ and $g(x) = \sqrt{x}$

3. $y = 2x - 4$, $y = x^3$ between $x = -2$ and $x = 2$

4. $g(x) = x + 1$ and $h(x) = 3 + 2x - x^2$

EXAM-STYLE QUESTION

5. Consider the function $f(x) = x^4 - x^2$.
   a. Find the x-intercepts.
   b. i. Find $f'(x)$.
      ii. Hence find the coordinates of the minimum and maximum points.
   c. i. Use your answers to parts **a** and **b** to sketch a graph of f.
      ii. Sketch a graph of $g(x) = 1 - x^2$ on the same axes.
   d. Write down an expression that gives the area of the region between f and g and find the area of the region.

In questions 6–9, sketch a graph of the region bounded by the given curves. Write down an expression that gives the area of the region. Find the area using your GDC.

6. $y = \ln x$ and $y = x - 2$

7. $f(x) = x^2 - 3x + 1$ and $g(x) = -x + 3$

8. $f(x) = e^x$ and $h(x) = 2 - x - x^2$

9. $y = \frac{x+2}{x-1}$ and $y = -\frac{1}{2}x + 6$

EXAM-STYLE QUESTION

10. Consider the functions $f(x) = x$ and $g(x) = 2\sqrt{x}$
    a. Sketch the graphs of f and g on the same axes.
    b. i. Write down an expression for the area of the region between f and g.
       ii. Find this area.
    c. The line $x = k$ divides the area of the region from part **b** in half.
       i. Write down an expression for half the area of the region from part **b**.
       ii. Find the value of k.

Now we look at cases where y_1 and y_2 are continuous on $a \le x \le b$, but y_1 is not greater than or equal to y_2 for all x in $a \le x \le b$. In this case you must find all the points of intersection and determine which curve is above the other in the intervals determined by the points of intersection.

Example 14

Write down an expression that gives the area of the region between $f(x) = 10x + x^2 - 3x^3$ and $g(x) = x^2 - 2x$. Find the area.

Answer

$10x + x^2 - 3x^3 = x^2 - 2x$
$x = -2, 0, 2$

$$\int_{-2}^{0} ((x^2 - 2x) - (10x + x^2 - 3x^3))\, dx$$

$$+ \int_{0}^{2} ((10x + x^2 - 3x^3) - (x^2 - 2x))\, dx$$

$= 24$

Find the points of intersection of f and g.
$g(x) = x^2 - 2x$ is greater than or equal to $f(x) = 10x + x^2 - 3x^3$ for $-2 \le x \le 0$, so on this interval the 'height of each rectangle' is represented by $(x^2 - 2x) - (10x + x^2 - 3x^3)$.
$f(x) = 10x + x^2 - 3x^3$ is greater than or equal to $g(x) = x^2 - 2x$ for $0 \le x \le 2$, so on this interval the 'height of each rectangle' is represented by $(10x + x^2 - 3x^3) - (x^2 - 2x)$.

Use a GDC to help find the x-coordinates of the points of intersection and determine which curve is above the other curve in the intervals formed by the points of intersection.

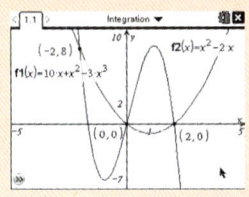

 GDC help on CD: Alternative demonstrations for the TI-84 Plus and Casio FX-9860GII GDCs are on the CD.

Exercise 9L

In questions 1–4, write down an expression to find the area of the region bounded by the two curves and then find the area.

1 $y = x^3 - 2x^2$ and $y = 2x^2 - 3x$

2 $f(x) = (x - 1)^3$ and $g(x) = x - 1$

3 $f(x) = xe^{-x^2}$ and $g(x) = x^3 - x$

4 $g(x) = -x^4 + 10x^2 - 9$ and $h(x) = x^4 - 9x^2$

EXAM-STYLE QUESTION

5 The curves shown in the figure are graphs of $f(x) = \frac{1}{4}x^2$, $g(x) = -x^2$ and $h(x) = 2x - 4$.
   a i Find the coordinates of point Q.
     ii Show that the line passing through points P and Q is tangent to $f(x) = \frac{1}{4}x^2$ at point Q.
   b i Find the coordinates of point P correct to 4 significant digits.
     ii Hence write down an expression for the area of the shaded region and then find the area.

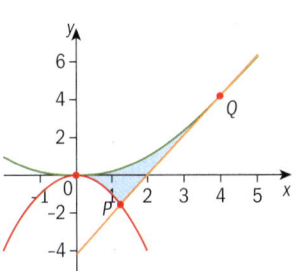

9.6 Volume of revolution

A **solid of revolution** is formed by rotating a plane figure about an **axis of revolution**.

First consider a rectangle perpendicular to the x-axis. Imagine rotating the rectangle 360° about the x-axis.

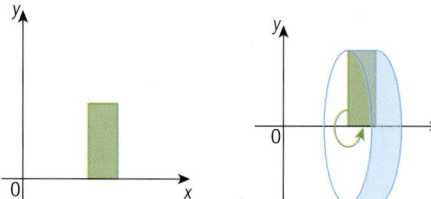

The solid that is formed is called a **disk**. The disk is cylindrical in shape.

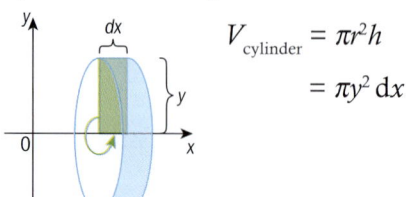

$$V_{cylinder} = \pi r^2 h$$
$$= \pi y^2 \, dx$$

> Solids of revolution are used in manufacturing many items such as pistons and crankshafts.
>
>
>
> ▲ Pistons
>
> ▲ Crankshaft

Investigation – volume of revolution

Consider the triangle formed by the line $f(x) = 0.5x$ and the x-axis between $x = 0$ and $x = 6$.

1 Copy and complete the table to give the dimensions and volume of each of the disks formed when the rectangles shown in the figure are rotated 360° about the x-axis. The last row in the table below is completed for you.

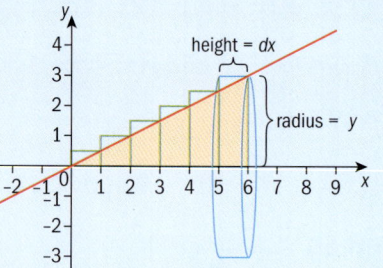

Interval	Radius	Height	Volume
$0 \leq x < 1$			
$1 \leq x < 2$			
$2 \leq x < 3$			
$3 \leq x < 4$			
$4 \leq x < 5$			
$5 \leq x < 6$	$f(6) = 3$	$6 - 5 = 1$	$\pi(3^2)(1) \approx 28.27$

2 Find the sum of the volumes of the six disks in question 1. Is this sum greater or less than the exact volume of the solid formed by rotating the triangle about the x-axis?

3 Write down a definite integral you think can be used to find the exact volume of the solid of revolution formed when the triangle is rotated about the x-axis. Evaluate the integral on a GDC and compare it to your estimate in question 2.

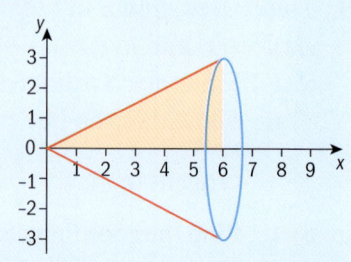

4 When the triangle is rotated about the x-axis the solid formed is a cone. Use a geometric formula to find the volume of the cone and compare it to the value of your definite integral in question 3.

→ If $y = f(x)$ is continuous on $a \leq x \leq b$ and the region bounded by $y = f(x)$ and the x-axis between $x = a$ and $x = b$ is rotated 360° about the x-axis then the volume of the solid formed is given by

$$V = \int_a^b \pi(f(x))^2\, dx \text{ or } \int_a^b \pi y^2\, dx.$$

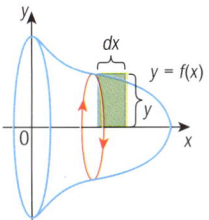

Radius of disk (height of 'representative rectangle') = y
Height of disk (width of 'representative rectangle') = dx
Volume of disk = $\pi r^2 h = \pi y^2 dx$
Sum of the volumes of an infinite number of disks from $x = a$ to $x = b$
and exact volume of the solid $\int_a^b \pi y^2\, dx.$

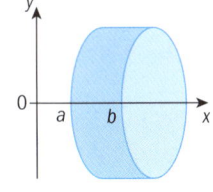

Example 15

Use a definite integral to find the volume of the solid formed when the region bounded by $f(x) = \sqrt{9 - x^2}$ and the x-axis is rotated 360° about the x-axis. Verify your answer using a geometric formula.

Answer

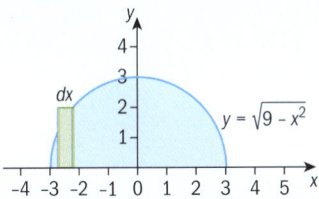

$$V = \int_a^b \pi y^2\, dx$$

$$= \int_{-3}^{3} \pi (\sqrt{9 - x^2})^2\, dx$$

$$\approx 113$$

To verify:

$$V = \frac{4}{3}\pi r^3 = \frac{4}{3}\pi(3^3)$$

$$= 36\pi$$

$$\approx 113$$

Sketching a graph and 'representative rectangle' is helpful.
Radius of disk is the height of representative rectangle, $\sqrt{9 - x^2}$.
Height of disk is the width of representative rectangle, dx.
The limits of integration are the x-intercepts, -3 and 3.

Use a GDC to evaluate the integral.

When the region is rotated about the x-axis, a sphere is formed.
Volume of sphere = $\frac{4}{3}\pi r^3$

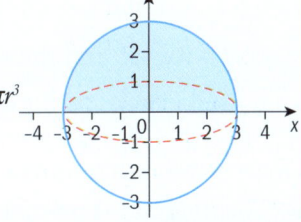

Exercise 9M

Use a definite integral to find the volume of the solid formed when the region bounded by the given curves is rotated 360° about the x-axis. Verify your answers using geometric formulae.

1 $f(x) = 4$ and the x-axis between $x = 0$ and $x = 5$

2 $f(x) = 6 - 2x$ and the x-axis between $x = 0$ and $x = 3$

3 $f(x) = \sqrt{4 - x^2}$ and the x-axis

4 $f(x) = \sqrt{16 - x^2}$ and the x-axis between $x = 0$ and $x = 4$

5 $f(x) = x$ and the x-axis between $x = 2$ and $x = 4$

> Ibn al-Haytham (965–1040), a mathematician who lived mainly in Egypt, is credited with calculating the integral of a function in order to find the volume of a paraboloid – the 3-D shape created by rotating a parabola about its axis of symmetry.

Example 16

Use a definite integral to find the volume of the solid formed when the region under the curve $y = x^2$ between $x = 0$ and $x = 2$ is rotated about the x-axis. Give your answer in terms of π.

Answer

$V = \int_a^b \pi y^2 \, dx$

$= \int_0^2 \pi (x^2)^2 \, dx$

$= \int_0^2 \pi x^4 \, dx$

$= \pi \left[\dfrac{1}{5} x^5 \right]_0^2$

$= \pi \left(\dfrac{1}{5}(2^5) - \dfrac{1}{5}(0^5) \right)$

$= \dfrac{32\pi}{5}$

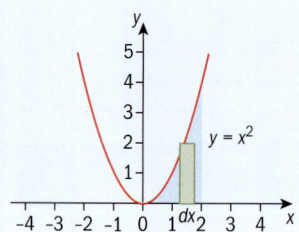

Sketching a graph and 'representative rectangle' is helpful.
Radius of disk is the height of representative rectangle, x^2.
Height of disk is the width of representative rectangle, dx.
The limits of integration are 0 and 2.

Exercise 9N

In questions 1–4, use a definite integral to find the volume of the solid formed by rotating the region bounded by the given curves about the x-axis.

1 $f(x) = x^3$ and the x-axis between $x = 1$ and $x = 2$

2 $y = x^2 + 1$ and the x-axis between $x = 0$ and $x = 1$

3 $f(x) = 3x - x^2$ and the x-axis

4 $y = \dfrac{1}{x}$ and the x-axis between $x = 1$ and $x = 4$

EXAM-STYLE QUESTION

5 The diagram shows part of the graph of $y = e^{\left(\frac{1}{4}x\right)}$. The shaded region between the graph of $y = e^{\left(\frac{1}{4}x\right)}$ and the x-axis from $x = 0$ to $x = \ln 4$ is rotated 360° about the x-axis.

a Write down a definite integral that represents the volume of the solid formed.

b This volume is equal to $k\pi$. Find the value of k.

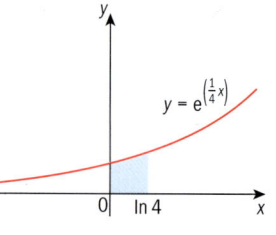

: EXAM-STYLE QUESTION
6 The shaded region in the diagram is bounded by $y = \dfrac{1}{\sqrt{x}}$, $x = 1$, $x = a$ and the x-axis. The shaded region is rotated 360° about the x-axis.
   a Write down a definite integral that represents the volume of the solid formed.
   b The volume of the solid formed is 3π. Find the value of a.

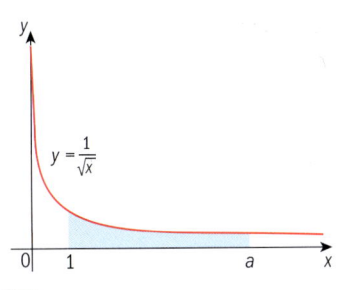

9.7 Definite integrals with linear motion and other problems

Extension material on CD:
Worksheet 9 - More volumes of solids of revolution

Another application of definite integrals is finding the change in a function over time.

Suppose the displacement function for a particle moving along a horizontal line is given by $s(t) = t^2 - 4t + 3$ for $t \geq 0$, where t is measured in seconds and s is measured in metres. The initial displacement of the particle, $s(0) = 0^2 - 4(0) + 3 = 3$, tells us that at time 0 seconds the particle is 3 metres to the right of the origin. $s(2) = 2^2 - 4(2) + 3 = -1$ tells us that at time 2 seconds the particle is 1 metre to the left of the origin.

Consider $\int_0^2 v(t)\,dt$. Since the antiderivative of velocity is displacement we have $\int_0^2 v(t)\,dt = [s(t)]_0^2 = s(2) - s(0) = -4$.

Recall that if displacement = $s(t)$, then velocity = $v(t) = s'(t)$ and acceleration = $a(t) = v'(t) = s''(t)$.

The displacement function tells us the distance and direction a particle is from an origin at any time t.

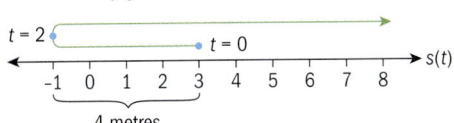

This gives us the change in displacement from time 0 to 2 seconds. It tells us that at 2 seconds the particle is 4 metres to the left of where it was at 0 seconds.

Note that $v(t) = 2t - 4$, and $v(t) = 0$ when $t = 2$. Velocity changes from negative to positive at $t = 2$, so the particle changes direction when $t = 2$.

$$\int_{t_1}^{t_2} v(t)\,dt = s(t_2) - s(t_1) \text{ is the change in displacement from } t_1 \text{ to } t_2.$$

Now consider $\int_0^5 v(t)\,dt = s(5) - s(0) = 8 - 3 = 5$. This tells us that at 5 seconds the particle is 5 metres to the right of where it was at 0 seconds.

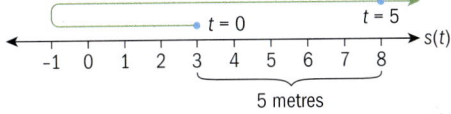

Note that the change in displacement of 5 metres is not the total distance traveled between 0 and 5 seconds. The total distance traveled is the total of 4 metres traveled to the left plus 9 metres traveled to the right or 13 metres as shown below.

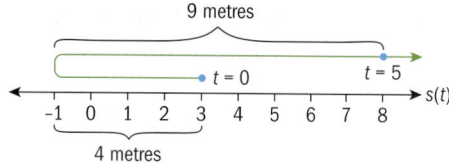

We will consider this in terms of the area under the curve of $v(t) = 2t - 4$.

Let the area of the triangle below the x-axis be A_1 and the area of the triangle above the x-axis be A_2. $\int_0^5 v(t)dt$ is the negative of A_1 plus A_2.

$$\int_0^5 |v(t)|\,dt = -A_1 + A_2 = -\frac{1}{2}(2)(4) + \frac{1}{2}(3)(6) = -4 + 9 = 5.$$

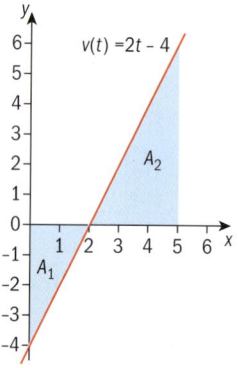

This gives us the displacement from time 0 to 5 seconds.

To find the total distance traveled from time 0 to 5 seconds we need the sum of the areas A_1 and A_2. We can find this by evaluating $\int_0^5 |v(t)|\,dt$.

$$\int_0^5 v(t)\,dt = A_1 + A_2 = \frac{1}{2}(2)(4) + \frac{1}{2}(3)(6) = 4 + 9 = 13$$

This gives us a total of 13 metres traveled from time 0 to 5 seconds.

|v(t)| means the absolute value or modulus of v(t).

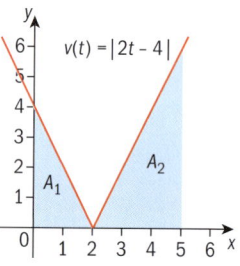

→ If v is the velocity function for a particle moving along a line, the **total distance** traveled from t_1 to t_2 is given by:

$$\text{distance} = \int_{t_1}^{t_2} |v(t)|\,dt.$$

Example 17

The displacement function for a particle moving along a horizontal line is given by $s(t) = 8 + 2t - t^2$ for $t \geq 0$, where t is measured in seconds and s is measured in metres.
  a Find the velocity of the particle at time t.
  b Find when the particle is moving right and when it is moving left.
  c Draw a motion diagram for the particle.
  d Write definite integrals to find the particle's change in displacement and the total distance traveled on the interval $0 \leq t \leq 4$. Use a GDC to evaluate the integrals and then use the motion diagram to verify the results.

▶ Continued on next page

Answers

a $v(t) = 2 - 2t$

b $2 - 2t = 0$
   $t = 1\,\text{s}$
   Moves right for $0 < t < 1$
   Moves left when $t > 1$

c $s(0) = 8$ and $s(1) = 9$

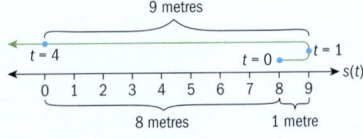

d Change in displacement
   $= \int_0^4 (2 - 2t)\,dt = -8\,\text{m}$

   Total distance $= \int_0^4 |2 - 2t|\,dt$
   $= 10\,\text{m}$

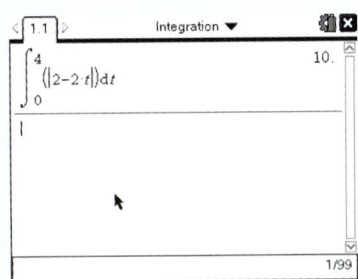

$v(t) = s'(t)$

Find when velocity equals zero.

The particle moves right when $v(t) > 0$ and left when $v(t) < 0$.

Find displacement at $t = 0$ and $t = 1$.

Change in displacement $= \int_{t_1}^{t_2} v(t)\,dt$

Total distance $= \int_{t_1}^{t_2} |v(t)|\,dt$

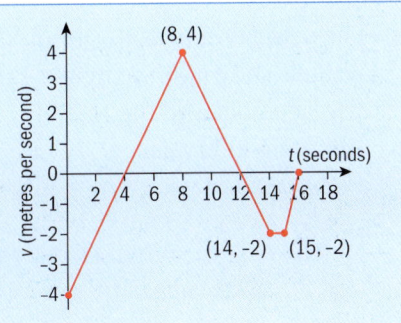

Show $s(4) = 0$ on the diagram. At 4 seconds the particle is 8 metres to the left of where it was at 0 seconds.
The particle traveled 1 metre to the right and 9 metres to the left for a total distance of 10 metres from time 0 to 4 seconds.

> **GDC help on CD:** *Alternative demonstrations for the TI-84 Plus and Casio FX-9860GII GDCs are on the CD.*

Example 18

The velocity function v, in m s^{-1}, of a particle moving along a line is shown in the figure. Find the particle's change in displacement and the total distance traveled on the interval $0 \leq t \leq 16$.

▶ Continued on next page

Answer

Let A_1, A_2 and A_3 be the areas of the two triangles and the trapezium.

Change in displacement

$$= \int_0^{16} v(t)dt$$

$$= -A_1 + A_2 - A_3$$

$$= -\frac{1}{2}(4)(4) + \frac{1}{2}(8)(4) - \frac{1}{2}(4+1)(2) = 3 \text{ m}$$

Total distance

$$= \int_0^{16} |v(t)| dt$$

$$= A_1 + A_2 + A_3$$

$$= \frac{1}{2}(4)(4) + \frac{1}{2}(8)(4) + \frac{1}{2}(4+1)(2)$$

$$= 29 \text{ m}$$

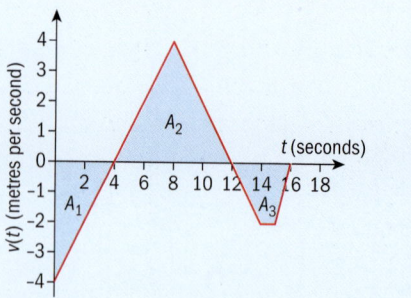
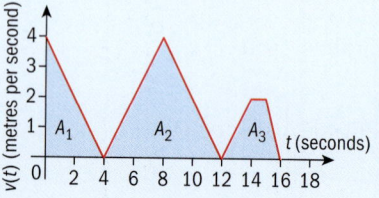

Exercise 9O

Each of questions 1–3 gives a displacement function and time interval, where t is measured in seconds and s is measured in metres.

   a Find the velocity of the particle at time t.
   b Draw a motion diagram for the particle.
   c Write definite integrals to find the particle's change in displacement and the total distance traveled on the given time interval.

Use a GDC to evaluate the integrals and then use the motion diagram to verify the results.

1 $s(t) = t^2 - 6t + 8$; $0 \le t \le 4$

2 $s(t) = \frac{1}{3}t^3 - 3t^2 + 8t$; $0 \le t \le 6$

3 $s(t) = (t-2)^3$; $0 \le t \le 4$

4   The velocity function v, in m s^{-1}, of a particle moving along a line is shown in the figure. Find the particle's change in displacement and the total distance traveled for each of the following intervals.

   a $2 \le t \le 12$
   b $0 \le t \le 5$
   c $0 \le t \le 12$

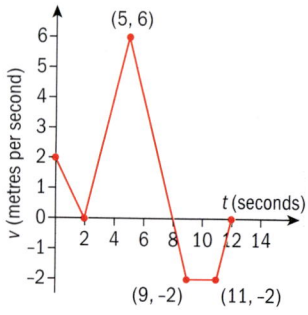

EXAM-STYLE QUESTIONS

5 The velocity, v, in m s^{-1} of a particle moving in a straight line is given by $v(t) = t^2 - 9$, where t is the time in seconds.
 a Find the acceleration of the particle at $t = 1$.
 b The initial displacement of the particle is 12 metres. Find an expression for s, the displacement, in terms of t.
 c Find the distance traveled between times 2 seconds and 8 seconds.

6 The velocity function, v, in m s^{-1} of a particle moving along a line is shown in the figure.
 a Find the acceleration when $t = 3$.
 b Write down the time interval(s) on which the particle is traveling to the right.
 c Find the total distance traveled for $0 \leq t \leq 16$.

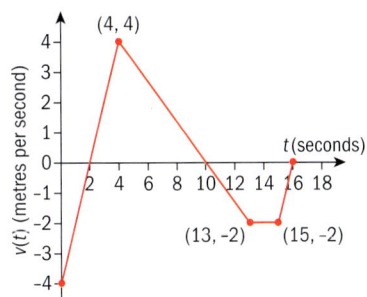

Definite integrals can be used in situations other than linear motion. We can use definite integrals to find the cumulative effect of any varying rate of change.

> The integral of a rate of change is the total change from t_1 to t_2:
> $$\int_{t_1}^{t_2} F'(t)\,dt = F(t_2) - F(t_1).$$

Example 19

A culture of bacteria is started with an initial population of 100 bacteria. The rate at which the number of bacteria changes over a one-month period can be modeled by the function $r(t) = e^{0.273t}$, where r is measured in bacteria per day.
Find the population of bacteria 20 days after the culture was started.

Answer

$r(t) = e^{0.273t}$ is a rate of change. It is the derivative of a function, say $R(t)$, that gives the number of bacteria at time t. Therefore

$$\int_0^{20} r(t)\,dt = R(20) - R(0)$$

is the change in the number of bacteria from day 0 to day 20. Since the initial population was 100 bacteria, the population after 20 days is $100 + \int_0^{20} e^{0.273t}\,dt$ or about 957 bacteria.

> Notice that the units show that the integral results in a number of bacteria.
> $$\int_0^{20} \underbrace{e^{0.273t}}_{\text{(bacteria per day)}} \underbrace{dt}_{\text{(days)}} \approx \underbrace{857}_{\text{bacteria}}$$
>
> *You could get the same result using the following longer method (see next page).*

▶ Continued on next page

$R(t) = \int e^{0.273t} \, dt$

$= \dfrac{1}{0.273} e^{0.273t} \, dt + C$

$= \dfrac{1000}{273} e^{0.273t} + C$

$100 = \dfrac{1000}{273} e^{0.273(0)} + C$

$C = 100 - \dfrac{1000}{273}$

$= \dfrac{26300}{273}$

$R(t) = \dfrac{1000}{273} e^{0.273t} + \dfrac{26300}{273}$

$R(20) = \dfrac{1000}{273} e^{0.273(20)} + \dfrac{26300}{273} \approx 957$

Find the function $R(t)$, such that $R'(t) = r(t)$. Recall that

$$\int e^{ax+b} \, dx = \dfrac{1}{a} e^{ax+b} + C.$$

Use the initial condition that $R(0) = 100$ to find C.

Find $R(20)$.

Notice how much more convenient it is to obtain the same result using

$100 + \int_0^{20} e^{0.273t} \, dt$

≈ 957.

Exercise 9P

Write an expression involving a definite integral that can be used to answer these. Use a GDC to evaluate the expression.

1 The rate of consumption of oil in a certain country from January 1, 2000 to January 1, 2010 (in billions of barrels per year) is modeled by the function $C'(t) = 18.4 e^{\frac{t}{20}}$, where t is the number of years since January 1, 2000.
Find the total consumption of oil over the 10-year period.

2 The number of spectators who enter a stadium per hour for a football game is modeled by the function $r(t) = 1375t^2 - t^3$ for $0 \le t \le 1.5$ hours. The function $r(t)$ is measured in people per hour. There are no spectators in the stadium when the gates open at $t = 0$ hours. The game begins at time $t = 1.5$ hours.
How many spectators are in the stadium when the game begins?

3 There is 36.5 cubic cm of snow on a driveway at midnight. From midnight to 8 a.m. snow accumulates on the driveway at a rate modeled by the function $s(t) = 5te^{(-0.01t^5 + 0.13t^3 - 0.38t^2 - 0.3t + 0.9)}$, where t is measured in hours and s in cm³.
How many cubic cm of snow are on the driveway at 8 a.m.?

4 Water begins leaking from a tank holding 4000 gallons of water. The rate at which it is leaking, measured in gallons per minute, can be modeled by the function $r(t) = -133\left(1 - \dfrac{t}{60}\right)$.

How much water is in the tank at the end of 20 minutes?

Review exercise

1. Find the indefinite integral.

   a $\int (4x^3 - 8x + 6)\,dx$ b $\int \sqrt[3]{x^4}\,dx$ c $\int \dfrac{3}{x^4}\,dx$

   d $\int \dfrac{5x^4 - 3x}{6x^2}\,dx$ e $\int e^{4x}\,dx$ f $\int x^2(x^3+1)^4\,dx$

   g $\int \dfrac{1}{2x+3}\,dx$ h $\int \dfrac{\ln x}{x}\,dx$ i $\int (3x^2+1)(6x)\,dx$

   j $\int \dfrac{2e^x}{e^x+3}\,dx$ k $\int 3\sqrt{2x-5}\,dx$ l $\int 2x\,e^{2x^2}\,dx$

2. Find the definite integral.

   a $\int_0^2 (3x^2 - 6)\,dx$ b $\int_4^{16} \dfrac{4}{\sqrt{t}}\,dt$ c $\int_1^{e^2} \dfrac{4}{x}\,dx$

   d $\int_0^1 6x\,e^{3x^2+3}\,dx$ e $\int_{-1}^1 (3x-1)^3\,dx$ f $\int_0^2 \dfrac{1}{2x+1}\,dx$

EXAM-STYLE QUESTIONS

3. The diagram shows part of the graph of $f(x) = x^2 - 1$. Regions A and B are shaded.

   a Write down an expression for the area of region B.
   b Calculate the area of region B.
   c Write down an expression for the total area of shaded regions A and B. (You need not evaluate the expression.)
   d Region B is rotated about the x-axis. Write down an expression for the volume of the solid formed. (You need not evaluate the expression.)

   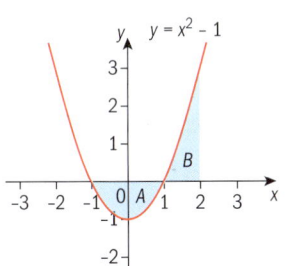

4. A curve with equation $y = f(x)$ passes through the point $(2, 6)$. Its gradient function is $f'(x) = 3x - 2$. Find the equation of the curve.

5. Given that $\int_1^5 f(x)\,dx = 20$, deduce the value of

   a $\int_1^5 \dfrac{1}{4}f(x)\,dx$;   b $\int_1^5 [f(x) + 2]\,dx$

6. A particle moves along a straight line so that its velocity, $v\,\text{m s}^{-1}$ at time t seconds is given by $v(t) = 4e^{2t} + 2$. When $t = 0$, the displacement, s, of the particle is 8 metres. Find an expression for s in terms of t.

7. Given $\int_1^k \dfrac{1}{2x-1}\,dx = \ln 5$, find the value of k.

Review exercise

EXAM-STYLE QUESTIONS

1. Find the volume of the solid formed when the region bounded by $f(x) = 4 - x^2$ and the x-axis is rotated $360°$ about the x-axis.

2. A particle moves along a horizontal line with velocity v m s^{-1} given by $v(t) = 2t^2 - 11t + 12$ where $t \geq 0$.
   a. Write down an expression for the acceleration, a m s^{-2}, in terms of t.
   b. The particle is moving to the left for $a < t < b$. Find the value of a and the value of b.
   c. Find the total distance the particle travels from time 2 seconds to time 5 seconds.

3. a. Find the equation of the tangent line to $f(x) = x^3 - 2$ at $x = -1$.
   b. The tangent line intersects $f(x) = x^3 - 2$ at a second point. Find the coordinates of this point.
   c. Graph f and the tangent line.
   d. Write an expression for the area enclosed by the graphs of f and the tangent line and then find the area.

CHAPTER 9 SUMMARY
Antiderivatives and the indefinite integral

- **Power rule:** $\int x^n \, dx = \dfrac{1}{n+1} x^{n+1} + C, \; n \neq -1$
- **Constant rule:** $\int k \, dx = kx + C$
- **Constant multiple rule:** $\int kf(x) \, dx = k \int f(x) \, dx$
- **Sum or difference rule:** $\int (f(x) \pm g(x)) \, dx = \int f(x) \, dx \pm \int g(x) \, dx$

More on indefinite integrals

- $\int \dfrac{1}{x} \, dx = \ln x + C, \; x > 0$
- $\int e^x \, dx = e^x + C$
- $\int (ax+b)^n \, dx = \dfrac{1}{a}\left(\dfrac{1}{n+1}(ax+b)^{n+1}\right) + C$
- $\int e^{ax+b} \, dx = \dfrac{1}{a} e^{ax+b} + C$
- $\int \dfrac{1}{ax+b} \, dx = \dfrac{1}{a} \ln(ax+b) + C, \; x > -\dfrac{b}{a}$

Continued on next page

Area and definite integrals

- When f is a non-negative function for $a \leq x \leq b$, $\int_a^b f(x)dx$ gives the area under the curve from $x = a$ to $x = b$.

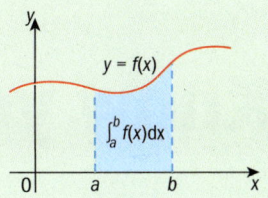

- **Some properties of definite integrals**

  1. $\int_a^b kf(x)\,dx = k\int_a^b f(x)\,dx$

  2. $\int_a^b (f(x) \pm g(x))\,dx = \int_a^b f(x)dx \pm \int_a^b g(x)dx$

  3. $\int_a^a f(x)\,dx = 0$

  4. $\int_a^b f(x)\,dx = -\int_b^a f(x)\,dx$

  5. $\int_a^b f(x)\,dx = \int_a^c f(x)\,dx + \int_c^b f(x)\,dx$

Fundamental Theorem of Calculus

If f is a continuous function on the interval $a \leq x \leq b$ and F is an antiderivative of f on $a \leq x \leq b$, then

$$\int_a^b f(x)\,dx = [F(x)]_a^b = F(b) - F(a)$$

Area between two curves

- If y_1 and y_2 are continuous on $a \leq x \leq b$ and $y_1 \geq y_2$ for all x in $a \leq x \leq b$, then the area between y_1 and y_2 from $x = a$ to $x = b$ is given by $\int_a^b (y_1 - y_2)\,dx$.

Volume of revolution

- If $y = f(x)$ is continuous on $a \leq x \leq b$ and the region bounded by $y = f(x)$ and the x-axis between $x = a$ and $x = b$ is rotated $360°$ about the x-axis then the volume of the solid formed is given by $\int_a^b \pi(f(x))^2$ or $\int_a^b \pi y^2 \, dx$.

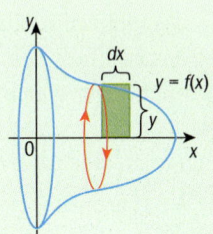

Definite integrals with linear motion and other problems

- $\int_{t_1}^{t_2} v(t)dt = s(t_2) - s(t_1)$ is the change in displacement from t_1 to t_2.

- If v is the velocity function for a particle moving along a line, the total distance traveled from t_1 to t_2 is given by: distance $= \int_{t_1}^{t_2} |v(t)|\,dt$.

Theory of knowledge

Know your limits!

The method of exhaustion

The ancient Greeks used concepts of calculus long before calculus was formalized. To find an estimate for the area of a circle of radius one, the ancient Greeks constructed regular inscribed and circumscribed polygons with increasing numbers of sides.

Let a_n be the areas of the regular polygons with n sides inscribed in a circle of radius one and A_n be the areas of the circumscribed polygons. The ancient Greeks found that both $\lim_{n\to\infty} A_n$ and $\lim_{n\to\infty} a_n$ were equal to π.

- What conclusion were they able to draw from these facts?
- Can you think of other applications of limits in real life?

Newton vs. Leibniz

The development of calculus was truly a culmination of centuries of work by mathematicians all over the world. The 17th-century mathematicians Isaac Newton (English) and Gottfried Wilhelm Leibniz (German) are recognized for the actual development of calculus. One of the most famous conflicts in mathematical history is the argument over which one of them invented or discovered calculus first and whether any plagiarism was involved.

- What are some possible consequences when people seek personal acclaim for their work?
- Suppose that Newton and Leibniz did develop calculus independently of one another. Would this offer support to the idea that calculus was discovered or that it was invented?

Although the debate was never fully resolved, today it is generally believed that Newton and Leibniz did develop calculus independently of one another. Modern-day calculus emerged in the 19th century, due to the efforts of mathematicians such as Augustin-Louis Cauchy (French), Bernhard Riemann (German), Karl Weierstrass (German), and others.

- Did the work of these mathematicians arise from the need to solve certain real-world problems or purely from intellectual curiosity?

Gabriel's horn

Consider the solid formed when the region bounded by $f(x) = \frac{1}{x}$, $x = 1$ and $x = a$, $a > 1$ is rotated about the x-axis. If $a \to \infty$, the solid is known as **Gabriel's horn**.

The volume of the solid of revolution about the x-axis is given by $\pi \int_a^b y^2 \, dx$. It can be shown that the surface area of the solid is given by $2\pi \int_a^b y\sqrt{1 + (y')^2} \, dx$.

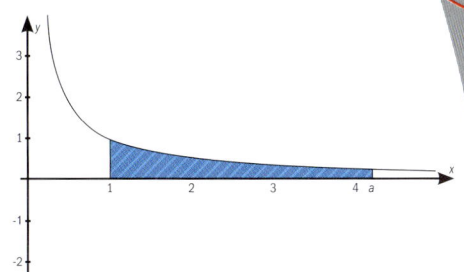

- Use a GDC to find, to four decimal places, the volume and surface area of the solid described above for the given values of a. Write them in a copy of the table. Then make a conjecture about what the volume and surface area approach as a approaches infinity.

a	Volume = $\pi \int_1^a \left(\frac{1}{x}\right)^2 dx$	Surface area = $2\pi \int_1^a \left[\frac{1}{x}\sqrt{1 + \frac{1}{x^4}}\right] dx$
10		
100		
1000		
10 000		
100 000		
1 000 000		
$a \to \infty$	Volume $\to$	Surface area $\to$

- Based on the results in your table, how much paint would it take to fill up Gabriel's horn?
- How much paint would it take to cover its surface?

Paradoxes

A result that defies logic is called a paradox. Gabriel's horn is one example of a paradox. Research some other examples of paradoxes.

10 Bivariate analysis

CHAPTER OBJECTIVES:

5.4 Linear correlation of bivariate data; Pearson's product–moment correlation coefficient *r*; scatter diagrams, lines of best fit; mathematical and contextual interpretation.

5.4 The equation of the regression line of *y* on *x*; use of the equation for prediction purposes.

Before you start

You should know how to:

1 Calculate simple positive exponents
   e.g. Evaluate 3^4.
   $3^4 = 3 \times 3 \times 3 \times 3 = 81$
   e.g. Evaluate $\left(\dfrac{2}{5}\right)^3$.
   $$\left(\dfrac{2}{5}\right)^3 = \dfrac{2^3}{5^3} = \dfrac{2 \times 2 \times 2}{5 \times 5 \times 5}$$
   $$= \dfrac{8}{125}$$

2 Convert numbers to exponential form
   e.g. Find n given $2^n = 8$.
   $2 \times 2 \times 2 = 8$
   $2^3 = 8$
   $n = 3$

Skills check

1 Evaluate:
   a 2^5
   b 3^3
   c 7^3
   d $\left(\dfrac{1}{2}\right)^7$
   e $\left(\dfrac{3}{4}\right)^4$
   f 0.001^3

2 State the value of n in the following equations:
   a $2^n = 16$
   b $3^n = 243$
   c $7^n = 343$
   d $5^n = 625$
   e $(-4)^n = -64$
   f $\left(\dfrac{1}{2}\right)^n = \dfrac{1}{8}$

In 1956, an Australian statistician, Oliver Lancaster, made the first convincing case for a link between exposure to sunlight and skin cancer. He observed that the rate of skin cancer in Australia among Caucasians was strongly correlated with latitude, and hence with amount of sunlight: the northern states had higher rates than the southern ones. This was well before the hole in the ozone layer! His discovery was the result of careful data collection and comparison of skin cancer rates.

In Chapter 5 we dealt with **univariate** analysis. We defined a **population**: it consists of all of the measurements of interest. A **sample** is a portion of the population.

Suppose that we are interested in studying the height x and weight y of adult males.

The sampling units are adult males and the **bivariate** data contains all of the pairs (x, y) of height and weight of the males in our sample.

Sampling unit	Variable(s)	Population
adult males	height	univariate
adult males	weight	univariate
adult males	height, weight	bivariate

→ Bivariate analysis is concerned with the relationships between pairs of variables (x, y) in a data set.

In this chapter we will look for associations between two sets of data using graphs, representing a relationship as an equation and using a scale to describe the strength of the relationship.

Investigation – leaning tower of Pisa

The bell tower of Pisa cathedral was built in 1178 and soon began leaning to one side – hence its name. The measurements below show the lean in tenths of a millimetre beyond 2.9 metres. So in 1975 the tower was leaning 2.9642 metres from the vertical.

Year	1975	1976	1977	1978	1979	1980	1981	1982	1983	1984	1985	1986	1987
Lean	642	644	656	667	673	688	696	698	713	717	725	742	757

Does it look like the lean of the tower is increasing with time?
If so, how fast is the lean increasing with time?
Is there evidence that the lean changes significantly with time?
Is there an approximate formula for calculating the lean?
Can you predict the lean in the future?

10.1 Scatter diagrams

One way to view data is by showing it on a **scatter diagram**.

→ **Scatter diagrams** (also called scatter plots) are used to investigate the possible relationship between two variables that both relate to the same 'event'.

Correlation is a way to measure how *associated* or *related* two variables are. The purpose of doing correlations is to allow us to make a prediction about one variable based on what we know about another variable.

Scatter diagrams are similar to line graphs in that they use horizontal and vertical axes to plot data points. However, they have a very specific purpose. A scatter diagram shows how much one variable affects another.

→ The relationship between two variables is called their **correlation**.

→ To draw a scatter diagram plot the (x, y) values from the data table as dots on a graph. The pattern formed by the dots can give us some indication of the correlation.

The **independent variable** should be on the horizontal axis with the **dependent variable** on the vertical axis.

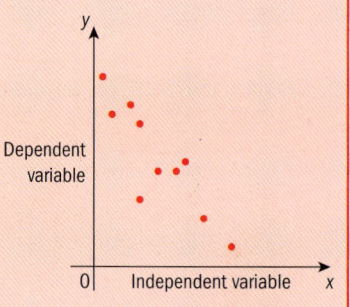

For the leaning tower of Pisa example, we think that the lean increases with time. Time is the **independent** variable. The lean depends on the time, so the amount of lean is the **dependent** variable.

→ A general upward trend in the pattern of dots shows **positive** correlation.

The value of the dependent variable increases as the value of the independent variable increases.

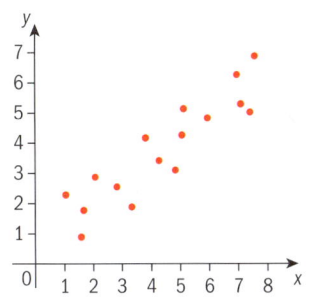

→ A general downward trend in the pattern of dots shows **negative** correlation.

The dependent variable decreases as the independent variable increases.

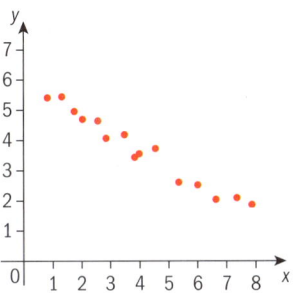

→ Scattered points with no trend may indicate correlation close to **zero**.

Scatter diagrams allow us to assess the strength of a correlation. Here are differing amounts of positive correlation:

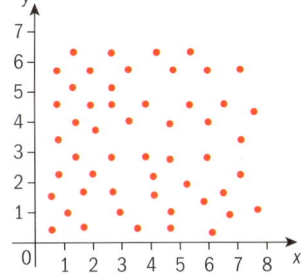

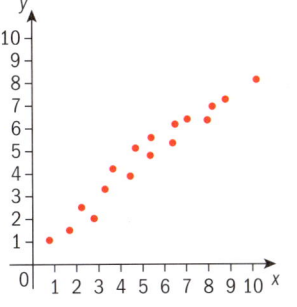

Strong positive correlation:
y increases as x increases

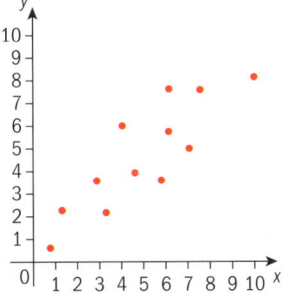

Moderate positive correlation

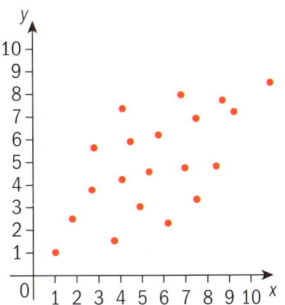

Weak positive correlation

Here are differing amounts of negative correlation:

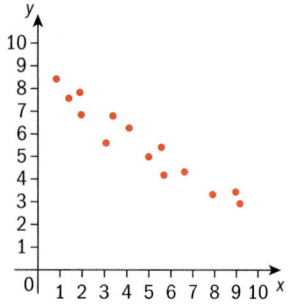

Strong negative correlation: y decreases as x increases

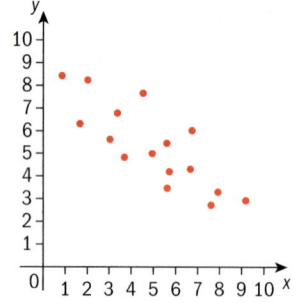

Moderate negative correlation

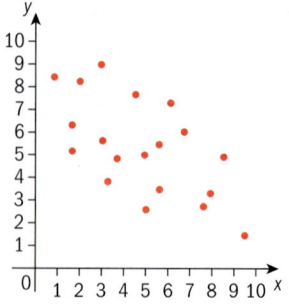

Weak negative correlation

Not all relationships are linear.

The points on this graph are approximately linear.

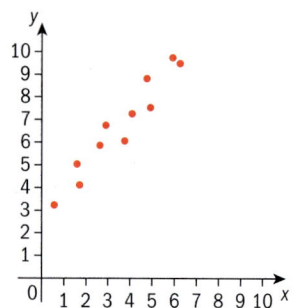

The points on this graph would be represented by a curve.
There is a **non-linear** relationship between the variables.

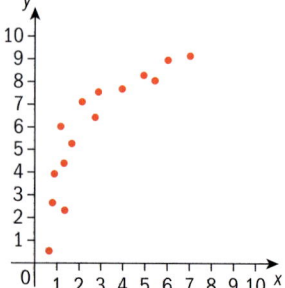

Causation

→ A correlation between two data sets does not necessarily mean that one causes the other.

Here is an example: the shoe sizes of grade school students and the students' vocabulary have a strong, positive correlation. In other words, the larger the shoe size, the larger the vocabulary the student has. Now it is easy to see that shoe size and vocabulary have nothing to do with each other, but they are highly correlated. The reason is that there is a **confounding factor**, age. The older grade school students will have larger shoe sizes and often a larger vocabulary.

You may wish to use 'causation versus correlation' as the stimulus for an exploration.

Example 1

a Represent this data on a scatter diagram.

x	1	2	3	4	4	6	6	6	7	8
y	1	3	3	5	6	7	5	6	8	9

b Is the relationship linear or non-linear?
c Describe the type and strength of the relationship.

Answers

a

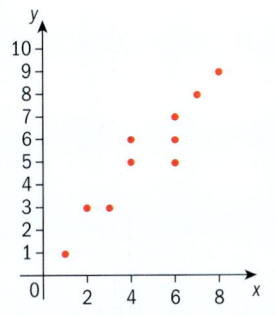

b This is a **linear** relationship.

c There is a **strong**, **positive** correlation.

Compare the scatter diagram with the examples earlier.

Exercise 10A

1 Describe the correlation shown by each of these scatter diagrams.

a

b

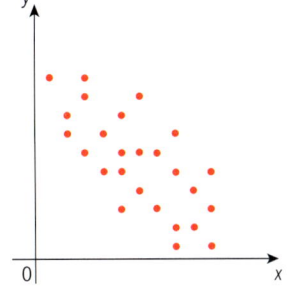

c

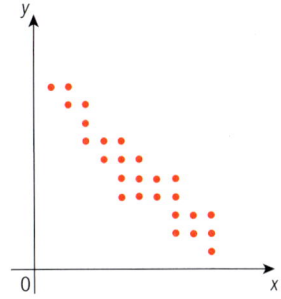

d

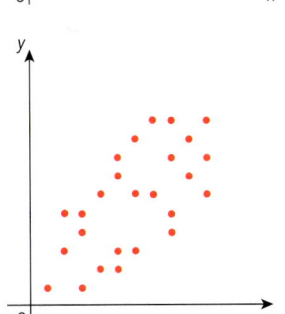

e

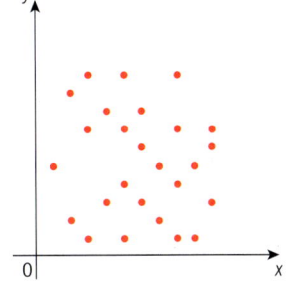

2 For the following data sets:
  i is the correlation positive, negative or is there no association
  ii is the relationship linear or non-linear
  iii is the association strong, moderate, weak or zero?

  a

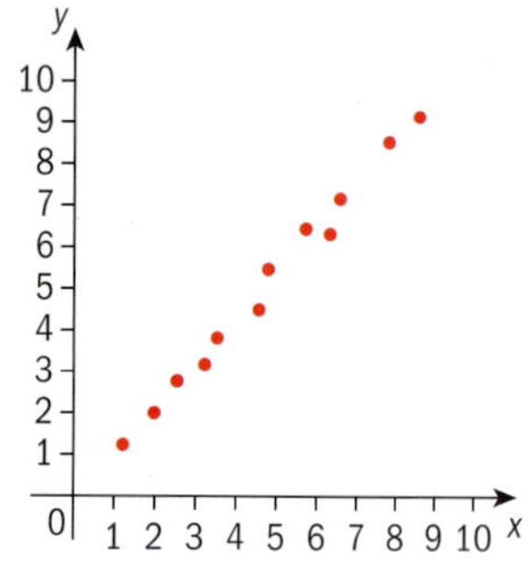

  b

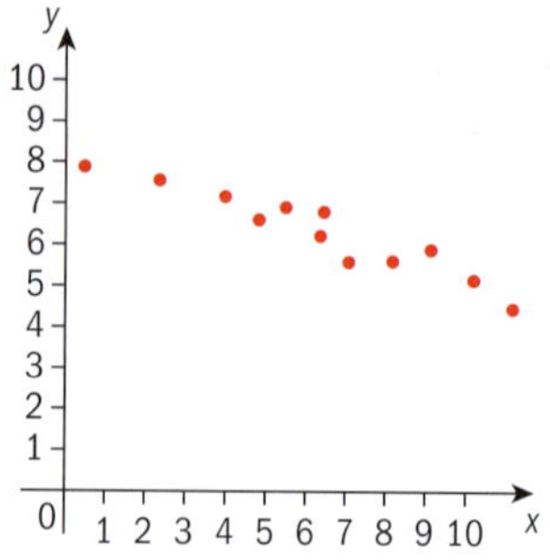

  c

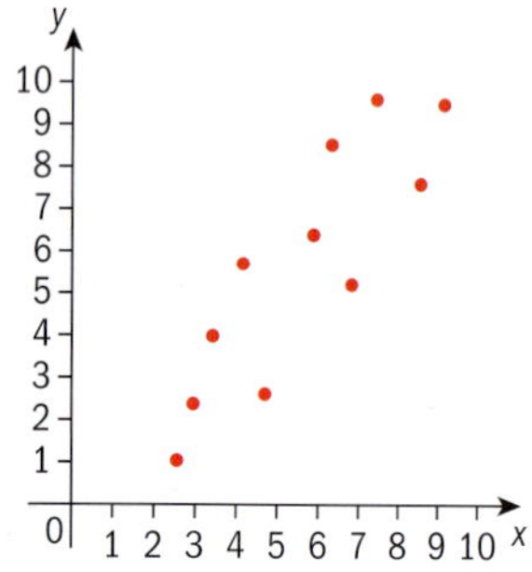

  d

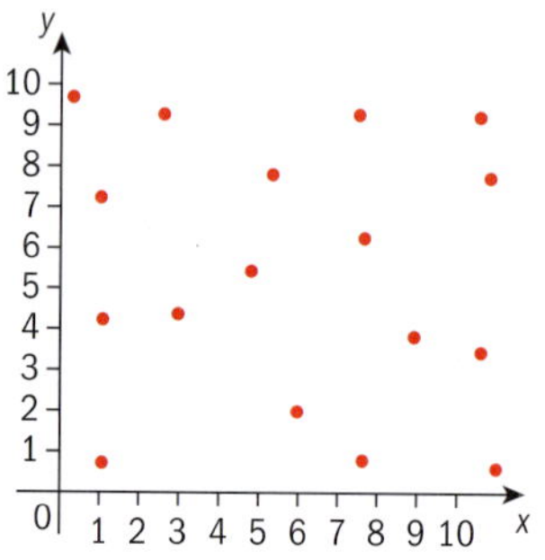

  e

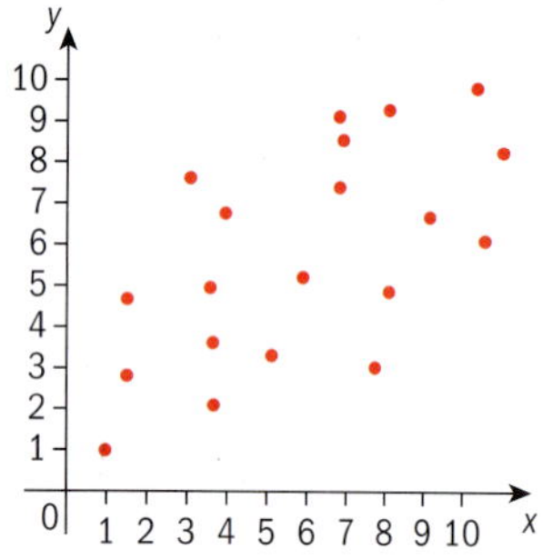

  f

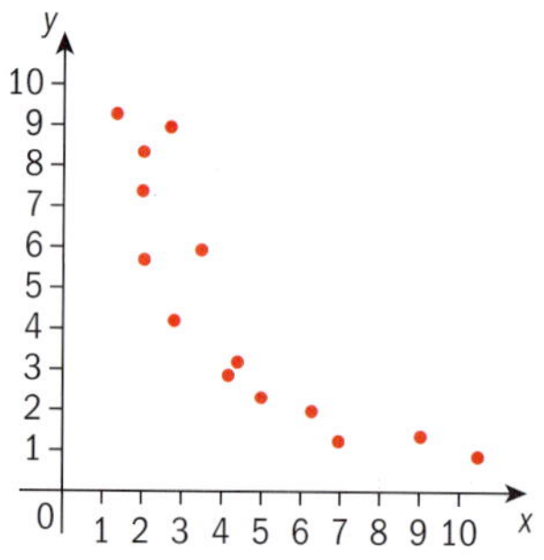

3 Copy and complete these sentences.
  a If the independent and dependent variables show a positive correlation then as the independent variable increases the dependent variable ....
  b If the independent and dependent variables show a negative correlation then as the independent variable increases the dependent variable ....

4 This table shows the rainfall in cm in Tennessee from 2000 to 2008.

Year	2000	2001	2002	2003	2004	2005	2006	2007	2008
Rainfall	42	51	39	44	31	33	30	28	21

   a Show this data on a scatter diagram.
   b Describe the correlation.
   c In general, what has happened to the rainfall since the year 2000?

5 This table shows a group of friends with their mathematics and science scores.

Friend	Tim	Ted	Tom	Tod	May	Ray	Kay	Jay
Mathematics	85	75	66	80	70	95	90	60
Science	75	65	40	72	55	88	80	40

   1 Draw a scatter diagram to represent this data.
   2 Describe the correlation in terms of strength, direction and form.

Investigation – leaning tower of Pisa (continued)

a Construct a scatter diagram for the data from the leaning tower of Pisa investigation at the start of this chapter.
b Describe the correlation.
c What is happening to the lean as the years increase?
d Research the latest developments on the efforts to save the leaning tower of Pisa. Comment on the dangers of extrapolation.

> **Extrapolation** means estimating a value at a point that is larger than (or smaller than) the data you have. Extrapolating here means assuming that the trend of the lean will remain the same.

10.2 The line of best fit

→ A **line of best fit** or trend line is drawn on a scatter diagram to find the direction of an association between two variables and to show the trend. This line of best fit can then be used to make predictions.

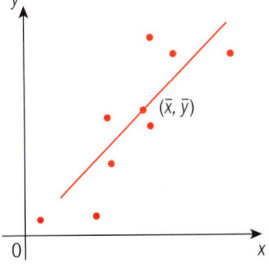

→ To draw a line of best fit by eye draw a line that will balance the number of points above the line with the number of points below the line. An improvement is to have a reference point: one point for the line to pass through. This is the **mean point** and is calculated by finding the mean of the x-values and the mean of the y-values.

> The mean point is written as $(\bar{x}, \bar{y})$.

Example 2

Is there a relationship between the grams of fat and the total calories in fast food?

Meal	Total fat (g)	Total calories
Hamburger	9	260
Cheeseburger	13	320
Quarter Pounder	21	420
Quarter Pounder with Cheese	30	530
Big Burger	31	560
Toasted Sandwich	31	550
Chicken Wings	34	590
Crispy Chicken	25	500
Fish Fillet	28	560
Grilled Chicken	20	440
Grilled Chicken Light	5	300

a Find the mean number of grams of fat.
b Find the mean number of calories.
c Construct a scatter diagram for this data.
d Plot the mean point on your scatter diagram and use it to draw a line of best fit.

Answers

a Mean grams of fat $= \dfrac{247}{11}$
$= 22.\dot{4}\dot{5}$

b Mean no. of calories $= \dfrac{5030}{11}$
$= 457.\dot{2}\dot{7}$

c and **d**

Mean grams of fat
$= \dfrac{\text{Total grams of fat}}{\text{Number of meals}}$

Mean no. of calories
$= \dfrac{\text{Total no. of calories}}{\text{Number of meals}}$

The line of best fit does not have to pass through (0, 0). It must pass through the mean point and have roughly the same number of data points either side of it.

Hence
$(\bar{x}, \bar{y}) = (22.\dot{4}\dot{5}, 457.\dot{2}\dot{7})$

A 'line of best fit' is also called a **regression line**. The British scientist and statistician Francis Galton (1822–1911) coined the term *regression* in the 19th century.

340 Bivariate analysis

Exercise 10B

1 The table below shows the relationship between the length and width of a mango tree leaf measured in mm.

Length	35	50	78	80	95	105	118	125	136	145
Width	25	30	38	50	36	42	52	48	58	62

   a Find the mean point.
   b Construct a scatter diagram and draw a line of best fit through your mean point.

2 The following table gives the heights and weights of ten sixteen-year-old students.

Name	Abe	Bill	Chavo	Dee	Eddie	Fah	Grace	Hanna	Ivy	Justin
Height (cm)	182	173	162	178	190	161	180	172	167	185
Weight (kg)	73	68	60	66	75	50	80	60	56	72

   a Find: **i** the mean height **ii** the mean weight.
   b Construct a scatter diagram and draw a line of best fit through your mean point.

3 The table below shows the number of hours spent studying and the increase in the students' grades in mathematics.

Hours studying	0	1	2	3	4	5	6	7	8
Increase in grade	−1	1	3	7	9	9	8	10	14

   a Find the mean point.
   b Construct a scatter diagram and draw a line of best fit through your mean point.
   c Describe the correlation.
   d What can you say about the effect of the number of hours spent studying mathematics and the increase in grade?

> What are the risks of extrapolation? You may wish to explore extrapolation in financial or climate models.

The equation of the line of best fit through the mean point

Raw data rarely fit a straight line exactly. Usually, you must be satisfied with rough predictions. Typically, you have a set of data whose scatter diagram appears to 'fit' a straight line, the line of best fit.

→ The equation of the line of best fit, also called the **regression line**, can be used for prediction purposes.

Example 3

Miss Lincy's 10 students' scores, out of 100, for their classwork and final exam are shown below.

Student	Ed	Craig	Uma	Phil	Jenny	James	Ron	Bill	Caroline	Steve
Classwork	95	66	88	75	90	82	50	45	80	84
Final	95	59	85	77	92	70	40	50	Abs	80

Caroline was absent for the final. Do not include her grades in finding the mean point.
a Find the mean classwork score.
b Find the mean final exam score.
c Construct a scatter diagram and draw a line of best fit through your mean point.
d Find the equation of the regression line.
e Use the equation of the regression line to estimate Caroline's score for the final exam.

Answers

a Mean classwork score = $\dfrac{\text{Classwork total}}{\text{Number of students}}$

   Mean classwork score = $\dfrac{675}{9} = 75$

b Mean final exam score = $\dfrac{\text{Final exam total}}{\text{Number of students}}$

   Mean final exam score = $\dfrac{648}{9} = 72$

c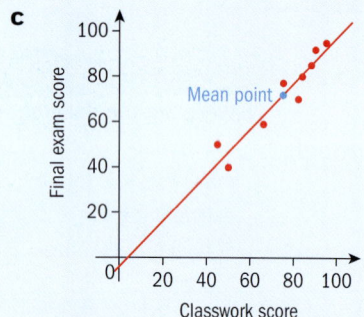

d Using the mean point and Uma's results, we have $(x_1, y_1) = (75, 72)$
   $(x_2, y_2) = (88, 85)$
   $m = \dfrac{85 - 72}{88 - 75} = 1$
   The equation of the line is:
   $y - 72 = 1(x - 75)$
   $y = x - 3$

   Use $m = \dfrac{y_2 - y_1}{x_2 - x_1}$ where (x_1, y_1) is the mean point and (x_2, y_2) is any point on the line. Use $y - y_1 = m(x - x_1)$ for the equation of the line.

e $y = 80 - 3 = 77$
   Caroline's exam score is estimated to be 77.

   Caroline's classwork score was 80. Let $x = 80$.

   Using the line of best fit to predict a data value within the range of the given data is called **interpolation**. It is generally more reliable than extrapolation.

Exercise 10C

EXAM-STYLE QUESTIONS

1 Tomato plants are at risk to a disease called blight. An agricultural scientist wishes to see how the temperature of the greenhouse affects the disease. She designs an experiment in which she monitors the percentage of diseased leaves occurring at different temperatures.

Temperature (x °F)	70	72	74	76	78	80
Percentage of diseased leaves (y)	12.3	9.5	7.7	6.1	4.3	2.3

a Draw a scatter diagram with a regression line passing through the mean point.
b Find the equation of the regression line.
c Use your equation to estimate the number of diseased leaves at 75 °F.

2 Market research of real estate investments reveals the following sales figures for new homes of different prices over the past year.

Price (thousands of £)	160	180	200	220	240	260	280
Sales of new homes this year	126	103	82	75	82	40	20

a Find the mean house price.
b Find the mean number of sales.
c Draw a scatter diagram with a regression line passing through the mean point.
d Find the equation of the regression line.
e Use your equation to estimate the number of new homes priced at £230 000 that were sold.

> Extension material on CD:
> Worksheet 10 - More bivariate analysis

Understanding the regression line

Example 4

> A study was done to investigate the relationship between the age x in years of a young person and the time y in minutes in which the child can run one kilometre. Data from children between the ages of 7 and 18 was collected. The equation of the regression line was found to be $y = 20 - \frac{1}{2}x$. Interpret the slope and y-intercept.
>
> **Answer**
>
> In the context of the question, we can say that, on average, as a child ages one year their time to run a kilometre goes down by 30 seconds (half a minute).
>
> For this question, the y-intercept is not relevant, since 0-year-old children cannot run one kilometre.
>
> The slope is $-\frac{1}{2}$. What this means is that for every increase of 1 in x there is a decrease of $\frac{1}{2}$ in y.
>
> The y-intercept is 20, which means that when x is 0, y is 20.

> The y-intercept is the height of the line when $x = 0$, and might not always have a meaning. Be careful with your interpretation of the intercept. Sometimes the value $x = 0$ is impossible or represents a dangerous extrapolation outside the range of the data.

Example 5

> A biologist wants to study the relationship between the number of trees x per hectare and the number of birds y per hectare.
> She calculates the equation of the regression line to be $y = 8 + 5.4x$.
> State the gradient and the y-intercept and interpret them.
>
> **Answer**
>
> The slope is 5.4. This means that for every additional tree, you can expect an average of 5.4 additional birds per hectare.
> The y-intercept is 8, which means that an area with no trees averages 8 birds per hectare.

Note that all these interpretations follow a pattern:
The **gradient** of the line is the amount by which y increases when x increases by 1 unit.

Exercise 10D

For each scenario, state what the slope and y-intercept are and interpret them if relevant. If not relevant, explain why.

1. A social science teacher has collected data on the number of days x per year a particular student plays sports and the number of hours y of homework that the same student does per week.
   She came up with the equation of the regression line $y = 40 - 0.3x$

2. A police chief wants to investigate the relationship between the number of times y a person has been convicted of a crime and the number of criminals x that person knows.
   The equation was found to be $y = 0.5 + 6x$

3. A doctor is researching the relationship between the number of packs of cigarettes x a person smokes per day and the number of days per year y the person is sick.
   The doctor comes up with the equation of the regression line $y = 7 + 2.4x$

4. A skateboard salesman wants to investigate the number of customers y that came to his shop each year x.
   The equation of the regression line is $y = -5 + 100x$

5. A group of mathematics and science teachers wished to compare their exam scores.
   The science score y and the mathematics score x gave the regression line $y = -10 + 0.8x$

10.3 Least squares regression

The term **regression** is used in statistics quite differently to other contexts. The method was first used to examine the relationship between the heights of fathers and sons. The two are related, of course, but the slope is less than 1.0. A tall father tends to have sons shorter than him; a short father tends to have sons taller than him. The height of sons regresses to (moves towards) the mean. The term 'regression' is now used for many sorts of curve fitting.

Let us visit the leaning tower of Pisa problem again. We know that there is a strong, positive correlation between the number of years and the lean of the tower. We can construct a scatter diagram to illustrate the data, find a mean point and draw a line of best fit (regression line) through the mean point. Inaccuracies occur because we only have one point to draw the line through and the line of best fit is then drawn 'by eye'.

There is another way to improve our line, involving **residuals**.

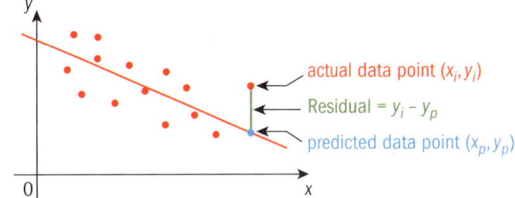

→ A **residual** is the vertical distance between a data point and the graph of a regression equation.

The residual is positive if the data point is above the graph.

The residual is negative if the data point is below the graph.

The residual is 0 only when the graph passes through the data point.

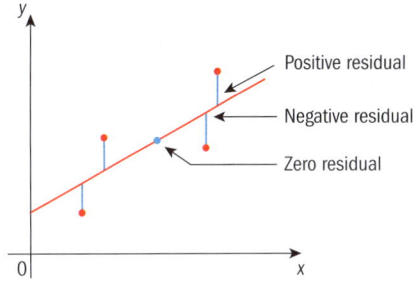

The equation of the regression line of y on x

The least squares regression line uses our previous formula $y - y_1 = m(x - x_1)$, but now uses the method of least squares to find a suitable value for the slope, m.

→ The least squares regression line is the one that has the smallest possible value for the sum of the squares of the residuals.

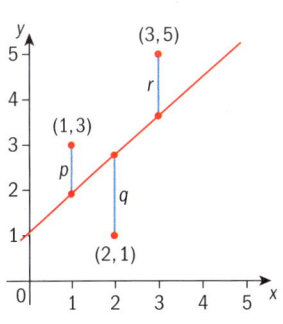

In the diagram we aim to make $p^2 + q^2 + r^2$ as close to zero as possible.

A rather complicated formula emerges:

> The formula for finding the gradient, or slope (m) of a regression line is:
> $$m = \frac{S_{xy}}{(S_x)^2}, \text{ where}$$
> $$S_{xy} = \sum xy - \frac{(\sum x)(\sum y)}{n} \text{ and}$$
> $$(S_x)^2 = \sum x^2 - \frac{(\sum x)^2}{n}$$

> The earliest form of regression was the method of least squares that was published by Legendre in 1805, and by Gauss four years later. Legendre and Gauss both applied the method to the problem of determining, from astronomical observations, the orbits of bodies about the Sun.

> Σ is the Greek letter 'S' and is used as an instruction to sum data. Σxy means the sum of all the xy values.

Example 6

Use the least squares regression formula to find the equation of the regression line through the points $(1, 3)$, $(2, 1)$, and $(3, 5)$ from the diagram on page 345.

Answer

$S_{xy} = \sum xy - \frac{(\sum x)(\sum y)}{n}$

$= 20 - \frac{6 \times 9}{3}$

$= 2$

$(S_x)^2 = \sum x^2 - \frac{(\sum x)^2}{n}$

$= 14 - \frac{6^2}{3}$

$= 2$

x	y	xy	x²
1	3	3	1
2	1	2	4
3	5	15	9
6	9	20	14

The terms in the formula

The sum of each column

> The line of regression of y on x, which can be used to estimate y given x.

The equation of the regression line is

$y - \bar{y} = \frac{S_{xy}}{(S_x)^2}(x - \bar{x})$

$y - 3 = \frac{2}{2}(x - 2)$

$y = x + 1$

The mean point $(\bar{x}, \bar{y})$ is $(2, 3)$.

Now you have seen how the formula for the equation of the regression line works, from now on you can use your GDC to find it.

> → You should use your GDC to find the equation of the regression line in examinations.

> See GDC Chapter 17, Sections 5.15 and 5.16.

Example 7

The table shows the distance in km and airfares in US dollars from Changi airport, Singapore, to twelve destinations.

Distance	Fare
576	178
370	138
612	94
1216	278
409	158
1502	258
946	198
998	188
189	98
787	179
210	138
737	98

a Use your GDC to sketch a scatter diagram of this data with the line of best fit.
b Write down the equation of your line of best fit.
c Use your equation to estimate the cost of a 1000 km flight.

Answers

a

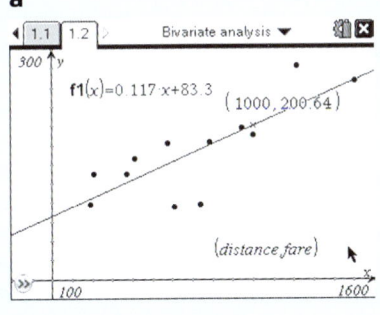

GDC help on CD: *Alternative demonstrations for the TI-84 Plus and Casio FX-9860GII GDCs are on the CD.*

b $y = 0.117x + 83.3$

You will usually have to round your answers to three significant figures.

c cost = $(0.117 \times 1000) + 83.3$
       = $200.30

Cost = \$(0.117 × distance + 83.3)
Dollars and cents – two decimal places.

 Exercise 10E

You should use your GDC for all of this exercise.

1 A patient is given medicine by a drip feed and its concentration in his blood is measured at hourly intervals. The doctors believe that a linear relationship will exist between the variables.

Time x (hours)	0	1	2	3	4	5	6
Concentration y	2.4	4.3	5.0	6.9	9.1	11.4	13.5

It would not be sensible to predict the concentration after 8 hours from this equation – we don't know whether the relationship will continue to be linear. The process of trying to predict a value from outside the range of your data is called **extrapolation**.

a Show the data on a scatter diagram with a line of best fit.
b Write down the equation of the line.
c Find the concentration of medicine in the blood after 3.5 hours.

2 The table below shows the value of Jai's car in thousands of Malaysian Ringgits for the first seven years after it was purchased.

Age (yrs)	0	1	2	3	4	5	6	7
Cost (MYR 1000)	30	25	21	19	18	15	12	10

  a Show the price of the car on a scatter diagram with a line of best fit.
  b Write down the equation of the regression line.
  c Estimate the cost of his car after $4\frac{1}{2}$ years.
  d Suppose Jai takes good care of the car. Explain why the equation cannot be used to estimate the cost of the car after 50 years.

3 The table below shows ten people who bought gym membership and the number of hours that they exercised in the past week.

Person	Nat	Nick	Nit	Noi	Nancy	Norm	Nada	Ned	New	Nat
Months of membership	7	8	9	1	5	12	2	10	4	6
Hours of exercise	5	3	5	10	5	3	8	2	8	7

  a Show the data on a scatter diagram with a line of best fit.
  b Find the equation of the regression line.
  c If Nino has been a member for 3 months, estimate how many hours he exercised last week.
  d Could you use your equation to estimate how many hours Naja exercised after 2 years membership? Explain why.

4 Sarah's parents are concerned that she seems short for her age. Their paediatrician has the following record of Sarah's height:

Age (months)	36	48	51	57	60
Height (cm)	86	90	91	94	95

A scatter diagram showed a strong positive association between age and height, and a least squares regression line was found to be HEIGHT = 71.95 + 0.3833 AGE. The doctor wants to predict Sarah's height when she is 50 years old if there is no intervention (growth hormone), and he uses the regression line for this prediction. Check the doctor's prediction, and then comment on this procedure.

5 Revisit the data from the leaning tower of Pisa.
  a Find the mean point.
  b Draw a scatter diagram with a regression line passing through the mean point.
  c Find the equation of the regression line.
  d Use your equation to estimate the lean in 1990.

10.4 Measuring correlation

Up to this point we have used a scatter diagram to see if there is a relationship (correlation) between two variables. We have determined it as positive or negative; zero if there is no correlation. We have also called the correlation weak, moderate or strong. We then found the equation of the regression line of y on x and used the line for prediction purposes.

Now we will seek to classify the strength of the correlation numerically. There are several scales that are in use; we will study a correlation coefficient developed by Karl Pearson.

Karl Pearson (1857–1936) founded the world's first university statistics department at University College London in 1911.

→ The **Pearson product–moment correlation coefficient** (denoted by r) is a measure of the correlation between two variables X and Y, giving a value between $+1$ and -1 inclusive. It is widely used in the sciences as a measure of the strength of **linear** dependence between two variables.

The regression coefficient is used to determine how nearly the points fall on a straight line, or how nearly linear they are. A perfect correlation will have a regression coefficient of $r = 1.000$. Normally in the physical sciences we would like to have a 'confidence level' of 0.01 or better. That means that a coefficient of $r = 0.990$ or higher gives us the confidence to say that a relationship is linear within a margin of 0.01%.

In the relationship between the variables is not linear, then this correlation coefficient does not adequately represent the strength of the relationship between the variables.

The r-value from Pearson's correlation coefficient indicates the strength of the relationship between two data sets.

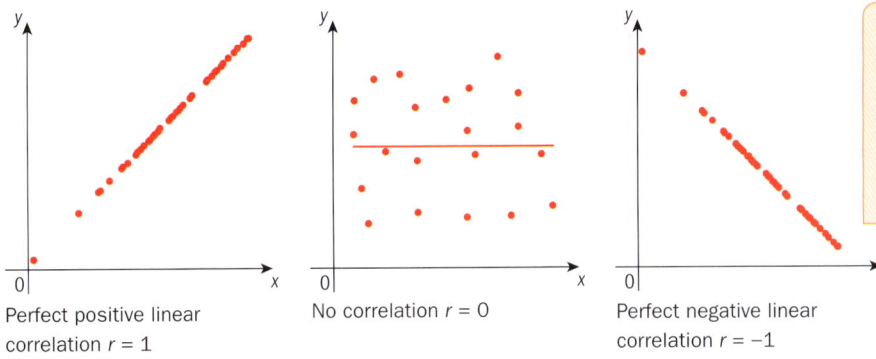

Perfect positive linear correlation $r = 1$

No correlation $r = 0$

Perfect negative linear correlation $r = -1$

Here are some more data sets and their r values:

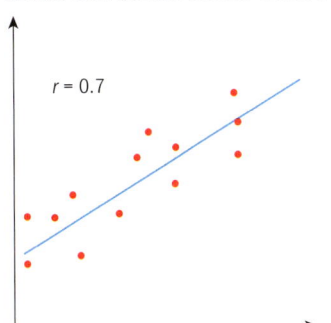

$r = 0.7$

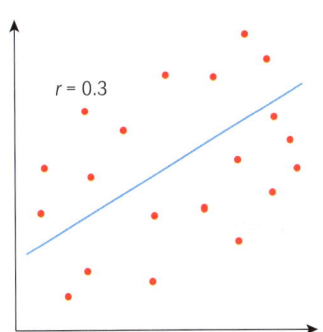

$r = 0.3$

Chapter 10 349

For negative correlation, the value of *r* is also negative.

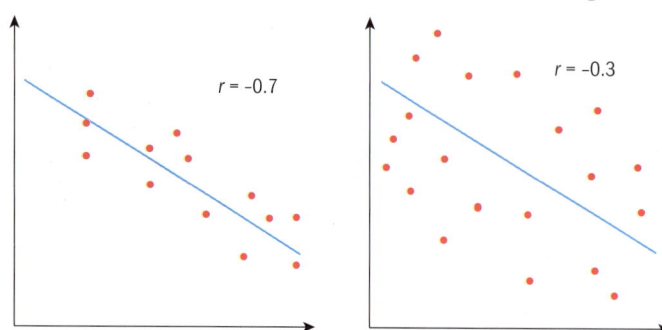

→ The formula for finding Pearson's correlation coefficient is

$$r = \frac{S_{xy}}{S_x S_y}$$

where

$$S_{xy} = \sum xy - \frac{(\sum x)(\sum y)}{n}, \quad S_x = \sqrt{\sum x^2 - \frac{(\sum x)^2}{n}} \text{ and}$$

$$S_y = \sqrt{\sum y^2 - \frac{(\sum y)^2}{n}}.$$

> You should recognise these formulae from the previous section.

→ A quick way to interpret the *r*-value is:

r-value	Correlation
$0 < |r| \leq 0.25$	Very weak
$0.25 < |r| \leq 0.5$	Weak
$0.5 < |r| \leq 0.75$	Moderate
$0.75 < |r| \leq 1$	Strong

Example 8

Sue wants to determine the strength of the correlation between the number of spoons of plant food she uses and the extra number of orchids grown from a plant. Use Pearson's correlation coefficient formula to interpret the relationship.

Plant	Spoons of plant food *x*	Increase in the number of orchids *y*
A	1	2
B	2	3
C	3	8
D	4	7

▶ Continued on next page

Answer

$$S_{xy} = \sum xy - \frac{(\sum x)(\sum y)}{n}$$

$$= 60 - \frac{10 \times 20}{4} = 10$$

$$S_x = \sqrt{\sum x^2 - \frac{(\sum x)^2}{n}}$$

$$= \sqrt{30 - \frac{10^2}{4}} = \sqrt{5}$$

$$S_y = \sqrt{\sum y^2 - \frac{(\sum y)^2}{n}}$$

$$= \sqrt{126 - \frac{20^2}{4}} = \sqrt{26}$$

$$r = \frac{S_{xy}}{S_x S_y} = \frac{10}{\sqrt{5}\sqrt{26}} \approx 0.877$$

A positive correlation means that as the number of spoons of plant food increases, the number of extra orchids increases.
The r-value of 0.877 indicates strong correlation.

Plant	x	y	xy	x^2	y^2
A	1	2	2	1	4
B	2	3	6	4	9
C	3	8	24	9	64
D	4	7	28	16	49
Total	10	20	60	30	126

GDC help on CD: *Alternative demonstrations for the TI-84 Plus and Casio FX-9860GII GDCs are on the CD.*

> You should use your GDC to calculate r in the examinations. We have shown the formula and table here to help you understand where the value comes from. See GDC Chapter 17, Section 5.16.

> Regression and correlation allow you to compare two different sets of data, to see if there may be a connection. For example, you might wish to explore the relationship between life expectancy and a country's GDP.

If two variables are correlated, we can predict one based on the other. For example, we know that IB scores and college achievement are positively correlated. So when college admission officials want to predict who is likely to succeed at their universities, they will choose students with high IB scores.

While the formulae appear complicated at first sight, making the table and evaluating the r-value are quite straightforward. From this point on we will be using technology to find the r-value.

> Which statistical methods would be useful for analyzing business performance?

Chapter 10

Exercise 10F

1 Nine students sat a French and a Spanish test. The table gives their results. Find the *r*-value and describe the correlation between the two sets of scores.

Subject	A	B	C	D	E	F	G	H	I
French	56	56	65	65	50	25	87	44	35
Spanish	87	91	85	91	75	28	92	66	58

2 A social psychologist thinks that there is a correlation between income and education. She found that people with higher income have more years of education. The results of her survey are shown below.

> You can also phrase it that people with more years of education have higher income.

Person	A	B	C	D	E	F	G	H	I	J
Income (thousand $)	125	100	40	35	41	29	35	24	50	60
Years of education	19	20	16	16	18	12	14	12	16	17

 a Find the *r*-value.
 b What can you say about the strength of the correlation?
 c What does the sign of the *r*-value indicate?

3 Does a car stop more slowly as it gets older? The table below shows the age (in months) of a car and its stopping distance (in metres) from a speed of 40 km h^{-1}.

Age (months)	9	15	24	30	38	46	53	60	64	76
Stopping distance (m)	28.4	29.3	37.6	36.2	36.5	35.3	36.2	44.1	44.8	47.2

 a Find the *r*-value.
 b What happens to the stopping distance as the car gets older?
 c Describe the strength of the correlation.

4 Kelly has always been told to stop chatting on the computer and focus on her studies. Kelly wants to see if there will be any effect on school grades first, and decides to survey 10 friends. Here are Kelly's results.

GPA	3.1	2.4	2.0	3.8	2.2	3.4	2.9	3.2	3.7	3.5
Chat time (h/week)	14	16	20	7	25	9	15	13	4	14

> An A grade is worth 4 points, a B is 3 points, a C is 2 points, a D 1 point, and an F is 0 points. The average grade is called the grade point average, GPA.

 a Find the *r*-value.
 b Describe the correlation.
 c Based on the survey, would Kelly's grade increase if the chat time decreased?

5 Mo had always been told to stop playing computer games and get on with some work, and so he decided to conduct a survey of 10 friends to see the effect on GPA. The results are in the table below.

GPA	2.7	3.8	1.5	3.6	2.2	3.8	2.0	1.9	2.5	3.0
Game time (h/week)	10	24	25	17	5	26	14	30	22	7

a Find the r-value.
b Describe the correlation.
c Based on the survey, would Mo's grade increase if his game time decreased?

6 Find and interpret the r-value for the leaning tower of Pisa data.

Extension material on CD:
Worksheet 10 - More bivariate analysis

Review exercise

1 Phrases **i**, **ii**, **iii**, **iv**, and **v** represent descriptions of the correlation between two variables:
  i High positive linear correlation
  ii Low positive linear correlation
  iii No correlation
  iv Low negative linear correlation
  v High negative linear correlation

Which phrase best represents the relationship between the two variables shown in each of the scatter diagrams below?

a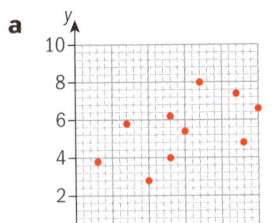

b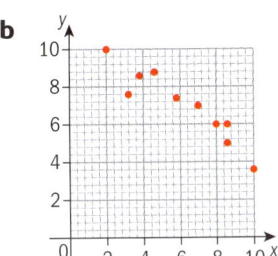

c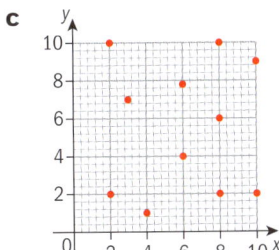

d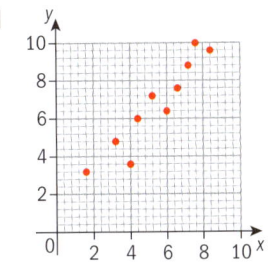

EXAM-STYLE QUESTIONS

2 The following table gives the amount of fuel in a car's fuel tank, and the number of kilometres traveled after filling the tank.

Distance traveled (km)	0	220	276	500	680	850
Amount of fuel in tank (litres)	55	43	30	24	10	6

a Copy the scatter diagram and plot the remaining points. The mean distance traveled $\bar{x}$ is 421 km, and the mean amount of fuel in the tank $\bar{y}$ is 28 litres. This point is plotted on the scatter diagram.

b Sketch the line of best fit through the mean point.

c A car traveled 350 km. Use your line of best fit to estimate the amount of fuel left in the tank.

3 This table shows the ages of ten policemen and the time that it took them to run 100 m.

Age	22	23	24	25	32	35	39	45	45	50
Time	10.9	11.1	10.8	12.0	11.2	12.1	12.6	13	12.7	13.6

a Plot the data on a scatter diagram.
b Find the mean age and the mean time.
c Draw a line of best fit through the mean point.
d How long would you expect a 30-year-old policeman to take to run the 100 m?

Review exercise

EXAM-STYLE QUESTIONS

1 The table shows the number of push-ups that David can do each minute for 6 minutes.

Minutes	1	2	3	4	5	6
Push-ups	7	8	5	3	2	2

a Show the points on a scatter diagram along with a line of best fit.
b What happens to the number of push-ups as the time increases?
c Find the equation of the regression line.
d Find the r-value and use it to describe the relationship.

2 The heights and weights of a sample of 11 students are:

Height (m) h	1.36	1.47	1.54	1.56	1.59	1.63	1.66	1.67	1.69	1.74	1.81
Weight (kg) w	52	50	67	62	69	74	59	87	77	73	67

a Write down the regression line of w on h.
b Use the regression line to estimate the weight of someone whose height is 1.6 m.

Bivariate analysis

EXAM-STYLE QUESTIONS

3 A psychologist wants to investigate the relationship between the IQ of a child and the IQ of their mother. She measures the IQ of a sample of 8 children and mothers:

Child's IQ x	87	91	94	98	103	108	111	123
Mother's IQ y	94	96	89	102	98	94	116	117

  a Write down the correlation coefficient between x and y.
  b Find the regression line of y on x.
  c Use the regression line to estimate the IQ of the mother of a child with an IQ of 100.

Using your answer to part **a**, explain how accurate you think this estimate is likely to be.

4 Eight students took two mathematics tests. We would like to know if we could predict the result of Test 2 from Test 1.
The percentage results are given below:

Test 1	54	72	32	68	55	80	45	77
Test 2	31	38	16	34	27	41	22	37

  a Plot your results on a scatter diagram.
  b Describe the correlation from your diagram.
  c Copy and complete the sentence 'Students with a high score on Test 1 tend to have a ……. score on Test 2'.
  d Find the equation of the line of best fit.
  e If another student achieved a score of 40 in Test 1, what can we predict that this student will get in Test 2?

5 The height of a pot plant was measured for the first 8 weeks from when it was bought:

Week x	0	1	2	3	4	5	6	7	8
Height (cm) y	23.5	25	26.5	27	28.5	31.5	34.5	36	37.5

  a Plot these pairs of values on a scatter diagram taking 1 cm to represent 1 week on the horizontal axis and 1 cm to represent 2 cm on the vertical axis.
  b Write down the value of the mean point.
  c Plot the mean point on the scatter diagram. Name it L.
  d **i** Write down the correlation coefficient, r, for these readings.
     ii Comment on this result.
  e Find the equation of the regression line of y on x.
  f Draw the line of regression on the scatter diagram.
  g Using your equation estimate the height of the plant after $4\frac{1}{2}$ weeks.
  h DJ uses the equation to claim that a plant would be 62.8 cm tall after 30 weeks. Comment on his claim.

EXAM-STYLE QUESTIONS

6 Personality researchers studied the behavior of a group of 10 teenagers. The researchers assessed a personality variable called 'agreeableness'. This is a measure of how nice that person is to be around. The questions ask about how cheerful, stubborn, polite, bossy and cooperative the person is. Each teenager's average of those items is reported.

They also created a measure of behavior problems. The youths reported on various behaviors in the last six months, including cheating, swearing, stealing and fighting. Each teenager's sum of behavior problems is reported.

Participant	Agreeableness factor	Behavior problems
George	4.3	5
Bill	3.0	22
Ronald	3.4	10
Jimmy	3.3	12
Gerald	2.9	23
Laura	4.0	21
Hilary	4.7	2
Nancy	2.4	35
Eleanor	2.9	12
Elizabeth	4.7	4

a Construct a scatter diagram and show the regression line.
b What happens as the agreeableness factor increases?
c Find the *r*-value.
d Describe the correlation.
e Copy and complete the sentence 'Teenagers who were more agreeable tended to have _____ behavior problems'.
f Write down the equation of the regression line.
g Michelle was absent for the behavior problems questions but scored 4.5 on her agreeableness. Estimate her score for the behavior problems.

7 Each day, a clothing factory recorded the number x of coats it produces and the total production cost y dollars. The results for nine days are shown in the following table.

x	26	44	65	43	50	31	68	46	57
y	400	582	784	625	699	448	870	537	724

a Write down the equation of the regression line of y on x.
Use your regression line as a model to answer the following:
b Interpret the meaning of **i** the gradient **ii** the *y*-intercept.
c Estimate the cost of producing 70 coats.
d The factory sells the boxes for $19.99 each. Find the smallest number of coats that the factory should produce in one day in order to make a profit.

CHAPTER 10 SUMMARY

- Bivariate analysis is concerned with the relationships between pairs of variables (x, y) in a data set.

Scatter diagrams

- **Scatter diagrams** (also called scatter plots) are used to investigate the possible relationship between two variables that both relate to the same 'event'.
- The relationship between two variables is called their **correlation**.
- To draw a scatter diagram plot the (x, y) values from the data table as dots on a graph. The pattern formed by the dots can give us some indication of the correlation.

  The **independent variable** should be on the horizontal axis with the **dependent variable** on the vertical axis.

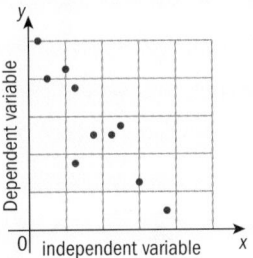

- A general upward trend in the pattern of dots shows **positive** correlation.
- A general downward trend in the pattern of dots shows **negative** correlation.
- Scattered points with no trend may indicate correlation close to **zero**.
- A correlation between two data sets does not necessarily mean that one causes the other.

The line of best fit

- A **line of best fit** or trend line is drawn on a scatter diagram to find the direction of an association between two variables and to show the trend. This line of best fit can then be used to make predictions.
  - If the line rises from left to right then there is a **positive** correlation.
  - If the line falls from left to right then there is a **negative** correlation.
  - Strong positive and negative correlations have data points very close to the line of best fit.
  - Weak positive and negative correlations have data points that are not clustered near or on the line of best fit.

Continued on next page

- To draw a line of best fit by eye draw a line that will balance the number of points above the line with the number of points below the line. An improvement is to have a reference point: one point for the line to pass through. This is the **mean point** and is calculated by finding the mean of the x-values and the mean of the y-values.

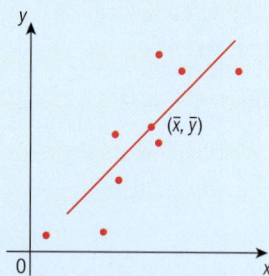

- The equation of the line of best fit, also called the **regression line**, can be used for prediction purposes.

Least squares regression

- A **residual** is the vertical distance between a data point and the graph of a regression equation.
- The least squares regression line is the one that has the smallest possible value for the sum of the squares of the residuals.
- The formula for finding the gradient, or slope (m) of a regression line is

$$m = \frac{S_{xy}}{(S_x)^2}, \text{ where}$$

$$S_{xy} = \sum xy - \frac{(\sum x)(\sum y)}{n} \text{ and } (S_x)^2 = \sum x^2 - \frac{(\sum x)^2}{n}$$

- You should use your GDC to find the equation of the regression line in examinations.

Continued on next page

Measuring correlation

- The **Pearson product–moment correlation coefficient** (denoted by r) is a measure of the correlation between two variables X and Y, giving a value between $+1$ and -1 inclusive. It is widely used in the sciences as a measure of the strength of **linear** dependence between two variables.
- The formula for finding Pearson's correlation coefficient is $r = \dfrac{S_{xy}}{S_x S_y}$

where

$$S_{xy} = \sum xy - \dfrac{(\sum x)(\sum y)}{n}, \quad S_x = \sqrt{\sum x^2 - \dfrac{(\sum x)^2}{n}} \text{ and } S_y = \sqrt{\sum y^2 - \dfrac{(\sum y)^2}{n}}.$$

- A quick way to interpret the r-value is:

r-value	Correlation
$0 < |r| \leq 0.25$	Very weak
$0.25 < |r| \leq 0.5$	Weak
$0.5 < |r| \leq 0.75$	Moderate
$0.75 < |r| \leq 1$	Strong

Theory of knowledge

Correlation or causation?

Correlation shows how closely two variables vary with each other. For example, as the value of one increases, so does the other.

Causation is when two variables directly affect each other. For example, the time you go to bed affects the number of hours you sleep.

- If we find a strong correlation between a baby's weight at birth and high achievement at age 24, should we suggest that pregnant women should try to boost their baby's birth weight, because heavier babies achieve more highly?

Sometimes a cause and effect are closely related, but not always. It is easy to assume that events that are closely correlated are also connected causally. But correlation between two events does not mean that one has caused the other.

For example, if your cat stays out all night and then gets sick, and this happens many times, it is likely that your cat gets sick because it stays out all night. But being outside all night may not cause the sickness. The cause is more likely to be a virus or bacteria.

A correlation between two variables is not necessarily proof of causation.

Experimental research investigates what happens when you change a variable, for example what happens to a liquid when you increase the temperature.

Correlated research does not change the variables. It observes the outcome of two events and offers statistical data as proof.

Correlation asks these questions:

- What relationship exists between the two variables?
- What connects or separates them from each other?

Which is causal and which is correlated?

- Bullying harms mental health.
- Stress of watching major sporting events can be hazardous to the heart.
- The temperature and the number of ice cream vendors out on that day.
- TV raises blood pressure in obese adults.
- Deep-voiced men have more children.
- Watching too much violence on television leads to people being violent in real life.
- Surgeons with video game skills perform better in simulated surgery.
- Swedish speakers are healthier than Dutch speakers.

Anscombe's Quartet

Anscombe's Quartet is a group of four data sets that provide a useful caution against applying individual statistical methods to data without first generating more evidence.

Francis Anscombe (1918–2001) British statistician.

- Find the mean of x, the mean of y, the variance of x and the variance of y and the r-value for each data set.

Set 1		Set 2		Set 3		Set 4	
x	y	x	y	x	y	x	y
4	4.26	4	3.1	4	5.39	8	6.58
5	5.68	5	4.74	5	5.73	8	5.76
6	7.24	6	6.13	6	6.08	8	7.71
7	4.82	7	7.26	7	6.42	8	8.84
8	6.95	8	8.14	8	6.77	8	8.47
9	8.81	9	8.77	9	7.11	8	7.04
10	8.04	10	9.14	10	7.46	8	5.25
11	8.33	11	9.26	11	7.81	8	5.56
12	10.84	12	9.13	12	8.15	8	7.91
13	7.58	13	8.74	13	12.74	8	6.89
14	9.96	14	8.1	14	8.84	19	12.5

- Write down what you think the graphs and their regression lines will look like.
- Using your GDC, sketch the graph of each set of points on a different graph.
- Draw the regression line on each graph.
- Explain what you notice.

11 Trigonometry

CHAPTER OBJECTIVES:

3.1 The circle: radian measures of angles; length of an arc; area of a sector

3.2 Definition of $\cos\theta$ and $\sin\theta$ in terms of the unit circle; definition of $\tan\theta$ as $\dfrac{\sin\theta}{\cos\theta}$; exact values of trigonometric ratios of $0, \dfrac{\pi}{6}, \dfrac{\pi}{4}, \dfrac{\pi}{3}, \dfrac{\pi}{2}$ and their multiples.

3.3 The Pythagorean identity $\cos^2\theta + \sin^2\theta = 1$

3.6 Solution of triangles; the cosine rule; the sine rule, including the ambiguous case; area of a triangle $\dfrac{1}{2}ab\sin C$; applications

Before you start

You should know how to:

1 Use properties of triangles, including Pythagoras' theorem.

e.g. Find the value of x in each diagram.

a $x° + 96° + 38° = 180°$
$x° = 180° - 96° - 38°$
$x = 46°$

b $\triangle ABC$ is isosceles, so $\angle A = \angle C$.
$\angle A + \angle B + \angle C = 180°$
$x° + 53° + x° = 180°$
$2x° = 180° - 53° = 127°$
$x = 63.5°$

c Using Pythagoras,
$x^2 = 6^2 + 9^2$
$x = \sqrt{6^2 + 9^2} = \sqrt{117}$
≈ 10.8

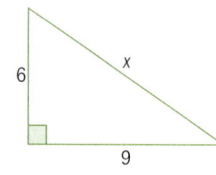

Skills check

1 Find the value of x in each diagram.

a, **b**

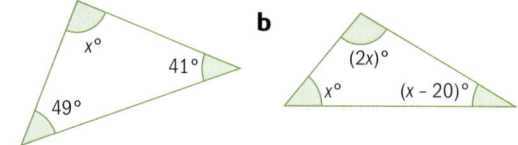

c, **d**

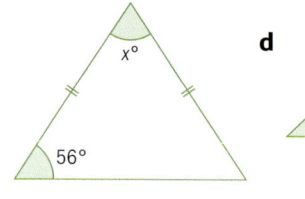

e, **f**

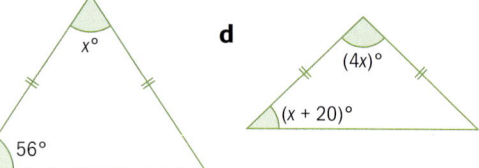

Sometimes we need to know dimensions (such as the height of a tree or a mountain or the width of a canyon) that we can't measure directly. Surveyors can find these dimensions using trigonometry and the method of triangulation.

For example, to find the distance across a canyon, a surveyor needs a point of reference on the far side of the canyon, such as a tree or rock formation. Then, standing on the near side, he measures accurately the distance between two known points, and also the angle formed by these points and the point on the far side. Using trigonometry, this is enough information to find the distance to the other side, without ever having to cross to the far side of the canyon.

Some mathematicians use the phrase 'measure of angle' instead of 'size of angle'.

11.1 Right-angled triangle trigonometry

This chapter starts by looking at the relationships between the sizes of angles and side lengths of right-angled triangles, and then extends to areas of triangles and real-life applications of trigonometry.

Some people say 'right triangle' instead of right-angled triangle.

Start by looking at the right-angled triangle with vertices at the points A, B and C. The angles at these vertices are called $\hat{A}$, $\hat{B}$ and $\hat{C}$, respectively.

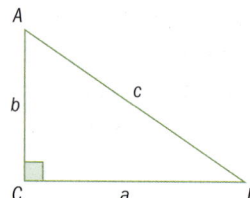

The side AB, the side opposite the right-angle, is called the **hypotenuse** of the right-angled triangle.

> Angles can be described in various ways. This triangle could be called △ABC the angle at A could be called $\hat{A}$; $B\hat{A}C$; $C\hat{A}B$; ∠BAC; ∠CAB. Angles can also be labeled with Greek letters like θ (theta).

In this triangle, notice that the side labeled a (side BC) is opposite $\hat{A}$, the side labeled b (side AC) is opposite $\hat{B}$, and the side labeled c (side AB) is opposite $\hat{C}$. It is convenient to identify sides in relation to their opposite angles.

Trigonometric ratios

Look at these two right-angled triangles.

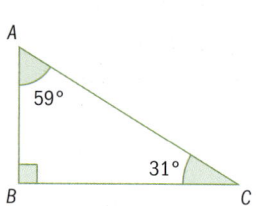 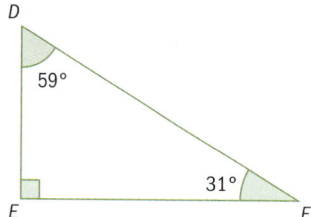

△ABC and △DEF each have angles measuring 59°, 31° and 90°. △DEF is larger than △ABC. Triangles with the same three angles are called **similar triangles**, and their corresponding sides are in the same proportions.

For △ABC and △DEF:

$\dfrac{AC}{AB} = \dfrac{DF}{DE}$, and $\dfrac{BC}{AB} = \dfrac{EF}{DE}$, and $\dfrac{BC}{AC} = \dfrac{EF}{DF}$

> Some textbooks call the two shortest sides of a right-angled triangle the **legs** of the triangle. The **hypotenuse** is the longest side of a right-angled triangle.

For any similar triangles, regardless how large or small they are, their sides will be in the same ratios. In other words, their corresponding sides will be in proportion to each other.

The fact that the sides of similar triangles form equal ratios helps us define the three trigonometric ratios – **sine**, **cosine** and **tangent**. These ratios vary according to the sizes of the angles in the right-angled triangles.

In any right-angled triangle
- the hypotenuse (often abbreviated to **h** or **H**) is the longest side and is opposite the right angle
- the side opposite the angle marked θ is called the **opposite** side (sometimes abbreviated to **o** or **O**)
- the side next to the angle θ is called the **adjacent** side (sometimes abbreviated to **a** or **A**)

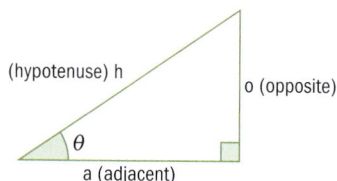

→ For any right-angled triangle with an angle θ:

$$\sin \theta = \frac{\text{opposite}}{\text{hypotenuse}} = \frac{O}{H}$$

$$\cos \theta = \frac{\text{adjacent}}{\text{hypotenuse}} = \frac{A}{H}$$

$$\tan \theta = \frac{\text{opposite}}{\text{adjacent}} = \frac{O}{A}$$

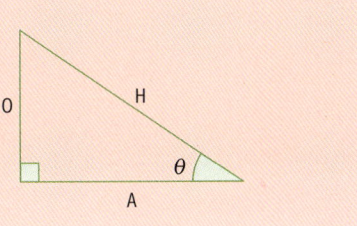

A mnemonic is a made-up word or phrase that helps you remember a list or a formula.
Remember these with the mnemonic SOH-CAH-TOA

Look at this right-angled triangle, with $\hat{A}$ highlighted.

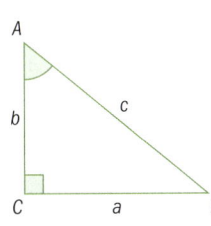

$$\sin A = \frac{BC}{AB} = \frac{a}{c}$$

$$\cos A = \frac{AC}{AB} = \frac{b}{c}$$

$$\tan A = \frac{BC}{AC} = \frac{a}{b}$$

We use the abbreviations sin, cos and tan for these trigonometric ratios.

You can use trigonometric ratios to find unknown side lengths and angles in right-angled triangles.

Relation between sine, cosine and tangent

In triangle ABC:

$\sin \theta = \dfrac{a}{c}$

$\cos \theta = \dfrac{b}{c}$

so $\dfrac{\sin \theta}{\cos \theta} = \dfrac{\frac{a}{c}}{\frac{b}{c}} = \dfrac{a}{b}$

but $\tan \theta = \dfrac{a}{b}$

so $\dfrac{\sin \theta}{\cos \theta} = \tan \theta$

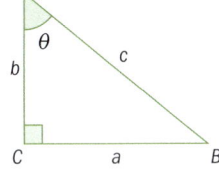

The astronomer Aryabhata, born in India in about 476 CE, believed that the Sun, planets and stars circled the Earth in different orbits. He began to invent trigonometry in order to calculate the distances from planets to the Earth.

→ $\tan \theta = \dfrac{\sin \theta}{\cos \theta}$

Example 1

For this triangle, find the length of side a.

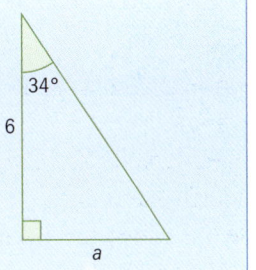

▶ Continued on next page

Although mathematicians have studied triangles for thousands of years, the term 'trigonometry' was first used in 1595 by Bartholomaeus Pitiscus (German, 1561–1613).

Chapter 11   365

Answer

$\tan 34° = \dfrac{\text{opposite}}{\text{adjacent}} = \dfrac{a}{6}$

$a = 6 \tan 34°$

$a = 6 \tan 34° \approx 4.05$

Use the tangent ratio.

The side opposite the angle of 34° is the opposite side, and the side adjacent to 34° has length 6. You can find the value of tan 34° using your GDC.

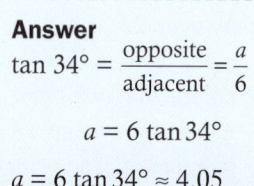

To enter tan press μ and then select tan.

Be sure you are in **degree mode**.

To change to degree mode press ⌂On and choose 5: Settings & Status 2:Settings 1:General Use the tab key to move to Angle and select Degree. Press enter and then select 4:Current

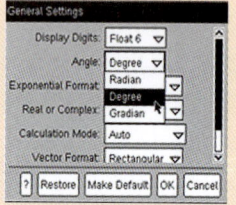

If you know the lengths of the sides of a right-angled triangle and you want to find the sizes of the angles of the triangle, you will need to use the inverse trigonometric functions $\sin^{-1}$, $\cos^{-1}$ and $\tan^{-1}$.

Example 2

Find the size of $\hat{B}$ in this triangle.

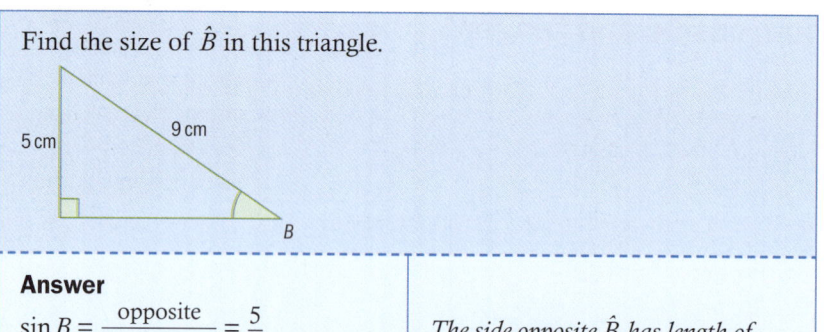

Answer

$\sin B = \dfrac{\text{opposite}}{\text{hypotenuse}} = \dfrac{5}{9}$

$\hat{B} = \sin^{-1}\left(\dfrac{5}{9}\right) \approx 33.7°$

The side opposite $\hat{B}$ has length of 5 cm and the hypotenuse has a length of 9 cm. Use the sine ratio.

To enter $\sin^{-1}$ press μ and then select $\sin^{-1}$.

GDC help on CD: Alternative demonstrations for the TI-84 Plus and Casio FX-9860GII GDCs are on the CD.

$\sin^{-1}$ is called 'arc sine,' $\cos^{-1}$ is 'arc cos' $\tan^{-1}$ is 'arc tan'

In this exercise, you will be **solving** right-angled triangles (finding missing angles and side lengths). Always make sure that your calculator is set in DEGREE mode.

Exercise 11A

For each question, use the diagram and the information given to find all the unknown angles and sides. All lengths are in cm. Give your answers correct to 3 significant figures where necessary.

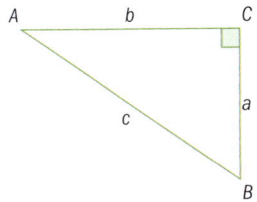

1. $a = 12, c = 20$
2. $b = 37, Â = 40°$
3. $c = 4.5, B̂ = 55°$
4. $b = 48, c = 60$
5. $a = 11, Â = 35°$
6. $a = 8.5, b = 9.7$
7. If $a = 2x$, $b = 5x - 1$ and $c = x^2 + 1$ ($x \in \mathbb{Z}$) find the value of x, and the angles $Â$ and $B̂$.

$x \in \mathbb{Z}$ means that x is an integer.

Special right-angled triangles

Look at this isosceles right-angled triangle.

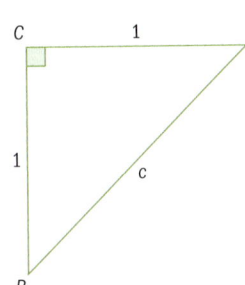

To solve the triangle, you need to find the length AB, and angles $Â$ and $B̂$.
Using Pythagoras' theorem
$1^2 + 1^2 = c^2$, so $c^2 = 2$, and $c = AB = \sqrt{2}$
Using the tangent ratio
$$\tan A = \frac{BC}{AC} = \frac{1}{1} = 1$$
$Â = \tan^{-1}(1) = 45°$
This is an isosceles triangle, so $Â = B̂$, and $B̂ = 45°$.

Now look at the values of the trigonometric ratios of this triangle.

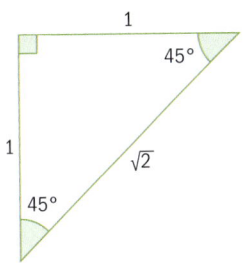

→ $\sin 45° = \dfrac{1}{\sqrt{2}} = \dfrac{\sqrt{2}}{2}$

$\cos 45° = \dfrac{1}{\sqrt{2}} = \dfrac{\sqrt{2}}{2}$

$\tan 45° = \dfrac{1}{1} = 1$

Now look at this right-angled triangle, which is half of an equilateral triangle.
To solve the triangle, you need to find BC, $Â$ and $B̂$.
Using Pythagoras' theorem gives
$1^2 + a^2 = 2^2$, so $a^2 = 3$, and $a = BC = \sqrt{3}$

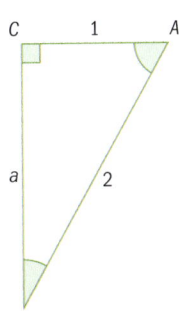

Using the cosine ratio,

$\cos A = \dfrac{AC}{AB} = \dfrac{1}{2}$

$\hat{A} = \cos^{-1}\left(\dfrac{1}{2}\right) = 60°$

$\hat{B} = 180° - 90° - 60° = 30°$

Here are the values of all the trigonometric ratios for this 30°–60°–90° triangle.

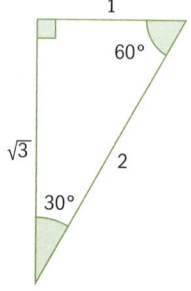

→ $\sin 30° = \dfrac{1}{2}$ $\quad\quad \sin 60° = \dfrac{\sqrt{3}}{2}$

$\cos 30° = \dfrac{\sqrt{3}}{2}$ $\quad\quad \cos 60° = \dfrac{1}{2}$

$\tan 30° = \dfrac{1}{\sqrt{3}} = \dfrac{\sqrt{3}}{3}$ $\quad \tan 60° = \dfrac{\sqrt{3}}{1} = \sqrt{3}$

Example 3

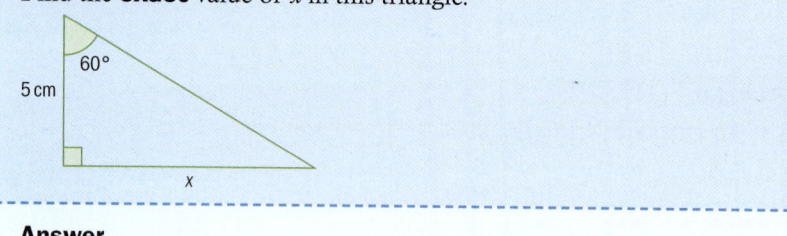

Find the **exact** value of x in this triangle.

Answer

$\tan 60° = \dfrac{x}{5} = \sqrt{3}$

$x = 5\sqrt{3}$ cm

When an **exact** answer is asked for you should leave the square root or radical in your answer and **not** change it to a rounded decimal. Square roots of numbers that are not perfect squares are called surds.

Exercise 11B

1 Use the diagram to solve each right-angled triangle. Give exact answers. Lengths are in cm.
   a $a = 12, c = 24$
   b $b = 9, \hat{A} = 45°$
   c $c = 4.5, \hat{B} = 60°$
   d $b = 6, c = 4\sqrt{3}$
   e $a = 5\sqrt{2}, c = 10$

'Solve' here means find all unknown sides and angles.

The diagram will not always be to scale

2 Find the exact values of x, y and z.

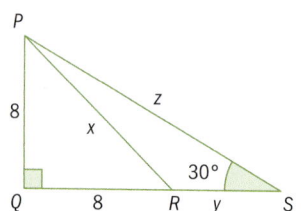

Trigonometry

3 △ABC has $\hat{A} = 60°$, $\hat{C} = 90°$, $BC = x + 2$, and $AB = x^2 - 4$.
   a Find the exact value of x.
   b Find the exact length of side AC.

4 Triangle ABC has $\hat{B} = 45°$, $\hat{C} = 90°$, $AC = 4x - 1$ and $BC = x^2 + 2$.
   a Find the value of x.
   b Find the exact length of side AB.

5 In the diagram find the values of w, x, y and z, to 1 dp. Lengths are in cm.

Sketch the triangle first.

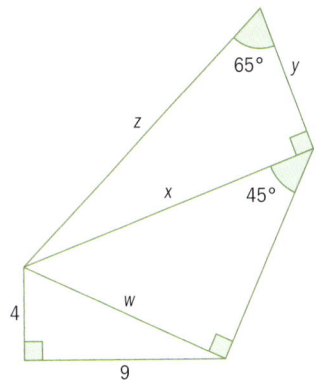

11.2 Applications of right-angled triangle trigonometry

In the last section, you found lengths and angles in right-angled triangles using sine, cosine and tangent. In this section, you will see how to apply these trigonometric ratios to solve problems in real-life situations.

Let's begin with some terminology.

→ The **angle of elevation** is the angle 'up' from horizontal.
   The **angle of depression** is the angle 'down' from horizontal.

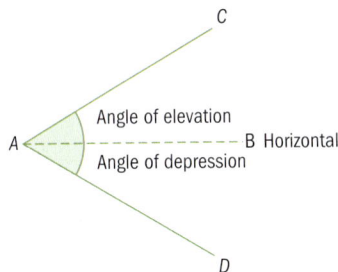

Example 4

An observer stands 100 m from the base of a building. The angle of elevation of the top of the building is 65°. How tall is the building, to the nearest metre?

▶ Continued on next page

Chapter 11   369

Answer

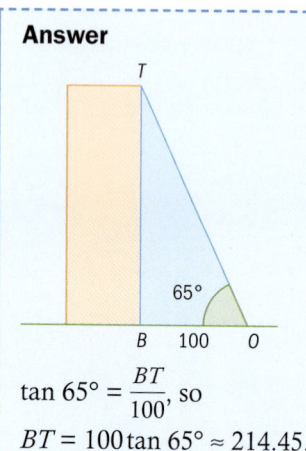

$\tan 65° = \dfrac{BT}{100}$, so
$BT = 100 \tan 65° \approx 214.45...$
The building is 214 metres tall, to the nearest metre.

Start by sketching a diagram. Let O represent the position of the observer on the ground B represent the base of the building and T represent the top of the building. Mark the 65° angle of elevation.

You are finding the height of the building, length BT.

You also need to solve problems using compass points and bearings.

→ The four cardinal **compass points** are north (N), south (S), east (E) and west (W).
Three-figure bearings give directions as angles measured clockwise from north.

When using **compass points** for direction, you will see notation such as:

N 40° E, which means 40° east of north.

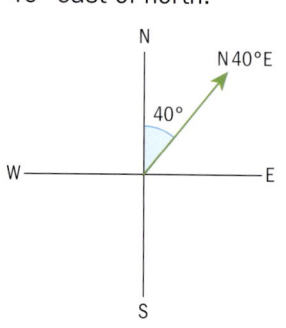

W 20°S, which means 20° south of west.

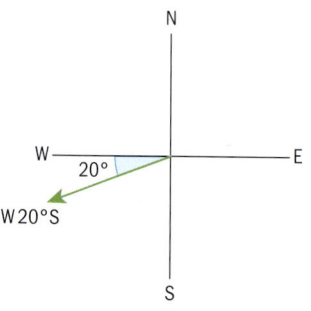

NW, which means 45° between north and west.

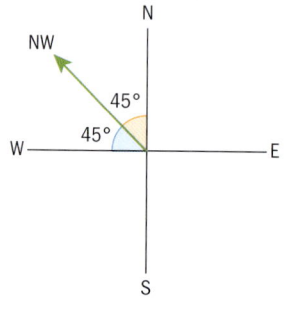

When using **bearings** for direction, you will see notation such as:

035° which means 35° clockwise from north.

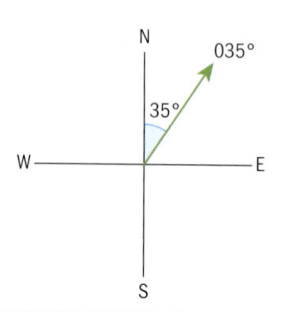

110°, which means 110° clockwise from north.

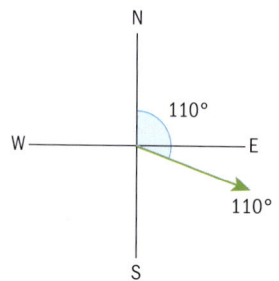

270°, which means 270° clockwise from north. Notice that a bearing of 270° is the same as 'due west'.

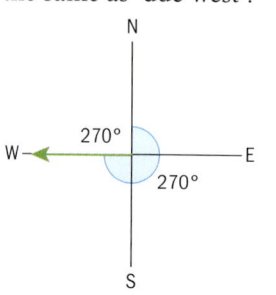

Example 5

Two ships leave dock at the same time.
Ship A sails due north for 30 km before dropping anchor.
Ship B sails on a bearing of 050° for 65 km before dropping anchor.
Find the distance between the ships when they are stationary, to the nearest kilometre.

Answer

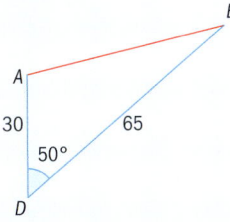

Sketch a diagram. Point D represents the dock from which the ships set sail. Ship A stops at A and ship B stops at B.
You need to find the length AB, the distance between the ships when they are stationary.

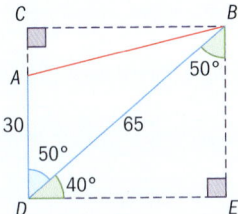

There are no right-angled triangles in the diagram, so draw them in. The hypotenuse of each right-angled triangle is the path of one of the ships. Add any angles you know from angle properties.

The angle DBE is found using the alternate angle property.

$\sin 40° = \dfrac{BE}{65}$

so $BE = 65 \sin 40° \approx 41.781...$

Find BE.

$\cos 40° = \dfrac{DE}{65}$

so $DE = 65 \cos 40° = 49.7928$

$BC = DE = 49.7928...$

$AC = BE - 30 = 11.7812...$

Find DE.

Store these values in your GDC.

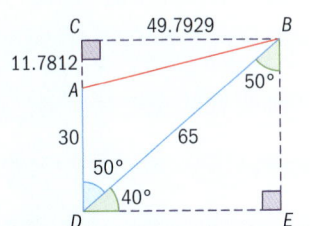

Add the new information to the diagram.

$AB^2 = (49.7929...)^2 + (11.7812...)^2$
so $AB = .1677...$
The ships are approximately 52 km apart, to the nearest km.

Use Pythagoras' theorem in △ABC. Use the values you stored.

Use exact values in the intermediate steps, and round only for the final answer.

Chapter 11   371

Exercise 11C

1. Isosceles triangle *ABC* has side *AC* = 10 cm and *AB* = *CB* = 15 cm, as shown.
   a. Find the height of the triangle.
   b. Find the sizes of $B\hat{A}C$ and $A\hat{B}C$.

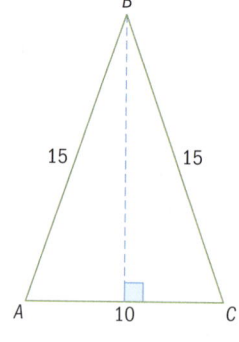

2. △*ABE* fits exactly inside the square *ABCD*, as shown. *BC* = 28 cm and *DE* = 8 cm.
   a. Find the lengths of segments *AE* and *BE*.
   b. Find the sizes of $A\hat{E}D$, $E\hat{B}A$ and $A\hat{E}B$.
   Give your answers correct to 3 sf.

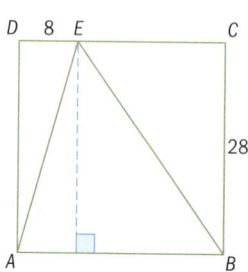

3. An observer standing on the top of a vertical cliff 120 m above sea level sees a ship in the water at an angle of depression of 9°. How far is the ship from the base of the cliff?

> If a diagram is not given with the question, start by drawing your own.

EXAM-STYLE QUESTION

4. A rectangle has length 25 mm and width 18 mm. Find the angles between the diagonals of the rectangle.

5. Anya walks 2 km due north, then turns and walks another 3 km in the direction N35°W. Find her distance and bearing from her starting point.

6. From a window 12 m above the ground in Building A, the angle of elevation of the top of Building B across the street is 40°. If the buildings are 70 m apart, what is the height of Building B?

EXAM-STYLE QUESTION

7. A ship leaves port and sails 35 km on a bearing of 047°. The ship then turns and sails 15 km on a bearing of 105°. How far, and on what bearing, must the ship sail to return directly to port?

8. Buildings X and Y are across the street from each other, 95 m apart. From a point on the roof of Building X, the angle of depression to the base of Building Y is 55° and the angle of elevation to the top of Building Y is 35°. How tall are the two buildings?

> It is a good idea to check your final answers to make sure that the shortest side is opposite the smallest angle and the longest side is opposite the largest angle.

9. Jacob is walking north along a straight road when he spots a tower in a field to his right on a bearing of 018°. After walking another 240 metres he notices the tower is now on a bearing of 066°. If he continues walking north, how close will he pass to the tower?

372 Trigonometry

10 From her position at ground level, Hayley notices that the angle of elevation of the top of a building is 40°. When she moves 20 metres closer to the building, the new angle of elevation is 55°. Find the height of the building.

> Unless the question tells you otherwise, assume the ground is level.

11 A car is traveling at a constant speed on a straight highway. A passenger in the car sees a bridge spanning the highway ahead at an angle of elevation of 5°. Ten seconds later, the angle of elevation of the bridge is 17°. How much more time will elapse before the car passes directly under the bridge?

12 The diagram shows a right rectangular prism, *ABCDEFGH*. *AD* = 24 cm, *DH* = 9 cm, and *HG* = 18 cm. Find these angles.
  a $H\hat{A}D$
  b $A\hat{B}E$
  c $H\hat{A}G$
  d $A\hat{G}D$

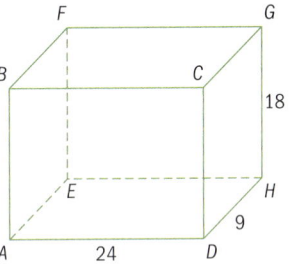

11.3 Using the coordinate axes in trigonometry

The angle θ in a Cartesian coordinate system has its vertex at the origin, as shown in the diagram. A positive angle is measured anticlockwise from the *x*-axis.

> Some people use 'counterclockwise' instead of anticlockwise.

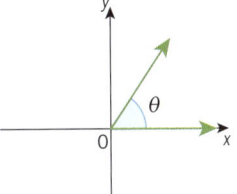

> In some textbooks the side of the angle that lies along the positive *x*-axis is called the **initial side**. The other side is called the **terminal side**. An angle like this, with its vertex at the origin and its initial side along the positive *x*-axis is said to be in **standard position**.

Here are positive three angles α, β and δ.

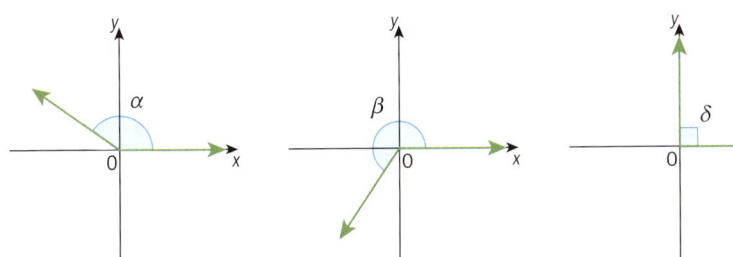

> The first four letters of the Greek alphabet are alpha α, beta β, gamma γ and delta δ.

This diagram shows a circle with equation $x^2 + y^2 = 1$
The center of the circle is at the origin and its radius is
one unit. This is called a **unit circle**.

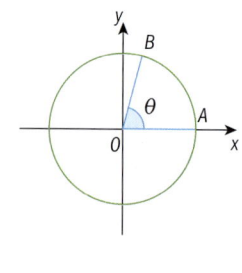

In the diagram, the angle θ is positive.
Now take a closer look at acute angles in
the first **quadrant** of the unit circle.
OA and OB are radii of the unit circle so
$OA = OB = 1$

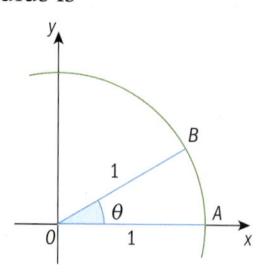

Next, use the acute angle θ to form a
right-angled triangle BOC.
Using the trigonometric ratios in $\triangle BOC$,
$\cos \theta = \dfrac{x}{1}$, so $x = \cos \theta$,
and $\sin \theta = \dfrac{y}{1}$, so $y = \sin \theta$.
So point B has coordinates $(\cos \theta, \sin \theta)$.

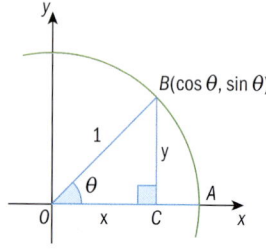

Example 6

Find the exact coordinates of point D,
then give these values to three
significant figures.

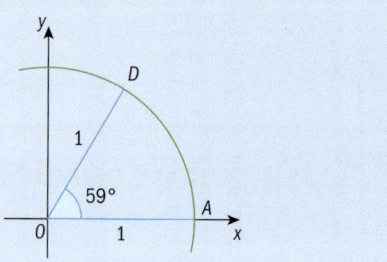

Answer

The exact coordinates of point D are $(\cos 59°, \sin 59°)$

To 3 sf the coordinates of D are $(0.515, 0.857)$

$A\hat{O}D$ is a positive angle.

Use your GDC to find the values of $\cos 59°$ and $\sin 59°$.

Example 7

In the diagram, find the exact coordinates of point P.

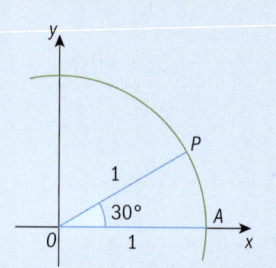

Answer

The exact coordinates of P are $\left(\dfrac{\sqrt{3}}{2}, \dfrac{1}{2}\right)$

$A\hat{O}P$ is in the first
quadrant. Therefore,
the coordinates of
point P are $(\cos 30°, \sin 30°)$.

See page 368 for the
exact values of sine
30° and cos 30°.

Trigonometry

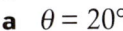

Exercise 11D

1 Use the diagram to find the coordinates of point P for each given value of θ. Give your answers to 3 significant figures.

  a θ = 20°
  b θ = 17°
  c θ = 60°
  d θ = 74°
  e θ = 90°

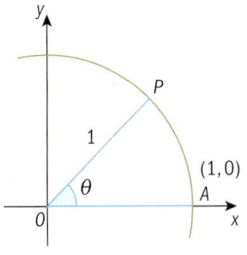

2 Use the diagram from question 1 to find the value of θ for the given coordinates of point P. Give your answers to the nearest degree.

  a P(0.408, 0.913)
  b P(0.155, 0.988)
  c P(0.707, 0.707)
  d P(0.970, 0.242)

> These coordinates have been rounded to 3 significant figures.

> The diagram will not always be to scale.

3 Use the diagram to find the area of △AOP for the given value of θ. Give your answers to 3 significant figures.

  a θ = 70°
  b θ = 38°
  c θ = 24°
  d θ = 30°

> The dashed line is the height of the triangle.

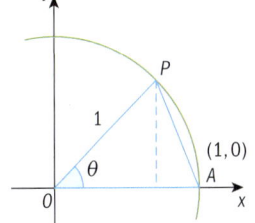

Now look at angles in the second quadrant. These angles are obtuse angles (between 90° and 180°). Here is an obtuse angle in the second quadrant in the unit circle.

When you are working with obtuse angles it is sometimes helpful to think of how they relate to angles in the first quadrant (acute angles).

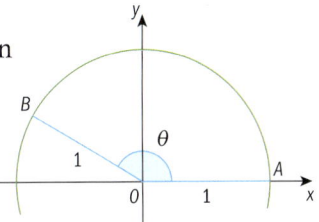

Investigation – obtuse angles

This diagram shows point B at a positive angle of 30° from OA, and point C at a positive angle of θ from OA.

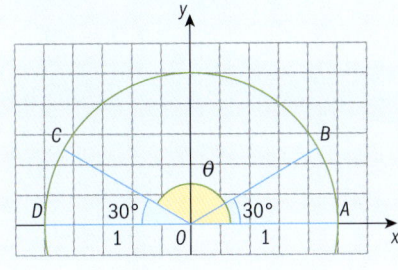

Find the value of θ.
What are the coordinates of point B?
Use the symmetry of the unit circle to write down the coordinates of point C.

▶ Continued on next page

Chapter 11   375

Now look at the triangles formed by the sides OB and OC and the x-axis.

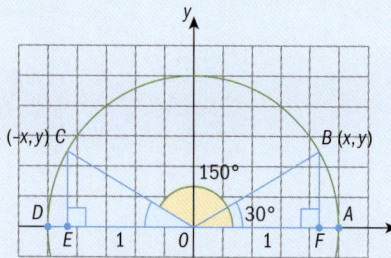

△EOC is congruent to △FOB. Both are 30°– 60°– 90° triangles with hypotenuse length 1. You can also see that if the coordinates of point B are (x, y), then the coordinates of point C are (–x, y).

The coordinates of B are (cos 30°, sin 30°), or $\left(\frac{\sqrt{3}}{2}, \frac{1}{2}\right)$.

So the coordinates of point C are (cos 150°, sin 150°), which are the same as the coordinates (–cos 30°, sin 30°), or $\left(-\frac{\sqrt{3}}{2}, \frac{1}{2}\right)$.

Draw diagrams showing each of these pairs of angles in the unit circle.
1 40° and 140°
2 25° and 155°
3 68° and 112°

Label the coordinates of the points where the non-horizontal sides meet the unit circle. What do you notice?

From the investigation you should now understand an important property of supplementary angles.

> For supplementary angles α and β, $\sin \alpha = \sin \beta$, and $\cos \alpha = -\cos \beta$

> For any angle θ, $\sin \theta = \sin(180° - \theta)$, and $\cos \theta = -\cos(180° - \theta)$

This property will be useful later on.

Supplementary angles add up to 180°.

You will see these properties illustrated graphically when you study graphs of sine and cosine functions in Chapter 13.

Exercise 11E

1 Use the diagram to find the coordinates of points B and C for the given values of θ. Give your answers to 3 significant figures.
   a $\theta = 30°$
   b $\theta = 57°$
   c $\theta = 45°$
   d $\theta = 13°$
   e $\theta = 85°$

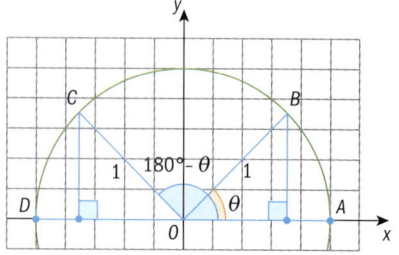

2 Use the diagram in question 1 to find the value of θ for each of the given positions of point C. Give your answers to the nearest tenth of a degree.

   a $C(-0.332, 0.943)$
   b $C(-0.955, 0.297)$
   c $C(-0.903, 0.429)$
   d $C(-0.769, 0.639)$

> These coordinates have been rounded to 3 significant figures.

3 Find the sine of each acute angle (to 4 sf), and state the obtuse angle that has the same sine.

   a $15°$
   b $36°$
   c $81°$
   d $64°$

4 Find one acute and one obtuse value for $\hat{A}$.

   a $\sin A = 0.871$
   b $\sin A = 0.436$
   c $\sin A = 0.504$
   d $\sin A = 0.5$

Next, look at the line with equation $y = mx$:

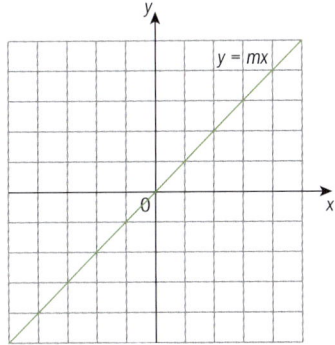

Any line with equation $y = mx$ has gradient m, and passes through the origin.

> This is a special form of the standard equation of a line $y = ax + b$ or $y = mx + c$

Now look at what happens when the line intersects the unit circle at point B in the first quadrant.

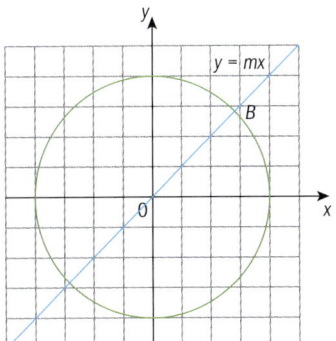

In the first quadrant the line forms an angle θ with the x-axis. A right-angled triangle is formed with segment OB (part of the line $y = mx$) as its hypotenuse.

This illustrates some important properties involving the right triangle and the line $y = mx$.

First, using Pythagoras' theorem gives $(\sin \theta)^2 + (\cos \theta)^2 = 1^2$. The usual way of writing $(\sin \theta)^2$ and $(\cos \theta)^2$ is $\sin^2 \theta$ and $\cos^2 \theta$, which gives

$$\sin^2 \theta + \cos^2 \theta = 1$$

Suppose you want to find the gradient of the line $y = mx$.

This line passes through the points $O(0, 0)$, and $B(\sin \theta, \cos \theta)$.

The gradient of a line $= \dfrac{y_2 - y_1}{x_2 - x_1}$

Here you can find the gradient, m, using the coordinates of points O and B

$$m = \dfrac{\sin \theta - 0}{\cos \theta - 0} = \dfrac{\sin \theta}{\cos \theta} = \tan \theta$$

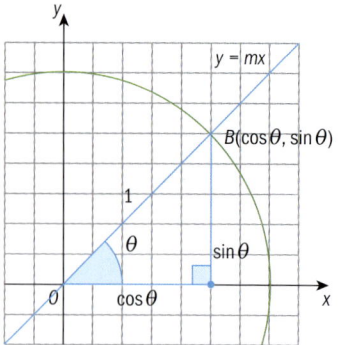

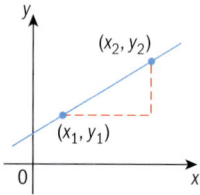

> These three properties are true for any angle θ
> 1. $\sin^2 \theta + \cos^2 \theta = 1$
> 2. $\tan \theta = \dfrac{\sin \theta}{\cos \theta}$
> 3. For any line $y = mx$ which forms an angle of θ with the x-axis, the value of m (the gradient of the line) is $\tan \theta$.

The gradient of a line is $\dfrac{\text{rise}}{\text{run}}$.

Property number 1 is also known as the Pythagorean Identity.

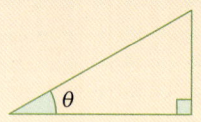

Property number 2 is often useful in calculations.

Example 8

Find the gradient of the line which forms a positive angle of 130° with the x-axis.

Answer

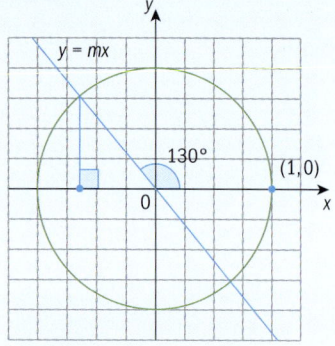

The gradient of the line is $\tan 130° \approx -1.19$.

Gradient $= \tan \theta$

You can find this value using your GDC.

Example 9

Find the gradient of the line shown in the diagram.

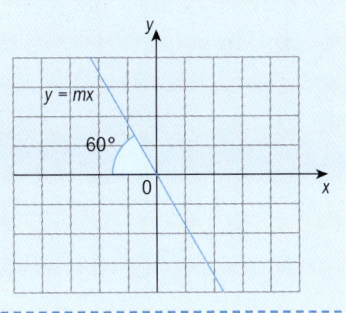

Answer

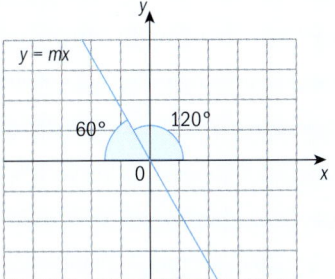

Find the 'standard position' angle formed by this line. The angle 60° is equivalent to a positive obtuse angle of 120°.
This line forms an angle of 120° in standard position.

The gradient of the line is

$$\tan 120° = \frac{\sin 120°}{\cos 120°} = \frac{\sin 60°}{\cos 60°} = \frac{\frac{\sqrt{3}}{2}}{-\frac{1}{2}}$$

$$= -\sqrt{3} = -1.73$$

Exercise 11F

1 Find the gradient of the line $y = mx$ in each diagram, giving your answers to three significant figures.

a

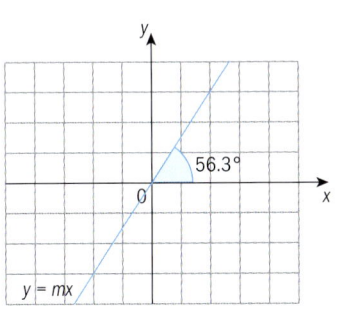

b

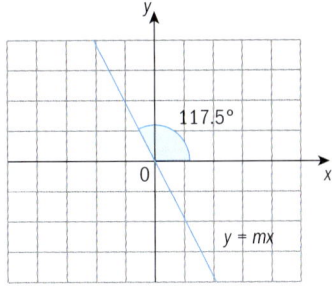

c

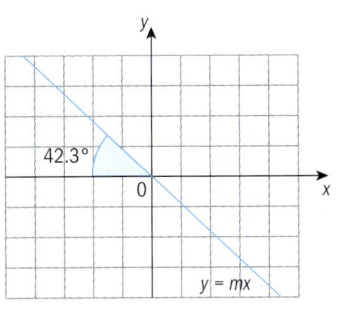

d

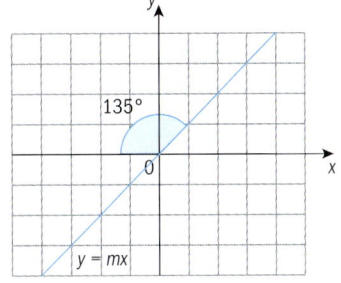

2 Find the equation of the line passing through the origin and point P. Find the value of θ to the nearest degree.

a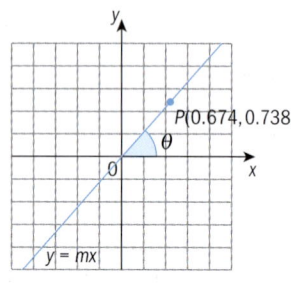

b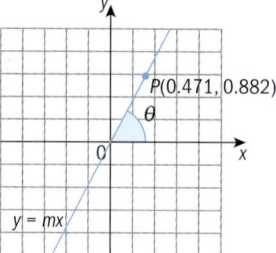

c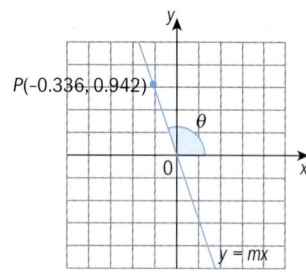

d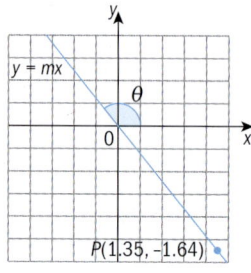

e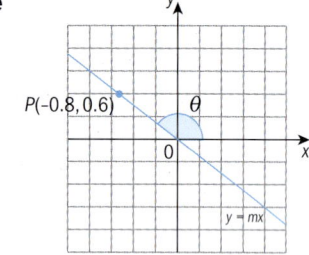

f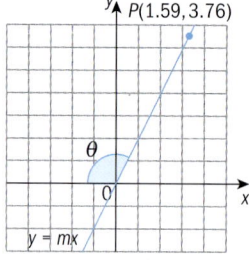

Extension material on CD:
Worksheet 11 - Angle sums and differences

11.4 The sine rule

You can use trigonometry to solve triangles that are not right-angled.

Look at $\triangle ABC$. The **altitude** (height), h, of the triangle is AD, perpendicular to BC.

In the right-angled triangle ABD

$$\sin B = \frac{h}{c}$$

This gives $h = c \sin B$

In the right-angled triangle ACD

$$\sin C = \frac{h}{b}$$

This gives $h = b \sin C$

Equate the values of h to give

$$c \sin B = b \sin C.$$

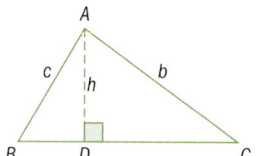

Rearranging this equation gives $\dfrac{\sin B}{b} = \dfrac{\sin C}{c}$

The ratios of the sine of each angle to the length of the opposite side are equal.

Now draw the altitude from B to side AC, and from C to AB and find the ratios $\dfrac{\sin A}{a} = \dfrac{\sin C}{c} = \dfrac{\sin B}{b}$ again. You should find that these ratios are equal, as before.

> **The sine rule**
>
> For any $\triangle ABC$, where a is the length of the side opposite $\hat{A}$, b is the length of the side opposite $\hat{B}$, and c is the length of the side opposite $\hat{C}$,
>
> $$\dfrac{\sin A}{a} = \dfrac{\sin B}{b} = \dfrac{\sin C}{c} \text{ or } \dfrac{a}{\sin A} = \dfrac{b}{\sin B} = \dfrac{c}{\sin C}$$

This formula is given in the Formula booklet that you use in the examination.

You can use the sine rule to solve triangles if you know the size of at least one angle and its opposite side, and one other measurement (the length of a side or the size of an angle).

Example 10

Find the missing angles and sides in this triangle, giving your answers to 3 sf.

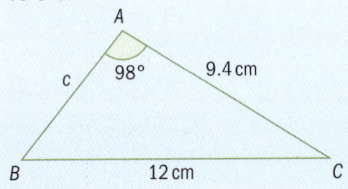

Be sure you are in **degree mode**.
To change to degree mode press 🏠 On and choose 5: Settings & Status |2: Settings |1: General
Use the tab key to move to Angle and select Degree. Press enter and then select 4: Current

Answer

Using the sine rule
$$\dfrac{\sin 98°}{12} = \dfrac{\sin B}{9.4}$$

so $\sin \hat{B} = \dfrac{9.4 \sin 98°}{12}$

$\hat{B} = 50.9° \,(3\,\text{sf})$

You need to find the angles $\hat{B}$ and $\hat{C}$, and the length c.

$\hat{C} = 180 - \hat{A} - \hat{B}$, so
$\hat{C} = 31.1305533...$
$\hat{C} = 31.1° \,(3\,\text{sf})$

The sum of the angles in any triangle is 180°.

$\dfrac{\sin 98°}{12} = \dfrac{\sin 31.13055...}{c}$

$c = \dfrac{12 \sin 31.13055...}{\sin 98°}$

$c = 6.26\,\text{cm} \,(3\,\text{sf})$

Use the sine rule once more to find c.

Don't round the intermediate steps, just the final values for $\hat{B}$, $\hat{C}$ and c.

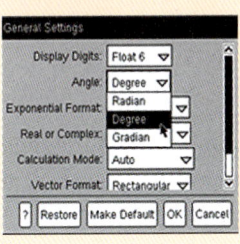

GDC help on CD: Alternative demonstrations for the TI-84 Plus and Casio FX-9860GII GDCs are on the CD.

In Example 10, the triangle with all measures labeled would look like this:

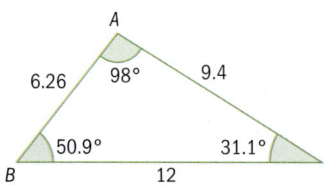

Always check your final answers to be sure the shortest side is opposite the smallest angle and the longest side is opposite the largest angle.

 ## Example 11

Find the missing angles and sides in this triangle, rounding your answers to 2 decimal places.

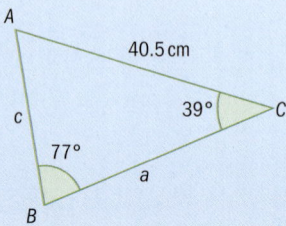

Answer

$\hat{A} = 180° − 77° − 39° = 64°$

$\dfrac{\sin 77°}{40.5} = \dfrac{\sin 64°}{a}$, $a = \dfrac{40.5 \sin 64°}{\sin 77°}$

so $a = 37.36$ cm (2 dp)

$\dfrac{\sin 77°}{40.5} = \dfrac{\sin 39°}{c}$

$c = \dfrac{40.5 \sin 39°}{\sin 77°}$

so $c = 26.16$ cm (2 dp)

You need to find angle $\hat{A}$, and the lengths a and c

Use the sine rule to find a and c.

Check: Shortest side (26.16) is opposite the smallest angle (39°). Longest side (40.5) is opposite the largest angle (77°).

 ## Example 12

A ship is sailing due north. The captain sees a lighthouse 10 km away on a bearing of 032°. Later, the captain observes that the bearing of the lighthouse is 132°. How far did the ship travel between these two observations?

Answer

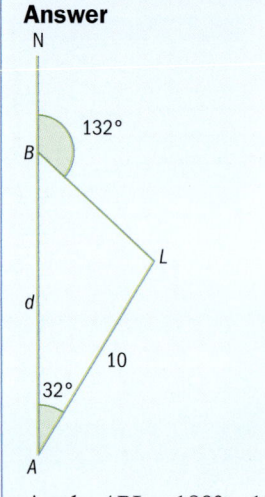

Angle $ABL = 180° − 132° = 48°$

Draw a diagram to model the situation.
A is the position from which the captain first spots the lighthouse, and B is his second position. L is the position of the lighthouse.
You have to find d, the distance the ship travels from point A to point B.

▶ Continued on next page

$\hat{L} = 180 - \hat{A} - \hat{B} = 100°$

$\dfrac{\sin 100°}{d} = \dfrac{\sin 48°}{10}$

$d = \dfrac{10 \sin 100°}{\sin 48°}$

$d = 13.251...$

The ship travels approximately 13.25 km between points A and B.

Use the sine rule to find d.

Give your answer to a sensible degree of accuracy.

> Ptolemy (c 90–168 CE), in his 13-volume work *Almagest*, wrote sine values for angles from 0° to 90°. He also included a theorem similar to the sine rule.

Exercise 11G

1 Solve each triangle ABC. Give your answers correct to 3 significant figures.

> 'Solve' a triangle means find all unknown sides and angles.

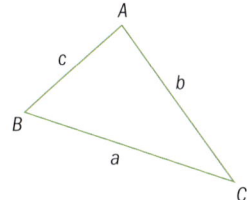

a $b = 24\,\text{cm}, \hat{A} = 47°, \hat{B} = 83°$ **b** $c = 2.5\,\text{cm}, \hat{A} = 40°, \hat{C} = 72°$
c $a = 4.5\,\text{cm}, b = 3.6\,\text{cm}, \hat{A} = 55°$ **d** $b = 60, \hat{B} = 15°, \hat{C} = 125°$
e $c = 5.8\,\text{cm}, \hat{A} = 27°, \hat{B} = 43°$

EXAM-STYLE QUESTION

2 An isosceles triangle has base 20 cm, and base angles of 68.2°, as shown. Use the sine rule to find the length of sides XY and XZ.

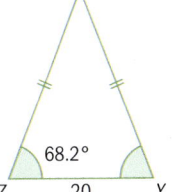

3 Julia sees a tree in a field S40°E from where she is standing. She then walks 2 km due south and notices that the tree is now S75°E from her new position. How far is the tree from both her first and second positions on the road?

4 Adam and Kevin are standing 35 metres apart, on opposite sides of a flagpole. From Adam's position, the angle of elevation of the top of the flagpole is 36°. From Kevin's position, the angle of elevation is 50°. How high is the flagpole?

Triangles are often used in architecture.
Left: The Hearst Tower in New York City is made up of isosceles triangles.
Right: A builder can strengthen a rectangular frame by making diagonal corners to form triangles.

A triangle is rigid – you cannot change its shape. Crossbars and struts give rigidity to a structure.

Investigation – ambiguous triangles

Try to draw triangle ABC, with $\hat{A} = 32°$, $a = 3$ cm, and $c = 5$ cm. You should find that there are actually two possible triangles that fit this description:

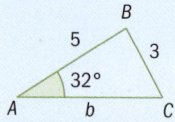 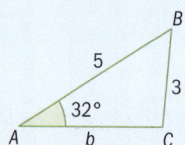

The given measurements do not describe a unique triangle.
1 Find the size of the angle C in each of the triangles (call them C_1 and C_2). What is the relationship between these two angles?
2 Using these angles for C, find angle B and the length AC for each triangle.
This is known as the **ambiguous case**, and it can sometimes happen when you are given two sides and a non-included angle of a triangle.

Example 13

In triangle ABC, $\hat{A} = 40°$, $a = 14$ cm, and $c = 20$ cm. Solve this triangle, giving all possible cases. Give your answers correct to 1 dp.

Answer

$\dfrac{\sin 40°}{14} = \dfrac{\sin C}{20}$

$\sin C = \dfrac{20 \sin 40°}{14}$

$\hat{C}_1 = 66.7°$

$\hat{C}_2 = 180° - 66.7°$, so $\hat{C}_2 = 113.3°$

$\hat{B}_1 = 180° - 40° - 66.7° = 73.3°$

$\hat{B}_2 = 180° - 40° - 113.3° = 26.7°$

Use your GDC in degree mode.

Round to 1 dp. Supplementary angles have the same sine value. The two possible values for C give two possible values for $\hat{B}$.

▶ Continued on next page

$$\frac{\sin 40°}{14} = \frac{\sin 73.3°}{b_1}$$

$$b_1 = \frac{14 \sin 73.3°}{\sin 140°}$$

$$b_1 = 20.9 \text{ cm}$$

$$\frac{\sin 40°}{14} = \frac{\sin 26.7}{b_2}$$

$$b_2 = \frac{14 \sin 26.7°}{\sin 40°}$$

$$b_2 = 9.8 \text{ cm}$$

And finally, find two values for b correct to 1 dp.

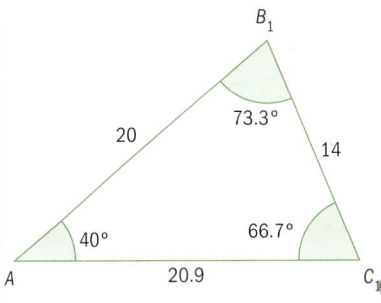

The ambiguous case does not occur every time you solve a triangle.

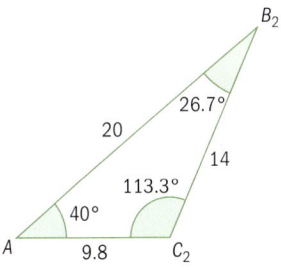

If you draw these triangles this is what you see.

→ There can be an ambiguous case when you use the sine rule if:
- you are given two sides and a non-included acute angle
- the side opposite the given acute angle is the shorter of the two given sides.

Exercise 11H

1 Use the given information to find the missing sides and angles in a triangle *ABC*. Give all possible solutions with answers to 1 dp. All lengths are in cm.

Some of these do not involve the ambiguous case.

  a $\hat{A} = 30°$, $a = 4$, and $c = 7$
  b $\hat{B} = 50°$, $b = 17$, and $c = 21$
  c $\hat{C} = 20°$, $b = 6.8$, and $c = 2.5$
  d $\hat{A} = 42°$, $a = 33$, and $c = 25$
  e $\hat{A} = 70°$, $a = 25$, and $b = 28$
  f $\hat{A} = 70°$, $a = 25$, and $b = 26$
  g $\hat{A} = 45°$, $a = 22$, and $b = 14$
  h $\hat{B} = 56°$, $b = 45$, and $c = 50$

2 Look at this diagram.
  a Find *BE*, *CE* and *DE*.
  b Find the sizes of angles $E\hat{A}B$, $B\hat{C}E$, $B\hat{C}D$, $B\hat{D}C$, $A\hat{B}D$ and $C\hat{B}D$.
  c Explain how this diagram relates to the ambiguous case of the sine rule.

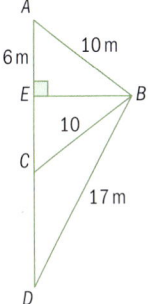

EXAM-STYLE QUESTION

3 A ship is sailing due west when the captain sees a lighthouse at a distance of 20 km on a bearing of 230°.
  a Draw a diagram to model this situation.
  b How far must the ship sail before the lighthouse is 16 km away?
  c How far must the ship sail beyond this point before the lighthouse is again at a distance of 16 km from the ship?
  d What is the bearing of the lighthouse from the ship the second time the two are 16 km apart?

11.5 The cosine rule

You cannot use the sine rule to solve triangles like these:

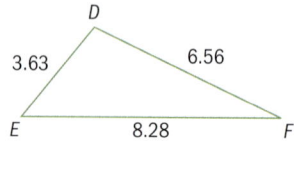
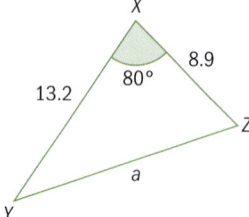

Consider the triangle ABC, with altitude h from A to side BC.

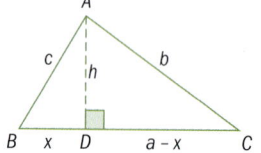

In triangle ACD, Pythagoras' theorem gives

$$b^2 = h^2 + (a-x)^2 = h^2 + a^2 - 2ax + x^2$$

In triangle ABD,

$$h^2 + x^2 = c^2$$

so $h^2 = c^2 - x^2$

Substitute for h^2 in the first equation to get

$$b^2 = c^2 - x^2 + a^2 - 2ax + x^2$$
$$= c^2 + a^2 - 2ax$$

In triangle ABD, $\cos B = \dfrac{x}{c}$, so $x = c \cos B$

By substituting for x, you get

$$b^2 = a^2 + c^2 - 2ac \cos B$$

This equation is one form of the **cosine rule**.

> **The cosine rule**
>
> For $\triangle ABC$, where a is the length of the side opposite $\hat{A}$, b is the length of the side opposite $\hat{B}$, and c is the length of the side opposite $\hat{C}$:
>
> $a^2 = b^2 + c^2 - 2bc \cos A$ or
> $b^2 = a^2 + c^2 - 2ac \cos B$ or
> $c^2 = a^2 + b^2 - 2ab \cos C$

You might also see $2bc \cos A$ written as $2bc \cdot \cos A$, where the dot means multiply. The cosine rule is in the Formula booklet.

Example 14

Find a and the missing angles in this triangle.

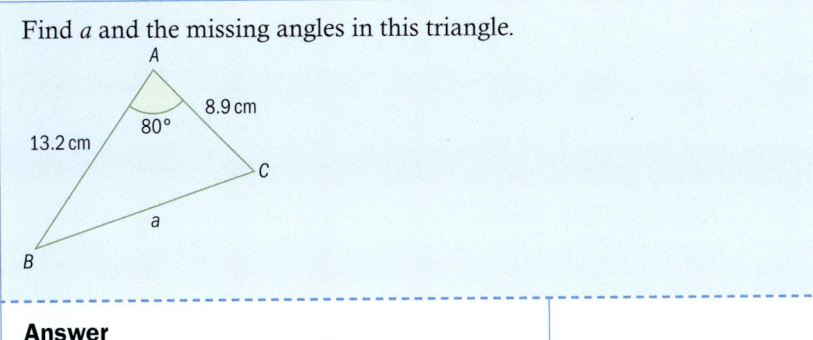

Answer

$a^2 = 13.2^2 + 8.9^2 - 2(13.2)(8.9)\cos 80°$ *Use the cosine rule.*
$a = \sqrt{13.2^2 + 8.9^2 - 2(13.2)(8.9)\cos 80°}$
$a = 14.6\,\text{cm}$

$\dfrac{\sin 80°}{a} = \dfrac{\sin B}{8.9}$ *Use the sine rule.*

$\sin B = \dfrac{8.9 \sin 80°}{14.6}$

so $\hat{B} = 36.9°$ (1dp)

$\hat{C} = 180° - 80° - 36.9° = 63.1°$

When you use the cosine rule to find angles, it is sometimes helpful to rearrange the formula like this:

→ **Cosine rule**

$\cos A = \dfrac{b^2 + c^2 - a^2}{2bc}$

$\cos B = \dfrac{a^2 + c^2 - b^2}{2ac}$

$\cos C = \dfrac{a^2 + b^2 - c^2}{2ab}$

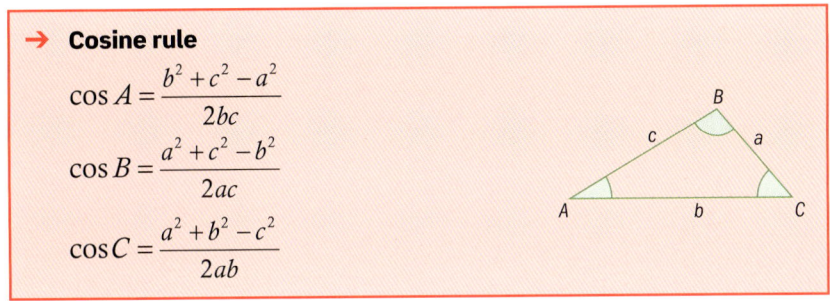

Example 15

Find angles A, B and C.

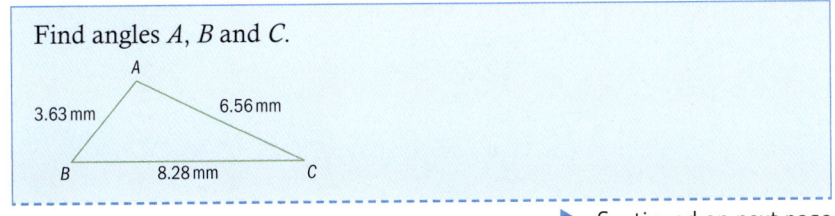

▶ Continued on next page

Answer

$$\cos A = \frac{(3.63)^2 + (6.56)^2 - (8.28)^2}{2(3.63)(6.56)}$$

Use the cosine rule.
$$\cos A = \frac{b^2 + c^2 - a^2}{2bc}$$

$$\hat{A} = \cos^{-1}\left(\frac{(3.63)^2 + (6.56)^2 - (8.28)^2}{2(3.63)(6.56)}\right)$$

$\hat{A} = 105°$ (3 sf)

$$\cos B = \frac{(3.63)^2 + (8.28)^2 - (6.56)^2}{2(3.63)(8.28)}$$

Cosine rule (You could use the sine rule here instead.)

$$\hat{B} = \cos^{-1}\left(\frac{(3.63)^2 + (8.28)^2 - (6.56)^2}{2(3.63)(8.28)}\right)$$

so $\hat{B} = 49.9°$

$\hat{C} = 180° - 105° - 49.9°$
  $= 25.1°$ (3 sf)

Now look again at Example 5 from Section 11.2. This problem can be solved more quickly using the cosine rule.

Example 16

Two ships leave dock at the same time. Ship A sails due north for 30 km before dropping anchor. Ship B sails on a bearing of 050° for 65 km before dropping anchor. Find the distance between the ships when they are stationary, to the nearest kilometre.

Answer

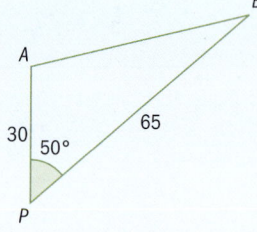

Draw a diagram.

$AB^2 = 30^2 + 65^2 - 2(30)(65) \times \cos 50°$

$AB = \sqrt{30^2 + 65^2 - 2(30)(65)\cos 50°}$
  $= 51.17$

The ships are 51 kilometres apart (to the nearest km).

Use the cosine rule:
$a^2 = b^2 + c^2 - 2bc \cos 50°$

Pythagoras' theorem is a special case of the cosine rule. See what happens to the equation when you use the cosine rule with an angle of 90°.

Exercise 11I

1. Use the given information to find all sides and angles in each triangle. Give your answers to 1 dp. All lengths are in metres.
   a $\hat{A} = 64°$, $b = 43$, and $c = 72$
   b $a = 20$, $b = 33$, and $c = 41$
   c $a = 3.6$, $b = 4.9$, and $c = 2.4$
   d $\hat{B} = 31°$, $a = 10$, and $c = 14$
   e $\hat{C} = 70°$, $a = 75$, and $b = 86$
   f $a = 45$, $b = 50$, and $c = 58$

   > There are many real-life applications of triangle trigonometry.

2. A hiker leaves camp and walks 5 km on a bearing of 058°. He stops for a break, then continues walking for another 8 km on a bearing of 103°. He stops again before heading straight back. How far must he walk to get back to camp?

EXAM-STYLE QUESTIONS

3. The diagonals of a parallelogram form an acute angle of 62°. The lengths of the diagonals are 6 cm and 9 cm.
   Find the lengths of the sides of the parallelogram.

4. Town B is 15 km away from Town A, in the direction N36°W. Town C is N27°E of town A, and towns A and C are 20 km apart. Find the distance between towns B and C.

5. Ship A leaves port and sails due east for 28 km. Ship B leaves from the same port and sails 49 km. The ships are then 36 km apart. On what bearing was ship B sailing?

6. The pyramid $ABCDE$ has a square base with sides 15 cm. Its other faces are **congruent** isosceles triangles with equal sides of 24 cm.
   Find these angles.
   a $A\hat{B}D$
   b $E\hat{D}C$
   c $E\hat{A}C$

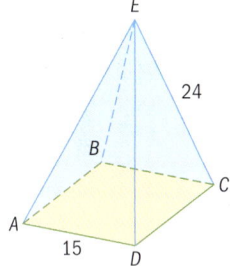

11.6 Area of a triangle

Look at triangle ABC with base b and height h.
You can find the area of the triangle using the formula:

$$\text{area} = \frac{1}{2}bh$$

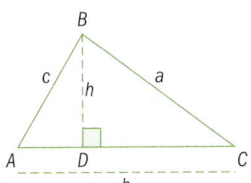

In $\triangle ADB$, $\sin A = \dfrac{h}{c}$, so $h = c \sin A$

Substituting for h in the area formula gives area $= \dfrac{1}{2} bc \sin A$.

Notice that in this formula you do not need to know the height of the triangle.

→ The area of any triangle *ABC* is given by the formula:

area = $\frac{1}{2}bc \sin A$ or area = $\frac{1}{2}ac \sin B$ or area = $\frac{1}{2}ab \sin C$

Example 17

a Find the area of triangle *ABC*.

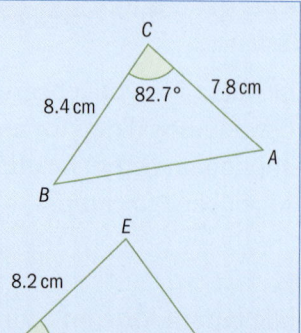

b The area of this triangle is 50 cm². Find angle θ.

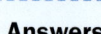

Answers

a Area = $\frac{1}{2}(8.4)(7.8)\sin 82.7°$

    = 32.5 cm² (3 sf)

Area = $\frac{1}{2}ab \sin C$

b $\frac{1}{2}(8.2)(13.7)\sin\theta = 50$

$\sin\theta = \dfrac{50}{\frac{1}{2}(8.2)(13.7)}$

$= \dfrac{100}{(8.2)(13.7)} = 0.8901...$

$\theta = \sin^{-1} 0.8901$

$= 62.9° (3\,sf)$

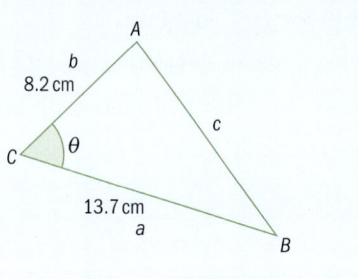

> In the first century CE, Hero (or Heron) of Alexandria developed a different method for finding the area of a triangle using only the lengths of the sides.

Exercise 11J

1 Find the area of each triangle. All lengths are in cm.

a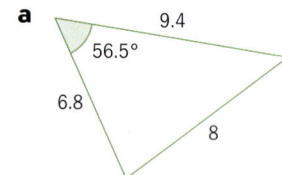

b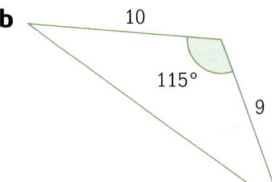

c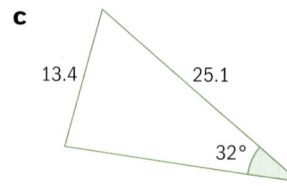

d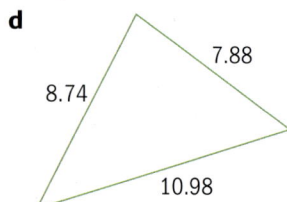

e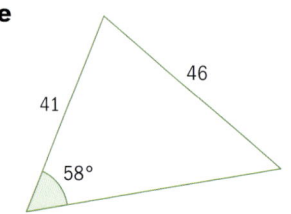

f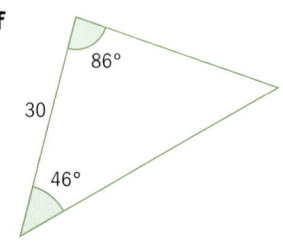

2 The triangle shown has an area of $100 \, \text{m}^2$.
Find the value of θ.

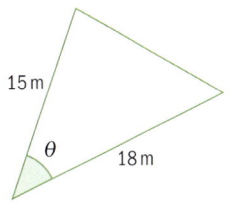

3 The triangle shown has an area of $324 \, \text{cm}^2$.
Find the value of x.

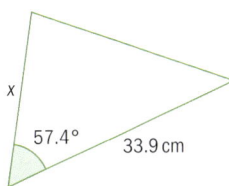

EXAM-STYLE QUESTIONS

4 **a** Find the largest angle in this triangle.
b Hence, find the area of the triangle.

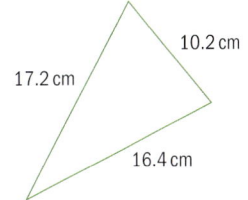

> The command term 'hence' tells you to use your answer from part **a** to help you answer part **b**.

5 The triangle shown has an area of $30 \, \text{cm}^2$.
Find the value of x.

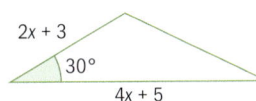

6 The area of a triangle is $20 \, \text{mm}^2$.
Two sides of the triangle are $8 \, \text{mm}$ and $11 \, \text{mm}$.
Find two possible lengths for the third side.

11.7 Radians, arcs and sectors

Angles can be measured in **radians** instead of degrees.

Why use radians?

One complete turn is $360°$, but the number 360 is a somewhat arbitrary measure. Radians, however, are directly related to measurements within a circle. In this section, you will see how radians are connected to *arc length* and *sector area*.

> The Babylonians believed that there were 360 days in a year and hence used $360°$ to represent one revolution.

One radian is defined as the size of the central angle **subtended** by an arc which is the same length as the radius of the circle.

Two radians is the size of the central angle subtended by an arc with a length equal to twice the radius of the circle.

> A central angle subtended by an arc is an angle with its vertex at the center of the circle and its sides passing through the endpoints of the arc.

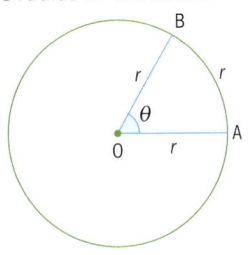

θ = 1 radian

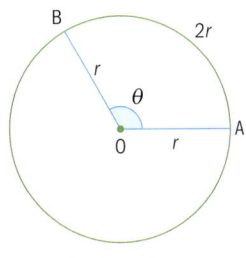

θ = 2 radians

One complete turn around the circle is subtended by an arc equal in length to the circumference of the circle.

$$\text{The circumference} = 2\pi r$$

Therefore, the angle which subtends the circumference of the circle is 2π radians.

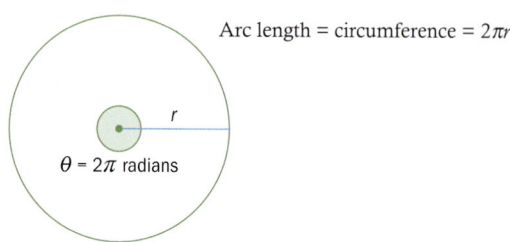

Any central angle in a circle is a fraction of 2π, so you can calculate the length of the arc the angle subtends as a fraction of the circumference.

→ Arc length $= \left(\dfrac{\theta}{2\pi}\right)(2\pi r) = r\theta$

where r is the radius and θ is the central angle measured in radians.

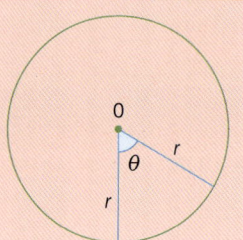

Similarly, the formula for the area of a circle is: area $= \pi r^2$
The area of a sector with a central angle θ will be a fraction of the area of the circle.

→ Area of sector $= \left(\dfrac{\theta}{2\pi}\right)(\pi r^2) = \dfrac{\theta r^2}{2}$

where r is the radius of the circle and θ is the central angle, in radians.

Example 18

a Find the length of the arc which subtends a central angle of 2.6 radians (see diagram) in a circle of radius 7 cm.
b Find the area of the sector.

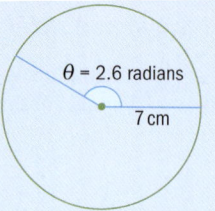

Answers
a Arc length $= 7(2.6) = 18.2$ cm   Arc length $= r\theta$
b Sector area $= \dfrac{2.6(7^2)}{2} = 63.7$ cm²   Sector area $= \dfrac{\theta r^2}{2}$

The abbreviation for radians is **rad**. In the example above, 2.6 radians could be written as 2.6 rad. If you see an angle with no units, such as 'sin 2.6', you can assume that the angle is 2.6 radians.

> Another way of writing angles in radians is 2.6^c where the c stands for circular measure.

Example 19

A circle has radius 2.5 mm. Find the size of the central angle subtended by an arc of length 9 mm.

Answer

$9 = 2.5\theta$ | Arc length = $r\theta$
$\theta = \dfrac{9}{2.5}$
$= 3.6 \text{ rad}$

Example 20

In this circle, arc $AB = 7.86$ cm and the area of sector $AOB = 23.58$ cm². Find the central angle θ and the radius r.

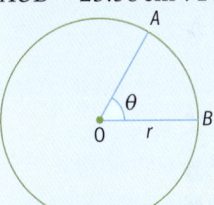

> Some farmers plant their crops in circular patterns. What other real-life applications are there for circles, arcs and sectors?

Answer

$23.58 = \dfrac{\theta r^2}{2}$, so $47.16 = \theta r^2$ | Sector area = $\dfrac{\theta r^2}{2}$

$7.86 = r\theta$, so $\theta = \dfrac{7.86}{r}$ | Arc length = $r\theta$

$47.16 = \dfrac{7.86}{r}(r^2) = 7.86r$, so | Substitute for θ in the previous equation.

$r = \dfrac{47.16}{7.86}$
$= 6 \text{ cm}$

$\theta = \dfrac{7.86}{6}$, so $\theta = 1.31 \text{ rad}$ | Use the result $\theta = \dfrac{7.86}{r}$

Exercise 11K

1 Find the length of the arc which subtends a central angle of 1.7 radians in a circle with radius 5.6 cm.

2 Find the length of the arc which subtends an angle of 3.25 radians at the center of a circle with diameter 24 cm.

Chapter 11

3 An angle θ is subtended by an arc of length 12.5 mm at the center of a circle. Find the value of θ if the circle has radius 2.5 mm.

4 An arc AB subtends an angle of 2.4 radians at the center O of a circle with radius 50 cm. Find the area and perimeter of sector AOB.

5 An arc WX subtends an angle of 5.1 radians at the center P of a circle with radius 3 cm. Find the area and perimeter of sector WPX.

EXAM-STYLE QUESTION

6 In the circle with center P the arc QR subtends an angle of θ at the center. If the length of arc QR is 27.2 cm and the area of sector PQR is 217.6 cm², find θ and the radius of the circle.

7 Circle O has radius 4 cm, and circle P has radius 6 cm. The centers of the circles are 8 cm apart. If the circles intersect at A and B, find the blue shaded area in the diagram.

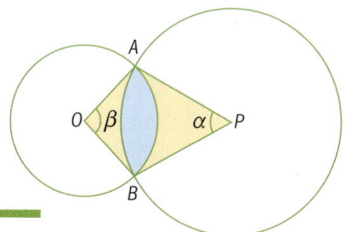

Degrees and radians

You have already seen that one full rotation around a circle gives a central angle of 2π, and you know that one full rotation is also equal to 360°. You can use this fact to convert between radians and degrees.

$360° = 2\pi$, so $180° = \pi$.

and $\dfrac{180°}{\pi} = 1$ radian

$1° = \dfrac{\pi}{180}$ radians

> Any angle which is a multiple of π is assumed to be measured in radians so you don't need to write 'rad'.

→ To convert degrees to radians multiply by $\dfrac{\pi}{180}$

→ To convert radians to degrees multiply by $\dfrac{180}{\pi}$

Example 21

a Convert these angles to radians: 30°, 45°, 60°
Give exact answers.
b Convert these angles to degrees: $\dfrac{2\pi}{5}$ rad, $\dfrac{\pi}{9}$ rad
Give exact answers.

> Exact radian values are written as multiples of π.

▶ Continued on next page

Answers

a $30° = 30\left(\dfrac{\pi}{180}\right) = \dfrac{30\pi}{180} = \dfrac{\pi}{6}$ *Multiply by* $\dfrac{\pi}{180}$

$45° = 45\left(\dfrac{\pi}{180}\right) = \dfrac{45\pi}{180} = \dfrac{\pi}{4}$

$60° = 60\left(\dfrac{\pi}{180}\right) = \dfrac{60\pi}{180} = \dfrac{\pi}{3}$

b $\dfrac{2\pi}{5} = \dfrac{2\pi}{5}\left(\dfrac{180}{\pi}\right) = 72°$ *Multiply by* $\dfrac{180}{\pi}$

$\dfrac{\pi}{9} = \dfrac{\pi}{9}\left(\dfrac{180}{\pi}\right) = 20°$

Example 22

a Convert these angles to radians: 43°, 70°, 136°
Give values to 3 significant figures.
b Convert these angles to degrees: 1 rad, 2.3 rad
Give values to one decimal place.

Answers

a $43° = 43\left(\dfrac{\pi}{180}\right) = \dfrac{43\pi}{180} = 0.750$ rad (3 sf)

$70° = 70\left(\dfrac{\pi}{180}\right) = \dfrac{70\pi}{180} = 1.22$ rad (3 sf)

$136° = 136\left(\dfrac{\pi}{180}\right) = \dfrac{136\pi}{180} = 2.37$ rad (3 sf)

b $1 \text{ rad} = 1\left(\dfrac{180}{\pi}\right) = 57.3°$ (1 dp)

$2.3 \text{ rad} = 2.3\left(\dfrac{180}{\pi}\right) = 131.8°$ (1 dp)

Exercise 11L

1 Convert these angles to radians.
Give exact values.
   a 75°   **b** 240°   **c** 80°   **d** 330°

2 Convert these angles to radians.
Give answers to 3 significant figures.
   a 56°   **b** 107°   **c** 324°   **d** 230°

3 Convert these angles to degrees.
Give exact values.
   a $\dfrac{5\pi}{6}$ **b** $\dfrac{5\pi}{3}$ **c** $\dfrac{3\pi}{2}$ **d** $\dfrac{5\pi}{4}$

4 Convert these angles to degrees.
Give answers to 3 significant figures.
   a 1.5 rad   **b** 0.36 rad   **c** 2.38 rad   **d** 3.59 rad

In Section 11.1, you looked at some 'special' angles in right-angled triangles: 30°, 45°, 60° and 90°. These angles, and their multiples, are often used in trigonometry and they can also be expressed in radians. It is helpful to remember these angles, so you do not have to do the conversion every time. The table shows some special angles in degrees and their equivalents in radians as multiples of π.

Angle in degrees	30°	45°	60°	90°	120°	135°	150°	180°	210°	225°
Angle in radians	$\frac{\pi}{6}$	$\frac{\pi}{4}$	$\frac{\pi}{3}$	$\frac{\pi}{2}$	$\frac{2\pi}{3}$	$\frac{3\pi}{4}$	$\frac{5\pi}{6}$	π	$\frac{7\pi}{6}$	$\frac{5\pi}{4}$

Angles which are multiples of 30°, 45°, 60° and 90° are usually written in exact radian form using π.

When you solve trigonometry problems you must be careful to note whether the angles are given in degrees or radians.

To find sine, cosine and tangent values for angles measured in radians, set your GDC to *radian mode*.

> To change to **radian mode** press ⇧On and choose 5:Settings & Status | 2:Settings | 1:General
> Use the tab key to move to Angle and select Radian.
> Press enter and then select 4:Current to return to the document

Example 23

The diagram shows a circle with center O and radius 5 cm.
Find the area of the shaded region, to 3 significant figures.

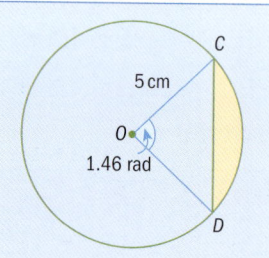

GDC help on CD: Alternative demonstrations for the TI-84 Plus and Casio FX-9860GII GDCs are on the CD.

Answer

Area of sector $OCD = \dfrac{(1.46)(5^2)}{2}$

$= 18.25 \text{ cm}^2$

Area of $\triangle OCD$

$= \dfrac{1}{2}(5)(5)\sin(1.46)$

$\approx 12.42335...$

Shaded area

$= 18.25 - \dfrac{1}{2}(5)(5)\sin(1.46)$

$\approx 5.83 \text{ cm}^2$ (3 sf)

Area of the shaded region = area of sector OCD – area of $\triangle OCD$

Area $= \dfrac{1}{2} ab \sin C$

Exercise 11M

1 Find the exact value of each trigonometric ratio.

   a $\sin \dfrac{\pi}{4}$ **b** $\cos \dfrac{2\pi}{3}$ **c** $\tan \dfrac{\pi}{6}$ **d** $\sin \dfrac{\pi}{3}$

2 Find the value of each trigonometric ratio, to 3 significant figures.

   a $\cos 0.47$ **b** $\sin 1.25$ **c** $\tan 2.3$ **d** $\cos 0.84$

EXAM-STYLE QUESTIONS

3 The diagram shows the circle, center A, radius 4.5 cm, and $B\hat{A}C = 1.3$ radians.

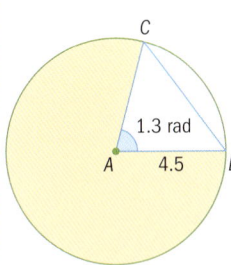

   a Find the area of $\triangle ABC$.
   b Find the length BC.
   c Find the area of the shaded region.

4 The diagram shows the circle, center O, with radius 3 m, $AB = 11$ and $A\hat{O}B = 0.94$ radians. Find the shaded area.

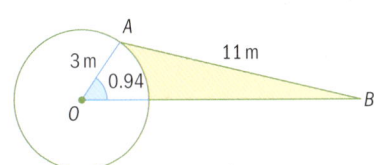

5 The diagram shows the circle, center O, with radius 6 cm, $QR = 11.2$ cm and $P\hat{O}Q = 1.25$ radians.

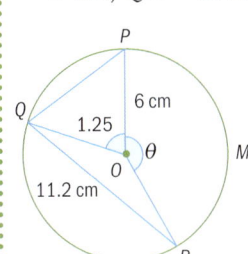

   a Find the area of $\triangle POQ$.
   b Find the area of $\triangle QOR$.
   c Find θ ($P\hat{O}R$).
   d Find the length of arc PMR.

Review exercise

1 In triangle ABC, $\hat{A} = \hat{B} = 45°$. The length of AC is 7 cm. Find the length of AB.

2 In triangle XYZ, $XY = 8$ cm, $XZ = 16$ cm, and $X\hat{Y}Z = 90°$.
   a Find $X\hat{Z}Y$.   **b** Find YZ.

EXAM-STYLE QUESTIONS

3 A straight line passes through the origin (0, 0) and through the point with coordinates (5, 2). The line forms an acute angle of θ with the x-axis.
Find the value of $\tan \theta$.

4 The diagram shows triangle XYZ, with $XZ = 4$ cm, $XY = 10$ cm, and $\hat{X} = 30°$
Find the area of triangle XYZ.

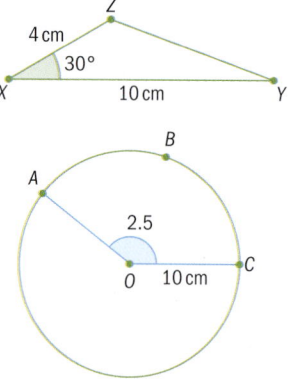

5 The diagram shows a circle, center O and radius 10 cm. $A\hat{O}C = 2.5$ radians.
   a Find the length of arc ABC.
   b Find the area of the shaded sector.

Review exercise

1 An observer standing 100 m from the base of a building sees the top of the building at an angle of elevation of 36°. How tall is the building?

2 The diagram shows part of a unit circle (radius 1 unit) with center O.
   a Angle $AOB = 32°$. Write down the coordinates of B.
   b Point C has coordinates (0.294, 0.956). Find angle AOC.
   c Angle $COD = 54°$. Find the coordinates of D.

EXAM-STYLE QUESTIONS

3 The diagram shows triangle XYZ, with $\hat{X} = 42.4°$, $\hat{Z} = 82.9°$ and $XY = 13.2$ cm.
   a Find $\hat{Y}$.
   b Find XZ.

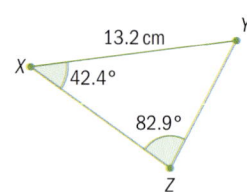

4 The diagram shows triangle PQR, with $\hat{Q} = 118°$, $PQ = 9.5$ m and $QR = 11.5$ m.
   a Find PR.
   b Find $\hat{P}$.

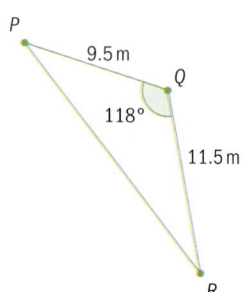

5 This diagram shows the triangle ABC, which has an area of 10 cm².
   a Find $A\hat{C}B$, given that it is an obtuse angle.
   b Find AB.

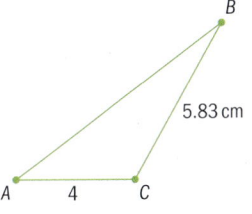

6 Two ships sail from the same port P at the same time.
   Ship A sails on a bearing of 050° for a distance of 24 km before droping anchor.
   Ship B sails on a bearing of 170° for a distance of 38 km before droping anchor.
   Find the distance between the two ships when they are stationary.

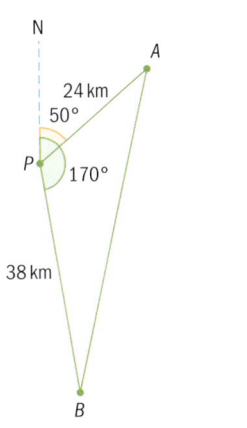

7 The diagram shows quadrilateral ABCD, with AB = 7 cm, BC = 9 cm, CD = 8 cm, and AD = 15 cm. Angle ACD = 82°, angle CAD = x°, and angle ABC = y°

   a Find the value of x.
   b Find AC.
   c Find the value of y.
   d Find the area of triangle ABC.

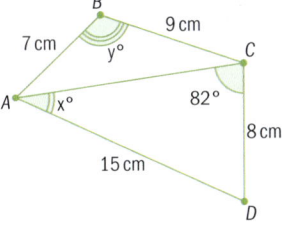

8 The diagram shows a circle with center A and radius 12 cm. Angle DAC = 0.93 radians, and angle BCA = 1.75 radians.
   a Find BC.
   b Find DB.
   c Find the length of arc DEC.
   d Find the perimeter of the region BDEC.

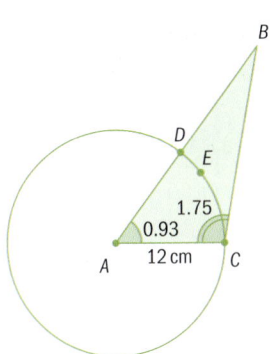

CHAPTER 11 SUMMARY
Right-angled triangle trigonometry

For any right-angled triangle with an angle θ:

- sine $\theta = \dfrac{\text{opposite}}{\text{hypotenuse}} = \dfrac{O}{H}$; cosine $\theta = \dfrac{\text{adjacent}}{\text{hypotenuse}} = \dfrac{A}{H}$;

  tangent $\theta = \dfrac{\text{opposite}}{\text{adjacent}} = \dfrac{O}{A}$

- $\tan \theta = \dfrac{\sin \theta}{\cos \theta}$

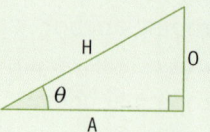

- The trigonometric ratios of 'special angles' are:

angle measure	sine	cosine	tangent
30°	$\dfrac{1}{2}$	$\dfrac{\sqrt{3}}{2}$	$\dfrac{1}{\sqrt{3}} = \dfrac{\sqrt{3}}{3}$
45°	$\dfrac{1}{\sqrt{2}} = \dfrac{\sqrt{2}}{2}$	$\dfrac{1}{\sqrt{2}} = \dfrac{\sqrt{2}}{2}$	$\dfrac{1}{1} = 1$
60°	$\dfrac{\sqrt{3}}{2}$	$\dfrac{1}{2}$	$\dfrac{\sqrt{3}}{1} = \sqrt{3}$

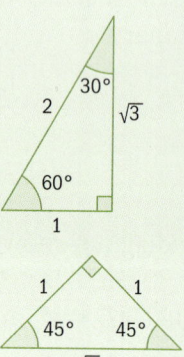

Applications of right-angled triangle trigonometry

- The **angle of elevation** is the angle 'up' from horizontal.
  The **angle of depression** is the angle 'down' from horizontal.
- The four cardinal **compass points** are north (N), south (S), east (E), and west (W).
  Three-figure bearings give directions as clockwise from north.

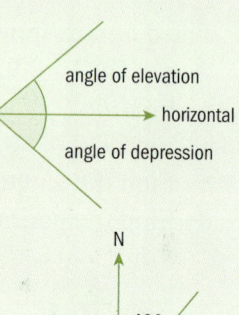

Using the coordinate axes in trigonometry

- For supplementary angles α and β, $\sin \alpha = \sin \beta$, and $\cos \alpha = -\cos \beta$
- For any angle θ, $\sin \theta = \sin(180° - \theta)$, and $\cos \theta = -\cos(180° - \theta)$
- For any angle $\theta°$, $\sin \theta = \sin(180° - \theta)$, and $\cos \theta = -\cos(180° - \theta)$
- These three properties are true for any angle θ:
  1. $\sin^2 \theta + \cos^2 \theta = 1$.
  2. $\tan \theta = \dfrac{\sin \theta}{\cos \theta}$
  3. For any line $y = mx$ which forms an angle of θ with the x-axis, the value of m (the gradient of the line) is $\tan \theta$.

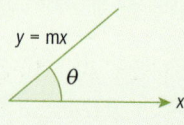

Continued on next page

The sine rule

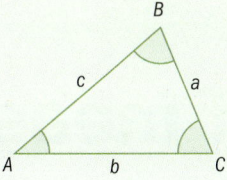

- For any $\triangle ABC$, a is the length of the side opposite $\hat{A}$, b is the length of the side opposite $\hat{B}$, and c is the length of the side opposite $\hat{C}$,

$$\frac{\sin A}{a} = \frac{\sin B}{b} = \frac{\sin C}{c} \text{ or } \frac{a}{\sin A} = \frac{b}{\sin B} = \frac{c}{\sin C}$$

- There can be an ambiguous case when you use the sine rule if:
  - you are given two sides and a non-included acute angle
  - the side opposite the given acute angle is the shorter of the two given sides.

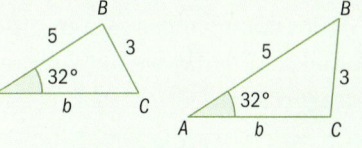

The cosine rule

- The cosine rule states that:
  $a^2 = b^2 + c^2 - 2bc \cos A$ or
  $b^2 = a^2 + c^2 - 2ac \cos B$ or
  $c^2 = a^2 + b^2 - 2ab \cos C$

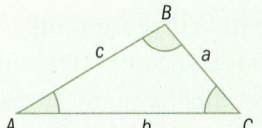

- $\cos A = \dfrac{b^2 + c^2 - a^2}{2bc}$

  $\cos B = \dfrac{a^2 + c^2 - b^2}{2ac}$

  $\cos C = \dfrac{a^2 + b^2 - c^2}{2ab}$

Area of a triangle

- The area of any triangle ABC is given by the formula:

$$\text{area} = \frac{1}{2} bc \sin A \text{ or area} = \frac{1}{2} ac \sin B \text{ or area} = \frac{1}{2} ab \sin C$$

Radians, arcs and sectors

For a sector with central angle θ radians in a circle of radius r:

- Arc length of sector $= r\theta$
- Area of sector $= \dfrac{\theta r^2}{2}$
- To convert degrees to radians, multiply by $\dfrac{\pi}{180}$
- To convert radians to degrees, multiply by $\dfrac{180}{\pi}$

Theory of knowledge

Units of measurement

Mathematics is often considered a 'universal language'. However, this language actually has many forms.

Angles can be measured in different units: degrees or radians. Why do we need more than one unit of measurement?

Actually, we don't need different units of measurement. But different forms of measurement have developed in different parts of the world and at different times.

The idea of 360° in a full circle is thought to date back thousands of years to the ancient Babylonians, who used a *sexagesimal* (base-60) number system. It may also be related to the fact that the orbit of the Earth about the Sun is close to 360 days.

There are other number systems as well as the decimal (base-10) system which we use.

Another important system is binary, which is base 2.

- Where is the binary system commonly used?

The Plimpton 322 tablet dates from Old Babylonian times, around 1800 BCE. Scholars have translated the cuneiform script into modern digits, and discovered that all the numbers are written in base 60.

The numbers are arranged in columns and show Pythagorean triples – the Babylonians were using these more than 1000 years before the time of Pythagoras.

- What is a Pythagorean triple?
- Why is this tablet called Plimpton 322?

- What do we measure in base 60?

The radian unit of measurement seems to be a much more sensible unit for angles, as radians are closely related to the measurements in a circle. Although this type of angle measurement had been used previously by mathematicians, the term 'radian' wasn't much used until the 1870s. Today, radian measurement is commonly used in geometry, trigonometry and calculus.

- How are radians related to the measurements in a circle?
- Who measures angles in **gradians**?

Angles are not the only area in which different units of measurement are common.
A look at the units of currency, distance and mass will show that the 'universal language of mathematics' may not be as universal as we might like to think.

▲ The term **radian** was used in papers written by James Thomson in the early 1870s in Belfast.

These road signs in the USA and France both tell a driver that the speed limit is 30, but give no units.

- Which speed is actually faster?

Would you rather be a millionaire in the US, UK or China?

6000 kg

7 tons (US)

11 000 pounds

- Which elephant is heaviest?

- Do you think it is possible for any language to be truly 'universal'?
- What sort of mathematical information has been sent into deep space, to perhaps communicate with other intelligent life-forms?

Chapter 11   403

12 Vectors

CHAPTER OBJECTIVES:

4.1 Vectors as displacements in the plane and in three dimensions; components of a vector; column representation the sum and difference of two vectors; the zero vector, the vector **–v**; multiplication by a scalar; magnitude of a vector; unit vectors; base vectors **i**, **j** and **k**; position vectors.

4.2 The scalar product of two vectors; perpendicular vectors; parallel vectors. Angle between vectors.

4.3 Vector equation of a line in two and three dimensions. The angle between two lines.

4.4 Coincident and parallel lines. Point of intersection of two lines. Determining whether two lines intersect.

Before you start

You should know how to:

1 Use coordinates in three dimensions.
   e.g. $OABCDEFG$ is a cube with sides 2 units. A lies on the x-axis, C lies on the y-axis and D lies on the z-axis. Write down the coordinates of A, B and F.
   A has coordinates $(2, 0, 0)$.
   B has coordinates $(2, 2, 0)$.
   F has coordinates $(2, 2, 2)$.

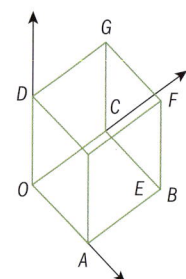

2 Use Pythagoras' theorem.
   e.g. Find the length of the hypotenuse, x, of a triangle with other sides 4 cm, 7 cm.
   $x^2 = 7^2 + 4^2 = 65$
   $x = \sqrt{65} = 8.06$ cm

Skills check

1 The cuboid, $OABCDEFG$ is such that OA has length 3 units, OC 4 units and OD 2 units. A lies on the x-axis, C on the y-axis and D on the z-axis.

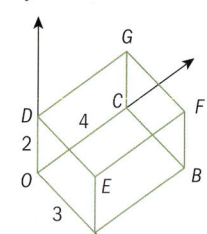

   Give the coordinates of
   a A b B
   c E d F
   e H, the midpoint of GF.

2 Find the length of the hypotenuse, x.

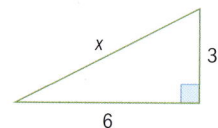

3 Use the cosine rule.
e.g. In triangle PQR, PQ = 6 cm, QR = 11 cm and $\hat{Q} = 95°$.
Calculate the length of PR.
$PR^2 = PQ^2 + QR^2 - 2PQ \times QR \times \cos 95°$
$= 6^2 + 11^2 - 2 \times 6 \times 11 \times \cos 95°$
$= 145.49...$
PR = 12.1 cm (3 sf)

3 a In triangle ABC, AB = 9 cm, BC = 15 cm and angle ABC = 110°. Calculate the length of AC. correct to the nearest cm.

b In triangle ABC, AB = 8.6 cm, BC = 3.1 cm and AC = 9.7 cm.

Diagram NOT accurately drawn

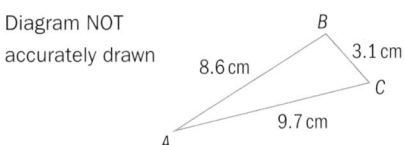

Calculate angle ABC to the nearest degree.

Some quantities can be described by a single number – they only require one piece of information. For example, normal body temperature is 37 °C, the length of the Amazon river is 6400 km, the density of water is 1000 kg m^{-3}. These quantities are determined by magnitude (size) alone and are called **scalars**.

However, some quantities require not only magnitude but also direction to completely define them. Such quantities are called **vectors**. If you wish to fly yourself from London to Paris and I tell you that the distance is 340 km, this piece of information is useless until I tell you what direction you need to travel in!

Vectors are used extensively in a branch of Physics called Mechanics. They are used to represent quantities such as displacement, force, weight, velocity and momentum. In Mathematics we are primarily interested in vectors as representations of displacements and velocities. The final exercise of this chapter has a number of questions where you will be able to see these applications in both two-dimensional and three-dimensional problems. This chapter deals with the basic concepts, vocabulary and notation of vectors and then leads on to the basic operations and geometry of vectors.

> You may wish to explore the role of vectors in Mechanics.

12.1 Vectors: basic concepts

If you travel 4 kilometres north and 3 kilometres east, how far have you traveled?

A simple question perhaps – but a question with two sensible answers:

- One answer to this question is to say that you have traveled 7 kilometres. This is the total **distance** that you have moved through (4 + 3 = 7 kilometres).

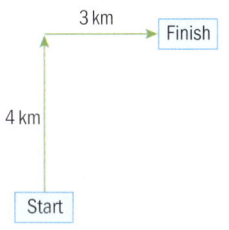

- A second answer to this question is to say that you have traveled 5 kilometres. This value has been found using Pythagoras' theorem ($\sqrt{4^2 + 3^2}$ = 5 kilometres). This value is called the **displacement**. It measures the difference between your initial position and your final position.

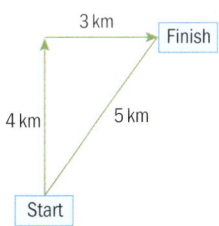

Vectors and scalars

→ A **vector** is a quantity that has **size** (magnitude) and **direction**. Examples of vectors are displacement and velocity.

→ A **scalar** is a quantity that has size but no direction. Examples of scalars are distance and speed.

As can be seen above, distance and displacement have different meanings. This is also true of velocity and speed. Speed refers to how fast something is traveling whereas velocity refers to the rate at which something changes its position.

For instance, if we think of a car that is traveling at 90 kilometres per hour then this is the car's speed.

If that car was traveling around a track where the starting line and the finish line were in the same place, then its velocity when it returns to the starting line would be zero.

If the same car was traveling down a straight road in a westerly direction, after one hour we would say that its velocity is 90 kilometres per hour west.

Representation of vectors

Vectors are represented using directed line segments where the length of the line represents the size of the vector quantity and the direction on the line (indicted by an arrow) shows the direction of the vector.

Consider the points $A(2, 3)$ and $B(5, 7)$ on the Cartesian plane:

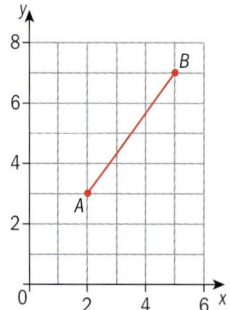

To describe the movement from A to B we could say 'move 3 units in the positive x direction and 4 units in the positive y direction'. The 3 is the called the **horizontal** (or x) **component**, the 4 is the **vertical** (or y) **component**. The direction of this movement and the length of the movement are both important. We can therefore use a vector to describe this.

This vector can be represented in a variety of ways:

In the diagram the line AB represents the vector $\overrightarrow{AB}$ where the arrow over the letters indicates the direction of the movement (from A to B). The components of the vector are here represented using **column vector form.**

$$\overrightarrow{AB} = \begin{pmatrix} 3 \\ 4 \end{pmatrix}$$

Vectors can also be represented using a lower case **bold** letter. For example we could use **a** to represent the vector $\overrightarrow{AB}$.

$$\mathbf{a} = \overrightarrow{AB} = \begin{pmatrix} 3 \\ 4 \end{pmatrix}$$

> In a column vector $\begin{pmatrix} x \\ y \end{pmatrix}$, the x represents a movement in the positive x direction and the y a movement in the positive y direction.

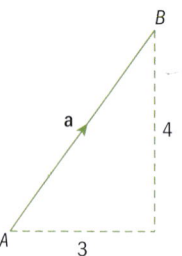

> Bold letters are difficult to write by hand so instead when writing you should underline the letter to show that it is a vector.
> So **a** is written by hand as <u>a</u>.

Finally the vector can be represented in **unit vector form**. We can write $\begin{pmatrix} 3 \\ 4 \end{pmatrix}$ as 3**i** + 4**j** where **i** and **j** are vectors of length 1 unit in the directions of *x* and *y* respectively. **i** and **j** are called base vectors.

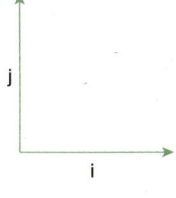

The vector 3**i** + 4**j** therefore means a movement of 3 units in the positive *x* direction and 4 in the positive *y* direction.

As well as objects that move along a flat surface in two dimensions, also think about objects that move around in three-dimensional space. We can represent a vector in three dimensions in a similar way as above but we introduce the letter **k** for the vector of length one unit in the *z*-direction.

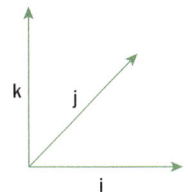

So now we have three components.

$\begin{pmatrix} 3 \\ -2 \\ 1 \end{pmatrix}$ = 3**i** − 2**j** + **k** would therefore represent a movement of 3 units in the positive *x*-direction, 2 units in the negative *y*-direction and 1 unit in the positive *z*-direction.

> → The unit vector in the direction of the *x*-axis is **i**.
>
> In two dimensions $\mathbf{i} = \begin{pmatrix} 1 \\ 0 \end{pmatrix}$ and in three dimensions $\mathbf{i} = \begin{pmatrix} 1 \\ 0 \\ 0 \end{pmatrix}$
>
> → The unit vector in the direction of the *y*-axis is **j**. In two dimensions $\mathbf{j} = \begin{pmatrix} 0 \\ 1 \end{pmatrix}$ and in three dimensions $\mathbf{j} = \begin{pmatrix} 0 \\ 1 \\ 0 \end{pmatrix}$
>
> → In three dimensions the unit vector in the direction of the *z*-axis is
>
> $\mathbf{k} = \begin{pmatrix} 0 \\ 0 \\ 1 \end{pmatrix}$
>
> The vectors **i**, **j**, **k** are called **base vectors**.

Example 1

a Write $\mathbf{a} = \begin{pmatrix} 6 \\ -7 \end{pmatrix}$ in unit vector form.

b Write −**i** + 5**k** in column vector form.

Answers

a **a** = 6**i** − 7**j**

b $\mathbf{b} = \begin{pmatrix} -1 \\ 0 \\ 5 \end{pmatrix}$ *Here the coefficient of the **j** component is zero.*

The magnitude of a vector

The **magnitude** of $\overrightarrow{AB}$ is the length of the vector and is denoted by $|\overrightarrow{AB}|$.

Magnitude is found by using Pythagoras' theorem.

If $\overrightarrow{AB} = \begin{pmatrix} 3 \\ 4 \end{pmatrix}$ then $|\overrightarrow{AB}| = \sqrt{3^2 + 4^2} = \sqrt{25} = 5$

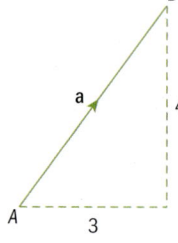

> Other names for magnitude are modulus, length, norm and size.

→ If $\overrightarrow{AB} = \begin{pmatrix} a \\ b \end{pmatrix} = a\mathbf{i} + b\mathbf{j}$ then $|\overrightarrow{AB}| = \sqrt{a^2 + b^2}$.

In three dimensions this becomes

→ If $\overrightarrow{AB} = \begin{pmatrix} a \\ b \\ c \end{pmatrix} = a\mathbf{i} + b\mathbf{j} + c\mathbf{k}$ then $|\overrightarrow{AB}| = \sqrt{a^2 + b^2 + c^2}$

Example 2

Find the magnitude of these vectors

a $\overrightarrow{OP} = \begin{pmatrix} -5 \\ 12 \end{pmatrix}$ **b** $\begin{pmatrix} 3 \\ -2 \\ 1 \end{pmatrix}$

Answers

a $|\overrightarrow{OP}| = \sqrt{(-5)^2 + 12^2} = \sqrt{169} = 13$

b $\left| \begin{pmatrix} 3 \\ -2 \\ 1 \end{pmatrix} \right| = \sqrt{3^2 + (-2)^2 + 1^2} = \sqrt{14} = 3.74 \text{ (3 sf)}$

> When physicists deal with problems of 'uniform acceleration' and 'free fall under gravity' they need to consider the magnitude and direction of the acceleration vector. You may wish to explore this concept further.

Exercise 12A

1 Write these in unit vector form.

  a $x = \begin{pmatrix} -2 \\ 3 \end{pmatrix}$ **b** $y = \begin{pmatrix} 0 \\ 7 \end{pmatrix}$ **c** $z = \begin{pmatrix} 1 \\ 1 \\ -1 \end{pmatrix}$

2 Write these in column vector form.

  a $\overrightarrow{AB} = 2\mathbf{i} + 3\mathbf{j}$ **b** $\overrightarrow{CD} = -\mathbf{i} + 6\mathbf{j} - \mathbf{k}$ **c** $\overrightarrow{EF} = \mathbf{k}$

3 Write the vectors **a**, **b**, **c, d** and **e** in the diagram in both unit vector form and column vector form.

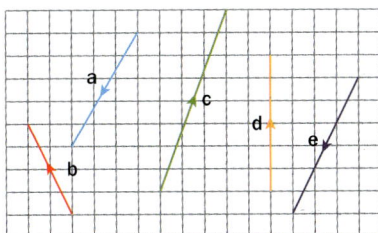

4 Find the magnitude of each vector.

a $\begin{pmatrix} 3 \\ 4 \end{pmatrix}$ b $\begin{pmatrix} 1 \\ -3 \end{pmatrix}$ c $2\mathbf{i} + 5\mathbf{j}$ d $\begin{pmatrix} 2.8 \\ 4.5 \end{pmatrix}$ e $2\mathbf{i} - 5\mathbf{j}$

5 Find the magnitude of each vector.

a $\begin{pmatrix} 3 \\ 2 \\ 5 \end{pmatrix}$ b $\begin{pmatrix} 4 \\ -1 \\ -3 \end{pmatrix}$ c $2\mathbf{i} + 2\mathbf{j} + \mathbf{k}$ d $\begin{pmatrix} -3 \\ 2 \\ 6 \end{pmatrix}$ e $\mathbf{j} - \mathbf{k}$

Equal, negative and parallel vectors

→ Two vectors are **equal** if they have the same direction and the same magnitude; their **i, j, k** components are equal too, and so their column vectors are equal.

Consider the following:

Vectors $\overrightarrow{AB}$ and $\overrightarrow{PQ}$ are pointing in the same direction (are parallel) and have the same magnitude. Therefore $\overrightarrow{AB} = \overrightarrow{PQ}$.

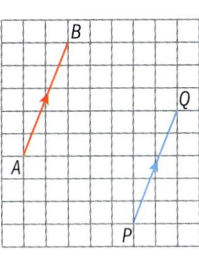

It does not matter where in the Cartesian plane these vectors are – they are still equal.

If two vectors are equal in length then their components will be the same. Here $\overrightarrow{AB} = \overrightarrow{PQ} = \begin{pmatrix} 2 \\ 5 \end{pmatrix}$

The two vectors $\overrightarrow{AB}$ and $\overrightarrow{MN}$ have the same magnitude but different directions.
So $\overrightarrow{AB} \neq \overrightarrow{MN}$.

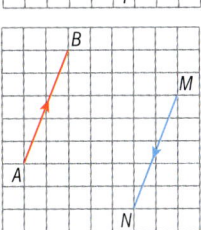

Here $\overrightarrow{AB} = \begin{pmatrix} 2 \\ 5 \end{pmatrix}$ and $\overrightarrow{MN} = \begin{pmatrix} -2 \\ -5 \end{pmatrix}$ and so $\overrightarrow{AB} = -\overrightarrow{MN}$.

$\overrightarrow{MN}$ is called the **negative vector**.

The direction of a vector is important, not just its length.

→ You can write $\overrightarrow{AB}$ as $-\overrightarrow{BA}$.

Vectors $\vec{AB}$, $\vec{CD}$ and $\vec{EF}$ are all **parallel** but have different magnitudes.

Here $\vec{AB} = \frac{1}{2}\vec{CD}$ and $\vec{AB} = 2\vec{EF}$.

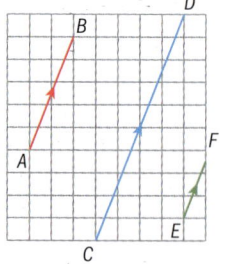

$\vec{AB} = \begin{pmatrix} 2 \\ 5 \end{pmatrix} = 2\mathbf{i} + 5\mathbf{j}$

$\vec{CD} = \begin{pmatrix} 4 \\ 10 \end{pmatrix} = 4\mathbf{i} + 10\mathbf{j}$

$\vec{EF} = \begin{pmatrix} 1 \\ 2.5 \end{pmatrix} = 1\mathbf{i} + 2.5\mathbf{j}$

Here $\vec{AB} = \frac{1}{2}\vec{CD}$ and $\vec{AB} = 2\vec{EF}$.

→ Two vectors are **parallel** if one is a scalar multiple of the other. So, $\vec{AB}$ and $\vec{RS}$ are parallel if $\vec{AB} = k\vec{RS}$ where k is a scalar quantity. This can also be written as $\mathbf{a} = k\mathbf{b}$.

Vectors $\vec{AB}$ and $\vec{GH}$ both have a magnitude of 29 but different directions. So $\vec{AB} \neq \vec{GH}$

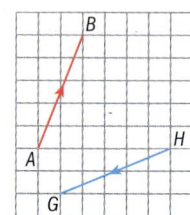

$\vec{AB} = \begin{pmatrix} 2 \\ 5 \end{pmatrix} = 2\mathbf{i} + 5\mathbf{j}$

$\vec{GH} = \begin{pmatrix} 5 \\ 2 \end{pmatrix} = -5\mathbf{i} + 2\mathbf{j}$

We cannot multiply $\vec{AB}$ by a scalar to get $\vec{GH}$.

Example 3

The diagram shows several vectors.

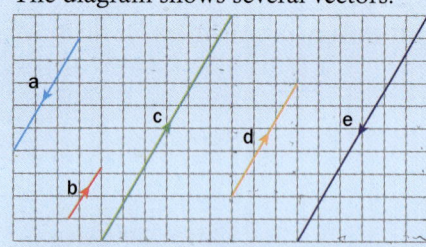

Write each of the other vectors in terms of the vector **a**.

Answer

From the diagram we can see the following:

$\mathbf{a} = \begin{pmatrix} -3 \\ -5 \end{pmatrix}$, $\mathbf{b} = \begin{pmatrix} 1.5 \\ 2.5 \end{pmatrix}$, $\mathbf{c} = \begin{pmatrix} 6 \\ 10 \end{pmatrix}$,

$\mathbf{d} = \begin{pmatrix} 3 \\ 5 \end{pmatrix}$, $\mathbf{e} = \begin{pmatrix} -6 \\ -10 \end{pmatrix}$,

▶ Continued on next page

therefore

$\mathbf{b} = -\frac{1}{2}\mathbf{a}$	**b** is parallel to **a**, in the opposite direction with half the magnitude
$\mathbf{c} = -2\mathbf{a}$	**c** is in the opposite direction to **a**, with twice the magnitude
$\mathbf{d} = -\mathbf{a}$	**d** is in the opposite direction to **a**, with the same magnitude
$\mathbf{e} = 2\mathbf{a}$	**e** is in the same direction as **a**, with twice the magnitude

Example 4

For what values of t and s are these two vectors parallel?
$\mathbf{m} = 3\mathbf{i} + t\mathbf{j} - 6\mathbf{k}$ and $\mathbf{n} = 9\mathbf{i} - 12\mathbf{j} + s\mathbf{k}$

Answer

For parallel vectors $\mathbf{m} = k\mathbf{n}$	
$3\mathbf{i} + t\mathbf{j} - 6\mathbf{k} = k(9\mathbf{i} - 12\mathbf{j} + s\mathbf{k})$	*Multiply out and equate coefficients.*
$3\mathbf{i} + t\mathbf{j} - 6\mathbf{k} = 9k\mathbf{i} - 12k\mathbf{j} + sk\mathbf{k}$	*From **i** components*
$3 = 9k$	
$k = \frac{1}{3}$	
So $t = -12 \times \frac{1}{3} = -4$	*From **j** components*
$-6 = s \times \frac{1}{3} \Rightarrow s = -18$	*From **k** components*

Exercise 12B

1 The diagram shows several vectors.

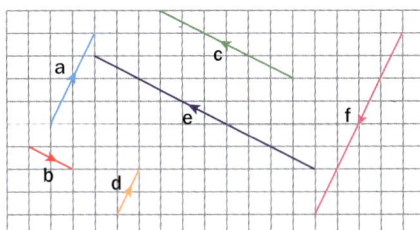

a Write each of the vectors **c**, **d**, **e** and **f** in terms of the vector **a** or **b**.

b How are **a** and **b** related?

2 Which of these vectors are parallel to $\mathbf{i} + 7\mathbf{j}$?

$\mathbf{a} = \begin{pmatrix} 0.1 \\ 0.7 \end{pmatrix}$
$\mathbf{b} = \begin{pmatrix} -1 \\ -7 \end{pmatrix}$
$\mathbf{c} = \begin{pmatrix} -0.05 \\ -0.03 \end{pmatrix}$

$\mathbf{d} = \begin{pmatrix} -10 \\ 70 \end{pmatrix}$
$\mathbf{e} = 60\mathbf{i} + 420\mathbf{j}$
$\mathbf{f} = 6\mathbf{i} - 42\mathbf{j}$

$\mathbf{g} = -\mathbf{i} + 7\mathbf{j}$

3 For what value of t are these two vectors parallel?
   a $r = 4i + tj$ and $s = 14i - 12j$
   b $a = \begin{pmatrix} t \\ -8 \end{pmatrix}$ and $b = \begin{pmatrix} 7 \\ -10 \end{pmatrix}$

4 For what values of t and s are these two vectors parallel?
   $v = ti - 5j + 8k$ and $w = 5i + j + sk$

5 In the cube $OABCDEFG$ the length of each edge is one unit.
   Express these vectors in terms of i, j and k.

   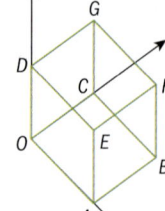

   a $\overrightarrow{OG}$
   b $\overrightarrow{BD}$
   c $\overrightarrow{AD}$
   d $\overrightarrow{OM}$ where M is the midpoint of GF.

6 Repeat question 5 given that $OABCDEFG$ is a cuboid where $OA = 5$ units, $OC = 4$ units and $OD = 3$ units.

Position vectors

Position vectors are vectors giving the position of a point, relative to a fixed origin, O.

The point P with coordinates $(-5, 12)$ has position vector $\overrightarrow{OP} = \begin{pmatrix} -5 \\ 12 \end{pmatrix} = -5i + 12j$.

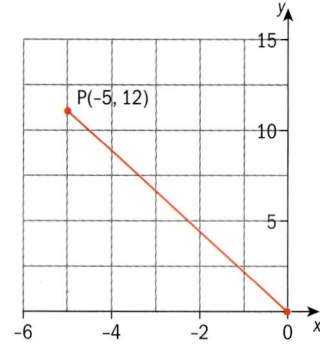

→ The point P with coordinates (x, y) has position vector
$\overrightarrow{OP} = \begin{pmatrix} x \\ y \end{pmatrix} = xi + yj$.

Resultant vectors

Consider the points $A(2, 3)$ and $B(6, 6)$.

The diagram shows the position vectors of A and B.

We can see that the vector $\overrightarrow{AB} = \begin{pmatrix} 4 \\ 3 \end{pmatrix}$

We can also see that to move from A to B we could describe this movement as either going directly from A to B or first going from A to O and then from O to B.

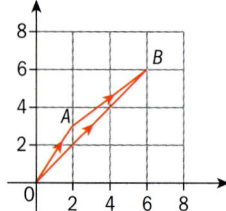

> Remember that $\overrightarrow{AB}$ should be written as a vector, **not** as a coordinate pair.

Thus we could write $\vec{AB} = \vec{AO} + \vec{OB}$.

The vector $\vec{AB}$ is called the resultant of the vectors $\vec{AO}$ and $\vec{OB}$.

Recall that $\vec{AO} = -\vec{OA}$,

and hence
$$\vec{AB} = -\vec{OA} + \vec{OB}$$
$$= \vec{OB} - \vec{OA}$$

> → To find the **resultant vector** $\vec{AB}$ between two points A and B we can subtract the position vector of A from the position vector of B.

Example 5

Points A and B have coordinates $(-3, 2, 0)$ and $(-4, 7, 5)$.
Find the vector $\vec{AB}$.

Answer

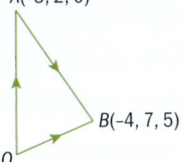

First we write down the position vectors $\vec{OA}$ and $\vec{OB}$.

$$\vec{OA} = \begin{pmatrix} -3 \\ 2 \\ 0 \end{pmatrix}$$

$$\vec{OB} = \begin{pmatrix} -4 \\ 7 \\ 5 \end{pmatrix}$$

$$\vec{AB} = \vec{OB} - \vec{OA} = \begin{pmatrix} -4 \\ 7 \\ 5 \end{pmatrix} - \begin{pmatrix} -3 \\ 2 \\ 0 \end{pmatrix} = \begin{pmatrix} -1 \\ 5 \\ 5 \end{pmatrix}$$

Similarly if we know a vector $\vec{PQ}$ and the vector $\vec{PR}$ then each of the points Q and R are given relative to point P.

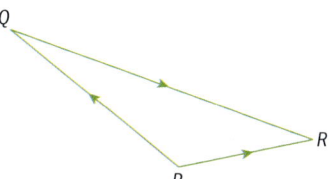

Now
$$\vec{QR} = \vec{QP} + \vec{PR}$$
$$= \vec{PR} - \vec{PQ}$$

Example 6

Given that $\overrightarrow{XY} = \begin{pmatrix} 2 \\ 1 \\ -3 \end{pmatrix}$ and $\overrightarrow{XZ} = \begin{pmatrix} 0 \\ -10 \\ -1 \end{pmatrix}$

Find the vectors a $\overrightarrow{YZ}$ b $\overrightarrow{ZY}$

Answers

a $\overrightarrow{YZ} = \overrightarrow{XZ} - \overrightarrow{XY} = \begin{pmatrix} 0 \\ -10 \\ -1 \end{pmatrix} - \begin{pmatrix} 2 \\ 1 \\ -3 \end{pmatrix} = \begin{pmatrix} -2 \\ -11 \\ 2 \end{pmatrix}$

b $\overrightarrow{ZY} = -\begin{pmatrix} -2 \\ -11 \\ 2 \end{pmatrix} = \begin{pmatrix} 2 \\ 11 \\ -2 \end{pmatrix}$

Exercise 12C

1. P has coordinates $(7, 4)$, Q has coordinates $(2, 3)$.
   Find the vectors $\overrightarrow{PQ}$ and $\overrightarrow{QP}$.

2. Point A has position vector $\begin{pmatrix} 5 \\ 1 \end{pmatrix}$, B has position vector $\begin{pmatrix} 1 \\ -3 \end{pmatrix}$ and C has position vector $\begin{pmatrix} -2 \\ 4 \end{pmatrix}$. Write these as column vectors:

   a $\overrightarrow{AB}$ b $\overrightarrow{BA}$ c $\overrightarrow{AC}$ d $\overrightarrow{CB}$

3. Write these vectors in $a\mathbf{i} + b\mathbf{j} + c\mathbf{k}$ form.
   a $\overrightarrow{OP}$ where P is $(2, -3, 5)$
   b the vector joining $(1, -5, 6)$ to the origin
   c the vector from $(2, -3, 5)$ to $(1, 2, -1)$
   d the vector from $(1, 2, -1)$ to $(2, -3, 5)$

4. $\overrightarrow{LN}\begin{pmatrix} 1 \\ -2 \\ 0 \end{pmatrix}$ and $\overrightarrow{NM}\begin{pmatrix} 4 \\ -2 \\ -3 \end{pmatrix}$. Find $\overrightarrow{LM}$.

5. Given that $\overrightarrow{TS} = 3\mathbf{i} + 4\mathbf{j} - \mathbf{k}$ and $\overrightarrow{TU} = \mathbf{i} - 4\mathbf{j} + 2\mathbf{k}$, find $\overrightarrow{US}$.

6. $\overrightarrow{AB} = \begin{pmatrix} 1 \\ y \\ -2 \end{pmatrix}$, $\overrightarrow{BC} = \begin{pmatrix} 2x \\ -3 \\ z \end{pmatrix}$ and $\overrightarrow{AC} = \begin{pmatrix} 1 \\ 4 \\ x+y \end{pmatrix}$.

   Find the values of the constants x, y and z.

The following example demonstrates how to show that three points are **collinear**.

> Collinear points all lie in a straight line.

Example 7

Show that the points A, B and C with position vectors $\mathbf{i} - 2\mathbf{j} + 3\mathbf{k}$, $-2\mathbf{i} + 3\mathbf{j} - \mathbf{k}$ and $4\mathbf{i} - 7\mathbf{j} + 7\mathbf{k}$ respectively are collinear.

Answer

$\vec{AB} = \vec{OB} - \vec{OA}$ $(-2-1)\mathbf{i} + (3-(-2))\mathbf{j} + (-1-3)\mathbf{k}$
$= -3\mathbf{i} + 5\mathbf{j} - 4\mathbf{k}$

$\vec{AC} = \vec{OC} - \vec{OA}$
$= (4-1)\mathbf{i} + (-7-(-2))\mathbf{j} + (7-3)\mathbf{k}$
$= 3\mathbf{i} - 5\mathbf{j} + 4\mathbf{k}$

$\vec{AB} = -\vec{AC}$

Hence $\vec{AB}$ and $\vec{AC}$ are parallel and, since they contain a common point A, points A, B and C must lie on the same line.

Start by finding the vector joining any two of the points, for example, $\vec{AB}$.

Now repeat using any two other points, for example $\vec{AC}$.

Note that we could have found $\vec{BC} = 6\mathbf{i} - 10\mathbf{j} + 8\mathbf{k}$ which is a scalar multiple of both $\vec{AB}$ and $\vec{AC}$, showing that AB and AC are both parallel to BC.

Exercise 12D

1 Show that the points A, B and C with position vectors $\mathbf{i} - 2\mathbf{j} + 3\mathbf{k}$, $-2\mathbf{i} + 3\mathbf{j} - \mathbf{k}$ and $4\mathbf{i} - 7\mathbf{j} + 7\mathbf{k}$ respectively are collinear.

EXAM-STYLE QUESTION

2 The points A, B and C have coordinates $(2, 3, -3)$, $(5, 1, 5)$ and $(8, -1, 13)$ respectively.
  a Find $\vec{AB}$.
  b Show that A, B and C are collinear.

3 Show that the points $P_1(1, 2, 4)$, $P_2(-2, 1, 4)$ and $P_3(-5, 0, 4)$ are collinear.
Given that P_4 is also collinear with P_1, P_2 and P_3 and the x-coordinate of P_4 is 2, find the y and z-coordinates.

4 The position vectors of A, B and C are given by $3\mathbf{i} + 4\mathbf{j}$, $x\mathbf{i}$, $\mathbf{i} - 2\mathbf{j}$ respectively. Find the value of x so that A, B and C are collinear and find the ratio $AB : BC$.

Distance between two points in space

→ If $A = (x_1, y_1, z_1)$ then $\mathbf{a} = \overrightarrow{OA} = x_1\mathbf{i} + y_1\mathbf{j} + z_1\mathbf{k}$
and if $B = (x_2, y_2, z_2)$ then $\mathbf{b} = \overrightarrow{OB} = x_2\mathbf{i} + y_2\mathbf{j} + z_2\mathbf{k}$

$\overrightarrow{AB} = \overrightarrow{AO} + \overrightarrow{OB}$
$= \overrightarrow{OB} - \overrightarrow{OA}$
$= \mathbf{b} - \mathbf{a}$
$= (x_2 - x_1)\mathbf{i} + (y_2 - y_1)\mathbf{j} + (z_2 - z_1)\mathbf{k}$

Distance $AB = \sqrt{(x_2 - x_1)^2 + (y_2 - y_1)^2 + (z_2 - z_1)^2}$

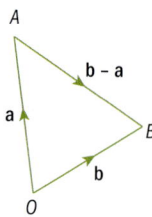

Example 8

Find the vector from $A(1, 3, 4)$ to $B(4, 2, 7)$ and hence determine the distance between the two points.

Answer

$\overrightarrow{OA} = \mathbf{i} + 3\mathbf{j} + 4\mathbf{k}$ and $\overrightarrow{OB} = 4\mathbf{i} + 2\mathbf{j} + 7\mathbf{k}$ *First write the points as position vectors.*

$\overrightarrow{AB} = \overrightarrow{OB} - \overrightarrow{OA}$
$= (4\mathbf{i} + 2\mathbf{j} + 7\mathbf{k}) - (\mathbf{i} + 3\mathbf{j} + 4\mathbf{k})$
$= 3\mathbf{i} - \mathbf{j} + 3\mathbf{k}$

Distance $= |\overrightarrow{AB}| \sqrt{(3)^2 + (-1)^2 + (3)^2}$
$= \sqrt{9 + 1 + 9} = \sqrt{19} = 4.36$ (3 sf)

Exercise 12E

1 Find the vector $\overrightarrow{AB}$ from $A(-1, 5, 1)$ to $B(4, 5, -1)$ and hence determine the distance between the two points.

EXAM-STYLE QUESTION

2 Point A has position vector $\begin{pmatrix} -5 \\ 2 \\ 4 \end{pmatrix}$, B has position vector $\begin{pmatrix} 6 \\ 0 \\ 6 \end{pmatrix}$ and C has position vector $\begin{pmatrix} 8 \\ 10 \\ 1 \end{pmatrix}$.

Show that triangle ABC is isosceles, and calculate the angle CAB.

3 If the position vector $\mathbf{a}$ of a point $(2, -3, t)$ is such that $|\mathbf{a}| = 7$, find two possible values of t.

EXAM-STYLE QUESTION

4 Given that $\mathbf{a} = x\mathbf{i} + 6\mathbf{j} - 2\mathbf{k}$ and $|\mathbf{a}| = 3x$, find two possible values of x.

5 $\mathbf{u} = \begin{pmatrix} a \\ -a \\ 2a \end{pmatrix}$, $\mathbf{v} = \begin{pmatrix} 2 \\ -4 \\ -2 \end{pmatrix}$. Given that $|\mathbf{u}| = |\mathbf{v}|$, find the possible values of a.

6 **a** and **b** are two vectors and $|\mathbf{a}| = 5$.
Find the value of $|\mathbf{a} + \mathbf{b}|$ when
  a $\mathbf{b} = 2\mathbf{a}$
  b $\mathbf{b} = -3\mathbf{a}$
  c **b** is perpendicular to **a** and $|\mathbf{b}| = 12$

Unit vectors

A **unit vector** is a vector of length 1 in a given direction.
To find a unit vector in the same direction as a vector **a** first find the length of the vector **a**, namely $|\mathbf{a}|$, and then multiply the vector **a** by $\frac{1}{|\mathbf{a}|}$. This vector will be in the same direction since it is a scalar multiple of **a** and will be one unit long since it is $\frac{1}{|\mathbf{a}|} \times$ the length of the original vector.

> → A vector of length 1 in the direction of **a** is found by using the formula $\frac{\mathbf{a}}{|\mathbf{a}|}$.

Using this method we can also find a vector of any length, say length k, in the direction of **a**. We would first find the unit vector and then multiple this by k.

> → A vector of length k in the direction of **a** is found by using the formula $k\frac{\mathbf{a}}{|\mathbf{a}|}$.

Example 9

> **a** Find the unit vector in the same direction as the vector $3\mathbf{i} + 4\mathbf{j}$
> **b** Find a vector of length 10 in the same direction as $\begin{pmatrix} 3 \\ -1 \end{pmatrix}$
>
> **Answers**
> **a** The vector $3\mathbf{i} + 4\mathbf{j}$ has length $\sqrt{3^2 + 4^2} = \sqrt{25} = 5$
>
> Therefore a vector of length 1 will be $\frac{1}{5}(3\mathbf{i} + 4\mathbf{j}) = \frac{3}{5}\mathbf{i} + \frac{4}{5}\mathbf{j}$
>
> **b** The vector $\begin{pmatrix} 3 \\ -1 \end{pmatrix}$ has length $\sqrt{10}$. The vector $\frac{1}{\sqrt{10}}\begin{pmatrix} 3 \\ -1 \end{pmatrix}$ has length 1.
>
> Therefore the vector of length 10 is $\frac{10}{\sqrt{10}}\begin{pmatrix} 3 \\ -1 \end{pmatrix}$
>
> This can be simplified if required:
>
> $\frac{10}{\sqrt{10}}\begin{pmatrix} 3 \\ -1 \end{pmatrix} = \frac{10\sqrt{10}}{\sqrt{10}\sqrt{10}}\begin{pmatrix} 3 \\ -1 \end{pmatrix} = \sqrt{10}\begin{pmatrix} 3 \\ -1 \end{pmatrix}$

Exercise 12F

1. Show that $\frac{3}{5}\mathbf{i} + \frac{4}{5}\mathbf{j}$ is a unit vector.

2. Show that $\frac{1}{3}\mathbf{i} - \frac{2}{3}\mathbf{j} + \frac{2}{3}\mathbf{k}$ is a unit vector.

3. Find a unit vector parallel to $4\mathbf{i} - 3\mathbf{j}$

4. Find a unit vector parallel to the vector $\begin{pmatrix} -1 \\ -5 \\ 4 \end{pmatrix}$

5. Find a unit vector in the direction of the vector between the points $P_1(1, 0, 1)$ and $P_2(3, 2, 0)$.

   > Show that the magnitude is 1.

6. $a\mathbf{i} + 2a\mathbf{j}$ is 1 unit long. Given that $a > 0$, find the value of a.

7. Find a vector of magnitude 5 that is parallel to $2\mathbf{i} - \mathbf{j}$.

   EXAM-STYLE QUESTION
8. Find a vector of magnitude 7 in the direction of $\begin{pmatrix} -1 \\ -3 \\ 2 \end{pmatrix}$.

9. Find a unit vector in the same direction as

   a $\begin{pmatrix} 2\cos\theta \\ 2\sin\theta \end{pmatrix}$ b $\begin{pmatrix} 1 \\ \tan\alpha \end{pmatrix}$

12.2 Addition and subtraction of vectors

Addition of vectors

Suppose we have two vectors $\mathbf{u} = \begin{pmatrix} 5 \\ 0 \end{pmatrix}$ and $\mathbf{v} = \begin{pmatrix} 3 \\ 4 \end{pmatrix}$

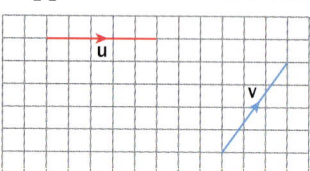

Now $\mathbf{u} + \mathbf{v}$ is interpreted geometrically as first 'move along vector $\mathbf{u}$' followed by 'move along vector $\mathbf{v}$'.

→ The resultant vector, **u** + **v**, is the third side of the triangle formed when **u** and **v** are placed next to each other head to tail.

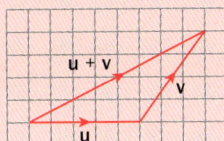

Notice also that vector addition is commutative since **u** + **v** = **v** + **u**. This gives rise to the parallelogram of vector addition.

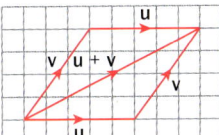

The resultant vector **u** + **v** in this case is $\begin{pmatrix} 8 \\ 4 \end{pmatrix}$.

Notice that this can easily be found arithmetically by adding the corresponding components together.

$$\mathbf{u} + \mathbf{v} = \begin{pmatrix} 5 \\ 0 \end{pmatrix} + \begin{pmatrix} 3 \\ 4 \end{pmatrix} = \begin{pmatrix} 5+3 \\ 0+4 \end{pmatrix} = \begin{pmatrix} 8 \\ 4 \end{pmatrix}$$

The word 'commute' means to exchange or switch.
In mathematics, the commutative property means you can switch the order without affecting the outcome. By considering the following calculations, which of addition, subtraction, multiplication and division would you say are commutative operations?
10 + 5 and 5 + 10
10 ÷ 5 and 5 ÷ 10
10 − 5 and 5 − 10
10 × 5 and 5 × 10

Subtraction of vectors

Again consider the two vectors $\mathbf{u} = \begin{pmatrix} 5 \\ 0 \end{pmatrix}$ and $\mathbf{v} = \begin{pmatrix} 3 \\ 4 \end{pmatrix}$.

u − **v** is interpreted geometrically as 'move along vector **u**' followed by 'move along negative **v**' or **u** + (−**v**).

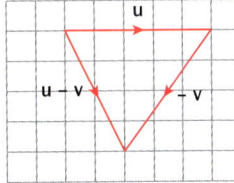

u − **v** = **u** + (− **v**)

The resultant vector here is **u** − **v** and in this specific case is $\begin{pmatrix} 2 \\ -4 \end{pmatrix}$.
Notice again that we can easily find this arithmetically by subtracting the corresponding components.

$$\mathbf{u} - \mathbf{v} = \begin{pmatrix} 5 \\ 0 \end{pmatrix} - \begin{pmatrix} 3 \\ 4 \end{pmatrix} = \begin{pmatrix} 5-3 \\ 0-4 \end{pmatrix} = \begin{pmatrix} 2 \\ -4 \end{pmatrix}$$

Subtraction is not commutative.
$\mathbf{v} - \mathbf{u} = \begin{pmatrix} -2 \\ 4 \end{pmatrix} \neq \mathbf{u} - \mathbf{v}$

→ Vectors are subtracted by adding a negative vector.

The zero vector

Consider the triangle *PQR*

$\vec{PQ} + \vec{QR} + \vec{RP}$ must be equal to zero as the overall journey results in a return to the starting point. This is written as $\vec{PQ} + \vec{QR} + \vec{RP} = \mathbf{0}$

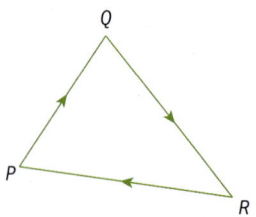

> The zero vector is in bold type to indicate that it is a vector.
>
> $\mathbf{0} = \begin{pmatrix} 0 \\ 0 \end{pmatrix}$ in two dimensions and $\begin{pmatrix} 0 \\ 0 \\ 0 \end{pmatrix}$ in three dimensions.

Equilibrium is the name for the state where a number of forces are in balance – their resultant is zero. You may wish to explore the concept of equilibrium further.

Example 10

Given that $\mathbf{a} = 2\mathbf{i} - 3\mathbf{j} + 3\mathbf{k}$ and $\mathbf{b} = 4\mathbf{i} - 2\mathbf{j} - \mathbf{k}$, find the vectors:
a $\mathbf{a} + \mathbf{b}$ **b** $\mathbf{b} - \mathbf{a}$ **c** $2\mathbf{b} - 3\mathbf{a}$

Answers

a $\mathbf{a} + \mathbf{b} = (2 + 4)\mathbf{i} + (-3 + (-2))\mathbf{j} + (3 + (-1))\mathbf{k}$
$= 6\mathbf{i} - 5\mathbf{j} + 2\mathbf{k}$

b $\mathbf{b} - \mathbf{a} = (4 - 2)\mathbf{i} + (-2 - (-3))\mathbf{j} + (-1 - 3)\mathbf{k}$
$= 2\mathbf{i} + \mathbf{j} - 4\mathbf{k}$

c $2\mathbf{b} - 3\mathbf{a} = (2(4) - 3(2))\mathbf{i} + (2(-2) - 3(-3))\mathbf{j} + (2(-1) - 3(3))\mathbf{k}$
$= 2\mathbf{i} + 5\mathbf{j} - 11\mathbf{k}$

Exercise 12G

1 Given that $\mathbf{a} = 2\mathbf{i} - \mathbf{j}$, $\mathbf{b} = 3\mathbf{i} + 2\mathbf{j}$, $\mathbf{c} = -\mathbf{i} + \mathbf{j}$ and $\mathbf{d} = 3\mathbf{i} + 3\mathbf{j}$, find these vectors.
 a $\mathbf{a} + \mathbf{b}$ **b** $\mathbf{b} + \mathbf{c}$ **c** $\mathbf{c} + \mathbf{d}$
 d $\mathbf{a} + \mathbf{b} + \mathbf{d}$ **e** $\mathbf{a} - \mathbf{b}$ **f** $\mathbf{d} - \mathbf{b} + \mathbf{a}$

2 Given $\mathbf{a} = \begin{pmatrix} 2 \\ -3 \end{pmatrix}$, $\mathbf{b} = \begin{pmatrix} -4 \\ 5 \end{pmatrix}$ and $\mathbf{c} = \begin{pmatrix} -5 \\ -3 \end{pmatrix}$, find these vectors.

 a $\mathbf{a} + \mathbf{b}$ **b** $\mathbf{b} - \mathbf{c}$ **c** $\frac{1}{2}(\mathbf{a} + \mathbf{c})$
 d $\mathbf{a} + 3\mathbf{b} - \mathbf{c}$ **e** $3\mathbf{c} - 2\mathbf{b} + 5\mathbf{a}$

3 Given that $\mathbf{a} = 3\mathbf{i} - \mathbf{j} - 2\mathbf{k}$ and $\mathbf{b} = 5\mathbf{i} - \mathbf{k}$, find these vectors.
  a $\mathbf{a} + \mathbf{b}$ **b** $\mathbf{b} - 2\mathbf{a}$
  c $2\mathbf{a} - \mathbf{b}$ **d** $4(\mathbf{a} - \mathbf{b}) + 2(\mathbf{b} + \mathbf{c})$

4 Given the vectors $\mathbf{p} = 3\mathbf{i} - 5\mathbf{j}$ and $\mathbf{q} = -\mathbf{i} + 4\mathbf{j}$, find the vectors $\mathbf{x}$, $\mathbf{y}$ and $\mathbf{z}$ where
  a $2\mathbf{x} - 3\mathbf{p} = \mathbf{q}$ **b** $4\mathbf{p} - 3\mathbf{y} = 7\mathbf{q}$ **c** $2\mathbf{p} + \mathbf{z} = 0$

5 The vectors $\mathbf{a}$ and $\mathbf{b}$ are such that $\mathbf{a} = \begin{pmatrix} x \\ x+y \end{pmatrix}$ and $\mathbf{b} = \begin{pmatrix} 6-y \\ -2x-3 \end{pmatrix}$.
Given that $\mathbf{a} = \mathbf{b}$, find the values of x and y.

6 The vectors $\mathbf{a}$ and $\mathbf{b}$ are such that $\begin{pmatrix} 3 \\ t \\ u \end{pmatrix}$ and $\begin{pmatrix} t-s \\ 3s \\ t+s \end{pmatrix}$.
Given that $3\mathbf{a} = 2\mathbf{b}$, find the values of s, t and u.

> The method of calculating the combined action of two or more forces by adding them is called the parallelogram law and has been known since the time of the Greek philosopher and polymath Aristotle (384–322 BCE). Dutch Mathematician Simon Stevin (1548–1620) used it in his treatise *Principles of the Art of Weighing* which led to a breakthrough in the development of Mechanics. It was not until around 1800 that Caspar Wessel (Danish-Norwegian, 1745–1818) and Jean-Robert Argand (Swiss, 1768–1822) started to formalize the general concept of a 'vector'.

Geometrical proofs

When you are not given specific vectors you can still use vector addition, subtraction and scalar multiples to deduce some geometrical results.

Example 11

In triangle OXY, A, B and C are the midpoints of OX, OY and XY respectively, $\overrightarrow{OX} = \mathbf{x}$ and $\overrightarrow{OY} = \mathbf{y}$.

a Find expressions for $\overrightarrow{OA}$, $\overrightarrow{OB}$, $\overrightarrow{XY}$, $\overrightarrow{OC}$ and $\overrightarrow{CO}$ in terms of $\mathbf{x}$ and $\mathbf{y}$.

b Find an expression for $\overrightarrow{AB}$ in terms of $\mathbf{x}$ and $\mathbf{y}$. What is the relationship between the line XY and the line AB?

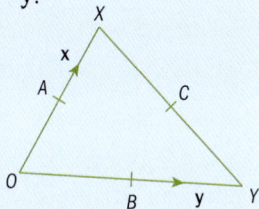

c P is the point such that $\overrightarrow{OP} = \overrightarrow{OX} + \frac{2}{3}\overrightarrow{XP}$. Find $\overrightarrow{OP}$.

d What can you conclude about the position of P?

Answers

a $\overrightarrow{OA} = \frac{1}{2}\overrightarrow{OX} = \frac{1}{2}\mathbf{x}$

$\overrightarrow{OB} = \frac{1}{2}\overrightarrow{OY} = \frac{1}{2}\mathbf{y}$

$\overrightarrow{XY} = \overrightarrow{XO} + \overrightarrow{OY} = -\mathbf{x} + \mathbf{y} = \mathbf{y} - \mathbf{x}$

Use information from the diagram.

Use vector addition.

▶ Continued on next page

$\vec{OC} = \vec{OX} + \vec{XC} = \mathbf{x} + \frac{1}{2}\vec{XY}$ Use vector addition.
From the diagram,
$= \mathbf{x} + \frac{1}{2}(\mathbf{y} - \mathbf{x})$ $\vec{XC} = \frac{1}{2}\vec{XY}.$

$= \mathbf{x} + \frac{1}{2}\mathbf{y} - \frac{1}{2}\mathbf{x}$

$= \frac{1}{2}\mathbf{x} + \frac{1}{2}\mathbf{y} = \frac{1}{2}(\mathbf{x} + \mathbf{y})$

$\vec{CO} = -\vec{OC} = -\frac{1}{2}(\mathbf{x} + \mathbf{y})$

b $\vec{AB} = \vec{AO} + \vec{OB} = -\frac{1}{2}\mathbf{x} + \frac{1}{2}\mathbf{y}$ $\vec{AO} = -\vec{OA}$

$= \frac{1}{2}(\mathbf{y} - \mathbf{x})$

Since $\vec{XY} = \mathbf{y} - \mathbf{x}$ and $\vec{AB} = \frac{1}{2}(\mathbf{y} - \mathbf{x})$
then the line AB is half the length of
XY and in the same direction as XY.
The lines are therefore parallel.

c $\vec{OP} = \vec{OX} + \frac{2}{3}\vec{XB}$

$= \mathbf{x} + \frac{2}{3}(\vec{XO} + \vec{OB})$ Use vector addition.
$\vec{XO} = -\vec{OX}$
$= \mathbf{x} + \frac{2}{3}(-\mathbf{x} + \frac{1}{2}\mathbf{y})$

$= \frac{1}{3}\mathbf{x} + \frac{1}{3}\mathbf{y}$

$= \frac{1}{3}(\mathbf{x} + \mathbf{y})$

So $OP : OC = 2 : 3$

d P lies $\frac{2}{3}$ of the way along the line OC.

Exercise 12H

1 In this triangle $OA = AP$, $BQ = 3OB$, N is the midpoint of PQ and $\mathbf{a} = \vec{OA}$, $\mathbf{b} = \vec{OB}$.
Show that
a $\vec{AP} = \mathbf{a}$
b $\vec{AB} = \mathbf{b} - \mathbf{a}$
c $\vec{PQ} = 4\mathbf{b} - 2\mathbf{a}$
d $\vec{PN} = 2\mathbf{b} - \mathbf{a}$
e $\vec{ON} = \mathbf{a} + 2\mathbf{b}$
f $\vec{AN} = 2\mathbf{b}$

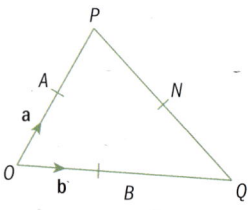

2 In this triangle $\mathbf{a} = \overrightarrow{OA}$, $\mathbf{b} = \overrightarrow{OB}$ and $AC:CB = 3:1$.
Show that
   a $\overrightarrow{AB} = \mathbf{b} - \mathbf{a}$
   b $\overrightarrow{AC} = \frac{3}{4}(\mathbf{b} - \mathbf{a})$
   c $\overrightarrow{CB} = \frac{1}{4}(\mathbf{b} - \mathbf{a})$
   d $\overrightarrow{OC} = \frac{1}{4}\mathbf{a} + \frac{4}{4}\mathbf{b}$

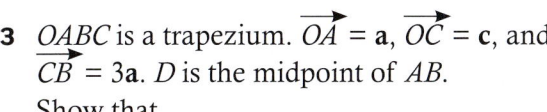

3 $OABC$ is a trapezium. $\overrightarrow{OA} = \mathbf{a}$, $\overrightarrow{OC} = \mathbf{c}$, and $\overrightarrow{CB} = 3\mathbf{a}$. D is the midpoint of AB.
Show that
   a $\overrightarrow{OB} = \mathbf{c} + 3\mathbf{a}$
   b $\overrightarrow{AB} = \mathbf{c} + 2\mathbf{a}$
   c $\overrightarrow{OD} = 2\mathbf{a} + \frac{1}{2}\mathbf{c}$
   d $\overrightarrow{OC} = 2\mathbf{a} - \frac{1}{2}\mathbf{c}$

4 $ABCDEF$ is a regular hexagon with center O. $\overrightarrow{FA} = \mathbf{a}$ and $\overrightarrow{FB} = \mathbf{b}$.
   a Express each of these in terms of $\mathbf{a}$ and/or $\mathbf{b}$.
      i $\overrightarrow{AB}$
      ii $\overrightarrow{FO}$
      iii $\overrightarrow{FC}$
      iv $\overrightarrow{BC}$
      v $\overrightarrow{FD}$

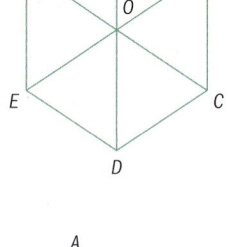

   b What geometrical facts can you deduce about the lines AB and FC?
   c Using vectors, determine whether FD and AC are parallel.

5 In the diagram $\overrightarrow{OA} = \mathbf{a}$ and $\overrightarrow{OB} = \mathbf{b}$. M is the midpoint of OA and P lies on AB such that $\overrightarrow{AP} = \frac{2}{3}\overrightarrow{AB}$.
Show that
   a $\overrightarrow{AB} = \mathbf{b} - \mathbf{a}$ and $\overrightarrow{AP} = \frac{2}{3}(\mathbf{b} - \mathbf{a})$

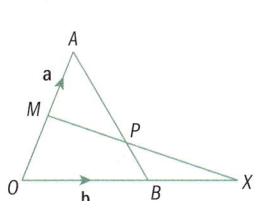

   b $\overrightarrow{MA} = \frac{1}{2}\mathbf{a}$ and $\overrightarrow{MP} = \frac{2}{3}\mathbf{b} - \frac{1}{6}\mathbf{a}$
   c If X is a point such that $OB = BX$, show that $\overrightarrow{MX} = 2\mathbf{b} - \mathbf{a}$.
   d Prove that MPX is a straight line.

12.3 Scalar product

We often need to find the angle θ between two vectors when solving problems.

Investigation – cosine rule

Consider two vectors $\overrightarrow{OA} = \mathbf{a} = 3\mathbf{i} + 4\mathbf{j}$
and $\overrightarrow{OB} = \mathbf{b} = 5\mathbf{i} + 12\mathbf{j}$.

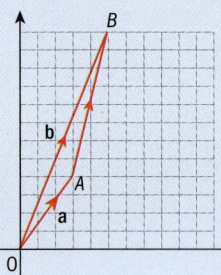

You are going to use the cosine rule to find the angle θ between the two vectors.

1 Find the vector $\overrightarrow{AB}$.

2 Find the lengths of OA, OB and AB ($|\overrightarrow{OA}|$, $|\overrightarrow{OB}|$ and $|\overrightarrow{AB}|$).

3 Recall the cosine rule and apply it to this situation
$$\cos\theta = \frac{|OA|^2 + |OB|^2 - |AB|^2}{2|OA| \times |OB|}$$

4 Find θ by finding $\cos^{-1}\left(\dfrac{|OA|^2 + |OB|^2 - |AB|^2}{2|OA| \times |OB|}\right)$

You should find that $\theta = 14.3°$.

Now repeat this process using $\overrightarrow{OA} = \mathbf{a} = a_1\mathbf{i} + a_2\mathbf{j}$ and $\overrightarrow{OB} = \mathbf{b} = b_1\mathbf{i} + b_2\mathbf{j}$.

At step 3 it is possible to simplify the expression you obtain to
$$\cos\theta = \frac{a_1 b_1 + a_2 b_2}{|\mathbf{a}||\mathbf{b}|}$$

Or alternatively $a_1 b_1 + a_2 b_2 = |\mathbf{a}||\mathbf{b}|\cos\theta$

$a_1 b_1 + a_2 b_2$, is called the **scalar product** of the two vectors $\mathbf{a} = a_1\mathbf{i} + a_2\mathbf{j}$ and $\mathbf{b} = b_1\mathbf{i} + b_2\mathbf{j}$.

> The scalar product is also known as the dot product.

It can be found by multiplying the coefficients of i together, the coefficients of j together and (if in three dimensions) the coefficients of k together and then adding them all up.

→ **Scalar product**
If $\mathbf{a} = a_1\mathbf{i} + a_2\mathbf{j}$ and $\mathbf{b} = b_1\mathbf{i} + b_2\mathbf{j}$ then $\mathbf{a} \cdot \mathbf{b} = a_1 b_1 + a_2 b_2$.
Similarly if $\mathbf{a} = a_1\mathbf{i} + a_2\mathbf{j} + a_3\mathbf{k}$ and $\mathbf{b} = b_1\mathbf{i} + b_2\mathbf{j} + b_3\mathbf{k}$ then
$\mathbf{a} \cdot \mathbf{b} = a_1 b_1 + a_2 b_2 + a_3 b_3$.

> Notice that the scalar product is commutative, that is $\mathbf{a} \cdot \mathbf{b} = \mathbf{b} \cdot \mathbf{a}$

→ The scalar product $\mathbf{a} \cdot \mathbf{b} = |\mathbf{a}||\mathbf{b}|\cos\theta$ where θ is the angle between the vectors.

Example 12

If $\mathbf{a} = \mathbf{i} + 4\mathbf{j} - 2\mathbf{k}$ and $\mathbf{b} = 2\mathbf{i} + 4\mathbf{j} + 6\mathbf{k}$, find $\mathbf{a} \cdot \mathbf{b}$.

Answer
$\mathbf{a} \cdot \mathbf{b} = (1 \times 2) + (4 \times 4) + (-2 \times 6)$
$= 2 + 16 - 12 = 6$

The result is a scalar number, not a vector.

You can also use your GDC to find the scalar product of two vectors.

Finding the angle between two vectors

If you do not know the angle θ between two vectors $\mathbf{a}$ and $\mathbf{b}$ then you can use

$\mathbf{a} \cdot \mathbf{b} = |\mathbf{a}||\mathbf{b}|\cos\theta$ or $\cos\theta = \dfrac{\mathbf{a} \cdot \mathbf{b}}{|\mathbf{a}||\mathbf{b}|}$

to find θ rather than resorting to the full cosine rule each time.

Example 13

Find the angle between $\mathbf{a}$ and $\mathbf{b}$ given that $\mathbf{a} = 3\mathbf{i} + 4\mathbf{j}$ and $\mathbf{b} = 5\mathbf{i} + 12\mathbf{j}$.

Answer
Using $\mathbf{a} \cdot \mathbf{b} = |\mathbf{a}||\mathbf{b}|\cos\theta$,
$\mathbf{a} \cdot \mathbf{b} = 3 \times 5 + 4 \times 12 = 63$
$|\mathbf{a}| = 5, |\mathbf{b}| = 13$
$|\mathbf{a}||\mathbf{b}|\cos\theta = 5 \times 13 \times \cos\theta$
$= 65\cos\theta$
$\Rightarrow 63 = 65\cos\theta$
$\cos\theta = \dfrac{63}{65}$
$\theta = \cos^{-1}\left(\dfrac{63}{65}\right)$
$= 14.25°$

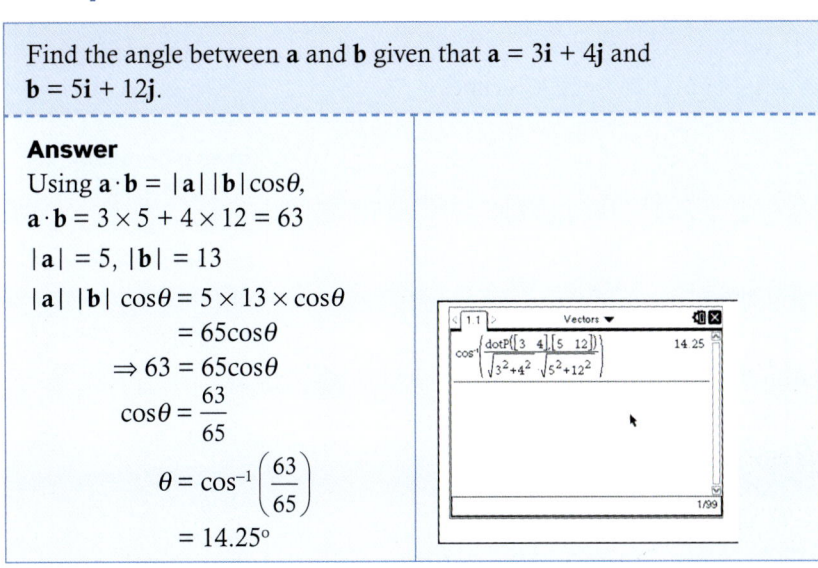

Special properties of the scalar product

Perpendicular vectors

An important fact is that two vectors are perpendicular if and only if their scalar product is zero.
This is because if $\theta = 90°$ then

$$\mathbf{a} \cdot \mathbf{b} = |\mathbf{a}||\mathbf{b}| \cos 90°$$
$$= |\mathbf{a}||\mathbf{b}| \times 0$$
$$= 0$$

> Perpendicular vectors are also called orthogonal.

→ For **perpendicular** vectors $\mathbf{a} \cdot \mathbf{b} = 0$

> Notice that since **i**, **j** and **k** are all perpendicular
> $\mathbf{i} \cdot \mathbf{j} = \mathbf{j} \cdot \mathbf{i} = \mathbf{i} \cdot \mathbf{k} = \mathbf{k} \cdot \mathbf{i} = \mathbf{j} \cdot \mathbf{k} = \mathbf{k} \cdot \mathbf{j} = 0$

Parallel vectors

If two vectors **a** and **b** are parallel then

$$\mathbf{a} \cdot \mathbf{b} = |\mathbf{a}||\mathbf{b}| \cos 0°$$
$$= |\mathbf{a}||\mathbf{b}|$$

→ For **parallel** vectors $\mathbf{a} \cdot \mathbf{b} = |\mathbf{a}||\mathbf{b}|$

Coincident vectors

Given a vector **a**

$$\mathbf{a} \cdot \mathbf{a} = |\mathbf{a}||\mathbf{a}| \cos 0°$$
$$= a^2$$

> Since **i** and **j** and **k** are all one unit in length
> $\mathbf{i} \cdot \mathbf{i} = \mathbf{j} \cdot \mathbf{j} = \mathbf{k} \cdot \mathbf{k} = 1$

→ For **coincident** vectors $\mathbf{a} \cdot \mathbf{a} = a^2$

Exercise 12I

1 Given that $\mathbf{a} = 2\mathbf{i} + 4\mathbf{j}$, $\mathbf{b} = \mathbf{i} - 5\mathbf{j}$ and $\mathbf{c} = -5\mathbf{i} - 2\mathbf{j}$, find
   a $\mathbf{a} \cdot \mathbf{b}$
   b $\mathbf{b} \cdot \mathbf{c}$
   c $\mathbf{a} \cdot \mathbf{a}$
   d $\mathbf{c} \cdot (\mathbf{a} + \mathbf{b})$
   e $(\mathbf{c} + \mathbf{a}) \cdot \mathbf{b}$

2 Given that $\mathbf{u} = \begin{pmatrix} -1 \\ 0 \\ 5 \end{pmatrix}$, $\mathbf{v} = \begin{pmatrix} 4 \\ -3 \\ -1 \end{pmatrix}$ and $\mathbf{w} = \begin{pmatrix} -1 \\ 3 \\ -6 \end{pmatrix}$, find
   a $\mathbf{u} \cdot \mathbf{v}$
   b $\mathbf{u} \cdot (\mathbf{v} - \mathbf{w})$
   c $\mathbf{u} \cdot \mathbf{v} - \mathbf{u} \cdot \mathbf{w}$
   d $2\mathbf{u} \cdot \mathbf{w}$
   e $(\mathbf{u} - \mathbf{v}) \cdot (\mathbf{u} + \mathbf{w})$

> In 1686 Newton published *Philosophiae Naturalis Principia Mathematica*, in which he detailed three **laws of motion**. In understanding and applying these we need to know how to resolve a force into two perpendicular directions and to find the resultant of forces that are perpendicular. You may wish to explore these laws further.

3 Determine whether these pairs of vectors are perpendicular, parallel or neither.

a $\mathbf{a} = 2\mathbf{i} + 4\mathbf{j}$ and $\mathbf{b} = 4\mathbf{i} - 2\mathbf{j}$

b $\mathbf{c} = \begin{pmatrix} 2 \\ 1 \end{pmatrix}$ and $\mathbf{d} = \begin{pmatrix} 1 \\ 2 \end{pmatrix}$

c $\mathbf{u} = \begin{pmatrix} -8 \\ 2 \\ 2 \end{pmatrix}$ and $\mathbf{v} = \begin{pmatrix} 4 \\ -1 \\ -1 \end{pmatrix}$

d $\mathbf{a} = 3\mathbf{i} - 2\mathbf{j} + \mathbf{k}$ and $\mathbf{b} = 3\mathbf{i} - 2\mathbf{j} - \mathbf{k}$

e $\overrightarrow{OX} = \begin{pmatrix} 1 \\ 0 \\ 0 \end{pmatrix}$ and $\overrightarrow{OZ} = \begin{pmatrix} 0 \\ 0 \\ 1 \end{pmatrix}$

f $\mathbf{n} = 2\mathbf{i} - 8\mathbf{j}$ and $\mathbf{m} = -\mathbf{i} + 4\mathbf{j}$

g $\overrightarrow{AB} = \begin{pmatrix} 2 \\ 2 \end{pmatrix}$ and $\overrightarrow{CD} = \begin{pmatrix} -1 \\ -1 \end{pmatrix}$

4 Find $(\mathbf{a} + 3\mathbf{b}) \cdot (2\mathbf{a} - \mathbf{b})$ if $\mathbf{a} = \mathbf{i} + \mathbf{j} + 2\mathbf{k}$ and $\mathbf{b} = 3\mathbf{i} + 2\mathbf{j} - \mathbf{k}$.

5 Given that $\mathbf{a} = 3\mathbf{i} - 5\mathbf{k}$, $\mathbf{b} = 2\mathbf{i} + 7\mathbf{j}$ and $\mathbf{c} = \mathbf{i} + \mathbf{j} + \mathbf{k}$, find the vector $\mathbf{d}$ such that $\mathbf{a} \cdot \mathbf{d} = -9$, $\mathbf{b} \cdot \mathbf{d} = 11$ and $\mathbf{c} \cdot \mathbf{d} = 6$.

6 Find the angle between the vectors $\mathbf{a}$ and $\mathbf{b}$ if $|\mathbf{a}| = \sqrt{3}$, $|\mathbf{b}| = 2$ and $\mathbf{a} \cdot \mathbf{b} = \sqrt{6}$.

7 Find the angles between these vectors, giving your answer in degrees to one decimal place.

a $\begin{pmatrix} 2 \\ -1 \end{pmatrix}$ and $\begin{pmatrix} 2 \\ 5 \end{pmatrix}$

b $\begin{pmatrix} 4 \\ 0 \end{pmatrix}$ and $\begin{pmatrix} -3 \\ 1 \end{pmatrix}$

c $2\mathbf{i} + 5\mathbf{j}$ and $2\mathbf{i} - 5\mathbf{j}$

EXAM-STYLE QUESTIONS

8 Consider the points $A(2,4)$, $B(1,9)$ and $C(3,2)$. Find

a $\overrightarrow{AB}$ and $\overrightarrow{AC}$

b $\overrightarrow{AB} \cdot \overrightarrow{AC}$

c the cosine of the angle between AB and AC.

9 Find the angles between these pairs of vectors.

a $\begin{pmatrix} -1 \\ 2 \\ 2 \end{pmatrix}$ and $\begin{pmatrix} 2 \\ -3 \\ 6 \end{pmatrix}$

b $\begin{pmatrix} 2 \\ 3 \\ 1 \end{pmatrix}$ and $\begin{pmatrix} 4 \\ -2 \\ -2 \end{pmatrix}$

c $2\mathbf{i} - 7\mathbf{j} + \mathbf{k}$ and $\mathbf{i} + \mathbf{j} - \mathbf{k}$

EXAM-STYLE QUESTION

10 Points *A*, *B* and *C* form a triangle. Their position vectors are $\begin{pmatrix} 1 \\ -1 \\ 4 \end{pmatrix}$, $\begin{pmatrix} 2 \\ 3 \\ 4 \end{pmatrix}$ and $\begin{pmatrix} 2 \\ -1 \\ -1 \end{pmatrix}$ respectively. Find

  a the lengths of the sides *AB* and *AC*
  b the exact value of the cosine of the angle *BAC*
  c the area of the triangle.

11 Find the angle between $\begin{pmatrix} 1 \\ 1 \\ 1 \end{pmatrix}$ and the *x*-axis.

EXAM-STYLE QUESTION

12 The position vectors *A* and *B* are $4\mathbf{i} + 4\mathbf{j} - 4\mathbf{k}$ and $\mathbf{i} + 2\mathbf{j} + 3\mathbf{k}$ respectively, relative to an origin *O*.
  a Show that *OA* and *OB* are perpendicular.
  b Find the length of *AB*.

13 Find λ if the vectors $2\mathbf{i} + \lambda\mathbf{j} + \mathbf{k}$ and $\mathbf{i} - 2\mathbf{j} + 3\mathbf{k}$ are perpendicular.

EXAM-STYLE QUESTIONS

14 Let $\mathbf{a} = 5\mathbf{i} - 3\mathbf{j} + 7\mathbf{k}$, $\mathbf{b} = \mathbf{i} + \mathbf{j} + \lambda\mathbf{k}$. Find λ such that $\mathbf{a} + \mathbf{b}$ is perpendicular to $\mathbf{a} - \mathbf{b}$.

15 Let $\mathbf{a} = \begin{pmatrix} p \\ 2 \\ -p \end{pmatrix}$ and $\mathbf{b} = \begin{pmatrix} 2 \\ -p \\ -3 \end{pmatrix}$.

Find the value of *p* such that $\mathbf{a} + \mathbf{b}$ and $\mathbf{a} - \mathbf{b}$ are perpendicular.

12.4 Vector equation of a line

Suppose that a straight line passes through a point *A* where *A* has a position vector **a** and that the line is parallel to a vector **b**.

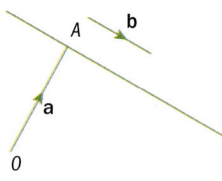

Now if we let *R* be any point on the line, then $\overrightarrow{AR}$ is parallel to **b**.

Any point *R* on the line *L* can be found by starting at the origin then moving through the vector **a** to reach the line. Now there must be some number *t* such that $\overrightarrow{AR} = t\mathbf{b}$.

Hence $\mathbf{r} = \overrightarrow{OR} = \overrightarrow{OA} + \overrightarrow{AR} = \mathbf{a} + t\mathbf{b}$

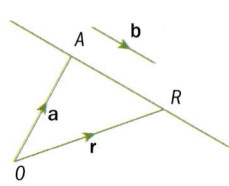

→ The **vector equation** of a line is given by **r** = **a** + *t***b** where **r** is the general position vector of a point on the line, **a** is a given position vector of a point on the line and **b** is a **direction vector** parallel to the line. *t* is called the parameter.

Example 14

a Find the vector equation of the line through $(1, -1, 3)$ and parallel to the vector $-\mathbf{i} + 3\mathbf{j} - \mathbf{k}$

b Find the vector equation of the line through the points $A(1, 0, -4)$ and $B(-2, 1, 1)$.

c Find the acute angle between these two lines.

Answers

a $\mathbf{a} = \mathbf{i} - \mathbf{j} + 3\mathbf{k}$ and $\mathbf{b} = -\mathbf{i} + 3\mathbf{j} - \mathbf{k}$

The vector equation is
$\mathbf{r} = (\mathbf{i} - \mathbf{j} + 3\mathbf{k}) + t(-\mathbf{i} + 3\mathbf{j} - \mathbf{k})$

b $\overrightarrow{OA} = \begin{pmatrix} 1 \\ 0 \\ -4 \end{pmatrix}$ and $\overrightarrow{OB} = \begin{pmatrix} -2 \\ 1 \\ 1 \end{pmatrix}$ *Write the position vectors of A and B.*

$\overrightarrow{AB} = \overrightarrow{OB} - \overrightarrow{OA} = \begin{pmatrix} -3 \\ 1 \\ 5 \end{pmatrix}$ *$\overrightarrow{AB}$ is in the direction of the line.*

Hence the equation of the line is

$\mathbf{r} = \begin{pmatrix} 1 \\ 0 \\ -4 \end{pmatrix} + t \begin{pmatrix} -3 \\ 1 \\ 5 \end{pmatrix}$

c The direction vectors are $\begin{pmatrix} -1 \\ 3 \\ -1 \end{pmatrix}$ and $\begin{pmatrix} -3 \\ 1 \\ 5 \end{pmatrix}$ *To find the angle between these two lines find the angle between their direction vectors.*

Using $\mathbf{a} \cdot \mathbf{b} = |\mathbf{a}||\mathbf{b}|\cos$

$-1 \times -3 + 3 \times 1 + -1 \times 5$

$= \sqrt{11} \times \sqrt{35} \cos\theta$

$\Rightarrow 1 = \sqrt{11}\sqrt{35} \cos\theta$

In the equation $\mathbf{r} = \mathbf{a} + t\mathbf{b}$, $\mathbf{b}$ is the direction vector.

$\cos\theta = \dfrac{1}{\sqrt{11}\sqrt{35}}$

$\theta = \cos^{-1}\left(\dfrac{1}{\sqrt{11}\sqrt{35}}\right)$

$= 87.1°$

Exercise 12J

1 Find the equation of the line parallel to vector **a** and passing through point B with position vector **b** as given.

a $\mathbf{a} = \begin{pmatrix} 3 \\ 2 \end{pmatrix}$ $\mathbf{b} = \begin{pmatrix} -1 \\ 2 \end{pmatrix}$

b $\mathbf{a} = \begin{pmatrix} 5 \\ -2 \end{pmatrix}$ $\mathbf{b} = \begin{pmatrix} -1 \\ 0 \end{pmatrix}$

c $\mathbf{a} = \begin{pmatrix} 3 \\ -2 \\ 8 \end{pmatrix}$ $\mathbf{b} = \begin{pmatrix} 3 \\ 1 \\ -2 \end{pmatrix}$

d $\mathbf{a} = 3\mathbf{i} - \mathbf{j} + \mathbf{k}$ $\mathbf{b} = 2\mathbf{j} - \mathbf{k}$

2 Find a vector equation of the line which passes through the two points.
  a $(4, 5)$ and $(3, -2)$
  b $(4, -2)$ and $(5, -2)$
  c $(3, 5, 2)$ and $(2, -4, 5)$
  d $(0, 0, 1)$ and $(1, -1, 0)$

3 Find an equation of the line perpendicular to vector **a** and passing through point B with position vector **b** as given.

a $\mathbf{a} = \begin{pmatrix} 3 \\ 2 \end{pmatrix}$ $\mathbf{b} = \begin{pmatrix} -1 \\ 6 \end{pmatrix}$

b $\mathbf{a} = \begin{pmatrix} 5 \\ -2 \end{pmatrix}$ $\mathbf{b} = \begin{pmatrix} -1 \\ 0 \end{pmatrix}$

c $\mathbf{a} = \begin{pmatrix} 3 \\ 0 \\ -1 \end{pmatrix}$ $\mathbf{b} = \begin{pmatrix} 4 \\ 2 \\ 1 \end{pmatrix}$

d $\mathbf{a} = \mathbf{i} - 3\mathbf{j} + 4\mathbf{k}$ $\mathbf{b} = 5\mathbf{k}$

4 Determine whether the given point lies on the given line.

a $(4, 5)$ $\mathbf{r} = \begin{pmatrix} 2 \\ 1 \end{pmatrix} + t \begin{pmatrix} 1 \\ 2 \end{pmatrix}$

b $(5, -2)$ $\mathbf{r} = \begin{pmatrix} 5 \\ 1 \end{pmatrix} + t \begin{pmatrix} 4 \\ -3 \end{pmatrix}$

c $(-3, 5, 1)$ $\mathbf{r} = \begin{pmatrix} -1 \\ 5 \\ -3 \end{pmatrix} + t \begin{pmatrix} 1 \\ 0 \\ -2 \end{pmatrix}$

d $(2, 1, 1)$ $\mathbf{r} = 2\mathbf{i} - \mathbf{j} - 3\mathbf{k} + t(-2\mathbf{j} - 3\mathbf{k})$

5 Find the equation of the line through the point $(2, 4, 5)$ in the direction $-2\mathbf{i} + 3\mathbf{j} + 8\mathbf{k}$.

Find p and q so that the point $(p, 10, q)$ lies on this line.

6 Find the vector equation of a vertical line passing through the point $(-6, 5)$.

7 Are the lines represented by these vector equations coincident, parallel or perpendicular, or none of these?

a $\mathbf{r}_1 = \begin{pmatrix} 3 \\ 4 \end{pmatrix} + s \begin{pmatrix} 2 \\ -1 \end{pmatrix}$ $\mathbf{r}_2 = \begin{pmatrix} -9 \\ 10 \end{pmatrix} + t \begin{pmatrix} -6 \\ 3 \end{pmatrix}$

b $\mathbf{r}_1 = \begin{pmatrix} 2 \\ 5 \end{pmatrix} + s \begin{pmatrix} -4 \\ 2 \end{pmatrix}$ $\mathbf{r}_2 = \begin{pmatrix} 2 \\ 1 \end{pmatrix} + t \begin{pmatrix} 1 \\ 2 \end{pmatrix}$

c $\mathbf{r}_1 = \begin{pmatrix} 5 \\ 1 \end{pmatrix} + s \begin{pmatrix} 4 \\ -3 \end{pmatrix}$ $\mathbf{r}_2 = \begin{pmatrix} 5 \\ 3 \end{pmatrix} + t \begin{pmatrix} 8 \\ -6 \end{pmatrix}$

d $\mathbf{r}_1 = \begin{pmatrix} 2 \\ 1 \end{pmatrix} + s \begin{pmatrix} 1 \\ 2 \end{pmatrix}$ $\mathbf{r}_2 = \begin{pmatrix} 2 \\ 2 \end{pmatrix} + t \begin{pmatrix} 1 \\ 1 \end{pmatrix}$

e $\mathbf{r}_1 = \begin{pmatrix} 5 \\ 7 \end{pmatrix} + s \begin{pmatrix} 4 \\ -3 \end{pmatrix}$ $\mathbf{r}_2 = \begin{pmatrix} 5 \\ 1 \end{pmatrix} + t \begin{pmatrix} 4 \\ 3 \end{pmatrix}$

8 Find the angle between these pairs of lines.

a $\mathbf{r} = \begin{pmatrix} 2 \\ 1 \\ 3 \end{pmatrix} + \mu \begin{pmatrix} 1 \\ 4 \\ 0 \end{pmatrix}$ and $\mathbf{r} = \begin{pmatrix} 6 \\ 10 \\ 4 \end{pmatrix} + \lambda \begin{pmatrix} 2 \\ 1 \\ 1 \end{pmatrix}$

b $\mathbf{r} = t \begin{pmatrix} 2 \\ 0 \\ -2 \end{pmatrix}$ and $\mathbf{r} = \begin{pmatrix} -4 \\ 7 \\ 2 \end{pmatrix} + t \begin{pmatrix} -1 \\ 3 \\ 1 \end{pmatrix}$

EXAM-STYLE QUESTIONS

9 The points A and B have coordinates $(-2, -3, -4)$ and $(-6, -7, -2)$ respectively. The line l_1 has equation $\mathbf{r} = \begin{pmatrix} -1 \\ -1 \\ 2 \end{pmatrix} + t \begin{pmatrix} 1 \\ 2 \\ 6 \end{pmatrix}$.

a Show that point A lies on l_1.

b Show that $\overrightarrow{AB}$ is perpendicular to l_1.

10 The figure shows a cuboid in which $OA = 2\,\text{m}$, $OC = 5\,\text{m}$ and $OD = 3\,\text{m}$.
Take O as the origin and unit vectors $\mathbf{i}$, $\mathbf{j}$ and $\mathbf{k}$ in the direction OA, OC and OD respectively.

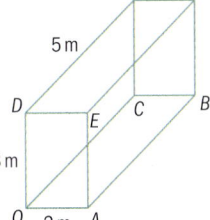

a Express these vectors in terms of the unit vectors.
  i $\overrightarrow{OF}$ **ii** $\overrightarrow{AG}$

b Calculate the value of
  i $|\overrightarrow{OF}|$ **ii** $|\overrightarrow{AG}|$
  iii Find the scalar product of $\overrightarrow{OF}$ and $\overrightarrow{AG}$.

c Hence find the angle between the diagonals OF and AG.

11 Relative to a fixed point O, the points A and B have position vectors $\mathbf{i} + 5\mathbf{j} - 2\mathbf{k}$ and $8\mathbf{i} - 3\mathbf{j} + 6\mathbf{k}$ respectively.
   a Find the vector $\overrightarrow{AB}$.
   b Find the cosine of angle OAB.
   c Show that, for all values of , the point P with position vector $(1 + 7\mu)\mathbf{i} + (5 - 8\mu)\mathbf{j} + (-2 + 8\mu)\mathbf{k}$ lies on the line through A and B.
   d Find the value of μ for which OP is perpendicular to AB.
   e Hence find the point on the foot of the perpendicular from O to AB.

Intersection point of two vectors

If you are given the vector equations of two different lines, you can work out where the lines cross.

Example 15

Two lines have equations $\mathbf{r}_1 = \begin{pmatrix} 3 \\ 0 \\ -1 \end{pmatrix} + s \begin{pmatrix} 1 \\ 1 \\ 1 \end{pmatrix}$ and $\mathbf{r}_2 = \begin{pmatrix} 6 \\ 2 \\ 0 \end{pmatrix} + t \begin{pmatrix} 0 \\ 4 \\ 8 \end{pmatrix}$.

Show that the lines intersect and find the coordinates of the point of intersection.

Answer

Two vectors are equal if their corresponding components are equal.

$\mathbf{r}_1 = \begin{pmatrix} x \\ y \\ z \end{pmatrix} = \begin{pmatrix} 3 \\ 0 \\ -1 \end{pmatrix} + s \begin{pmatrix} 1 \\ 1 \\ 1 \end{pmatrix} \Rightarrow \begin{matrix} x = 3 + s \\ y = s \\ z = -1 + s \end{matrix}$

$\mathbf{r}_2 = \begin{pmatrix} x \\ y \\ z \end{pmatrix} = \begin{pmatrix} 6 \\ 2 \\ 0 \end{pmatrix} + t \begin{pmatrix} 0 \\ 4 \\ 8 \end{pmatrix} \Rightarrow \begin{matrix} x = 6 \\ y = 2 + 4t \\ z = 8t \end{matrix}$

$3 + s = 6$ (1)
$s = 2 + 4t$ (2)
$-1 + s = 8t$ (3)

Equation (1) gives $s = 3$
Substituting $s = 3$ into equation (2):
$3 = 2 + 4t$ so $t = \frac{1}{4}$
Substituting $s = 3$ into equation (3):
$-1 + 3 = 8t$ so $t = \frac{1}{4}$
Since the value of s and the value of t are consistent for all three equations the two lines must intersect.

r_1 and r_2 intersect if there is a value of t and a value of s such that $r_1 = r_2$.

Equate components and solve the resulting simultaneous equations.

In three dimensions, two lines will either
1. **intersect** – if the value of the variables is consistent in all three equations
2. **be parallel** – they will have direction vectors that are scalar multiples of each other
3. **be skew** – if the lines are not parallel and the values are not consistent so the lines do not intersect.

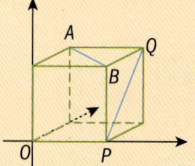

AB and PQ are skew – they never meet.

▶ Continued on next page

Substituting $s = 3$ into $\mathbf{r}_1$:

$$\mathbf{r}_1 = \begin{pmatrix} x \\ y \\ z \end{pmatrix} = \begin{pmatrix} 3 \\ 0 \\ -1 \end{pmatrix} + 3\begin{pmatrix} 1 \\ 1 \\ 1 \end{pmatrix}$$

$x = 3 + 3 = 6$
$y = 0 + 3 = 3$
$z = -1 + 3 = 2$

Therefore the coordinates of the point of intersection are $(6, 3, 2)$.

$$\mathbf{r}_2 = \begin{pmatrix} x \\ y \\ z \end{pmatrix} = \begin{pmatrix} 6 \\ 2 \\ 0 \end{pmatrix} + \frac{1}{4}\begin{pmatrix} 0 \\ 4 \\ 8 \end{pmatrix}$$

$x = 6$
$y = 2 + 4\left(\dfrac{1}{4}\right) = 3$
$z = 0 + 8\left(\dfrac{1}{4}\right) = 2$

To find the point of intersection substitute the value of s into $\mathbf{r}_1$ to give the position vector of the point of in intersection.

Alternatively we could substitute the value of t into $\mathbf{r}_2$

This gives the same coordinates and is a useful way of checking the answer.

Exercise 12K

1. Find the coordinates of the point where $\mathbf{r}_1 = 4\mathbf{i} + 2\mathbf{j} + \lambda(2\mathbf{i} - 4\mathbf{j})$ intercepts $\mathbf{r}_2 = 11\mathbf{i} + 16\mathbf{j} + \mu(\mathbf{i} + 2\mathbf{j})$.

2. The vector equations of two lines are given by $\mathbf{r}_1 = \begin{pmatrix} 4 \\ -2 \end{pmatrix} + s\begin{pmatrix} 8 \\ 2 \end{pmatrix}$ and $\mathbf{r}_2 = \begin{pmatrix} 6 \\ -3 \end{pmatrix} + t\begin{pmatrix} 9 \\ 6 \end{pmatrix}$. The lines intersect at the point P. Find the position vector of P.

EXAM-STYLE QUESTIONS

3. The line l_1 has equation

$$\mathbf{r} = \begin{pmatrix} 5 \\ -1 \\ 2 \end{pmatrix} + t\begin{pmatrix} 2 \\ 1 \\ -1 \end{pmatrix}$$

The line l_2 has equation

$$\mathbf{r} = \begin{pmatrix} 3 \\ -2 \\ -4 \end{pmatrix} + s\begin{pmatrix} 2 \\ 1 \\ 2 \end{pmatrix}$$

Show that the lines l_1 and l_2 intersect, and find the coordinates of the point of intersection.

4 Find where the lines with equations $\mathbf{r}_1 = \mathbf{i} + \mathbf{j} + t(3\mathbf{i} - \mathbf{j})$ and $\mathbf{r}_2 = -\mathbf{i} + s\mathbf{j}$ intersect.

5 Show that the two straight lines $\mathbf{r}_1 = \begin{pmatrix} 3 \\ 0 \\ 5 \end{pmatrix} + t \begin{pmatrix} -1 \\ 1 \\ 2 \end{pmatrix}$ and $\mathbf{r}_2 = \begin{pmatrix} 1 \\ 4 \\ 0 \end{pmatrix} + s \begin{pmatrix} 1 \\ 1 \\ 1 \end{pmatrix}$ are skew.

6 The lines L and M have vector equations
$L: \mathbf{l} = 3\mathbf{i} - 2\mathbf{j} + 5\mathbf{k} + s(-\mathbf{i} + 3\mathbf{j} - 5\mathbf{k})$

$M: \mathbf{m} = 14\mathbf{i} - 20\mathbf{j} + 6\mathbf{k} + t(3\mathbf{i} - 4\mathbf{j} - 3\mathbf{k})$
  a Show that the lines L and M meet and find the position vector of their point of intersection.
  b Show that the lines L and M are perpendicular.

7 The line L has equation $\mathbf{r} = \begin{pmatrix} 6 \\ 9 \\ 3 \end{pmatrix} + t \begin{pmatrix} 1 \\ 2 \\ -2 \end{pmatrix}$.

The point A has coordinates $(5, 7, a)$, where a is a constant.

The point B has coordinates $(b, 13, -1)$, where b is a constant.

The points A and B lie on the line L.

  a Find the values of a and b.
     Point P lies on L such that OP is perpendicular to L.
  b Find the coordinates of P.
  c Hence find the exact distance OP.

8 The points A and B have position vectors $\mathbf{a} = 2\mathbf{i} - \mathbf{j} + 2\mathbf{k}$ and $\mathbf{b} = 3\mathbf{i} - 2\mathbf{j} - \mathbf{k}$ respectively, relative to a fixed origin O.
  a Determine a vector equation of the line L_1, passing through A and B.

The line L_2 has vector equation $\mathbf{r} = 7\mathbf{i} + 3\mathbf{k} + s(2\mathbf{i} + \mathbf{j} + 2\mathbf{k})$.
  b Show that the lines L_1 and L_2 intersect and find the position vector of the point of intersection C.
  c Find the length of the line AC.
  d Find, to the nearest degree, the acute angle between the lines L_1 and L_2.

Extension material on CD:
Worksheet 12 - Equation of a line in three dimensions

12.5 Applications of vectors

Vectors are applicable to real-life situations that include vector quantities such as displacements and velocities.

Example 16

> The position vector of a boat, A, t hours after it leaves a harbour is given by $\mathbf{r}_1 = t \begin{pmatrix} 30 \\ 15 \end{pmatrix}$. A second boat, B, is passing near the harbour. Its position vector at time t is given by $\mathbf{r}_2 = \begin{pmatrix} 50 \\ 5 \end{pmatrix} + t \begin{pmatrix} 10 \\ 10 \end{pmatrix}$.
>
> a How far apart are the two boats at the time the first boat leaves the harbour?
> b How fast is each boat traveling?
> c Are the boats in danger of colliding if one of the boats does not change course?

Answers

a At $t = 0$ boat A is at the 'origin' with position vector $\begin{pmatrix} 0 \\ 0 \end{pmatrix}$ and boat B has position vector $\begin{pmatrix} 50 \\ 5 \end{pmatrix}$ therefore the distance between them is $\sqrt{50^2 + 5^5} = \sqrt{2525} = 50.2$ km.

b The speed of the boats is found by calculating the magnitude of their direction vectors – this is each boat's velocity vector.

For boat A the vector that it will pass through in one hour is $\begin{pmatrix} 30 \\ 15 \end{pmatrix}$ which has length $\sqrt{30^2 + 15^2} = \sqrt{1125} = 33.5$ km.
Therefore boat A has a speed of 33.5 km h^{-1}.
For boat B the vector that it will pass through in one hour is $\begin{pmatrix} 10 \\ 10 \end{pmatrix}$ which has length $\sqrt{10^2 + 10^2} = \sqrt{200} = 14.1$ km.
Therefore boat B has a speed of 14.1 km h^{-1}.

c For the boats to collide there would need to be a value of t such that the position vectors of the two boats are the same.
x components: $30t = 50 + 10t \Rightarrow t = 2.5$ h
y components: $15t = 5 + 10t$ $t = 1$ h
Therefore the boats will not collide.

Exercise 12L

1 The position vector of ship S is 30 km north and 60 km east.
The position vector of buoy B is 20 km north and 45 km east.
What is
  a the position of the ship relative to the buoy
  b the exact distance from the ship to the buoy?

2 A particle P is at the origin O at time $t = 0$. The particle moves with constant velocity and arrives at the point Q with position vector $\begin{pmatrix} x \\ y \end{pmatrix} = \begin{pmatrix} 20 \\ -8 \end{pmatrix}$ m 4 seconds later. Find

  a the velocity of P
  b the position of P if it continues moving past this point with the same velocity for 6 more seconds.

Another particle T is moving with constant velocity $(12\mathbf{i} - 5\mathbf{j})\,\mathrm{m\,s^{-1}}$. It passes through the point A whose position vector is $(4\mathbf{i} - \mathbf{j})$ m at $t = 0$.

  c Find the speed of the particle.
  d Find the distance of T from O when $t = 3$ s.
  e Will the two particles collide?

3 In this question distances are given in kilometres and time in hours. A unit vector represents a displacement of 1 km.
At 3 p.m. a man is standing on the top of a cliff looking out to sea and observing two ships traveling. Ship A's position relative to a point on the shore is given by $3\mathbf{i} + 3\mathbf{j}$ and it is traveling with a velocity of $4\mathbf{i} + 3\mathbf{j}$. Ship B's position is given by $4\mathbf{i} + 3\mathbf{j}$ and it is traveling with a velocity of $3\mathbf{i} + 3\mathbf{j}$. Find

  a the time at which the two ships will collide if one does not change course
  b the point at which the ships will collide.

EXAM-STYLE QUESTION

4 The position of two helicopters X and Y at time t seconds are given by the formulae

$$\mathbf{r}_x = \begin{pmatrix} 11 \\ 3 \\ -3 \end{pmatrix} + t \begin{pmatrix} 1 \\ -1 \\ 4 \end{pmatrix} \text{ and } \mathbf{r}_y = \begin{pmatrix} 1 \\ -7 \\ -2 \end{pmatrix} + t \begin{pmatrix} 2 \\ 1 \\ 9 \end{pmatrix} \text{ respectively.}$$

Distances are given in metres.

  a Find the speed of the two helicopters.
  b Show that the two helicopters do not meet.
  c Find the distance between the helicopters when $t = 10$.

Review exercise

1 Prove using a vector method that the points $A(1, 2, 3)$, $B(-2, 3, 5)$ and $C(7, 0, -1)$ are collinear.

EXAM-STYLE QUESTION

2 Show the points A, B and C with position vectors $5\mathbf{i} - \mathbf{j} + 6\mathbf{k}$, $2\mathbf{i} + 2\mathbf{j}$ and $-3\mathbf{i} - 5\mathbf{j} + 8\mathbf{k}$ respectively form a right-angled triangle.

EXAM-STYLE QUESTION

3 Given that $\mathbf{a} = \begin{pmatrix} 5 \\ -1 \\ -3 \end{pmatrix}$ and $\mathbf{b} = \begin{pmatrix} 1 \\ 3 \\ -5 \end{pmatrix}$, show that the vectors $\mathbf{a} + \mathbf{b}$ and $\mathbf{a} - \mathbf{b}$ are perpendicular.

4 Two lines with equations $\mathbf{r}_1 = \begin{pmatrix} 0 \\ 6 \\ -1 \end{pmatrix} + s\begin{pmatrix} 7 \\ 3 \\ 1 \end{pmatrix}$ and $\mathbf{r}_2 = \begin{pmatrix} 3 \\ 1 \\ 2 \end{pmatrix} + t\begin{pmatrix} 2 \\ 4 \\ -1 \end{pmatrix}$ intersect at the point P. Find the coordinates of P.

EXAM-STYLE QUESTIONS

5 A triangle has its vertices at $A(-2, 4)$, $B(1, 7)$ and $C(-3, 2)$.
   a Find $\overrightarrow{AB}$ and $\overrightarrow{AC}$.
   b Find $\overrightarrow{AB} \cdot \overrightarrow{AC}$.
   c Show that $\cos B\hat{A}C = \dfrac{3}{\sqrt{2}\sqrt{5}}$.

6 Two lines L_1 and L_2 are given by $\mathbf{r}_1 = \begin{pmatrix} 6 \\ 2 \\ -3 \end{pmatrix} + s\begin{pmatrix} -2 \\ 2 \\ 1 \end{pmatrix}$ and $\mathbf{r}_2 = \begin{pmatrix} 0 \\ -12 \\ 7 \end{pmatrix} + t\begin{pmatrix} -1 \\ 11 \\ -3 \end{pmatrix}$.

   a P is the point on L_1 when $s = 4$. Find the position vector of P.
   b Show that P is also on L_2.

7 The line L_1 has vector equation $\mathbf{r} = \begin{pmatrix} 2 \\ -3 \\ -3 \end{pmatrix} + t\begin{pmatrix} 1 \\ 3 \\ 2 \end{pmatrix}$.

   L_2 is parallel to L_1 and passes through the point B(2, 2, 4).

   a Write down a vector equation for L_2 in the form $\mathbf{r} = \mathbf{a} + s\mathbf{b}$.

   A third line L_3 is perpendicular to L_1 and is represented by $\mathbf{r} = \begin{pmatrix} 3 \\ 11 \\ 7 \end{pmatrix} + q\begin{pmatrix} 7 \\ x \\ 1 \end{pmatrix}$.

   b Show that $x = -3$.
   c Find the coordinates of the point C, the intersection of L_1 and L_3.
   d Find $\overrightarrow{BC}$.
   e Find $|\overrightarrow{BC}|$ in the form $a\sqrt{b}$ where a and b are integers to be found.

EXAM-STYLE QUESTION

8 (In this question distances are measured in km and time in hours.) At noon a lighthouse keeper observes two ships A and B.

Ship A's position at time t is given by $r_1 = \begin{pmatrix} -4 \\ 3 \end{pmatrix} + \lambda \begin{pmatrix} 4 \\ 17 \end{pmatrix}$.

Ship B's position at time t is given by $r_2 = \begin{pmatrix} 4 \\ 9 \end{pmatrix} + \mu \begin{pmatrix} -12 \\ 5 \end{pmatrix}$.

a Show that A and B will collide, and find the time when this will occur and the position vector of the point of collision.

In order to prevent collision, at 12:15 ship A changes its direction to $\begin{pmatrix} 16 \\ 17 \end{pmatrix}$.

b Find the distance between A and B at 12:30.

Review exercise

1 Find the size of the angle between the two vectors $\begin{pmatrix} 3 \\ 5 \end{pmatrix}$ and $\begin{pmatrix} 2 \\ -4 \end{pmatrix}$. Give your answer to the nearest degree.

EXAM-STYLE QUESTIONS

2 The vertices of the triangle PQR are defined by the position vectors

$\overrightarrow{OP} = \begin{pmatrix} 3 \\ -2 \\ 1 \end{pmatrix}$, $\overrightarrow{OQ} = \begin{pmatrix} 3 \\ -1 \\ 0 \end{pmatrix}$ and $\overrightarrow{OR} = \begin{pmatrix} 2 \\ -1 \\ 5 \end{pmatrix}$. Find

a $\overrightarrow{QR}$ and $\overrightarrow{QP}$ **b** $P\hat{Q}R$ **c** the area of triangle PQR.

3 A tent OABCDE is a triangular prism with a constant cross-section that is an equilateral triangle with sides of 2 m. The tent is 4 m long. The base OADC is horizontal. Support poles are to be laid along the diagonals BC and BD.

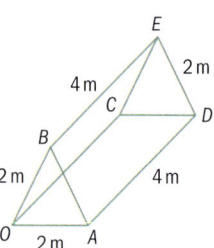

Take O as the origin and unit vectors **i** and **j** in the directions of OA and OC respectively, **k** is a unit vector vertically upwards.

a Express these vectors in terms of the unit vectors **i**, **j** and **k**.
   i $\overrightarrow{OC}$ **ii** $\overrightarrow{OB}$ **iii** $\overrightarrow{OD}$

b Hence find vectors $\overrightarrow{BC}$ and $\overrightarrow{BD}$.

c Calculate the values of
   i $|\overrightarrow{BC}|$ **ii** $|\overrightarrow{BD}|$
   iii the scalar product of $\overrightarrow{BC}$ and $\overrightarrow{BD}$.

d Hence find the angle between the support poles.

4 Given that $\mathbf{a} = x\mathbf{i} + (x-2)\mathbf{j} + \mathbf{k}$ and $\mathbf{b} = x^2\mathbf{i} - 2x\mathbf{j} - 12x\mathbf{k}$ where x is a scalar variable, find

a the values of x for which **a** and **b** are perpendicular

b the angle between **a** and **b** when $x = -1$.

EXAM-STYLE QUESTIONS

5 The points P and Q have position vectors $\begin{pmatrix} 1 \\ -1 \\ 3 \end{pmatrix}$ and $\begin{pmatrix} 1 \\ 5 \\ 5 \end{pmatrix}$ respectively, with respect to an origin O.

   a Show that $\overrightarrow{OP}$ is perpendicular to $\overrightarrow{PQ}$.

   b Write down the vector equation of the line L_1, which passes through P and Q.

   The line L_2 has equation $\mathbf{r} = \begin{pmatrix} 2 \\ -1 \\ 2 \end{pmatrix} + \mu \begin{pmatrix} 1 \\ -3 \\ -2 \end{pmatrix}$.

   c Show that the lines L_1 and L_2 intersect and find the position vector of their point of intersection.

   d Calculate, to the nearest degree, the acute angle between the lines L_1 and L_2.

6 All distances in this question are in metres and time is in seconds.

   An insect is flying at a constant height. At time $t = 0$, the insect is at point A with coordinates $(0, 0, 6)$. Two seconds later the insect is at point B with coordinates $(6, -2, 6)$.

   a Find vector $\overrightarrow{AB}$.

   The insect continues to fly in the same direction at the same speed.

   b Show that the position vector of the insect at time t is given by
   $\begin{pmatrix} x \\ y \\ z \end{pmatrix} = \begin{pmatrix} 0 \\ 0 \\ 6 \end{pmatrix} + t \begin{pmatrix} 3 \\ -1 \\ 0 \end{pmatrix}$.

   At time $t = 0$, a bird takes off from the ground. The position vector of the bird at time t is given by $\begin{pmatrix} x \\ y \\ z \end{pmatrix} = \begin{pmatrix} 36 \\ 18 \\ 0 \end{pmatrix} + t \begin{pmatrix} -3 \\ -4 \\ 1 \end{pmatrix}$.

   c Write down the coordinates of the starting position of the bird.

   d Find the speed of the bird.

   The bird reaches the insect at point C.

   e Find the time the bird takes to reach the insect.

   f Find the coordinates of C.

CHAPTER 12 SUMMARY

Vector: basic concepts

- A **vector** is a quantity that has **size** (magnitude) and **direction**.
  Examples of vectors are displacement and velocity.
- A **scalar** is a quantity that has size but no direction.
  Examples of scalars are distance and speed.
- The unit vector in the direction of the x-axis is $\mathbf{i}$.

  In two dimensions $\mathbf{i} = \begin{pmatrix} 1 \\ 0 \end{pmatrix}$ and in three dimensions $\mathbf{i} = \begin{pmatrix} 1 \\ 0 \\ 0 \end{pmatrix}$

- The unit vector in the direction of the y-axis is $\mathbf{j}$.

  In two dimensions $\mathbf{j} = \begin{pmatrix} 0 \\ 1 \end{pmatrix}$ and in three dimensions $\mathbf{j} = \begin{pmatrix} 0 \\ 1 \\ 0 \end{pmatrix}$

- In three dimensions the unit vector in the direction of the z-axis is $\mathbf{k}$, where

  $\mathbf{k} = \begin{pmatrix} 0 \\ 0 \\ 1 \end{pmatrix}$

- The vectors $\mathbf{i}, \mathbf{j}, \mathbf{k}$ are called **base vectors**.
- If $\overrightarrow{AB} = \begin{pmatrix} a \\ b \end{pmatrix} = a\mathbf{i} + b\mathbf{j}$ then $|\overrightarrow{AB}| = \sqrt{a^2 + b^2}$.

- If $\overrightarrow{AB} = \begin{pmatrix} a \\ b \\ c \end{pmatrix} = a\mathbf{i} + b\mathbf{j} + c\mathbf{k}$ then $|\overrightarrow{AB}| = \sqrt{a^2 + b^2 + c^2}$.

- Two vectors are **equal** if they have the same direction and the same magnitude; their $\mathbf{i}, \mathbf{j}, \mathbf{k}$ components are equal too, and so their column vectors are equal.
- You can write $\overrightarrow{AB}$ as $-\overrightarrow{BA}$.
- Two vectors are **parallel** if one is a scalar multiple of the other.
  So, $\overrightarrow{AB}$ and $\overrightarrow{RS}$ are parallel if $\overrightarrow{AB} = k\overrightarrow{RS}$ where k is a scalar quantity.
  This can also be written as $\mathbf{a} = k\mathbf{b}$.
- The point with coordinates (x, y) has **position vector** $\overrightarrow{OP} = \begin{pmatrix} x \\ y \end{pmatrix} = x\mathbf{i} + y\mathbf{j}$.
- To find the **resultant vector** $\overrightarrow{AB}$ between two points A and B, subtract the position vector of A from the position vector of B.

Continued on next page

- If $A = (x_1, y_1, z_1)$ then $\mathbf{a} = \overrightarrow{OA} = x_1\mathbf{i} + y_1\mathbf{j} + z_1\mathbf{k}$
  and if $B = (x_2, y_2, z_2)$ then $\mathbf{b} = \overrightarrow{OB} = x_2\mathbf{i} + y_2\mathbf{j} + z_2\mathbf{k}$
  $$\overrightarrow{AB} = \overrightarrow{AO} + \overrightarrow{OB}$$
  $$= \overrightarrow{OB} - \overrightarrow{OA}$$
  $$= \mathbf{b} - \mathbf{a}$$
  $$= (x_2 - x_1)\mathbf{i} + (y_2 - y_1)\mathbf{j} + (z_2 - z_1)\mathbf{k}$$
  Distance $AB = \sqrt{(x_2 - x_1)^2 + (y_2 - y_1)^2 + (z_2 - z_1)^2}$

- A vector of length 1 in the direction of $\mathbf{a}$ is found by using the formula $\dfrac{\mathbf{a}}{|\mathbf{a}|}$.

- A vector of length k in the direction of $\mathbf{a}$ is found by using the formula $k\dfrac{\mathbf{a}}{|\mathbf{a}|}$.

Addition and subtraction of vectors

- The resultant vector, $\mathbf{u} + \mathbf{v}$, is the third side of the triangle formed when $\mathbf{u}$ and $\mathbf{v}$ are placed next to each other head to tail.

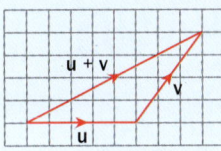

- Vectors are subtracted by adding a negative vector.

Scalar product

- **Scalar product**
  If $\mathbf{a} = a_1\mathbf{i} + a_2\mathbf{j}$ and $\mathbf{b} = b_1\mathbf{i} + b_2\mathbf{j}$ then $\mathbf{a} \cdot \mathbf{b} = a_1b_1 + a_2b_2$.
  Similarly if $\mathbf{a} = a_1\mathbf{i} + a_2\mathbf{j} + a_3\mathbf{k}$ and $\mathbf{b} = b_1\mathbf{i} + b_2\mathbf{j} + b_3\mathbf{k}$ then
  $\mathbf{a} \cdot \mathbf{b} = a_1b_1 + a_2b_2 + a_3b_3$.
- The **scalar product** $\mathbf{a} \cdot \mathbf{b} = |\mathbf{a}||\mathbf{b}|\cos\theta$ where θ is the angle between the vectors.
- For **perpendicular** vectors $\mathbf{a} \cdot \mathbf{b} = 0$.
- For **parallel** vectors $\mathbf{a} \cdot \mathbf{b} = |\mathbf{a}||\mathbf{b}|$.
- For **coincident** vectors $\mathbf{a} \cdot \mathbf{a} = a^2$.

Vector equation of a line

- The **vector equation** of a line is $\mathbf{r} = \mathbf{a} + t\mathbf{b}$ where $\mathbf{r}$ is the general position vector of a point on the line, $\mathbf{a}$ is a given position vector of a point on the line and $\mathbf{b}$ is a **direction vector** parallel to the line. t is called the parameter.

Theory of knowledge

Separate or connected?

Mathematics is often separated into different topics, or fields of knowledge.
- List the different fields of mathematics you can think of.
- Why do humans feel the need to categorize and compartmentalize knowledge?

Algebra and Geometry

In this chapter you represented vectors geometrically and used them to prove geometrical properties. You also used vector algebra to describe and generalize geometrical properties.
- So where do vectors fit – in Algebra or Geometry?

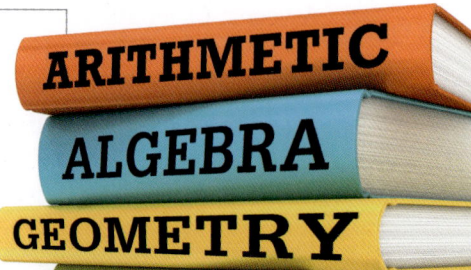

Making connections

Making connections between seemingly different mathematical domains develops understanding. The French mathematician René Descartes (1596–1650) was one of the first to use algebra to solve geometry problems. His key development was Cartesian – or coordinate – geometry.

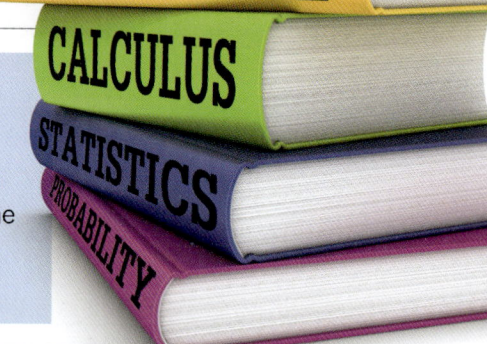

"As long as algebra and geometry have been separated, their progress have been slow and their uses limited, but when these two sciences have been united, they have lent each mutual forces, and have marched together towards perfection."

Joseph Louis Lagrange, 1736–1813, French mathematician

Proving Pythagoras's theorem

Here is a right-angled triangle.
To prove Pythagoras' theorem we need to show that $a^2 + b^2 = c^2$

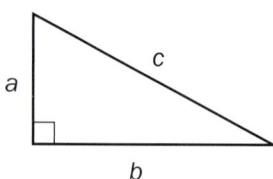

You can see the links between algebra and geometry when they are used to tackle the same problem.

Geometric Proof

Draw and cut out four triangles identical to this one.

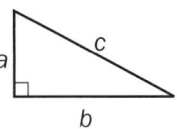

Arrange them to make a square with side lengths $a + b$ like this:

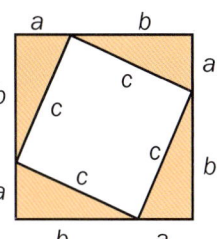

- What is the area of the white square in the center?

Rearrange the triangles to make another square with the same side length, like this:

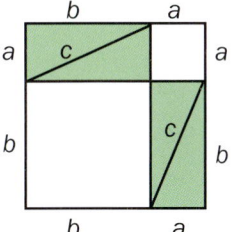

- What are the areas of the two white squares?

The area of the central square in the first diagram must be equal to the sum of the areas of the two squares in the second diagram. That is $c^2 = a^2 + b^2$

Algebraic Proof

Use the same diagram, but look at the triangles instead of the squares.

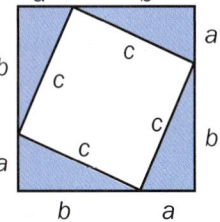

- Use these two methods for finding the area of the large square with side lengths $a + b$.

    Method 1. Square the side lengths: $(a + b)^2$

    Method 2. Calculate the area of the four congruent triangles and add this to c^2, the area of the square.

Methods 1 and 2 both give expressions for the area of the large square.

Equating these gives $b^2 + 2ab + a^2 = 2ab + c^2 \Rightarrow a^2 + b^2 = c^2$

Vector Proof

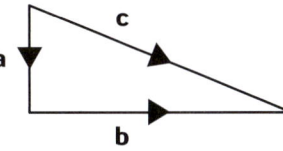

Represent the sides of the right-angled triangle by vectors **a**, **b** and **c**.

Because they form a triangle,   **a** + **b** = **c**

So   (**a** + **b**)·(**a** + **b**) = **c**·**c**

Expanding this gives   **a**·**a** + **a**·**b** + **b**·**a** + **b**·**b** = **c**·**c**

a·**b** = **b**·**a** = 0, because **a** and **b** are perpendicular

So   **a**·**a** + **b**·**b** = **c**·**c**

or $a^2 + b^2 = c^2$

- Which method of proof did you prefer?
- Which was the easiest?
- Which was the most beautiful?

13 Circular functions

CHAPTER OBJECTIVES

3.2 Definition of $\cos\theta$ and $\sin\theta$ in terms of the unit circle; definition of $\tan\theta$ as $\dfrac{\sin\theta}{\cos\theta}$

3.2 Exact values of trigonometric ratios of $0, \dfrac{\pi}{6}, \dfrac{\pi}{4}, \dfrac{\pi}{3}, \dfrac{\pi}{2}$ and their multiples

3.3 The Pythagorean identity $\cos^2\theta + \sin^2\theta = 1$

3.3 Double-angle identities for sine and cosine

3.3 Relationship between trigonometric ratios

3.4 The circular functions $\sin x$, $\cos x$ and $\tan x$

3.4 Composite functions of the form $f(x) = a\sin(b(x+c)) + d$

3.4 Transformations

3.4 Applications of trigonometric functions

3.5 Solving trigonometric equations in a finite interval, both graphically and analytically

Before you start

You should know how to:

1. Find the exact values of certain trigonometric ratios.
   e.g. Find the exact value of $\sin 30°$.
   $\sin 30° = 0.5$
   e.g. Find the exact value of $\tan\left(\dfrac{3\pi}{4}\right)$.
   $\tan\left(\dfrac{3\pi}{4}\right) = -1$

2. Work with the graphing functions of your GDC.
   e.g. Use the graphing functions of your GDC to find the x-intercepts of the graph of $f(x) = x^3 - 3x^2 + 2$.
   $x \approx -0.732, 1, 2.73$
   e.g. Use the graphing functions of your GDC to solve the equation $4x^2 - 7 = 2\ln x$. $x \approx 0.0303, 1.38$

Skills check

1. Find the exact value of
   a $\sin 45°$ b $\tan 60°$
   c $\cos 150°$ d $\sin 225°$

2. Find the exact value of
   a $\sin\dfrac{2\pi}{3}$ b $\tan\dfrac{3\pi}{4}$
   c $\cos\pi$ d $\sin\dfrac{7\pi}{6}$

3. Use the graphing functions of your GDC to find the x-intercepts of the graph of each function.
   a $f(x) = 2x^3 - x + 5$ b $f(x) = \ln(x^2 - 3)$

4. Use the graphing functions of your GDC to solve each equation.
   a $x^3 - 5x = \sqrt{x+1}$ b $x^4 = 3 - x^2$

The London Eye, on the south bank of the River Thames, was opened to the public in the year 2000. Each of the 32 capsules can carry up to 25 people. It is a major tourist attraction, and has an average 3.5 million visitors each year.

The Eye makes approximately one revolution every 30 minutes. It is about 135 metres tall at the highest point. A passenger in one of the capsules travels in a circle in a complete revolution. The passenger's height above the boarding platform can be modeled by the function

$$h(t) = 67.5\cos\left(\frac{2\pi}{30}(t-15)\right) + 67.5,$$

where h is the height in metres, and t is the time in minutes after a passenger boards the capsule. This is an example of a circular function, which you will study in this chapter.

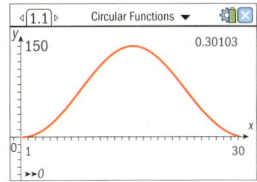

▲ This is the graph of the function which models the passenger's height above the boarding platform.

13.1 Using the unit circle

In this section, we will continue to work with the unit circle.

→ The unit circle has its center at the origin $(0, 0)$ and a radius length of 1 unit. The terminal side of any angle θ in standard position will meet the unit circle at a point with coordinates $(\cos\theta, \sin\theta)$.

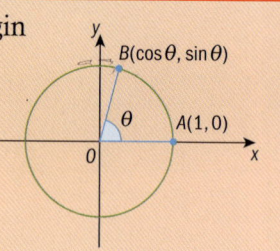

Remember that the unit circle has equation $x^2 + y^2 = 1$.

In this diagram, $A\hat{O}B$ (θ) is in standard position. The point A has coordinates $(1, 0)$, and the point B has coordinates $(\cos\theta, \sin\theta)$.

Look at some angles in standard position in the unit circle. If the angle θ opens in a counterclockwise direction (from the positive x-axis), then θ is positive. These angles can be measured in degrees or in radians.

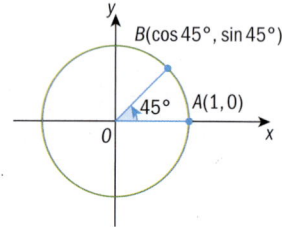
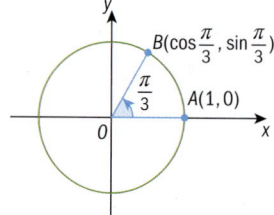

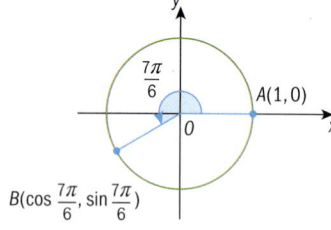
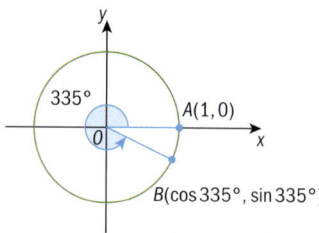

If the angle θ opens in a clockwise direction (from the positive x-axis), then θ is negative.

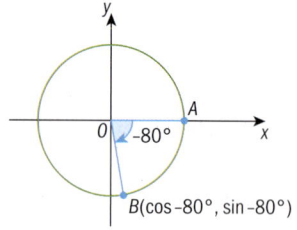
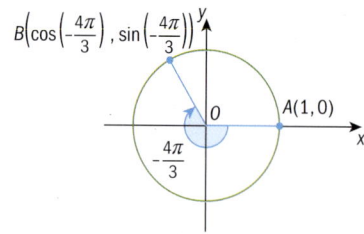

448 Circular functions

If we know the sine and cosine values for an angle, we can give numerical coordinates to the point where the angle meets the unit circle.

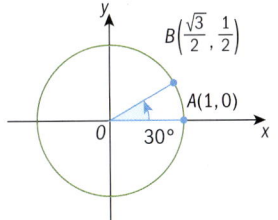
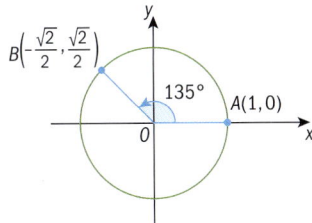

Investigation – sine, cosine and tangent on the unit circle

You can also use the unit circle to help you understand the sine and cosine values of angles whose terminal sides lie on the x- and y-axes.

Sketch each angle in standard position on the unit circle. Use your sketch (not your GDC) to help you determine the sine, cosine and tangent of each angle.

Angles in degrees:

1 90°
2 180°
3 270°
4 360°
5 −90°
6 −180°

Angles in radians:

7 0
8 $\dfrac{\pi}{2}$
9 π
10 $\dfrac{3\pi}{2}$
11 $-\dfrac{3\pi}{2}$
12 4π

In Chapter 11, you used right triangles to help you find the exact values of sine, cosine and tangent for 30°, 45° and 60°. You will now extend this to include other special angles in degrees and radians.

Angle measure degrees, radians	Sine	Cosine	Tangent
0°, 0 radians	0	1	0
30°, $\dfrac{\pi}{6}$	$\dfrac{1}{2}$	$\dfrac{\sqrt{3}}{2}$	$\dfrac{1}{\sqrt{3}} = \dfrac{\sqrt{3}}{3}$
45°, $\dfrac{\pi}{4}$	$\dfrac{1}{\sqrt{2}} = \dfrac{\sqrt{2}}{2}$	$\dfrac{1}{\sqrt{2}} = \dfrac{\sqrt{2}}{2}$	$\dfrac{1}{1} = 1$
60°, $\dfrac{\pi}{3}$	$\dfrac{\sqrt{3}}{2}$	$\dfrac{1}{2}$	$\dfrac{\sqrt{3}}{1} = \sqrt{3}$
90°, $\dfrac{\pi}{2}$	1	0	undefined

It is important for you to remember these values, as you will be required to know them without using your GDC.

In Chapter 11, you discovered that supplementary angles have the same sine value. You also found that these angles have opposite cosine values.

For example, $\sin 30° = \sin 150°$, and $\cos 150° = -\cos 30°$.

Now, you will use the unit circle to find other angles with 'related' trigonometric values.

Consider angles in each quadrant that form the same angle with the x-axis. As the coordinates on the unit circle represent the cosine and sine values, you can see that angles in different quadrants have related sine and cosine values.

For angles in the second quadrant, the sine is positive, and the cosine is negative.

For angles in the first quadrant, the sine and cosine values are both positive.

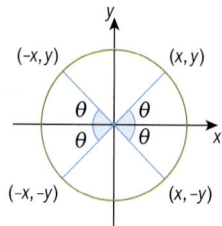

For angles in the third quadrant, the sine and cosine values are both negative.

For angles in the fourth quadrant, the cosine is positive, and the sine is negative.

→ For any angle θ, $\tan \theta = \dfrac{\sin \theta}{\cos \theta}$, where $\cos \theta \neq 0$.

It follows that, for angles in the first and third quadrants, the tangent will be positive, and for angles in the second and fourth quadrants, the tangent will be negative.

Example 1

Find three other angles with the same value as:
a sine 35°
b cosine 35°
c tangent 35°.

Answers

a To find angles with the same sine:

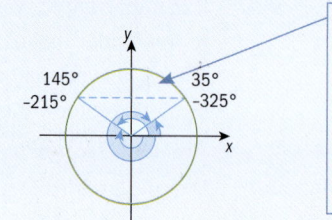

Angles with the same sine values meet the unit circle at points with the same y-coordinates.

To find angles with the same sine values, draw a horizontal line across the unit circle.

$\sin 35° = \sin 145° = \sin(-215°) = \sin(-325°)$

These angles all form a 35° angle with the x-axis.

▶ Continued on next page

b To find angles with the same cosine:

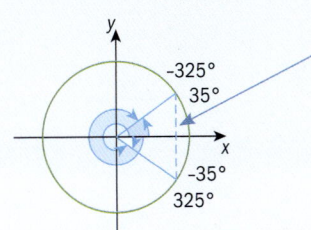

Angles with the same cosine values meet the unit circle at points with the same x-coordinates.

To find angles with the same cosine values, draw a vertical line across the unit circle.

> These angles all form a 35° angle with the x-axis.

$\cos 35° = \cos 325° = \cos(-35°) = \cos(-325°)$

c To find angles with the same tangent:

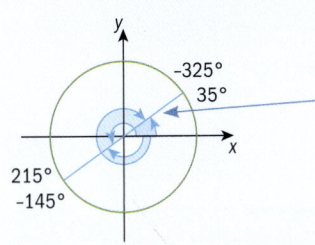

Tangent values are positive in the first and third quadrants.

To find angles with the same tangent values, draw a line through the origin of the unit circle.

> These angles all form a 35° angle with the x-axis.

$\tan 35° = \tan 215° = \tan(-145°) = \tan(-325°)$

This last example helps illustrate some useful properties.

→ For any angle θ:
$\sin \theta = \sin(180 - \theta)$
$\cos \theta = \cos(-\theta)$
$\tan \theta = \tan(180 + \theta)$

Exercise 13A

1 Sketch each angle in standard position on the unit circle.
   a 75°   **b** 110°   **c** 250°   **d** 330°
   e −100°   **f** −270°   **g** −180°   **h** 40°

2 Sketch each angle in standard position on the unit circle.

> These angles are measured in radians.

   a $\dfrac{\pi}{6}$ **b** $\dfrac{5\pi}{3}$ **c** $\dfrac{\pi}{2}$ **d** $\dfrac{11\pi}{6}$

   e $-\dfrac{\pi}{3}$ **f** $-\dfrac{5\pi}{6}$ **g** -2π **h** 3

3 Find three other angles (in degrees) with the same sine as the given angle.
   a 60°   **b** 200°   **c** −75°   **d** 115°

4 Find three other angles (in degrees) with the same cosine as the given angle.
   a 35°   **b** 130°   **c** 295°   **d** −240°

5 Find three other angles (in degrees) with the same tangent as the given angle.
  a 50°   **b** 100°   **c** 220°   **d** −25°

6 Find three other angles (in radians) with the same sine as the given angle.
  a $\dfrac{\pi}{3}$ **b** $\dfrac{5\pi}{4}$ **c** 4.1 rad   **d** −3 rad

7 Find three other angles (in radians) with the same cosine as the given angle.
  a $\dfrac{\pi}{6}$ **b** 1 rad   **c** 2.5 rad   **d** $-\dfrac{3\pi}{5}$

8 Find three other angles (in radians) with the same tangent as the given angle.
  a $\dfrac{\pi}{4}$ **b** 1.3 rad   **c** $\dfrac{5\pi}{7}$ **d** −5 rad

Example 2

Given that $\sin 50° = 0.766$ (to 3 significant figures), find the values of
a $\cos 50°$ **b** $\cos 130°$ **c** $\sin 230°$ **d** $\cos(-50°)$

Answers

a $\sin^2 50° + \cos^2 50° = 1$

   $(0.766)^2 + \cos^2 50° = 1$
   $\cos^2 50° = 1 - (0.766)^2$
   $\cos 50° = \sqrt{1-(0.766)^2}$
   $\cos 50° = \pm 0.643$ (3 sf)

Use $\sin^2 \theta + \cos^2 \theta = 1$, the 'Pythagorean identity' found in Section 11.3.

Substitute $\sin 50° = 0.766$, then solve for $\cos \theta$.

b
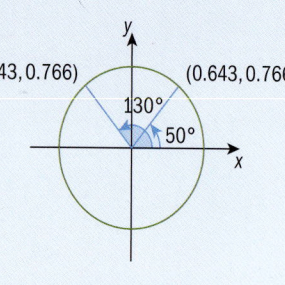

$\cos 130° = -0.643$ (3 sf)

It is a good strategy to make a sketch of the angles on a unit circle. This makes the relationships between the angle easier to see.

▶ Continued on next page

c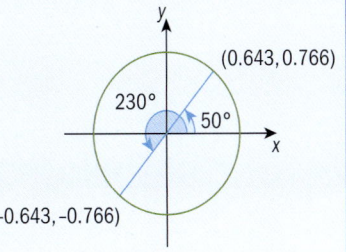

*Use similar sketches to help answer parts **c** and **d**.*

sin 230° = −0.766

These related angles all make an angle of 50° with the x-axis.

d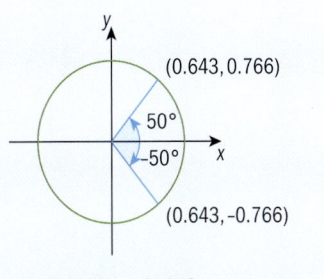

cos (−50°) = 0.643

Exercise 13B

1 Given that sin 70° = 0.940 and cos 70° = 0.342 (to 3 sf), find each value.
   a sin 110°   **b** cos (−70°)   **c** cos 250°   **d** sin 290°

EXAM-STYLE QUESTIONS

2 Given that $\sin \frac{\pi}{6} = \frac{1}{2}$ and $\cos \frac{\pi}{6} = \frac{\sqrt{3}}{2}$, find each value.
   a $\sin \frac{7\pi}{6}$ **b** $\cos \frac{5\pi}{6}$ **c** $\sin \left(-\frac{\pi}{6}\right)$ **d** $\cos \left(-\frac{11\pi}{6}\right)$

3 Given that sin A = 0.8 and cos A = 0.6, find each value.
   a sin (180° − A)   **b** cos (−A)   **c** cos (360° − A)
   d sin (180° + A)   **e** tan A **f** tan (−A)
   g sin (360° − A)   **h** tan (180° + A)

4 Given that sin θ = a and cos θ = b, find each value in terms of a and b.
   a tan θ **b** sin ($\pi - \theta$)   **c** cos ($\pi + \theta$)   **d** tan ($\pi + \theta$)
   e sin ($\pi + \theta$)   **f** cos (−θ)   **g** sin ($2\pi - \theta$)   **h** cos ($\theta - \pi$)

13.2 Solving equations using the unit circle

Suppose we want to solve an equation such as $\sin x = \dfrac{1}{2}$.

We know that $\sin 30° = \dfrac{1}{2}$, but we also know that

$\sin 150° = \dfrac{1}{2}$, $\sin \dfrac{\pi}{6} = \dfrac{1}{2}$, and $\sin\left(-\dfrac{7\pi}{6}\right) = \dfrac{1}{2}$.

So what is the value of x in the equation $\sin x = \dfrac{1}{2}$?

In fact, there are an infinite number of values we could substitute for x, so we need more information about the values of x that we are looking for. We need to know two things:

- Is x measured in degrees or radians?
- What is the domain?

Now suppose we want to solve the equation $\sin x = \dfrac{1}{2}$, for $-360° \leq x \leq 360°$. There are two positions on the unit circle for which $\sin x = \dfrac{1}{2}$, so we will find the angles at those positions which are within our domain $-360° \leq x \leq 360°$.

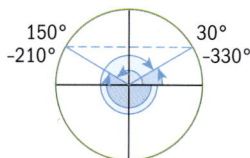

The equation has four solutions within the given domain.
$x = -330°, -210°, 30°, 150°$

Example 3

Solve the equation $\cos x = -\dfrac{\sqrt{2}}{2}$, $-2\pi \leq x \leq 2\pi$.

Answer

$\dfrac{3\pi}{4}, \dfrac{-5\pi}{4}$

$\dfrac{-3\pi}{4}, \dfrac{5\pi}{4}$

$x = -\dfrac{5\pi}{4}, -\dfrac{3\pi}{4}, \dfrac{3\pi}{4}, \dfrac{5\pi}{4}$

You know $\cos\left(\dfrac{3\pi}{4}\right) = -\dfrac{\sqrt{2}}{2}$.

Draw a vertical line to help find the other position on the unit circle with this same cosine value.
Once you have found both positions on the unit circle, find all the angles within the domain that have their terminal sides at these positions.

Example 4

Solve the equation $\tan x = \sqrt{3}$, $0 \leq x \leq 720°$.

Answer

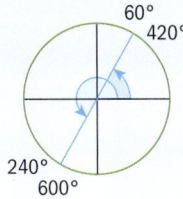

$x = 60°, 240°, 420°, 600°$

$\tan 60° = \sqrt{3}$.
Draw a line through the origin to find the other position on the unit circle with this same tangent value. You can find the angles 420° and 600° by making another rotation around the unit circle.

Exercise 13C

1 Solve each equation for $-360° \leq x \leq 360°$.

  a $\sin x = \dfrac{\sqrt{3}}{2}$ **b** $\cos x = -\dfrac{1}{2}$ **c** $\tan x = 1$

  d $\sin x = 0$ **e** $\cos^2 x = \dfrac{1}{2}$ **f** $\tan^2 x = \dfrac{1}{3}$

2 Solve each equation for $-2\pi \leq \theta \leq 2\pi$.

  a $\sin\theta = \dfrac{1}{2}$ **b** $\tan\theta = 0$ **c** $\cos\theta = \dfrac{\sqrt{3}}{2}$

  d $\sin\theta = -1$ **e** $2\tan^2\theta = 6$ **f** $\sin\theta = \cos\theta$

3 Solve each equation for $-180° \leq \theta \leq 720°$.

  a $\cos\theta = 1$ **b** $\sin\theta = -\dfrac{\sqrt{2}}{2}$

  c $\sin\theta = -\cos\theta$ **d** $3\tan^2 x - 1 = 8$

4 Solve each equation for $-\pi \leq x \leq \pi$.

  a $\sin x = 1$ **b** $2\sin x + 3 = 2$

  c $10\sin^2 x = 5$ **d** $4\cos^2 x + 2 = 5$

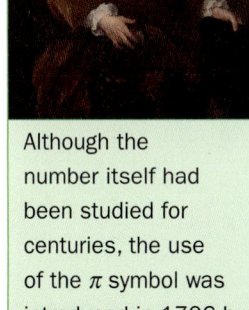

Although the number itself had been studied for centuries, the use of the π symbol was introduced in 1706 by William Jones (Welsh, 1675–1749).

Example 5

Solve the equation $\sin(2x) = \dfrac{\sqrt{2}}{2}$, $0° \leq x \leq 360°$.

Answer

If $0° \leq x \leq 360°$, then $0° \leq 2x \leq 720°$

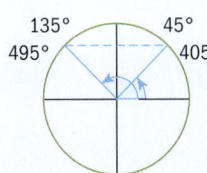

$2x = 45°, 135°, 405°, 495°$
$x = 22.5°, 67.5°, 202.5°, 247.5°$

We know that $\sin 45° = \sin 135° = \dfrac{\sqrt{2}}{2}$.
To find the other angles, make another rotation around the unit circle.

These angles represent the value of $2x$, not the value of x.

Example 6

> Solve the equation $2\sin^2 x + 5\sin x - 3 = 0$, $0 \leq x \leq 2\pi$.

Answers

$2\sin^2 x + 5\sin x - 3 = 0$	*This is a 'quadratic-type' equation.*
$(2\sin x - 1)(\sin x + 3) = 0$	*Solve by factorizing.*
$\sin x = \dfrac{1}{2}$ or $\sin x = -3$	*The value of sine cannot be less than -1, so we can disregard $\sin x = -3$.*
$x = \dfrac{\pi}{6}, \dfrac{5\pi}{6}$	

Exercise 13D

1 Solve each equation for $-180° \leq x \leq 180°$.

   a $\cos(2x) = \dfrac{\sqrt{3}}{2}$ **b** $6\sin(2x) - 2 = 1$

   c $\sin\left(\dfrac{x}{2}\right) - \cos\left(\dfrac{x}{2}\right) = 0$ **d** $\sin^2\left(\dfrac{x}{3}\right) = 3\cos^2\left(\dfrac{x}{3}\right)$

2 Solve each equation for $-\pi \leq \theta \leq \pi$.

   a $\sin(2\theta) = -\dfrac{1}{2}$ **b** $\tan(3\theta) = 1$

   c $\cos^2\left(\dfrac{\theta}{2}\right) = \dfrac{1}{2}$ **d** $\sin^2\left(\dfrac{2\theta}{3}\right) = 1$

> **EXAM-STYLE QUESTION**
> **3** Solve each equation for $0 \leq \theta \leq 2\pi$.
>   **a** $2\cos^2 x - 5\cos x - 3 = 0$ **b** $2\sin^2 x + 3\sin x + 1 = 0$
>   **c** $\tan^2 x + 2\tan x + 1 = 0$ **d** $\sin^2 x = 6\sin x - 5$

13.3 Trigonometric identities

In this section, we will look at special kinds of equations called **identities**. You are already familiar with one important trigonometric identity, $\sin^2 x + \cos^2 x = 1$.

This equation is an identity, because it is true for ALL values of x.

Another identity with which you are familiar is $\tan x = \dfrac{\sin x}{\cos x}$, the definition of tangent, which is also true for every value of x.

Double-angle identity for cosine

The diagram shows the angles θ and $-\theta$ drawn in standard position in the unit circle.

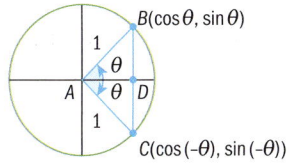

The length of segment CD is equal to the length of segment BD, and we have $BD = CD = \sin\theta$.

$$BC = BD + CD, \text{ so } BC = 2\sin\theta. \qquad [1]$$

We can see that $\angle BAC = 2\theta$. The length of segment BC can be found using the cosine rule in $\triangle ABC$:

$$BC^2 = AB^2 + AC^2 - 2(AB)(AC)\cos(2\theta)$$
$$BC^2 = 1^2 + 1^2 - 2(1)(1)\cos(2\theta) = 2 - 2\cos(2\theta)$$
$$BC = \sqrt{2 - 2\cos(2\theta)} \qquad [2]$$

Now we have two expressions for BC.

If we put [1] and [2] equal, we find

$$2\sin\theta = \sqrt{2 - 2\cos(2\theta)}.$$

Squaring both sides gives us $4\sin^2\theta = 2 - 2\cos(2\theta)$.
Rearranging this equation gives us $2\cos(2\theta) = 2 - 4\sin^2\theta$.
Finally, we divide by 2 to get $\cos(2\theta) = 1 - 2\sin^2\theta$.

> → The equation $\cos(2\theta) = 1 - 2\sin 2\theta$ is an **identity**, as it is true for all values of θ.

We will use this identity to help us find other identities.

We know $\sin^2\theta + \cos^2\theta = 1$, so $\sin^2\theta = 1 - \cos^2\theta$.
Using substitution, we get $\cos(2\theta) = 1 - 2(1 - \cos^2\theta)$.
Rearranging this equation gives us

$$\cos(2\theta) = 2\cos^2\theta - 1.$$

We can substitute $\sin^2\theta + \cos^2\theta = 1$ into this equation to get $2\cos(2\theta) = 2\cos^2\theta - (\sin^2\theta + \cos^2\theta)$, which gives us

$$\cos(2\theta) = \cos^2\theta - \sin^2\theta.$$

The three equations we have just found are:

> → The **double-angle identities** for cosine:
> $$\cos(2\theta) = 1 - 2\sin^2\theta$$
> $$\qquad\quad = 2\cos^2\theta - 1$$
> $$\qquad\quad = \cos^2\theta - \sin^2\theta$$

Double-angle identity for sine

Now we will find a double-angle identity for sine.

We know that $\sin^2(2\theta) + \cos^2(2\theta) = 1$, so

$$\cos^2(2\theta) = 1 - \sin^2(2\theta). \qquad [1]$$

From the double-angle identity for cosine,

$$\cos(2\theta) = 1 - 2\sin^2\theta$$
$$\cos^2(2\theta) = (1 - 2\sin^2\theta)^2 \qquad [2]$$
$$1 - \sin^2(2\theta) = (1 - 2\sin^2\theta)^2 \quad \text{Equate [1] and [2]}$$
$$1 - \sin^2(2\theta) = 1 - 4\sin^2\theta + 4\sin^4\theta$$
$$4\sin^2\theta - 4\sin^4\theta = \sin^2(2\theta)$$
$$4\sin^2\theta(1 - \sin^2\theta) = \sin^2(2\theta) \quad 1 - \sin^2\theta = \cos^2\theta$$
$$4\sin^2\theta \cos^2\theta = \sin^2(2\theta)$$
$$2\sin\theta \cos\theta = \sin(2\theta) \quad \text{Take square roots of both sides}$$

→ The double-angle identity for sine is $\sin(2\theta) = 2\sin\theta \cos\theta$

Example 7

Given that $\sin x = \dfrac{3}{4}$, and $0° < x < 90°$, find the exact values of

a $\cos x$
b $\sin(2x)$
c $\cos(2x)$
d $\tan(2x)$.

Answers

a $\sin^2 x + \cos^2 x = 1$ *Pythagorean identity.*

$\left(\dfrac{3}{4}\right)^2 + \cos^2 x = 1$ *Substitute the value of $\sin x$.*

$\cos^2 x = 1 - \dfrac{9}{16} = \dfrac{7}{16}$

$\cos x = \dfrac{\sqrt{7}}{4}$ *Take the square root of $\dfrac{7}{16}$.*

Remember, if x is an acute angle, the cosine must be positive.

b $\sin(2x) = 2\sin x \cos x$ *Double-angle identity.*

$\sin(2x) = 2\left(\dfrac{3}{4}\right)\left(\dfrac{\sqrt{7}}{4}\right)$ *Substitute the value of $\sin x$ and $\cos x$.*

$\sin(2x) = \dfrac{3\sqrt{7}}{8}$

▶ Continued on next page

c $\cos(2x) = 1 - 2\sin^2 x$	Use a double-angle identity.
$\cos(2x) = 1 - 2\left(\dfrac{3}{4}\right)^2 = 1 - \dfrac{9}{8}$	Substitute the value of $\sin x$.
$\cos(2x) = -\dfrac{1}{8}$	
d $\tan(2x) = \dfrac{\sin(2x)}{\cos(2x)}$	Definition of tangent.
$\tan(2x) = \dfrac{\left(\dfrac{3\sqrt{7}}{8}\right)}{\left(-\dfrac{1}{8}\right)}$	Substitute the values of $\sin(2x)$ and $\cos(2x)$.
$= \left(\dfrac{3\sqrt{7}}{8}\right)\left(-\dfrac{8}{1}\right)$	
$\tan(2x) = -3\sqrt{7}$	

You could use any of the three identities for $\cos(2x)$.

Example 8

Given that $\cos\theta = \dfrac{4}{5}$, and $\dfrac{3\pi}{2} < \theta < 2\pi$, find the exact values of

a $\sin\theta$ **b** $\cos(2\theta)$.

Answers

a $\sin^2\theta + \cos^2\theta = 1$	Pythagorean identity.
$\sin^2\theta + \left(\dfrac{4}{5}\right)^2 = 1$	Substitute the value of $\cos\theta$.
$\sin^2\theta = 1 - \dfrac{16}{25} = \dfrac{9}{25}$	
$\sin\theta = -\dfrac{3}{5}$	Take the square root of $\dfrac{9}{25}$.
b $\cos(2\theta) = 2\cos^2\theta - 1$	Use a double-angle identity.
$\cos(2\theta) = 2\left(\dfrac{4}{5}\right)^2 - 1 = \dfrac{32}{25} - 1$	Substitute the value of $\cos\theta$.
$\cos(2\theta) = \dfrac{7}{25}$	

Remember, if $\dfrac{3\pi}{2} < \theta < 2\pi$, the angle will be in the fourth quadrant. The cosine is positive but the sine is negative.

Notice that, in Example 8, we could find the values of $\sin\theta$ and $\cos(2\theta)$ without ever finding the measure of the angle θ.

Exercise 13E

EXAM-STYLE QUESTIONS

1. Given that $\sin\theta = \dfrac{5}{6}$, and $0° < \theta < 90°$, find the exact value of each.

   a $\sin(2\theta)$ b $\cos(2\theta)$ c $\tan(2\theta)$

2. Given that $\cos x = -\dfrac{2}{3}$, and $90° < x < 180°$, find each value.

   a $\sin(2x)$ b $\cos(2x)$ c $\tan(2x)$

3. Given that $\cos\theta = \dfrac{5}{6}$, and $0 < \theta < \pi$, find each value.

   a $\tan\theta$ b $\sin(2\theta)$ c $\cos(2\theta)$ d $\tan(2\theta)$

4. Given that $\sin x = -\dfrac{1}{8}$, and $180° < x < 270°$, find each value.

   a $\sin(2x)$ b $\cos(2x)$ c $\tan(2x)$ d $\sin(4x)$

EXAM-STYLE QUESTION

5. Given that $\tan\theta = \dfrac{3}{4}$, and $0 < \theta < \pi$, find each value.

   a $\sin\theta$ b $\cos\theta$ c $\sin(2\theta)$ d $\cos(2\theta)$

6. Given that $\sin(2x) = \dfrac{24}{25}$, and $\dfrac{\pi}{4} < x < \dfrac{\pi}{2}$, find each value.

   a $\cos(2x)$ b $\tan(2x)$ c $\sin(4x)$ d $\cos(4x)$

7. Given that $\tan x = \dfrac{a}{b}$, and $0° < x < 90°$, find each value in terms of a and b.

   a $\sin x$ b $\cos x$ c $\sin(2x)$ d $\cos(2x)$

> You should be able to answer all of these questions without finding the size of the angle.

You can also use identities when working with equations.

Example 9

Solve the equation $\sin(2x) = \sin x$ for $0° \leq x \leq 360°$.
Do not use your GDC.

Answer

$\sin(2x) = \sin x$
$2(\sin x)(\cos x) = \sin x$ *Use double-angle identity.*
$2(\sin x)(\cos x) - \sin x = 0$ *Rearrange.*
$(\sin x)(2\cos x - 1) = 0$ *Factorize.*
$\sin x = 0$ or $2\cos x - 1 = 0$
If $\sin x = 0$, then $x = 0°, 180°, 360°$.
If $2\cos x - 1 = 0$, then $\cos x = \dfrac{1}{2}$,
so $x = 60°, 300°$.
$x = 0°, 60°, 180°, 300°, 360°$

> There are more trigonometric identities. What are they? What identities are used in other areas of mathematics?

Example 10

Prove that $(1 + \tan^2 x) \cos(2x) = 1 - \tan^2 x$.

Answer

$(1 + \tan^2 x) \times \cos(2x) = 1 - \tan^2 x$	
$\left(1 + \dfrac{\sin^2 x}{\cos^2 x}\right)(2\cos^2 x - 1) = 1 - \dfrac{\sin^2 x}{\cos^2 x}$	Rewrite using $\sin x$ and $\cos x$.
$2\cos^2 x - 1 + 2\sin^2 x - \dfrac{\sin^2 x}{\cos^2 x} = 1 - \dfrac{\sin^2 x}{\cos^2 x}$	Multiply through on the left side of the equation.
$2\cos^2 x + 2\sin^2 x = 2$	Simplify.
$\sin^2 x + \cos^2 x = 1$	Divide by 2.

In Example 10, we ended up with a known identity, which is true for all values of x. Therefore, the original equation is true for all values of x and is also an identity, though it is not one you must learn.

When you show equations to be true in this way, it is called 'proving identities'.

Exercise 13F

1 Solve each equation for $0° \leq x \leq 180°$.
   a $\sin(2x) = \cos x$
   b $\sin(2x) = \cos(2x)$
   c $(\sin x + \cos x)^2 = 0$
   d $\cos^2 x = \dfrac{1}{2}$

2 Solve each equation for $-180° \leq \theta \leq 180°$.
   a $2\sin x \cos x = \dfrac{\sqrt{3}}{2}$
   b $\sin x(1 - \sin x) = \cos^2 x$
   c $\cos^2 x = \dfrac{1}{2} + \sin^2 x$
   d $\cos(2x) = \sin x$

3 Solve each equation for $0 \leq x \leq \pi$.
   a $\tan x = \sin x$
   b $2\cos^2 x - 1 = \dfrac{1}{\sqrt{2}}$
   c $\cos(2x) = \cos x$
   d $\sin(4x) = \sin(2x)$

4 Solve each equation for $0 \leq \theta \leq \pi$.
   a $(\sin(2x) + \cos(2x))^2 = 2$
   b $\sin x - 1 = \cos^2 x$
   c $\cos^2 x = \cos(2x)$
   d $2\sin^2 x = 1$

5 Prove each identity.
   a $(\sin x + \cos x)^2 = 1 + \sin(2x)$
   b $\dfrac{1}{\cos \theta} = \sin \theta \tan \theta + \cos \theta$
   c $\dfrac{1 - \cos^2(2x)}{2 \sin x \cos x} = 2 \sin x \cos x$
   d $\cos \theta + \sin \theta = \dfrac{1 - 2\sin^2 \theta}{\cos \theta - \sin \theta}$
   e $\cos^4 x - \sin^4 x = \cos(2x)$

: **EXAM-STYLE QUESTIONS**

6 The expression $2\sin 3x \cos 3x$ can be written in the form $\sin kx$. Find the value of k.

7 The expression $\cos 4x$ can be written in the form $1 - b\sin^2 x \cos^2 x$. Find the value of b.

13.4 Graphing circular functions

In previous sections, we have used the unit circle to find relationships between different angles and their sine, cosine and tangent values. In this section, you will see how these values can be used to help you understand the trigonometric functions $y = \sin x$, $y = \cos x$, and $y = \tan x$. You will also practice graphing these functions with the GDC to help you solve equations.

Sine and cosine functions

You already know the exact sine values for many angles, as seen in the table.

Angle measure (x) degrees, radians	Sine value ($\sin x$)	Angle measure (x) degrees, radians	Sine value ($\sin x$)
0°, 0 radians	0	210°, $\frac{7\pi}{6}$	$-\frac{1}{2}$
30°, $\frac{\pi}{6}$	$\frac{1}{2}$	225°, $\frac{5\pi}{4}$	$-\frac{1}{\sqrt{2}} = -\frac{\sqrt{2}}{2}$
45°, $\frac{\pi}{4}$	$\frac{1}{\sqrt{2}} = \frac{\sqrt{2}}{2}$	240°, $\frac{4\pi}{3}$	$-\frac{\sqrt{3}}{2}$
60°, $\frac{\pi}{3}$	$\frac{\sqrt{3}}{2}$	270°, $\frac{3\pi}{2}$	-1
90°, $\frac{\pi}{2}$	1	300°, $\frac{5\pi}{3}$	$-\frac{\sqrt{3}}{2}$
120°, $\frac{2\pi}{3}$	$\frac{\sqrt{3}}{2}$	315°, $\frac{7\pi}{4}$	$-\frac{1}{\sqrt{2}} = -\frac{\sqrt{2}}{2}$
135°, $\frac{3\pi}{4}$	$\frac{1}{\sqrt{2}} = \frac{\sqrt{2}}{2}$	330°, $\frac{11\pi}{6}$	$-\frac{1}{2}$
150°, $\frac{5\pi}{6}$	$\frac{1}{2}$	360°, 2π	0
180°, π	0		

If we let $y = \sin x$, we can plot these values as coordinates on a graph.

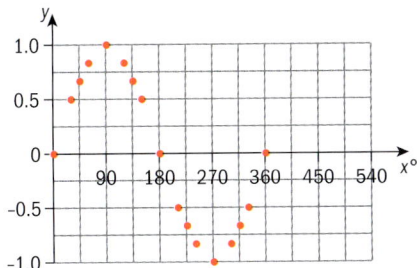

▼ Graphing the equation $y = \sin x$ on this same set of axes, we see this:

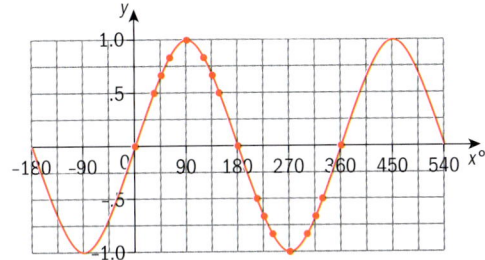

▼ If we measure x in radians, the graph has the same shape.

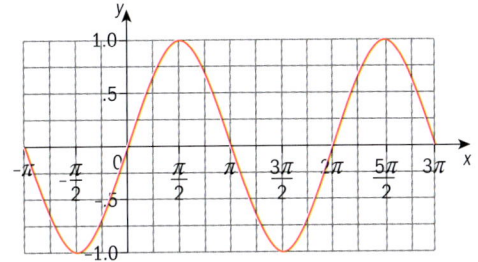

We can see that the graph of the function $y = \sin x$ is generating the same sine values that we found using the unit circle.

Similarly, if we let $y = \cos x$, we can plot the cosine values we know, along with the graph of the function $y = \cos x$.

▼ $y = \cos x$, with x measured in degrees:

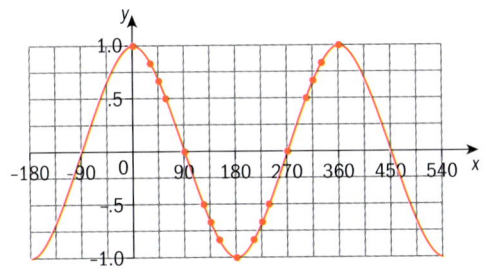

▼ $y = \cos x$, with x measured in radians:

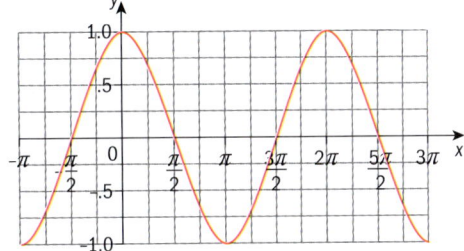

→ If we compare the sine and cosine functions, we see many similarities.

- The curves are the same size and shape, only their horizontal positions on the axes differ. The sine curve passes through the origin (0, 0), and the cosine curve passes through the point (0, 1).
- The functions are **periodic**, which means that they repeat the same cycle of values over and over. The **period**, or length of one cycle, is 360° or 2π. This means that if you look at two points whose x-coordinates are 360° (or 2π) apart, the y-coordinates of those two points would be the same.
- Both functions have a maximum value of 1 and a minimum value of -1. Each of these functions has an **amplitude** of 1. The amplitude is the difference between the horizontal axis of the wave ($y = 0$, in this case) to a maximum or minimum value ($y = 1$ or $y = -1$, in this case). We can also say that the amplitude is one-half the vertical distance from a maximum to a minimum.

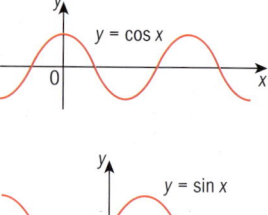
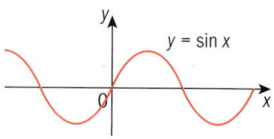

We can use the graphs of $y = \sin x$ and $y = \cos x$ to help us to solve equations, much as we used the unit circle to help us solve equations earlier in this chapter.

Consider the equation $\sin x = \dfrac{1}{2}$, $-360° \leq x \leq 360°$.

By graphing a horizontal line $y = \dfrac{1}{2}$ on the same set of axes as $y = \sin x$, we can see that there are four points where $\sin x = \dfrac{1}{2}$.

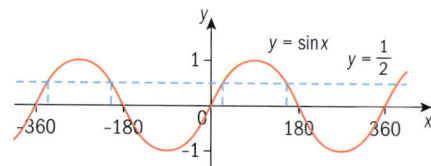

These points correspond to the values $x = -330°, -210°, 30°, 150°$.

Example 11

Solve the equation $\cos\theta = 0.4$, $-360° \leq \theta \leq 360°$.
Give your answers to the nearest tenth.

Answer

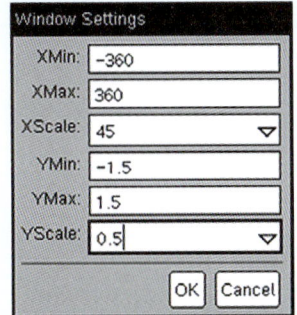

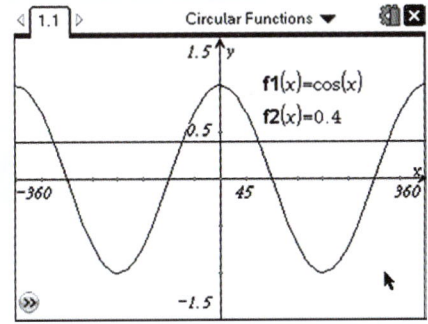

Enter $y = \cos x$ and $y = 0.4$ into the GDC, and set an appropriate window to view the graph. Be sure your GDC is in DEGREE mode!

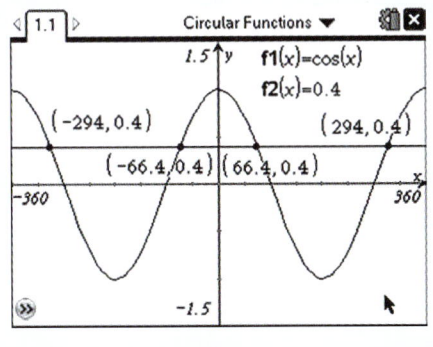

There are four intersection points within this domain, so the equation will have four solutions. Use 6:Analyze Graph | 4:Intersection to find these intersection points.

$\theta = -293.6°, -66.4°, 66.4°, 293.6°$

The GDC can be very helpful in solving equations with sine and cosine functions.

To change to **degree mode** press ⌂On and choose 5:Settings & Status | 2:Settings | 2:Graphs & Geometry Use the Tab key to move to Graphing Angle and select Degree. Press Enter and then select 4:Current to return to the document.

GDC help on CD: *Alternative demonstrations for the TI-84 Plus and Casio FX-9860GII GDCs are on the CD.*

Example 12

Solve the equation $\sin x = 0.25x - 0.3$, $-2\pi \leq x \leq 2\pi$.
Give your answers to three significant figures.

Angle measures are in radians

Answer

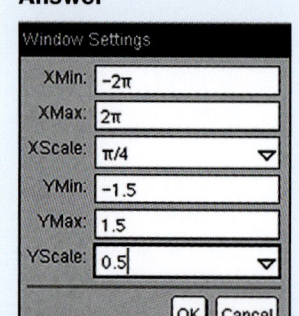

GDC help on CD: *Alternative demonstrations for the TI-84 Plus and Casio FX-9860GII GDCs are on the CD.*

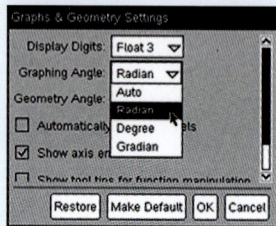

To change to **radian mode** press ⌂ On and choose 5:Settings & Status | 2:Settings | 2:Graphs & Geometry Use the Tab key to move to Graphing Angle and select Radian. Press Enter and then select 4:Current to return to the document.

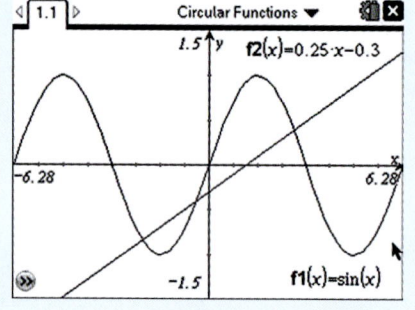

Enter $y = \sin x$ and $y = 0.25x - 0.3$ into the GDC, and set an appropriate window to view the graph. Be sure your GDC is in RADIAN mode!

There are three intersection points within this domain, so the equation will have three solutions. Use 6:Analyze Graph | 4: Intersection to find these intersection points.

$x = -2.15, -0.416, 2.75$

Exercise 13G

In questions 1 to 4, use your GDC to solve each equation.
Give your answers to the nearest degree.

1. $\sin x = \dfrac{1}{4}$, $-360° \le x \le 360°$

2. $\cos\theta = \sqrt{0.8}$, $-180° \le \theta \le 360°$

3. $\sin\theta = -0.9$, $0° \le \theta \le 360°$

4. $\sin x = \cos(x - 20)$, $0° \le x \le 540°$

In questions 5 to 8, use your GDC to solve each equation.
Give your answers to three significant figures.

5. $\sin\theta = \sqrt{\dfrac{2}{3}}$, $-2\pi \le x \le 2\pi$

6. $\cos\theta = -\dfrac{1}{e^2}$, $-\pi \le x \le 2\pi$

7. $\cos x = -x$, $-\pi \le x \le 2\pi$

8. $\sin x = x^2 - 1$, $-2\pi \le x \le 2\pi$

Tangent function

Investigation – graphing tan x

For the sine and cosine functions, we began with values for sinx and cosx that we already knew.

Now try a similar approach to the function $y = \tan x$.

1. List the tangent values for the angles:
   0°, ±30°, ±45°, ±60°, 120°, 135°, 150°, 180°, 210°, 225°, 240°, 300°, 315°, 330°, 360°.
2. On a piece of grid paper, plot these values as points. Let the x-axis represent the angle (measured in degrees), and let the y-axis represent the value of tanx.
3. Why are there no tangent values for the angles ±90° or 270°? What feature do you sometimes have on the graph of a function for values that do not exist?
4. Connect the points on your grid paper to sketch the graph of $y = \tan x$.
5. Graph the function $y = \tan x$ on your GDC, and compare it to your sketch.
   Are your graphs similar?

If you had been using radians, rather than degrees, the graph of the tangent function would look like this:

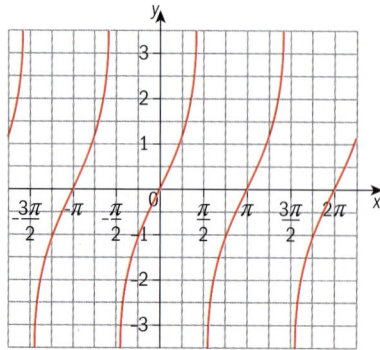

→ Like the sine and cosine functions, the tangent function is **periodic**. There are vertical asymptotes at values of x where the function does not exist. The same cycle of values repeats between each pair of vertical asymptotes.

The period of the tangent function is 180° (or π radians). Unlike the sine and cosine functions, the tangent function does not have an amplitude. It has no maximum or minimum values.

Example 13

Solve the equation $\tan\theta = 1 - x$, $-2\pi \leq \theta \leq 2\pi$.
Give your answers to three significant figures.

Answer

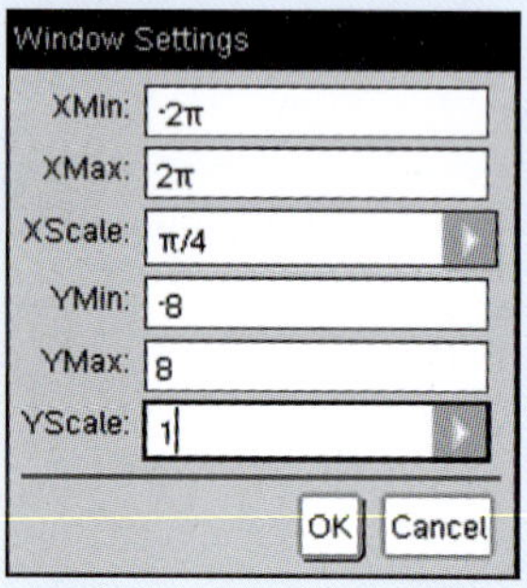

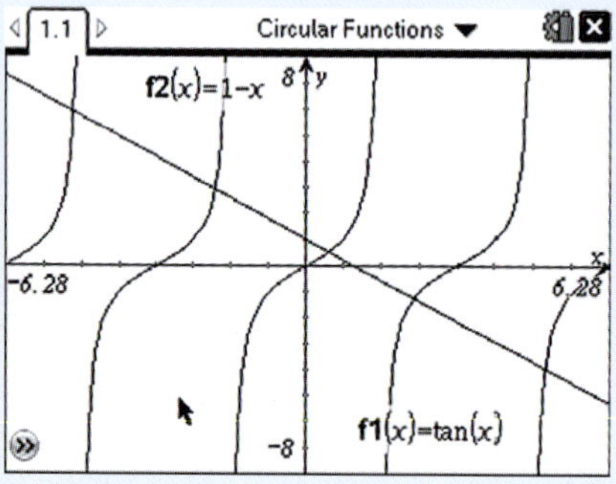

Be sure your GDC is in RADIAN mode!

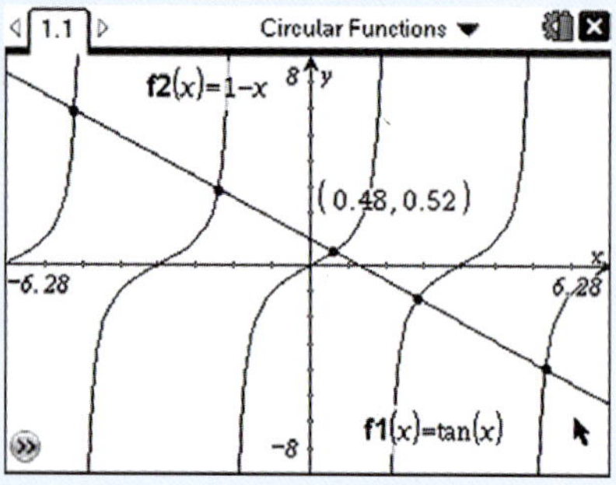

There are five intersection points within this domain, so the equation will have five solutions.

$\theta = -4.88, -1.90, 0.480, 2.25, 4.96$

GDC help on CD: *Alternative demonstrations for the TI-84 Plus and Casio FX-9860GII GDCs are on the CD.*

Exercise 13H

In questions 1 to 4, use your GDC to solve each equation.
Give your answers to the nearest degree.

1 $\tan x = 2, -360° \leq x \leq 360°$

2 $\tan \theta = \sqrt{11}, -180° \leq \theta \leq 360°$

3 $\tan \theta = -1.5, 0° \leq \theta \leq 360°$

4 $\tan x = \cos x, 0° \leq x \leq 720°$

In questions 5 to 8, use your GDC to solve each equation.
Give your answers to three significant figures.

5 $\tan \theta = \dfrac{3}{7}, -2\pi \leq x \leq 2\pi$

6 $\tan \theta = \pi, -\pi \leq \theta \leq \pi$

7 $\tan x = 2x - 3, 0 \leq x \leq 2\pi$

8 $\tan x = 4 - x^2, -2\pi \leq x \leq 2\pi$

13.5 Translations and stretches of trigonometric functions

Investigation – transformations of sinx and cosx

Using your GDC in radian mode, graph the functions $y = \cos x$ and $y = \cos\left(x - \dfrac{\pi}{2}\right)$ on the same axes.

What do you notice about the graphs of these two functions?

What do they have in common?

Describe how the graphs are different, and try to explain why this is happening.

Now, repeat this process for each of these pairs of functions.

1 $y = \sin x$ and $y = \sin x + 3$
2 $y = \cos x$ and $y = 2\cos x$
3 $y = \cos x$ and $y = \cos(2x)$
4 $y = \sin x$ and $y = \sin\left(x + \dfrac{\pi}{3}\right)$
5 $y = \sin x$ and $y = \cos\left(x - \dfrac{\pi}{2}\right)$

In the last section, we looked at the basic trigonometric functions $y = \sin x$, $y = \cos x$ and $y = \tan x$. Now we will study transformations of these functions.

Let's begin by looking at the graphs of the sine and cosine functions, and reviewing some vocabulary relating to these functions.

> You need to be very familiar with the features of the basic sine and cosine curves.

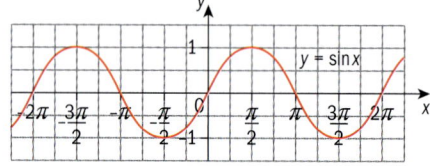
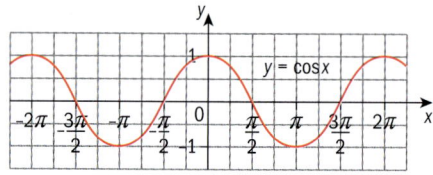

These functions have a **period** of 2π (or 360°, if we were graphing the functions in degrees rather than radians).

These functions have an **amplitude** of 1.

The graphs of these functions can be transformed in the same way that you transformed the graphs of other functions earlier in this book; see Chapter 1.

Translations

→ The function $y = \sin(x) + d$ is a **vertical translation** of the standard sine curve.
The curve shifts up if d is positive, down if d is negative.

The function $y = \sin(x - c)$ is a **horizontal translation** of the standard cosine curve. The curve shifts to the right if c is positive, left if c is negative.

> A horizontal translation is also known as a 'phase shift'.

It is important to note that a translation does not change the period or the amplitude of a trigonometric function.

▼ This graph shows a vertical translation. The sine curve has been shifted up 2 units. The green arrow shows the direction of the translation.

▼ This graph shows a horizontal translation. The sine curve has been shifted $\dfrac{\pi}{2}$ units to the right. The green arrow shows the direction of the translation.

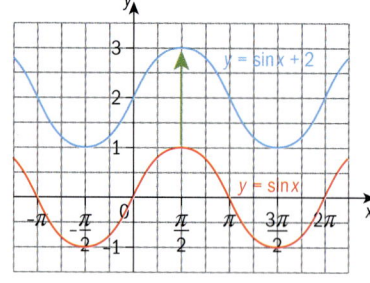
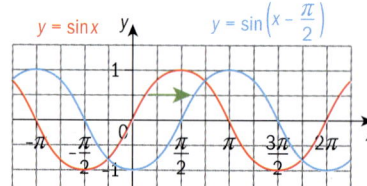

Circular functions

> The function $y = \cos(x) + d$ is a **vertical translation** of the standard cosine curve.
> The curve shifts up if d is positive, down if d is negative.
>
> The function $y = \cos(x - c)$ is a **horizontal translation** of the standard cosine curve. The curve shifts to the right if c is positive, left if c is negative.

As with the sine curve, a translation does not change the period or the amplitude of the cosine function.

▼ This graph shows a vertical translation. The cosine curve has been shifted down 3 units. The green arrow shows the direction of the translation.

▼ The graph below shows a horizontal translation. The cosine curve has been shifted $\frac{3\pi}{4}$ units to the left.

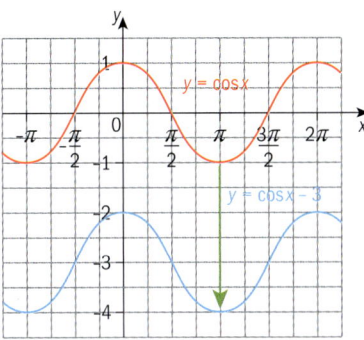

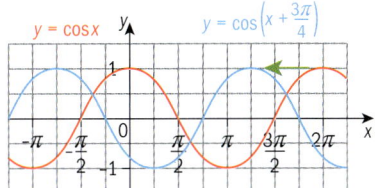

Now consider the graph of the tangent function.

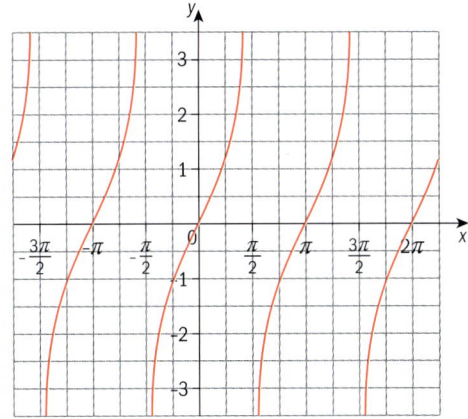

Remember that this function has a period of π (or 180°). It has no amplitude, because there are no maximum or minimum points.

There are vertical asymptotes at $x = \pm\frac{\pi}{2}, \pm\frac{3\pi}{2}$, etc. (or at $x = \pm 90°$, $x = \pm 270°$, etc).

As with the sine and cosine functions, vertical and horizontal translations do not change the period of the tangent function.

We can combine horizontal and vertical translations by looking at equations in the form $y = \sin(x - c) + d$, $y = \cos(x - c) + d$, and $y = \tan(x - c) + d$.

Example 14

Sketch the graph of $y = \sin x$.
On the same set of axes, sketch the graph of:

a $y = \sin x + 1$ b $y = \sin\left(x - \dfrac{2\pi}{3}\right)$ c $y = \sin\left(x - \dfrac{2\pi}{3}\right) + 1$

Answers

a $y = \sin x + 1$

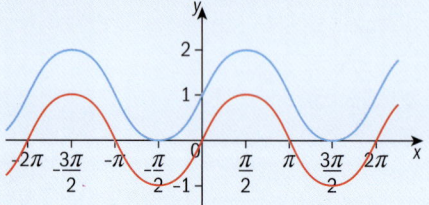

The basic sine curve is shown in red, the translated function is shown in blue. This is a vertical shift of 1 unit upward.

b $y = \sin\left(x - \dfrac{2\pi}{3}\right)$

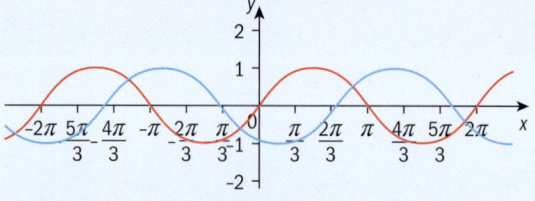

Again, the basic sine curve is shown in red, the new function is shown in blue. This is a horizontal shift of $\dfrac{2\pi}{3}$ units to the right.

c $y = \sin\left(x - \dfrac{2\pi}{3}\right) + 1$

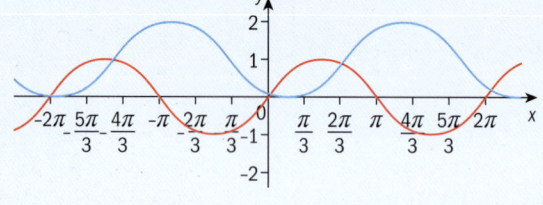

This is a combination of the translations from parts a and b, with the new function shown in blue. The basic sine curve (shown in red) has been shifted $\dfrac{2\pi}{3}$ units to the right, and 1 unit up.

Example 15

Write an equation for each function, as directed.
a Write a sine equation.

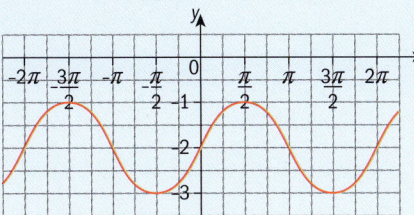

b Write a cosine equation.

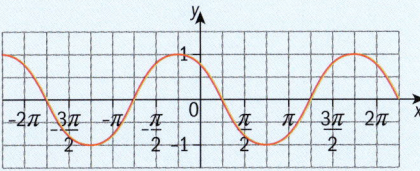

c Write one sine and one cosine equation.

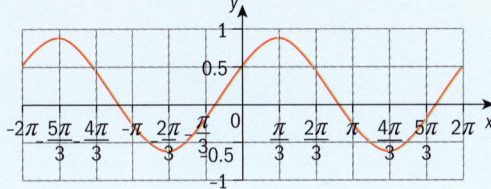

Answers

a $y = \sin x - 2$

You can see this is a sine curve with a maximum value of −1 and a minimum value of −3. It has been shifted down 2 units.

b $y = \cos\left(x + \dfrac{\pi}{4}\right)$

You can see this is a cosine curve which has been shifted to the left by $\dfrac{\pi}{4}$.

c $y = \cos\left(x - \dfrac{\pi}{3}\right) + 0.5$

or

$y = \sin\left(x - \dfrac{\pi}{6}\right) + 0.5$

You can see this as a cosine curve which has been shifted to the right by $\dfrac{\pi}{3}$, and up 0.5.

You might also view this as a sine curve which has been shifted to the left by $\dfrac{\pi}{6}$, and up 0.5.

> Because the shapes of the sine and cosine curves are so similar, there may be many correct equations for the graph of a sine or cosine function.

Exercise 13I

For questions 1 to 8, sketch the graph of the function for $-2\pi \leq x \leq 2\pi$.

1. $y = \sin x - 5$
2. $y = \cos x + 2$
3. $y = \tan\left(x - \dfrac{\pi}{4}\right)$
4. $y = \sin\left(x + \dfrac{\pi}{3}\right)$
5. $y = \cos(x - \pi)$
6. $y = \sin\left(x + \dfrac{\pi}{4}\right) - 2$
7. $y = \cos\left(x + \dfrac{2\pi}{3}\right) - 1.5$
8. $y = \tan\left(x + \dfrac{\pi}{2}\right) + 4$

For questions 9 to 12, write an equation for the function shown.

9.
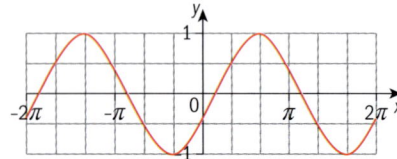

10.
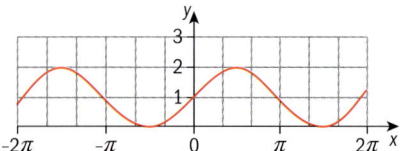

11.
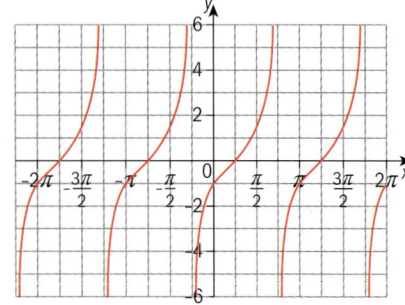

12.
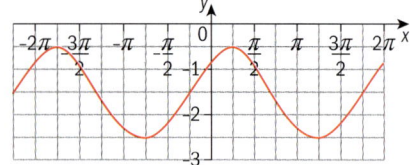

Vertical stretches

> → The functions $y = a\sin x$ and $y = a\cos x$ are **vertical stretches** of the sine and cosine functions. When the graph of a function undergoes a vertical stretch, every y-value in the original function is multiplied by the value of a.
> If $|a| > 1$, the function will appear to stretch away from the x-axis.
>
> If $0 < |a| < 1$, the function will appear to compress closer to the x-axis.
> If a is negative, the function will also be reflected over the x-axis.
> With a vertical stretch, the **amplitude** of the sine or cosine function will change from 1 to $|a|$. The period of the function will not change.

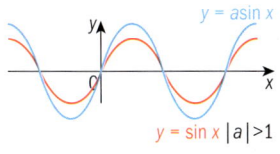

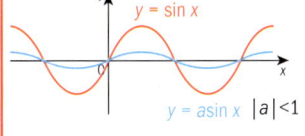

In the graph below, the sine curve has been stretched vertically by a factor of 3. The maximum values are at $y = 3$, and the minimum values are at $y = -3$. The amplitude of the new function is 3.

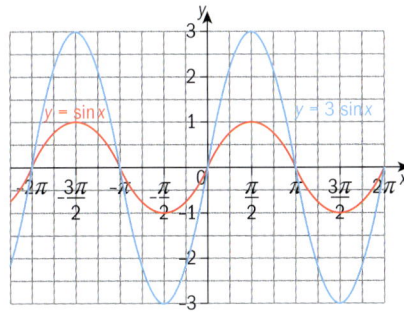

The next graph shows a vertical stretch that is also a reflection about the x-axis. All of the y-values in the standard cosine curve have been multiplied by -0.5.
The maximum values are at $y = 0.5$, and the minimum values are at $y = -0.5$

The amplitude of the new function is 0.5.

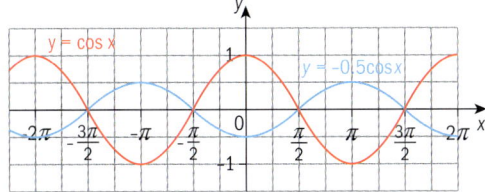

Example 16

Sketch the graph of $y = \cos x$.
On the same set of axes, sketch the graph of:
a $y = 0.25\cos x$ **b** $y = -2\cos x$

Answers

a $y = 0.25\cos x$

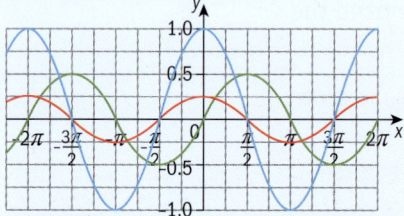

The basic cosine curve is shown in blue, the new function is shown in red. This is a vertical stretch of stretch factor 0.25.

b $y = -2\cos x$

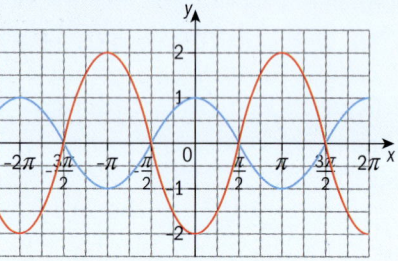

Again, the basic cosine curve is shown in blue, the new function is shown in red.
Every y-value in the blue function has been multiplied by −2 to give the red function.

Horizontal stretches

→ The functions $y = \sin(bx)$, $y = \cos(bx)$ and $y = \tan(bx)$ represent horizontal stretches of the sine, cosine and tangent functions. When the graph of a function undergoes a horizontal stretch, every x-value in the original function is multiplied by $\frac{1}{b}$.

We could also say that every x-value in the original function is *divided* by b.

Multiplying (or dividing) the x-values by a number in this way changes the **period** of a trigonometric function.

- If $|b| > 1$, the period will be shorter, and the function will appear to compress toward the y-axis.
- If $0 < |b| < 1$, the period will be longer, and the function will appear to stretch away from the y-axis.
- If b is negative, the function will also be reflected over the y-axis.

When a sine or cosine function undergoes a horizontal stretch, the period of the function will change from 2π to $\frac{2\pi}{|b|}$, or from 360° to $\frac{360°}{|b|}$.

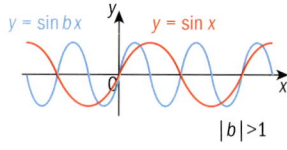

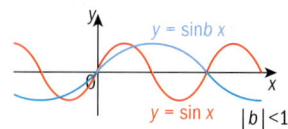

▼ In the graph below, the sine curve in blue has a period of π.

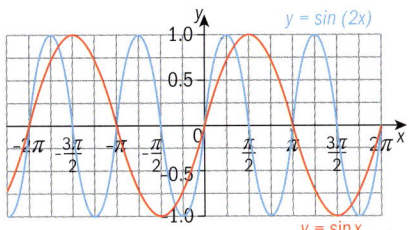

▼ In the graph below, the sine curve in blue has a period of 4π. The function has also been reflected about the y-axis.

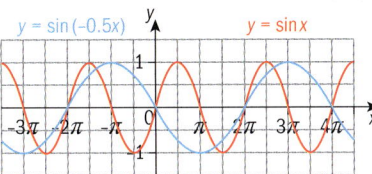

→ For a function in the form $y = \tan(bx)$, the period will change from π to $\dfrac{\pi}{|b|}$, or from $180°$ to $\dfrac{180°}{|b|}$.

The graph to the right shows the function $y = \tan(0.5x)$. The period of this function is 2π.

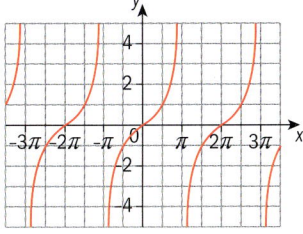

Example 17

Sketch the graph of: **a** $y = \sin(0.5x)$ **b** $y = \tan(2x)$ **c** $y = 2\cos(3x)$

Answers

a $y = \sin(0.5x)$

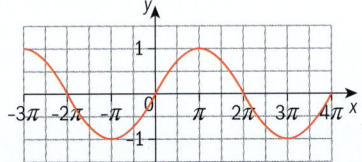

The period of this function is $\dfrac{2\pi}{0.5}$, or 4π.

b $y = \tan(2x)$

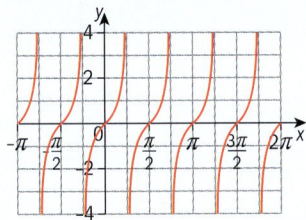

The period of this function is $\dfrac{\pi}{2}$.

c $y = 2\cos(3x)$

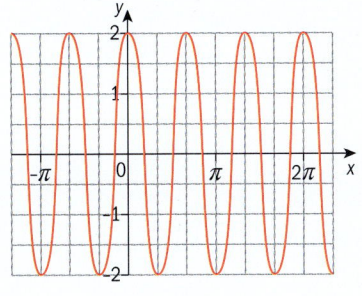

The period of this function is $\dfrac{2\pi}{3}$. The amplitude is 2.

Chapter 13

Exercise 13J

For questions 1 to 8, sketch the graph of the function from $-2\pi \le x \le 2\pi$.

1. $y = 0.5 \sin x$
2. $y = -4 \cos x$
3. $y = \tan\left(\dfrac{2}{3}x\right)$
4. $y = \sin(-2x)$
5. $y = 2\cos\left(\dfrac{3}{2}x\right)$
6. $y = 3\sin(3x)$
7. $y = -2.5\sin(0.5x)$
8. $y = -\cos\left(-\dfrac{1}{3}x\right)$

For questions 9 to 12, write an equation for the function shown.

9
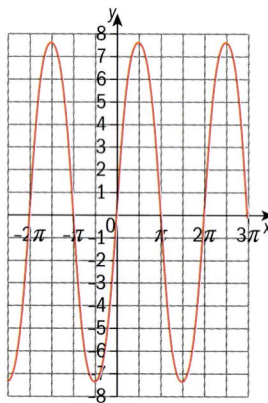

10
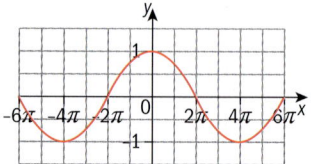

11
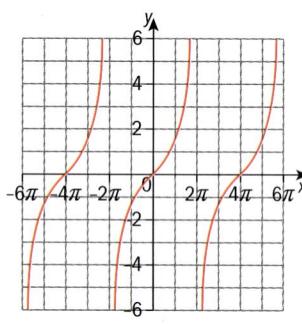

12
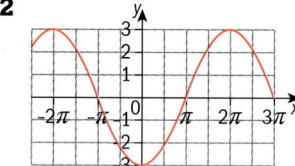

13.6 Combined transformations with sine and cosine functions

In this section, we will be looking at functions of the form $y = a\sin(b(x - c)) + d$ and $y = a\cos(b(x - c)) + d$.

For functions of this type, there are four possible transformations happening:

- a represents a vertical stretch. The amplitude of the sine or cosine function will be equal to $|a|$.
- b represents a horizontal stretch, which affects the period of the function. The period of the sine or cosine function will be equal to $\dfrac{2\pi}{|b|}$.

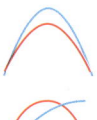

478 Circular functions

- c represents a horizontal translation, or shift. The function will shift to the right if c is positive or to the left if c is negative.
- d represents a vertical translation, or shift. The functions shifts up if d is positive or down if d is negative.

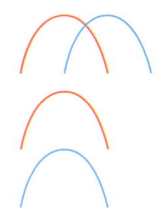

The function $y = 2\sin\left(\frac{1}{2}\left(x + \frac{\pi}{3}\right)\right) + 1$ is shown in blue on the same axes as the basic sine curve (shown in red).

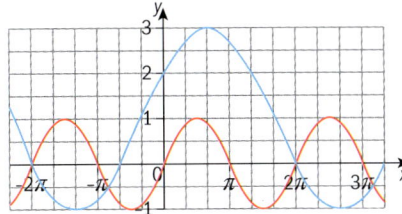

This function has an amplitude of 2 and a period of 4π. The function $y = \sin x$ has undergone four transformations to become the function in blue. There have been two changes to the y-values and two changes to the x-values.

- There has been a vertical stretch of scale factor 2 and a vertical translation of 1. All the y-values in the standard sine function have been multiplied by 2, then increased by 1.
- There has been a horizontal stretch of scale factor 2 and a horizontal translation of $-\frac{\pi}{3}$. All the x-values in the original sine function have been multiplied by 2 (divided by $\frac{1}{2}$), then decreased by $\frac{\pi}{3}$.

The function $y = 3\cos\left(-2\left(x - \frac{\pi}{4}\right)\right)$ is shown in blue on the same axes as the basic cosine curve (shown in red).

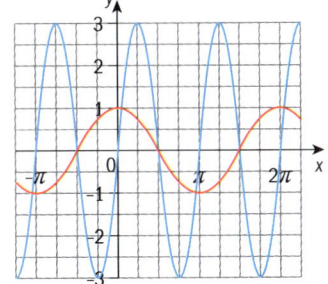

This function has an amplitude of 3 and a period of π. The function $y = \cos x$ has undergone four transformations to become the function in blue.

- There has been a vertical stretch of scale factor 3. All the y-values in the standard cosine function have been multiplied by 3.
- There has been a horizontal stretch of scale factor $\frac{1}{2}$, a reflection about the y-axis, and a horizontal translation of $\frac{\pi}{4}$. All the x-values in the original cosine function have been divided by -2, then increased by $\frac{\pi}{4}$.

When sketching functions like these by hand, it is best to take a step-by-step approach.

Example 18

Sketch the graph of the function $y = 5\cos\left(\dfrac{2}{3}(x+\pi)\right) - 2$.

Answers

This function will have an amplitude of 5 and a vertical shift of -2.

The maximum and minimum values of the function will be 3 and -7, respectively.

The horizontal axis of the wave will be $y = -2$, which is the vertical translation.

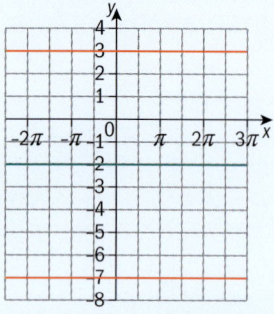

Mark these maximum and minimum values (shown in red) and the axis of the wave (shown in green).

These will be helpful guidelines when graphing the function.

$$\text{Period} = \dfrac{2\pi}{b} = \dfrac{2\pi}{\left(\dfrac{2}{3}\right)} = (2\pi)\left(\dfrac{3}{2}\right) = 3\pi$$

This function will have a period of 3π and a horizontal shift of $-\pi$.

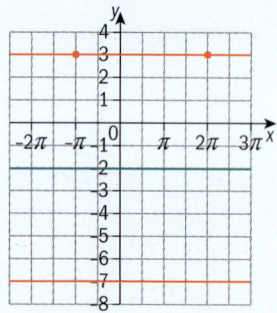

The standard cosine curve has a maximum where $x = 0$, so this function will have a maximum where $x = -\pi$.

As the period is 3π, there will be another maximum 3π units to the right, where $x = 2\pi$. Mark these maximum points on the red line.

Use your knowledge of the features of the cosine curve to mark other points, such as the minimum and the points on the axis of the wave.

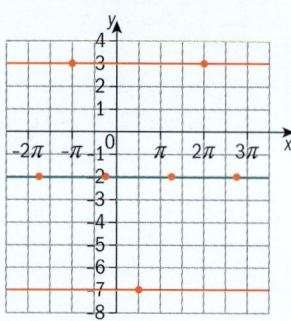

Midway between two maximum values, there is a minimum value.

Midway between the maximum and minimum values, there will be points on the green (horizontal) axis.

▶ Continued on next page

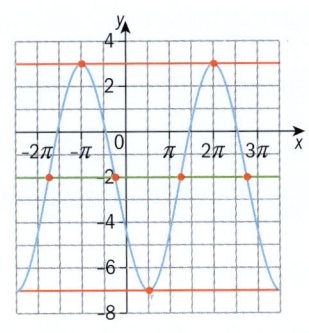

Connect these points and sketch the function.

You may want to erase the guidelines when your sketch is complete.

Example 19

Find the amplitude and period, then write one sine equation and one cosine equation for the function shown in the diagram.

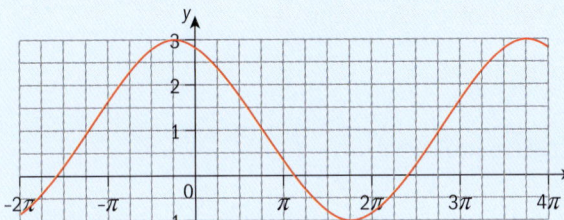

Answer

The amplitude is $\dfrac{3-(-1)}{2} = 2$

The vertical shift is $\dfrac{3+(-1)}{2} = 1$

The period is 4π

$y = 2\sin\left(\dfrac{1}{2}\left(x + \dfrac{5\pi}{4}\right)\right) + 1$

The amplitude is one-half the difference between the maximum and the minimum value.

The period is the horizontal distance it takes for the function to complete one cycle. The easiest way to find this is by looking at the horizontal distance from a maximum point to a maximum point or from a minimum point to a minimum point.

For a sine function, the horizontal translation is found by looking at the x-coordinate of a point on the horizontal axis of the wave, with a positive gradient. This corresponds to the point (0, 0) on the standard sine curve.

In this function, one such point is $\left(-\dfrac{5\pi}{4}, 1\right)$, so the horizontal translation is $-\dfrac{5\pi}{4}$.

▶ Continued on next page

$$y = 2\cos\left(\frac{1}{2}\left(x + \frac{\pi}{4}\right)\right) + 1$$

The sine and cosine function have the same amplitude, period, and vertical translation. For a cosine function, you can find the horizontal translation by looking at the x-coordinate of a maximum point on the curve. This corresponds to the point (0, 1), which is a maximum on the standard cosine curve. In this function, one such point is $\left(-\frac{\pi}{4}, 1\right)$, so the horizontal translation is $-\frac{\pi}{4}$.

Remember, there may be more than one possible correct equation that can be written for a given sine or cosine function.

→ For sine and cosine functions of the same curve, the horizontal translations will differ by one-fourth the period of the function.

Exercise 13K

For questions 1 to 4, write one sine and one cosine equation for the function.

1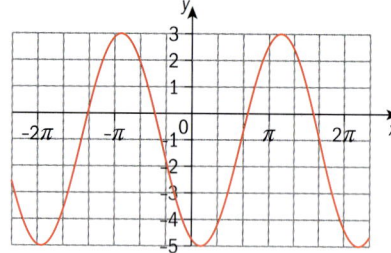

2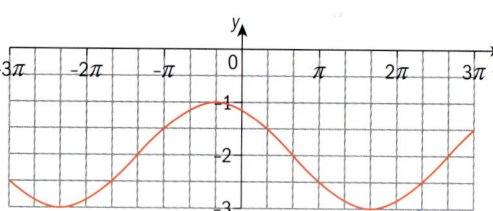

3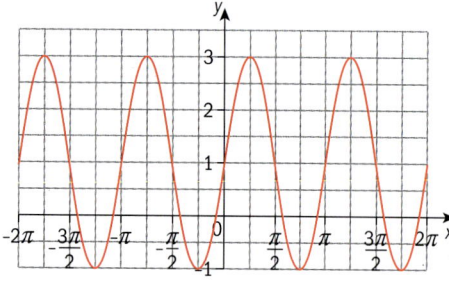

4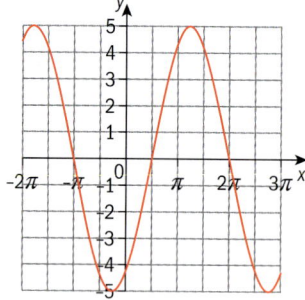

For questions 5 to 10, make a neat sketch of the function over at least one full cycle.

5 $y = 3\cos\left(\frac{1}{3}\left(x - \frac{\pi}{6}\right)\right) + 2$

6 $y = -\sin\left(-2\left(x + \frac{\pi}{4}\right)\right) - 1$

7 $y = 1.5\cos\left(3\left(x + \frac{\pi}{2}\right)\right)$

8 $y = -2\cos\left(\frac{1}{2}x\right) + 4$

482 Circular functions

13.7 Modeling with sine and cosine functions

Many real-life situations can be modeled using sine and cosine functions. Examples include tide heights, sunrise times, and average temperatures. In this section, we will use our knowledge of transformations to look at how sine and cosine functions can be used to model data.

→ To create a cosine function model for data you need to know:
- the amplitude of the function
- the vertical translation
- the horizontal translation
- the period.

The sine function has the same amplitude, vertical translation and period but its horizontal translation is one-fourth of the period to the left of the cosine curve.

Example 20

Create a model for this data, which shows the depth of the water measured off a buoy in the ocean over an 18-hour period, starting at midnight.

Time	0:00	2:00	4:00	6:00	8:00	10:00	12:00	14:00	16:00	18:00
Water depth (m)	6.7	8.3	9.1	8.1	6.4	5.6	6.7	8.4	9.2	8.2

Answer
Enter data into lists (label these *time* and *depth*), then plot the data on the GDC. The independent variable, time, will be on the *x*-axis, and the water depth will be the dependent variable on the *y*-axis.

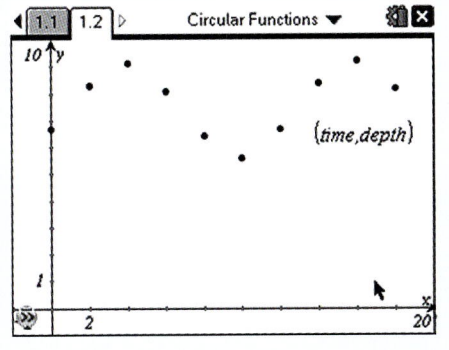

From the graph, the minimum value is 5.6 metres, which occurs at 10:00. The maximum value is 9.2 metres. Use these values to estimate the amplitude.

> Be sure your GDC is in RADIANS mode.

GDC help on CD: *Alternative demonstrations for the TI-84 Plus and Casio FX-9860GII GDCs are on the CD.*

▶ Continued on next page

The data is clearly periodic, with the water height rising and falling in an apparent pattern.

Now try to find a trigonometric function to model this data.

To develop the model, estimate the amplitude, period, and vertical and horizontal translations of the function.

The amplitude is one-half the vertical distance between the maximum and minimum values.

$$\text{Estimated amplitude} = \frac{9.2 - 5.6}{2} = 1.8 \text{ metres}$$

The vertical translation is the value half-way between the maximum and minimum values.

$$\text{Vertical translation} = \frac{9.2 + 5.6}{2} = 7.4$$

The period is the horizontal distance the function takes to complete one cycle. The maximum values are at 4:00 and 16:00, so estimate the period to be 12 hours.

Finally, estimate the horizontal translation. To create a cosine model for the data, the easiest way is to look for a maximum point. The plotted points seem to have maximum point where $x = 4$ and where $x = 16$. Use either of these x-values for the horizontal translation.

Substitute these estimates into the equation $y = a\cos(b(x - c)) + d$

$$y = 1.8\cos\left(\frac{2\pi}{12}(x - 4)\right) + 7.4.$$

Enter this equation into the GDC and graph the function on the same axes as the data points.

> You can also find the vertical translation by subtracting the amplitude from a maximum value, or by adding the amplitude to a minimum value.

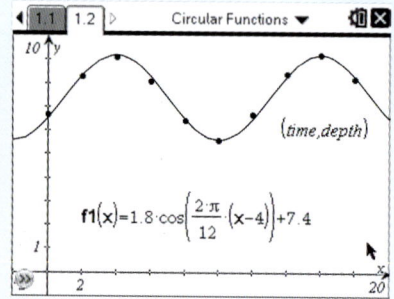

> **GDC help on CD:** *Alternative demonstrations for the TI-84 Plus and Casio FX-9860GII GDCs are on the CD.*

The function appears to be a very good model for the data. You could try to make minor adjustments to get a 'better fit'.

> You could create a sine function instead. Try it – you should get
> $$y = 1.8\sin\left(\frac{2\pi}{12}(x - 1)\right) + 7.4$$

Example 21

The following set of data can be modeled by the function
$y = a\cos(b(x - c)) + d$.

x	1	2	3	4	5	6	7	8	9	10	11
y	4	7.6	9.4	7.6	4	2.2	4	7.6	9.4	7.6	4

a Use the data to estimate the period, amplitude, and vertical and horizontal translations.
b Write the cosine function which models the data.
c Graph the function on the same axes as the data points.
d Use the regression function on your GDC to get a sine model for the data, and graph this function on the same axes as the data points.

Answers

a Amplitude $= \dfrac{9.4 - 2.2}{2} = 3.6$

Vertical translation $= \dfrac{9.4 + 2.2}{2} = 5.8$

Horizontal translation $= 3$
Period $= 9 - 3 = 6$

b $y = 3.6\cos\left(\dfrac{2\pi}{6}(x-3)\right) + 5.8$

c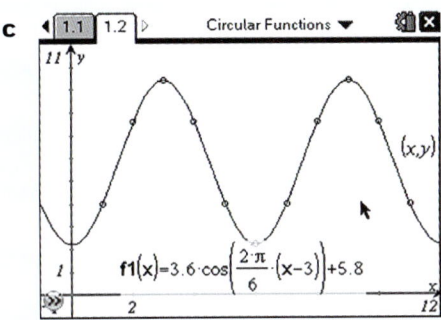

d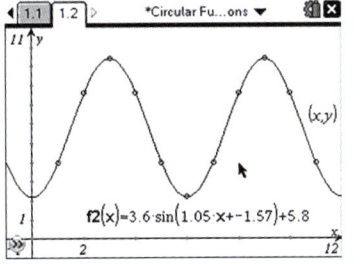

> Be sure your GDC is in RADIAN mode!

> Use the Sine Regression function under the STAT CALC menu. Be sure to tell the GDC which lists contain the data (x, y).

> **GDC help on CD:** *Alternative demonstrations for the TI-84 Plus and Casio FX-9860GII GDCs are on the CD.*

Exercise 13L

For each set of data,
a Use the data to estimate the period, amplitude, and vertical and horizontal translations.
b Write a cosine function in the form $y = a\cos(b(x - c)) + d$ to model the data.
c Graph the function on the same axes as the data points.
d Use the regression function on your GDC to get a sine model for the data, and graph this function on the same axes as the data points.

> What real life situations can be modeled by periodic functions? what adjustments might need to be made to account for fluctuations in the data?

1

x	0	0.5	1	1.5	2	2.5	3	3.5	4	4.5	5
y	11.8	8.5	2.2	5.5	11.8	8.5	2.2	5.5	11.8	8.5	2.2

2

x	5	10	15	20	25	30	35	40	45	50	55
y	12.5	9.3	12.5	18.9	21.9	18.9	12.5	9.3	12.5	18.9	21.9

3

x	2	4	6	8	10	12	14	16	18	20	22
y	1.8	2.1	1.8	1.3	0.7	0.5	0.7	1.3	1.8	2.1	1.8

When you have a function to model data, you can use that function to make predictions.

Example 22

The function
$$h(t) = 67.5\cos\left(\frac{2\pi}{30}(t - 15)\right) + 67.5$$
can be used to model the height of a passenger above the boarding platform on the London Eye.
a Use this function to estimate the height of a passenger above the platform
   i 8 minutes after boarding
   ii 19 minutes after boarding.
b Use this function to estimate how long it takes for a passenger to first reach a height of 100 m.

Answers

a i 8 minutes after boarding: | Substitute $t = 8$ in the function.

$$h(8) = 67.5\cos\left(\frac{2\pi}{30}(8 - 15)\right) + 67.5 \approx 74.6$$

The passenger is approximately 74.6 metres above the platform.

▶ Continued on next page

    ii 19 minutes after boarding:
$$h(19) = 67.5\cos\left(\frac{2\pi}{30}(19-15)\right) + 67.5 \approx 112.7$$

    The passenger is approximately 112.7 metres above the platform.

b $h(t) = 67.5\cos\left(\frac{2\pi}{30}(t-15)\right) + 67.5 = 100$ *Set the function equal to 100, for the height.*

  $t \approx 9.90$ minutes

Exercise 13M

EXAM-STYLE QUESTIONS

1 The depth of the water at the end of a pier can be estimated by the function $d(t) = 5.6\sin(0.5236(t-2.5)) + 14.9$, where d is the depth of the water in metres, and t is the number of hours after midnight.
  a What is the period of this function?
  b Estimate the depth of the water at midnight.
  c Estimate the depth of the water at 14:00.
  d At what time will the water first reach its highest depth?

2 The average high temperature in a city can be modeled by the function $T(d) = 17.5\cos(0.0172(d-187)) + 12.5$, where T is the temperature in degrees Celsius, and d is the day of the year (1 Jan = 1, 14 Jan = 14, etc,...)
  a What is the expected high temperature in this city on the first day of February?
  b What is the highest expected temperature, and on what day does it occur?
  c How many days each year are expected to remain below freezing?

3 A Ferris wheel at an amusement park reaches a maximum height of 46 metres and a minimum height of 1 metre. It takes 20 minutes for the wheel to make one full rotation.
  a If a child gets on the Ferris wheel when $t = 0$, how high will he be after riding for 10 minutes?
  b Write a sine function to model the height of the child t minutes after boarding the Ferris wheel.
  c How high is the child if he has been riding for 3 minutes?
  d For what length of time will the child be higher than 40 metres?

EXAM-STYLE QUESTION

4 The owner of an ice cream shop tracks his annual sales and discovers he sells a minimum of 5 gallons of ice cream on the first day of January, and a maximum of 37 gallons on the first day of July.
  a Assuming the annual sales can be modeled by a cosine function, create an equation to model this situation. Let x represent the month.
  b How many gallons would he expect to sell on the first day of April?
  c During what month would he expect to sell 30 gallons of ice cream in one day?

Extension material on CD:
Worksheet 13 - Modeling temperatures project

Review exercise

1 Given that $\cos 70° = 0.342$ (to three significant figures), find the values of
  a $\cos 110°$
  b $\cos 250°$
  c $\cos(-290°)$

2 Given that $\sin 40° = 0.643$ (to three significant figures), find the values of
  a $\sin 140°$
  b $\sin 320°$
  c $\sin(-140°)$

3 Solve each equation for $-360° \leq x \leq 360°$.
  a $\cos x = -\dfrac{1}{2}$
  b $\tan x = \dfrac{1}{\sqrt{3}}$
  c $2\sin^2 x - \sin x = 1$

EXAM-STYLE QUESTIONS

4 Solve the equation $\sin 2x + \sin x = 0$, for $0 \leq x \leq \pi$.

5 The graph of f, for $0 \leq x \leq 9$, is shown.
  a Given that the function can be written in the form
    $f(x) = a\sin(b(x-c)) + d$,
    i find the values of a, c and d.
    ii explain why $b = \dfrac{\pi}{4}$.
  b Write down the interval for which $f(x) > 6$.

The graph has a maximum at (6, 11) and a minimum at (2, 1).

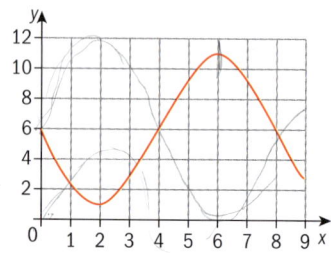

Circular functions

6 Given that $\cos x = \dfrac{2}{5}$, and x is an acute angle, find

   a $\sin x$ **b** $\tan x$ **c** $\sin 2x$

7 Sketch the graph of the function $f(x) = 3\cos\left(\dfrac{2\pi}{5}(x+1)\right) - 2$, for $-3 \leq x \leq 5$.

Review exercise

1 Solve each equation for $-180° \leq x \leq 360°$.

   a $\sin x = 0.75$ **b** $\cos x = -0.63$ **c** $\tan x = -2.8$

2 Solve each equation for $-2\pi \leq \theta \leq 2\pi$.

   a $2\sin\theta = \cos\left(\dfrac{\theta}{3}\right)$ **b** $\cos x = 3x - 1$ **c** $2\tan\left(\dfrac{x}{5}\right) = 4x - x^3$

3 The graph of f, for $0 \leq x \leq 7$, is shown.

   a Given that the function can be written in the form $f(x) = a\cos bx + c$, find the values of a, b and c.

   b Write down the solutions to the equation $f(x) = 1$.

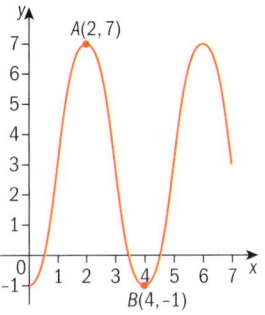

4 The depth of the water at the end of a fishing pier is given by the function $D(t) = P\sin\left(\dfrac{\pi}{6}(t - Q)\right) + 10$, where D is the depth of the water in metres, and t is the number of hours after midnight.

Low tide occurs at 4:00, when the depth of the water is 6 m, and high tide occurs at 10:00, when the depth of the water is 14 m.

   a Find the values of P and of Q.
   b Sketch a graph of the function D, for $0 \leq t \leq 24$.
   c At what time does the water first reach a height of 8 metres?
   d Fishing is prohibited when the water depth is less than 8 metres. How many hours each day is fishing prohibited?

5 The longest day of the year in a city is 21 June, with 15 hours of daylight. The shortest day of the year is 21 December, with 9.35 hours of daylight.
The number of hours of daylight can be modeled by the function $h(x) = A\sin 0.0172(x - 86) + B$, where x is the day of the year (i.e. $x = 1$ on 1 Jan).

   a Find the value of A and of B.
   b How many hours of daylight would you expect on 1 Feb?

CHAPTER 13 SUMMARY

Using the unit circle

- The unit circle has its center at the origin (0, 0) and a radius length of 1 unit. The terminal side of any angle θ in standard position will meet the unit circle at a point with coordinates $(\cos\theta, \sin\theta)$.

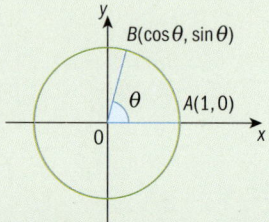

- For any angle θ, $\tan\theta = \dfrac{\sin\theta}{\cos\theta}$, where $\cos\theta \neq 0$.
- For any angle θ:
  - $\sin\theta = \sin(180 - \theta)$
  - $\cos\theta = \cos(-\theta)$
  - $\tan\theta = \tan(180 + \theta)$

Trigonometric identities

- The equation $\cos(2\theta) = 1 - 2\sin^2\theta$ is an **identity**, as it is true for all values of θ.
- The double-angle identity for cosine are:
$$\cos(2\theta) = 1 - 2\sin^2\theta$$
$$= 2\cos^2\theta - 1$$
$$= \cos^2\theta - \sin^2\theta$$
- The double-angle identity for sine is $\sin(2\theta) = 2\sin\theta\cos\theta$

Graphing circular functions

- The sine and cosine functions have graphs of the same size and shape but different horizontal positions on the axes. The functions are **periodic** with period 360° or 2π. Both functions have a maximum value of 1 and a minimum value of −1, and an **amplitude** of 1.
- Like the sine and cosine functions, the tangent function is **periodic**. There are vertical asymptotes at values of x where the function does not exist. The same cycle of values repeats between each pair of vertical asymptotes.
- The period of the tangent function is 180° (or π radians). Unlike the sine and cosine functions, the tangent function does not have an amplitude. It has no maximum or minimum values.
- The function $y = \sin(x) + d$ is a **vertical translation** of the standard sine curve. The curve shifts up if d is positive, down if d is negative.

Continued on next page

- The function $y = \sin(x - c)$ is a **horizontal translation** of the standard cosine curve. The curve shifts to the right if c is positive, left if c is negative.
- The function $y = \cos(x) + d$ is a **vertical translation** of the standard cosine curve. The curve shifts up if d is positive, down if d is negative.
- The function $y = \cos(x - c)$ is a **horizontal translation** of the standard cosine curve. The curve shifts to the right if c is positive, left if c is negative.
- The functions $y = a\sin x$ and $y = a\cos x$ are **vertical stretches** of the sine and cosine functions. When the graph of a function undergoes a vertical stretch, every y-value in the original function is multiplied by the value of a.
  With a vertical stretch, the **amplitude** of the sine or cosine function will change from 1 to $|a|$. The period of the function will not change.
- The functions $y = \sin(bx)$, $y = \cos(bx)$ and $y = \tan(bx)$ represent **horizontal stretches** of the sine, cosine and tangent functions. When the graph of a function undergoes a horizontal stretch, every x-value in the original function is multiplied by $\frac{1}{b}$.
- When a sine or cosine function undergoes a horizontal stretch, the period of the function will change from 2π to $\frac{2\pi}{|b|}$, or from $360°$ to $\frac{360°}{|b|}$.
- For a function in the form $y = \tan(bx)$, the period will change from π to $\frac{\pi}{|b|}$, or from $180°$ to $\frac{180°}{|b|}$.

Combined transformations with sine and cosine functions

- For sine and cosine functions of the same curve, the horizontal translations will differ by one-fourth the period of the function.

Modeling with sine and cosine functions

- To create a cosine function model for data you need to know:
  - the amplitude of the function
  - the vertical translation
  - the horizontal translation
  - the period.
- The sine function has the same amplitude, vertical translation and period but its horizontal translation is one-fourth of the period to the left of the cosine curve.

Theory of knowledge

Pure vs. applied mathematics

Mathematics is often categorized into 'pure' and 'applied'. What is the difference between the two areas?

You often get this type of question in trigonometry:

A weight is suspended on a spring, as shown. When the weight is pulled down and released, it will oscillate up and down.

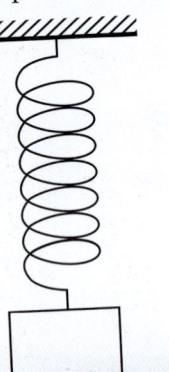

If we let $h = 0$ represent the resting height of the weight, the oscillation height at time t seconds is given by $h(t) = a \sin(b(t - c))$

The weight is pulled down 5cm and completes one full oscillation every two seconds. Find the values of a, b and c.

Ignore the effects of friction and air resistance.

This question is an example of pure mathematics. If we ignore the effects of friction and air resistance, the weight should oscillate indefinitely. But in real life, the oscillations will reduce until the weight comes to rest.

- What is the point of studying pure mathematics problems like this, when the results are unrealistic in real life?
- Should we only study applied mathematics, which could have some practical use?

"As far as the laws of mathematics refer to reality, they are not certain; and as far as they are certain, they do not refer to reality."

Albert Einstein, *Sidelights on Relativity*

Pure mathematicians study mathematics for its own sake, without having any application in mind. Applied mathematicians use mathematics to help them research, model and solve problems in other areas of knowledge, for example physics, economics, computer science and engineering.

> There are 10 types of people in this world: those who understand binary and those who don't.

Applications of pure mathematics

Practical applications are often found for pure mathematics – sometimes many years after the first ideas were formulated.

1 Modern computers use the binary (base 2) number system. German mathematician Gottfried Wilhelm Leibniz (1646–1716) wrote about this number system, which uses only 0s and 1s, in the early 1700s. When he was studying this 'pure mathematics' idea, he didn't know how it would be used 300 years later.

2 George Boole, an English mathematician, developed his Boolean Logic system in the 1850s. This system was later used in digital electronics.

3 In physics, elementary particles were discovered through arguments involving the beauty, symmetry or elegance of the underlying mathematics.

Perhaps all 'pure' mathematics will be used to model some aspect of real life one day.

> "Physics is mathematical not because we know so much about the physical world, but because we know so little; it is only its mathematical properties that we can discover."
>
> Bertrand Russell, English mathematician and philosopher (1872–1970)

▶ George Boole (1815–64)

Pure mathematics in applications

Studying applied mathematics has led to the development of entirely new mathematical disciplines, such as statistics and game theory.

- Can we model the real world with mathematics because we create mathematics to mirror the world, or because the world is intrinsically mathematical?
- What does this tell us about the relationship between the natural sciences, mathematics and the natural world?
- Is mathematics invented or discovered?

Chapter 13   493

14 Calculus with trigonometric functions

CHAPTER OBJECTIVES:

6.1 Tangents and normals, and their equations

6.2 Derivative of $\sin x$, $\cos x$, $\tan x$, including differentiation of a sum and a real multiple, chain rule, product and quotient rules, second derivative of these functions

6.3 Local maximum and minimum points, points of inflexion, graphs of functions, including the relationship between graphs of f, f' and f''

6.4 Indefinite integral of $\sin x$ and $\cos x$, including composites with the linear function $ax + b$, integration by inspection, or substitution of the form $\int f(g(x))g'(x)dx$

6.5 Anti-differentiation with a boundary condition to determine the constant term, definite integrals, areas between the curve and the x-axis, areas between curves, volumes of revolution about the x-axis

6.6 Kinematic problems involving displacement s, velocity v and acceleration a, total distance traveled

Before you start

You should know how to:

1 Find the exact value of trigonometric functions for unit circle values.

2 Use trigonometric identities to solve equations.
e.g. Solve $\cos 2x = -\cos x$ for $0 \leq x \leq 2\pi$.
$\cos 2x = -\cos x$
$2\cos^2 x - 1 = -\cos x$
$2\cos^2 x + \cos x - 1 = 0$
$(2\cos x - 1)(\cos x + 1) = 0$
$\cos x = \dfrac{1}{2}$ or $\cos x = -1$
$x = \dfrac{\pi}{3}, \dfrac{5\pi}{3}, \pi$

3 Use the product, quotient and chain rules to find derivatives.
e.g. Find the derivative of $f(x) = x^2 \ln x$.
$f(x) = x^2 \ln x$
$f'(x) = x^2 \left(\dfrac{1}{x}\right) + (\ln x)(2x) = x + 2x \ln x$

Skills check

1 Find the exact value of

a $\cos \dfrac{7\pi}{4}$ **b** $\sin \dfrac{3\pi}{2}$

c $\tan \dfrac{11\pi}{6}$ **d** $\sin \dfrac{4\pi}{3}$

2 Solve each equation for $0 \leq x \leq 2\pi$.

a $1 + \tan x = \sin^2 x + \cos^2 x$

b $\sin 2x - \cos x = 0$

c $\sin^2 x = 1 + \cos x$

3 Find the derivative of

a $f(x) = 2x^3 e^x$

b $f(x) = x \ln(x^2)$

c $f(x) = \dfrac{x-5}{x^2+4}$

d $f(x) = \dfrac{\ln x}{x}$

At *The Chocolate Factory* in Ghirardelli Square, San Francisco, California, a vat of milk chocolate is stirred by a stirrer blade which is driven by a wheel that pushes the blade back and forth across the bottom of the vat.

In the chocolate stirrer, the periodic circular motion of the wheel is translated into periodic linear motion of the blade. The diagram shows the mechanism. One end of a rod is attached to a wheel or crankshaft and the other end of the rod is attached to the stirrer blade inside the vat. As the wheel turns, the rod pushes the stirrer blade backwards and forwards across the vat. The distance between the center of the wheel and the stirrer blade can be modeled by a function like this: $d(\theta) = 2\cos\theta + \sqrt{25 - 4\sin^2\theta}$, where d is the distance in metres and θ is the angle of rotation of the wheel in radians.

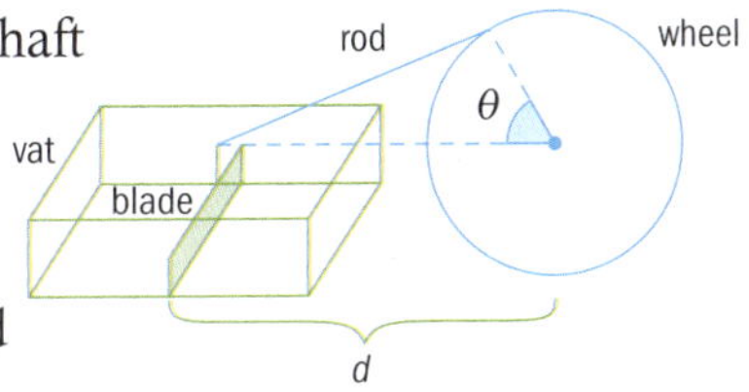

To find the angle of rotation when the blade is the shortest distance from the center of the wheel, you would use the derivative of $d(\theta)$.

Many real world phenomena, such as heart rhythms, movement of hands on a clock, the tides and circular motion, have periodic behavior – they follow a pattern that recurs at regular intervals. Periodic behavior can be modeled by trigonometric functions – sine, cosine and tangent – which are periodic functions. You can see from their graphs that each function's values recur.

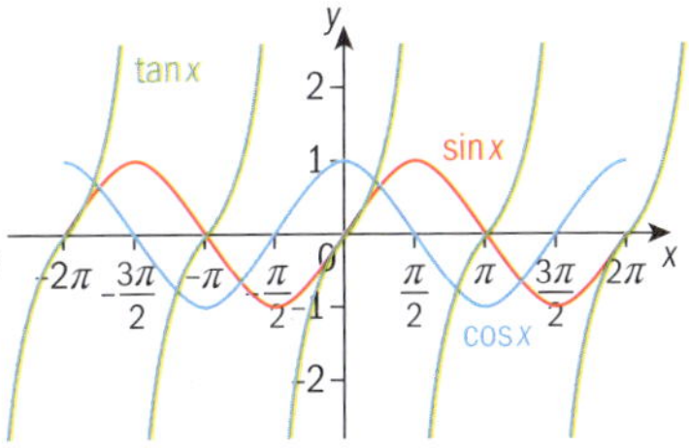

In this chapter you will find derivatives of the sine, cosine and tangent functions and integrate sine and cosine functions, in order to investigate the behavior of periodic functions like this one.

Chapter 14

14.1 Derivatives of trigonometric functions

In Chapter 7 you met these properties of derivatives, where c is a constant real number.

Constant rule: $\dfrac{d}{dx}[c] = 0$

Constant multiple rule: $\dfrac{d}{dx}[cf(x)] = cf'(x)$

Sum or difference rule: $\dfrac{d}{dx}[f(x) \pm g(x)] = f'(x) \pm g'(x)$

Product rule: $\dfrac{d}{dx}[f(x) \cdot g(x)] = f(x) \cdot g'(x) + g(x) \cdot f'(x)$

Quotient rule: $\dfrac{d}{dx}\left[\dfrac{f(x)}{g(x)}\right] = \dfrac{g(x) \cdot f'(x) - f(x) \cdot g'(x)}{[g(x)]^2}$, $g(x) \neq 0$

Chain rule: $\dfrac{d}{dx}[f(g(x))] = f'(g(x)) \cdot g'(x)$

Investigation: The derivative of sine

Here is the graph of $f(x) = \sin x$ for $-2\pi \leq x \leq 2\pi$. Use it to answer the questions below.

1 There are four values of x in $-2\pi \leq x \leq 2\pi$ where the gradient of the tangent line to $f(x) = \sin x$ is equal to zero. What are they?
Use these values to plot four points that are on the graph of the derivative of f against x.

> The gradient of a horizontal line is 0. So at the values of x where the tangent lines to f are horizontal the derivative of f is equal to 0.

2 List the intervals of $-2\pi \leq x \leq 2\pi$ where the graph of $f(x) = \sin x$ is increasing and those where it is decreasing. When f is increasing, what is true about the sign of the derivative of f? When f is decreasing, what is true about the sign of the derivative of f? Use this information and the points you plotted in question 1 to make a possible sketch of the graph of the derivative of f.

3 Use a GDC to graph the derivative of $f(x) = \sin x$ in the interval $-2\pi \leq x \leq 2\pi$. Be sure your GDC is in radian mode. Compare the graph of the derivative on your GDC to the one you drew in question 2. Adjust your drawing if necessary.

> Enter the graph:
>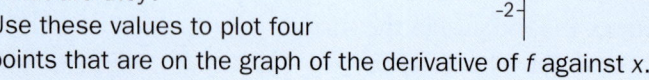

4 Make a conjecture based on the graph of the derivative of $f(x) = \sin x$. What function do you think is the derivative of sine?

5 Verify your conjecture numerically with a GDC by comparing the table of values for the function you graphed in question 3 and the function you chose in question 4.

In the investigation, you should have found that $\frac{d}{dx}(\sin x) = \cos x$. Now consider the derivative of $f(x) = \cos x$.

If you translate the graph of sine to the left $\frac{\pi}{2}$, you get the graph of cosine. So $f(x) = \cos x = \sin\left(x + \frac{\pi}{2}\right)$.

> You can examine a geometric justification of this fact in the TOK section at the end of this chapter.

Hence, $\frac{d}{dx}(\cos x) = \frac{d}{dx}\left[\sin\left(x + \frac{\pi}{2}\right)\right]$

$= \left[\cos\left(x + \frac{\pi}{2}\right)\right](1)$

$= \cos\left(x + \frac{\pi}{2}\right)$

> Using the chain rule:
> $\frac{d}{dx}\left[\sin\left(x + \frac{\pi}{2}\right)\right]$
> $= \left[\cos\left(x + \frac{\pi}{2}\right)\right]\left[\frac{d}{dx}\left(x + \frac{\pi}{2}\right)\right]$
> $= \left[\cos\left(x + \frac{\pi}{2}\right)\right][1]$

If you translate the graph of cosine to the left $\frac{\pi}{2}$, you get a reflection of the graph of sine in the x-axis. So

$f(x) = \cos\left(x + \frac{\pi}{2}\right) = -\sin x.$

Therefore we conclude that $\frac{d}{dx}(\cos x) = \cos\left(x + \frac{\pi}{2}\right) = -\sin x$

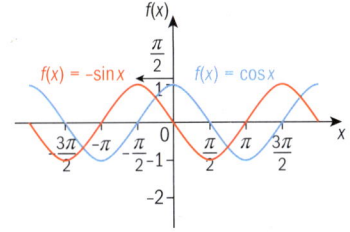

Finally consider the derivative of $f(x) = \tan x$. We know that

$f(x) = \tan x = \frac{\sin x}{\cos x}$, where $\cos x \neq 0$.

So, $\frac{d}{dx}(\tan x) = \frac{d}{dx}\left(\frac{\sin x}{\cos x}\right)$

$= \frac{\cos x(\cos x) - \sin x(-\sin x)}{(\cos x)^2}$

$= \frac{\cos^2 + \sin^2 x}{\cos^2 x}$

$= \frac{1}{\cos^2 x}, \cos x \neq 0$

> Apply the quotient rule.

> Use the identity $\cos^2 \theta + \sin^2 \theta = 1$ to simplify the numerator.

→ **Derivatives of sine, cos and tan:**

$f(x) = \sin x \Rightarrow f'(x) = \cos x$

$f(x) = \cos x \Rightarrow f'(x) = -\sin x$

$f(x) = \tan x \Rightarrow f'(x) = \frac{1}{\cos^2 x}, \cos x \neq 0$

Example 1

Find the derivative of each function.

a $f(x) = \sin x + \cos x$

b $y = \cos(t^2)$

c $y = \dfrac{1}{\tan x}$

d $f(x) = \sin^3(2x)$

Answers

a $f(x) = \sin x + \cos x$
$f'(x) = \cos x - \sin x$ — Take the derivative of each term.

b $y = \cos(t^2)$
$y' = \underbrace{[-\sin(t^2)]}_{\text{derivative of outside function with respect to inside function}} \cdot \underbrace{[2t]}_{\text{derivative of inside function with respect to } t}$

Apply the chain rule, where the outside function is $u(t) = \cos t$ and the inside function is $v(t) = t^2$.

$= -2t \sin(t^2)$

c $y = \dfrac{1}{\tan x}$

$= (\tan x)^{-1}$ — Rewrite using rational exponents.

$y' = -1(\tan x)^{-2}\left(\dfrac{1}{\cos^2 x}\right)$

$= -\dfrac{1}{\tan^2 x \cos^2 x}$ or $-\dfrac{1}{\sin^2 x}$

Apply the chain rule, where the outside function is $u(x) = x^{-1}$ and the inside function is $v(x) = \tan x$.

d $f(x) = \sin^3(2x)$
$= (\sin(2x))^3$
$f'(x) = 3(\sin(2x))^2 (\cos(2x))(2)$
$= 6\sin^2(2x)\cos(2x)$

Apply the chain rule twice. First the outside function is $u(x) = x^3$ and the inside function is $v(x) = \sin(2x)$. Then when finding the derivative of $\sin(2x)$, the outside function is $u(x) = \sin x$ and the inside function is $v(x) = 2x$.

> In the 17th and 18th centuries the development of mechanical devices shifted the field of trigonometry from its original connection to triangles to modeling periodic motion. Joseph Fourier (1768–1830), a French mathematician and physicist, found that almost any periodic function, such as the vibration of a violin string or the movement of the pendulum on a clock, could be expressed as an infinite sum of sine and cosine functions.

> The oscillation of a spring and the movement of a pendulum are examples of simple harmonic motion. How are trigonometric functions and calculus used to model this motion?

Exercise 14A

In questions 1–10, find the derivative of each function.

1 $f(x) = 3\sin x - 2\cos x$

2 $y = \tan(3x)$

3 $y = \dfrac{2}{\sin x}$

4 $s(t) = \cos^2 t$

5 $f(x) = \sin \sqrt{x}$

6 $y = \tan^2 x$

7 $y = \cos \dfrac{x}{2} + \sin(4x)$

8 $f(x) = \dfrac{1}{\cos(2x)}$

9 $y = \dfrac{4}{\sin^2(\pi x)}$

10 $f(x) = \sin(\sin x)$

EXAM-STYLE QUESTIONS

11 Differentiate with respect to x.

    a $\tan(x^3)$ **b** $\cos^4 x$

12 A function has the equation $y = \sin(3x - 4)$.

    a Find $\dfrac{dy}{dx}$.   **b** Find $\dfrac{d^2y}{dx^2}$.

Example 2

Find the equations of the tangent line and the normal line to the curve $f(x) = \cos 3x$ at the point where $x = \dfrac{\pi}{9}$.

Answer

$f\left(\dfrac{\pi}{9}\right) = \cos\left(3\left(\dfrac{\pi}{9}\right)\right)$

$= \cos\dfrac{\pi}{3} = \dfrac{1}{2}$

Evaluate the function f at $x = \dfrac{\pi}{9}$ to find the point of tangency.

The point of tangency is $\left(\dfrac{\pi}{9}, \dfrac{1}{2}\right)$.

$f'(x) = -3\sin(3x)$

$f'\left(\dfrac{\pi}{9}\right) = -3\sin\left(3\left(\dfrac{\pi}{9}\right)\right)$

Find the derivative of f and evaluate it at $x = \dfrac{\pi}{9}$ to find the slope of the tangent line.

> 'Slope' is another word for 'gradient'.

$= -3\sin\left(\dfrac{\pi}{3}\right) = -3\left(\dfrac{\sqrt{3}}{2}\right)$

$= -\dfrac{3\sqrt{3}}{2}$

The slope of the tangent line at $x = \dfrac{\pi}{9}$ is $-\dfrac{3\sqrt{3}}{2}$.

The slope of the normal line at $x = \dfrac{\pi}{9}$ is $\dfrac{2}{3\sqrt{3}}$ or $\dfrac{2\sqrt{3}}{9}$.

The normal line is perpendicular to the tangent line, so the slopes are negative reciprocals.

Tangent line: $y - \dfrac{1}{2} = -\dfrac{3\sqrt{3}}{2}\left(x - \dfrac{\pi}{9}\right)$

Normal line: $y - \dfrac{1}{2} = \dfrac{2\sqrt{3}}{9}\left(x - \dfrac{\pi}{9}\right)$

Use the point–slope equation for a line, $y - y_1 = m(x - x_1)$, to write the equations.

Exercise 14B

In questions 1 and 2, find the equations of the tangent line and the normal line to the curve at the given value of x.

1 $f(x) = \sin x - \cos x$; $x = \dfrac{\pi}{2}$

2 $f(x) = 2\tan x$; $x = \dfrac{\pi}{4}$

EXAM-STYLE QUESTIONS

3. The point $P\left(\dfrac{\pi}{2}, 0\right)$ lies on the graph of $y = \sin(2x)$.
   Find the gradient of the tangent to the curve at P.

4. Let $f(x) = \cos(2x)$.
   a. Write down the value of $f\left(\dfrac{\pi}{3}\right)$.
   b. Find $f'(x)$.
   c. Find the equation of the tangent line to f at $x = \dfrac{\pi}{3}$.

5. Consider the function $f(x) = 3\sin x$ for $0 \le x \le 2\pi$.
   Find the value(s) of x for which the tangent lines to the graph of f are parallel to the line $y = \dfrac{3}{2}x + 4$.

14.2 More practice with derivatives

You now know the derivatives of these functions:

$\dfrac{d}{dx}[x^n] = nx^{n-1}, n \ne 1$

$\dfrac{d}{dx}[\sin x] = \cos x$

$\dfrac{d}{dx}[e^x] = e^x$

$\dfrac{d}{dx}[\cos x] = -\sin x$

$\dfrac{d}{dx}[\ln x] = \dfrac{1}{x}, x > 0$

$\dfrac{d}{dx}[\tan x] = \dfrac{1}{\cos^2 x}, \cos x \ne 0$

Using these facts and the rules stated at the beginning of Section 14.1, you can find the derivatives of a wide variety of functions.

> Most phenomena in the sciences, engineering, business and other fields can be modeled by an **elementary function**. An elementary function is a function that is algebraic, transcendental, or a sum, difference, product, quotient or composition of algebraic and transcendental functions.
> **Algebraic functions**
> - Polynomials
> - Rational functions
> - Functions involving radicals
>
> **Transcendental functions**
> (cannot be expressed as sums, difference, products, quotients and radicals involving x^n)
> - Logarithmic functions
> - Exponential functions
> - Trigonometric functions
> - Inverse trigonometric functions
>
> With the exception of the inverse trigonometric function, you can now differentiate almost any elementary function.

Example 3

Find the derivative of each function.
a. $f(x) = 4e^{2x} + \sin(3x + 2)$
b. $y = e^x \sin x$
c. $y = \cos^3 x \sin x$
d. $s(t) = \ln(\sin t)$

Answers

a. $f(x) = 4e^{2x} + \sin(3x + 2)$
   $f'(x) = 4(e^{2x})(2) + [\cos(3x+2)](3)$
   $ = 8e^{2x} + 3\cos(3x + 2)$

   Use the constant, multiple and chain rules to differentiate the first term and the chain rule to differentiate the second term.

b. $y = e^x \sin x$
   $y' = e^x(\cos x) + \sin x (e^x)$
   $ = e^x(\cos x + \sin x)$

   Use the product rule.

c. $y = \cos^3 x \sin x$
   $ = (\cos x)^3 \sin x$
   $y' = (\cos x)^3 (\cos x) + \sin x (3(\cos x)^2)(-\sin x)$
   $ = \cos^4 x - 3\cos^2 x \sin^2 x$

   Use the product rule and apply the chain rule when finding the derivative of $(\cos x)^3$.

d. $s(t) = \ln(\sin t)$
   $s'(t) = \dfrac{1}{\sin t}(\cos t) = \dfrac{\cos t}{\sin t}$ or $\dfrac{1}{\tan t}$

   Apply the chain rule.

Exercise 14C

In questions 1–10, find the derivative of each function.

1. $f(x) = 6\cos\left(2x - \dfrac{\pi}{3}\right) + 3x$
2. $y = \dfrac{\sin x}{1 + \cos x}$
3. $f(x) = xe^x - e^x$
4. $s(t) = \dfrac{1}{2}e^{\sin 2t}$
5. $f(x) = e^x(\sin x - \cos x)$
6. $s(t) = t \tan t$
7. $y = e^{3x} \cos 4x$
8. $y = \sqrt{\tan 2x}$
9. $f(x) = (\ln x)(\cos x)$
10. $f(x) = \ln(\cos x)$

EXAM-STYLE QUESTIONS

11. **a** Let $f(x) = \ln(3x^2)$. Write down $f'(x)$.
    b Let $g(x) = \sin\dfrac{x}{2}$. Write down $g'(x)$.
    c Let $h(x) = \ln(3x^2)\sin\dfrac{x}{2}$. Find $h'(x)$.

12. Given that $f(x) = \dfrac{\sin x}{1 + \cos^2 x}$ and $f'(x) = \dfrac{\cos x(1 + a\cos^2 x + b\sin^2 x)}{(1 + \cos^2 x)^2}$, find a and b.

You can use the first and second derivatives of a function to analyze the graph of a function.

> See Chapter 7, Section 7.

Example 4

Consider the function $f(x) = \sin x + \cos x$ for $0 \leq x \leq 2\pi$. Analyze it without using a GDC.
a Find the x- and y-intercepts.
b Find the intervals on which f is increasing and decreasing and the relative extreme points.
c Find the intervals on which f is concave up and concave down and the inflexion points.
d Use the information from parts **a** to **c** to sketch the graph of f.

Answers

a $\sin x + \cos x = 0$
$\sin x = -\cos x$
$x = \dfrac{3\pi}{4}, \dfrac{7\pi}{4}$
x-intercepts: $\dfrac{3\pi}{4}$ and $\dfrac{7\pi}{4}$.

$f(0) = \sin 0 + \cos 0$
$= 0 + 1$
$= 1$
y-intercept: 1

To find the x-intercept, set the function equal to 0 and solve for x. Use your knowledge of unit circle values to find the solutions.

To find the y-intercept, evaluate the function when $x = 0$.

▶ Continued on next page

b $f(x) = \sin x + \cos x$
$f'(x) = \cos x - \sin x$
$\cos x - \sin x = 0$
$\cos x = \sin x$
$f'(x) = 0$ at $x = \dfrac{\pi}{4}, \dfrac{5\pi}{4}$

Increasing: $0 < x < \dfrac{\pi}{4}$ and $\dfrac{5\pi}{4} < x < 2\pi$

Decreasing: $\dfrac{\pi}{4} < x < \dfrac{5\pi}{4}$

Relative maximum point: $\left(\dfrac{\pi}{4}, \sqrt{2}\right)$

Relative minimum point: $\left(\dfrac{5\pi}{4}, -\sqrt{2}\right)$

Find the derivative of f and find where $f'(x) = 0$.
Make a sign diagram for f'.
f is increasing when f' is positive and decreasing when f' is negative.

f'(x) + $\quad$ − $\quad$ +
0 $\quad \dfrac{\pi}{4} \quad \dfrac{5\pi}{4} \quad 2\pi$

The first derivative test tells us that relative extrema occur when the first derivative changes sign.
Evaluate f at $x = \dfrac{\pi}{4}$ and $\dfrac{5\pi}{4}$ to find the maximum and minimum values.

c $f''(x) = -\sin x - \cos x$
$-\sin x - \cos x = 0$
$-\sin x = \cos x$
$x = \dfrac{3\pi}{4}, \dfrac{7\pi}{4}$

Concave up: $\dfrac{3\pi}{4} < x < \dfrac{7\pi}{4}$

Concave down: $0 < x < \dfrac{3\pi}{4}$ and $\dfrac{7\pi}{4} < x < 2\pi$

Inflexion points: $\left(\dfrac{3\pi}{4}, 0\right)$ and $\left(\dfrac{7\pi}{4}, 0\right)$

Find the second derivative of f and find where $f''(x) = 0$.
Make a sign diagram for f''.
f is concave up when f'' is positive and concave down when f'' is negative.

f''(x) $\quad$ − $\quad$ + $\quad$ −
0 $\quad \dfrac{3\pi}{4} \quad \dfrac{7\pi}{4} \quad 2\pi$

Inflexion points occur when the second derivative changes sign. Evaluate f at $x = \dfrac{3\pi}{4}$ and $\dfrac{7\pi}{4}$ to find the y-coordinates of the inflexion points.

d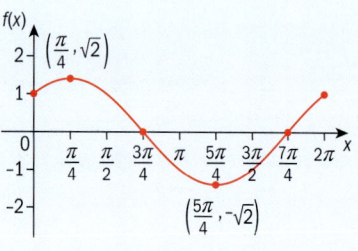

Derivatives are useful for finding both relative extrema and absolute extrema on a closed interval.

> Absolute extrema are sometimes called 'global extrema'.

Example 5

a Show how to use the second derivative test to find the x-coordinates of the relative extrema of $f(x) = \ln x + \sin x$ on $0 \leq x \leq 2\pi$.

b Find the global extrema of the function $f(x) = x + \sin(x^2)$ on the closed interval $0 \leq x \leq \pi$.

Answers

a $f(x) = \ln x + \sin x$

$f'(x) = \dfrac{1}{x} + \cos x$

$\dfrac{1}{x} + \cos x = 0$

$x \approx 2.04, 4.48$

$f''(x) = -\dfrac{1}{x^2} - \sin x$

$f''(2.04) \approx -1.11 < 0 \Rightarrow$ relative maximum at $x = 2.07$

$f''(4.48) \approx 0.925 > 0 \Rightarrow$ relative minimum at $x = 4.49$

Find the first derivative and set it equal to zero to find the critical numbers. Use a GDC to solve.

Find the second derivative and evaluate each of the critical numbers from the first derivative in the second derivative. $f'' > 0$ implies a relative minimum and $f''(x) < 0$ implies a relative maximum.

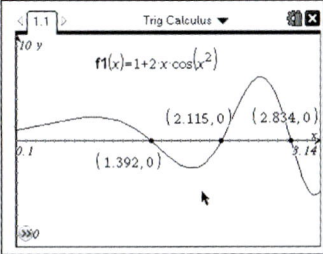

b $f(x) = x + \sin(x^2)$

$f'(x) = 1 + 2x\cos(x^2)$

$1 + 2x\cos(x^2) = 0$

$x = 1.392, 2.115, 2.834$

$f(0) = 0$
$f(1.392) \approx 2.33$
$f(2.115) \approx 1.14$
$f(2.834) \approx 3.82$
$f(\pi) \approx 2.71$

The maximum is 3.82 and the minimum is 0.

Find the first derivative. Set it equal to zero to find the critical numbers. Use a GDC to solve. Evaluate f at the endpoints of the interval and each of the critical numbers from the first derivative. The largest value is the global maximum and the smallest is the minimum.

 GDC help on CD: Alternative demonstrations for the TI-84 Plus and Casio FX-9860GII GDCs are on the CD.

Exercise 14D

Do not use a GDC for questions 1–5.

For questions 1 and 2, find any relative minimum points and relative maximum points of the function on the given interval.

1 $f(x) = \sqrt{3}\sin x + \cos x, \ 0 \leq x \leq 2\pi$

2 $f(x) = 2\sin x + \cos 2x, \ 0 \leq x \leq 2\pi$

For questions 3–4, find the intervals on which the function is increasing, decreasing, concave up and concave down. Find any relative minimum points, relative maximum points and inflexion points. Use this information to sketch a graph of the function.

3 $f(x) = \sqrt{\sin x}, \ 0 \leq x \leq \pi$

4 $f(x) = \cos^2(2x), \ 0 \leq x \leq \pi$

EXAM-STYLE QUESTIONS

5 Let $f(x) = \cos 2x + \cos^2 x$.
  a Show that $f'(x) = -3\sin 2x$.
  b f has one relative minimum point on the interval $0 \leq x \leq \pi$. Find the coordinates of this point.
  c Find $f''(x)$.
  d Find the coordinates of the inflexion point(s) of f on the interval $0 \leq x \leq \pi$.

You may use a GDC for questions 6–8.

6 Let $f(x) = \pi + x \sin x$.
  a i Find $f'(x)$.
   ii $f''(x)$ can be expressed in the form $ax \sin x + b \cos x$. Find a and b.
  b i Solve the equation $f'(x) = 0$ for $0 \leq x \leq 2\pi$.
   ii Hence use $f''(x)$ to identify the x-coordinate of any relative minimum points and any relative maximum points of f for $0 \leq x \leq 2\pi$.

7 Let $f(x) = x^2 \cos x$.
  a Find $f'(x)$.
  b Hence find the global extrema of $f(x) = x^2 \cos x$ on the interval $0 \leq x \leq 5$.

8 The photograph shows the machine that stirs chocolate in *The Chocolate Factory* in Ghirardelli Square, San Francisco. A vat of milk chocolate is stirred by a stirrer blade that is driven by a wheel that pushes the blade back and forth across the bottom of the vat. Suppose that the distance between the center of the wheel and the stirrer blade can be modeled by the function

$$d(\theta) = 2\cos\theta + \sqrt{25 - 4\sin^2\theta}$$

where d is the distance in metres and θ is the angle of rotation of the wheel in radians.
  a Find $d'(\theta)$.
  b Sketch a graph of $d'(\theta)$ for $0 \leq \theta \leq 2\pi$, and label the coordinates of all x-intercepts and relative minimum and maximum points.
  c i Explain how to use the graph of $d'(\theta)$ to determine the angle of rotation when the blade is at the shortest distance from the center of the wheel. What is this angle and this distance?
   ii At which angle(s) of rotation is the distance between the center of the wheel and the blade changing the fastest? Explain how you determine your answer.

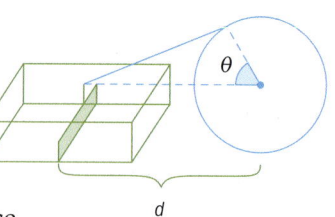

14.3 Integral of sine and cosine

You met these integration rules in Chapter 9.

Power rule: $\int x^n \, dx = \dfrac{1}{n+1} x^{n+1} + C, \, n \neq 1$

Constant rule: $\int k \, dx = kx + C$

Constant multiple rule: $\int kf(x) \, dx = k \int f(x) \, dx$

Sum or difference rule: $\int (f(x) \pm g(x)) \, dx = \int f(x) \, dx \pm \int g(x) \, dx$

Integrals of $\dfrac{1}{x}$ and e^x: $\int \dfrac{1}{x} \, dx = \ln x + C, \, x > 0$

$$\int e^x \, dx = e^x + C$$

Integral with linear composition:

$$\int f(ax+b) \, dx = \dfrac{1}{a} F(ax+b) + C, \text{ where } F'(x) = f(x).$$

These integrals result directly from the derivatives of sine and cosine.

→ **Integrals of sine and cosine**

$$\int \sin x \, dx = -\cos x + C \qquad \int \cos x \, dx = \sin x + C$$

Check:

$\dfrac{d}{dx}(-\cos x) =$
$-(-\sin x) = \sin x$

$\dfrac{d}{dx}(\sin x) = \cos x$

The integrals of the composition of sine or cosine with a linear function are:

→ $\int \sin(ax+b) \, dx = -\dfrac{1}{a} \cos(ax+b) + C$

$\int \cos(ax+b) \, dx = \dfrac{1}{a} \sin(ax+b) + C$

You can use the substitution method to find some integrals or perhaps recognize when you have an integral of the form

$$\int f(g(x)) g'(x) \, dx.$$

Example 6

Find the integrals.

a $\int 3\sin x \, dx$ **b** $\int \cos(4x-6) \, dx$

c $\int e^x \sin(e^x) \, dx$ **d** $\int x^3 \cos(3x^4) \, dx$

Answers

a $\int 3\sin x \, dx = 3\int \sin x \, dx$ — *Use the constant multiple rule and then integrate sine.*
$= 3(-\cos x) + C$
$= -3\cos x + C$

b $\int \cos(4x-6) \, dx = \dfrac{1}{4}\sin(4x-6) + C$ $\quad \int \cos(ax+b) \, dx = \dfrac{1}{a}\sin(ax+b) + C$

c $\int e^x \sin(e^x) \, dx = \int \left(\dfrac{du}{dx}\right) \sin u \, dx$ — *Recognize this as the form*
$\int f(g(x))g'(x) \, dx$ *and write down the answer*
$= \int \sin u \, du$
or let $u = e^x$ and then use $\dfrac{du}{dx} = e^x$. Simplify, integrate
$= -\cos u + C$
and substitute e^x for u.
$= -\cos e^x + C$

d $\int x^3 \cos(3x^4) \, dx = \int \dfrac{1}{12}\left(\dfrac{du}{dx}\right) \times \cos u \, dx$ — *Let $u = 3x^4$ and then $\dfrac{du}{dx} = 12x^3$, so $\dfrac{1}{12}\left(\dfrac{du}{dx}\right) = x^3$.*

$= \dfrac{1}{12}\int \cos u \, du$ — *Simplify and integrate.*

$= \dfrac{1}{12}\sin u + C$

$= \dfrac{1}{12}\sin(3x^4) + C$ — *Substitute $3x^4$ for u.*

Exercise 14E

Find the integrals in questions 1–10.

1 $\int (2\cos x + 3\sin x) \, dx$

2 $\int \left(x^2 + \cos\left(\dfrac{1}{3}x\right)\right) dx$

3 $\int \pi \sin(\pi x) \, dx$

4 $\int \sin(2x+3) \, dx$

5 $\int 20x^3 \cos(5x^4) \, dx$

6 $\int (2x-1)\cos(4x^2-4x) \, dx$

7 $\int \dfrac{e^{\tan(3x)}}{\cos^2(3x)} \, dx$

8 $\int \dfrac{\cos(\ln x)}{x} \, dx$

9 $\int \cos x \sin^2 x \, dx$

10 $\int \dfrac{\sin x}{\cos x} \, dx$, for $\cos x > 0$

EXAM-STYLE QUESTIONS

11 Let $f(x) = e^{\sin x} \cos x$.

    a Find $f'(x)$.

    b Write down $\int f(x)\,dx$.

12 Let $f(x) = \ln(\cos x)$.

    a Show that $f'(x) = -\tan x$.

    b Hence find $\int \tan x \ln(\cos x)\,dx$.

You can use the **Fundamental Theorem of Calculus** to evaluate definite integrals:

> See Section 9.4.

$$\int_a^b f(x)\,dx = [F(x)]_a^b = F(b) - F(a),\text{ where } F \text{ is an antiderivative of } f.$$

Example 7

Evaluate the definite integral without a GDC to get the exact value.
Then check your answer by evaluating the definite integral on a GDC.

a $\displaystyle\int_0^{\frac{\pi}{4}} 2\cos x\,dx$ **b** $\displaystyle\int_{\frac{\pi}{4}}^{\frac{\pi}{2}} \sin(2x)\cos^3(2x)\,dx$

Answers

a $\displaystyle\int_0^{\frac{\pi}{4}} 2\cos x\,dx = 2\int_0^{\frac{\pi}{4}} \cos x\,dx$ *Apply the Fundamental Theorem of Calculus.*

$\qquad = 2[\sin x]_0^{\frac{\pi}{4}}$

$\qquad = 2\left(\sin\dfrac{\pi}{4} - \sin 0\right)$

$\qquad = 2\left(\dfrac{\sqrt{2}}{2} - 0\right)$ *Use unit circle values to evaluate.*

$\qquad = \sqrt{2}$

Using a GDC:

$\displaystyle\int_0^{\frac{\pi}{4}} 2\cos x\,dx \approx 1.41$ and since

$\sqrt{2} \approx 1.41$, our answer is verified.

Look at the investigation in Section 9.3 if you need to review how to enter a definite integral into your calculator.

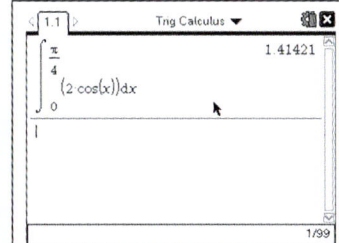

GDC help on CD: *Alternative screenshots for the TI-84 Plus and Casio FX-9860GII GDCs are on the CD.*

▶ Continued on next page

b $\int_{\frac{\pi}{4}}^{\frac{\pi}{2}} \sin(2x)\cos^3(2x)\,dx$ 　　Let $u = \cos(2x)$ and $\frac{du}{dx} = -2\sin(2x)$.

$= \int_{x=\frac{\pi}{4}}^{x=\frac{\pi}{2}} -\frac{1}{2}\left(\frac{du}{dx}\right) u^3\,dx$ 　　Substitute $-\frac{1}{2}\left(\frac{du}{dx}\right)$ for $\sin(2x)$ and u^3 for $\cos^3(2x)$.

$= -\frac{1}{2} \int_{u=0}^{u=-1} u^3\,du$ 　　When $x = \frac{\pi}{4}$, $u = \cos\left(2\left(\frac{\pi}{4}\right)\right) = \cos\frac{\pi}{2} = 0$

$= -\frac{1}{2}\left[\frac{1}{4}u^4\right]_0^{-1}$ 　　When $x = \frac{\pi}{2}$, $u = \cos\left(2\left(\frac{\pi}{2}\right)\right) = \cos\pi = -1$

$= -\frac{1}{8}((-1)^4 - 0)$ 　　Then apply the Fundamental Theorem of Calculus.

$= -\frac{1}{8}$

Using a GDC: $\int_{\frac{\pi}{4}}^{\frac{\pi}{2}} \sin(2x)\cos^3(2x)\,dx = -0.125$ 　　*Evaluate the definite integral on your GDC.*

and since $-\frac{1}{8} = -0.125$, our answer is verified.

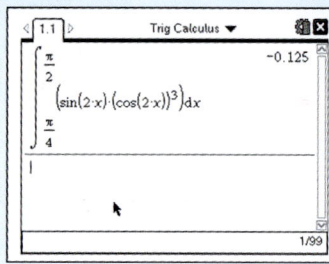

 GDC help on CD: *Alternative screenshots for the TI-84 Plus and Casio FX-9860GII GDCs are on the CD.*

 ## Exercise 14F

Evaluate the definite integral without a GDC to get the exact value.
Then check your answer by evaluating the definite integral on a GDC.

1 $\int_{-\frac{\pi}{3}}^{\frac{\pi}{3}} \cos x\,dx$ 　　　　**2** $\int_0^{\pi} (2\sin x + \sin 2x)\,dx$

3 $\int_0^{\frac{\pi}{2}} \cos\left(\frac{2}{3}x\right) dx$ 　　**4** $\int_{\ln\frac{\pi}{4}}^{\ln\frac{\pi}{3}} e^x \cos(e^x)\,dx$

You can use definite integrals to find area and volume.

→ When the area bounded by the curve $y = f(x)$, the x-axis and the lines $x = a$ and $x = b$ is rotated 360° about the x-axis, the volume of the solid formed is $\int_a^b \pi y^2\,dx$.

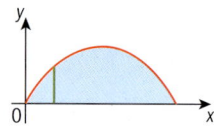

Example 8

A portion of the graph of $f(x) = x \sin x$ is shown in the diagram.
a Find the area of the shaded region.
b Write down the integral representing the volume of the solid formed when the shaded region is rotated 360° about the x-axis.
Hence, find the volume of the solid.

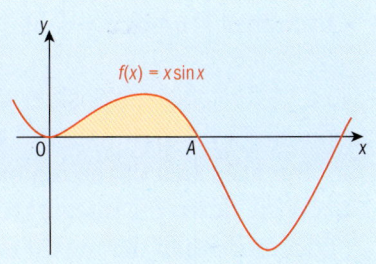

Answers

a $\quad x \sin x = 0$
$x = 0$ or $\sin x = 0$
$x = 0, \pi$
$\int_0^\pi x(\sin x)\,dx \approx 3.14$

Set the function equal to 0 to find the x-coordinates of O and A.

Set up the definite integral and evaluate on a GDC. Notice that the area of this region happens to be π.

b $\quad \pi \int_0^\pi [x(\sin x)]^2\,dx \approx 13.8$

Use $\int_a^b \pi y^2\,dx$ to set up the definite integral and evaluate on a GDC.

You can also find the area between two curves.

→ If $y_1 \geq y_2$ for all x in $a \leq x \leq b$, then $\int_a^b (y_1 - y_2)\,dx$ is the area between the two curves.

Example 9

Find the area of the region in quadrant 1 that is bounded by the curves $y = 0.4x$ and $y = \sin x$.

Answer

Area = $\int_0^{2.125} (\sin(x)) - 0.4x\,dx$

≈ 0.623

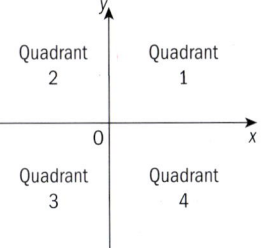

Use a GDC to help sketch a graph and find the points of intersection where $\sin x = 0.4x$.

The area is equal to $\int_a^b (y_1 - y_2)\,dx$ where $a = 0$ and $b \approx 2.125$.
Since $\sin x \geq 0.4x$ for $0 \leq x \leq 2.125$, choose $y_1 = \sin x$ and $y_2 = 0.4x$.

Exercise 14G

In questions 1–2, a region is bounded by the given curves. Use a definite integral to find the area of the region.

1 $y = x \sin x$ and $y = 2x - 6$ in quadrant 1

2 $y = x^2 - 2$ and $y = x + \cos x$

EXAM-STYLE QUESTIONS

3 Given that $\int_0^k \cos x \, dx = \frac{1}{2}$ and $0 \le k \le \frac{\pi}{2}$, find the **exact** value of k.

4 Let $f(x) = \tan \sqrt{x}$. Consider the region in the first quadrant bounded by f, the x-axis and the line $x = 2$.
  a Find the area of the region.
  b Write down the integral representing the volume of the solid formed when the region is rotated 360° about the x-axis. Hence find the volume of the solid.

5 The graph represents the function $f(x) = a \sin(bx)$.
  a Find the values of a and b.
  b Hence find the area of the shaded region.

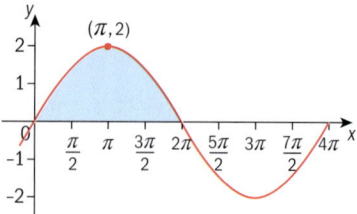

6 The diagram shows part of the graph of $y = \cos x + \sin 2x$. Regions A and B are shaded.
  a i $y = \cos x + \sin 2x$ can be written as $y = \cos x(c + d \sin x)$. Find the values of c and d.
    ii Hence find the **exact** values of the two x-intercepts shown in the diagram.
  b i Find the area of region A.
    ii Find the total area of the shaded regions.
  c Find the volume of the solid formed when region A is rotated 360° about the x-axis.

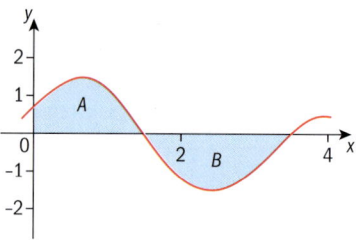

14.4 Revisiting linear motion

> **Extension material on CD:**
> *Worksheet 14 - More trigonometric derivatives and integrals*

Derivatives and integrals are used in kinematics problems involving motion along a straight line.

Suppose that an object is moving along a straight line and that its position from an origin at any time t is given by the displacement function $s(t)$. We then have the following relationships.

Displacement function = $s(t)$

Velocity $v(t) = \dfrac{ds}{dt} = s'(t)$

Acceleration $a(t) = \dfrac{dv}{dt} = v'(t)$ or $s''(t)$

Total distance traveled from time t_1 to t_2 = $\int_{t_1}^{t_2} |v(t)| \, dt$

We will now look at some examples where the linear motion is modeled by trigonometric functions.

> **Remember that…**
> Initially ⇒ at time 0
> At rest ⇒ $v(t) = 0$
> Initially at rest
>   ⇒ $v(0) = 0$
> Moving right or up
>   ⇒ $v(t) > 0$
> Moving left or down
>   ⇒ $v(t) < 0$
> Speed = |velocity|

Example 10

A particle moves along a horizontal line. The particle's displacement, in metres, from an origin O is given by $s(t) = 5 - 2\cos 3t$ for time t seconds.
a Find the particle's velocity and acceleration at any time t.
b Find the particle's initial displacement, velocity and acceleration.
c Find when the particle is moving to the right, to the left and stopped during the time $0 \leq t \leq \pi$.
d Write down a definite integral that represents the total distance traveled for $0 \leq t \leq \pi$ seconds and use a GDC to find the distance.

Answers

a $v(t) = 0 - 2(-\sin 3t)(3)$ $v(t) = s'(t)$
  $= 6\sin 3t$
  $a(t) = 6(\cos 3t)(3)$ $a(t) = v'(t)$
  $= 18\cos 3t$

b $s(0) = 5 - 2\cos(3(0))$ *Evaluate each function at $t = 0$*
  $= 5 - 2(1) = 3\,\text{m}$
  $v(0) = 6\sin(3(0))$
  $= 6(0) = 0\,\text{m s}^{-1}$
  $a(0) = 18\cos(3(0))$
  $= 18(1) = 18\,\text{m s}^{-2}$

c $v(t) = 0$ *The particle is at rest when $v(t) = 0$. The particle moves right when $v(t) > 0$ and left when $v(t) < 0$. A sign diagram is helpful to analyze the motion.*
  $6\sin 3t = 0$
  $\sin 3t = 0$
  $3t = 0, \pi, 2\pi, 3\pi$
  $t = 0, \dfrac{\pi}{3}, \dfrac{2\pi}{3}, \pi$

  The particle is at rest at $0, \dfrac{\pi}{3}, \dfrac{2\pi}{3}$ and π seconds.

  The particle moves right for
  $0 < t < \dfrac{\pi}{3}$ and $\dfrac{2\pi}{3} < t < \pi$
  seconds and left for
  $\dfrac{\pi}{3} < t < \dfrac{2\pi}{3}$ seconds.

d $\displaystyle\int_0^\pi |6\sin 3t|\, dt = 12\,\text{m}$ *The total distance traveled from time t_1 to t_2 is $\displaystyle\int_{t_1}^{t_2} |v(t)|\, dt$. Use a GDC to evaluate the integral.*

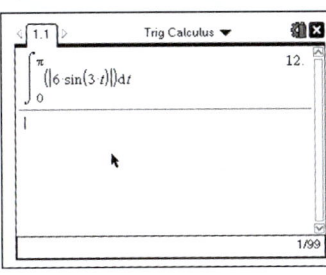

GDC help on CD: Alternative screenshots for the TI-84 Plus and Casio FX-9860GII GDCs are on the CD.

Example 11

A particle moves along a straight line so that its velocity, v m s^{-1} at time t seconds is given by
$v(t) = 5\sin t \cos^2 t$.

a Find the speed of the particle when $t = \dfrac{5\pi}{6}$ seconds.

b When $t = 0$, the displacement, s, of the particle is 3 m.
Find an expression for s in terms of t.

c Find an expression for the acceleration, a, of the particle in terms of t.

Answers

a $v\left(\dfrac{5\pi}{6}\right) = 5\sin\left(\dfrac{5\pi}{6}\right)\cos^2\left(\dfrac{5\pi}{6}\right)$

$= 5\left(\dfrac{1}{2}\right)\left(\dfrac{\sqrt{3}}{2}\right)^2$

$= \dfrac{15}{8}$

Speed $= \left|\dfrac{15}{8}\right| = \dfrac{15}{8}$ m s^{-1}

Velocity has both magnitude and direction, and speed is the magnitude of velocity. Therefore speed = |velocity|.

b $\int 5\sin t \cos^2 t \, dt = \int 5\left(-\dfrac{du}{dt}\right) u^2 \, dt$

$= -5\int u^2 \, du$

$= -5\left(\dfrac{1}{3}u^3\right) + C$

$s(t) = -\dfrac{5}{3}\cos^3 t + C$

$3 = -\dfrac{5}{3}\cos^3(0) + C$

$3 = -\dfrac{5}{3}(1) + C$

$C = \dfrac{14}{3}$

So $s(t) = -\dfrac{5}{3}\cos^3 t + \dfrac{14}{3}$

Integrate velocity to get displacement. Using substitution let $u = \cos t$, then $\dfrac{du}{dt} = -\sin t$ so $-\dfrac{du}{dt} = \sin t$.

Use the fact that $s(0) = 3$ to find C.

c $a(t) = v'(t)$

$= 5\sin t [2(\cos t)(-\sin t)] + \cos^2 t (5\cos t)$

$= -10\sin^2 t \cos t + 5\cos^3 t$

Use the product rule and the chain rule to find the derivative of velocity.

Exercise 14H

Do not use a GDC for question 1–3.

EXAM-STYLE QUESTION

1 A particle moves along a straight line so that its displacement s in metres from an origin O is given by $s(t) = e^t \sin t$ for time t seconds.

  a Write down an expression for the velocity, v, in terms of t.

  b Write down an expression for the acceleration, a, in terms of t.

2 A particle moves along a straight line. The particle's displacement, in metres, from an origin O is given by $s(t) = 1 - 2\sin t$ for time t seconds.
  a Calculate the velocity when $t = 0$.
  b Calculate the value of t for $0 < t < \pi$ when the velocity is zero.
  c Calculate the displacement of the particle from O when the velocity is zero.

3 The velocity v m s^{-1} of a moving body along a horizontal line at time t seconds is given by $v(t) = e^{\sin t} \cos t$.
  a i Find when the particle is at rest during the interval $0 \leq t \leq 2\pi$.
  ii Find when the particle is moving left during the interval $0 \leq t \leq 2\pi$.
  b Find the body's acceleration a in terms of t.
  c The initial displacement s is 4 metres. Find an expression for s in terms of t.

You may use a GDC for questions 4–6.

4 An object starts by moving from a fixed point O. Its velocity v m s^{-1} after t seconds is given by $v(t) = 4\sin t + 3\cos t$, $t \geq 0$. Let d be the displacement from O when $t = 4$.
  a Write down an integral which represents d.
  b Calculate the value of d.

5 A particle moves with a velocity v m s^{-1} given by
$$v(t) = -(t+1)\sin\left(\frac{t^2}{2}\right) \text{ where } t \geq 0.$$
  a i Find the acceleration at time 1.5 seconds.
  ii A particle is speeding up when velocity and acceleration have the same sign and slowing down when the signs are different. Determine whether the particle is speeding up or slowing down at time 1.5 seconds.
  b Find all the times in the interval $0 < t < 4$ that the particle changes direction.
  c Find the total distance traveled by the particle during the time $0 < t < 4$.

6 The velocity, v, in m s^{-1} of a particle moving in a straight line is given by $v(t) = e^{2\sin t} - 1$, where t is the time in seconds for $0 \leq t \leq 12$.
  a Find the acceleration of the particle at $t = 1$.
  b i Sketch a graph of $v(t) = e^{2\sin t} - 1$ for $0 \leq t \leq 12$.
  ii Determine the value(s) of t, for $0 \leq t \leq 12$ where the particle has a velocity of 5 m s^{-1}.
  iii At time $t = 0$ the particle is at the origin. Use the graph of velocity to explain whether or not the particle returns to the origin in the interval $0 \leq t \leq 12$.
  c Find the distance traveled in the 12 seconds.

Review exercise

1. Find the derivative of
   a. $f(x) = \cos(1 - 2x)$
   b. $y = \sin^3 x$
   c. $s(t) = e^{\tan t}$
   d. $f(x) = \sqrt{\sin x^2}$
   e. $f(x) = x^2 \cos x$
   f. $y = \ln(\tan x)$
   g. $f(x) = (\ln x)(\sin x)$
   h. $y = 2 \sin x \cos x$

2. Find the integral of
   a. $\int (4x^3 - \sin x)\,dx$
   b. $\int \cos(3x)\,dx$
   c. $\int \sin(4x+1)\,dx$
   d. $\int x \cos(2x^2)\,dx$
   e. $\int \dfrac{\sin(2t+1)}{\cos^2(2t+1)}\,dt$
   f. $\int \dfrac{\sin(\ln x)}{x}\,dx$
   g. $\int x e^{\sin x^2} \cos x^2\,dx$
   h. $\int \dfrac{6\cos x}{(2+\sin x)^2}\,dx$

3. Evaluate the definite integral of
   a. $\int_{-\frac{\pi}{3}}^{\frac{\pi}{3}} \sin x\,dx$
   b. $\int_0^{\pi} (1 + \sin x)\,dx$
   c. $\int_0^{\pi} (\sin x + \cos 2x)\,dx$
   d. $\int_0^{\frac{\pi}{2}} 5 \sin^{\frac{3}{2}} x \cos x\,dx$

EXAM-STYLE QUESTIONS

4. Find the equation of the normal to the curve with equation $y = \cos(3x - 6)$ at the point $(2, 1)$.

5. Find the coordinates of the point on the graph of $y = \sin\left(\dfrac{x}{2}\right)$, $0 \leq x \leq \pi$, at which the tangent is parallel to the line $y = \dfrac{1}{4}x + 3$.

6. A curve with equation $y = f(x)$ passes through the point $(0, 2)$. Its gradient function is $f'(x) = x - \sin x$. Find the equation of the curve.

7. The graph represents the function $f(x) = p \sin(x) + q$, $p, q \in \mathbb{N}$. Find
   a. the values of p and q
   b. the area of the shaded region.

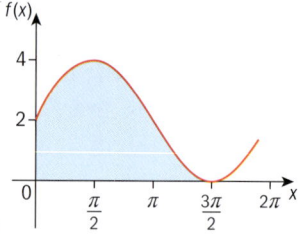

Review exercise

1. A region is bounded by the given curves. Use a definite integral to find the area of the region.
   a. $y = 2\cos^2 x + \cos x + 1$, $x = 0$, $x = 2$ and the x-axis
   b. $y = \sqrt{2 \sin x}$ and $y = 0.5x$

2 A region is bounded by the given curves. Use a definite integral to find the volume of the solid formed when the region is rotated 360° about the x-axis.
  a $y = \sin x$ and the x-axis for $0 \leq x \leq \pi$
  b $y = e^{\cos x}$, $x = 0$ and $x = 2\pi$

EXAM-STYLE QUESTIONS

3 The area under the curve $y = \cos x$ between $x = 0$ and $x = k$, where $0 < k < \frac{\pi}{2}$, is 0.942. Find the value of k.

4 Let $s(t) = 2e^{\cos(5t)} - 4$.
  a **i** Find $s'(t)$.
  ii Show that $s''(t) = 50 e^{\cos(5t)} (\sin^2(5t) - \cos(5t))$.
  iii Hence verify that s has a relative minimum at $t = \frac{\pi}{5}$.
  b s is the displacement function for a particle moving along a straight line, where s is measured in metres and t is in seconds. Find the total distance traveled by the particle from $t = 0$ to $t = 2$ seconds.

CHAPTER 14 SUMMARY
Derivatives of trigonometric functions

- Derivatives of sine, cos and tan:
  $f(x) = \sin x \Rightarrow f'(x) = \cos x$
  $f(x) = \cos x \Rightarrow f'(x) = -\sin x$
  $f(x) = \tan x \Rightarrow f'(x) = \frac{1}{\cos^2 x}$, $\cos x \neq 0$

Integral of sine and cosine

- Integrals of sine and cosine:
  $\int \sin x \, dx = -\cos x + C$
  $\int \cos x \, dx = \sin x + C$
- $\int \sin(ax+b) \, dx = -\frac{1}{a} \cos(ax+b) + C$
  $\int \cos(ax+b) \, dx = \frac{1}{a} \sin(ax+b) + C$
- When the area bounded by the curve $y = f(x)$, the x-axis and the lines $x = a$ and $x = b$ is rotated 360° about the x-axis, the volume of the solid formed is $\int_a^b \pi y^2 \, dx$.
- If $y_1 \geq y_2$ for all x in $a \leq x \leq b$, then $\int_a^b (y_1 - y_2) \, dx$ is the area between the two curves.

Theory of knowledge

From conjecture to proof

The investigation into the derivative of sine in Chapter 14 graphed the derivative of $\sin x$, which led to the conjecture that $\frac{d}{dx}(\sin x) = \cos x$. This was tested with several values and found to be true for these values.

- Does this prove that $\frac{d}{dx}(\sin x) = \cos x$?

Follow these steps to find the derivative of sine using geometry.

- For each step, are you using inductive or deductive reasoning?

STEP ONE:

Here is a unit circle. $Q\hat{O}P = h$ radians.

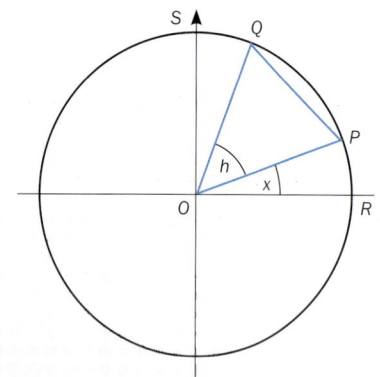

- How you do know that $\triangle QOP$ is isosceles?
- Hence, why is $O\hat{Q}P$ equal to $\frac{\pi - h}{2}$ radians?
- And why is arc QP equal to h?

STEP TWO:

- As h approaches zero, how does the length of arc QP compare to the length of segment QP?

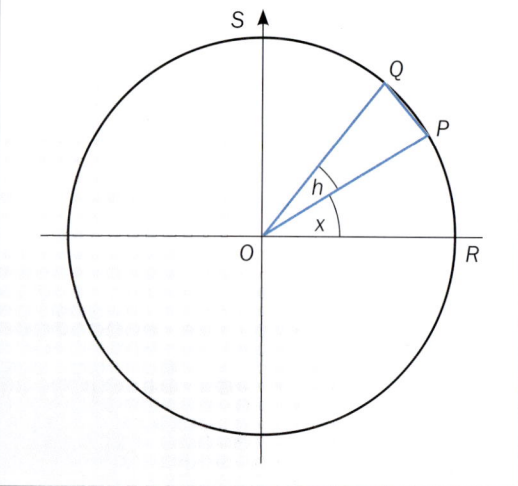

STEP THREE:

- Why is $S\hat{O}Q$ equal to $\frac{\pi}{2} - h - x$?
- Find a line segment parallel to SO.
- Hence, why is $O\hat{Q}A$ also equal to $\frac{\pi}{2} - h - x$?
- Use $O\hat{Q}P$ and $O\hat{Q}A$ to explain why $A\hat{Q}P = \frac{h}{2} + x$.

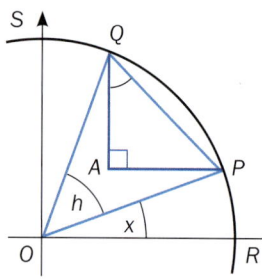

STEP FOUR:

- Why does QA equal $\sin(x+h) - \sin x$?

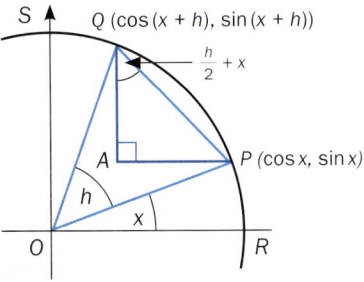

STEP FIVE:

Now show that $\dfrac{d}{dx}(\sin x) = \cos x$.

Explain each line of working:

$$\begin{aligned}\frac{d}{dx}(\sin x) &= \lim_{h \to 0} \frac{\sin(x+h) - \sin x}{h} \\ &= \lim_{h \to 0} \frac{QA}{\text{arc } QP} \\ &= \lim_{h \to 0} \frac{QA}{QP} \\ &= \lim_{h \to 0} \left[\cos\left(\frac{h}{2} + x\right)\right] \\ &= \cos x\end{aligned}$$

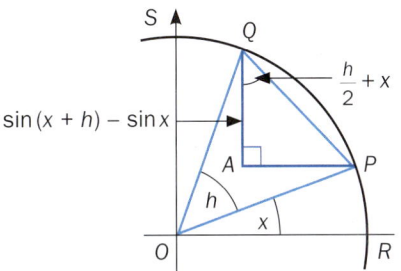

- Which type of reasoning did you use to show that $\dfrac{d}{dx}(\sin x) = \cos x$? Deductive or inductive?
- Explain your answer. Give an example of the other type of reasoning.
- Does this prove that $\dfrac{d}{dx}(\sin x) = \cos x$?

> "Every meaningful mathematical statement can also be expressed in plain language. Many plain-language statements of mathematical expressions would fill several pages, while to express them in mathematical notation might take as little as one line. One of the ways to achieve this remarkable compression is to use symbols to stand for statements, instructions and so on."
>
> Lancelot Hogben (1895–1975)
> English scientist

Mathematical symbols

The underlying concepts of calculus come from investigating limits at infinity. English mathematician John Wallis is credited with introducing the symbol ∞ for infinity.

- Could calculus have developed without the use of mathematical symbols?

▶ John Wallis (1616–1703)

15 Probability distributions

CHAPTER OBJECTIVES:

5.7 Concept of discrete random variables and their probability distributions.
Expected value (mean), E(X) for discrete data.
Applications
5.8 Binomial distribution and its mean and variance.
5.9 Normal distribution and curves.
Standardization of normal variables (z-values)
Properties of the normal distribution

Before you start

You should know how to:

1 Calculate the mean of a set of numbers
e.g. Calculate the mean of this frequency distribution of x.

x	0	1	2	3
Frequency	3	6	9	2

$$\bar{x} = \frac{\sum fx}{\sum f} = \frac{(0 \times 3)+(1 \times 6)+(2 \times 9)+(3 \times 2)}{3+6+9+2}$$
$$= \frac{30}{20} = 1.5$$

2 Use $\binom{n}{r}$ notation

e.g. Evaluate $\binom{5}{2}$

$\binom{5}{2} = \frac{5!}{2!3!} = \frac{5 \times 4}{2} = 10$

3 Solve equations
e.g. Solve the equation $\frac{4}{x} = 3$

$\frac{4}{x} = 3 \quad 4 = 3x \quad x = \frac{4}{3}$

Skills check

1 Calculate the mean of these frequency distributions of x:

a
x	3	4	5	6	7	8
Frequency	3	5	7	9	6	2

b
x	10	12	15	17	20
Frequency	3	10	15	9	2

Repeat question 1 above using your GDC.

2 Evaluate

a $\binom{6}{2}$ b $\binom{8}{5}$ c $\binom{9}{6}(0.3)^3(0.7)^6$

3 Solve these equations

a $\frac{5.5}{x} = 3.2$

b $\frac{x-2.5}{1.2} = 0.4$

c $\frac{9-x}{0.2} = 1.6$

The 2010 Soccer World Cup produced an unlikely celebrity. Paul the octopus correctly predicted the results of 12 out of 14 football matches between 2008 and 2010! Paul lived in a tank in the Sea Life Center in Oberhausen, Germany, and became internationally famous after his feeding behavior was used to predict the winners of a series of soccer matches. Two boxes, each containing a mussel and marked with the flag of one of the national teams in an upcoming match were placed in his tank. His choice of which mussel to eat first was interpreted as predicting that the country with that flag would win.

Paul was right 86% of the time!

This chapter will look at situations like this and how to determine the probability of an event *if it were entirely due to chance*. Perhaps, though, Paul really was able to predict the results of soccer matches!

> Why do people want to believe that someone or something (like an octopus) can predict the future when, rationally, predicting the future seems to be illogical?

15.1 Random variables

→ A **random variable** is a quantity whose value depends on chance.

Random variables are represented by capital letters.

Here are some examples of random variables:
X = the number of sixes obtained when a dice is rolled 3 times.
B = the number of babies in a pregnancy
M = the mass of crisps in a packet
T = the time taken for a runner to complete 100 m.

There are two basic types of random variables:

Discrete random variables – These have a finite or countable number of possible values (e.g. X and B above).

Continuous random variables – These can take on any value in some interval (e.g. M and T above).

Consider the discrete random variable X, the number of sixes obtained when a dice is rolled 3 times. You can write $P(X = x)$ to represent 'the probability that the number of sixes is x' where x can take the values 0, 1, 2 and 3.

> A discrete random variable does not necessarily need to take just positive integer values (e.g. shoe sizes of a set of students could have possible values of …4, 4.5, 5, 5.5, 6, 6.5,…).

> Use capital letters for random variables. Use lower case letters for the actual measured values.

The first random number table was published by Leonard Tippett, an English statistician, in 1927. Tippett took numbers 'at random' from census registers. By 1939 a set of 100 000 digits was published by Maurice Kendall and Bernard Babington Smith using a specialized machine operated by a human. To use such a list, decide the starting point, the number of digits, and the direction (up, down, left, right, diagonal, etc.) before selecting the numbers. For example, starting at the 15th number, on the 5th row going backwards gives 22, 40, 20, 44, 62, …
Most computers and calculators can now be used to generate random numbers. These are in fact pseudo-random numbers: that is, they are numbers which are generated by a mathematical formula but which have the appearance of random numbers.

73735	45963	78134	63873
02965	58303	90708	20025
98859	23851	27965	62394
33666	62570	64775	78428
81666	26440	20422	05720
15838	47174	76866	14330
89793	34378	08730	56522
78155	22466	81978	57323
16381	66207	11698	99314
75002	80827	53867	37797
99982	27601	62686	44711
84543	87442	50033	14021
77757	54043	46176	42391
80871	32792	87989	72248
30500	28220	12444	71840

Probability distributions of discrete random variables

→ A **probability distribution** for a discrete random variable is a list of each possible value of the random variable and the probability that each outcome occurs.

Example 1

Let X be the random variable that represents the number of sixes obtained when a fair dice is rolled three times. Tabulate the probability distribution for X.

Answer

X can take the values 0, 1, 2 and 3

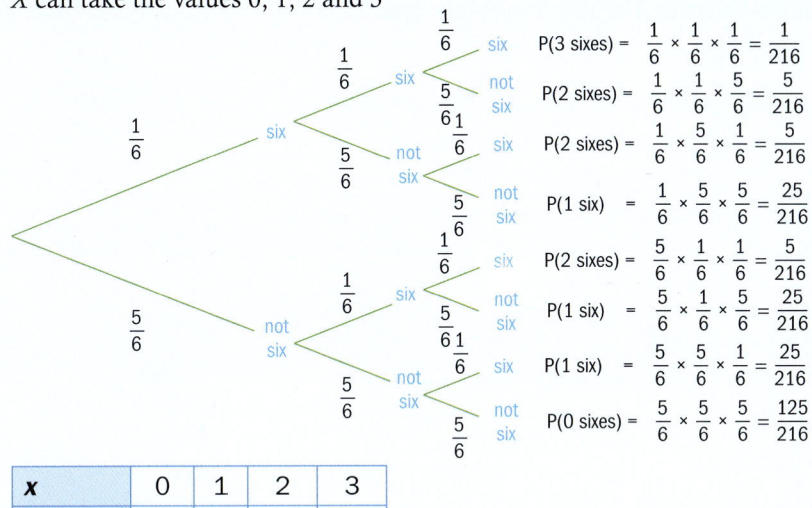

Use a tree diagram to find the values of $P(X = 0)$, $P(X = 1)$, $P(X = 2)$ and $P(X = 3)$

x	0	1	2	3
$P(X = x)$	$\frac{125}{216}$	$\frac{25}{72}$	$\frac{5}{72}$	$\frac{1}{216}$

Write the probabilities in a table.

Notice that in the example the sum of the probabilities is

$$\frac{125}{216} + \frac{25}{72} + \frac{5}{72} + \frac{1}{216} = 1$$

Sometimes $P(X = x)$ is replaced with just $P(x)$ or P_x – these mean the same thing.

→ For any random variable X
$0 \leq P(X = x) \leq 1$ $\quad \sum P(X = x) = 1$

$0 \leq P(X = x) \leq 1$ means that a probability must always be between 0 and 1.
$\sum P(X = x) = 1$ means that the sum of the probabilities will always be 1.

Example 2

The random variable X has the probability distribution

x	1	2	3	4	5
$P(X = x)$	$7c$	$5c$	$4c$	$3c$	c

a Find the value of c **b** Find $P(X \geq 4)$

Answers

a $7c + 5c + 4c + 3c + c = 1$
$20c = 1$
$c = \frac{1}{20}$

Using $\sum P(X = x) = 1$
Solve for c

b $P(X \geq 4) = P(X = 4) + P(X = 5)$
$= \frac{3}{20} + \frac{1}{20} = \frac{4}{20} = \frac{1}{5}$

The solutions of lots of examination questions start with the fact that the total probability must add up to 1.

Exercise 15A

1. Decide whether each random variable is continuous or discrete:
   a A is 'the age in completed years of the next person to call me on my phone'.
   b B is 'the length of the next banana I buy when shopping'.
   c C is 'how many cats I will see before the first white one'.
   d D is 'the diameter of the donuts in the cafeteria'.

2. Tabulate the probability distribution for each random variable:
   a the sum of the faces when two ordinary dice are thrown
   b the number of sixes obtained when two ordinary dice are thrown
   c the smaller or equal number when two ordinary dice are thrown
   d the product of the faces when two ordinary dice are thrown.

3. A fair six-sided dice has a '1' on one face, a '2' on two of its faces and a '3' on the remaining three faces.
   The dice is thrown twice. T is the random variable 'the total score thrown'. Find
   a the probability distribution of T
   b the probability that the total score is more than 4.

   > A fair dice means a dice that is equally likely to land on any of the faces.

4. A board game is played by moving a counter S squares forward at a time, following this rule:
   A fair six-sided dice is thrown once. If the number shown is even, S is half that number.
   If the number shown is odd, S is twice the number shown on the dice.
   a Write out a table showing the possible values of S and their probabilities.
   b What is the probability that in a single go in the game the counter moves more than 2 spaces?

5. The random variable X has the probability distribution

x	1	2	3	4
P(X = x)	$\frac{1}{3}$	$\frac{1}{3}$	c	c

   a Find the value of c.
   b Find $P(1 < X < 4)$.

EXAM-STYLE QUESTION

6. The probability distribution of a random variable Y is given by
   $P(Y = y) = cy^3$ for $y = 1, 2, 3$
   Given that c is a constant, find the value of c.

   > In question 6, $P(Y = y) = cy^3$ This is called a probability function for Y. You can use it to find the probability at various values of the random variable Y.

EXAM-STYLE QUESTIONS

7 The random variable X has this probability distribution.

x	−1	0	1	2
$P(X = x)$	$2k$	$4k^2$	$6k^2$	k

Find the value of k.

8 The random variable X has the probability distribution given by
$$P(X = x) = k\left(\frac{1}{3}\right)^{x-1} \text{ for } x = 1, 2, 3, 4 \text{ and } k \text{ is a constant.}$$
Find the exact value of k.

9 The discrete random variable X can take only the values 0, 1, 2, 3, 4, 5. The probability distribution of X is given by
$$P(X = 0) = P(X = 1) = P(X = 2) = a$$
$$P(X = 3) = P(X = 4) = P(X = 5) = b$$
$$P(X \geq 2) = 3P(X < 2)$$
where a and b are constants.
a Determine the values of a and b.
b Determine the probability that the sum of two independent observations from this distribution exceeds 7.

10 The discrete random variables A and B are independent and have these distributions:

A	1	2	3
$P(A = a)$	$\frac{1}{3}$	$\frac{1}{3}$	$\frac{1}{3}$

B	1	2	3
$P(B = b)$	$\frac{1}{6}$	$\frac{2}{3}$	$\frac{1}{6}$

The random variable C is the sum of one observation from A and one observation from B.
a Show that $P(C = 3) = \frac{5}{18}$.
b Tabulate the probability distribution for C.

Expectation

The **mean** or **expected value** of a random variable X is the average value that we should expect for X over many trials of the experiment.

The mean or expected value of a random variable X is represented by $E(X)$.

> Expectation is actually the mean of the underlying distribution (the parent population). It is often denoted by μ.

Investigation – dice scores

Two dice are rolled together and the difference, D, between the scores on the dice is noted.

1. Copy and complete the probability distribution for D.

d	0	1	2	3	4	5
$P(D=d)$		$\frac{10}{36}$				

2. This experiment is repeated 36 times. Copy and complete the following table to show the expected frequency of obtaining each of the different values of d.

d	0	1	2	3	4	5
Expected frequency		10				

> You may find it helpful to draw a sample space diagram like the ones in Chapter 3.

3. Calculate the mean of this frequency distribution.
4. The original experiment is repeated 100 times. Repeat questions 2 and 3 for this situation:

d	0	1	2	3	4	5
Expected frequency		$\frac{250}{9}$				

5. What do you notice?
6. What would the mean be if the experiment were repeated 10 times? Or 1000 times? Or just once?

We would expect the mean to be the same in each case. Therefore we can find the mean or expected value of the random variable D by just multiplying each value of d by its respective probability (the equivalent of conducting the experiment just once).

→ The expected value of a random variable X is
$E(X) = \sum x\, P(X = x)$.

Example 3

Here is the probability distribution from Example 1:

x	0	1	2	3
$P(X=x)$	$\frac{125}{216}$	$\frac{25}{72}$	$\frac{5}{72}$	$\frac{1}{216}$

What is the expected value of X?

Answer

Using the formula:

$E(X) = \left(0 \times \frac{125}{216}\right) + \left(1 \times \frac{25}{72}\right)$
$\qquad + \left(2 \times \frac{5}{72}\right) + \left(3 \times \frac{1}{216}\right)$

$E(X) = \frac{1}{2}$

Use $E(X) = \sum x\, P(X = x)$

Therefore if 3 dice are rolled a large number of times, you should expect the mean number of sixes to be 0.5.

▶ Continued on next page

Using a GDC:

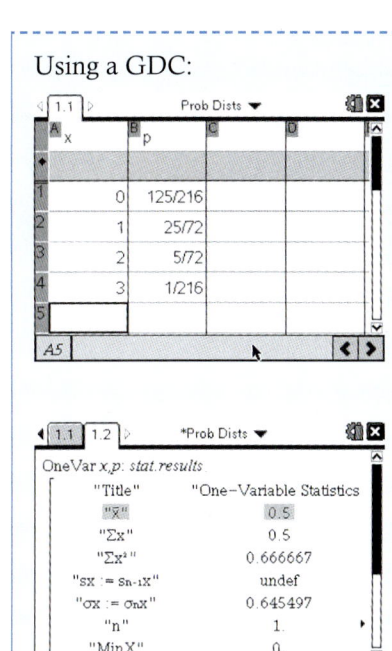

Enter the list of possible x-values in x and the set of corresponding probability values $P(X = x)$ in p.

GDC help on CD: *Alternative demonstrations for the TI-84 Plus and Casio FX-9860GII GDCs are on the CD.*

Now use One-Variable Statistics as when finding the mean of a data set. Use x as the X1 List and p as the Frequency List.

Note that the expected value of X does not need to be a value of X that is actually obtainable.

For more on entering data, see Chapter 17 Sections 5.1 and 5.2

$E(X) = \bar{x} = 0.5$

Exercise 15B

1 When throwing a standard six-sided dice, let X be the random variable defined by $X =$ the square of the score shown on the dice. What is the expectation of X?

EXAM-STYLE QUESTION

2 The random variable Z has probability distribution

z	2	3	5	7	11
P(Z = z)	$\frac{1}{6}$	$\frac{1}{6}$	$\frac{1}{6}$	x	y

and $E(Z) = 5\frac{2}{3}$

Find x and y.

3 A 'Fibonacci dice' is unbiased, six-sided and labeled with these numbers: 1, 2, 3, 5, 8, 13.
What is the expected score when the dice is rolled?

4 The discrete random variable X has probability distribution
$p(x) = \frac{x}{36}$ for $x = 1, 2, 3, \ldots, 8$
Find $E(X)$.

EXAM-STYLE QUESTIONS

5 For the discrete random variable P, the probability distribution is given by
$$P(X = x) = \begin{cases} kx & x = 1, 2, 3, 4, 5 \\ k(10-x) & x = 6, 7, 8, 9 \end{cases}$$
Find
a the value of the constant k **b** $E(X)$

6 a Copy and complete, in terms of k, this probability distribution for a discrete random variable, X:

x	1	2	3
$P(X = x)$	0.2	1–k	

b What range of values can k take? Give your answer in the form $a \leq k \leq b$, $a, b \in Q$
c Find in terms of k the mean of the distribution.

7 X is a discrete random variable which can only take the three values 1, 2 and 4.
It is known that $P(X = 2) = 0.3$ and that the mean of the distribution is 2.8.
Find $P(X = 1)$.

8 Ten balls are in a bag. They are all identical sizes but two of them are red and the rest are blue. Balls are picked out at random from the bag and are not replaced.
Let R be the number of balls drawn out up to and including the first red one.
a List the possible values of R and their associated probabilities.
b Calculate the mean value of R.
c What is the most likely value of R?

9 Ten balls are in a bag as in question 8 above. Again, balls are picked at random, but, instead, each ball is replaced before the next is drawn.
a Show that the probability that the first red is drawn out on the second go is $\frac{4}{25}$.
b Calculate the probability that the first red is drawn after the third go.
c Derive a formula to find the probability that the first red is drawn on the nth go.
d What is the most likely value of R?

EXAM-STYLE QUESTION

10 An instant lottery ticket is purchased for $2. The possible prizes are $0, $2, $20, $200 and $1000. Let Z be the random variable representing the amount won on the ticket. Z has the distribution:

z	0	2	20	200	1000
P(Z)		0.2	0.05	0.001	0.0001

a Determine $P(Z = 0)$.
b Determine $E(Z)$ and interpret its meaning.
c How much should you expect to gain or lose on average per ticket?

15.2 The binomial distribution

Definition of a binomial distribution

Investigation – the binomial quiz

Here are five questions. The answer to each question is either 'true' or 'false'. Write down the answer to each question.
You may need to guess the answer to some of them!

1 The bulletproof vest was created by a woman.
2 On average, the 'eye muscles' move 300 000 times a day.
3 Bowling was first played in Italy.
4 It took Leonardo da Vinci ten days to paint Mona Lisa's lips.
5 Selachophobia is the fear of light flashes.

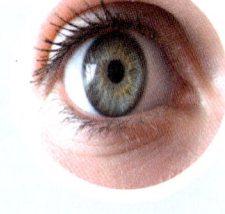

Now look in the Answers at the back of the book to find the correct answers to these five questions. How many did you get right?
Is this a good score? How many would you expect to get right if you were just guessing the answer to every question?

To pass this 'test' you need to get 3 correct out of 5.
What is the probability that you get exactly 3 right out of 5?

→ The essential elements of a binomial distribution are:
  • There is a fixed number (n) of trials.
  • Each trial has only two possible outcomes – a 'success' or a 'failure'.
  • The probability of a success (p) is constant from trial to trial.
  • Trials are independent of each other.

• In the quiz above there are 5 trials.
• Here, success is getting the question right and failure is getting the question wrong.
• In this case the probability of success is 0.5 assuming you obtained every answer by guessing.
• If you get the answer to one question right, it does not mean you are more or less likely to get the answer to the next question right.

The outcomes of a **binomial experiment** and the corresponding probabilities of these outcomes are called a **binomial distribution**.

The **binomial distribution** describes the behavior of a discrete variable X if the conditions above apply.

> → The parameters that define a unique binomial distribution are the values of n (the number of trials) and p (the probability of a success). Any binomial distribution is represented as $X \sim B(n, p)$

Consider this problem, which you first met in Chapter 3: determine the probability of getting exactly two heads in three tosses of a biased coin for which $P(\text{head}) = \frac{2}{3}$.

You could use a tree diagram to help you answer this question.

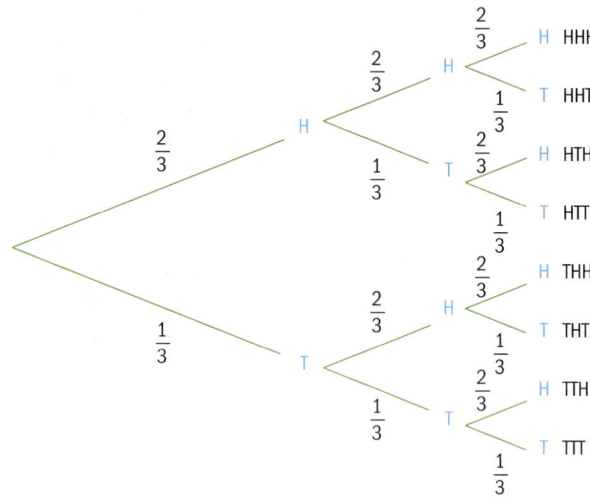

$P(\text{two heads in three tosses}) = P(HHT) + P(HTH) + P(THH)$

Each of the three probabilities are the same.

$P(HHT) = P(HTH) = P(THH) = \left(\frac{2}{3}\right)^2 \left(\frac{1}{3}\right) = \frac{4}{27}$

And so $P(\text{two heads in three tosses}) = 3\left(\frac{2}{3}\right)^2 \left(\frac{1}{3}\right) = \frac{12}{27} = \frac{4}{9}$

However, you should only use a tree diagram if the number of trials, n, is small.

What if you were asked to find the probability of obtaining exactly two heads in six tosses of this coin? The tree diagram for this question would be too large, so we will look for a formula.

> We often use a theoretical distribution, such as the binomial distribution, to describe a random variable that occurs in real life. This process is called *modeling* and enables us to carry out calculations. If the theoretical distribution matches the real-life variable perfectly, then the model is perfect. However, this is usually not the case. Generally the results of any calculations will not necessarily give a completely accurate description of the real-life situation. Does this make them any less useful?

First note that the conditions for a binomial distribution have been met:

• There is a fixed number (*n*) of trials.	In this case there are six trials.
• Each trial has two possible outcomes – a 'success' or a 'failure'.	Here a success is getting a head and a failure is getting a tail.
• The probability of a success (*p*) is constant from trial to trial.	The probability of a success is $\frac{2}{3}$ each time the coin is tossed.
• Trials are independent of each other.	Getting a head on one trial will not affect the outcome of the next trial.

One combination of Hs and Ts that will produce 2 heads and 4 tails is HHTTTT

And P(HHTTTT) = $\left(\frac{2}{3}\right)^2 \left(\frac{1}{3}\right)^4 = \frac{4}{729} (= 0.00548...)$

And every possible combination of 2 Hs and 4 Ts will have the same probability.

But how many combinations are there?

$\binom{n}{r}$ represents the number of ways of choosing *r* items out of *n* items.

The number of combinations of 6 items that have 2 Hs and 4 Ts is therefore $\binom{6}{2} = \binom{6}{4} = 15$

You can use your GDC to calculate $\binom{6}{2}$

> The most usual error when calculating a binomial probability is to forget that if there are exactly *r* successes, there must also be *n* – *r* failures.

> For more on the binomial expansion, see Chapter 6.

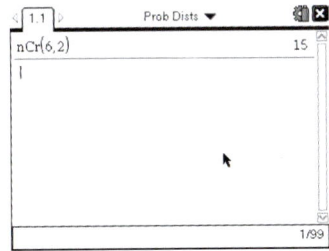

GDC help on CD: Alternative demonstrations for the TI-84 Plus and Casio FX-9860GII GDCs are on the CD.

Instead, you could use the formula $\binom{6}{2} = \frac{6!}{2!4!} = \frac{6 \times 5}{2} = 15$

or

the 3rd entry on the 6th row of Pascal's triangle:

1   6   ⑮   15   6   1

Therefore

P(2 heads in 6 tosses) = $\binom{6}{2}\left(\frac{2}{3}\right)^2\left(\frac{1}{3}\right)^4 = 15 \times \frac{4}{729} = \frac{20}{243} = 0.0823$ (3 sf)

Generalizing this method gives the binomial distribution function:

> If X is binomially distributed, $X \sim B(n, p)$, then the probability of obtaining r successes out of n independent trials, when p is the probability of success for each trial, is
>
> $$P(X = r) = \binom{n}{r} p^r (1-p)^{n-r}$$
>
> This is often shortened to
>
> $$P(X = r) = \binom{n}{r} p^r q^{n-r} \text{ where } q = 1 - p$$

Example 4

X is binomially distributed with 6 trials and a probability of success equal to $\frac{1}{5}$ at each attempt. What is the probability of

a exactly four successes
b at least one success?
c three or fewer successes?

Answers

By hand:

a $P(X = 4) = \binom{6}{4}\left(\frac{1}{5}\right)^4\left(\frac{4}{5}\right)^2$

$= 15 \times \frac{1}{625} \times \frac{16}{25}$

$= \frac{48}{3125}$

$= 0.01536$

$= 0.0154 \, (3 \, \text{sf})$

b $1 - \left(\frac{4}{5}\right)^6$

$= 1 - \frac{4096}{15\,625}$

$= \frac{11529}{15\,625}$

$= 0.738 \, (3\,\text{sf})$

c $P(X \le 3) = 0.983$

You can rewrite the question as

If $X \sim B\left(6, \frac{1}{5}\right)$, what is

a $P(X = 4)$
b $P(X \ge 1)$
c $P(X \le 3)$

Use $P(X = r) = \binom{n}{r} p^r q^{n-r}$

For $P(X \ge 1)$ it is quicker to calculate $1 - P(X = 0)$ than to calculate $P(X = 1) + P(X = 2) + \dots + P(X = 6)$

$P(X \le 3) = P(X = 0) + P(X = 1) + P(X = 2) + P(X = 3)$
Use your GDC for this calculation (see the following).

▶ Continued on next page

> It is easy to confuse $P(X < r)$ and $P(X \le r)$ so read the questions carefully.

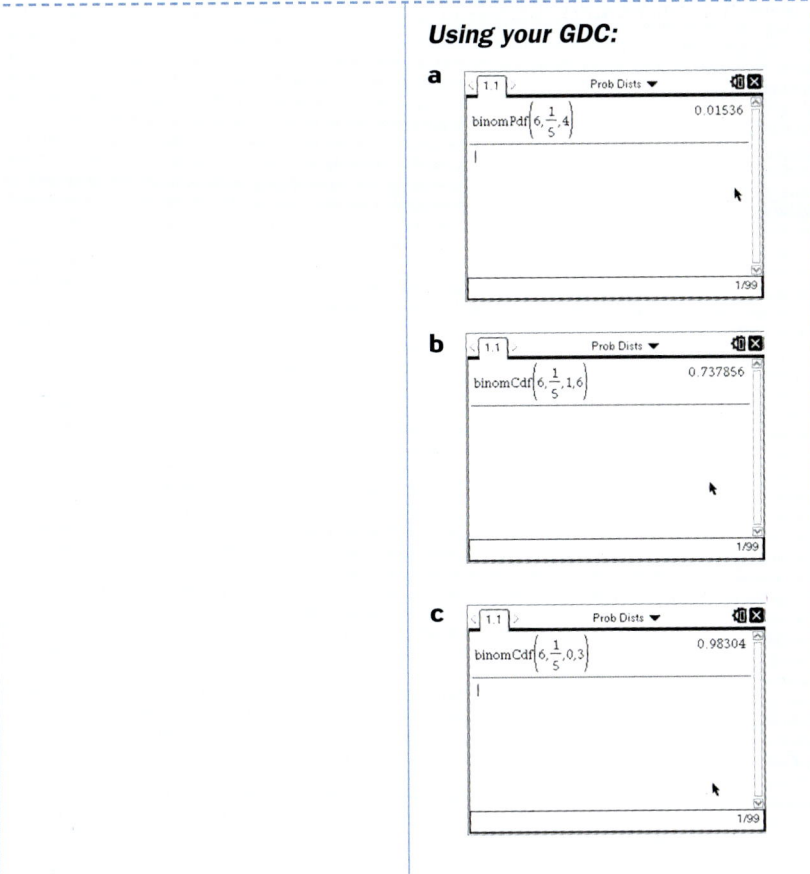

Exercise 15C

1 X is binomially distributed with 4 trials and a probability of success equal to $\frac{1}{2}$ on each trial.

Without a calculator determine the probability of

a $P(X = 1)$ **b** $P(X < 1)$
c $P(X \leq 1)$ **d** $P(X \geq 1)$

2 If $X \sim B\left(6, \frac{1}{3}\right)$ find to 3 significant figures

a $P(X = 2)$ **b** $P(X < 2)$
c $P(X \leq 2)$ **d** $P(X \geq 2)$

In question 2 **b**, **c**, and **d** use Binomcdf instead of Binompdf on the calculator, as you are calculating a cumulative probability.

3 If X is binomially distributed with 8 trials and a probability of success equal to $\frac{2}{7}$ at each attempt, what is the probability of

a exactly 5 successes **b** less than 5 successes
c more than 5 successes **d** at least one success?

Example 5

The probability that I get a bus to work on any morning is 0.4. What is the probability that in a working week of five days I will get a bus only twice?

Answer
By hand:
Let X be the number of days I get a bus.
$X \sim B(5, 0.4)$

$$P(X = 2) = \binom{5}{2}(0.4)^2 (0.6)^3$$
$$= 10 \times 0.16 \times 0.216$$
$$= 0.3456$$
$$= 0.346 \, (3\,\text{sf})$$

Can you see why this is a binomial situation?

We require P(X = 2)

Using your GDC:

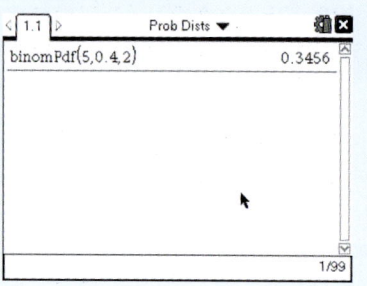

See Chapter 17, Section 5.12

GDC help on CD: *Alternative demonstrations for the TI-84 Plus and Casio FX-9860GII GDCs are on the CD.*

Example 6

When administering a drug it was known that 80% of people using it were cured. The testing program administered the drug to two groups of 10 patients.
What is the probability that all 10 patients were cured in both groups?

Answer

Let X be 'the number of patients cured in a group of 10'.
$P(X = 10) = 0.8^{10} = 0.10737 \ldots$
$[P(X = 10)]^2 = (0.10737\ldots)^2$
$= 0.0115 \, (3\,\text{sf})$

Multiply the probabilities P(X = 10) and P(X = 10) because the two events (the patients being cured in each group) are independent. So for two groups of 10 patients, the probability all are cured is [P(X = 10)]²

Assume that X is binomial since there are two outcomes: a success is a cure and a failure is 'not a cure'. Assume that the trial results from patient to patient are independent. The fixed probability of success is 0.8.

Exercise 15D

1. A regular tetrahedron has three white faces and one red face. It is rolled four times and the color of the bottom face is noted. What is the most likely number of times that the red face will end downwards? What is the probability of this value occurring?

2. The probability that a marksman scores a bull's eye when he shoots at a target is 0.55.
   Find the probability that in eight attempts
   a he **hits** the bull five times
   b he **misses** the bull at least five times.

EXAM-STYLE QUESTIONS

3. A factory has four machines making the same type of component. The probability that any machine will produce a substandard component is 0.01. What is the probability that, in a sample of four components from each machine,
   a none will be faulty   b exactly 13 will be not be faulty
   c at least two will be faulty?

4. The probability that a telephone line is engaged at a company switchboard is 0.25. If the switchboard has 10 lines, find the probability that
   a one half of the lines are engaged
   b at least three lines are free (to 4 significant figures)

5. The probability that Nicole goes to bed at 7:30 on a given day is 0.4. Calculate the probability that on five consecutive days she goes to bed at 7:30 on at most three days.

6. In an examination hall, it is known that 15% of desks are wobbly.
   a What is the probability that, in a row of six desks, more than one will be wobbly?
   b What is the probability that exactly one will be wobbly in a row of six desks?

7. In the mass production of computer processors it is found that 5% are defective. Processors are selected at random and put into packs of 15.
   a A packet is selected at random. Find the probability that it will contain
      i   three defective processors   ii no defective processors
      iii at least two defective processors
   b Two packets are selected at random. Find the probability that there are
      i   no defective processors in either packet
      ii at least two defective processors in either packet
      iii no defective processors in one packet and at least two in the other.

Example 7

A box contains a large number of carnations of which one-quarter are red. The rest are white. Carnations are picked at random from the box. How many flowers must be picked so that the probability that there is at least one red carnation among them is greater than 0.95?

Answer

Let X be the random variable 'the number of red carnations'.
$X \sim B(n, 0.25)$

$\frac{1}{4}$ are red, so $P(\text{red}) = 0.25$

$P(X \geq 1) = 1 - P(X = 0)$
$\qquad\quad\; = 1 - (0.75)^n$

$1 - (0.75)^n > 0.95$
$\quad\; 0.05 > (0.75)^n$

We require $P(X \geq 1) > 0.95$

$\log 0.05 > n \log 0.75$
and so
$n \log 0.75 < \log 0.05$
$\qquad n > \dfrac{\log 0.05}{\log 0.75}$

Solve the inequality for n.

> When you divide by a negative amount, the inequality reverses.

$\qquad n > 10.4$

The least value of n is 11.

At least 11 carnations must be picked out of the box to ensure that the probability that there is at least one red carnation among them is greater than 0.95.

Exercise 15E

1. If $X \sim B(n, 0.6)$ and $P(X < 1) = 0.0256$, find n.

2. 1% of fuses in a large box of fuses are faulty. What is the largest sample size that can be taken if the probability that there are no faulty fuses in the sample must be greater than 0.5?

3. If $X \sim B(n, 0.2)$ and $P(X \geq 1) > 0.75$, find the least possible value of n.

4. The probability that Anna scores a penalty goal in a hockey competition is 0.3. Find the least number of attempts that she would need to take if the probability that a goal is scored at least once is greater than 0.95.

5. How many times must an unbiased coin be tossed so that the probability that at least one tail will occur is at least 0.99?

Expectation of a binomial distribution

Think of the example of the biased coin where $P(H) = \frac{2}{3}$.
If you toss the coin 3 times how many times would you expect to get a head?

Intuitively the answer is 2.

This is the same as calculating $3 \times \frac{2}{3} = 2$

> → For the binomial distribution where $X \sim B(n, p)$, the expectation of X, $E(X) = np$.

The proof of this formula is not on the standard level syllabus.

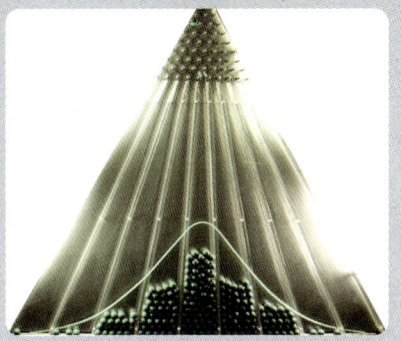

The Galton Board, quincunx or bean machine, is a device for statistical experiments named after English scientist Sir Francis Galton. It consists of an upright board with evenly spaced nails driven into its upper half, where the nails are arranged in staggered order, and a lower half divided into a number of evenly-spaced rectangular slots. In the middle of the top, there is a funnel into which balls can be poured. The funnel is directly above the top nail so that each ball falls directly on to this nail.
Each time a ball hits one of the nails, it can bounce right or left with equal probability.
This process therefore gives rise to a binomial distribution in the heights of heaps of balls in the slots at the bottom.
If the number of balls is sufficiently large then the distribution of the heights of the ball heaps will approximate a normal distribution (See Section 15.3). You may wish to investigate further why this is.

Example 8

A biased dice is thrown 30 times and the number of sixes seen is 8. The dice is then thrown a further 12 times. Find the expected number of sixes for these 12 throws.

Answer

$X \sim B(12, p)$ where $p = \frac{8}{30} = \frac{4}{15}$

$E(X) = np = 12 \times \frac{4}{15} = 3.2$

Let X be 'the number of sixes seen in 12 throws'.

You may wish to conduct your own binominal experiment and explore how close your results are to the expected binomial results.

Exercise 15F

1 a A fair coin is tossed 40 times. Find the expected number of heads.
  b A fair dice is rolled 40 times. Find the expected number of sixes obtained.
  c A card is drawn from a pack of 52 cards, noted and returned. 13 of these cards are labeled as Hearts. This is repeated 40 times. Find the expected number of Hearts.

EXAM-STYLE QUESTIONS

2 X is a random variable such that $X \sim B(n, p)$. Given that the mean of the distribution is 10 and $p = 0.4$, find n.

3 A multiple choice test has 15 questions and four possible answers for each one with only one correct answer per question. Assume a student guesses each answer.
If X is 'the number of questions answered correctly' give:
  a the distribution of X
  b the mean of X
  c the probability that a student will achieve the pass mark of 10 or more purely by guessing.

4 100 families each with three children are found to have these numbers of girls:

Number of girls	0	1	2	3
Frequency	13	34	40	13

  a Find the probability that a single baby born is a girl.
  b Using your value from **a** calculate the number of families with three children, in a sample of 100, you would expect to have two girls.

Variance of a binomial distribution

Chapter 8 introduced the concept of the variance of a set of data, as a measure of dispersion.

> The proof of the variance formula is not on the Standard Level syllabus.

The formula for the variance of the binomial distribution is given in the formula booklet:

> → If $X \sim B(n, p)$ then $\text{Var}(X) = npq$ where $q = p - 1$

Thinking about the original example of the biased coin where $P(H) = \frac{2}{3}$, if you toss the coin 3 times you expect to get a head 2 times. However, obviously this will not happen every time. If you repeat this experiment many times you will sometimes get 0, 1 and 3 heads.

Using the formula for variance,

$$\text{Variance} = npq = 3 \times \frac{2}{3} \times \frac{1}{3} = \frac{2}{3}$$

In general

> You can find the standard deviation, σ, by taking the square root of the variance.

> → For the binomial distribution where $X \sim B(n, p)$
>   • expectation of X, $E(X) = np$
>   • variance of X, $\text{Var}(X) = npq$ where $q = 1 - p$.

> The expected value of X, $E(X)$, is also called the mean, μ, so $E(X) = \mu$.

Example 9

In a large company, 40% of the workers travel to work on public transport.
A random sample of 15 workers is selected.
Find the expected number of workers in this sample that travel to work on public transport, and the standard deviation.

Answer

Let W be the number of workers who travel to work on public transport.	$n = 15, p = 0.4$
$W \sim B(15, 0.4)$	
$E(W) = 15 \times 0.4 = 6$	$E(W) = np$
$Var(W) = 15 \times 0.4 \times 0.6 = 3.6$	$Var(W) = npq$
Standard deviation is $\sqrt{3.6} = 1.90 (3\,sf)$	Standard deviation is square root of variance

Exercise 15G

1 If $X \sim B\left(0, \dfrac{1}{4}\right)$, calculate the mean and variance of X.

2 Find the mean and standard deviation of the binomial distribution $B(12, 0.6)$.

3 A fair coin is tossed 40 times. Find the mean and standard deviation of the number of heads.

4 An unbiased dice is thrown 10 times. Let X be the number of sixes obtained. Find
   a the expected number of sixes
   b $Var(X)$
   c $P(X < \mu)$.

EXAM-STYLE QUESTION

5 A frequent flyer finds that she is delayed at a particular airport once in every 5 trips, on average. One year she uses the airport on 22 occasions. Using a binomial model, find
   a the expected number of journeys that will be delayed at that airport
   b the variance
   c the probability that she is delayed on fewer than 4 occasions.

6 At the local athletics club, the expected number of people that can run 100 metres in under 13 seconds is 4.5 and the variance is 3.15.
Find the probability that at least 3 people can run 100 m in under 13 seconds.

EXAM-STYLE QUESTION

7 X is a random variable such that $X \sim B(n, p)$. Given that the mean of the distribution is 7.8 and $p = 0.3$ find:
  a n
  b the variance of X.

8 For a random variable $X \sim B(n, p)$, $E(X) = 9.6$ and $Var(X) = 1.92$.
Find the possible values of n and p.
Hence calculate $P(X = 6)$ for each possible pair.

15.3 The normal distribution

Investigation – normal distribution

Collect data from around 50 students in your school for one of these categories: height, weight, maximum hand span, length of foot, circumference of wrist.
1 Draw a histogram of the data
2 Where is the peak of the histogram?
3 Is the histogram roughly symmetrical?
4 Join the midpoints of the tops of the bars of your histograms with a curve.

Your histogram from the investigation is probably roughly symmetrical and the curve is bell-shaped with the majority of measurements around a central value.

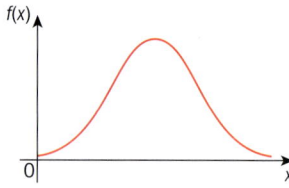

If more measurements were taken, a histogram plotted and the midpoints of the tops of the bars joined with a curve, then it would become more symmetrical and bell-shaped until it would look like the curve shown. This is a **normal distribution**.

The normal distribution is probably the most important distribution in statistics, since it is a suitable model for many naturally occurring variables. These include the physical attributes of people, animals and plants, and even mass-produced items from factories. The distribution could also be applied as an approximation of, for example, student exam scores, times to complete a piece of work, reaction times, or IQ scores.
In each case

- the curve is bell-shaped
- it is symmetrical about the mean (μ)
- the mean, mode and median are the same.

The Gaussian Curve
The normal curve is also called the "Gaussian curve" after the German mathematician Carl Friedrich Gauss (1777–1855). Gauss used the normal curve to analyze astronomical data in 1809. In Germany, the portrait of Gauss, the normal curve and its probability function appeared on the old 10-Deutschmark note.
   Although Gauss played an important role in its history, the French statisticians Abraham de Moivre (1667–1754) and Pierre-Simon Laplace (1749–1827) were involved in much of the early work. De Moivre developed the normal curve mathematically in 1733 as an approximation to the binomial distribution, although his paper on this was not discovered until 1924 by Karl Pearson. Laplace used the normal curve in 1783 to describe the distribution of errors, and in 1810 he proved an essential statistical theorem called the Central Limit Theorem.

The characteristics of any normal distribution

There is no single normal curve, but a **family of curves**, each one defined by its mean, μ, and standard deviation, σ.

→ If a random variable, X, has a normal distribution with mean μ and standard deviation σ, this is written $X \sim N(\mu, \sigma^2)$

Remember that the mean, μ, is the average, and the standard deviation, σ, is a measure of spread.

μ and σ are called the **parameters** of the distribution.

The mean is the central point of the distribution and the standard deviation describes the spread of the distribution. The higher the standard deviation, the wider the normal curve will be.

Note that in the expression $X \sim N(\mu, \sigma^2)$, σ^2 is the variance. Remember that the variance is the standard deviation squared.

These three graphs show $X_1 \sim N(5, 2^2)$, $X_2 \sim N(10, 2^2)$ and $X_3 \sim N(15, 2^2)$. The standard deviations are all the same, so the curves are all the same width but $\mu_1 < \mu_2 < \mu_3$.

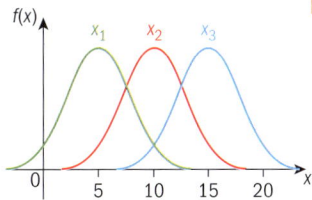

These three graphs show $X_1 \sim N(5, 1^2)$, $X_2 \sim N(5, 2^2)$ and $X_3 \sim N(5, 3^2)$. Here the means are all the same and all the curves are centered around this but $\sigma_1 < \sigma_2 < \sigma_3$ so curve X_1 is narrower than X_2, and X_2 is narrower than X_3.

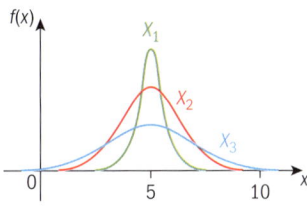

The curves may have different means and/or different standard deviations but they all have the same characteristics.

The area beneath the normal distribution curve

No matter what the values of μ and σ are for a normal probability distribution, the total area under the curve is always the same and equal to 1. We can therefore consider partial areas under the curve as representing probabilities.

So in this normal distribution we could find the probability $P(X < 5)$ by finding the shaded area on the diagram.

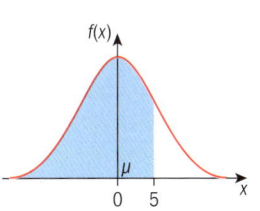

Unfortunately the probability function (the equation of the curve) for the normal distribution is very complicated and difficult to use.

$$f(X) = \frac{1}{\sqrt{2\pi}\sigma} e^{\left(\frac{-(X-\mu)^2}{2\sigma^2}\right)} \quad -\infty < X < \infty$$

It would be too hard for us to use integration to find areas under this curve! However, there are other methods we can use.

The standard normal distribution

The **standard normal distribution** is the normal distribution where $\mu = 0$ and $\sigma = 1$. The random variable is called Z. It uses 'z-values' to describe the number of standard deviations any value is away from the mean.

→ The standard normal distribution is written $Z \sim N(0, 1)$

You can use your GDC to calculate the areas under the curve of $Z \sim N(0, 1)$ for values between a and b and hence $P(a < Z < b)$.

> Note that the $P(Z = a) = 0$. You can think of this as a line having no width and therefore no area. This means that
> $P(a < Z < b) =$
> $P(a \leq Z \leq b) =$
> $P(a < Z \leq b) =$
> $P(a \leq Z < b)$

Example 10

Given that $Z \sim N(0, 1)$, find
a $P(-2 < Z < 1)$
b $P(Z < 1)$
c $P(Z > -1.5)$
d $P(Z < 0)$
e $P(|Z| > 0.8)$

Answers

a $P(-2 < Z < 1) = 0.819$

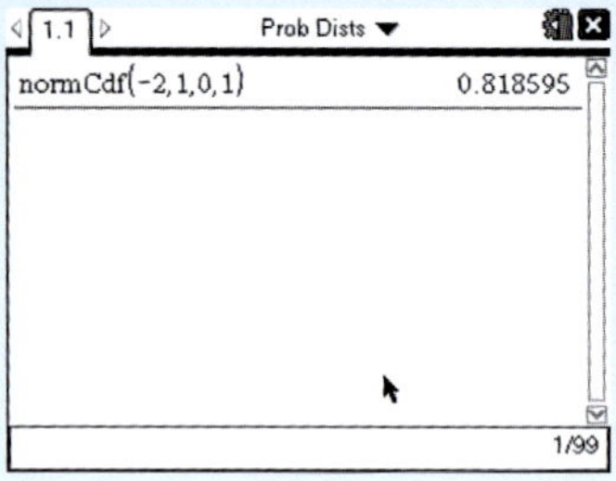

normCdf(−2,1,0,1)   0.818595

Using the distribution menu on your GDC, choose normCdf and enter the values in this order: lower limit, upper limit, mean, standard deviation.

GDC help on CD: *Alternative demonstrations for the TI-84 Plus and Casio FX-9860GII GDCs are on the CD.*

b $P(Z < 1) = 0.841$

normCdf(−9.E999,1,0,1)   0.841345

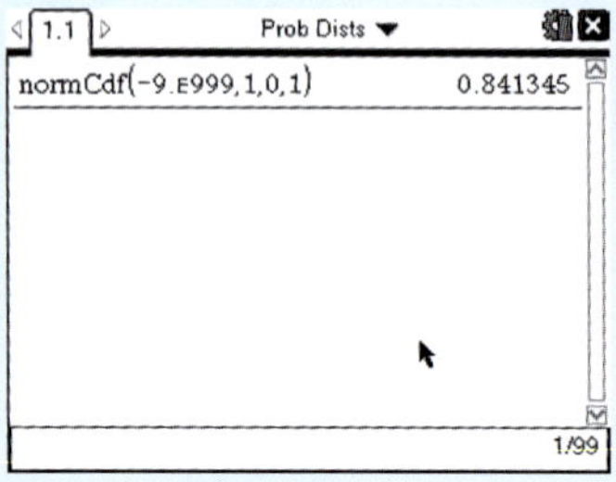

Enter the lower limit as a very small negative number,
-9×10^{999}

c $P(Z > -1.5) = 0.933$

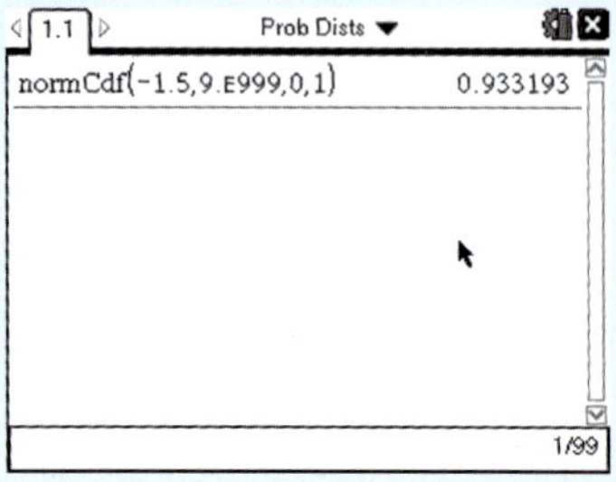

normCdf(−1.5,9.E999,0,1)   0.933193

Enter the upper limit as a very large number, 9×10^{999}

▶ Continued on next page

d P(Z < 0) = 0.5

Here you do not need to use the calculator because the graph is symmetrical about the mean.

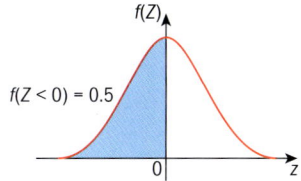

e P(|Z| > 0.8) = 1 − 0.576
= 0.424

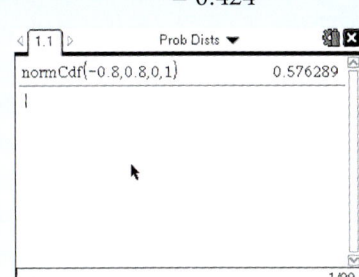

*|Z| > 0.8 means
−0.8 < Z < 0.8*

See Chapter 17, Section 5.13.

Exercise 15H

1 Given that $Z \sim N(0, 1)$ find
  a P(−1 < Z < 1)   **b** P(−2 < Z < 2)   **c** P(−3 < Z < 3)

2 Find the area under the standard normal curve:
  a between 1 and 2 standard deviations from the mean
  b between 0.5 and 1.5 standard deviations from the mean.

3 Find the area under the curve which is more than:
  a 1 standard deviation above the mean
  b 2.4 standard deviations above the mean.

4 Find the area under the curve that is less than
  a 1 standard deviation below the mean
  b 1.75 standard deviations below the mean.

5 Given that $Z \sim N(0, 1)$ use the GDC to find
  a P(Z < 0.65)   **b** P(Z > 0.72)   **c** P(Z ≥ 1.8)
  d P(Z > −2)   **e** P(Z ≤ −0.28)

6 Given that $Z \sim N(0, 1)$ use the GDC to find
  a P(0.2 < Z < 1.2)   **b** P(−2 < Z ≤ 0.3)   **c** P(−1.3 ≤ Z ≤ −0.3)

7 Given that $Z \sim N(0, 1)$ use the GDC to find
  a P(|Z| < 0.4)   **b** P(|Z| > 1.24)

In Question 1 of Exercise 15H, you found the probability that Z lies within one standard deviation of the mean, two standard deviations of the mean and three standard deviations of the mean respectively.

You can see that most of the data for a normal distribution will lie within three standard deviations of the mean.

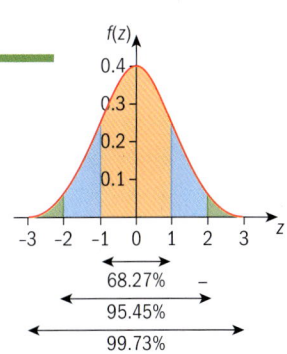

Probabilities for other normal distributions

Clearly, however, very few real-life variables are distributed like the standard normal distribution (with a mean of 0 and a standard deviation of 1). But you can transform any normal distribution $X \sim N(\mu, \sigma^2)$ to the standard normal distribution, because all normal distributions have the same basic shape but are merely shifts in location and spread.

To transform any given value of x on $X \sim N(\mu, \sigma^2)$ to its equivalent z-value on $Z \sim N(0, 1)$ use the formula

$$z = \frac{x - \mu}{\sigma}$$

You can then use your GDC to find the required probability.

→ If $X \sim N(\mu, \sigma^2)$ then the transformed random variable $Z = \dfrac{X - \mu}{\sigma}$ has a standard normal distribution.

Example 11

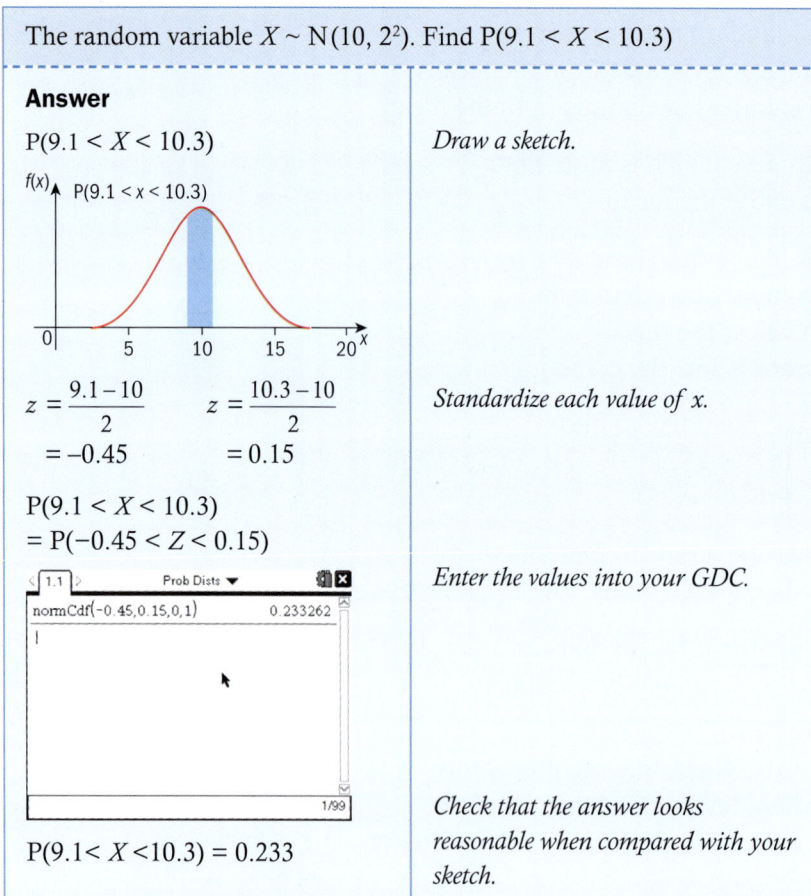

The random variable $X \sim N(10, 2^2)$. Find $P(9.1 < X < 10.3)$.

Answer

$P(9.1 < X < 10.3)$ — *Draw a sketch.*

$z = \dfrac{9.1 - 10}{2} = -0.45$ $z = \dfrac{10.3 - 10}{2} = 0.15$ — *Standardize each value of x.*

$P(9.1 < X < 10.3)$
$= P(-0.45 < Z < 0.15)$

normCdf(−0.45,0.15,0,1) 0.233262 — *Enter the values into your GDC.*

$P(9.1 < X < 10.3) = 0.233$ — *Check that the answer looks reasonable when compared with your sketch.*

GDC help on CD: *Alternative demonstrations for the TI-84 Plus and Casio FX-9860GII GDCs are on the CD.*

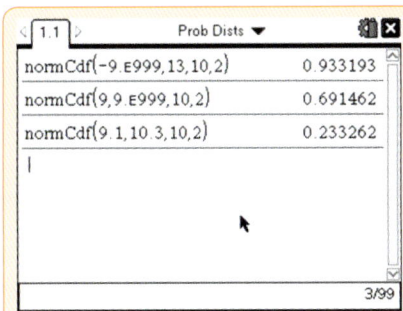 You can also find these solutions directly using the GDC. Without using the standardization formula, this is the quickest and most efficient method of answering this question. But it is important to know the method of standardization.
Enter lower limit, upper limit, mean = 10, standard deviation = 2.

Exercise 15I

1 The random variable $X \sim N(14, 5^2)$.
   Find
   a $P(X < 16)$ **b** $P(X > 9)$ **c** $P(9 \leq X < 12)$ **d** $P(X < 14)$

2 The random variable $X \sim N(48, 81)$.
   Find
   a $P(X < 52)$ **b** $P(X \geq 42)$ **c** $P(37 < X < 47)$

3 The random variable $X \sim N(3.15, 0.02^2)$.
   Find
   a $P(X < 3.2)$ **b** $P(X \geq 3.11)$ **c** $P(3.1 < X < 3.15)$

Example 12

Eggs laid by a chicken are known to have the mass normally distributed, with mean 55 g and standard deviation 2.5 g.
What is the probability that
a an egg weighs more than 59 g   **b** an egg is smaller than 53 g
c an egg is between 52 and 54 g?

Answer

$W \sim N(55, 2.5^2)$

Sketch first.
Mean = 55
$3 \times \sigma = 3 \times 2.5 = 7.5$

Enter the value in your GDC: lower limit, upper limit, mean = 55, standard deviation = 2.5

a $P(W > 59) = 0.0548$ (3 sf)
b $P(W < 53) = 0.212$ (3 sf)
c $P(52 < W < 54) = 0.230$ (3 sf)

 GDC help on CD: Alternative demonstrations for the TI-84 Plus and Casio FX-9860GII GDCs are on the CD.

Exercise 15J

EXAM-STYLE QUESTIONS

1. Households in Portugal spend an average of €100 per week on groceries with a standard deviation of €20. Assuming that the distribution of grocery expenditure follows a normal distribution, what is the probability of a household spending:
   a less than €130 per week
   b more than €90 per week
   c between €80 and €125 per week.

2. A machine produces bolts with diameters distributed normally with a mean of 4 mm and a standard deviation of 0.25 mm. Bolts are measured accurately and any which are smaller than 3.5 mm or bigger than 4.5 mm are rejected. Out of a batch of 500 bolts how many would be acceptable?

3. The length of time patients have to wait in Dr. Barrett's waiting room is known to be normally distributed with mean 14 minutes and standard deviation 4 minutes.
   a Find the probability that I will have to wait more than 20 minutes to see the doctor.
   b What proportion of patients wait less than 10 minutes?

4. Packets of 'Flakey flakes' breakfast cereal are said to contain 550 g. The production of packets of 'Flakey flakes' is such that the masses are normally distributed with a mean of 551.3 g, and standard deviation 15 g. What proportion of packets will contain more than the stated mass?

5. The mass of packets of washing powder is normally distributed with a mean of 500 g and standard deviation of 20 g.
   a Find the probability that a packet chosen at random has a mass of less than 475 g.
   b Three packets are chosen at random. What is the probability that all packets have a mass which is less than 475 g?

The inverse normal distribution

Here you need to find the value in the data that has a given cumulative probability. For example, a company fills cartons of juice to a nominal value of 150 ml. 5% of cartons are rejected for containing too little juice. The owner of the company may wish to find the cut-off point for the minimum volume of a carton.

You can use your GDC to help find this value. The calculator has a function called Inverse Normal which will do this. In these examples we will return to the standard normal distribution $Z \sim N(0, 1)$

See Chapter 17, Section 5.14.

Example 13

Given that $Z \sim N(0, 1)$ use your GDC to find a
a $P(Z < a) = 0.877$ **b** $P(Z > a) = 0.2$ **c** $P(-a < Z < a) = 0.42$

Answers

a

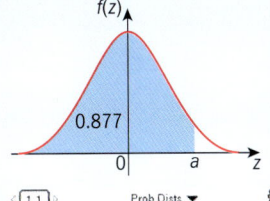

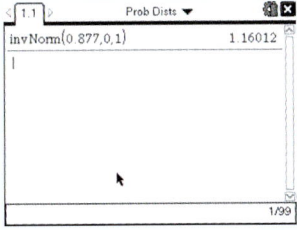

Draw a sketch.

$P(Z < a) = 0.877$
$a = 1.16$ (3 sf)

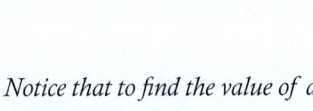

GDC help on CD: *Alternative demonstrations for the TI-84 Plus and Casio FX-9860GII GDCs are on the CD.*

b

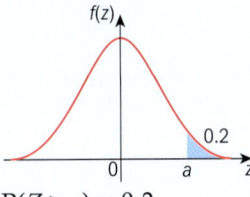 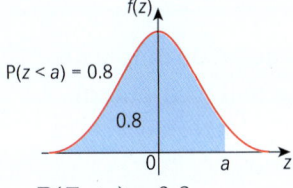

$P(Z > a) = 0.2$ $P(Z < a) = 0.8$

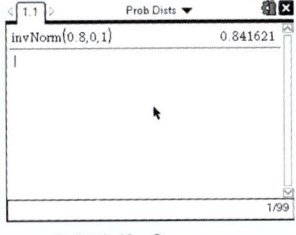

Notice that to find the value of a for $P(Z > a) = 0.2$ you can more easily find a for $P(Z < a) = 0.8$

$a = 0.842$ (3 sf)

c $P(-a < Z < a) = 0.42$

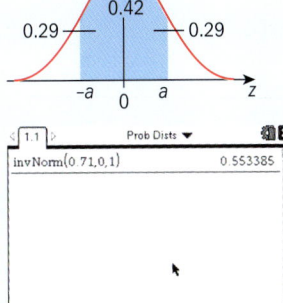

The areas either side of the shaded region are the same size and equal to $\frac{1}{2}(1 - 0.42) = 0.29$. Hence $P(Z < a) = 1 - 0.29 = 0.71$

$a = 0.533$ (3 sf)

Exercise 15K

1 Find *a* such that:
   a $P(Z < a) = 0.922$
   b $P(Z > a) = 0.342$
   c $P(Z > a) = 0.005$

2 Find *a* such that:
   a $P(1 < Z < a) = 0.12$
   b $P(a < Z < 1.6) = 0.787$
   c $P(a < Z < -0.3) = 0.182$

3 Find *a* such that:
   a $P(-a < Z < a) = 0.3$
   b $P(|Z| > a) = 0.1096$

4 Find the values of *a* shown in these diagrams:

a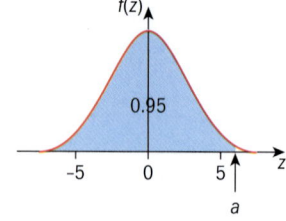

b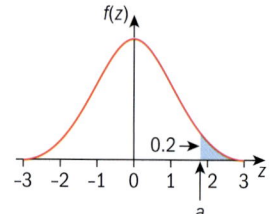

Once again, however, we are more likely to be dealing with distributions that are not the standard normal distribution.

Example 14

Given that $X \sim N(15, 3^2)$ determine x where $P(X < x) = 0.75$

Answer

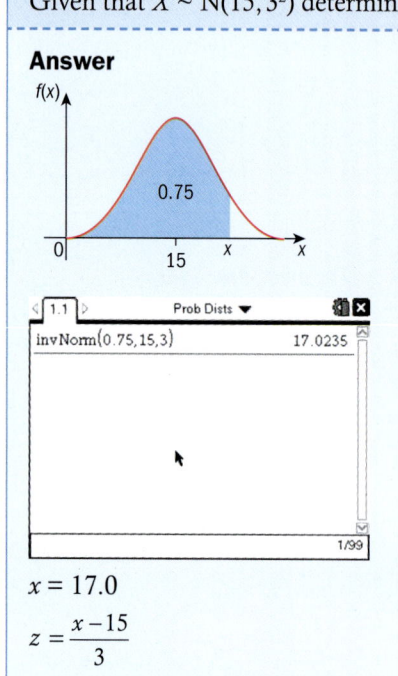

$x = 17.0$

$z = \dfrac{x - 15}{3}$

Draw a sketch to show the value of x required.

This question is best done on the GDC. In invNorm enter x, mean, standard deviation.

GDC help on CD: *Alternative demonstrations for the TI-84 Plus and Casio FX-9860GII GDCs are on the CD.*

You could also answer the question by first standardizing the value of x.

▶ Continued on next page

$P(X < x) = 0.75$

$P\left(Z < \dfrac{x-15}{3}\right) = 0.75$

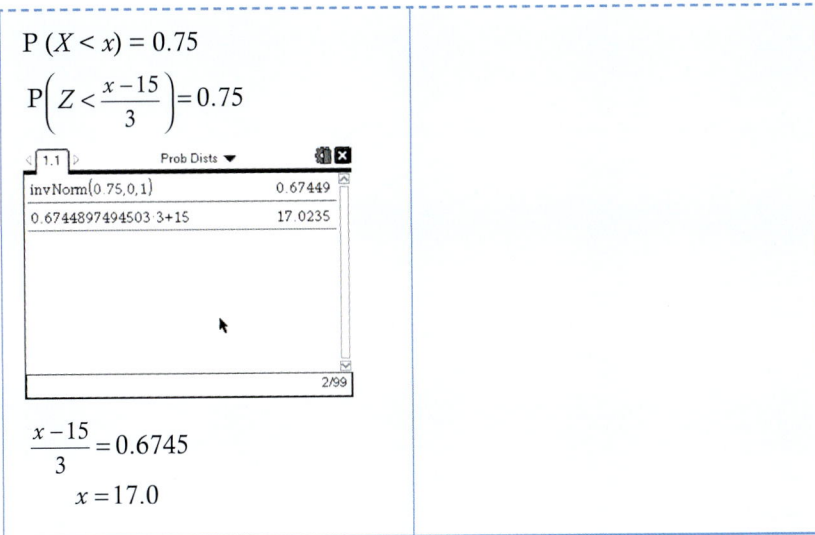

$\dfrac{x-15}{3} = 0.6745$

$x = 17.0$

Example 15

Cartons of juice are such that their volumes are normally distributed with a mean of 150 ml and a standard deviation of 5 ml.
5% of cartons are rejected for containing too little juice.
Find the minimum volume, to the nearest ml, that a carton must contain if it is to be accepted.

Answer

Let V be the volume of a carton.
$V \sim N(150, 5^2)$
$P(V < m) = 0.05$

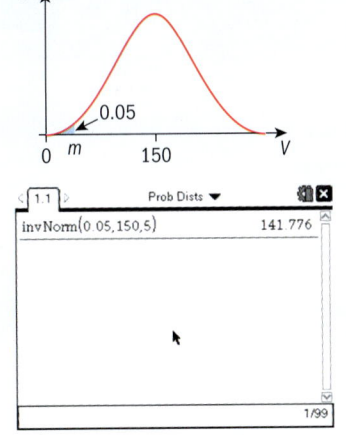

The minimum volume is 142 ml to the nearest ml.

Let m be the minimum volume that a carton must have to be accepted.

Draw a sketch.

GDC help on CD: Alternative demonstrations for the TI-84 Plus and Casio FX-9860GII GDCs are on the CD.

Exercise 15L

1. $X \sim N(5.5, 0.2^2)$ and $P(X > a) = 0.235$ Find the value of a.

2. The mass, M, of a randomly chosen tin of dog food is such that $M \sim N(420, 10^2)$. Find
   a the first quartile
   b the 90th percentile.

$\frac{1}{4}$ of the values are less than the first quartile.

EXAM-STYLE QUESTION

3. Regulations in a country insist that all mineral bottles that claim to contain 500 ml must have at least that amount. 'Yummy Cola' has a machine for filling bottles, which puts an average of 502 ml into each bottle with a standard deviation of 1.6 ml and follows a normal distribution.
   a An inspector randomly selects a bottle of 'Yummy Cola'. What is the probability that it will break the regulations?
   b What proportion of bottles will contain between 500 ml and 505 ml?
   c 95% of bottles contain between a ml and b ml of liquid where a and b are symmetrical about the mean. What are a and b?

4. The masses of lettuce sold at a hypermarket are normally distributed with a mean mass of 550 g and standard deviation of 25 g.
   a If a lettuce is chosen at random, find the probability that its mass lies between 520 g and 570 g.
   b Find the mass exceeded by 10% of the lettuces.

5. The marks of 500 candidates in an examination are normally distributed with a mean of 55 marks and a standard deviation of 15 marks.
   a If 5% of the candidates obtain a distinction by scoring d marks or more, find the value of d.
   b If 10% of students fail by scoring f marks or less, find the value of f.

You may also be given cumulative probabilities and asked to find either the mean (if σ is known) or the standard deviation (if μ is known) or both.

Extension material on CD:
Worksheet 15 - Normal distribution as an approximation to a binomial distribution

Example 16

Sacks of potatoes with mean weight 5 kg are packed by an automatic loader. In a test it was found that 10% of bags were over 5.2 kg. Use this information to find the standard deviation of the process.

Answer

Let M be the mass of potatoes in a sack.
$M \sim N(5, \sigma^2)$

10% (0.1) of bags were over 5.2 kg.

▶ Continued on next page

$P(M > 5.2) = 0.1$

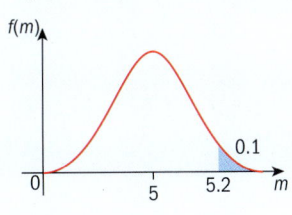

Draw a sketch.

$Z = \dfrac{5.2 - 5}{\sigma} = \dfrac{0.2}{\sigma}$

Standardize.

$P\left(Z > \dfrac{0.2}{\sigma}\right) = 0.1$

or $P\left(Z < \dfrac{0.2}{\sigma}\right) = 0.9$

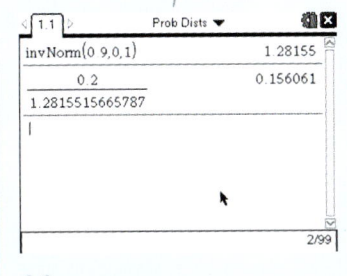

From the GDC
$P(Z < 1.28155\ldots) = 0.9$

GDC help on CD: *Alternative demonstrations for the TI-84 Plus and Casio FX-9860GII GDCs are on the CD.*

$\dfrac{0.2}{\sigma} = 1.28155\ldots$

$\sigma = 0.156$ (3 sf)

Example 17

A manufacturer does not know the mean and standard deviation of the diameters of ball bearings she is producing. However a sieving system rejects all ball bearings larger than 2.4 cm and those under 1.8 cm in diameter. It is found that 8% of the ball bearings are rejected as too small and 5.5% as too big. What is the mean and standard deviation of the ball bearings produced?

Answer

Let D be the diameters of ball bearings produced.
$D \sim N(\mu, \sigma^2)$
$P(D < 1.8) = 0.08$
$P(D > 2.4) = 0.055$

We know that 8% are too small, and 5.5% are too big.

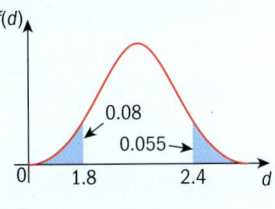

Draw a sketch.

$\dfrac{1.8 - \mu}{\sigma}$ and $\dfrac{2.4 - \mu}{\sigma}$

Standardize each value.

$P\left(Z < \dfrac{1.8 - \mu}{\sigma}\right) = 0.08$

From the first expression.

▶ Continued on next page

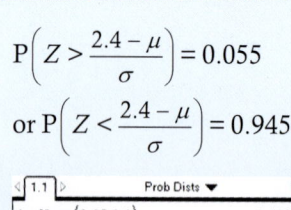

$P\left(Z > \dfrac{2.4 - \mu}{\sigma}\right) = 0.055$

or $P\left(Z < \dfrac{2.4 - \mu}{\sigma}\right) = 0.945$

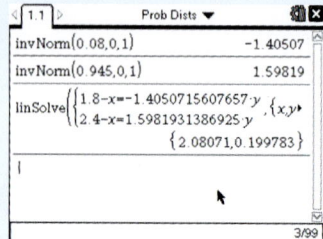

$\dfrac{1.8 - \mu}{\sigma} = -1.40507\ldots$ and

$\dfrac{2.4 - \mu}{\sigma} = 1.59819\ldots$

$\mu = 2.08$ and $\sigma = 0.200$

From the second expression.

$1 - 0.005 = 0.945$

From the GDC we know that
$P(Z < -1.40507\ldots) = 0.08$ *and*
$P(Z < 1.59819\ldots) = 0.945$

Solve simultaneously for μ and σ.

GDC help on CD: Alternative demonstrations for the TI-84 Plus and Casio FX-9860GII GDCs are on the CD.

Exercise 15M

1. $X \sim N(30, \sigma^2)$ and $P(X > 40) = 0.115$. Find the value of σ.

2. $X \sim N(\mu, 4^2)$ and $P(X < 20.5) = 0.9$. Find the value of μ.

3. $X \sim N(\mu, \sigma^2)$. Given that $P(X > 58.39) = 0.0217$ and $P(X < 41.82) = 0.0287$, find μ and σ.

4. A random variable X is normally distributed with mean μ and standard deviation σ, such that $P(X < 89) = 0.90$ and $P(X < 94) = 0.95$. Find μ and σ.

EXAM-STYLE QUESTIONS

5. The mean height of children of a certain age is 136 cm. 12% of children have a height of 145 cm or more. Find the standard deviation of the heights.

6. The standard deviation of masses of loaves of bread is 20 g. Only 1% of loaves weigh less than 500 g. Find the mean mass of the loaves.

7. The masses of cauliflowers are normally distributed with mean 0.85 kg. 74% of cauliflowers have mass less than 1.1 kg. Find:
   a. the standard deviation of cauliflowers' masses
   b. the percentage of cauliflowers with mass greater than 1 kg.

8. The lengths of nails are normally distributed with mean μ and standard deviation 7 cm. If 2.5% of the nails measure more than 68 cm find the value of μ.

9. A roll of wrapping paper is sold as '3 m long'. It is found that actually only 35% of rolls are over 3 m long and that the average length of the rolls of wrapping paper is 2.9 m. Find the value of the standard deviation of the lengths of rolls of wrapping paper, assuming that the lengths of rolls follow a normal distribution.

> Lambert Adolphe Jacques Quételet (1796–1874), a Flemish scientist, was the first to apply the normal distribution to human characteristics. He noted that characteristics such as height, weight, and strength were normally distributed.

EXAM-STYLE QUESTIONS

10 It is suspected that the scores in a test are normally distributed. 30% of students score less than 108 marks on the test, and 20% score more than 154 marks.
  a Find the mean and standard deviation of the scores, if they are normally distributed.
  b 60% of students score more than 117 marks. Does this fact appear to be reasonably consistent with the idea that the scores are normally distributed as above?

11 Due to variations in manufacturing, the length of wool in a randomly chosen ball of wool can be modeled by a normal distribution. Find the mean and standard deviation given that 95% of balls of wool have lengths exceeding 495 m and 99% have lengths exceeding 490 m.

Review exercise

EXAM-STYLE QUESTIONS

1 The table shows the probability distribution of a discrete random variable X.
  a Find the value of k.
  b Find the expected value of X.

x	−2	−1	0	1	2
$P(X = x)$	0.3	$\frac{1}{k}$	$\frac{2}{k}$	0.1	0.1

2 The probability distribution of a discrete random variable X is defined by $P(X = x) = cx(6 - x)$, $x = 1, 2, 3, 4, 5$.
  a Find the value of c.   **b** Find $E(X)$.

3 In a game a player rolls a biased tetrahedral (four-faced) dice. The probability of each possible score is shown. Find the probability of a total score of six after two rolls.

Score	1	2	3	4
Probability	$\frac{1}{4}$	$\frac{1}{4}$	$\frac{1}{8}$	x

4 A game involves spinning two spinners. One is numbered 1, 2, 3, 4. The other is numbered 2, 2, 4, 4. Each spinner is spun once and the number on each is recorded.
Let P be the product of the numbers on the spinners
  a Write down all the possible values for P.
  b Find the probability of each value of P.
  c What is the expected value of P?
  d A mathematician determines the amount of pocket money to give his son each week by getting him to play the game on Monday morning. If the son spins and the product is greater than 10 then he gets £10. Otherwise he gets £5. How much in total will the boy expect to get after 10 weeks of playing the game?

EXAM-STYLE QUESTION

5 In a train, $\frac{1}{3}$ of the passengers are listening to music. Five passengers are chosen at random. Find the probability that exactly three are listening to music.

Chapter 15

6 When a boy plays a game at a fair, the probability that he wins a prize is 0.1. He plays the game twice. Let X denote the total number of prizes that he wins. Assuming that the games are independent, find E(X).

7 Let X be normally distributed with mean 75 and standard deviation 5.
   a Given that P($X < 65$) = P($X > a$), find the value of a.
   b Given that P($65 < X < a$) = 0.954, Find P($X > a$).

Review exercise

1 Three dice are thrown. If a 1 or a 6 is rolled somewhere on these three dice you will be paid $1, but if neither is rolled you will pay $5. You play the game.
   a What is the probability you will win $1?
   b Copy and complete the table showing the probability distribution of X, 'the number of dollars won in a game'.

x	−5	1
P(X = x)		

   c How much would you expect to gain (or lose) in
      i 1 game   ii 9 games?

EXAM-STYLE QUESTION

2 I like 30% of the songs on my friends MP3 player. If I choose eight songs at random
   a find the probability that I like exactly three songs
   b find the probability that I like at least three songs.

3 Find the probability of throwing three sixes twice in five throws of six dice.

EXAM-STYLE QUESTION

4 In a large school one person in five is left-handed.
   a A random sample of 10 people is taken. Find the probability that
      i exactly four will be left-handed
      ii more than half will be left-handed.
   b Find the most likely number of left-handed people in the sample of 10 people.
   c How large must a random sample be if the probability that it contains at least one left-handed person is to be greater than 0.95?

5 Z is the standardized normal random variable with mean 0 and variance 1. Find the value of a such that P($|Z| \leq a$) = 0.85.

6 The results of a test given to a group of students are normally distributed with a mean of 71. 85% of students have scores of less than 80.
   a Find the standard deviation of the scores.

   To pass the test a student must score more than 65.
   b Find the probability that a student chosen at random passes the test.

EXAM-STYLE QUESTIONS

7 The lifespans of certain batteries are normally distributed. It is found that 15% of batteries last less than 30 hours and 10% of batteries last more than 50 hours. Find the mean and standard deviation of the lifespans of the batteries.

8 The time taken for Samuel to get to school each morning is normally distributed with a mean of μ minutes and a standard deviation of 2 minutes.
The probability that the journey takes more than 35 minutes is 0.2.
  a Find the value of μ.

Samuel should be at school at 08:45 each morning and so on five consecutive days he sets out at 08:10.
  b Find the probability that he arrives before 08:45 on all five days.
  c Find the probability he is late on at least two days.

CHAPTER 15 SUMMARY
Random variables

- A **random variable** is a quantity whose value depends on chance.
- A **probability distribution** for a discrete random variable is a list of each possible value of the random variable, with the probability that the outcome occurs.
- For any random variable X $0 \leq P(X = x) \leq 1$ $\sum P(X = x) = 1$
- The **expected value** of a random variable X is $E(X) = \sum x \, P(X = x)$.

The binomial distribution

The essential elements of a binomial distribution are:

- There is a fixed number (n) of trials.
- Each trial has only two possible outcomes – a 'success' or a 'failure'.
- The probability of a success (p) is constant from trial to trial.
- Trials are independent of each other.

- A binomial distribution of the random variable X is written as $X \sim B(n, p)$.
- The probability of obtaining r successes out of n independent trials, when p is the probability of success for each trial, is
$$P(X = r) = \binom{n}{r} p^r (1-p)^{n-r}$$
- For the binomial distribution where $X \sim B(n, p)$, the expectation of X, $E(X) = np$. The variance of X, $Var(X) = npq$ where $q = 1 - p$.

The normal distribution

- If a random variable, X, has a normal distribution with mean μ and standard deviation σ, this is written $X \sim N(\mu, \sigma^2)$
- The standard normal distribution is written $Z \sim N(0, 1)$
- If $X \sim N(\mu, \sigma^2)$ then the transformed random variable $Z = \dfrac{X - \mu}{\sigma}$ has a standard normal distribution.

Theory of knowledge

Statistics of human behavior

Social scientists use statistics to study human behavior. Some variables can be measured fairly easily by collecting and analyzing data, for example population, income, birth rates and mortality rates.

The United Nations and World Health Organization collect data like this, and use it to help plan aid and development programmes.

- Is it possible to reduce all human activity to a set of statistical data?
- Are there characteristics of human behavior that cannot be measured?

Dynamic data

Hans Rosling is Professor of International Health at the Karolinska Institute in Sweden, where he teaches courses on global health. These courses involve looking at lots of data from different countries on fertility, life expectancy, child mortality and other aspects of health. To help make this data easier to understand, he converts the statistics into moving graphs, which show dynamically how the data change over time. He set up Gapminder to develop the software, which is available free.

His innovative and entertaining approach has provided startling insights into global poverty and international health. You can see him and his graphs in action on YouTube, or on the Gapminder website.

Human behavior experiments

Social scientists and psychologists often set up experiments to collect data on human behavior in as scientific a way as possible.

Here are some of the problems associated with this type of data collection. Find an example for each one.

People's behavior may change when they are being studied.

- Research the 'Hawthorne Effect'.

Repeatability

Scientific experiments need to be 'repeatable'. This means that another group of researchers (who have no connection with the first group) who carry out the same experiment should get the same results. Each repetition of the experiment that gives the same results helps to confirm the theory.

- Does repetition **prove** the theory?

Controllability

In a science experiment in a laboratory, you can control the conditions so that you can study only one variable (for example, the effect of heat on a substance). All the variables that could affect the outcome of the experiment are controlled.

- Is it possible to control all the factors in human research?

To carry out a controlled investigation into the effects of television violence on child behavior, you would have to isolate the children from all other factors that could influence their behavior.

People lie!

A study that asked American teenagers how much TV they watched, found the average to be 7 hours a week. But evidence from other sources suggests the national average is 7 hours per day.

- Why do you think the teenagers in the study lied about how much TV they watch?

Self-fulfilling prophecies

People may (subconsciously perhaps) behave in the way that is expected of them. When a Minister for Transport predicts there will be queues at the petrol pumps, this could cause people to go to the petrol stations and cause long queues.

- How might your behavior change if you knew your interactions with your fellow students were being studied?

Hidden agenda

The researchers may have a reason to hope for a particular outcome.

- Who would you trust to carry out an unbiased survey into the effects of passive smoking?
  - A major tobacco manufacturer?
  - An anti-smoking campaign?
  - A company that manufactures drugs for lung cancer?

Prejudice and leading questions

Prejudice can creep in when questions are worded so that the answers will support the researcher's theory. Leading questions encourage a particular answer.

- Why are these questions 'leading'?
  - Do you agree that children today get too little exercise?
  - How much do you think unemployment will rise?

16 Exploration

CHAPTER OBJECTIVES:

As part of your Mathematics SL course, you need to write an exploration, which will be assessed and counts as 20% of your final grade. This chapter gives you advice on planning your exploration, hints and tips to help you get a good grade by making sure your exploration satisfies the assessment criteria, as well as suggestions on choosing a topic and getting started on your exploration.

16.1 About the exploration

The exploration is an opportunity for you to show that you can apply mathematics to an area that interests you.

You should aim to spend:

10 hours of class time	10 hours of your own time
<ul><li>Discussing the assessment criteria</li><li>Discussing suitable topics/titles</li><li>Discussing your progress with your teacher</li></ul>	<ul><li>Planning your exploration, doing research to help select an appropriate topic</li><li>Researching, collecting and organising your data and/or information</li><li>Applying mathematical processes:<ul><li>Ensuring that all of your results are derived using logical deductive reasoning</li><li>Ensuring that your proofs (when necessary) are coherent and correct</li></ul></li><li>Demonstrating mathematical communication and presentation:<ul><li>Checking that your notation and terminology are consistently correct</li><li>Adding diagrams, graphs or charts where necessary</li><li>Making sure your exploration is clearly structured and reads well</li></ul></li></ul>

Your school will set you deadlines for submitting a draft and the final piece of work.

If you do not submit an exploration then you receive a grade of "N" for Mathematics SL, which means you will not receive your IB diploma.

> Every candidate taking Mathematics SL **must** submit an exploration.
> Ensure that you know your school's deadlines and keep to them.

16.2 Internal assessment criteria

Your exploration will be assessed by your teacher, against given criteria.
It will then be externally moderated by the IB using the same assessment criteria.
The final mark for each exploration is the sum of the scores for each criterion.
The maximum possible final mark is 20.
This is 20% of your final grade for Mathematics SL.
A good exploration should be clear and easily understood by one of your peers, and self-explanatory all the way through.
The criteria are split into five areas, A to E:

Criterion A	Communication
Criterion B	Mathematical presentation
Criterion C	Personal engagement
Criterion D	Reflection
Criterion E	Use of mathematics

> These criteria are explained in more detail, with tips on how to ensure your exploration satisfies them.
> Make sure you understand these criteria and consult them frequently when writing your exploration.

Criterion A: Communication

This criterion assesses the organization, coherence, conciseness and completeness of the exploration.

Achievement level	Descriptor
0	The exploration does not reach the standard described by the descriptors below.
1	The exploration has some coherence.
2	The exploration has some coherence and shows some organization.
3	The exploration is coherent and well organized.
4	The exploration is coherent, well organized, concise and complete.

Your exploration

To get a good mark for Criterion A: Communication

✓ A well organized exploration should have
- An **introduction** in which you should discuss the context of the exploration
- A **rationale** which should include an explanation of why you chose this topic
- A description of the **aim** of the exploration which should be clearly identifiable
- A **conclusion**.

✓ A coherent exploration is logically developed and easy to follow.

✓ Your exploration should 'read well'.

✓ Any graphs, tables and diagrams that you use should accompany the work in the appropriate place and not be attached as appendices to the document.

✓ A concise exploration is one that focuses on the aim and avoids irrelevancies.

✓ A complete exploration is one in which all steps are clearly explained without detracting from its conciseness.

✓ It is essential that references are cited where appropriate, i.e.,
- Your exploration should contain footnotes as appropriate. For example, if you are using a quote from a publication, a formula from a mathematics book, etc., put the source of the quote in a footnote.
- Your exploration should contain a bibliography as appropriate. This can be in an appendix at the end. List any books you use, any websites you consult, etc.

Criterion B: Mathematical presentation

This criterion assesses to what extent you are able to:

- use appropriate mathematical language (notation, symbols, terminology)
- define key terms, where required
- use multiple forms of mathematical representation such as formulae, diagrams, tables, charts, graphs and models, where appropriate.

Achievement level	Descriptor
0	The exploration does not reach the standard described by the descriptors below.
1	There is some appropriate mathematical presentation.
2	The mathematical presentation is mostly appropriate.
3	The mathematical presentation is appropriate throughout.

Your exploration

To get a good mark for Criterion B: Mathematical communication

✓ You are expected to use appropriate mathematical language when communicating mathematical ideas, reasoning and findings.

✓ You are encouraged to choose and use appropriate ICT tools such as graphic display calculators, mathematical software, spreadsheets, databases, drawing and word-processing software, as appropriate, to enhance mathematical communication.

✓ You should define key terms, where required.

✓ You should express your results to an appropriate degree of accuracy, when appropriate.

✓ You should always include scales and labels if you use a graph. Tables should have appropriate headings.

✓ Variables should be explicitly defined.

✓ Do not use calculator or computer notation. For example, use 2^x and not 2^x; use × not *; use 0.028 and not 2.8E-2.

Criterion C: Personal engagement

This criterion assesses the extent to which you engage with the exploration and make it your own.

Achievement level	Descriptor
0	The exploration does not reach the standard described by the descriptors below.
1	There is evidence of limited or superficial personal engagement.
2	There is evidence of some personal engagement.
3	There is evidence of significant personal engagement.
4	There is abundant evidence of outstanding personal engagement.

Your exploration

To get a good mark for Criterion C: Personal engagement

✓ You should choose a topic for your exploration that you are interested in as it will be easier to display personal engagement.

✓ You can demonstrate personal engagement by using some of the following different attributes and skills. These include:
- Thinking and working independently
- Thinking creatively
- Addressing your personal interests
- Presenting mathematical ideas in your own way
- Asking questions, making conjectures and investigating mathematical ideas
- Looking for and creating mathematical models for real-world situations
- Considering historical and global perspectives
- Exploring unfamiliar mathematics.

Criterion D: Reflection

This criterion assesses how you review, analyze and evaluate the exploration.

Achievement level	Descriptor
0	The exploration does not reach the standard described by the descriptors below.
1	There is evidence of limited or superficial reflection.
2	There is evidence of meaningful reflection.
3	There is substantial evidence of critical reflection.

Your exploration

To get a good mark for Criterion D: Reflection

✓ Although reflection may be seen in the conclusion to the exploration, it may also be found throughout the exploration.

✓ You can show reflection in your exploration by
- Discussing the implications of your results
- Considering the significance of your findings and results
- Stating possible limitations and/or extensions to your results
- Making links to different fields and/or areas of mathematics.

Criterion E: Use of mathematics

This criterion assesses to what extent you use mathematics in your exploration.

Achievement level	Descriptor
0	The exploration does not reach the standard described by the descriptors below.
1	Some relevant mathematics is used.
2	Some relevant mathematics is used. Limited understanding is demonstrated.
3	Relevant mathematics commensurate with the level of the course is used. Limited understanding is demonstrated.
4	Relevant mathematics commensurate with the level of the course is used. The mathematics explored is partially correct. Some knowledge and understanding are demonstrated.
5	Relevant mathematics commensurate with the level of the course is used. The mathematics explored is mostly correct. Good knowledge and understanding are demonstrated.
6	Relevant mathematics commensurate with the level of the course is used. The mathematics explored is correct. Thorough knowledge and understanding are demonstrated.

Your exploration

To get a good mark for Criterion E: Use of mathematics

✓ You are expected to produce work that is commensurate with the level of the course you are studying. The mathematics you explore should either be part of the syllabus, or at a similar level (or beyond).

✓ You should ensure that the mathematics involved is not completely based on mathematics listed in the prior learning.

✓ **If the level of mathematics is not commensurate with the level of the course you can only get a maximum of two marks for this criterion.**

✓ You need to demonstrate that within your exploration that you fully understand the mathematics used.

16.3 How the exploration is marked

Once you have submitted the final version of your exploration, your teacher will mark it. The teacher looks at each criterion in turn, starting from the lowest grade. As soon as your exploration fails to meet one of the grade descriptors, then the mark for that criterion is set.

The teacher submits these marks to the International Baccalaureate, via a special website. A sample of your school's explorations is selected automatically from the marks that are entered and this sample is sent to an external moderator to be checked. This person moderates the explorations according to the assessment criteria and checks that your teacher has marked the explorations accurately.

If your teacher has applied the criteria to the exploration too severely then your school's exploration marks may be increased.

If your teacher has applied the exploration criteria too leniently then your school's exploration marks may be decreased.

16.4 Academic Honesty

This is extremely important in all your work. Make sure that you have read and are familiar with the IB Academic Honesty document.

> Your teacher or IB Diploma Programme coordinator will be able to give you this document.

Academic Honesty means:

- that your work is authentic
- that your work is your own intellectual property
- that you conduct yourself properly in written examinations
- that any work taken from another source is properly cited.

Authentic work:

- is work based on your own original ideas
- can draw on the work and ideas of others, but this must be fully acknowledged (e.g. in footnotes and a bibliography)
- must use your own language and expression – for both written and oral assignments
- must acknowledge all sources fully and appropriately (e.g. in a bibliography).

Malpractice

The IB defines **malpractice** as 'behavior that results in, or may result in, the candidate or any other candidate gaining an unfair advantage in one or more assessment components'.

Malpractice includes:

- plagiarism – copying from others' work, published or otherwise
- collusion – working secretly with at least one other person in order to gain an undue advantage. This includes having someone else write your exploration, and passing it off as your own
- duplication of work
- any other behavior that gains an unfair advantage.

> 'Plagiarism' is a word derived from Latin, meaning 'to kidnap'.

Advice to schools:

- A school-wide policy must be in place to promote Academic Honesty
- All candidates must clearly understand this policy
- All subject areas must promote the policy
- Candidates must be clearly aware of the penalties for academic dishonesty
- Schools must enforce penalties, if incurred.

Acknowledging sources

Remember to acknowledge all your sources. Both teachers and moderators can usually tell when a project has been plagiarised. Many schools use computer software to check for plagiarism. If you are found guilty of plagiarism then you will not receive your diploma. It is not worth taking the risk.

> You will find a definition of plagiarism in the Academic Honesty document.

16.5 Record keeping

Throughout the course, it would be a good idea to keep an exploration journal, either manually or online. Keeping a journal will help you to focus your search for a topic, and also remind you of deadlines.

> Keeping a journal while you write your exploration will also help you to demonstrate its academic honesty.

If you use a journal for Theory of knowledge you will probably appreciate how much help it is when writing your essays. In the same way, keeping a journal for your exploration will be a great assistance in focusing your efforts.

- Make notes of any books or websites you use, as you go along, so you can include them in your bibliography.
- There are different ways of referencing books, websites, etc. Make sure that you use the style advised by your school and **be consistent**.
- Keep a record of your actions so that you can show your teacher how much time you are spending on your exploration. Include any meetings you may have with your teacher about your exploration.
- Remember to follow your teacher's advice and meet the school's deadlines.
- The teacher is there to help you – so do not be afraid to ask for guidance. The more focused your questions are, the better guidance your teacher can give you.

16.6 Choosing a topic

You need to choose a topic that interests you, because then you will enjoy working on your exploration, you will put more effort into the exploration, and you will be able to demonstrate authentic personal engagement more effectively. You should discuss the topic with your teacher before you put too much time and effort into writing your exploration.

> Your teacher might give your class a set of stimuli – general areas from which you could choose a topic. Alternatively they might encourage you to find your own topic based on your interests and level of mathematical competence.

Each chapter of this book suggests some ideas for explorations, which could be starting points for you to choose a topic.

These questions may help you to find a topic for your exploration:

- What areas of the syllabus am I enjoying the most?
- What areas of the syllabus am I performing best in?
- Which mathematical skills are my strengths?
- Do I prefer pure mathematics, or applied problems and modeling?
- Have I discovered, either through reading or the media, mathematical areas outside of the syllabus that I find interesting?
- What career do I eventually want to enter, and what mathematics is important in this field?
- What are my own special interests or hobbies? Where is the mathematics in these areas?

Mind map

One way of choosing a topic is to start with a general area of interest and create a mind map. This can lead to some interesting ideas on applications of mathematics to explore.

The mind map below shows how the broad topic 'Geography' can lead to suggestions for explorations into such diverse topics as the spread of disease, earthquakes or global warming.

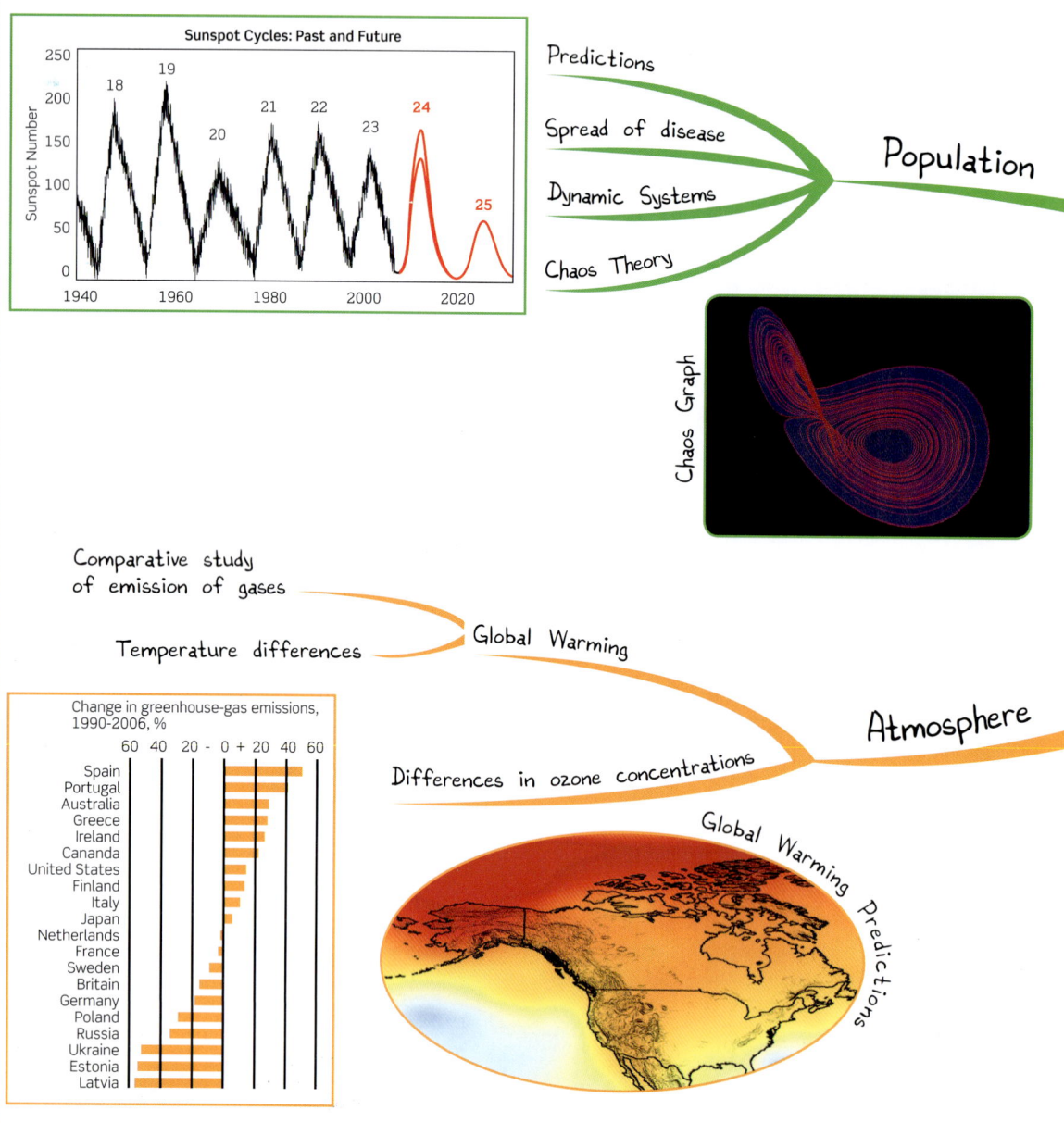

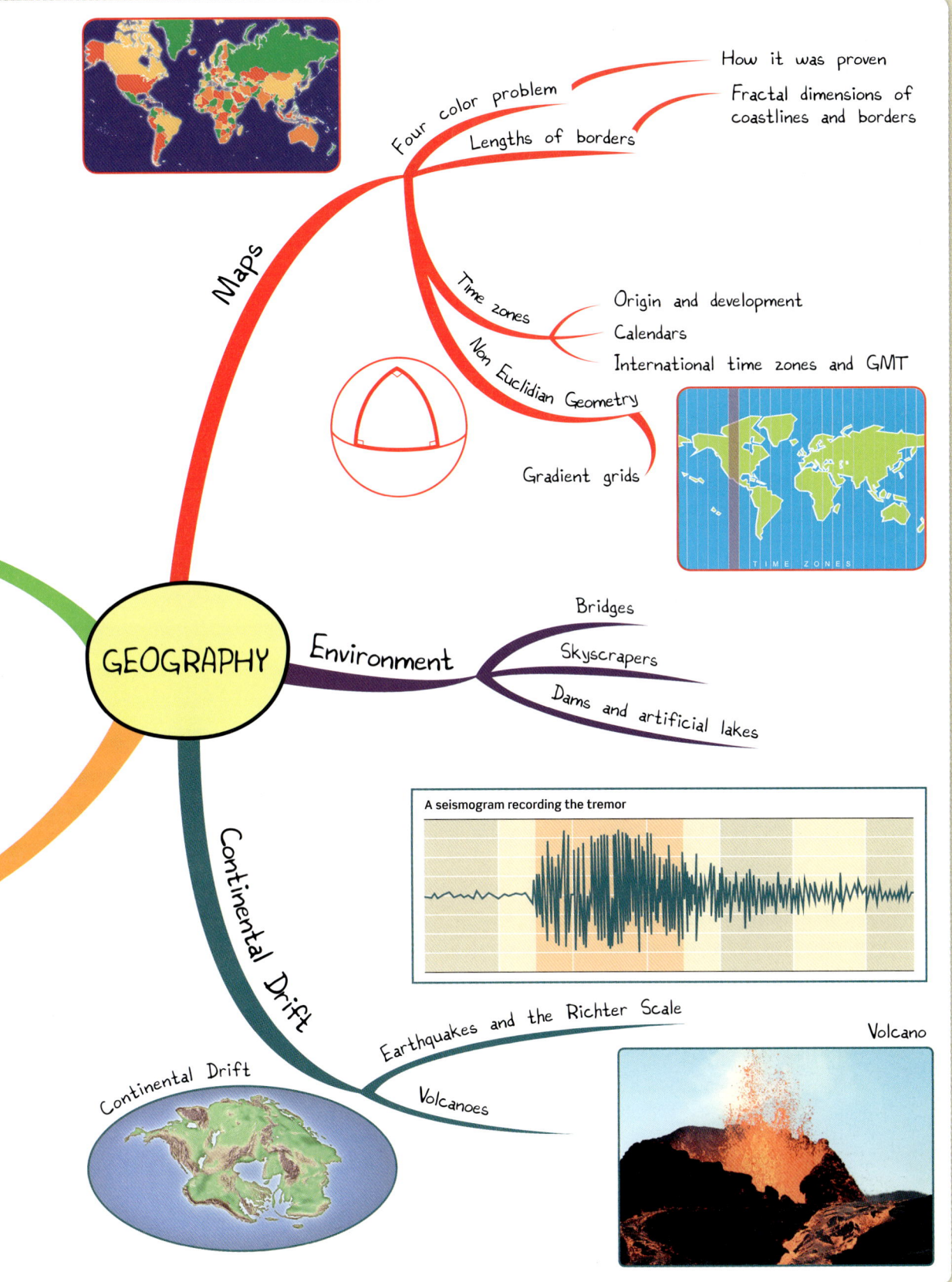

16.7 Getting started

Once you have chosen your topic, the next step is to do some research. The purpose of the research is to determine the suitability of your topic.

Do not limit your research to the internet. Your school library will have books on mathematics that are interesting and related to a variety of different fields.

These questions will help you to decide if your chosen topic is suitable.

- What areas of mathematics are contained in my topic?
- Which of these areas are accessible to me or are part of the syllabus?
- Is there mathematics outside the syllabus that I would have to learn in order to complete the exploration successfully? Am I capable of doing this?
- Can I show personal engagement in my topic, and how?
- Can I limit my work to the recommended length of 6 to 12 pages if I choose this topic?

If your original choice of topic is not suitable, has your research suggested another, better topic? Otherwise, could you either widen out or narrow down your topic to make it more suitable for the exploration?

Once you think you have a workable topic, write a brief outline covering:

- why you chose this topic
- how your topic relates to mathematics
- the mathematical areas in your topic, e.g. algebra, geometry, trigonometry, calculus, probability and statistics, etc.
- the key mathematical concepts in your topic, e.g. areas of irregular shapes, curve fitting, modeling data, etc.
- the mathematical skills you will need, e.g. writing formal proofs, integration, operations with complex numbers, graphing piecewise functions, etc.

- any mathematics outside the syllabus that you will need
- possible technology and software that can help in the design of your exploration and in doing the mathematics
- key mathematical terminology and notation required in your topic.

Now you are ready to start writing the topic in detail.

Remember that your fellow students (your peers) should be able to read and understand your exploration. You could ask one of your classmates to read your work and comment on any parts which are unclear, so you can improve them.

> Make sure you keep every internal deadline that your teacher assigns. In this way, you will receive feedback in time for you to be able to complete your exploration successfully.

17 Using a graphic display calculator

CHAPTER OBJECTIVES:
This chapter shows you how to use your graphic display calculator (GDC) to solve the different types of problems that you will meet in your course. You should not work through the whole of the chapter – it is simply here for reference purposes. When you are working on problems in the mathematical chapters, you can refer to this chapter for extra help with your GDC if you need it.

GDC instructions on CD: *The instructions in this chapter are for the TI-Nspire model. Instructions for the same techniques using the TI-84 Plus and the Casio FX-9860GII are available on the CD.*

Chapter contents

1 Functions
- 1.1 Graphing linear functions 572
- **Finding information about the graph**
- 1.2 Finding a zero 572
- 1.3 Finding the gradient (slope) of a line 573
- **Simultaneous equations**
- 1.4 Solving simultaneous equations graphically 574
- 1.5 Solving simultaneous linear equations 576
- **Quadratic functions**
- 1.6 Drawing a quadratic graph 577
- 1.7 Solving quadratic equations 578
- 1.8 Finding a local minimum or maximum point 579
- **Exponential functions**
- 1.9 Drawing an exponential graph 583
- 1.10 Finding a horizontal asymptote 584

Logarithmic functions
- 1.11 Evaluating logarithms 585
- 1.12 Finding an inverse function 585
- 1.13 Drawing a logarithmic graph 588
- **Trigonometric functions**
- 1.14 Degrees and radians 589
- 1.15 Drawing trigonometric graphs 590
- **More complicated functions**
- 1.16 Solving a combined quadratic and exponential equation 591
- **Modeling**
- 1.17 Using sinusoidal regression 592
- 1.18 Using transformations to model a quadratic function 594
- 1.19 Using sliders to model an exponential function 596

2 Differential calculus
- **Finding gradients, tangents and maximum and minimum points**
- 2.1 Finding the gradient at a point 598
- 2.2 Drawing a tangent to a curve 599

2.3	Finding maximum and minimum points	600	
Derivatives			
2.4	Finding a numerical derivative	602	
2.5	Graphing a numerical derivative	603	
2.6	Using the second derivative	605	
3	**Integral calculus**		
3.1	Finding the value of an indefinite integral	606	
3.2	Finding the area under a curve	607	
4	**Vectors**		
4.1	Calculating a scalar product	608	
4.2	Calculating the angle between two vectors	610	
5	**Statistics and probability**		
Entering data			
5.1	Entering lists of data	612	
5.2	Entering data from a frequency table	612	
Drawing charts			
5.3	Drawing a frequency histogram from a list	613	
5.4	Drawing a frequency histogram from a frequency table	614	
5.5	Drawing a box and whisker diagram from a list	614	
5.6	Drawing a box and whisker diagram from a frequency table	616	
Calculating statistics			
5.7	Calculating statistics from a list	617	
5.8	Calculating statistics from a frequency table	618	
5.9	Calculating the interquartile range	619	
5.10	Using statistics	620	
Calculating binomial probabilities			
5.11	Use of nCr	621	
5.12	Calculating binomial probabilities	622	
Calculating normal probabilities			
5.13	Calculating normal probabilities from X-values	624	
5.14	Calculating X-values from normal probabilities	625	
Scatter diagrams, linear regression and the correlation coefficient			
5.15	Scatter diagrams using a Data & Statistics page	627	
5.16	Scatter diagrams using a Graphs page	629	

Before you start

You should know:

- Important keys on the keyboard: On, Off, menu, esc, tab, ctrl, shift, enter, del
- The home screen
- Opening new documents, adding new pages, changing settings
- Moving between pages in a document
- Panning and grabbing axes to change a window in a Graphs page
- Change window settings in a Graphs page
- Using zoom tools in a Graphs page
- Using trace in a Graphs page
- Setting the number of significant figures or decimal places

1 Functions

1.1 Graphing linear functions

Example 1

Draw the graph of the function $y = 2x + 1$	
Open a new document and add a Graphs page. The entry line is displayed at the bottom of the work area. The default graph type is Function, so the form '$f1(x)=$' is displayed. The default axes are $-10 \leq x \leq 10$ and $-6.67 \leq y \leq 6.67$. Type $2x + 1$ and press enter .	
The graph of $y = 2x + 1$ is now displayed and labeled on the screen.	

Finding information about the graph

Your GDC can give you a lot of information about the graph of a function, such as the coordinates of points of interest and the gradient (slope).

1.2 Finding a zero

The x-intercept is known as a *zero* of the function.

> At the x-intercept, $y = 0$.

Example 2

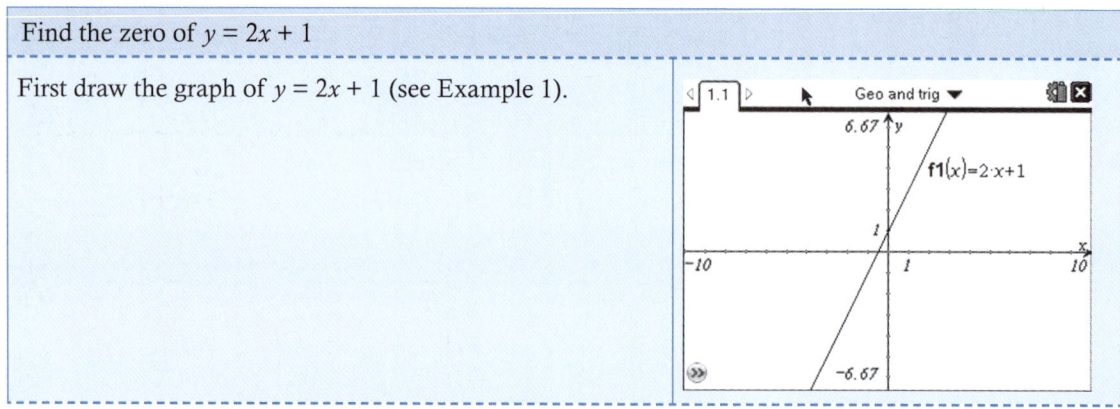

Continued on next page

Press menu 6:Analyze Graph | 1:Zero Press enter To find the zero you need to give the lower and upper bounds of a region that includes the zero. The GDC shows a line and asks you to set the lower bound. Move the line using the touchpad and choose a position to the left of the zero. Click the touchpad.	
The GDC shows another line and asks you to set the upper bound. Use the touchpad to move the line so that the region between the upper and lower bounds contains the zero. When the region contains the zero, the calculator will display the word 'zero' in a box. Click the touchpad.	
The GDC displays the zero of the function $y = 2x + 1$ at the point $(-0.5, 0)$.	

1.3 Finding the gradient (slope) of a line

The correct mathematical notation for gradient (slope) is $\frac{dy}{dx}$, and this is how the GDC denotes gradient.

Example 3

Find the gradient of $y = 2x + 1$
First draw the graph of $y = 2x + 1$ (see Example 1).

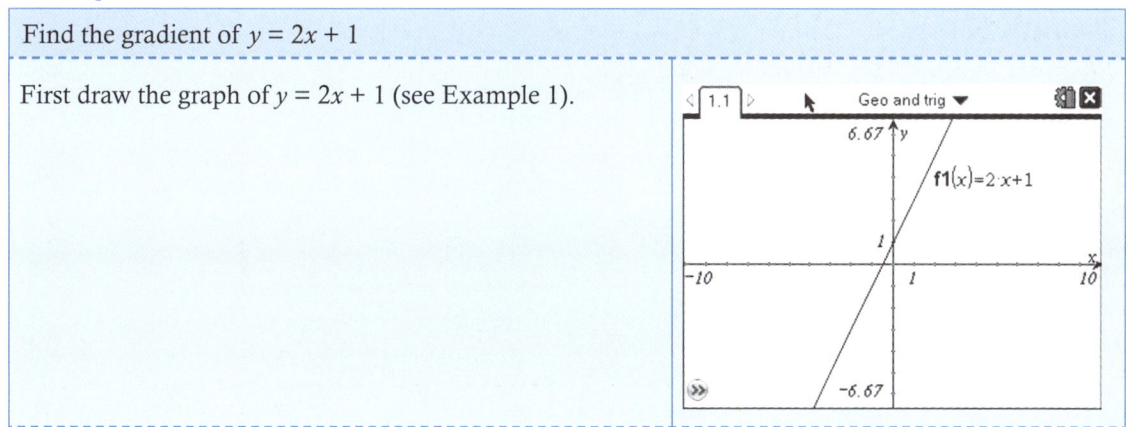

▶ Continued on next page

Press menu 6:Analyze Graph | 5: $\frac{dy}{dx}$ Press enter Use the touchpad to select a point on the line. Click the touchpad.	
The point you selected is now displayed together with the gradient of the line at that point. The gradient (slope) is 2.	
With the open-hand symbol showing, click the touchpad again. The hand is now grasping the point. Move the point along the line using the touchpad. This confirms that the gradient (slope) of $y = 2x + 1$ at every point on the line is 2.	

Simultaneous equations

1.4 Solving simultaneous equations graphically

To solve simultaneous equations graphically you draw the straight lines and then find their point of intersection. The coordinates of the point of intersection give you the solutions x and y.

> For solving simultaneous equations using a non-graphical method, see section 1.5.

Example 4

Use a graphical method to solve the simultaneous equations
$2x + y = 10$
$x - y = 2$

First rewrite both equations in the form '$y =$'.

$2x + y = 10$ $x - y = 2$
   $y = 10 - 2x$ $-y = 2 - x$
                               $y = x - 2$

> The GDC will only draw the graphs of functions that are expressed explicitly, '$y =$' as a function of x. If the equations are written in a different form, you need to rearrange them before using your GDC to solve them.

▶ Continued on next page

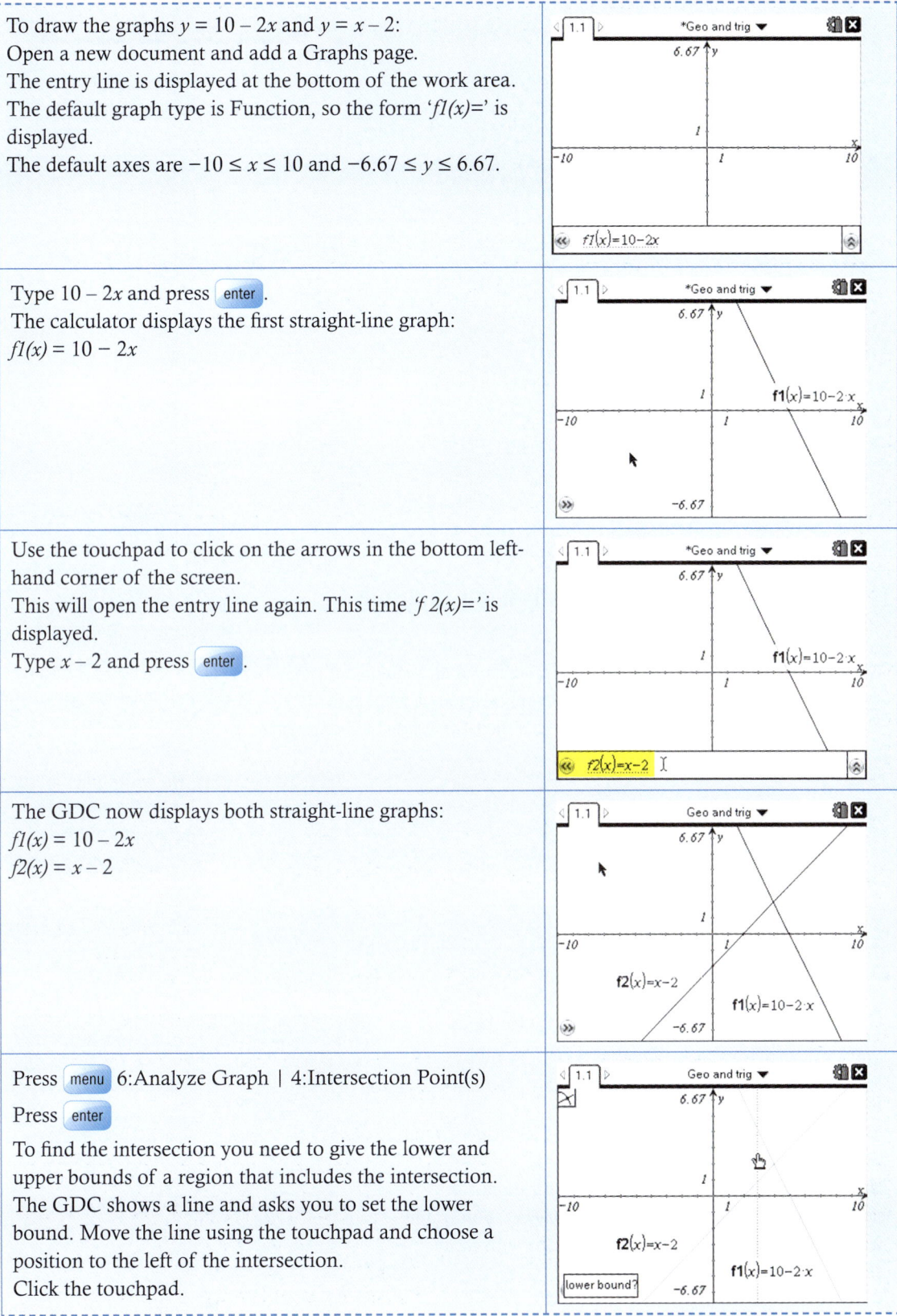

To draw the graphs $y = 10 - 2x$ and $y = x - 2$:
Open a new document and add a Graphs page.
The entry line is displayed at the bottom of the work area.
The default graph type is Function, so the form '$f1(x)=$' is displayed.
The default axes are $-10 \leq x \leq 10$ and $-6.67 \leq y \leq 6.67$.

Type $10 - 2x$ and press enter.
The calculator displays the first straight-line graph:
$f1(x) = 10 - 2x$

Use the touchpad to click on the arrows in the bottom left-hand corner of the screen.
This will open the entry line again. This time '$f2(x)=$' is displayed.
Type $x - 2$ and press enter.

The GDC now displays both straight-line graphs:
$f1(x) = 10 - 2x$
$f2(x) = x - 2$

Press menu 6:Analyze Graph | 4:Intersection Point(s)
Press enter
To find the intersection you need to give the lower and upper bounds of a region that includes the intersection. The GDC shows a line and asks you to set the lower bound. Move the line using the touchpad and choose a position to the left of the intersection.
Click the touchpad.

▶ Continued on next page

The GDC shows another line and asks you to set the upper bound.
Use the touchpad to move the line so that the region between the upper and lower bounds contains the intersection.
When the region contains the intersection, the calculator will display the word 'intersection' in a box.
Click the touchpad.

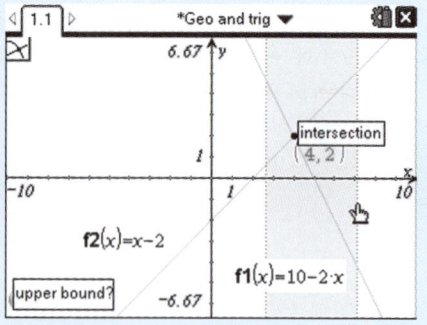

The calculator displays the intersection of the two straight lines at the point (4, 2).
The solution is $x = 4$, $y = 2$.

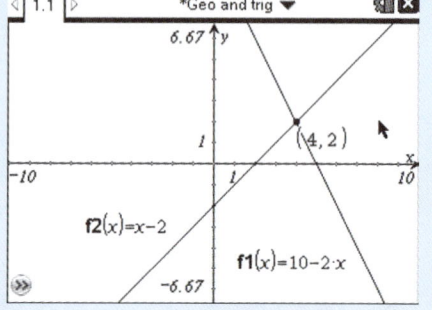

1.5 Solving simultaneous linear equations

When solving simultaneous equations in an examination, you do not need to show any method of solution. You should simply write out the equations in the correct form and then give the solutions. The GDC will do all the working for you.

You do not need the equations to be written in any particular format to use the linear equation solver, as long as they are both *linear*, that is, neither equation contains x^2 or higher order terms.

Example 5

Solve the equations:
$2x + y = 10$
$x - y = 2$

Open a new document and add a Calculator page.
Press [menu] 3:Algebra | 2:Solve Systems of Linear Equations…
Press [enter]
You will see this dialogue box, showing 2 equations and two variables, x and y.
Note: This is how you will use the linear equation solver in your examinations. In your project, you might want to solve a more complicated system with more equations and more variables.

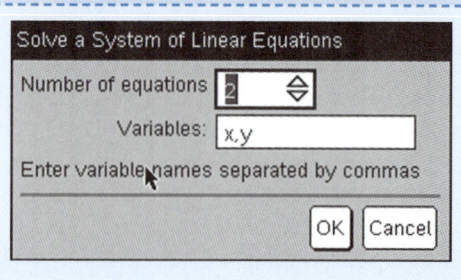

▶ Continued on next page

Press `enter` and you will see the template on the right. Type the two equations into the template, using the arrow keys ▲▼ to move within the template. Press `enter` and the GDC will solve the equations, giving the solutions in the form $\{x, y\}$.	
The solutions are $x = 4$, $y = 2$.	

Quadratic functions

1.6 Drawing a quadratic graph

Example 6

Draw the graph of $y = x^2 - 2x + 3$ and display using suitable axes.
Open a new document and add a Graphs page. The entry line is displayed at the bottom of the work area. The default graph type is Function, so the form '$f1(x) =$' is displayed. The default axes are $-10 \leq x \leq 10$ and $-6.67 \leq y \leq 6.67$.
Type $x^2 - 2x + 3$ and press `enter`. The calculator displays the curve with the default axes.
Pan the axes to get a better view of the curve. For help with panning, see your GDC manual.

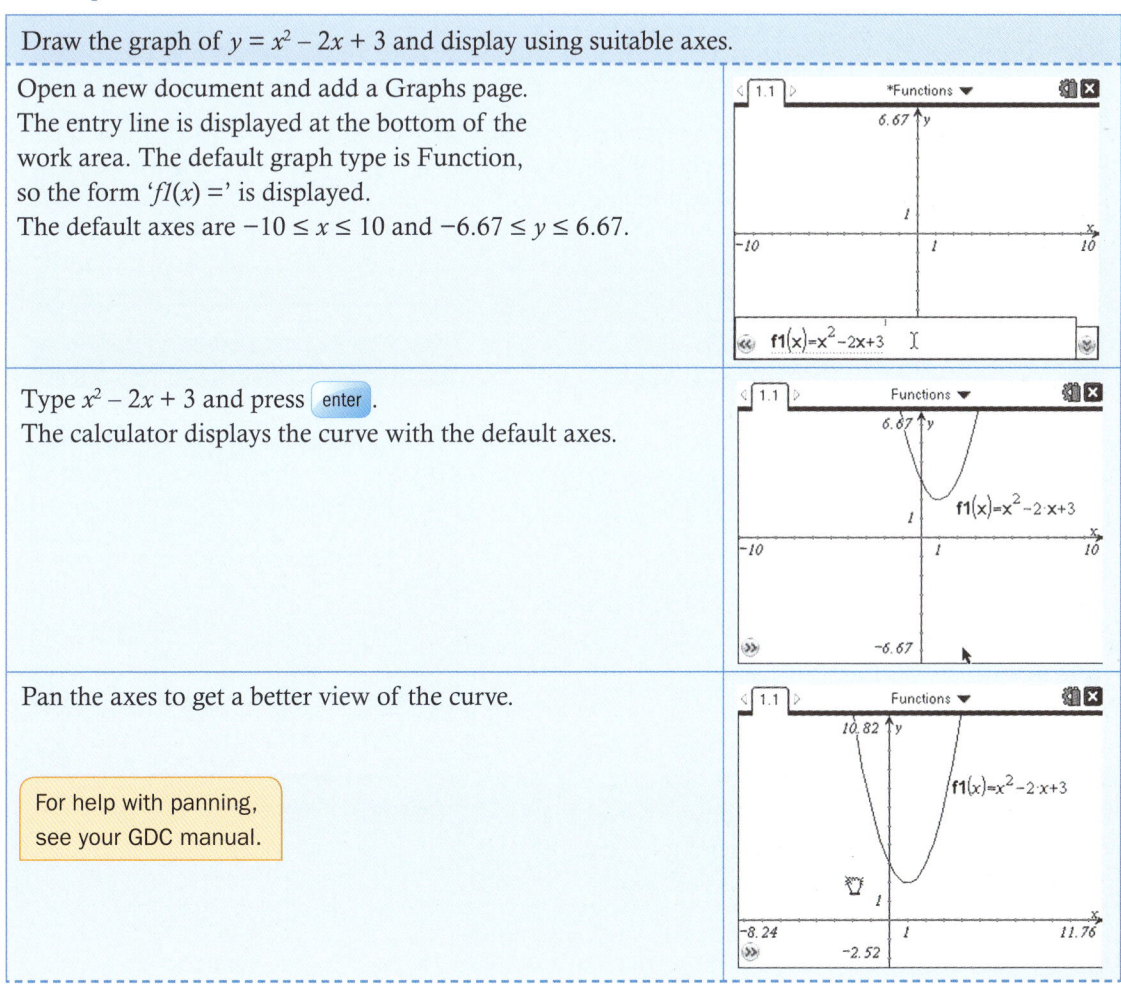

▶ Continued on next page

Chapter 17

Grab the *x*-axis and change it to make the quadratic curve fit the screen better.

For help with changing axes, see your GDC manual.

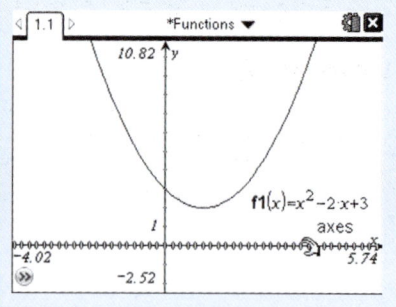

1.7 Solving quadratic equations

When solving quadratic equations in an examination, you do not need to show any method of solution. You should simply write out the equations in the correct form and then give the solutions. The GDC will do all the working for you.

Example 7

Solve $3x^2 - 4x - 2 = 0$

Press menu 3:Algebra | 3:Polynomial Tools | 1:Find Roots of a Polynomial...
Press enter
You will see this dialogue box, showing a polynomial of degree 2 (a quadratic equation) with real roots. You do not need to change anything.
Press enter

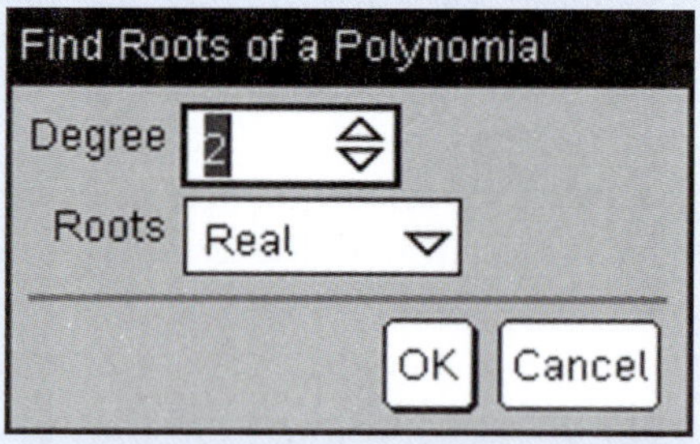

Another dialogue box opens for you to enter the equation. The general form of the quadratic equation is $a_2x^2 + a_1x + a_0 = 0$, so enter the coefficients in a_2, a_1 **and** a_0.
Here, $a_2 = 3$, $a_1 = -4$ and $a_0 = -2$. Be sure to use the (-) key to enter the negative values. Use the tab key to move around the dialogue box.
Press enter and the GDC will solve the equation, giving the roots in the form $\{x, y\}$.

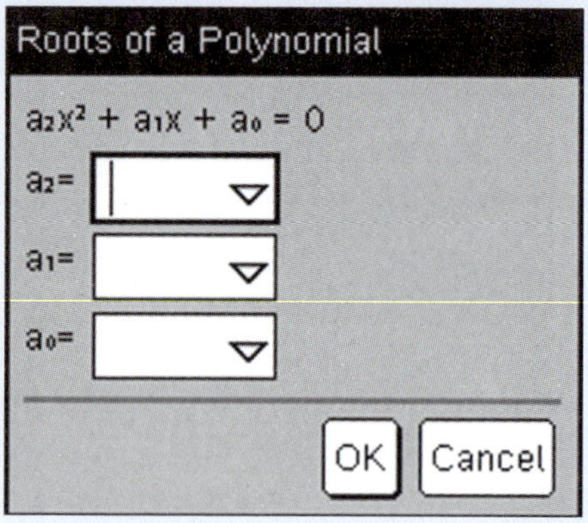

The solutions are $x = -0.387$ or $x = 1.72$ (to 3 sf).

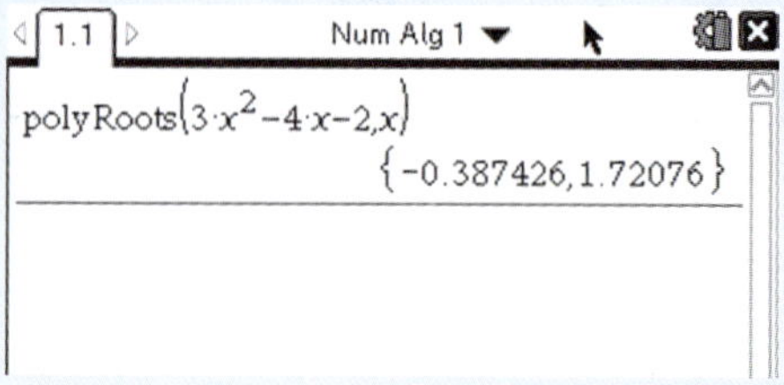

578 Using a graphic display calculator

1.8 Finding a local minimum or maximum point

Example 8

Find the minimum point on the graph of $y = x^2 - 2x + 3$

First draw the graph of $y = x^2 - 2x + 3$ (see Example 6).

Method 1: Using a table
You can look at the graph **and** a table of the values by using a split screen.
Press `menu` 2:View | 9:Show Table
(or simply press `ctrl` `T`)
The minimum value shown in the table is 2 when $x = 1$.

Look more closely at the values of the function around $x = 1$.
Change the settings in the table.
Choose any cell and press `menu` 5:Table | 5:Edit Table Settings…
Set Table Start to 0.98 and Table Step to 0.01.
Press `enter`

The table shows that the function has larger values at points around (1, 2). We can conclude that the point (1, 92) is a local minimum on the curve.

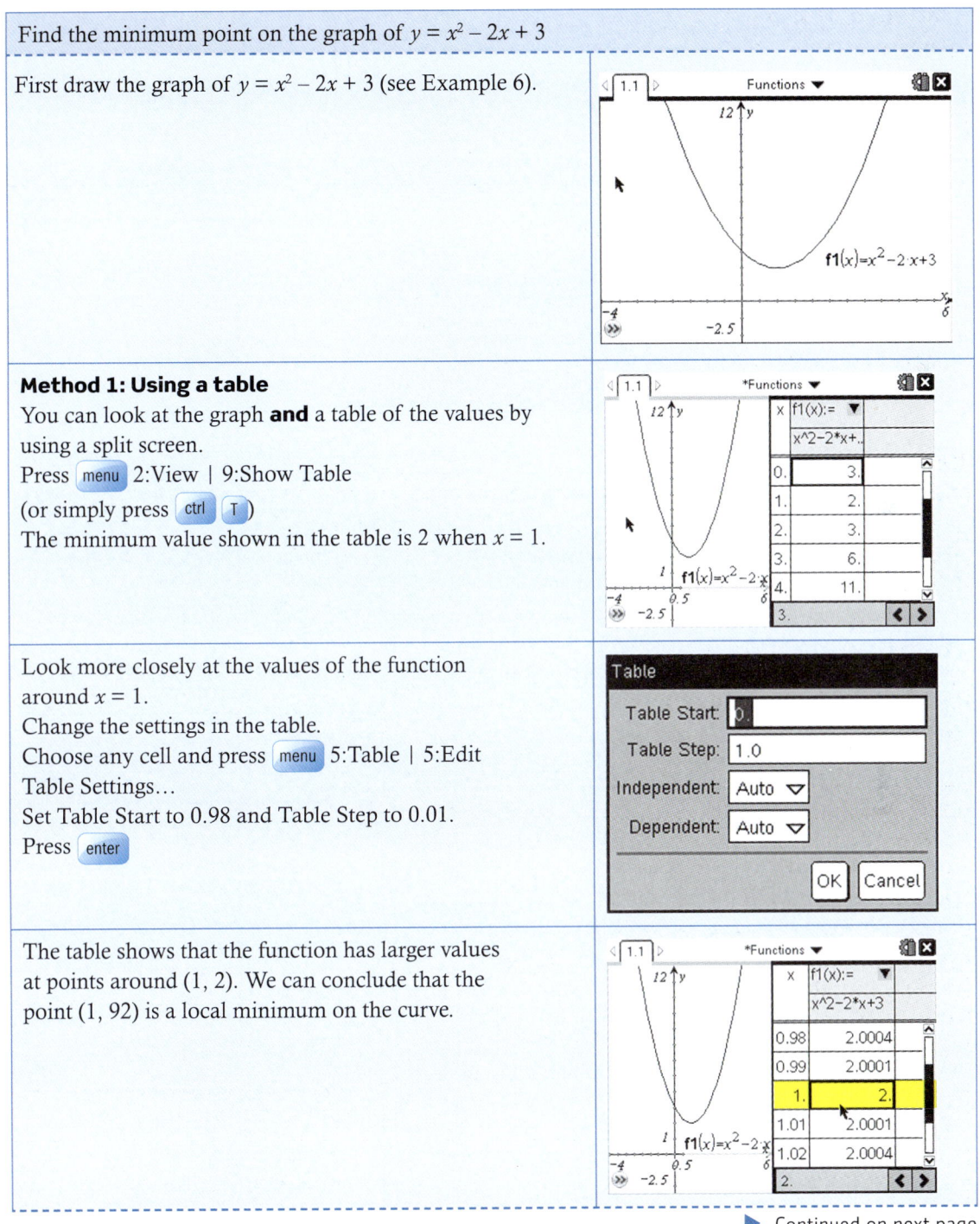

▶ Continued on next page

Chapter 17

Method 2: Using the minimum function

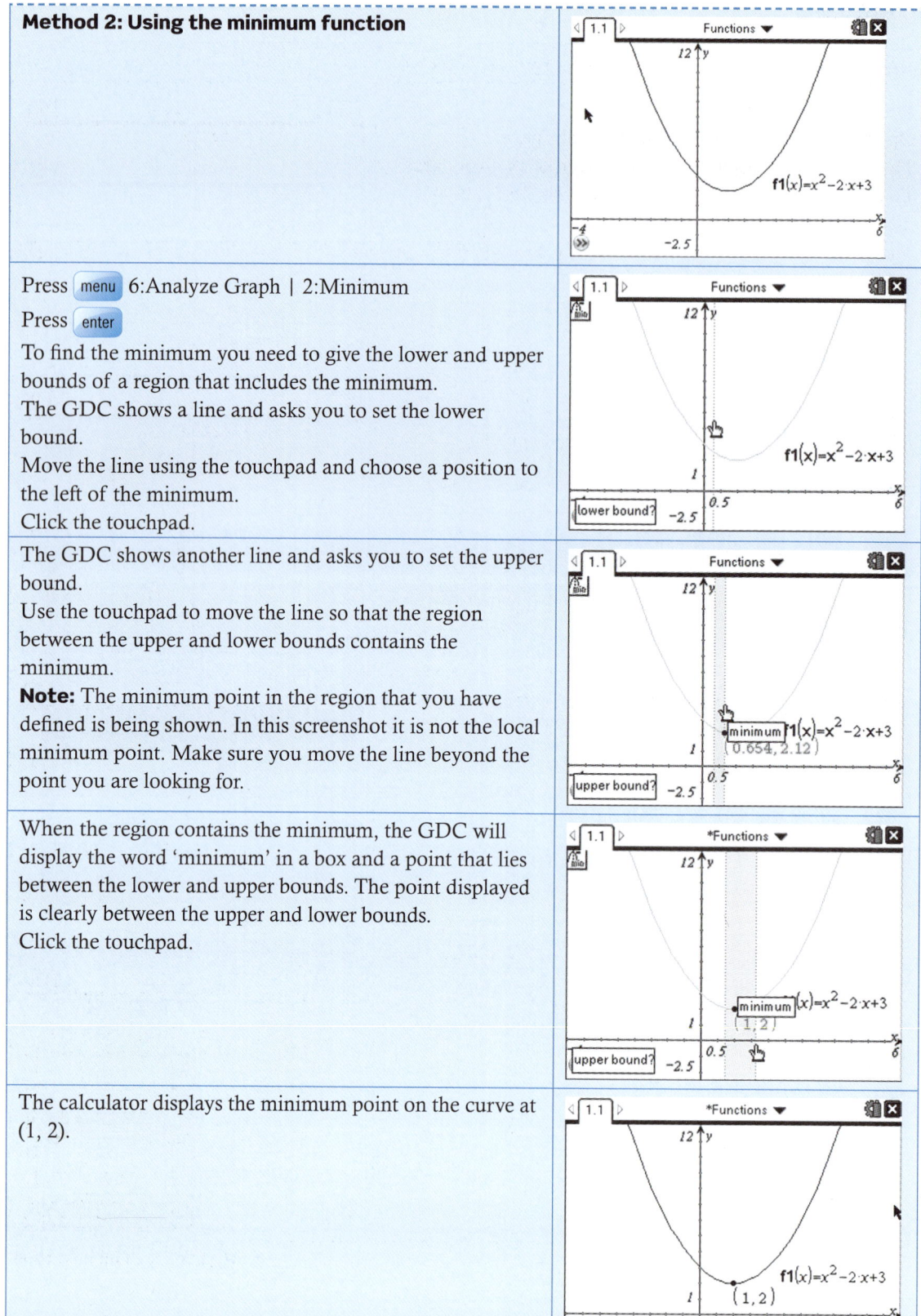

Press menu 6:Analyze Graph | 2:Minimum
Press enter
To find the minimum you need to give the lower and upper bounds of a region that includes the minimum.
The GDC shows a line and asks you to set the lower bound.
Move the line using the touchpad and choose a position to the left of the minimum.
Click the touchpad.

The GDC shows another line and asks you to set the upper bound.
Use the touchpad to move the line so that the region between the upper and lower bounds contains the minimum.
Note: The minimum point in the region that you have defined is being shown. In this screenshot it is not the local minimum point. Make sure you move the line beyond the point you are looking for.

When the region contains the minimum, the GDC will display the word 'minimum' in a box and a point that lies between the lower and upper bounds. The point displayed is clearly between the upper and lower bounds.
Click the touchpad.

The calculator displays the minimum point on the curve at (1, 2).

Example 9

Find the maximum point on the graph of $y = -x^2 + 3x - 4$

First draw the graph of $y = -x^2 + 3x - 4$:
Open a new document and add a Graphs page.
The entry line is displayed at the bottom of the work area.
The default graph type is Function, so the form '$f1(x)=$' is displayed.
The default axes are $-10 \leq x \leq 10$ and $-6.67 \leq y \leq 6.67$.

Type $-x^2 + 3x - 4$ and press enter.
The GDC displays the curve with the default axes.

Pan the axes to get a better view of the curve.
Grab the x-axis and change it to make the quadratic curve fit the screen better.

> For help with panning or changing axes, see your GDC manual.

Method 1: Using a table
You can look at the graph **and** a table of the values by using a split screen.
Press menu 2:View | 9:Show Table
(or simply press ctrl T)
The maximum value shown in the table is -2 when $x = 1$ and $x = 2$.

▶ Continued on next page

Chapter 17   581

Look more closely at the values of the function between $x = 1$ and $x = 2$. Change the settings in the table. Choose any cell and press `menu` 5:Table | 5:Edit Table Settings… Set Table Start to 1.0 and Table Step to 0.1. Press `enter`	Table Table Start: 0. Table Step: 1.0 Independent: Auto ▼ Dependent: Auto ▼ OK Cancel
Scroll down the table and you can see that the function has its largest value at $(1.5, -1.75)$. We can conclude that the point $(1.5, -1.75)$ is a local maximum on the curve.	$f_1(x) := -x^2 + 3*x - 4$ 1.3 −1.79 1.4 −1.76 1.5 −1.75 1.6 −1.76 1.7 −1.79 −1.75
Method 2: Using the maximum function	$f_1(x) = -x^2 + 3 \cdot x - 4$
Press `menu` 6:Analyze Graph | 3:Maximum Press `enter` To find the maximum you need to give the lower and upper bounds of a region that includes the maximum. The GDC shows a line and asks you to set the lower bound. Move the line using the touchpad and choose a position to the left of the maximum. Click the touchpad.	lower bound? 0.7 $f_1(x) = -x^2 + 3 \cdot x - 4$
The GDC shows another line and asks you to set the upper bound. Use the touchpad to move the line so that the region between the upper and lower bounds contains the maximum. **Note:** The maximum point in the region that you have defined is being shown. In this screenshot it is not the local maximum point. Make sure you move the line beyond the point you are looking for.	maximum (1.16, −1.87) $f_1(x) = -x^2 + 3 \cdot x - 4$ upper bound? 0.7

▶ Continued on next page

When the region contains the maximum, the GDC will display the word 'maximum' in a box and a point that lies between the lower and upper bounds. The point displayed is clearly between the upper and lower bounds. Click the touchpad.	
The GDC displays the maximum point on the curve at (1.5, −1.75).	

Exponential functions

1.9 Drawing an exponential graph

Example 10

Draw the graph of $y = 3^x + 2$	
Open a new document and add a Graphs page. The entry line is displayed at the bottom of the work area. The default graph type is Function, so the form '$f1(x) =$' is displayed. The default axes are $-10 \leq x \leq 10$ and $-6.67 \leq y \leq 6.67$.	
Type $y = 3^x + 2$ and press enter. (**Note:** Type 3 ^ x ▶ to enter 3^x. The ▶ returns you to the baseline from the exponent.) The GDC displays the curve with the default axes.	

▶ Continued on next page

Pan the axes to get a better view of the curve.

For help with panning, see your GDC manual.

Grab the *x*-axis and change it to make the exponential curve fit the screen better.

For help with changing axes, see your GDC manual.

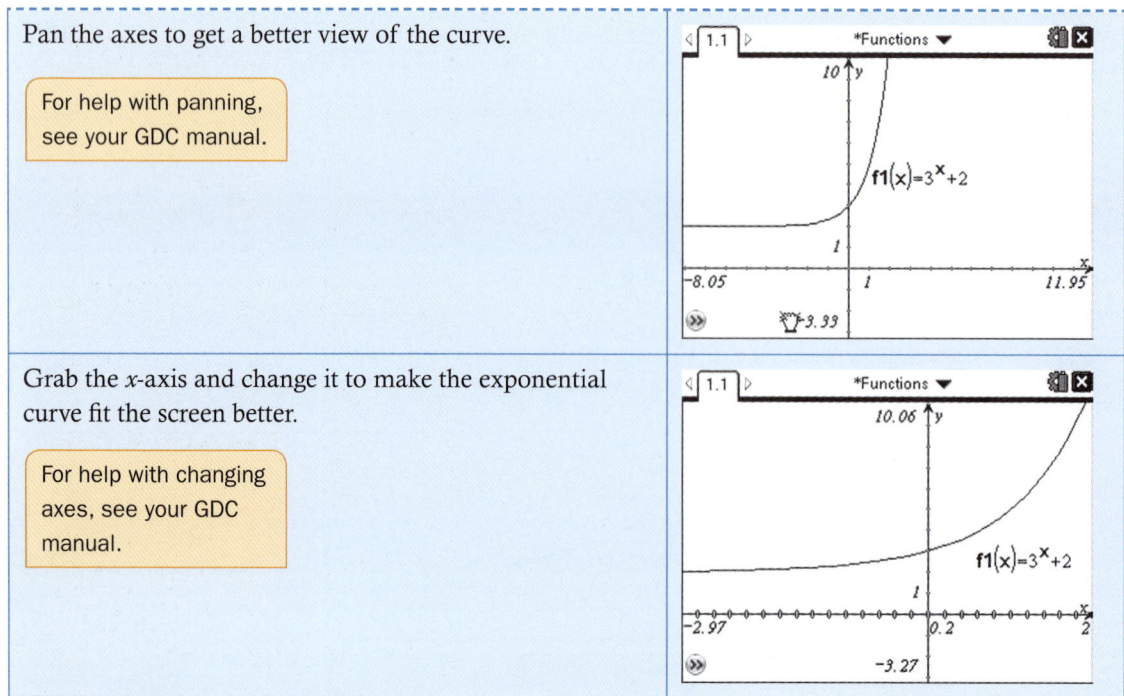

1.10 Finding a horizontal asymptote

Example 11

Find the horizontal asymptote to the graph of $y = 3^x + 2$

First draw the graph of $y = 3^x + 2$ (see Example 10).

You can look at the graph **and** a table of the values by using a split screen.

Press menu 2:View | 9:Show Table

(or simply press ctrl T)
The values of the function are clearly decreasing as $x \to 0$.

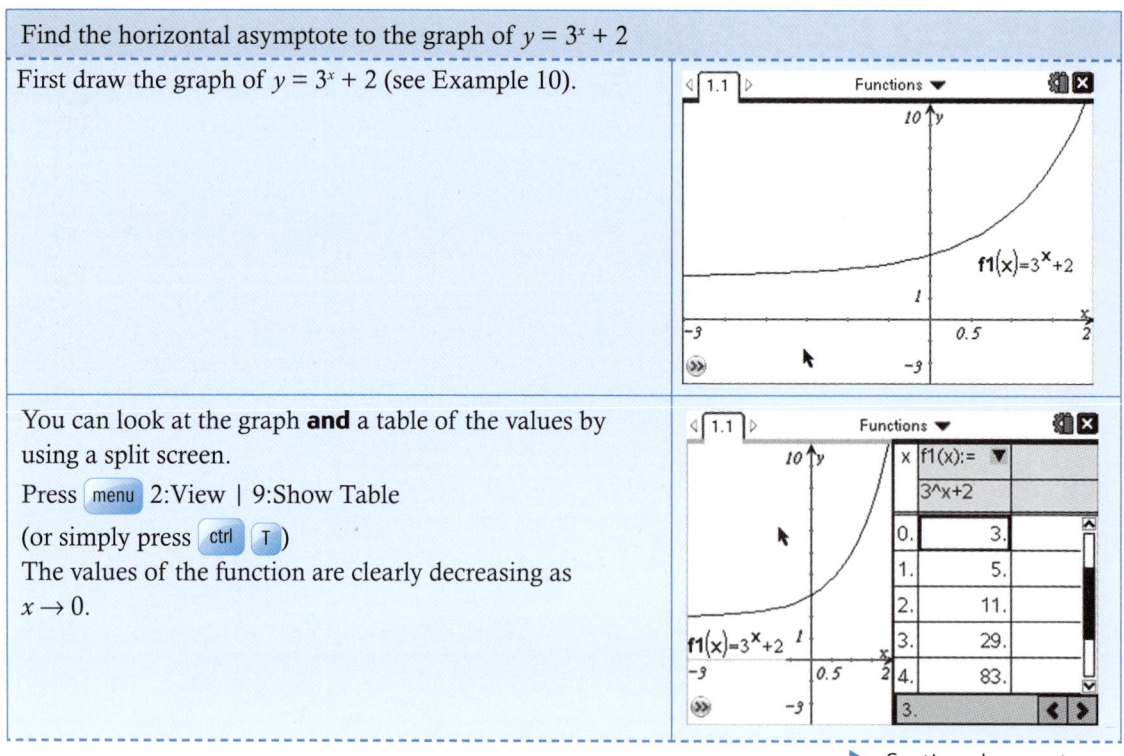

▶ Continued on next page

584 Using a graphic display calculator

Press and hold ▲ to scroll up the table.
The table shows that as the values of x get smaller, $f1(x)$ approaches 2.

Eventually, the value of $f1(x)$ reaches 2. On closer inspection, you can see, at the bottom of the screen, that the actual value of $f1(x)$ is 2.000 001 881 6...
We can say that $f1(x) \to 2$ as $x \to -\infty$.
The line $x = 2$ is a horizontal asymptote to the curve $y = 3^x + 2$.

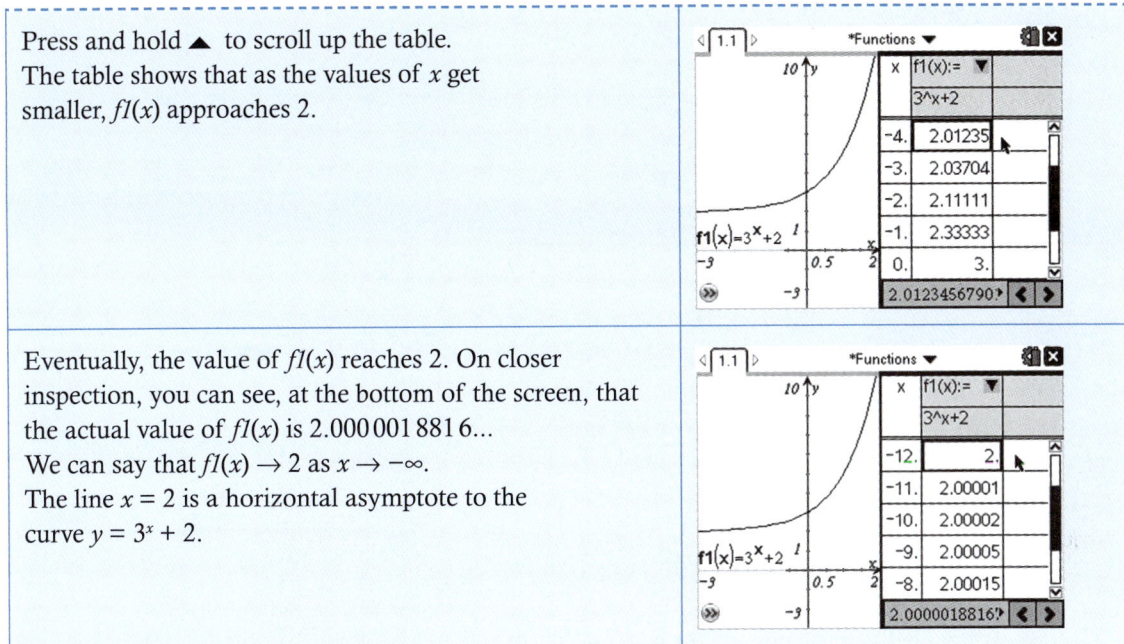

Logarithmic functions

1.11 Evaluating logarithms

Example 12

Evaluate $\log_{10} 3.95$, $\ln 10.2$ and $\log_5 2$.

Open a new document and add a Calculator page.
Press ctrl log to open the log template.
Enter the base and the argument then press enter del

For natural logarithms it is possible to use the same method, with the base equal to e, but it is far less time consuming to press ctrl ln .

Note that the GDC will evaluate logarithms with any base without having to use the change of base formula.

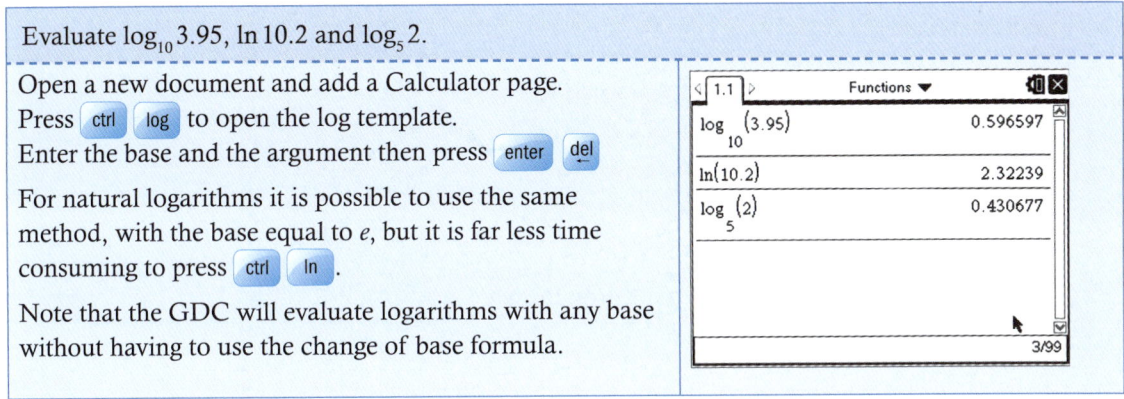

1.12 Finding an inverse function

The inverse of a function can be found by interchanging the x and y values. Geometrically this can be done by reflecting points in the line $y = x$.

Example 13

Show that the inverse of the function $y = 10^x$ is $y = \log_{10} x$ by reflecting $y = 10^x$ in the line $y = x$.

Open a new document and add a Graphs page.
First we will draw the line $y = x$. So that it can be recognised the axis of reflection, it has to be drawn and not plotted as a function.

Press `menu` 7: Points & Lines | 1: Point

Then type `(` `1` `enter` `1` `enter` then `(` `4` `enter` `4` `enter` `esc`

This will plot the points $(1, 1)$ and $(4, 4)$, which both lie on the line $y = x$.

Press `menu` 7: Points & Lines | 4: Line

Select both the points you have plotted and draw a line through them.

Press `esc` to exit the drawing function.

Click in the entry line at the bottom of the work area. The default graph type is Function, so the form "$f1(x)=$" is displayed.

Type $10\wedge x$ and press `enter`.

The calculator displays the function with the default axes, $-10 \leq x \leq 10$ and $-6.67 \leq y \leq 6.67$.

▶ Continued on next page

Press menu 7: Points & Lines | 2: Point On Select the curve with the touchpad (you will see that it is highlighted when it is selected). You can place a point anywhere on the curve.	
Press menu B: Transformation | 2: Reflection Use the touch pad to select the point that you just placed on the curve and then the line $y = x$. Press esc when you have finished. You should see the reflected image of the point in the line $y = x$.	
Press menu A: Construction | 6: Locus Use the touch pad to select each of the points. The calculator will display the locus of the reflection as the point moves along the curve.	
Click in the entry line at the bottom of the work area. "$f2(x)=$" is displayed. Type $\log_{10}(x)$ and press enter. The reflected curve and the logarithmic function coincide, showing that $y = \log_{10} x$ is inverse of the function $y = 10^x$.	

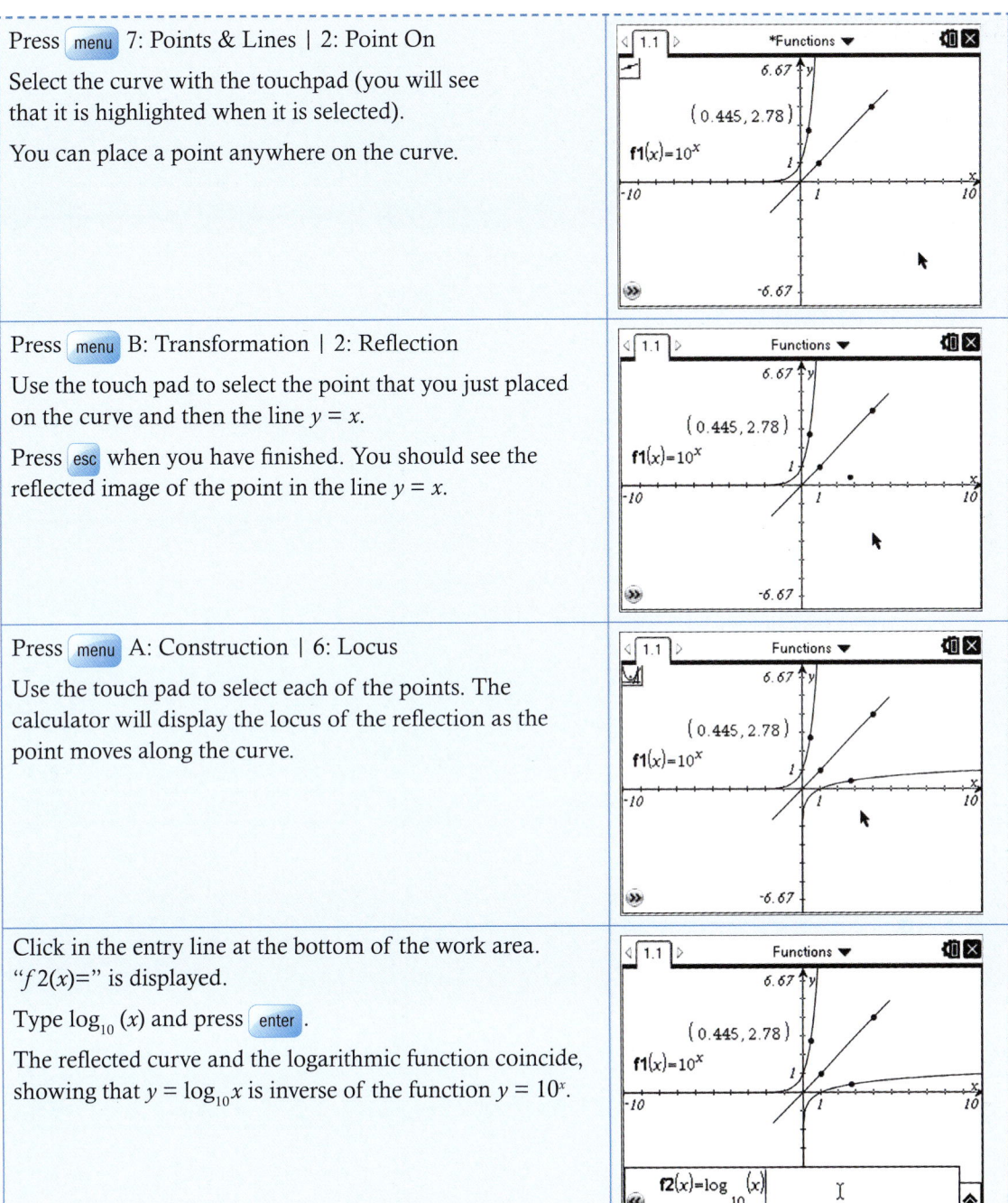

1.13 Drawing a logarithmic graph

Example 14

Draw the graph of $y = 2\log_{10} x + 3$.

Open a new document and add a Graphs page. The entry line is displayed at the bottom of the work area. The default graph type is Function, so the form "$f1(x)=$" is displayed. The default axes are $-10 \leq x \leq 10$ and $-6.67 \leq y \leq 6.67$.	
Type $2\log_{10}(x) + 3$ and press enter. (**Note:** Type 2 ctrl log and enter 10 as the base of the logarithm. Enter x in the argument section of the template, use the ▶ to move beyond the brackets to enter +3) The calculator displays the curve with the default axes.	
Pan the axes to get a better view of the curve.	
Grab the x-axis and change it to make the logarithmic curve fit the screen better.	

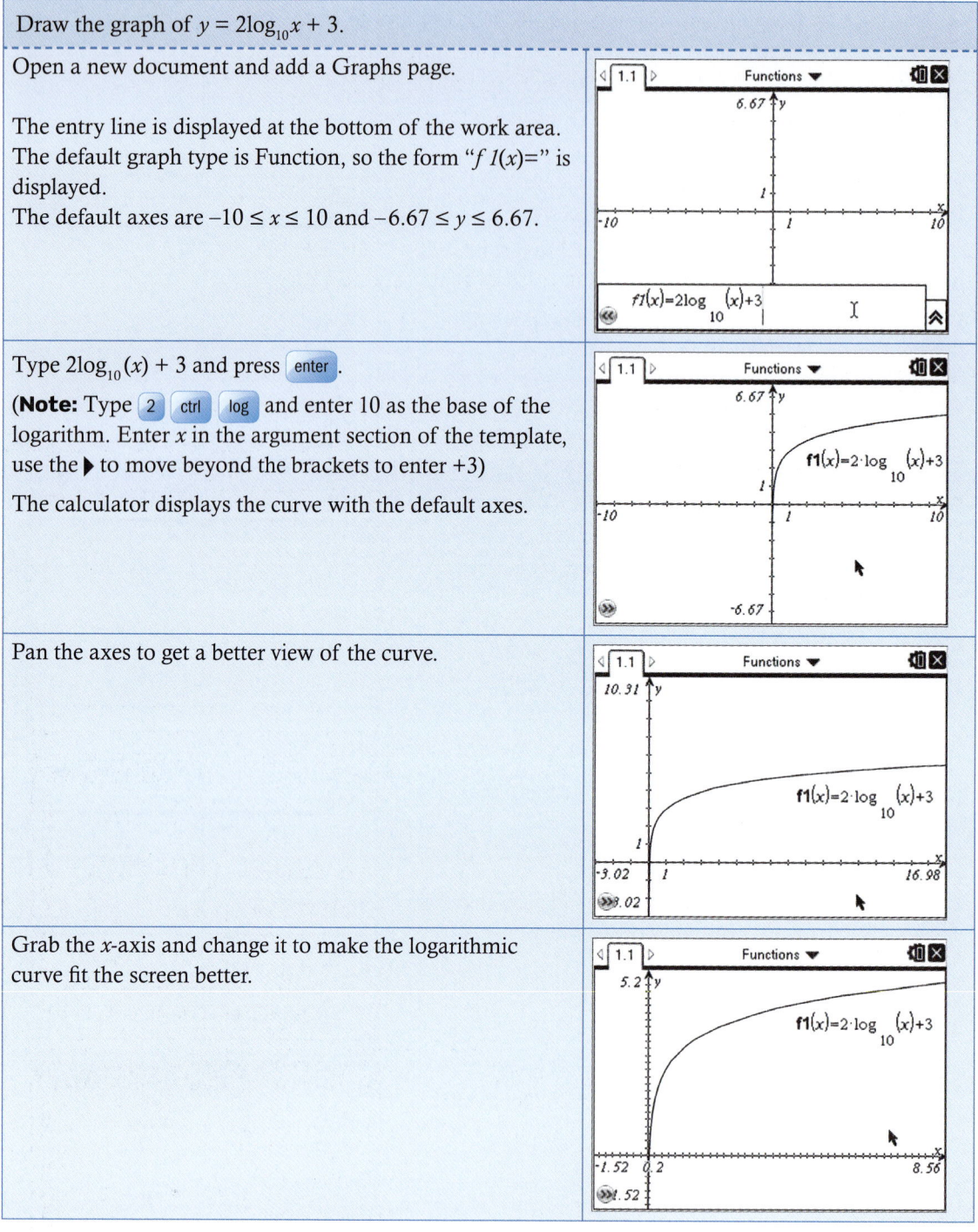

Trigonometric functions

1.14 Degrees and radians

Work in trigonometry will be carried out either in degrees or radians. It is important, therefore, to be able to check which mode the calculator is in and to be able to switch back and forth. On the TI-Nspire, there are three separate settings to make: general, graphing and geometry. The defaults for general and graphing are radians and for geometry the default is degrees. Geometry is only used for drawing plane geometrical figures. Normally the two important settings are general and graphing. General refers to the angle used in calculations and graphing is for drawing trigonometric graphs.

Example 15

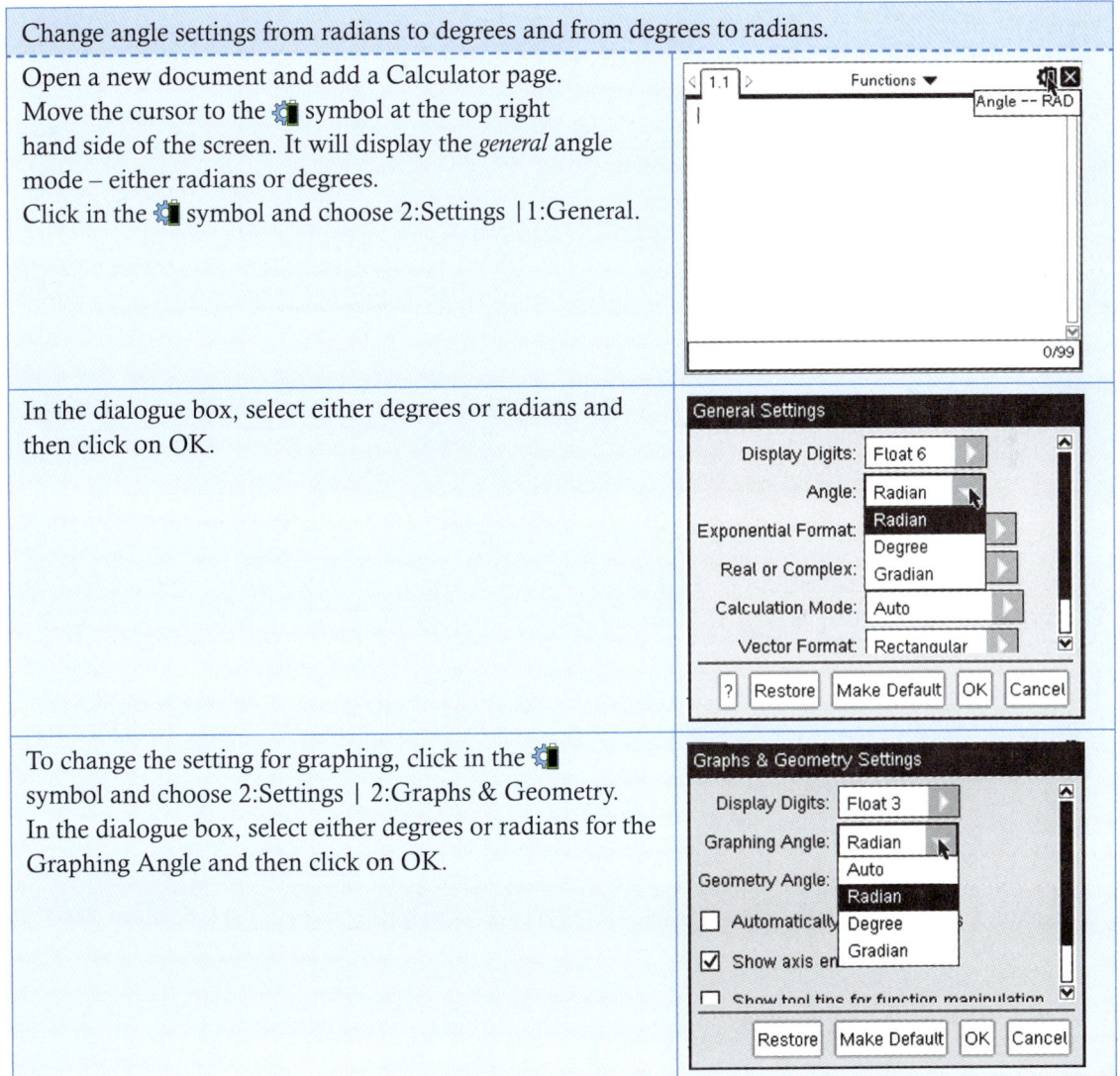

1.15 Drawing trigonometric graphs

Example 16

Draw the graph of $y = 2\sin\left(x + \dfrac{\pi}{6}\right) + 1$.

Open a new document and add a Graphs page.
Press `menu` 4:Window / Zoom | 8:Zoom - Trig
The entry line is displayed at the bottom of the work area.
The default graph type is Function, so the form "$f1(x)=$" is displayed.
The default axes are $-6.28 \leq x \leq 6.28$ and $-4.19 \leq y \leq 4.19$.
These are the basic axes for graphing trigonometric graphs with x between -2π and 2π. If the calculator is in degree mode, the x-axis will be between -360 and 360.

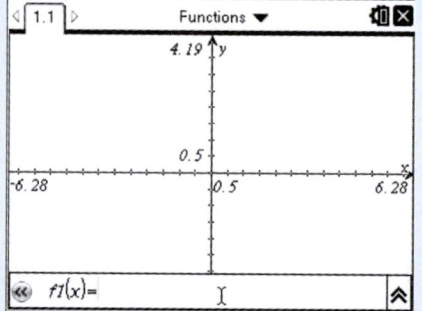

Type $y = 2\sin\left(x + \dfrac{\pi}{6}\right) + 1$ and press `enter`.

To enter sin, press `trig` and choose sin from the dialogue box.

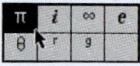

To enter π, press `π▸` and choose π from the dialogue box.

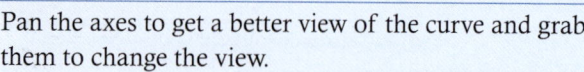

Pan the axes to get a better view of the curve and grab them to change the view.

It is also useful to change the x-axis scale to a multiple of π, such as $\dfrac{\pi}{6}$ as this will often show the positions of intercepts and turning points more clearly.
Change the scale by pressing `menu` 4:Window / Zoom | 1:Window Settings
XScale: `pi/6`
Type pi/6 in the dialogue box for XScale.

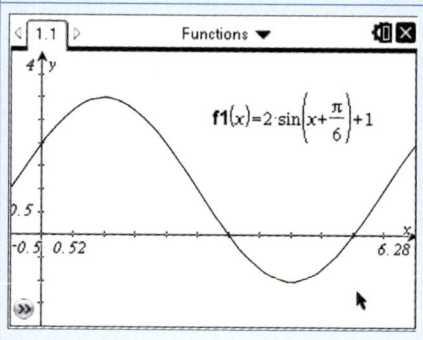

More complicated functions

1.16 Solving a combined quadratic and exponential equation

> Follow the same GDC procedure when solving simultaneous equations graphically or solving a combined quadratic and exponential equation. See Examples 4 and 17.

Example 17

Solve the equation $x^2 - 2x + 3 = 3 \cdot 2^{-x} + 4$

> To solve the equation, find the point of intersection of the quadratic function $f1(x) = x^2 - 2x + 3$ with the exponential function $f2(x) = 3 \cdot 2^{-x} + 4$.

To draw the graphs $f1(x) = x^2 - 2x + 3$ and $f2(x) = 3 \cdot 2^{-x} + 4$:
Open a new document and add a Graphs page.
The entry line is displayed at the bottom of the work area.
The default graph type is Function, so the form '$f1(x)=$' is displayed.
The default axes are $-10 \leq x \leq 10$ and $-6.67 \leq y \leq 6.67$.

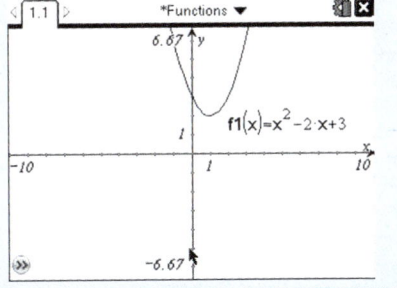

Type $x^2 - 2x + 3$ and press enter.
The GDC displays the first curve:
$f1(x) = x^2 - 2x + 3$

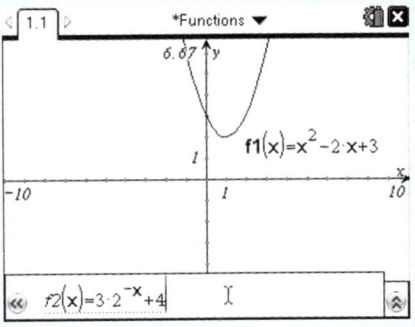

Use the touchpad to click on the arrows in the bottom left-hand corner of the screen.
This will open the entry line again. This time '$f2(x)=$' is displayed.
Type $3 \cdot 2^{-x} + 4$ and press enter.

The GDC displays both curves:
$f1(x) = x^2 - 2x + 3$
$f2(x) = 3 \cdot 2^{-x} + 4$

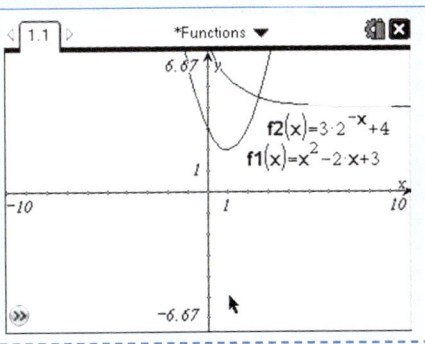

▶ Continued on next page

Pan the axes to get a better view of the curves. For help with panning, see your GDC manual.	
Press menu 6:Analyze Graph | 4:Intersection Point(s) Press enter To find the intersection you need to give the lower and upper bounds of a region that includes the intersection. The GDC shows a line and asks you to set the lower bound. Move the line using the touchpad and choose a position to the left of the intersection. Click the touchpad.	
The GDC shows another line and asks you to set the upper bound. Use the touchpad to move the line so that the region between the upper and lower bounds contains the intersection. When the region contains the intersection, the calculator will display the word 'intersection' in a box. Click the touchpad.	
The GDC displays the intersection of the two curves at the point (2.58, 4.5). The solution is $x = 2.58$.	

Modeling

1.17 Using sinusoidal regression

Note: the notation $\sin^2 x$, $\cos^2 x$, $\tan^2 x$, ... is a mathematical convention that has little algebraic meaning. To enter these functions on the GDC, you *should* enter $(\sin(x))^2$, etc. However, the calculator will conveniently interpret $\sin(x)^2$ and translate it as $(\sin(x))^2$.

Example 18

It is known that the following data can be modeled using a sine curve.

x	0	1	2	3	4	5	6	7
y	6.9	9.4	7.9	6.7	9.2	8.3	6.5	8.9

Use sine regression to find a function to model this data.

Open a new document and add a Lists & Spreadsheet page.
Type 'x' in the first cell and 'y' in the cell to its right.
Type the numbers from the x-list in the first column and those from the y-list in the second.
Use the ▼ ▲ ◄► keys to navigate around the spreadsheet.

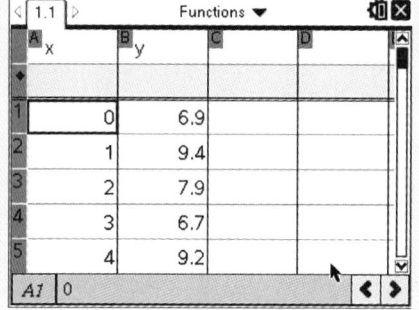

Press ⌂ On and add a new graphs page to your document.

Press menu 3:Graph Type | 4:Scatter Plot

Press enter

The entry line is displayed at the bottom of the work area.
Scatter plot type is displayed.
Enter the names of the lists, x and y, into the scatter plot function
Use the tab key to move from x to y.
Press enter del

Adjust your window settings to show your data and the x- and y-axes.
You now have a scatter plot of x against y.

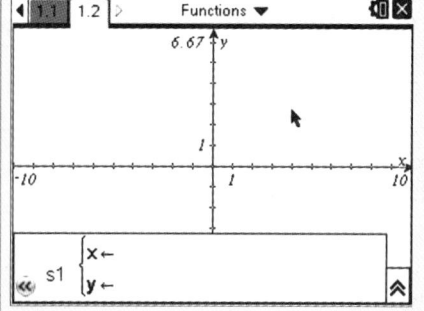

Press ctrl ◄ to return to the Lists & Spreadsheet page.
Select an empty cell and press menu 4:Statistics | Stat Calculations | C:Sinusoidal Regression...
Press enter
From the drop down menus choose 'x' for X List and 'y' for Y List. You should press tab to move between the fields.
Press enter

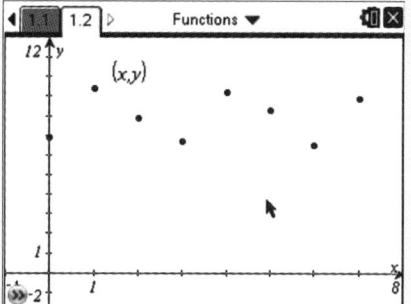

▶ Continued on next page

On screen, you will see the result of the sinusoidal regression in lists next to the lists for x and y.
The equation is in the form $y = a\sin(bx + c) + d$ and you will see the values of a, b, c and d displayed separately.
The equation of the sinusoidal regression line is
$y = 1.51\sin(2.00x - 0.80) + 7.99$

Press ctrl ▶ to return to the Graphs page.
Using the touchpad, click on ›› to open the entry line at the bottom of the work area.
You will see that the equation of the regression line has been pasted into $f1(x)$.
Press enter

The regression line is now shown on the graph.

1.18 Using transformations to model a quadratic function

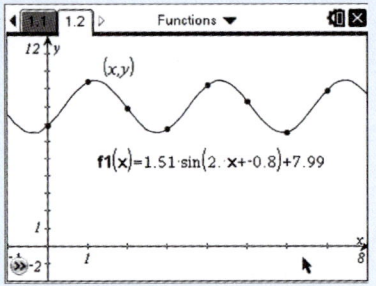

You can also model a linear function by finding the equation of the least squares regression line (see section 5.15).

Example 19

This data is approximately connected by a quadratic function.

x	−2	−1	0	1	2	3	4
y	9.1	0.2	−4.8	−5.9	−3.1	4.0	15.0

Find a function that fits the data.

Transform a basic quadratic curve to find an equation to fit some quadratic data.

Open a new document and add a Lists & Spreadsheet page.
Enter the data in two lists:
Type 'x' in the first cell and 'y' in the cell to its right.
Enter the x-values in the first column and the y-values in the second. Remember to use (−) to enter a negative number.
Use the ▼▲◀▶ keys to navigate around the spreadsheet.

▶ Continued on next page

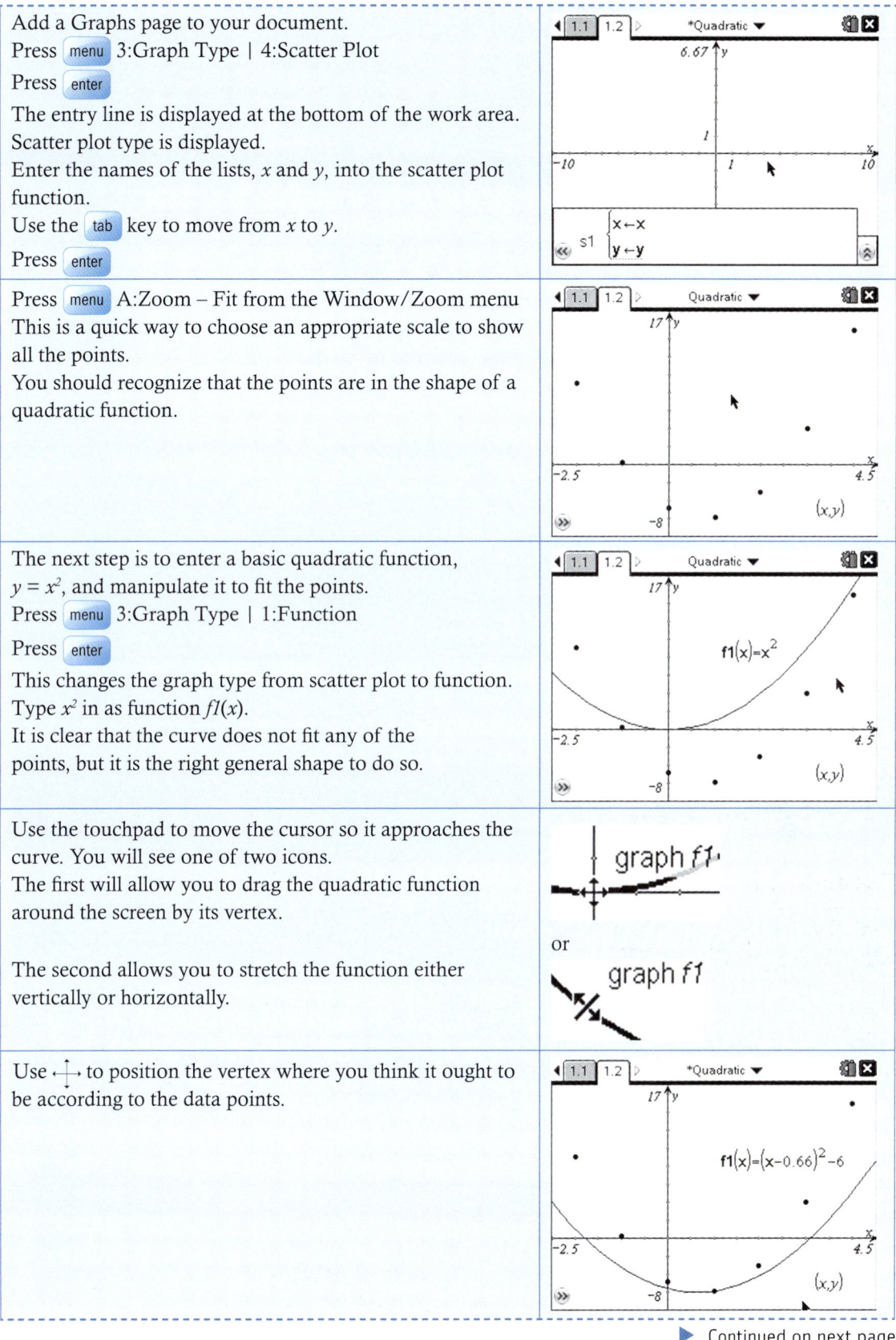

Add a Graphs page to your document.
Press menu 3:Graph Type | 4:Scatter Plot
Press enter
The entry line is displayed at the bottom of the work area. Scatter plot type is displayed.
Enter the names of the lists, x and y, into the scatter plot function.
Use the tab key to move from x to y.
Press enter

Press menu A:Zoom – Fit from the Window/Zoom menu
This is a quick way to choose an appropriate scale to show all the points.
You should recognize that the points are in the shape of a quadratic function.

The next step is to enter a basic quadratic function, $y = x^2$, and manipulate it to fit the points.
Press menu 3:Graph Type | 1:Function
Press enter
This changes the graph type from scatter plot to function.
Type x^2 in as function $f1(x)$.
It is clear that the curve does not fit any of the points, but it is the right general shape to do so.

Use the touchpad to move the cursor so it approaches the curve. You will see one of two icons.
The first will allow you to drag the quadratic function around the screen by its vertex.

The second allows you to stretch the function either vertically or horizontally.

Use ↔ to position the vertex where you think it ought to be according to the data points.

▶ Continued on next page

Chapter 17

Use ✕ to adjust the stretch of the curve.
Make some final fine adjustments using both the tools until you have a good fit to the data points.
The equation of the function that fits the data is:
$f1(x) = 2(x - 0.75)^2 - 6.11$

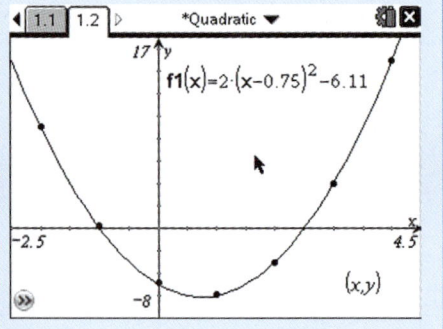

1.19 Using sliders to model an exponential function

Example 20

In general, an exponential function has the form $y = ka^x + c$.
For this data, it is known that the value of a is 1.5, so $y = k(1.5)^x + c$.

x	−3	−2	−1	0	1	2	3	4	5	6	7	8
y	3.1	3.2	3.3	3.5	3.8	4.1	4.7	5.5	6.8	8.7	11.5	15.8

Find the values of the constants k and c.

Open a new document and add a Lists & Spreadsheet page.
Enter the data in two lists:
Type '*x*' in the first cell and '*y*' in the cell to its right.
Enter the *x*-values in the first column and the *y*-values in the second. Remember to use (−) to enter a negative number.
Use the ▼▲◄► keys to navigate around the spreadsheet.

Add a Graphs page to your document.
Press [menu] 3:Graph Type | 4:Scatter Plot
Press [enter]
The entry line is displayed at the bottom of the work area.
Scatter plot type is displayed.
Enter the names of the lists, *x* and *y*, into the scatter plot function.
Use the [tab] key to move from *x* to *y*.
Press [enter]

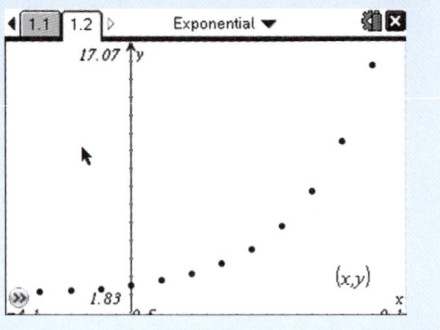

▶ Continued on next page

Adjust the window settings to fit the data and to display the axes clearly.	
Press menu 1:Actions | A:Insert Slider Position the slider somewhere where it is not in the way and change the name of the constant to k. Repeat and add a second slider for c. For help with sliders, see your GDC manual.	
Press menu 3:Graph Type | 1:Function Press enter This changes the graph type from scatter plot to function. Type $k.(1.5)^x + c$ in as function $f1(x)$.	
Try adjusting the sliders. You can get the curve closer to the points but they are not sufficiently adjustable to get a good fit.	

You can change the slider settings by selecting the slider, pressing ctrl menu and selecting 1:Settings.
Change the default values for k to:
    Minimum 0
    Maximum 2
    Step Size 0.1
Change the default values for c to:
    Minimum 0
    Maximum 4
    Step Size 0.1	

▶ Continued on next page

You can now adjust the sliders to get a much better fit to the curve.
The screen shows the value of k is 0.5 and c is 3.
So the best fit for the equation of the function is approximately $y = 0.5(1.5)^x + 3$.

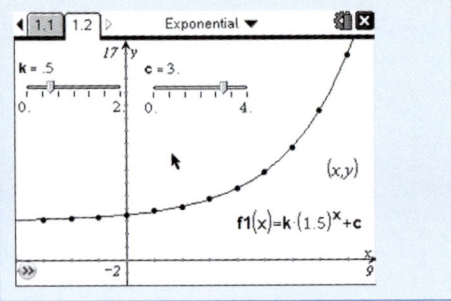

2 Differential calculus

Finding gradients, tangents and maximum and minimum points

2.1 Finding the gradient at a point

Example 21

Find the gradient of the cubic function $y = x^3 - 2x^2 - 6x + 5$

Open a new document and add a Graphs page.
The entry line is displayed at the bottom of the work area.
The default graph type is Function, so the form '$f1(x)=$' is displayed.
The default axes are $-10 \leq x \leq 10$ and $-6.67 \leq y \leq 6.67$.
Type $x^3 - 2x^2 - 6x + 5$ and press [enter].
(**Note:** Type [x] [^] [3] [▶] to enter x^3. The ▶ returns you to the baseline from the exponent.)

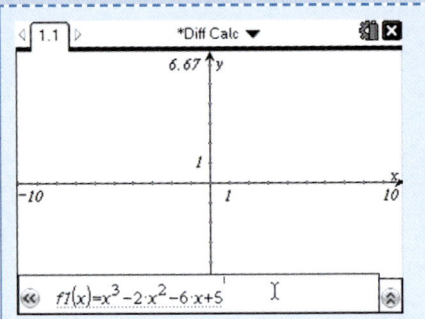

Pan the axes to get a better view of the curve and then grab the x- and y-axes to fit the curve to the window.

For help with panning and changing axes, see your GDC manual.

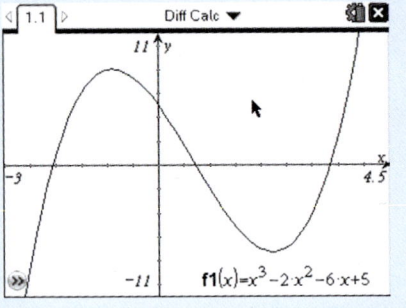

Press [menu] 6:Analyze Graph | 5:$\dfrac{dy}{dx}$
Press [enter]
Using the touchpad, move the 👆 towards the curve. As it approaches the curve, it turns to ∅ and displays the numerical value of the gradient.
Press [enter] to attach a point on the curve.

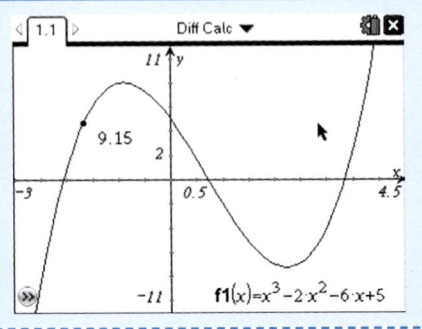

▶ Continued on next page

Use the touchpad to move the icon to the point. You can move the point along the curve and observe how the gradient changes as the point moves. Here, gradient at point = 9.31.	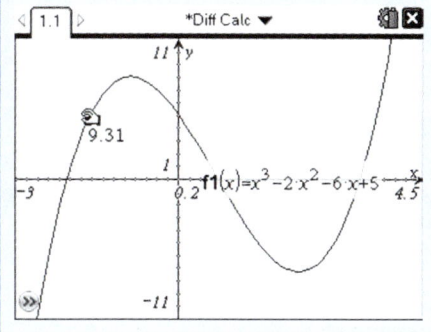

2.2 Drawing a tangent to a curve

Example 22

Draw a tangent to the curve $y = x^3 - 2x^2 - 6x + 5$	
First draw the graph of $y = x^3 - 2x^2 - 6x + 5$ (see Example 21).	
Press menu 7:Points & Lines | 7:Tangent Press enter Using the touchpad, move the ↖ towards the curve. As it approaches the curve, it turns to ☝. Press enter The cursor changes to ⬚ and displays 'point on'. Choose a point where you want to draw a tangent and press enter .	
You can move the point that the tangent line is attached to with the touchpad.	

▶ Continued on next page

Chapter 17 599

Use the touchpad to drag the arrows at each end of the tangent line to extend it.
Press ctrl menu with the tangent line selected – move to the arrow at the end and look for the word 'line'.
Choose 7:Coordinates and Equations
Click on the line to display the equation of the tangent: $y = -2.83x + 5.97$.
Click on the point to display the coordinates of the point: $(-0.559, 7.55)$.

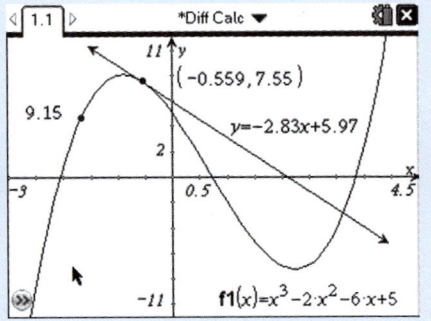

2.3 Finding maximum and minimum points

Example 23

Find the local maximum and local minimum points on the cubic curve:
$y = x^3 - 2x^2 - 6x + 5$

First draw the graph of $y = x^3 - 2x^2 - 6x + 5$ (see Example 21).

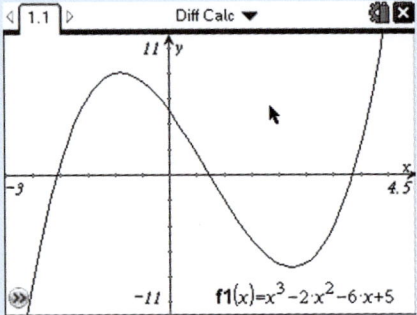

Press menu 6:Analyze Graph | 2:Minimum

Press enter
To find the minimum you need to give the lower and upper bounds of a region that includes the minimum.
The GDC shows a line and asks you to set the lower bound. Move the line using the touchpad and choose a position to the left of the minimum.
Click the touchpad.

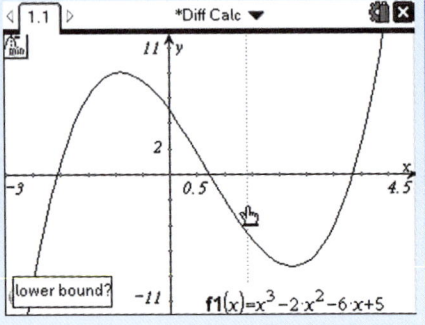

▶ Continued on next page

The GDC shows another line and asks you to set the upper bound. Use the touchpad to move the line so that the region between the upper and lower bounds contains the minimum. **Note:** The minimum point in the region that you have defined is being shown. In this screenshot it is not the local minimum point. Make sure you move the line beyond the point you are looking for.	
When the region contains the minimum, the GDC will display the word 'minimum' in a box and a point that lies between the lower and upper bounds. The point displayed is clearly between the upper and lower bounds. Click the touchpad.	
The GDC displays the local minimum at the point $(2.23, -7.24)$.	
Press menu 6:Analyze Graph | 3:Maximum to find the local maximum point on the curve in exactly the same way. The maximum point is $(-0.897, 8.05)$.	

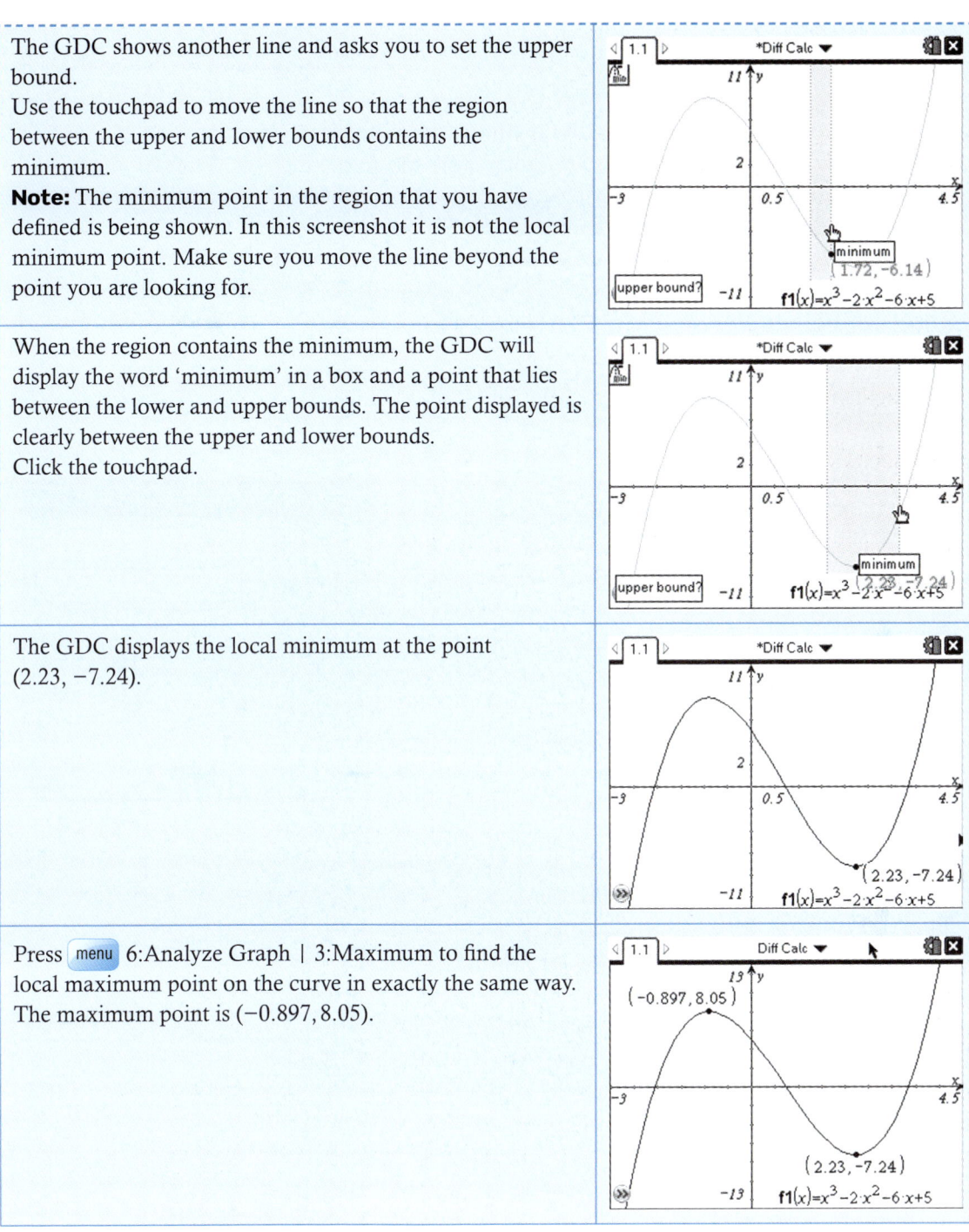

Derivatives

2.4 Finding a numerical derivative

Using the calculator it is possible to find the numerical value of any derivative for any value of x. The calculator will not, however, differentiate a function algebraically. This is equivalent to finding the gradient at a point graphically (see Section 2.1 example 21).

Example 24

If $y = \dfrac{x+3}{x}$, evaluate $\dfrac{dy}{dx}\Big|_{x=2}$

Open a new document and add a Calculator page.
Press menu 4:Calculus | 1:Numerical Derivative at a Point...
Leave the variable as x and the Derivative as 1st Derivative. Change the Value to the value of x at which you wish to evaluate the derivative, in this case $x = 2$.

Enter the function in the template.
Press enter

The calculator shows that the value of the first derivative of $y = \left(\dfrac{x+3}{x}\right)$ is $-\dfrac{3}{4}$ when $x = 2$.

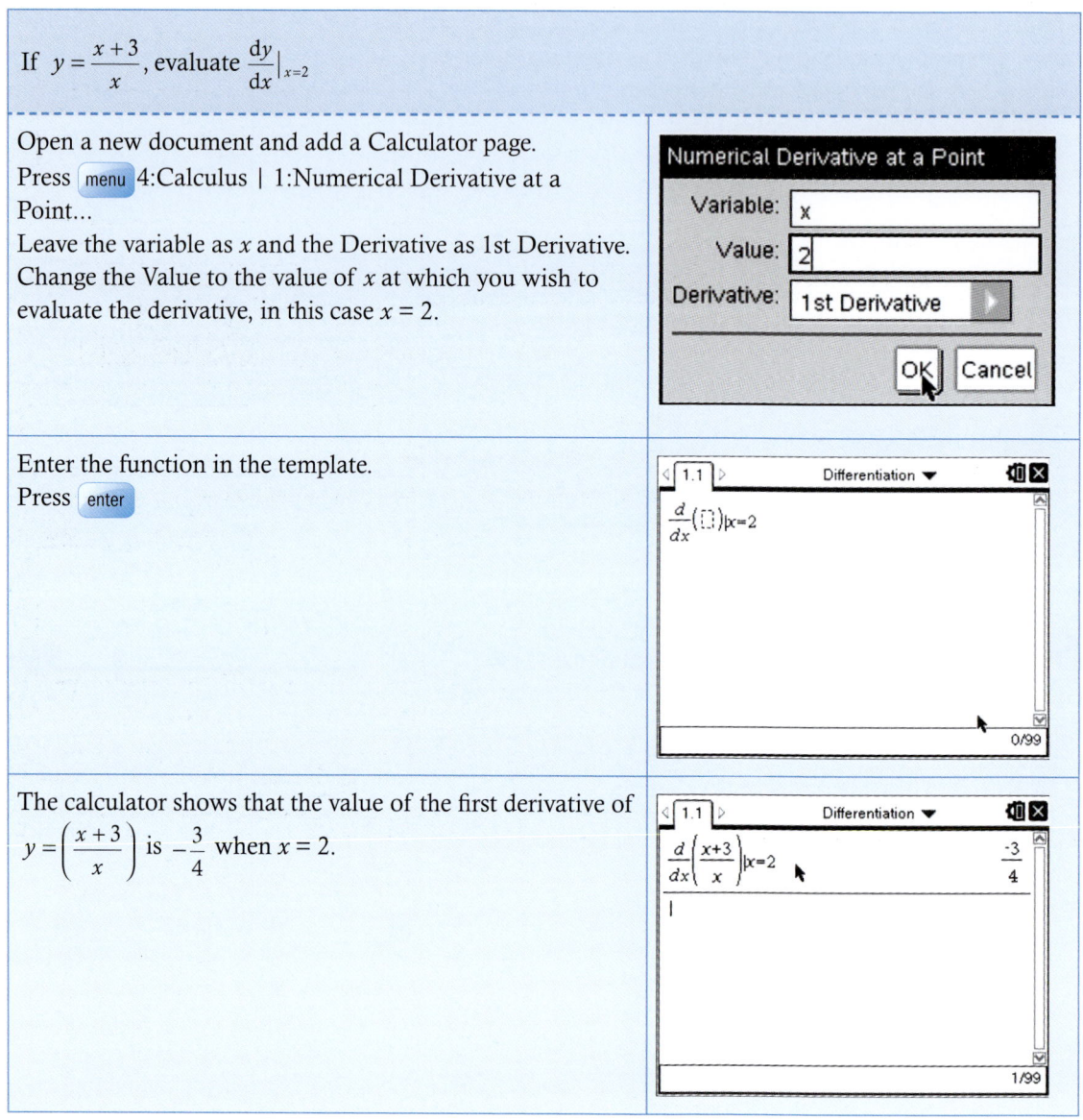

2.5 Graphing a numerical derivative

Although the calculator can only evaluate a numerical derivative at a point, it will graph the gradient function for all values of x.

Example 25

If $y = \dfrac{x+3}{x}$, draw the graph of $\dfrac{dy}{dx}$.

Open a new document and add a Graph page. The entry line is displayed at the bottom of the work area. The default graph type is Function, so the form "$f1(x)=$" is displayed. The default axes are $-10 \leq x \leq 10$ and $-6.67 \leq y \leq 6.67$.	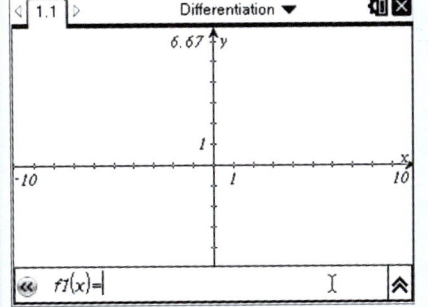
Press the templates button marked 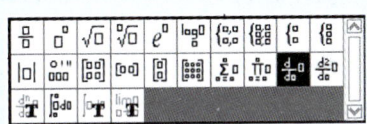 and choose the numerical derivative.	
In the template enter x and the function $\dfrac{x+3}{x}$. Press enter	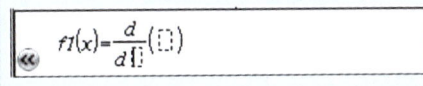
The calculator displays the graph of the numerical derivative function of $y = \dfrac{x+3}{x}$.	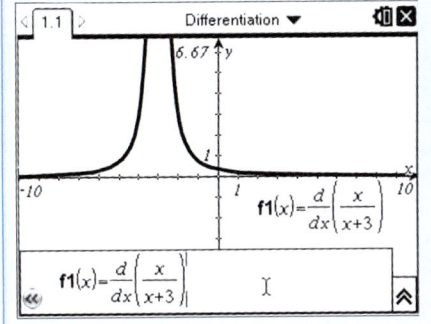

Example 26

Find the values of x on the curve $y = \dfrac{x^3}{3} + x^2 - 5x + 1$ where the gradient is 3.

Open a new document and add a Graphs page. The entry line is displayed at the bottom of the work area. The default graph type is Function, so the form "$f1(x) =$" is displayed. The default axes are $-10 \leq x \leq 10$ and $-6.67 \leq y \leq 6.67$.	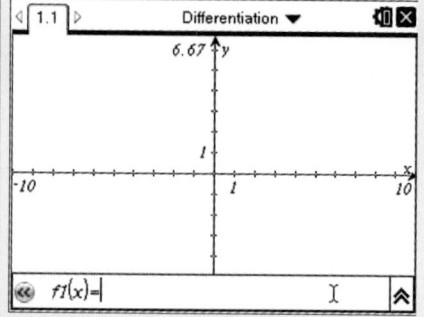
Press the templates button marked [⋅↕] and choose the numerical derivative.	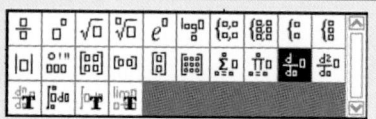
In the template enter x and the function $\dfrac{x^3}{3} + x^2 - 5x + 1$. Press `enter`.	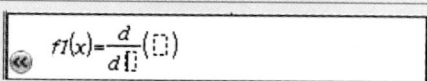
The calculator displays the graph of the numerical derivative function of $y = \dfrac{x^3}{3} + x^2 - 5x + 1$.	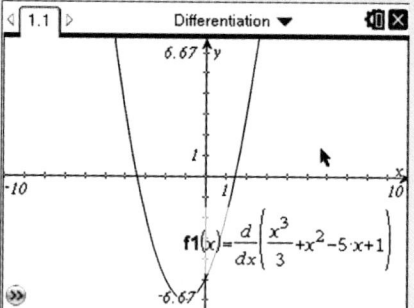
Using the touchpad, click on ›› to open the entry line at the bottom of the work area. Enter the function $f2(x) = 3$ Press `enter` The calculator now displays the curve and the line $y = 3$.	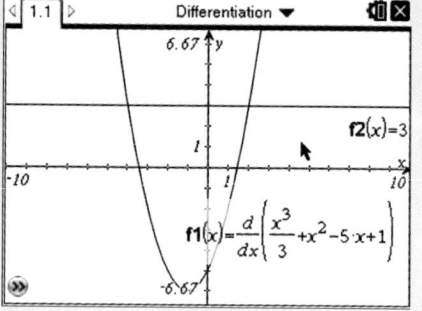
Press `menu` 7:Points & Lines | 3:Intersection Point(s) Using the touchpad, select graph $f1$ and graph $f2$. The calculator displays the coordinates of the intersection points of the gradient function and the line $y = 3$. The curve has gradient 3 when $x = -4$ and $x = 2$	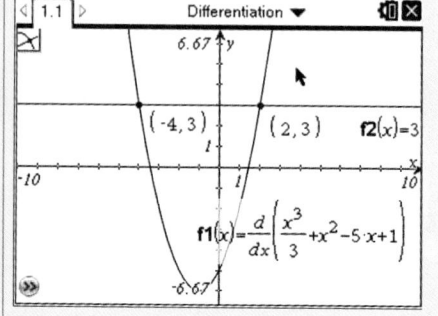

2.6 Using the second derivative

The calculator can find first and second derivatives. The second derivative can be used to determine whether a point is a maximum or minimum point.

Example 27

Find the stationary points on the curve $f(x) = x^4 - 4x^3$ and determine their nature.

$f(x) = x^4 - 4x^3$
$f'(x) = 4x^3 - 12x^2$
At stationary points
$f'(x) = 0$
$4x^3 - 12x^2 = 0$
$4x^2 - (x-3) = 0$
Therefore $x = 0$ or $x = 3$

Use the calculator to find the coordinates of the points and to determine their nature.
Open a new document and add a Calculator page.
Define the function $f1(x)$
Type F 1 (X) ctrl := and type the function.
Evaluate the function when $x = 0$ and $x = 3$

The stationary points are at $(0, 0)$ and $(3, -27)$

Press menu 4:Calculus | 1:Numerical Derivative at a Point...
Leave the variable as x and choose 2nd Derivative. Change the Value to the value of x at which you wish to evaluate the derivative, in this case $x = 0$ (and $x = 3$).

Enter $f1(x)$ in the template as the function.

Repeat for the second derivative when $x = 3$

(Note: you can cut and past the expression and change the 0 to 3)

In this case we are not certain what the nature of the stationary point is at $(0, 0)$ but the point $(3, -27)$ is a minimum because $f''(x) > 0$

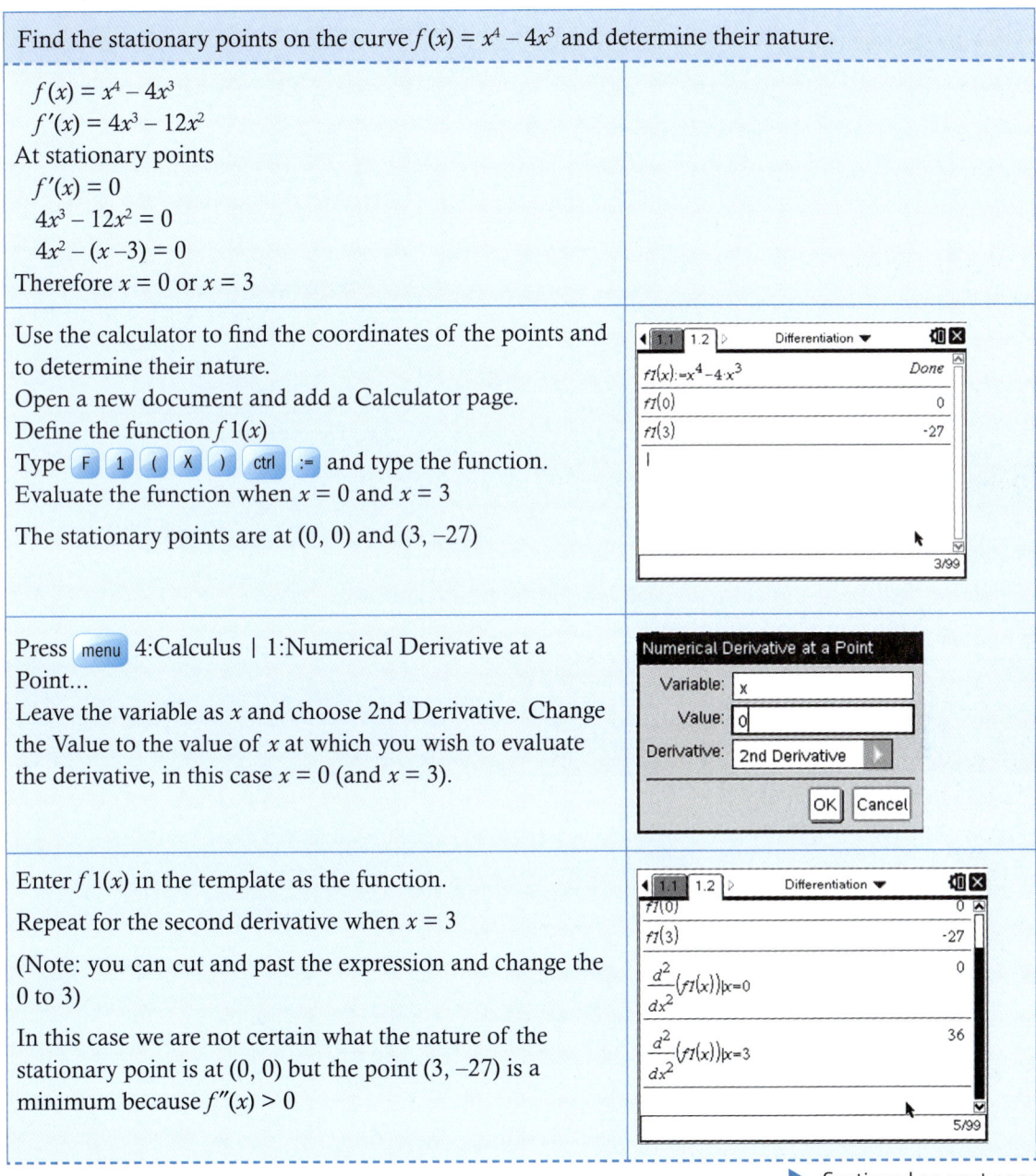

▶ Continued on next page

Evaluate $f'(x)$ either side of $x = 0$.

In this case using $x = -0.01$ and $x = 0.01$

The gradient is negative either side of the stationary point. Hence (0, 0) is a negative point of inflexion.

The graph on the right illustrates the curve, the minimum at (3, −27) and the point of inflexion at (0, 0).

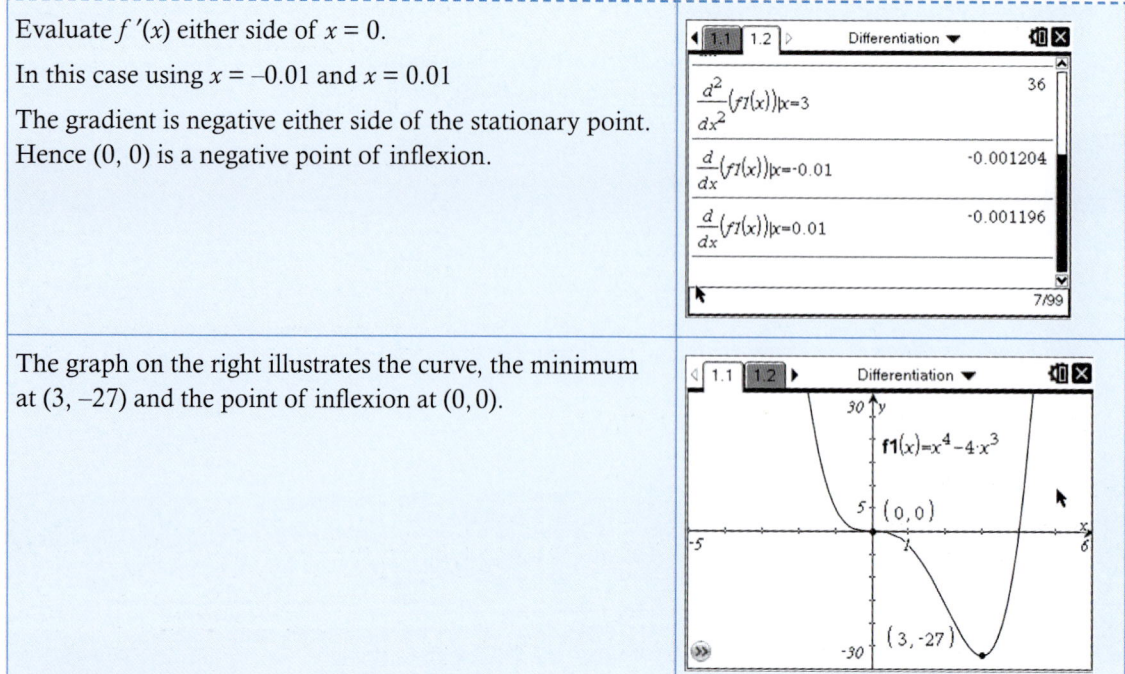

3 Integral calculus

The calculator can find the values of definite integrals either on a calculator page or graphically. The calculator method is quicker, but the graphical method is clearer and shows discontinuities, negative areas and other anomalies that can arise.

3.1 Finding the value of an indefinite integral

Example 28

Evaluate $\int \left(x - \dfrac{3}{\sqrt{x}} \right) dx$

Open a new document and add a Calculator page.
Press menu 4:Calculus | 1:Numerical Integral...
Enter the upper and lower limits, the function and x in the template.
Use the ▼▲◀▶ keys to navigate around the template.
In this example you will also use templates to enter the rational function and the square root.

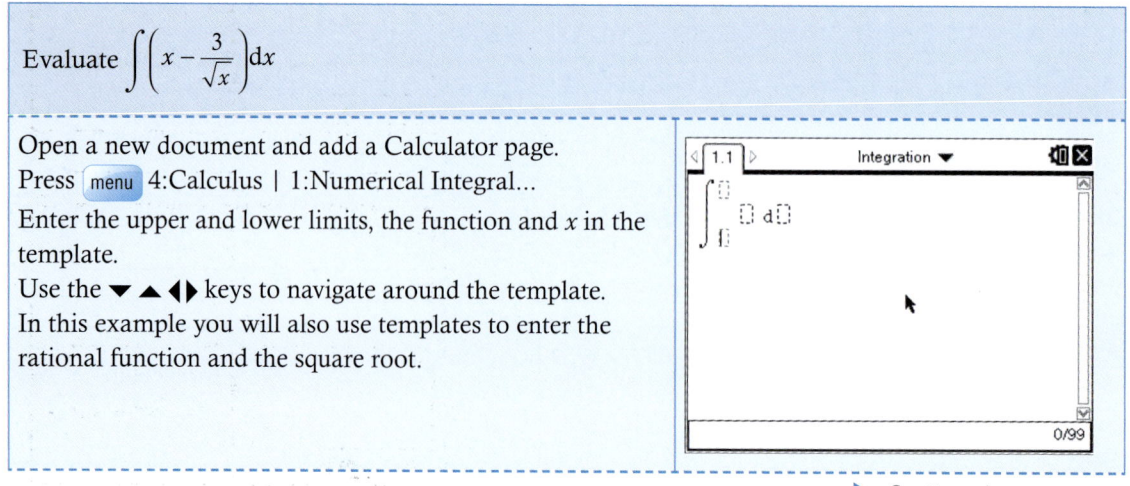

▶ Continued on next page

606   Using a graphic display calculator

The value of the integral is 21.5 (to 3 sf)

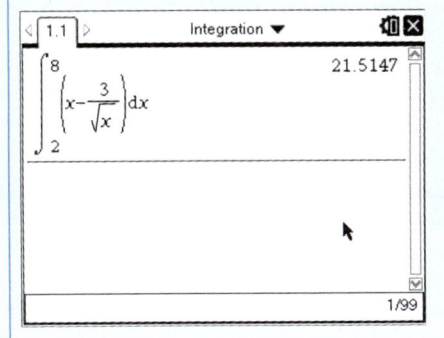

3.2 Finding the area under a curve

Example 29

Find the area bounded by the curve $y = 3x^2 - 5$, the x-axis and the lines $x = -1$ and $x = 1$.

Open a new document and add a Graphs page.
The entry line is displayed at the bottom of the work area.
The default graph type is Function, so the form "$f1(x)=$" is displayed.

The default axes are $-10 \leq x \leq 10$ and $-6.67 \leq y \leq 6.67$.
Type the function $3x^2 - 5$
Press enter

Press menu 6:Analyze Graph | 6:Integral

The calculator prompts you to enter the lower limit for the integral. There are several ways to do this.

You can click manually. This is not very accurate, however, and you will need to add the coordinates of the point you entered and edit them to obtain an accurate figure.

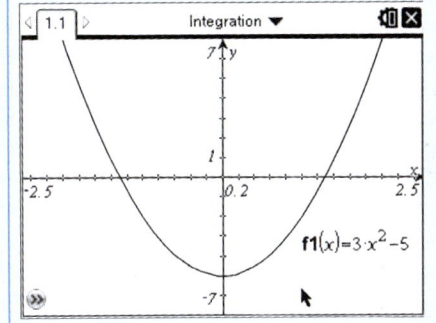

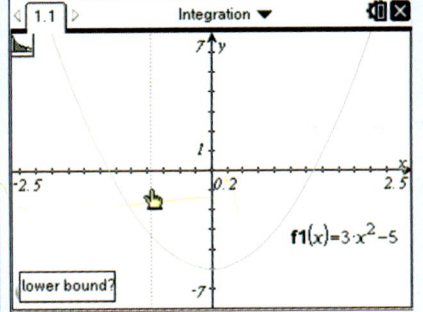

You can use the points on the axis.
Here the scale was set to 0.2, so the point $(-1, 0)$ can be selected as shown.

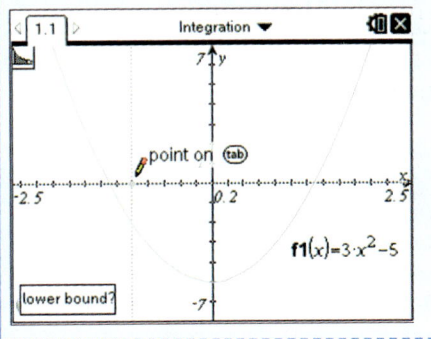

▶ Continued on next page

You can enter the point with the keyboard.
Enter a left bracket (and then type (-) 1 and press enter
There is no need to complete the coordinates.

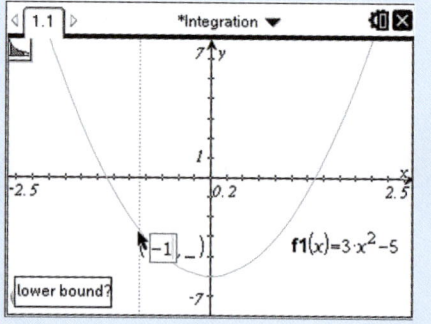

Repeat for the upper limit.

The calculator displays a changing value for the area.

Using one of the methods above, select a point where the value of x is 1.

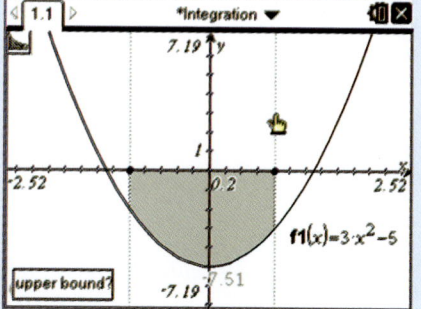

Repeat for the upper limit.

The area found is shaded and the value of the integral (−8) is shown on the screen.

Note: since the area lies below the x-axis in this case, the integral is negative.

The required area is 8.

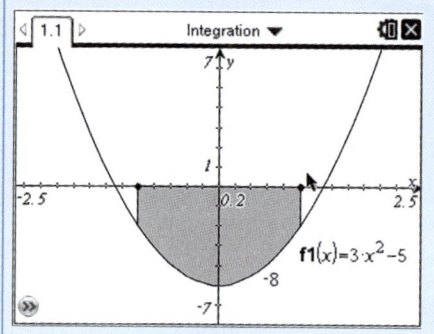

4 Vectors

4.1 Calculating a scalar product

Example 30

Evaluate the scalar products:

a $\begin{pmatrix} 1 \\ 3 \end{pmatrix} \cdot \begin{pmatrix} -3 \\ 4 \end{pmatrix}$

b $\begin{pmatrix} 1 \\ -1 \\ 4 \end{pmatrix} \cdot \begin{pmatrix} 3 \\ 2 \\ -1 \end{pmatrix}$

▶ Continued on next page

a Open a new document and add a Calculator page.

Press b 7: Matrix & Vector | C: Vector | 3: Dot Product

(or type DOTP()).

Press t and choose the 2 × 1 column vector template.

Enter the vector type, and enter the second vector.

Press

$$\begin{pmatrix} 1 \\ 3 \end{pmatrix} \cdot \begin{pmatrix} -3 \\ 4 \end{pmatrix} = 9$$

b Press b 7: Matrix & Vector | C: Vector | 3: Dot Product
Press t and choose the matrix template

Choose 3 rows and 1 column and then click on OK.

Enter the vector type, and enter the second vector.

Press

$$\begin{pmatrix} 1 \\ -1 \\ 4 \end{pmatrix} \cdot \begin{pmatrix} 3 \\ 2 \\ -1 \end{pmatrix} = -3$$

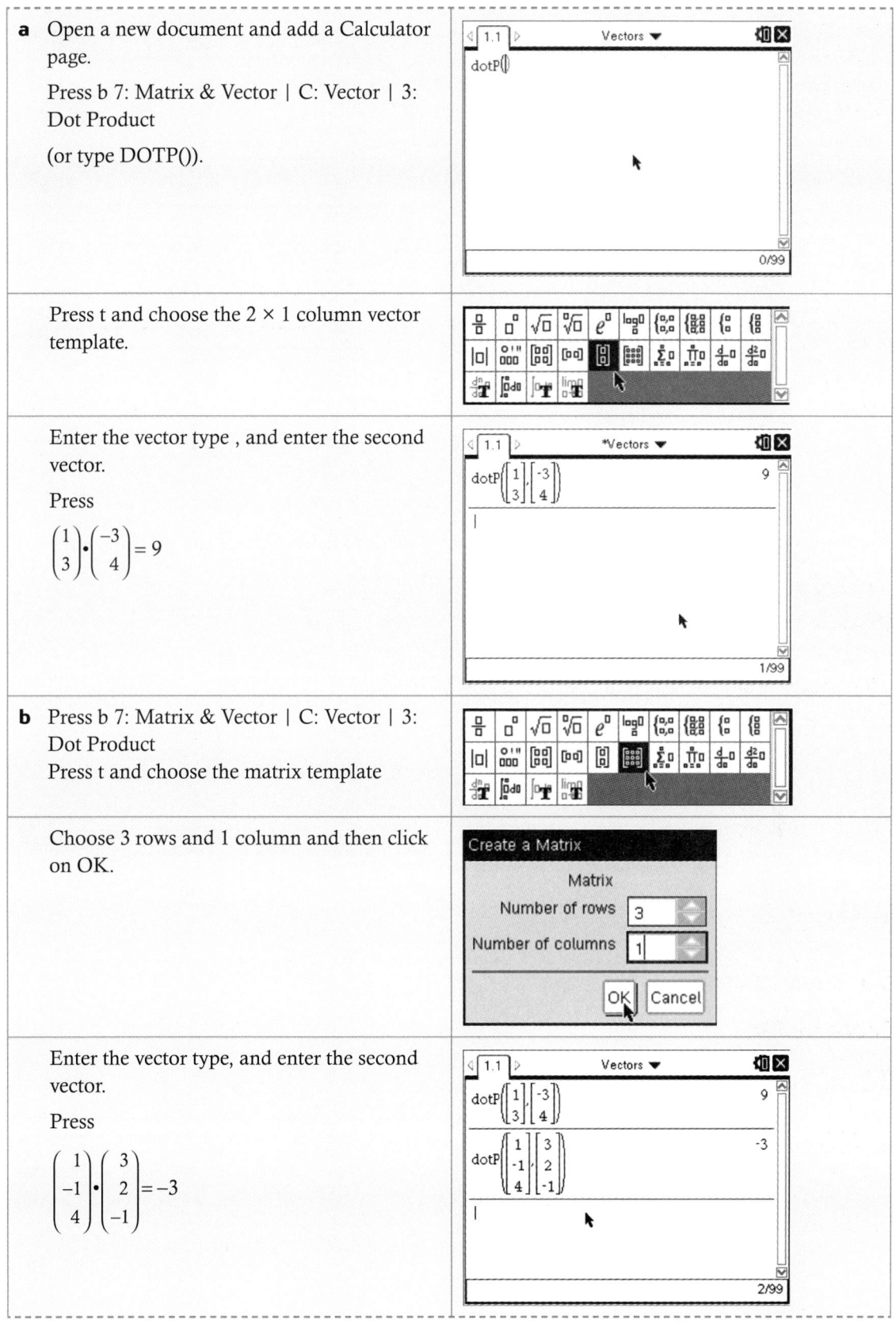

▶ Continued on next page

You can also enter vectors as rows by typing them in directly instead of using the templates. Separate the values in the vector with commas. When you press enter, the GDC changes the entry line and calculates the result.	*Vectors* dotp([1,3],[-3,4]) *Vectors* dotP([1 3],[-3 4]) 9
This method can be quicker, especially with 3 × 1 vectors.	Vectors dotP([1 -1 4],[3 2 -1]) -3

4.2 Calculating the angle between two vectors

The angle θ between two vectors **a** and **b**, can be calculated using the formula

$$\theta = \arccos\left(\frac{\mathbf{a} \cdot \mathbf{b}}{|\mathbf{a}||\mathbf{b}|}\right)$$

Example 31

Calculate the angle between $2\mathbf{i} + 3\mathbf{j}$ and $3\mathbf{i} - \mathbf{j}$

Open a new document and add a Calculator page.

Move the cursor to the ⚙ symbol at the top right-hand side of the screen. It will display the general angle mode – either radians or degrees.

Click in the ⚙ symbol and choose 2:Settings | 1:General.

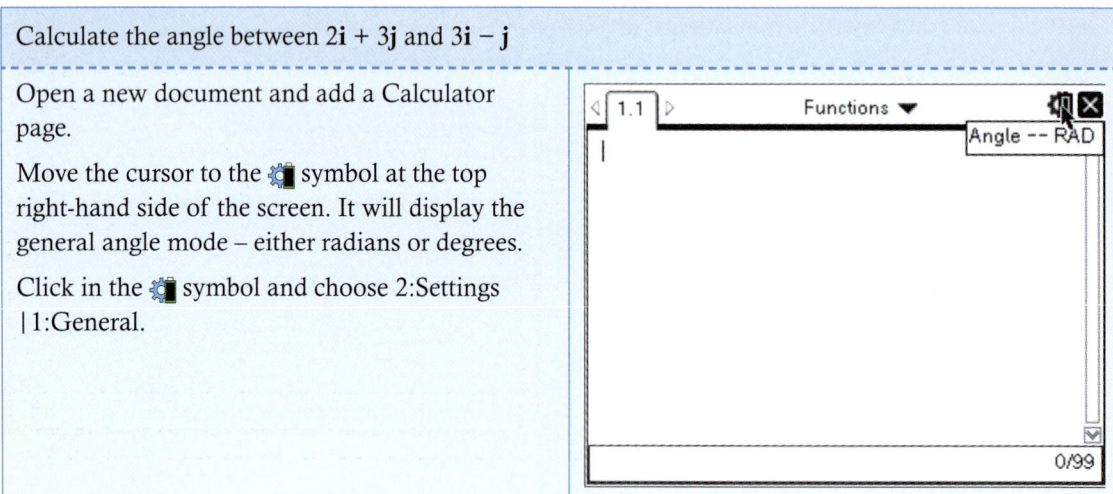

▶ Continued on next page

In the dialogue box, select either degrees or radians (according to the units you need your answer in) and then click on OK.	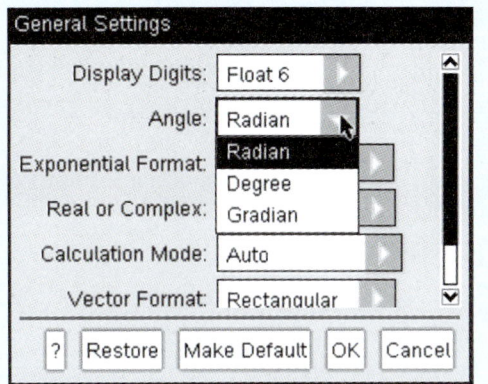
Press μ and choose $\cos^{-1}$ from the menu.	
Enter the values in the formula as shown, using the fraction template and the 2 × 1 column vector template. To calculate the magnitudes of the vectors use the formula $\lvert a\mathbf{i} + b\mathbf{j}\rvert = \sqrt{a^2 + b^2}$	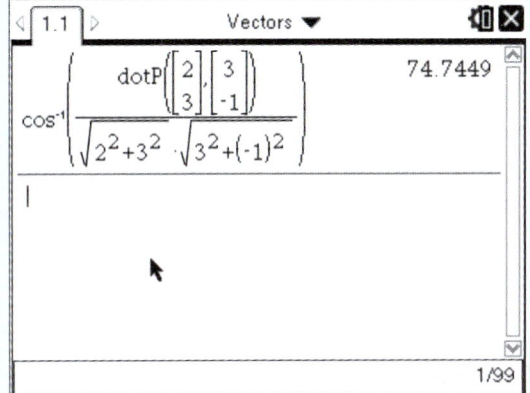
As an alternative to using the formula for the magnitude of a vector, you can use the norm function. Press b 7: Matrix & Vector | 7: Norms | 1: Norm or simply typing norm(Instead of retyping the vectors, you can use /C and /V to cut and paste.	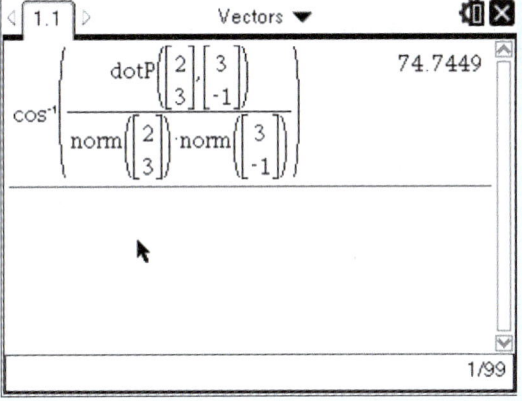

5 Statistics and probability

You can use your GDC to draw charts to represent data and to calculate basic statistics such as mean, median, etc. Before you can do this, you need to enter the data into a list or spreadsheet. This is done in a Lists & Spreadsheet page in your document.

Entering data

There are two ways of entering data: as a list or as a frequency table.

5.1 Entering lists of data

Example 32

Enter the data in the list
1, 1, 3, 9, 2

Open a new document and add a Lists & Spreadsheet page.
Type 'data' in the first cell.
Type the numbers from the list in the first column.
Press enter or ▼ after each number to move down to the next cell.
Note: The word 'data' is a label that will be used later when you want to create a chart or do some calculations with this data. You can use any letter or name to label the list.

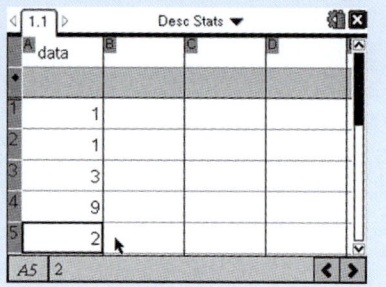

5.2 Entering data from a frequency table

Example 33

Enter the data in a table

Number	1	2	3	4	5
Frequency	3	4	6	5	2

Add a new Lists & Spreadsheet page to your document.
To label the columns, type 'number' in the first cell and 'freq' in the cell to its right.
Enter the numbers in the first column and the frequencies in the second.
Use the 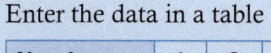 keys to navigate around the spreadsheet.

Drawing charts

You can draw charts from a list or from a frequency table.

5.3 Drawing a frequency histogram from a list

Example 34

Draw a frequency histogram for this data:
1, 1, 3, 9, 2

Enter the data in a list called 'data' (see Example 32).
Add a new Data & Statistics page to your document.
Note: You do not need to worry about what this screen shows.

Click at the bottom of the screen where it says 'Click to add variable', choose 'data' from the list and press `enter`.

The first chart you will see is a dot plot of your data.

Press `menu` 1:Plot Type | 3:Histogram
Press `enter`
You should now see a frequency histogram for the data in the list.

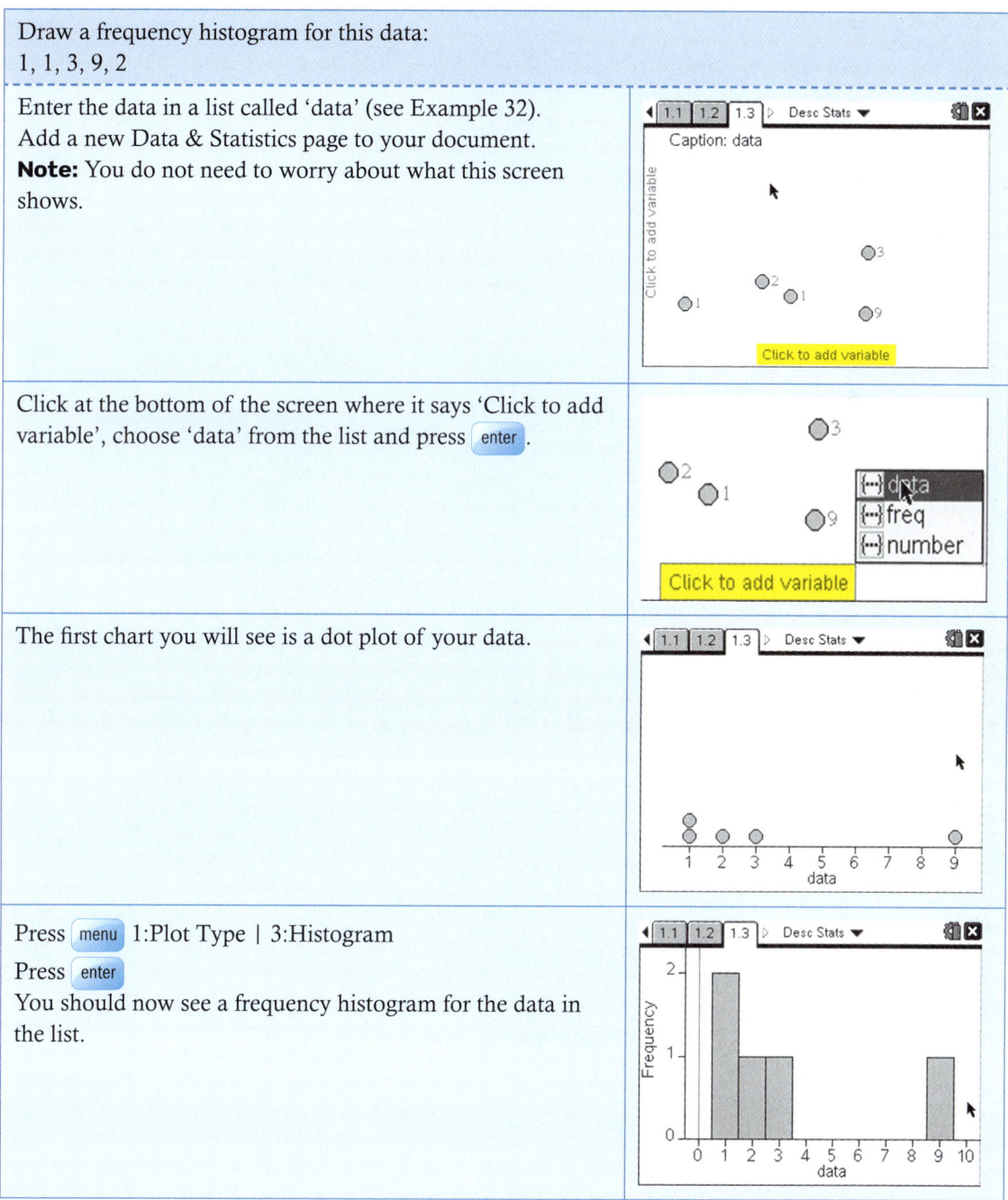

5.4 Drawing a frequency histogram from a frequency table

Example 35

Draw a frequency histogram for this data:

Number	1	2	3	4	5
Frequency	3	4	6	5	2

Enter the data in lists called 'number' and 'freq' (see Example 33).
Add a new Data & Statistics page to your document.
Note: You do not need to worry about what this screen shows.

Press menu 2:Plot Properties | 5:Add X Variable with Frequency
Press enter
You will see this dialogue box.
From the drop-down menus, choose 'number' for the Data List and 'freq' for the Frequency List.
Press enter

You should now see a frequency histogram for the data in the table.

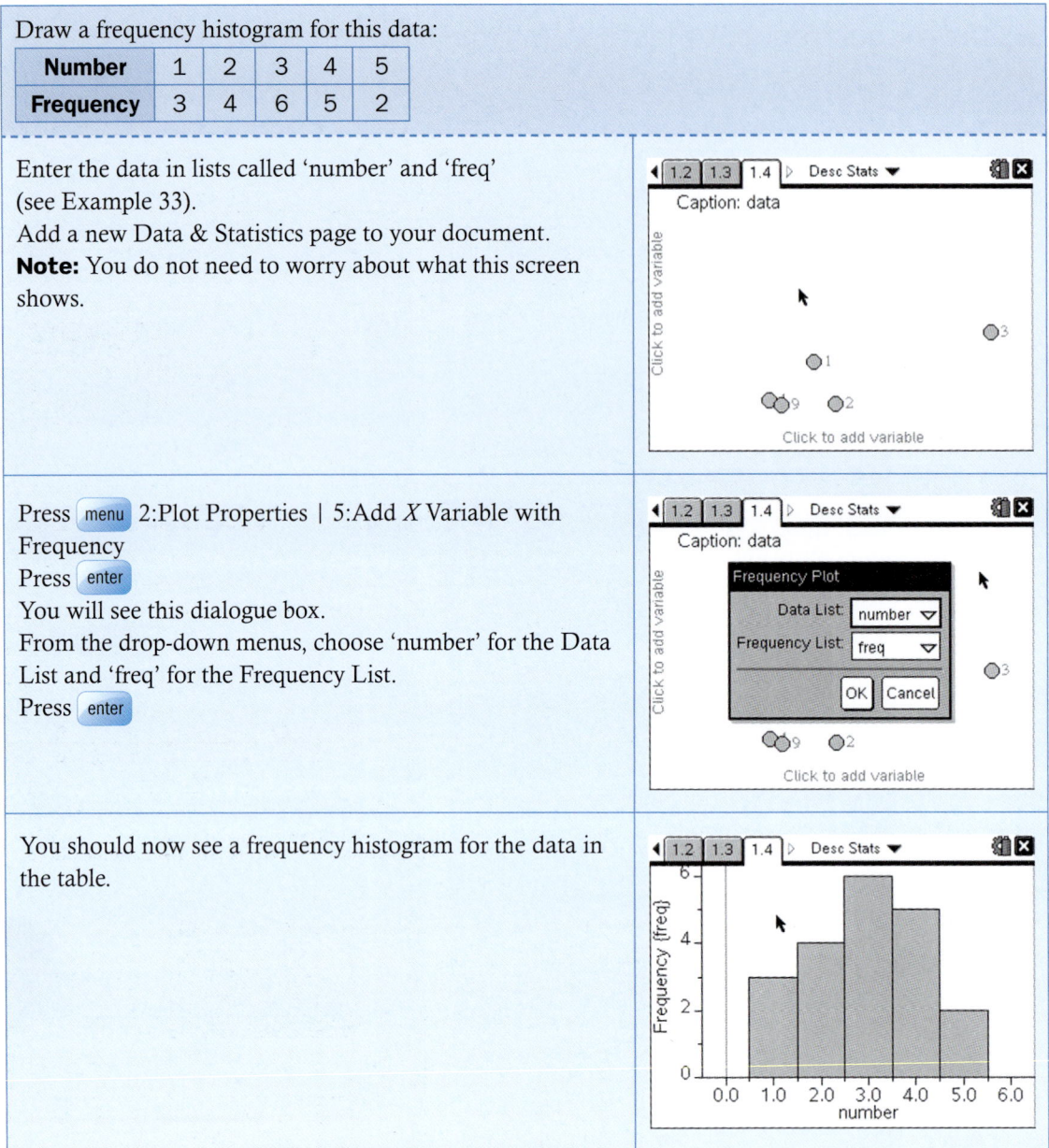

5.5 Drawing a box and whisker diagram from a list

Example 36

Draw a box and whisker diagram for this data:
1, 1, 3, 9, 2

▶ Continued on next page

Enter the data in a list called 'data' (see Example 32). Add a new Data & Statistics page to your document. **Note:** You do not need to worry about what this screen shows.	
Click at the bottom of the screen where it says 'Click to add variable', choose 'data' from the list and press enter.	
The first chart you will see is a dot plot of your data.	
Press menu 1:Plot Type | 2:Box Plot Press enter You should now see a box plot (box and whisker diagram) for the data in the list.	
Move the cursor over the plot and you will see the quartiles, Q_1 and Q_3, the median, and the maximum and minimum values.	

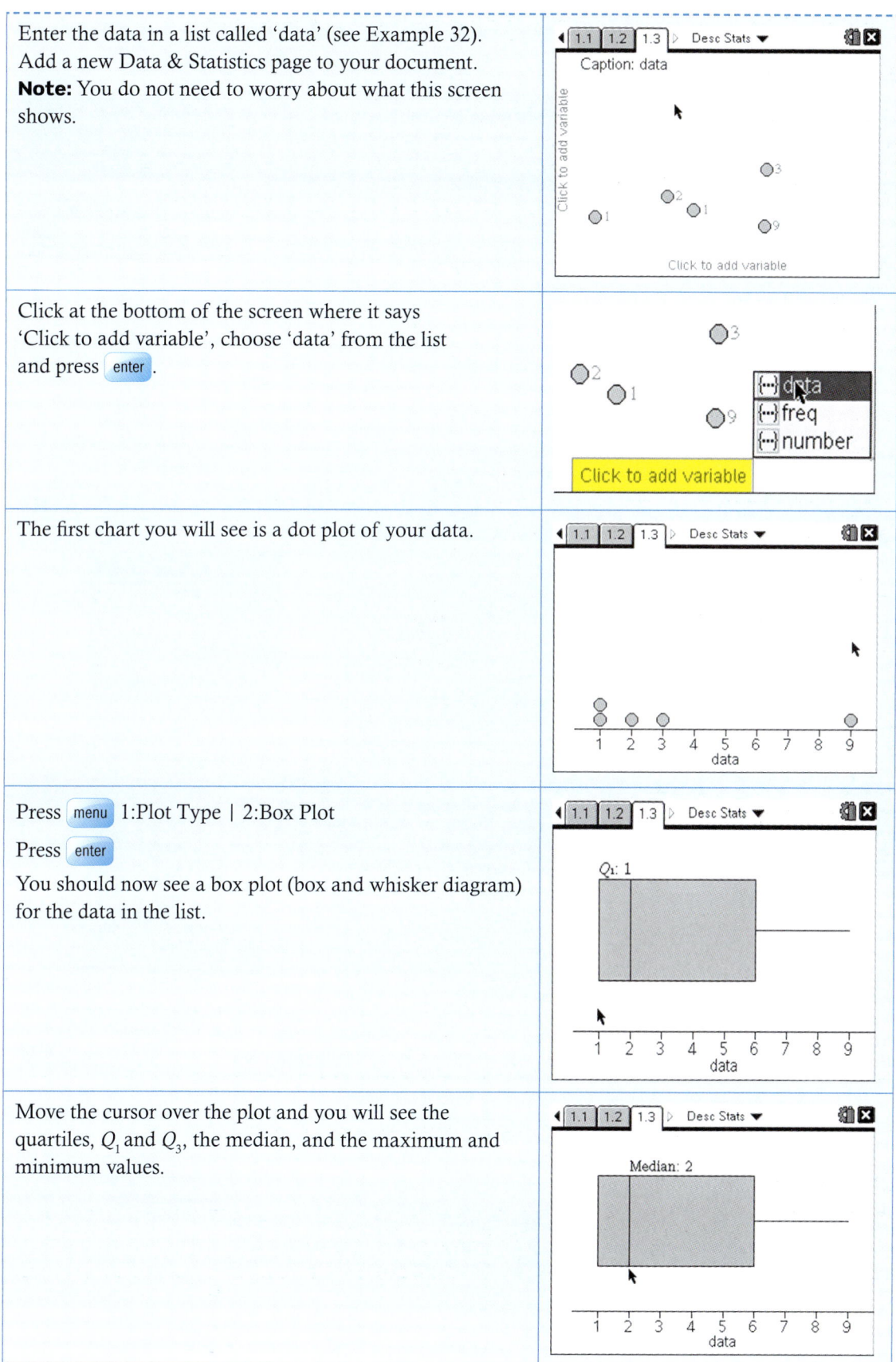

5.6 Drawing a box and whisker diagram from a frequency table

Example 37

Draw a box and whisker diagram for this data:

Number	1	2	3	4	5
Frequency	3	4	6	5	2

Enter the data in lists called 'number' and 'freq' (see Example 33).
Add a new Data & Statistics page to your document.
Note: You do not need to worry about what this screen shows.

Press menu 2:Plot Properties | 5:Add *X* Variable with Frequency
Press enter
You will see this dialogue box.

From the drop-down menus, choose 'number' for the Data List and 'freq' for the Frequency List.
Press enter

You should now see a frequency histogram.

Press menu 1:Plot Type | 2:Box Plot
Press enter
You should now see a box plot (box and whisker diagram) for the data in the table.

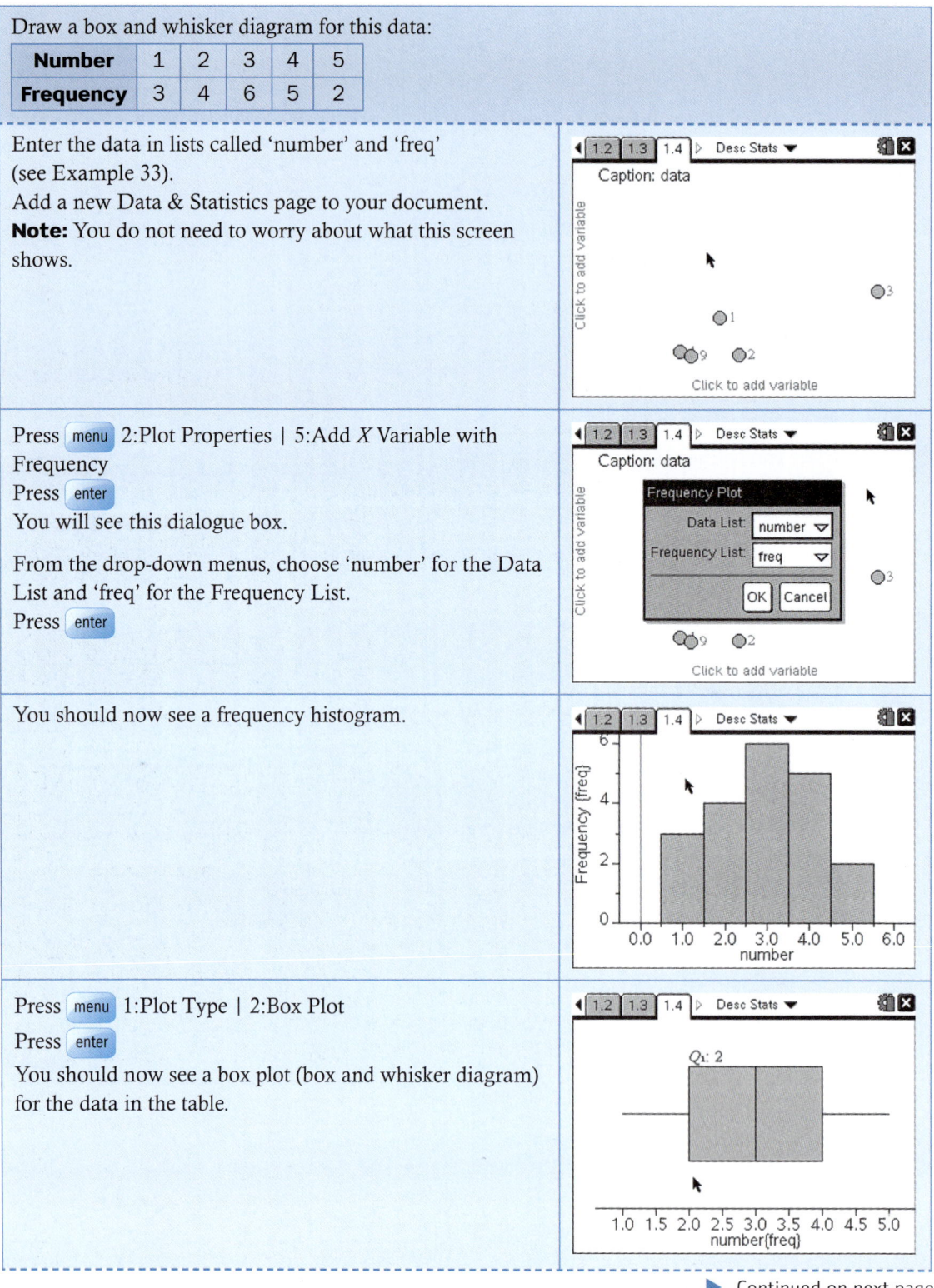

▶ Continued on next page

Move the cursor over the plot and you will see the quartiles, Q_1 and Q_3, the median, and the maximum and minimum values.

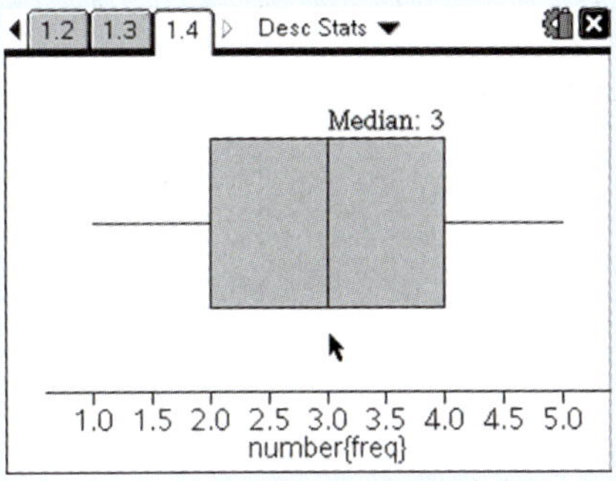

Calculating statistics

You can calculate statistics such as mean, median, etc. from a list, or from a frequency table.

> Mean, median, range, quartiles, standard deviation, etc. are called **summary statistics**.

5.7 Calculating statistics from a list

Example 38

Calculate the summary statistics for this data: 1, 1, 3, 9, 2

Enter the data in a list called 'data' (see Example 32).
Add a new Calculator page to your document.
Press menu 6:Statistics | 1:Stat Calculations | 1:One-Var Statistics…
Press enter
This opens a dialogue box.
Leave the number of lists as 1 and press enter.

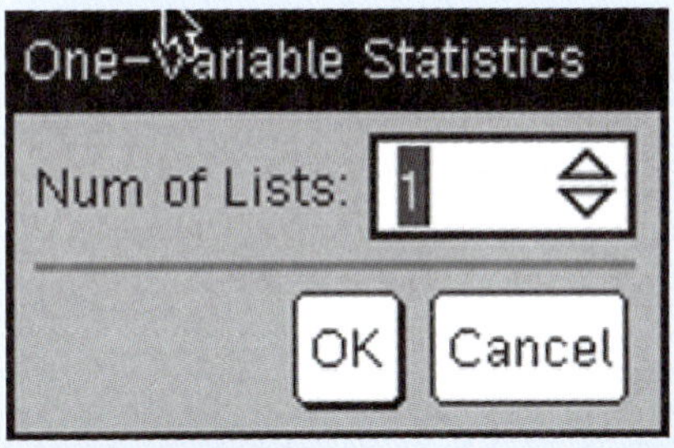

This opens another dialogue box.
Choose 'data' from the drop-down menu for $X1$ List and leave the Frequency List as 1.
Press enter

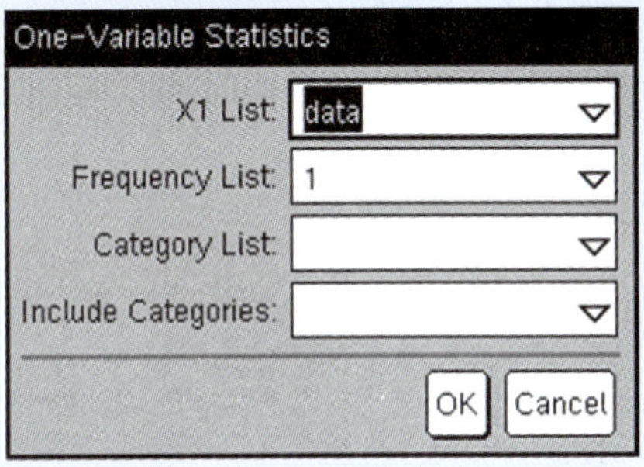

The information shown will not fit on a single screen.
You can scroll up and down to see it all.
The statistics calculated for the data are:

mean	$\bar{x}$
sum	Σx
sum of squares	Σx^2
sample standard deviation	s_x
population standard deviation	σ_x

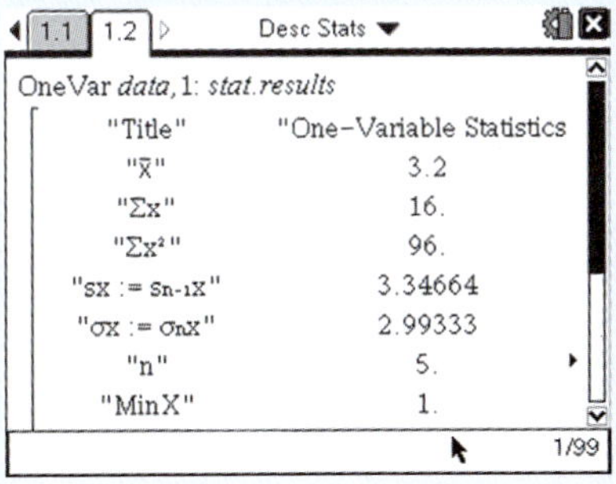

▶ Continued on next page

Chapter 17

number	n	
minimum value	Min$X$	
lower quartile	$Q_1 X$	
median	Median$X$	
upper quartile	$Q_3 X$	
maximum value	Max$X$	
sum of squared deviations from the mean	SSX	

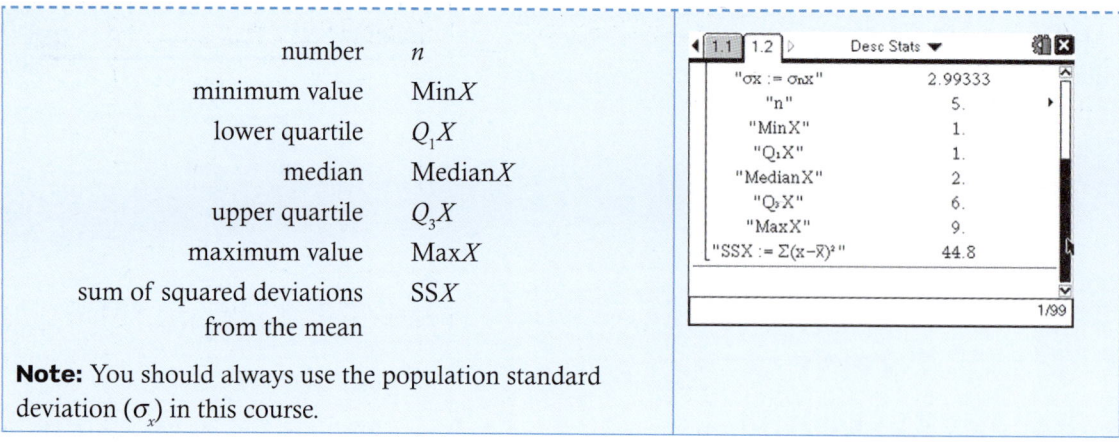

Note: You should always use the population standard deviation (σ_x) in this course.

5.8 Calculating statistics from a frequency table

Example 39

Calculate the summary statistics for this data:

Number	1	2	3	4	5
Frequency	3	4	6	5	2

Enter the data in lists called 'number' and 'freq' (see Example 33).
Add a new Calculator page to your document.
Press menu 6:Statistics | 1:Stat Calculations | 1:One-Var Statistics…
Press enter
This opens a dialogue box.
Leave the number of lists as 1 and press enter.

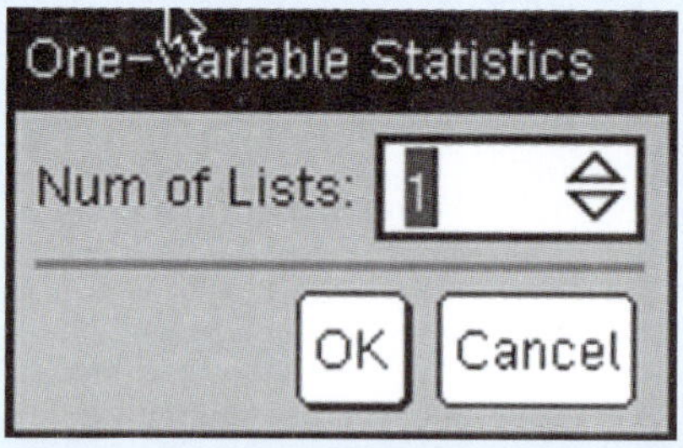

This opens another dialogue box.
From the drop-down menus, choose 'number' for X1 List and 'freq' for the Frequency List.
Press enter

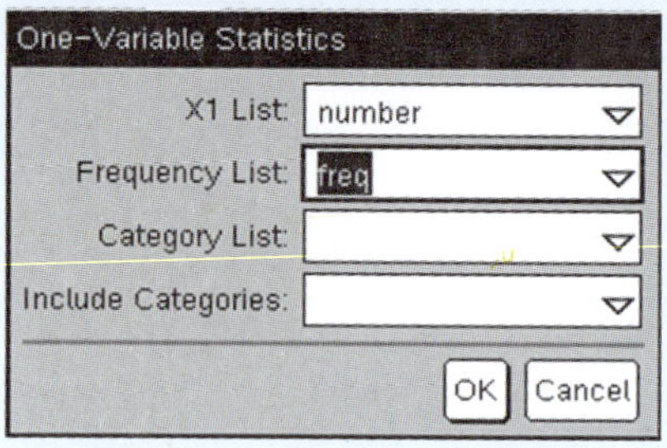

The information shown will not fit on a single screen.
You can scroll up and down to see it all.
The statistics calculated for the data are:

mean	$\bar{x}$	
sum	Σx	
sum of squares	Σx^2	
sample standard deviation	s_x	

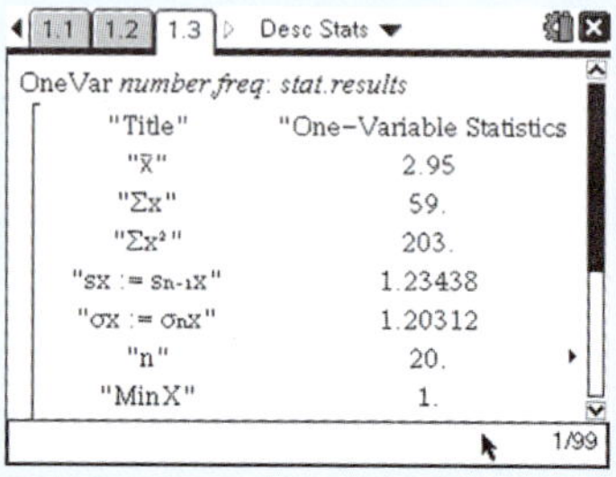

▶ Continued on next page

The information shown will not fit on a single screen.
You can scroll up and down to see it all.
The statistics calculated for the data are:

population standard deviation	σ_x
number	n
minimum value	$\text{Min}X$
lower quartile	$Q_1 X$
median	$\text{Median}X$
upper quartile	$Q_3 X$
maximum value	$\text{Max}X$
sum of squared deviations from the mean	SSX

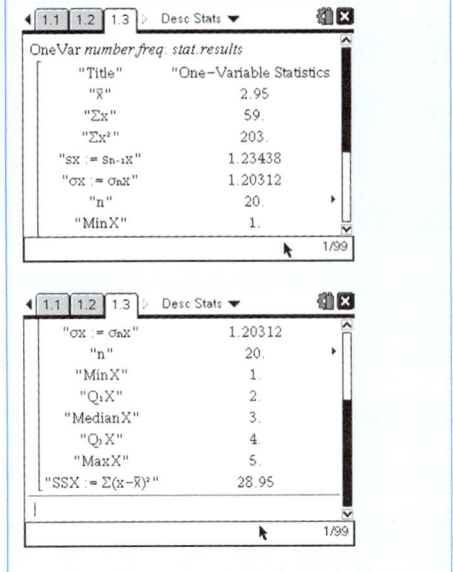

Note: You should always use the population standard deviation (σ_x) in this course.

5.9 Calculating the interquartile range

The interquartile range is the difference between the upper and lower quartiles ($Q_3 - Q_1$).

Example 40

Calculate the interquartile range for this data:

Number	1	2	3	4	5
Frequency	3	4	6	5	2

First calculate the summary statistics for this data (see Example 38).

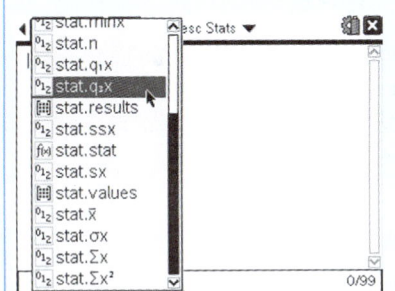

The values of the summary statistics are stored after One-Variable Statistics have been calculated and remain stored until the next time they are calculated.

Add a new Calculator page to your document.
Press `var`
A dialogue box will appear with the names of the statistical variables.
Scroll down to stat.$q_3 x$ using the touchpad, or the ▼▲ keys, and then press `enter`.

Type `(-)` and press `var` again.
Scroll down to stat.$q_1 x$ using the touchpad, or the ▼▲ keys, and then press `enter`.

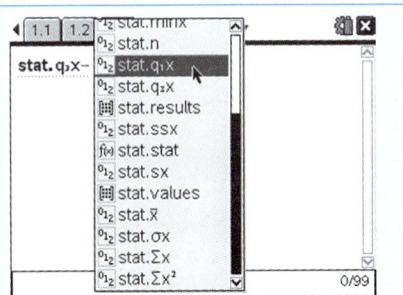

▶ Continued on next page

Press enter again.
The calculator now displays the result:
Interquartile range = $Q_3 - Q_1 = 3$

5.10 Using statistics

Example 41

Calculate $\bar{x} + \sigma_x$ for this data:

Number	1	2	3	4	5
Frequency	3	4	6	5	2

The calculator stores the values you calculate in One-Variable Statistics so that you can access them in other calculations. The values are stored until you do another One-Variable Statistics calculation.

First calculate the summary statistics for this data (see Example 38).
Add a new Calculator page to your document.
Press var
A dialogue box will appear with the names of the statistical variables.
Scroll down to stat.$\bar{x}$ using the touchpad, or the ▼ ▲ keys, and then press enter.

Type + and press var again.
Scroll down to stat.σx using the touchpad, or the ▼ ▲ keys, and then press enter.

Press enter again.
The calculator now displays the result:
$\bar{x} + \sigma_x = 4.15$ (to 3 sf)

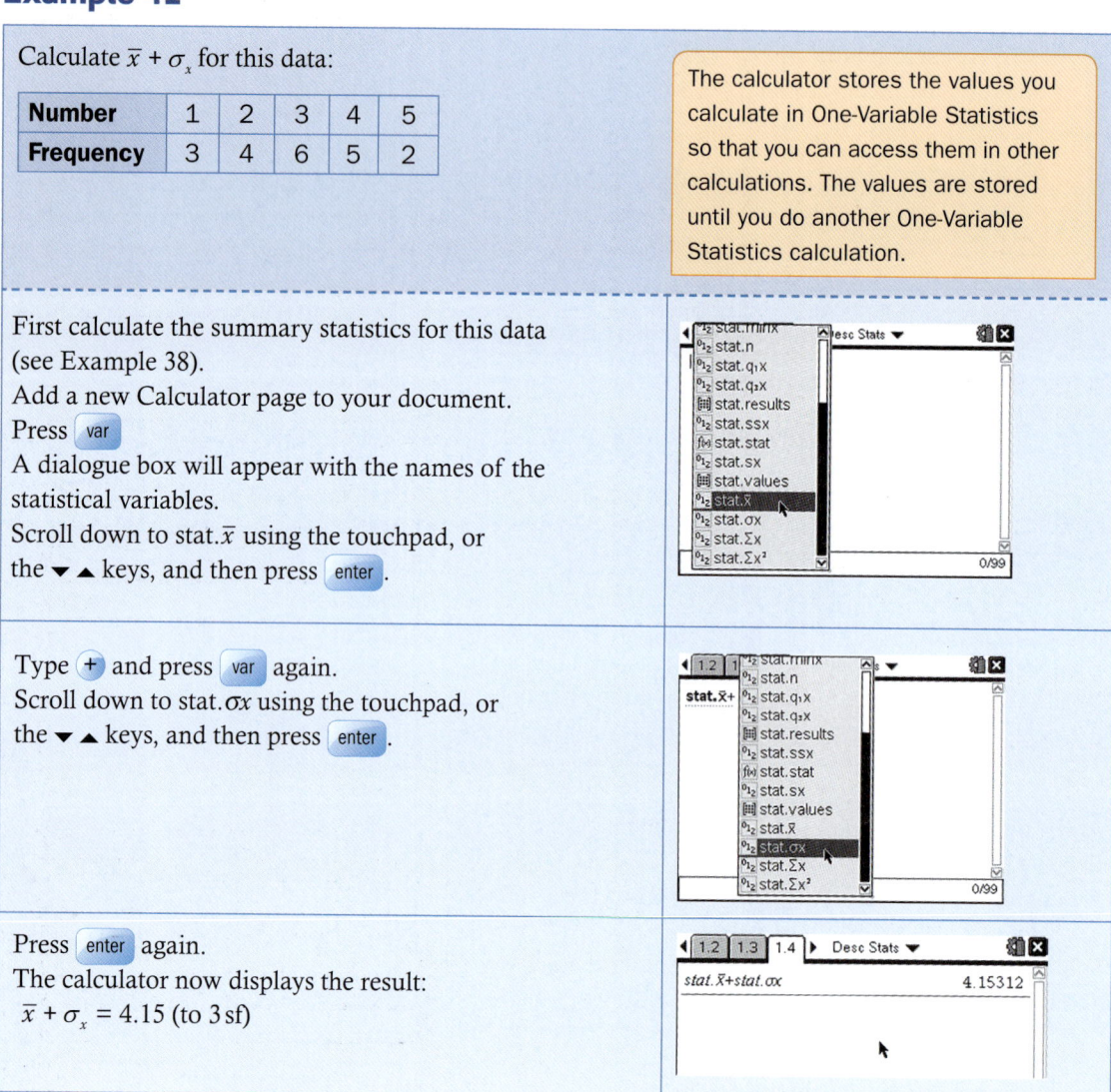

620 Using a graphic display calculator

Calculating binomial probabilities

5.11 The use of nCr

Example 42

Find the value of $\binom{8}{3}$ (or $_8C_3$)

Open a new document and add a Calculator page.
Press `menu` 5:Probability | 3:Combinations
Alternatively you can just type `N` `C` `R` `(`.
There is no need to worry about upper or lower case, the calculator recognises the key sequence and translates it accordingly.
Type 8,3
Press `enter`

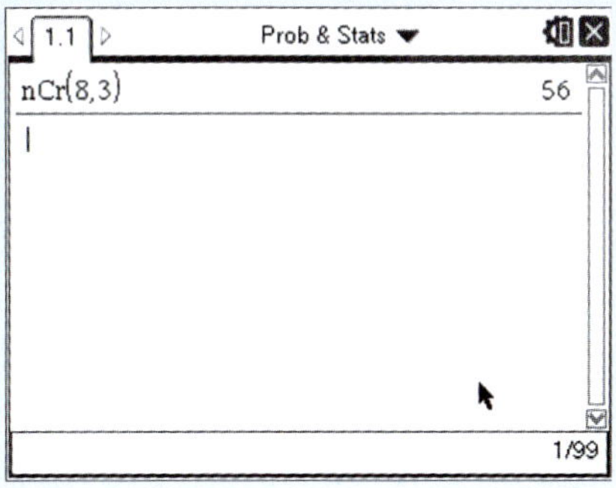

nCr(8,3)   56

Example 43

List the values of $\binom{4}{r}$ for $r = 0, 1, 2, 3, 4$

Open a new document and add a Calculator page.
Type `F` `1` `(` `X` `)` `ctrl` `:=`
Press `menu` 5:Probability | 3:Combinations
Alternatively you can just type `N` `C` `R` `(`.
There is no need to worry about upper or lower case, the calculator recognises the key sequence and translates it accordingly.
Type 4, x
Press `enter`

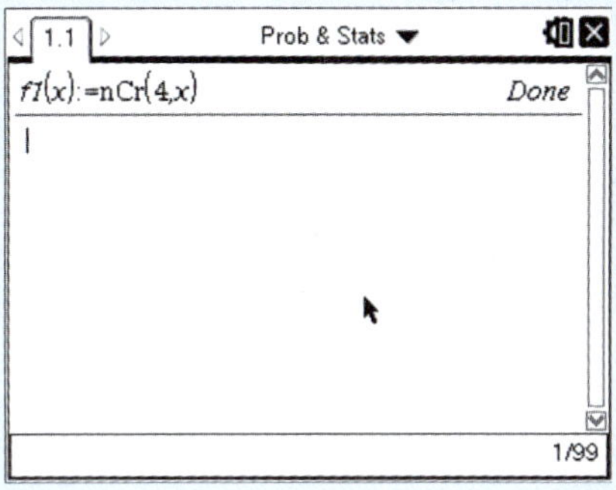

$f1(x):=nCr(4,x)$ Done

Press `⌂ On` and add a new Lists and Spreadsheet page to your document.

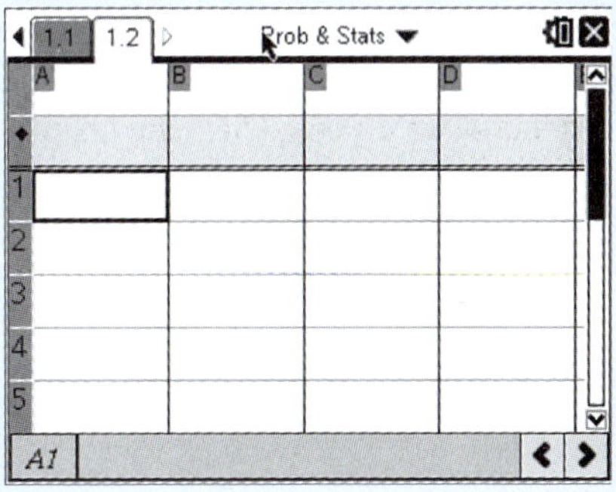

▶ Continued on next page

Chapter 17 621

Press `ctrl` `T` to switch from spreadsheet view to table view.
Press `enter` to display the function $f1(x)$

The table shows that

$\binom{4}{0}=1, \binom{4}{0}=1, \binom{4}{1}=4, \binom{4}{2}=6, \binom{4}{3}=4$ and $\binom{4}{4}=1$

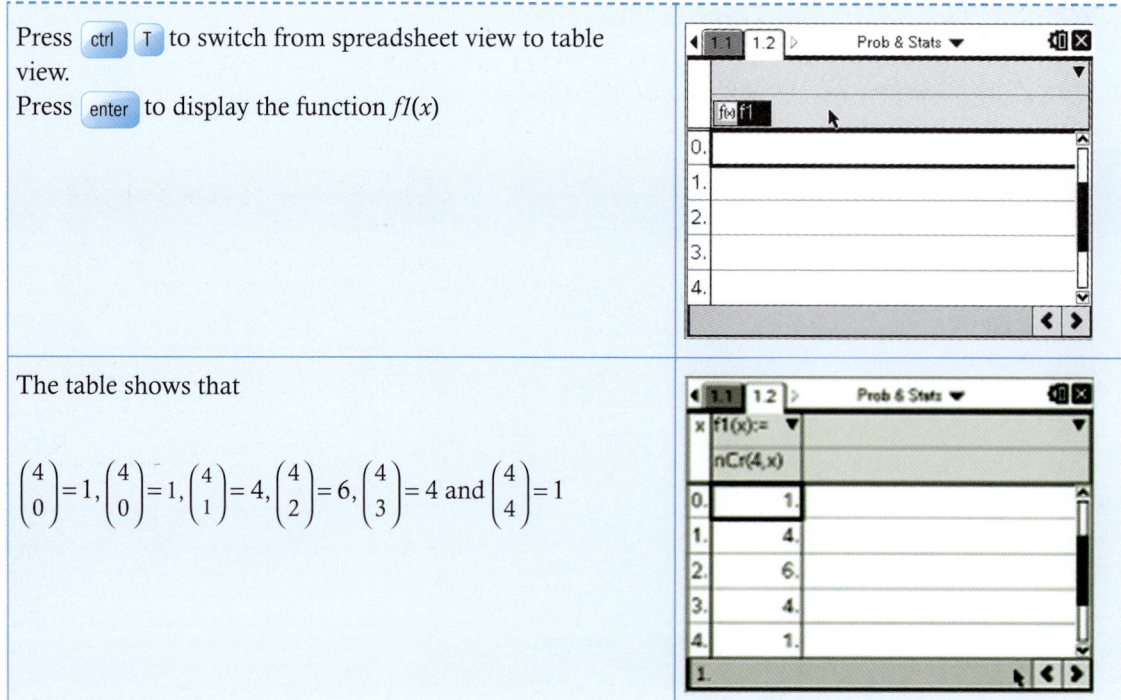

5.12 Calculating binomial probabilities

Example 44

X is a discrete random variable and $X \sim B(9, 0.75)$
Calculate $P(X = 5)$

$$P(x=5)=\binom{9}{5}0.75^5\,0.25^4$$

The calculator can find this value directly

Open a new document and add a Calculator page.
Press `menu` 5:Probability | 3:Probability | 5:Distributions | D:Binomial Pdf...
Enter the number of trials, probability of success and the X value.
Click on OK

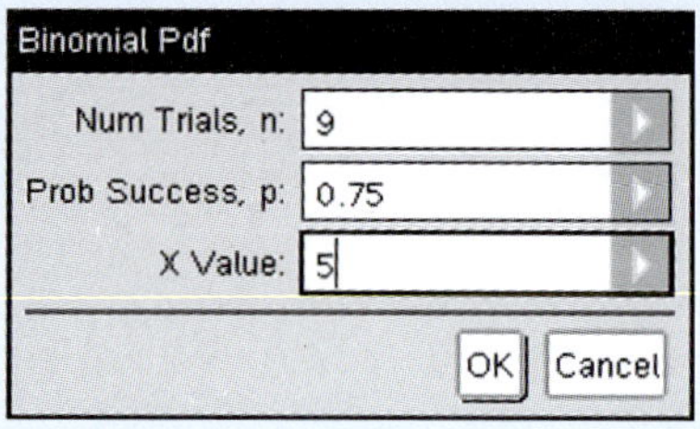

The calculator shows that
$P(X = 5) = 0.117$ (to 3 sf)

You can also type the function straight in without using the dialogue box.

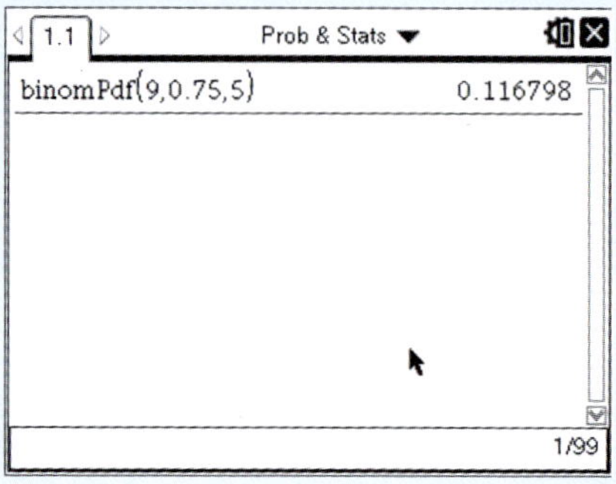

Example 45

X is a discrete random variable and $X \sim B(7, 0.3)$
Calculate the probabilities that X takes the values $\{0, 1, 2, 3, 4, 5, 6, 7\}$

Open a new document and add a Calculator page. Press menu 5:Probability | 3:Probability | 5:Distributions | D:Binomial Pdf... Enter the number of trials, probability of success and leave the X value blank. Click on OK	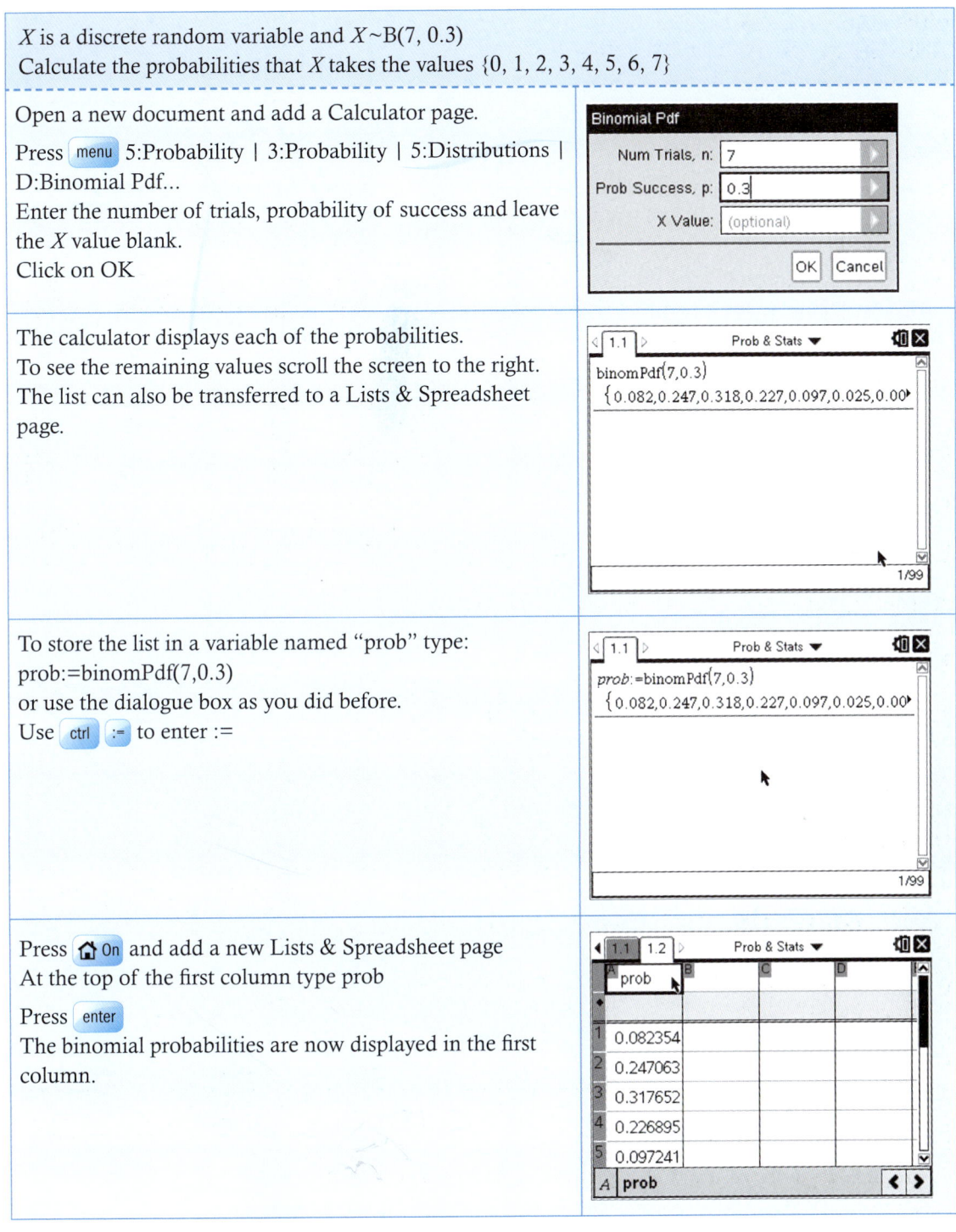
The calculator displays each of the probabilities. To see the remaining values scroll the screen to the right. The list can also be transferred to a Lists & Spreadsheet page.	
To store the list in a variable named "prob" type: prob:=binomPdf(7,0.3) or use the dialogue box as you did before. Use ctrl := to enter :=	
Press On and add a new Lists & Spreadsheet page At the top of the first column type prob Press enter The binomial probabilities are now displayed in the first column.	

Example 46

X is a discrete random variable and $X \sim B(20, 0.45)$
Calculate
a the probability that X is less than or equal to 10
b the probability that X lies between 5 and 15 inclusive
c the probability that X is greater than 11

Open a new document and add a Calculator page.
Press menu 5:Probability | 3:Probability | 5:Distributions | E:Binomial Cdf
Enter the number of trials and the probability of success
The lower bound in this case is 0 and the upper bound is 10.
Click on OK

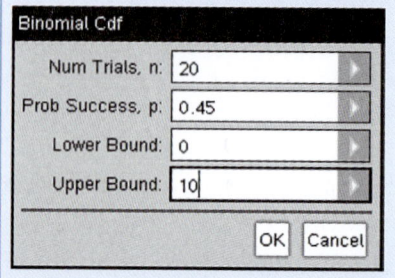

a $P(X \leq 10) = 0.751$ (to 3 sf)
b $P(5 \leq X \leq 15) = 0.980$ (to 3 sf)
c $P(X > 11) = 0.131$ (to 3 sf)
Note: the lower bound is 12 here.

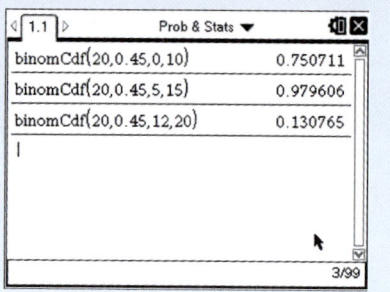

Calculating normal probabilities

5.13 Calculating normal probabilities from X-values

Example 47

A random variable X is normally distributed with a mean of 195 and a standard deviation of 20, or $X \sim N(195, 20^2)$. Calculate
a the probability that X is less than 190
b the probability that X is greater than 194
c the probability that X lies between 187 and 196.

Open a new document and add a Calculator page.
Press menu 5:Probability | 5:Distributions | 2:Normal Cdf

Press enter
You need to enter the values Lower Bound, Upper Bound, μ and σ in the dialogue box.
For the Lower Bound, enter -9×10^{999} as $-9\text{E}999$. This is the smallest number that can be entered in the GDC, so it is used in place of $-\infty$. To enter the E, you need to press the key marked EE.

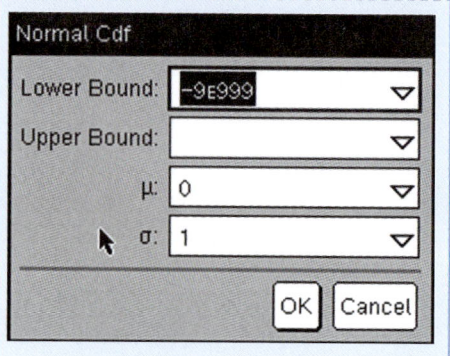

▶ Continued on next page

a $P(X < 190)$
Leave the Lower Bound as $-9\text{E}999$.
Change the Upper Bound to 190.
Change μ to 195 and σ to 20.
$P(X < 190) = 0.401$ (to 3 sf)

b $P(X > 194)$
Change the Lower Bound to 194.
For the Upper Bound, enter 9×10^{999} as 9E999. This is the largest number that can be entered in the GDC, so it is used instead of $+\infty$. Leave μ as 195 and σ as 20.
$P(X > 194) = 0.520$ (to 3 sf)

c $P(187 < X < 196)$
Change the Lower Bound to 187 and the Upper Bound to 196; leave μ as 195 and σ as 20.
$P(187 < X < 196) = 0.175$ (to 3 sf)

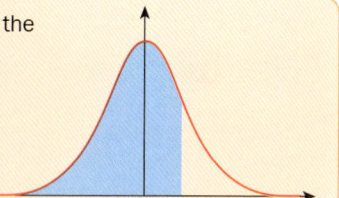

> It can be quicker to type the function directly into the calculator, without using the menus and the wizard, but there are a lot of parameters to remember for the function normCdf.

5.14 Calculating *X*-values from normal probabilities

> When using the inverse normal function (invNorm), make sure that you find the probability on the correct side of the normal curve. The areas are always the lower tail, that is, they are of the form $P(X < x)$ (see Example 48).

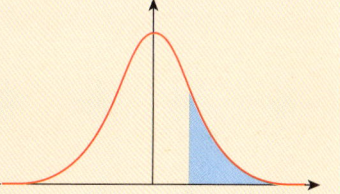

> If you are given the upper tail, $P(X > x)$, you must first subtract the probability from 1 to before you can use invNorm (see Example 49).

Example 48

A random variable X is normally distributed with a mean of 75 and a standard deviation of 12, or $X \sim N(75, 12^2)$. If $P(X < x) = 0.4$, find the value of x.

> You are given a *lower*-tail probability, so you can find $P(X < x)$ directly.

Open a new document and add a Calculator page.
Press menu 5:Probability | 5:Distributions | 3:Inverse Normal...
Press enter
Enter the probability (area = 0.4), mean ($\mu = 75$) and standard deviation ($\sigma = 12$) in the dialogue box.

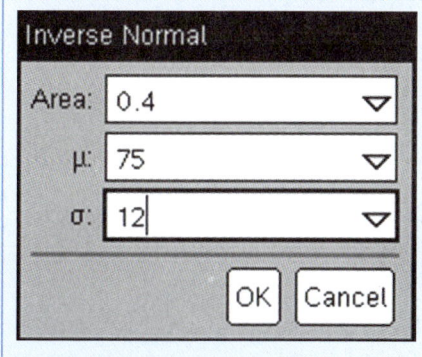

> It can be quicker to type the function directly into the calculator, without using the menus and the wizard, but there are a lot of parameters to remember for the function invNorm.

▶ Continued on next page

So, if P(X < x) = 0.4 then x = 72.0 (to 3 sf).

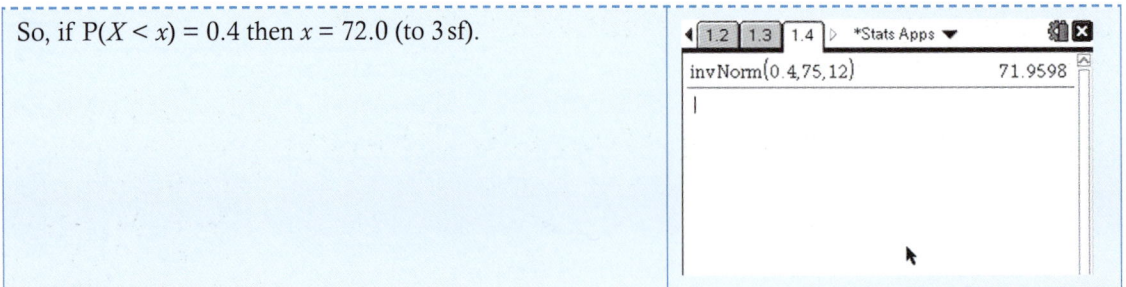

Example 49

A random variable X is normally distributed with a mean of 75 and a standard deviation of 12, or $X \sim N(75, 12^2)$.
If P(X > x) = 0.2, find the value of x.

You are given an *upper*-tail probability, so you must first find P(X < x) = 1 − 0.2 = 0.8. You can now use the invNorm function as before.

Open a new document and add a Calculator page.
Press menu 5:Probability | 5:Distributions | 3:Inverse Normal...
Press enter
Enter the probability (area = 0.8), mean ($\mu = 75$) and standard deviation ($\sigma = 12$) in the dialogue box.

So, if P(X > x) = 0.2 then x = 85.1 (to 3 sf).

This sketch of a normal distribution curve shows the value of x and the probabilities for Example 49.

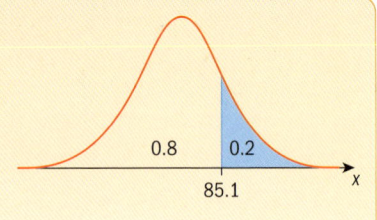

Using a graphic display calculator

Scatter diagrams, linear regression and the correlation coefficient

5.15 Scatter diagrams using a Data & Statistics page

Using a Data & Statistics page is a quick way to draw scatter graphs and find the equation of a regression line.

> For Pearson's product–moment correlation coefficient, see section 5.16, Scatter diagrams using a Graphs page.

Example 50

This data is approximately connected by a linear function.

x	1.0	2.1	2.4	3.7	5.0
y	4.0	5.6	9.8	10.6	14.7

Find the equation of the least squares regression line for y on x.
Use the equation to predict the value of y when $x = 3.0$.

Open a new document and add a Lists & Spreadsheet page. Enter the data in two lists: Type 'x' in the first cell and 'y' in the cell to its right. Enter the x-values in the first column and the y-values in the second. Use the ▼ ▲ ◀▶ keys to navigate around the spreadsheet.	

| Press **On** and add a new Data & Statistics page. **Note**: You do not need to worry about what this screen shows. | 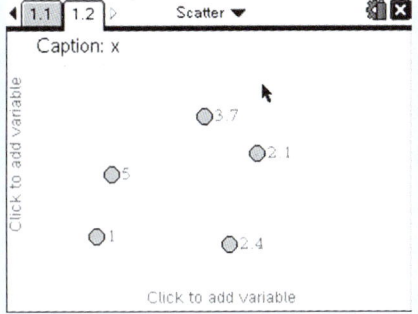 |

| Click at the bottom of the screen where it says 'Click to add variable', choose 'x' from the list and press **enter**. | 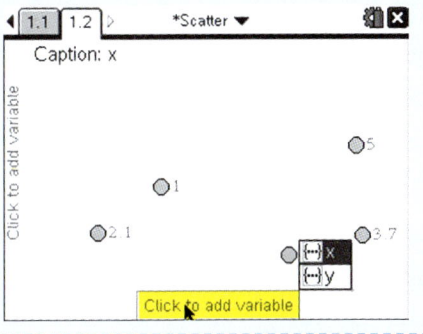 |

▶ Continued on next page

Chapter 17 627

You now have a dot plot of the *x*-values. Move the ▸ near to the side of the screen on the left. The message 'Click to add variable' reappears. Click on the message, choose '*y*' from the list and press enter.	
You now have a scatter graph of *y* against *x*.	
Press menu 4:Analyze | 6:Regression | 1:Show Linear(*mx* + *b*) Press enter You will see the least squares regression line for *y* on *x* and its equation: $y = 2.6282x + 1.47591$	
If you click the ▸ away from the line, it will no longer be selected and the equation disappears.	

▶ Continued on next page

Press `menu` 4:Analyze | 7:Residuals | 1:Show Residual Squares
Press `enter`
The squares on the screen represent the squared deviations of the y-values of the data from the regression line.

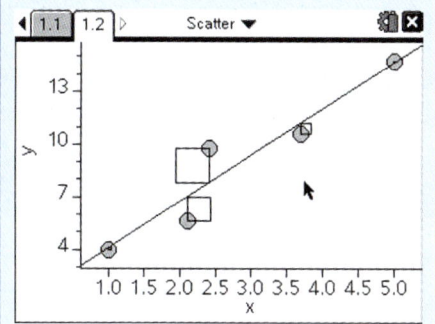

Move the ▸ towards the regression line. When it becomes a ✋, click the touchpad.
You now see the equation of the least squares regression line for y on x and the sum of squares.
The sum of squares is related to Pearson's product–moment correlation coefficient.
Press `menu` 4:Analyze | 7:Residuals | 1:Hide Residual Squares
Press `enter`

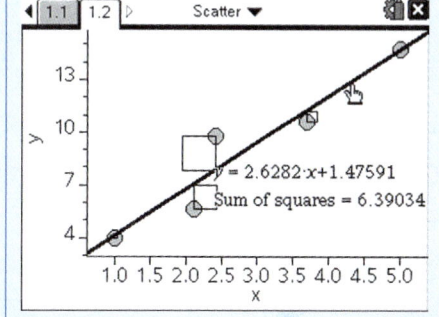

Press `menu` 4:Analyze | A:Graph Trace
Press `enter`
Use the ▶ ◀ keys to move the trace along the line.
It is not possible to move the trace point to an exact value, so get as close to $x = 3$ as you can.
From the graph, $y \approx 9.4$ when $x = 3.0$.

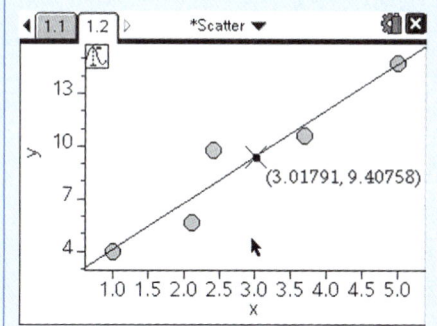

5.16 Scatter diagrams using a Graphs page

Using a Graphs page takes a little longer than the Data & Statistics page, but you will get more detailed information about the data such as Pearson's product–moment correlation coefficient.

Example 51

This data is approximately connected by a linear function.

x	1.0	2.1	2.4	3.7	5.0
y	4.0	5.6	9.8	10.6	14.7

a Find the equation of the least squares regression line for y on x.
b Find Pearson's product–moment correlation coefficient.
c Predict the value of y when $x = 3.0$.

This is the same data as in Example 50.

▶ Continued on next page

Open a new document and add a Lists & Spreadsheet page. Enter the data in two lists: Type 'x' in the first cell and 'y' in the cell to its right. Enter the x-values in the first column and the y-values in the second. Use the ▼▲◀▶ keys to navigate around the spreadsheet.	
Press [On] and add a new Graphs page to your document. Press [menu] 3:Graph Type | 4:Scatter Plot Press [enter] The entry line is displayed at the bottom of the work area. Scatter plot type is displayed. Enter the names of the lists, x and y, into the scatter plot function. Use the [tab] key to move from x to y. Press [enter]	
Adjust your window settings to show the data and the x- and y-axes. You now have a scatter plot of x against y.	
Press [ctrl] ◀ to return to the Lists & Spreadsheet page. Press [menu] 4:Statistics | 1:Stat Calculations | 3:Linear Regression ($mx + b$) Press [enter] From the drop-down menus, choose 'x' for X List and 'y' for Y List. You should press [tab] to move between the fields. Press [enter]	

▶ Continued on next page

On the screen, you will see the result of the linear regression in lists next to the lists for x and y.
The values of m (2.6282) and b (1.475 91) are shown separately.

a The equation of the least squares regression line for y on x is $y = 2.6282x + 1.47591$.

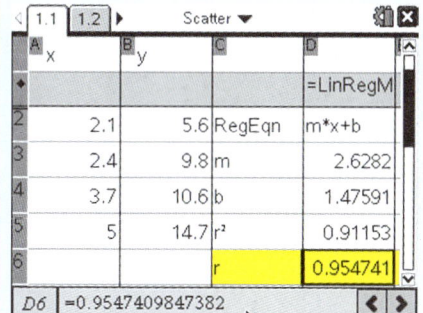

Scroll down the table to see the value of Pearson's product–moment correlation coefficient, given by r.

b Pearson's product–moment correlation coefficient, $r = 0.954\,741$.

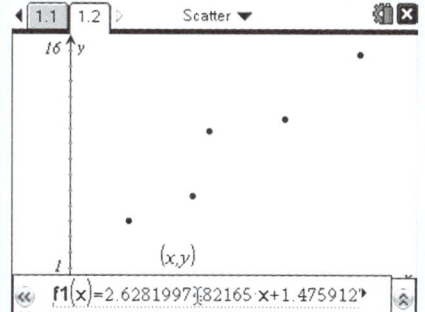

Press ctrl ▶ to return to the Graphs page.
Using the touchpad, click on ◀◀ to open the entry line at the bottom of the work area.
You will see that the equation of the regression line has been pasted into $f1(x)$.

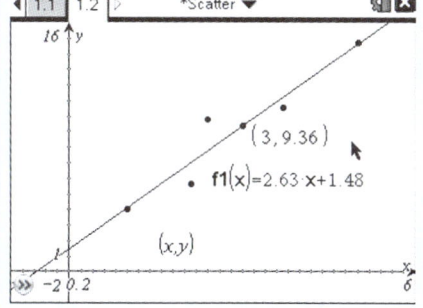

Press enter
The regression line is now shown on the graph.
Use the trace function menu 5:Trace | 1:Graph Trace to find the point where x is 3.0.
Using the ▶ ◀ keys, move the trace point close, then edit the x-coordinate and change it to exactly 3.0.

c When $x = 3.0$, $y = 9.36$.

18 Prior learning

CHAPTER OBJECTIVES:
This chapter contains a number of short topics that you should know before starting the course. You do not need to work through the whole of this chapter in one go. For example, before you start work on an algebra chapter in the book, make sure you have covered the algebra prior learning in this chapter.

The IB Standard Level examination questions will expect you to know all the topics in this chapter. Make sure you have covered them all.

Chapter contents

1 **Number**
 1.1 Calculation 633
 1.2 Simplifying expressions involving roots 634
 1.3 Primes, factors and multiples 637
 1.4 Fractions and decimals 638
 1.5 Percentages 640
 1.6 Ratio and proportion 643
 1.7 The unitary method 645
 1.8 Number systems 646
 1.9 Rounding and estimation 648
 1.10 Standard form 650
 1.11 Sets 651

2 **Algebra**
 2.1 Expanding brackets and factorization 657
 2.2 Formulae 662
 2.3 Solving linear equations 664
 2.4 Simultaneous linear equations 666
 2.5 Exponential expressions 667
 2.6 Solving inequalities 668
 2.7 Absolute value 669
 2.8 Adding and subtracting algebraic fractions 670

3 **Geometry**
 3.1 Pythagoras' theorem 673
 3.2 Geometric transformations 674
 3.3 Congruence 676
 3.4 Similarity 678
 3.5 Points, lines, planes and angles 682
 3.6 Two-dimensional shapes 683
 3.7 Circle definitions and properties 684
 3.8 Perimeter 685
 3.9 Area 686
 3.10 Volumes and surface areas of 3-dimensional shapes 688
 3.11 Coordinate geometry 692

4 **Statistics**
 4.1 Statistical graphs 699
 4.2 Data analysis 703

1 Number

1.1 Calculation

There are several versions of the rules for the order of operations. They all amount to the same thing:

- Brackets or parentheses are calculated first.
- Next come exponents, indices or orders.
- Then multiplication and division, in order from left to right.
- Finally additions and subtractions.

A fraction line or the line above a square root counts as a bracket too.

Your GDC follows the rules, so if you enter a calculation correctly you should get the correct answers.

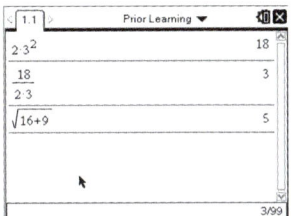

GDC help on CD: *Alternative screenshots for the TI-84 Plus and Casio FX-9860GII GDCs are on the CD.*

BEDMAS:	Brackets, exponents, division, multiplication, addition, subtraction.
BIDMAS:	Brackets, indices, division, multiplication, addition, subtraction.
BEMDAS:	Brackets, exponents, multiplication, division, addition, subtraction.
BODMAS:	Brackets, orders, division, multiplication, addition, subtraction.
BOMDAS:	Brackets, orders, multiplication, division, addition, subtraction.
PEMDAS:	Parentheses, exponents, multiplication, division, addition, subtraction.

Simple calculators, like the ones on phones, do not always follow the calculation rules.

The GDC shows divisions as fractions, which makes the order of operations clearer.

Example 1

a Evaluate $\dfrac{11 + (-1)^2}{4 - (3 - 5)}$

$= \dfrac{11 + 1}{4 - (-2)}$ *brackets first*

$= \dfrac{12}{6}$ *simplify numerator and denominator*

$= 2$

b Evaluate $\dfrac{-3 + \sqrt{9 - 8}}{4}$

$= \dfrac{-3 + \sqrt{1}}{4}$ *simplify the terms inside the square root*

$= \dfrac{-3 + 1}{4}$ *evaluate the root*

$= \dfrac{-2}{4}$ *simplify the numerator and denominator*

$= -\dfrac{1}{2}$

▶ Continued on next page

On your GDC you can either use templates for the fractions and roots or you can use brackets.

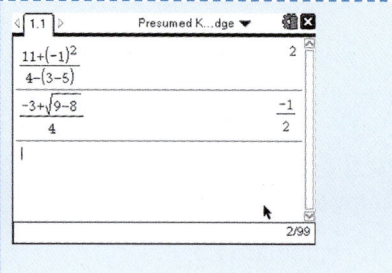

GDC help on CD: *Alternative demonstrations for the TI-84 Plus and Casio FX-9860GII GDCs are on the CD.*

Exercise 1A

Do the questions by hand first, then check your answers with your GDC.

1 Calculate
   a $12 - 5 + 4$
   b $6 \div 3 \times 5$
   c $4 + 2 \times 3 - 2$
   d $8 - 6 \div 3 \times 2$
   e $4 + (3 - 2)$
   f $(7 + 2) \div 3$
   g $(1 + 4) \times (8 - 4)$
   h $1 - 3 + 5 \times (2 - 1)$

2 Find
   a $\dfrac{6 + 9}{4 - 1}$
   b $\dfrac{2 \times 9}{3 \times 4}$
   c $\dfrac{2 - (3 + 4)}{4 \times (2 - 3)}$
   d $\dfrac{6 \times 5 \times 4}{3 \times 2 - 1}$

3 Determine
   a $3 \times (-2)^2$
   b $2^2 \times 3^3 \times 5$
   c $4 \times (5 - 3)^2$
   d $(-3)^2 - 2^2$

4 Calculate
   a $\sqrt{3^2 + 4^2}$
   b $\left(\sqrt{4}\right)^3$
   c $\sqrt{4^3}$
   d $\sqrt{2 + \sqrt{2 + 2}}$

5 Find
   a $\sqrt{\dfrac{13^2 - (3^2 + 4^2)}{2 \times 18}}$
   b $2\sqrt{\dfrac{3 + 5^2}{7}}$
   c $2(3^2 - 4(-2)) - (2 - \sqrt{7 - 3})$

1.2 Simplifying expressions involving roots

$\sqrt{2}$, $2 - \sqrt{3}$, $2\sqrt{5}$, $\dfrac{\sqrt{3}}{3}$, are **irrational numbers** that involve square roots. They are called **surds**.

In calculations, you can use approximate decimals for these types of irrational number, but for more accurate results you can use surds.

Surds are written in their **simplest form** when:
- there is no surd in the denominator
- the smallest possible whole number is under the $\sqrt{}$ sign.

If a question asks for an exact value, it means leave your answer in surd form.

According to some historians Pythagoras was so disturbed by the concept of $\sqrt{2}$ being irrational that it eventually killed him.

→ **Rules of surds**

$\left(\sqrt{a}\right)^2 = a \qquad \sqrt{a \times b} = \sqrt{a} \times \sqrt{b} \qquad \sqrt{\dfrac{a}{b}} = \dfrac{\sqrt{a}}{\sqrt{b}}$

Example 2

Simplify

a $\dfrac{4}{\sqrt{5}}$ **b** $\dfrac{3}{\sqrt{3}}$

Answers

a $\dfrac{4}{\sqrt{5}} = \dfrac{4}{\sqrt{5}} \times \dfrac{\sqrt{5}}{\sqrt{5}}$ *Multiply top and bottom by $\sqrt{5}$*

$= \dfrac{4\sqrt{5}}{\left(\sqrt{5}\right)^2}$

$= \dfrac{4\sqrt{5}}{5}$

b $\dfrac{3}{\sqrt{3}} = \dfrac{3}{\sqrt{3}} \times \dfrac{\sqrt{5}}{\sqrt{3}}$ *Multiply top and bottom by $\sqrt{3}$*

$= \dfrac{3\sqrt{3}}{\left(\sqrt{3}\right)^2}$

$= \dfrac{3\sqrt{3}}{3}$ *Cancel*

$= \sqrt{3}$

Example 3

Simplify

a $\sqrt{20}$ **b** $\sqrt{8} - \sqrt{18}$

Answers

a $\sqrt{20} = \sqrt{4} \times \sqrt{5} = 2\sqrt{5}$ $\sqrt{a \times b} = \sqrt{a} \times \sqrt{b}$

b $\sqrt{8} - \sqrt{18} = \sqrt{4 \times 2} - \sqrt{9 \times 2}$ *Look for square numbers that divide into 8 and 18. Use these to write 8 and 18 as products*

$= 2\sqrt{2} - 3\sqrt{3}$ *Use $\sqrt{a \times b} = \sqrt{a} \times \sqrt{b}$*

$= -\sqrt{2}$

Example 4

Expand the brackets and simplify $(1+\sqrt{2})(1-\sqrt{2})$

Answer

$(1+\sqrt{2})(1-\sqrt{2}) = 1 - \sqrt{2} + \sqrt{2} - \left(\sqrt{2}\right)^2$ $(a+b)(c+d)$
$= ac + ad + bc + bd$

$= 1 - 2$

$= -1$

Example 5

Rewrite the fraction $\dfrac{1}{(1+\sqrt{3})}$ without surds in the denominator

Answer

$\dfrac{1}{(1+\sqrt{3})} = \dfrac{1}{(1+\sqrt{3})} \times \dfrac{(1-\sqrt{3})}{(1-\sqrt{3})}$ *Multiply top and bottom by $1-\sqrt{3}$*

$= \dfrac{1-\sqrt{3}}{1-3} = \dfrac{1-\sqrt{3}}{-2}$

Exercise 1B

1 Simplify

  a $\dfrac{1}{\sqrt{2}}$ **b** $\dfrac{6}{\sqrt{3}}$ **c** $\dfrac{5}{\sqrt{5}}$ **d** $\dfrac{10\sqrt{2}}{\sqrt{5}}$ **e** $\sqrt{\dfrac{2}{5}}$

2 Simplify

  a $\sqrt{12}$ **b** $\sqrt{75}$ **c** $\sqrt{72}$

  d $3\sqrt{8}$ **e** $5\sqrt{27}$

3 Simplify

  a $\sqrt{3} \times \sqrt{12}$ **b** $\sqrt{3} \times \sqrt{27}$ **c** $\sqrt{24} \times \sqrt{32}$

  d $2\sqrt{3} \times 3\sqrt{2}$ **e** $3\sqrt{5} \times 5\sqrt{75}$

4 Simplify

  a $3\sqrt{5}+2\sqrt{5}$ **b** $5\sqrt{2}-3\sqrt{2}$ **c** $2\sqrt{3}+\sqrt{12}$

  d $\sqrt{2}-\sqrt{8}$ **e** $\sqrt{12}-2\sqrt{3}$

5 Expand and simplify

  a $(3+\sqrt{2})^2$ **b** $(\sqrt{2}+\sqrt{3})^2$ **c** $(3+\sqrt{2})(1-\sqrt{2})$

  d $(4+\sqrt{3})(1-\sqrt{2})$ **e** $(2+\sqrt{2})(2-\sqrt{2})$

6 Simplify

  a $\dfrac{1+\sqrt{3}}{\sqrt{7}}$ **b** $\dfrac{1}{1-2\sqrt{3}}$ **c** $\dfrac{\sqrt{5}}{1+\sqrt{5}}$ **d** $\dfrac{4+\sqrt{2}}{3-2\sqrt{2}}$

7 Write these without a surd on the denominator. Simplify as much as possible.

  a $\dfrac{2}{\sqrt{3}}+3\sqrt{3}$ **b** $\dfrac{\sqrt{3}}{2}+\dfrac{5}{\sqrt{3}}$ **c** $\sqrt{20}+\dfrac{2}{\sqrt{5}}$

1.3 Primes, factors and multiples

A **prime** number is an integer, greater than 1, that is not a multiple of any other number apart from 1 and itself.

Example 6

List all the factors of 42.	
Answer $42 = 1 \times 42$, $42 = 2 \times 21$, $42 = 3 \times 14$, $42 = 6 \times 7$ The factors of 42 are 1, 2, 3, 6, 7, 14, 21 and 42.	*Write 42 as a product of two numbers every way you can.*

> In 2009, the largest known prime was a 12 978 189 digit number.
> Prime numbers have become big business because they are used in cryptography.

Example 7

Write the number 24 as a product of prime factors.	
Answer $2\overline{)24}$ $\quad 24 = 2 \times 2 \times 2 \times 3$ $2\overline{)12}$ $\qquad\quad = 2^3 \times 3$ $2\overline{)6}$ $3\overline{)3}$ $\quad 1$	*Begin dividing by the smallest prime number. Repeat until you reach an answer of 1.*

Example 8

Find the **lowest common multiple** (LCM) of 12 and 15.	
Answer The multiples of 12 are 12, 24, 36, 48, 60, 72, 84, 96, 108, 120, 132, 144...	
The multiples of 15 are 15, 30, 45, 60, 75, 90, 105, 120, 135...	
The common multiples are 60, 120... The LCM is 60.	*List all the multiples until you find some in both lists. The LCM is the smallest number in each of the lists.*

Example 9

Find the **highest common factor** (HCF) of 36 and 54.

Answer

$2\overline{)36}$ $36 = 2 \times 2 \times 3 \times 3$ $2\overline{)54}$ $54 = 2 \times 3 \times 3 \times 3$
$2\overline{)18}$ $3\overline{)27}$
$3\overline{)9}$ $3\overline{)9}$
$3\overline{)3}$ $3\overline{)3}$
$\phantom{2\overline{)}}1$ $\phantom{2\overline{)}}1$

The HCF of 36 and 54 is $2 \times 3 \times 3 = 18$

> Write each numbers as a product of prime factors. Find the product of all the factors that are common to both numbers.

Exercise 1C

1. List all the factors of
   a 18   b 27   c 30   d 28   e 78

2. Write as products of prime factors.
   a 36   b 60   c 54   d 32   e 112

3. Find the LCM of
   a 8 and 20
   b 6, 10 and 16

4. Find the HCF of
   a 56 and 48
   b 36, 54 and 90

1.4 Fractions and decimals

There are two types of fraction:
- **common** fractions (often just called 'fractions')
  like $\frac{4}{5}$ $\frac{\text{numerator}}{\text{denominator}}$
- **decimal** fractions (often just called 'decimals') like 0.125.

Fractions can be:

proper like $\frac{2}{3}$, where the numerator is less than the denominator

improper like $\frac{4}{3}$ where the numerator is greater than the denominator

mixed numbers like $6\frac{7}{8}$.

Fractions where the numerator and denominator have no common factor are in their **lowest terms**.

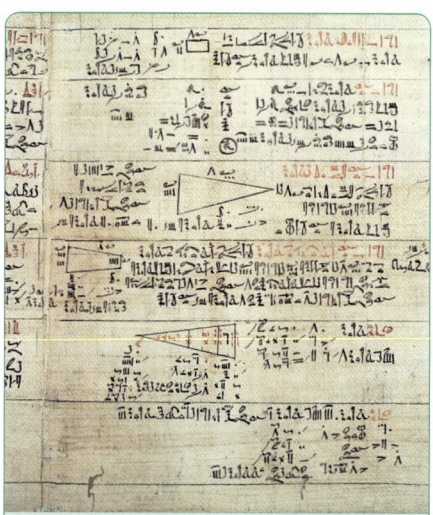

The Rhind Papyrus from ancient Egypt in around 1600 BCE shows calculations using fractions. Egyptians used **unit** fractions so for $\frac{4}{5}$ they would write $\frac{1}{2} + \frac{1}{4} + \frac{1}{20}$. This is not generally regarded as a very helpful way of writing fractions.

$\dfrac{1}{3}$ and $\dfrac{4}{12}$ are **equivalent** fractions.

0.675 is a **terminating** decimal.

0.32... or $0.\overline{32}$ or $0.\dot{3}\dot{2}$ are different ways of writing the **recurring** decimal 0.3232 323 232...

Non-terminating, non-recurring decimals are **irrational** numbers, like π or $\sqrt{2}$.

> $\pi \approx$ 3.14159265358979323846264
> 33832795028841971693993775...
> $\sqrt{2} \approx$ 1.41421356237309504880168
> 87242096980785696718537...
> They do not terminate and there are no repeating patterns in the digits.

Using a GDC, you can either enter a fraction using the fraction template $\dfrac{\square}{\square}$ or by using the divide key ÷. Take care – you will sometimes need to use brackets.

Example 10

a Evaluate

$\dfrac{1}{2} + \dfrac{3}{8} \times \dfrac{4}{9}$ × before +

$= \dfrac{1}{2} + \dfrac{1}{6}$

$= \dfrac{4}{6}$ *simplify*

$= \dfrac{2}{3}$

b Evaluate

$\dfrac{\dfrac{1}{2} + \dfrac{1}{3}}{\dfrac{1}{2} \times \dfrac{1}{3}}$ *evaluate numerator and denominator first*

$= \dfrac{\dfrac{5}{6}}{\dfrac{1}{6}}$

$= 5$

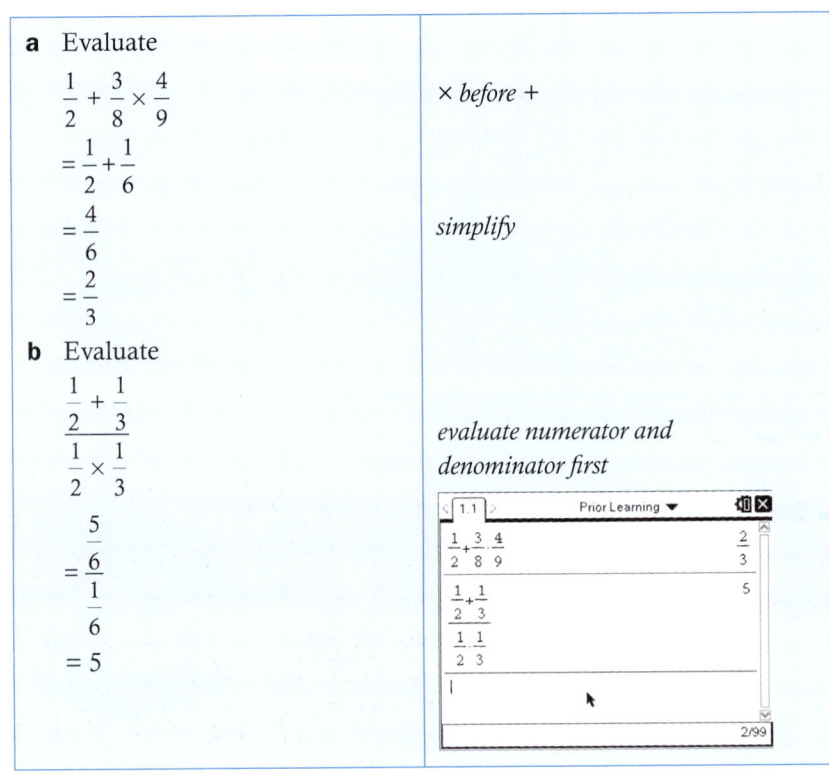

GDC help on CD: *Alternative demonstrations for the TI-84 Plus and Casio FX-9860GII GDCs are on the CD.*

Example 11

a Convert $\dfrac{7}{16}$ to a decimal.   **b** Write $3\dfrac{7}{8}$ as an improper fraction.

Answers

a $\dfrac{7}{16} = 0.4375$

b $3\dfrac{7}{8} = \dfrac{24}{8} + \dfrac{7}{8}$

$= \dfrac{31}{8}$

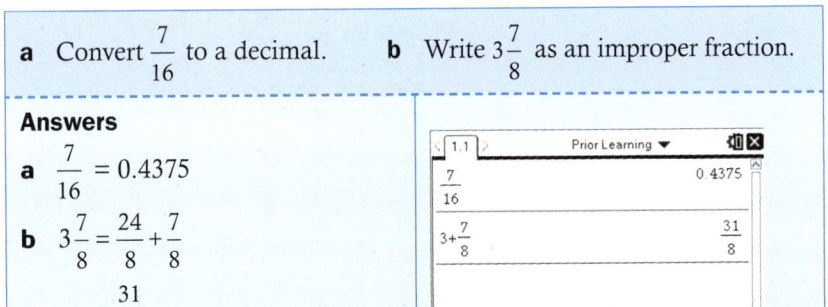

GDC help on CD: *Alternative demonstrations for the TI-84 Plus and Casio FX-9860GII GDCs are on the CD.*

Exercise 1D

1 Calculate

a $\dfrac{1}{2} + \dfrac{3}{4} \times \dfrac{5}{9}$ b $\dfrac{2}{3} \div \dfrac{5}{6} \times 1\dfrac{1}{3}$

c $\sqrt{\left(\dfrac{3}{5}\right)^2 + \left(\dfrac{4}{5}\right)^2}$ d $\dfrac{1 - \left(\dfrac{2}{3}\right)^5}{1 - \dfrac{2}{3}}$

2 Write the following fractions in their lowest terms.

a $\dfrac{16}{36}$ b $\dfrac{35}{100}$ c $\dfrac{34}{51}$ d $\dfrac{125}{200}$

> There are some useful tools for working with fractions. Look in menu 2:Number.

3 Write these mixed numbers as improper fractions.

a $3\dfrac{3}{5}$ b $3\dfrac{1}{7}$ c $23\dfrac{1}{4}$ d $2\dfrac{23}{72}$

4 Write these improper fractions as mixed numbers.

a $\dfrac{32}{7}$ b $\dfrac{100}{3}$ c $\dfrac{17}{4}$ d $\dfrac{162}{11}$

> To convert a fraction to a decimal, divide the numerator by the denominator. Pressing ctrl ≈ will give the result as a decimal instead of a fraction.

5 Convert to decimals.

a $\dfrac{8}{25}$ b $\dfrac{5}{7}$ c $3\dfrac{4}{5}$ d $\dfrac{45}{17}$

1.5 Percentages

A percentage is a way of expressing a fraction or a ratio as part of a hundred.

For example 25% means 25 parts out of 100.

As a fraction, $25\% = \dfrac{25}{100} = \dfrac{1}{4}$.

As a decimal, $25\% = 0.25$.

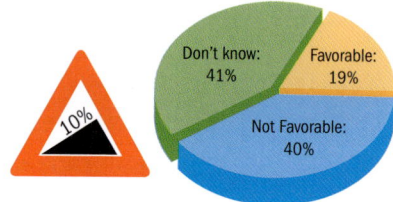

Example 12

Lara's mark in her maths test was 25 out of 40. What was her mark as a percentage?	
Answer	
$\dfrac{25}{40} \times 100 = 62.5\%$	*Write the mark as a fraction.* *Multiply by 100.* *Use your GDC.*

Example 13

There are 80 students taking the IB in a school. 15% take Maths Standard level. How many students is this?

Answer
Method 1
$\frac{15}{100} \times 80 = 12$ *Write the percentage as a fraction out of a hundred and then multiply by 80.*

Method 2
$15\% = 0.15$ *Write the percentage as a decimal.*
$0.15 \times 80 = 12$ *Multiply by 80.*

International currencies
Questions in the Mathematics Standard Level examination may use international currencies.
For example: Swiss franc (CHF); US dollar (USD); British pound sterling (GBP); euro (EUR); Japanese yen (JPY) and Australian dollar (AUD).

Exercise 1E

1. Write as percentages
    a. 13 students from a class of 25
    b. 14 marks out of 20

2. Find the value of
    a. 7% of 32 CHF
    b. $4\frac{1}{2}$% of 12.00 GBP
    c. 25% of 750.28 EUR
    d. 130% of 8000 JPY

> 7% = 0.07

Percentage increase and decrease

Consider an increase of 35%.
The new value after the increase will be 135% of the original value.
So to increase an amount by 35%, find 135% of the amount.
Multiply by $\frac{135}{100}$ or 1.35.
Now consider a decrease of 15%.
After a 15% decrease, the new value will be 85% of the original. So to decrease an amount by 15% find 85%. Multiply by $\frac{85}{100}$ or by 0.85.

Example 14

a The manager of a shop increases the prices of CDs by 12%.
A CD originally cost 11.60 CHF.
What will it cost after the increase?

b The cost of a plane ticket is decreased by 8%.
The original price was 880 GBP. What is the new price?

c The rent for an apartment has increased from 2700 EUR to 3645 EUR per month.
What is the percentage increase?

After an 8% decrease, the amount will be 92% of its original value.

After a 12% increase, the amount will be 112% of its original value.

Answers

a $11.60 \times 1.12 = 12.99$ CHF
(to the nearest 0.01 CHF)

b $880 \times 0.92 = 809.60$ GBP

c **Method 1**
The increase is $3645 - 2700$
$= 945$ EUR

The percentage increase is $\dfrac{945}{2700} \times 100 = 35\%$

Method 2
$\dfrac{3645}{2700} = 1.35 = 135\%$
Percentage increase is 35%.

Calculate the new price as a percentage of the old price.

Find the increase. Work out the increase as a percentage of the original amount.

Percentage increase = $\dfrac{\text{actual increase}}{\text{original amount}} \times 100\%$

Example 15

In a shop, an item's price is given as 44 AUD, **including** tax.
The tax rate is 10%.
What was the price without the tax?

Answer
Call the original price x.
After tax has been added, the price will be $1.10x$.
Hence $1.10x = 44$
$x = 44 \div 1.10$
$= 40$
The price without tax is 40 AUD.

110% = 1.10
Solve for x.
Divide both sides by 1.10.

Exercise 1F

1 In the UK, prices of some goods include a government tax called VAT, which is at 20%.
A TV costs 480 GBP before VAT. How much will it cost including VAT?

2 In a sale in a shop in Tokyo, a dress that was priced at 17 000 JPY is reduced by 12.5%. What is the sale price?

3 The cost of a weekly train ticket goes up from 120 GBP to 128.40 GBP. What is the percentage increase?

4 Between 2004 and 2005, oil production in Australia fell from 731 000 to 537 500 barrels per day. What was the percentage decrease in the production?

5 Between 2005 and 2009 the population of Venezuela increased by 7%. The population was 28 400 000 in 2009. What was it in 2005 (to the nearest 100 000)?

6 An item appears in a sale marked as 15% off with a price tag of 27.20 USD. What was the original price before discount?

7 The rate of GST (goods and service tax) that is charged on items sold in shops was increased from 17% to 20%. What would the price increase be on an item that costs 20 GBP before tax?

8 A waiter mistakenly adds a 10% service tax onto the cost of a meal which was 50.00 AUD. He then reduces the price by 10%. Is the price now the same as it started? If not, what was the percentage change from the original price?

1.6 Ratio and proportion

The **ratio** of two numbers r and s is $r:s$. It is equivalent to the fraction $\frac{r}{s}$. Like the fraction, it can be written in its lowest terms.

For example, $6:12$ is equivalent to $1:2$ (dividing both numbers in the ratio by 6).

> When you write a ratio in its lowest terms, both numbers in the ratio should be positive whole numbers.

In a **unitary ratio**, one of the terms is 1.

For example $1:4.5$ or $25:1$.

If two quantities a and b are in **proportion**, then the ratio $a:b$ is fixed.

> When you write a unitary ratio, you can use decimals.

We also write $a \propto b$ (a is proportional to b).

Example 16

> 200 tickets were sold for a school dance. 75 were bought by boys and the rest by girls. Write down the ratio of boys to girls at the dance, in its lowest terms.
>
> **Answer**
> The number of girls is $200 - 75 = 125$
> The ratio of boys to girls is $75:125 = 3:5$

> Always give the ratio in its lowest terms.

Map scales are often written as a ratio. A scale of 1 : 50 000 means that 1 cm on the map represents 50 000 cm = 0.5 km on the earth.

Example 17

An old English map was made to the scale of 1 inch to a mile. Write this scale as a ratio.

Answer
1 mile = 1760 × 3 × 12
= 63 360 inches
The ratio of the map is 1 : 63 360

Always make sure that the units in ratios match each other.

12 inches = 1 foot
3 feet = 1 yard
1760 yards = 1 mile

Example 18

Three children, aged 8, 12 and 15 win a prize of 140 USD. They decide to share the prize money in the ratio of their ages. How much does each receive?

Answer
140 USD is divided in the ratio 8 : 12 : 15.
This is a total of 8 + 12 + 15
= 35 parts.
140 ÷ 35 = 4 USD
8 × 4 = 32, 12 × 4 = 48 and
15 × 4 = 60
The children receive 32 USD, 48 USD and 60 USD.

Divide the money into 35 parts. One part is 4 USD.

Exercise 1G

1 Aspect ratio is the ratio of an image's width to its height.
   A photograph is 17.5 cm wide by 14 cm high. What is its aspect ratio, in its lowest terms?

2 Gender ratio is expressed as the ratio of men to women in the form $n : 100$. Based on the figures for 2008, the gender ratio of the world was 102 : 100. In Japan, there were 62 million men and 65.2 million women in 2008. What was the gender ratio in Japan?

3 Ryoka was absent for a total of 21 days during a school year of 32 weeks. What is the ratio of the number of days that she was absent to the number of possible days she could have spent at the school during the year in its simplest terms? (A school week is 5 days.)

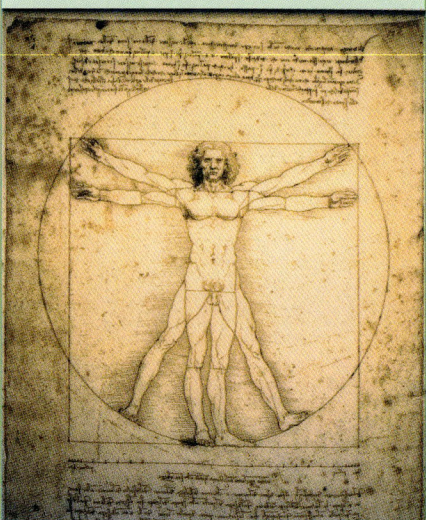

Leonardo da Vinci drew this famous drawing of Vitruvian Man around 1487. The drawing is based on ideal human proportions described by the ancient Roman architect Vitruvius.

4 A model airplane has a wingspan of 15.6 cm. The model is built to a scale of 1 : 72. What was the wingspan of a full-sized airplane (in metres)?

5 On a map, a road measures 1.5 cm. The actual road is 3 km long. What is the scale of the map and how long would a footpath that is 800 m long be on the map?

6 A joint collection is made for two charities and it is agreed that the proceeds should be split in the ratio 5 : 3 between an animal charity and one for sick children. 72 USD is collected. How much is donated to the two charities?

7 For a bake sale, a group of students decide to make brownies, chocolate chip cookies and flapjacks in the ratio 5 : 3 : 2. They plan to make 150 items all together. How many of each will they need to make?

1.7 The unitary method

In the unitary method, you begin by finding the value of **one** part or item.

Example 19

A wheelbarrow full of concrete is made by mixing together 6 spades of gravel, 4 spades of sand, 2 spades of cement and water as required. When there are only 3 spades of sand left, what quantities of the other ingredients will be required to make concrete?

Answer
The ratio   gravel : sand : cement
          is   6 : 4 : 2
          or $\frac{6}{4} : \frac{4}{4} : \frac{2}{4}$
              $= \frac{3}{2} : 1 : \frac{1}{2} = \frac{9}{2} : 3 : \frac{3}{2}$

Hence the mixture requires $4\frac{1}{2}$ spades of gravel to 3 spades of sand to $1\frac{1}{2}$ spades of cement.

> Since the value you want to change is the sand, make sand equal to 1 by dividing through by 4. Then multiply through by 3 to make the quantity of sand equal to 3.

Exercise 1H

1. Josh, Jarrod and Se Jung invested 5000 USD, 7000 USD and 4000 USD to start up a company. In the first year, they make a profit of 24 000 USD which they share in the ratio of the money they invested. How much do they each receive?

2. Amy is taking a maths test. She notices that there are three questions worth 12, 18 and 20 marks. The test lasts one hour and fifteen minutes. She decides to allocate the time she spends on each question in the ratio of the marks. How long does she spend on each question?

1.8 Number systems

Throughout this course, you will be working with **real numbers**. There are two types of real numbers – rational numbers and irrational numbers.

> → **Rational numbers** are numbers that can be written in the form $\frac{a}{b}$, where a and b are both integers, and $b \neq 0$.

$\frac{2}{5}$, $-\frac{17}{8}$, 0.41, $1.\dot{3}$, and 9 are rational numbers.

$\frac{2}{5}$ and $-\frac{17}{8}$ are written in the form $\frac{a}{b}$.

0.41 can be written in the form $\frac{a}{b}$, because $0.41 = \frac{41}{100}$.

$1.\dot{3}$ can be written in the form $\frac{a}{b}$, because $1.\dot{3} = \frac{4}{3}$.

9 can be written in the form $\frac{a}{b}$, because $9 = \frac{9}{1}$.

Repeating or terminating decimals can be written as fractions, so they are rational numbers.

Within the set of rational numbers are sets of numbers called **natural numbers** (0, 1, 2, 3, ...) and **integers** (−4, −3, −1, 0, 1, 2, 3, ...).

$\mathbb{R}$ represents the set of real numbers, $\mathbb{Q}$ rational numbers, $\mathbb{N}$ natural numbers, and $\mathbb{Z}$ integers.

> → **Irrational numbers** are real numbers that can be written as decimals that never terminate or repeat.

$\sqrt{3}$, π, e, and $\sqrt{117}$ are irrational numbers.

$\sqrt{3} = 1.7320508...$ $\pi = 3.14159265...$

$e = 2.7182818...$ $\sqrt{117} = 10.8166538...$

Example 20

Classify each of these real numbers as rational or irrational.

$0.75, -2, \sqrt{37}, \sqrt{25}, 0, \dfrac{2\pi}{3}$

Answer

0.75 is a rational number	0.75 can be written in the form $\dfrac{3}{4}$, and -2 can be written as $-\dfrac{2}{1}$.
-2 is a rational number	
$\sqrt{37}$ is an irrational number	$\sqrt{37} = 6.08276...$ This decimal does not repeat or terminate.
$\sqrt{25}$ is a rational number	$\sqrt{25}$ is a rational number, since it is equal to 5.
0 is a rational number	
$\dfrac{2\pi}{3}$ is an irrational number	Even though it is written in fractional form, $\dfrac{2\pi}{3}$ is not a rational number. Multiples of π are irrational.

Example 21

Write the rational number $0.8\dot{3}$ in the form $\dfrac{a}{b}$.

Answer

Let $x = 0.8\dot{3}$.

$100x = 83.\dot{3}$, and $10x = 8.\dot{3}$.

$100x - 10x = 83.\dot{3} - 8.\dot{3}$

$90x = 75$

$x = \dfrac{75}{90} = \dfrac{5}{6}$

Multiply by powers of 10 to change the position of the decimal point.

Subtracting these values cancels out the repeating 3s.

Exercise 1I

1 Classify each of these real numbers as rational or irrational.

   a 83   **b** $\dfrac{4}{9}$ **c** $\dfrac{2\pi}{3}$ **d** -0.96

   e $-0.4\dot{5}$ **f** e^5 **g** $-4\sqrt{81}$ **h** $\dfrac{\sqrt{5}}{7}$

   i $1.24\dot{7}$ **j** $\sqrt{18}$

2 Which of the numbers from question 1 are:
   a integers
   b natural numbers?

3 Write each rational number from question 1 in the form $\dfrac{a}{b}$, where a and b are integers, and $b \neq 0$.

Properties of real numbers

Real number arithmetic uses three important properties.

Commutative property

→ When adding or multiplying two or more numbers, the order does not matter.

These properties may seem like common sense, but you should think about when you can or can't use them.

For example:
- $a + b = b + a$
- $15 + 7 = 7 + 15$
- $xy = yx$
- $3(8) = 8(3)$

Addition and multiplication are commutative. Subtraction and division are not.

Associative property

→ When adding or multiplying three or more numbers, you can group the numbers in different ways for the calculation without changing their order.

The commutative and associative properties do not work for subtraction.
- $20 - 7 \neq 7 - 20$
- $(18 - 9) - 3 \neq 18 - (9 - 3)$

For example:
- $a + b + c = (a + b) + c = a + (b + c)$
- $5 + 9 + 16 = (5 + 9) + 16 = 5 + (9 + 16)$
- $xyz = (xy)z = x(yz)$
- $6 \times 4 \times 10 = (6 \times 4) \times 10 = 6 \times (4 \times 10)$

Use BIDMAS – calculate the value in the brackets first.

Distributive property

→ $a(b + c) = ab + ac$ and $a(b - c) = ab - ac$.

*We use this when expanding brackets in algebra or simplifying multiplication.
For example $5 \times 32 = (5 \times 30) + (5 \times 2)$*

1.9 Rounding and estimation

To round to a given number of **decimal places**:
- Look at the figure in the next decimal place.
- If this figure is less than 5, round down.
- If this figure is 5 or more, round up.

An exam question might tell you to give your answer to two decimal places, for example.

To round to a given number of **significant figures**:

- For any number, read from left to right and ignore the decimal point.
- The first significant figure is the first non-zero digit, the second significant figure is the next digit (which can be zero or otherwise), and so on.

3	5	.	2	7	1		0	.	5	3	9
1st	2nd		3rd	4th	5th				1st	2nd	3rd
sf	sf		sf	sf	sf				sf	sf	sf

> In IB exams, unless otherwise stated in the question, all numerical answers must be given exactly or correct to three significant figures.

Example 22

Write the number 8.0426579 to
a 2 decimal places   b 1 significant figure   c 1 decimal place
d 4 decimal places   e 6 significant figures

Answers
a 8.04   8.04*2* next digit less than 5 so round down
b 8   8.*0* next digit less than 5 so round down
c 8.0   8.0*4* next digit less than 5 so round down
d 8.0427   8.0426*5* next digit 5 so round up
e 8.04266   8.04265*7* next digit greater than 5 so round up

> When a question asks for a number of decimal places, write them down even if some of the values are zero.

Example 23

Round 42536 to 3 significant figures.

Answer
42500   425*3*6 next digit (3) less than 5 so round down.
              Replace any other digits before the decimal place with zeros.

Estimation

To **estimate** the value of a calculation, write all the numbers to one significant figure.

For example, to estimate the value of 197.2 ÷ 3.97, calculate 200 ÷ 4 = 50

> Estimating the answer to a calculation first gives you an idea of the answer to expect. If your GDC gives you a very different answer, you can then check if you keyed in the values correctly.

Exercise 1J

1 Write each number to the nearest number given in the brackets.
    a 2177 (ten)   b 439 (hundred)   c 3532 (thousand)
    d 20.73 (unit)   e 12.58 (unit)

2 Write down each number correct to the number of decimal places given in the brackets.
  a 0.6942 (2)   **b** 28.75 (1)   **c** 0.9999 (2)
  d 77.984561 (3)   **e** 0.05876 (2)

3 Write down each number in question 1 correct to 2 significant figures.

4 Write down each number in question 2 correct to 3 significant figures.

5 Write each fraction as a decimal to 3 significant figures.
  a $\dfrac{2}{3}$ **b** $\dfrac{3}{46}$ **c** $\dfrac{5}{13}$

> Use your GDC to convert each fraction to a decimal.

6 Write down an estimate for the value of the following calculations
  a $54.04 \div 9.89$ **b** $\dfrac{2.8 \times 3.79}{1.84}$ **c** $\dfrac{7.08 - 0.7556}{(8.67)^2}$

7 Use your GDC to evaluate each part of question 6 to 3 significant figures.

1.10 Standard form

Very large and very small numbers can be written in standard form as

$A \times 10^n$ where n is an integer and $1 \leq A < 10$

- First write the number with the place values adjusted so that it is between 1 and 10.
- Then work out the value of the index, n, the number of columns the digits have moved.

> For example, 37300 is 3.73×10^4 in standard form.

Example 24

Write **a** 89 445 **b** 0.000 000 065 in standard form	
Answers	
a $89\,445 = 8.9445 \times 10^4$	Write 89 445 as 8.9445×10^n The digits have moved 4 places to the right so $n = 4$.
b $0.000\,000\,065 = 6.5 \times 10^{-8}$	Write 6.5×10^n The digits have moved 8 places to the left, so $n = -8$

Exercise 1K

1 Write these in standard form
    a 1475
    b 231000
    c 2.8 billion
    d 0.35×10^6
    e 73.5×10^5

> 1 billion = 1 thousand million

2 Write these as ordinary numbers
    a 6.25×10^4
    b 4.2×10^8
    c 3.554×10^2

3 Write these in standard form
    a 0.0001232
    b 0.00004515
    c 0.617
    d 0.75×10^{-5}
    e 34.9×10^{-5}

4 Write these as ordinary numbers
    a 3.5×10^{-7}
    b 8.9×10^{-8}
    c 1.253×10^{-2}

5 Light travels about 3×10^5 metres per second. Find the time it takes to travel 1 metre. Give your answer in standard form.

1.11 Sets

A **set** is a group of items. We generally use a capital letter to name a set, and the brackets { } to enclose the items of the set.
For example, if P is the set of all the prime numbers less than 20, then P = {2, 3, 5, 7, 11, 13, 17, 19}.

> The curly brackets used for sets are sometimes called "braces".

Each item in the set is called an **element** of the set.

- The symbol $\in$ means "is an element of".
  For example, $3 \in P$ means "3 is an element of the set P".
- The symbol $\notin$ means "is not an element of".
  For example, $8 \notin P$ means "8 is not an element of the set P".

We use a lower-case *n* for the number of items in a set.
The set P has 8 elements, so $n(P) = 8$.

If the number of items in a set is zero, then that set is an **empty set**, or *null set*. We represent the empty set with empty brackets, { }, or with the symbol $\varnothing$.

A set which contains all relevant items is called the **universal set**, and is represented by the letter U. In some cases, the universal set can be assumed. For example, a common universal set is "all real numbers".

> The universal set can also be thought of as the *reference set*.

Set builder notation

To fully specify a set, you can use this notation:
$$A = \{x \mid x \in \mathbb{Z}, 10 < x < 15\}$$

- A is the set of all values of x
- such that x is an integer
- greater than 10 and less than 15

The elements of this set are A = {11, 12, 13, 14}.

Example 25

Write the elements of each set, and give the number of items in each set.
a B, the set of all multiples of 5 that are less than 30.
b $T = \{x \mid x \in \mathbb{N}, x \geq 7\}$

Answers
a B = {5, 10, 15, 20, 25}   *This set is **infinite**, which means it goes on forever.*
   n(B) = 5
b T = {7, 8, 9, 10, 11, ...}   *We cannot count the number of items in the set.*

> You can use ellipses (3 dots ...) to show that a series continues.

Exercise 1L

1 List the elements in each set.
   a A, the set of all the factors of 72.
   b B, the set of all the prime factors of 72.
   c C, the set of all even prime numbers.
   d D, the set of all the even multiples of 7.
   e $E = \{x \mid x \in \mathbb{Z}, |x| < 4\}$
   f $F = \{x \mid x \in \mathbb{N}, x \geq 20\}$
   g G, the set of all prime numbers that are multiples of 4.

2 State the number of items in each of the sets from question 1.

Subsets, intersections, and unions

→ We say that a set B is a **subset** of set A if all the elements of set B are also elements of set A.

> The symbol ⊆ means "is a subset of".

Let A = {1, 2, 3, 4, 5, 6}, and B = {2, 3, 4}.
Since B is a subset of A, we write B ⊆ A.

There are many other subsets of A, such as {1, 3, 5, 6}, {2, 5}, {4},

> $\mathbb{Z} \subseteq \mathbb{R}$, since all integers belong to the set of real numbers.

and even the empty set {}, as well as the set {1, 2, 3, 4, 5, 6} itself.

> → A set C is called a **proper subset** of set A if C is a subset of A, but has fewer elements than A.

For example, C = {2, 5} is a proper subset of the set A = {1, 2, 3, 4, 5, 6}. We write this as C ⊂ A.

The symbol ⊂ means "is a subset of".

> → Two sets that share elements in common have an **intersection**. We use the symbol ∩ to represent the intersection of two sets.

For example, let D = {2, 4, 6, 8, 10} and E = {1, 2, 3, 4, 5}. Both sets contain the elements 2 and 4, so D ∩ E = {2, 4}.

They only have 2 and 4 in common.

> → The **union** of two sets is the set of all the elements of both sets.
> We use the symbol ∪ to represent the union of two sets.

For example, if D = {2, 4, 6, 8, 10} and E = {1, 2, 3, 4, 5}, the union of these sets is D ∪ E = {1, 2, 3, 4, 5, 6, 8, 10}.

Numbers which appear in both sets should only be listed once.

Example 26

Let A = {the odd natural numbers less than 16} and B = {x | x is a factor of 15}.
a List the elements of each set.
b Is B a subset of A? Explain.
c Give the intersection and the union of sets A and B.

Answers

a A = {1, 3, 5, 7, 9, 11, 13, 15} *You could write B ⊆ A*
   B = {1, 3, 5, 15}
b Yes, B is a subset of A.   *B is also a proper subset of A.*
   All the elements of B are   *You could write B ⊂ A.*
   elements of A.
   You could write B ⊆ A.
c A ∩ B = {1, 3, 5, 15}   *These numbers are elements of both sets.*
   A ∪ B = {1, 3, 5, 7, 9, 11, 13, 15}   *This set includes all the elements of A and all the elements of B, once only.*

There are two types of sets which have no intersections.

> → **Disjoint sets** contain no elements in common.
>
> For example, if A = {2, 4, 6, 8}, and B = {1, 3, 5, 7}, A and B are disjoint sets. We write A ∩ B = {}, or A ∩ B = ∅.

Chapter 18   653

> → Sets are **complements** if they have no elements in common, and they contain *all* the elements of U between them.

For example, let U = {all positive integers}, and
A = { 2, 4, 6, 8, 10, ...}.
The complement of A is the set {1, 3, 5, 7, 9, ...}.
We write A′ = {1, 3, 5, 7, 9, ...}. Together, sets A and A′ contain all the positive integers, but they have no elements in common.

The complement of a set A is written A′, called "A prime".

Example 27

Let U = {multiples of 5} and M = {10, 20, 30, ...}. What is the complement of M?	
Answer M′ = { 5, 15, 25, ...}	*Since M contains all the even multiples of 5, M′ must contain all the odd multiples of 5.* *Together, $M \cup M' = U$.*

Exercise 1M

1 Let A = {1, 2, 3, 4, 5, 6}, and let B = {4, 5}.
   a Is B a subset of A? Explain.
   b Are the sets A and B disjoint? Explain.
   c List the intersection of sets A and B.
   d List the union of sets A and B.

2 Let A = {$x \mid x$ is a factor of 36} and B = {$x \mid x$ is a factor of 15}.
   a List the elements of each set.
   b Is B a subset of A? Explain.
   c Are the sets A and B disjoint? Explain.
   d List the intersection of sets A and B.
   e List the union of sets A and B.

3 Let A = {$x \mid x \in \mathbb{Z}, x > 16$} and B = {$x \mid x$ is a multiple of 20}.
   a List the elements of each set.
   b Is B a subset of A? Explain.
   c Are the sets A and B disjoint? Explain.
   d List the intersection of sets A and B.
   e List the union of sets A and B.

4 Let U = {positive integers} and D = { $x \mid x$ is a multiple of 3}.
   List the elements of the complement of D.

5 Let U = {multiples of 10}, and let B = {10, 20, 30}. List the elements of B′

6 Give two sets A and B such that
  a A ∩ B = {}
  b A ∩ B = {4, 7, 10}
  c A ∪ B = {1, 2, 3, 4, 5}
  d $n(A ∩ B) = 2$
  e $n(A ∪ B) = 8$
  f $n(A ∪ B) = 7$ and $n(A ∩ B) = 3$
  g B ⊆ A and $n(A ∩ B) = 3$

Sets related to number lines and inequalities

Subsets of the set of real numbers can be represented as intervals on a **real number line**. These intervals can also be expressed using set notation and inequalities.

Example 28

Write each interval using set notation and inequalities.

a
b

Answers

a $\{x \mid x \in \mathbb{R}, x \geq -1\}$ The numbers greater than −1 are shaded on the number line. The solid circle at −1 tells us that −1 is included.

b $\{x \mid x \in \mathbb{R}, -3 < x < 1\}$ The numbers between −3 and 1 are shaded on the number line. The open circles at −3 and 1 tell us that −3 and 1 are not included.

Example 29

Shade the number line to indicate the interval of real numbers given by the set.
  a $\{x \mid x \in \mathbb{R}, x < 2\}$ **b** $\{x \mid x \in \mathbb{R}, 0 < x \leq 4\}$

Answers

a $\{x \mid x \in \mathbb{R}, x < 2\}$ 2 is not included, so use an open circle at 2.

b $\{x \mid x \in \mathbb{R}, 0 < x \leq 4\}$
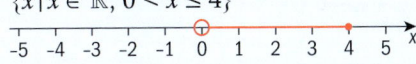
Draw the line between 0 and 4.
0 is not included, so use an open circle.
4 is included, so use a filled circle.

Exercise 1N

1 Write each interval using set notation and inequalities.

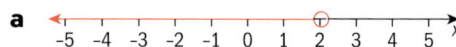

a
b
c
d

2 Shade the number line to indicate the interval of real numbers given by the set.

a $\{x \mid x \in \mathbb{R}, x \leq 0\}$
b $\{x \mid x \in \mathbb{R}, -3 \leq x < 2\}$
c $\{x \mid x \in \mathbb{R}, x > -1\}$
d $\{x \mid x \in \mathbb{R}, -5 < x < 1\}$

Mappings

You can show the mathematical **relations** between two sets in several different ways.

Example 30

Each member of $\{x \mid x \in \mathbb{R}, -4 < x < 4\}$ is mapped to its square. Express this relation as:
a a mapping diagram
b a table
c a set of ordered pairs
d a graph.

Answers

a Input   Output

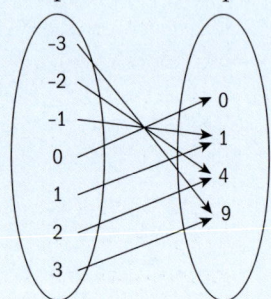

Write the integers −3, −2, −1, 0, 1, 2 and 3 in the input set. Write the squares of the input values, 0, 1, 4 and 9, in the output set.

Draw arrows to map each input value to an output value.

b

x	−3	−2	−1	0	1	2	3
y	9	4	1	0	1	4	9

Use the variable x for the input values and the variable y for the output values.

c {(−3, 9), (−2, 4), (−1, 1), (0, 0), (1, 1), (2, 4), (3, 9)}

Write each input value as the first member of an ordered pair and its corresponding output value as the second member.

▶ Continued on next page

d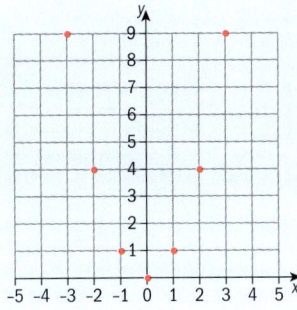

Represent each input value on the horizontal axis and each output value is represented on the vertical axis.

Exercise 10

Express each relation as:
- **a** a mapping diagram
- **b** a table
- **c** a set of ordered pairs
- **d** a graph.

1 Each member of $\{x \mid x \in \mathbb{N}, x \leq 5\}$ is mapped to 2 more than the number.

2 Each member of $\{x \mid x \in \mathbb{Z}, -4 < x < 4\}$ is mapped to the absolute value of the number.

2 Algebra

2.1 Expanding brackets and factorization

The **distributive law** is used to expand brackets and factorize expressions.
$$a(b + c) = ab + ac$$

> The word **algebra** comes from the title of a book Hisab al-jabr w'al-muqabala written by Abu Ja'far Muhammad ibn Musa Al-Khwarizmi in Baghdad around 800 AD. It is regarded as the first book to be written about algebra.

Example 31

Expand $2y(3x + 5y - z)$

Answer
$2y(3x + 5y - z) = 2y(3x) + 2y(5y) + 2y(-z)$
$\qquad\qquad\qquad\quad = 6xy + 10y^2 - 2yz$

> Two other laws used in algebra are the **commutative law**
> $ab = ba$
> and the **associative law** $(ab)c = a(bc)$.

Chapter 18

Example 32

Factorize $6x^2y - 9xy + 12xz^2$

Answer
$6x^2y - 9xy + 12xz^2 = 3x(2xy - 3y + 4z^2)$

Look for a common factor. Write this outside the bracket. Find the terms inside the bracket by dividing each term by the common factor.

Exercise 2A

1. Expand
   a. $3x(x-2)$
   b. $\dfrac{x}{y}(x^2y - y^2 + x)$
   c. $a(b - 2c) + b(2a + b)$

2. Factorize
   a. $3pq - 6p^2q^3r$
   b. $12ac^2 + 15bc - 3c^2$
   c. $2a^2bc + 3ab^2c - 5abc^2$

Products resulting in quadratic expressions

The product of two **binomials**, such as $x + a$ and $x + b$, results in a **quadratic expression**.

$(x + a)(x + b) = (x + a)x + (x + a)b = x^2 + ax + bx + ab = x^2 + (a + b)x + ab$

Here is a shorter method to find the product of two binomials.

$(x + a)(x + b)$ — First terms, Last terms, Inner terms, Outer terms

= **F**irst terms + **O**uter terms + **I**nner terms + **L**ast terms

$= x^2 + bx + ax + ab$

$= x^2 + (a + b)x + ab$

Example 33

Find each product.
a. $(x + 2)(x + 5)$
b. $(x + 6)(x - 4)$
c. $(2x - 3)(3x + 1)$

Answers

a. $(x + 2)(x + 5) = x^2 + 5x + 2x + 10$
$ = x^2 + 7x + 10$

b. $(x + 6)(x - 4) = x^2 - 4x + 6x - 24$
$ = x^2 + 2x - 24$

▶ Continued on next page

c $(2x - 3)(3x + 1) = 6x^2 + 2x - 9x - 3$
$= 6x^2 - 7x - 3$

Exercise 2B

Find each product and simplify your answer.

1 $(x + 7)(x - 4)$ **2** $(x - 3)(x - 2)$ **3** $(3x - 4)(x + 2)$

4 $(2x - 5)(3x + 2)$ **5** $(3x + 2)(3x + 1)$

→ Consider the following special products.
$(x + a)^2 = (x + a)(x + a) = x^2 + ax + ax + a^2 = x^2 + 2ax + a^2$
$(x - a)^2 = (x - a)(x - a) = x^2 - ax - ax + a^2 = x^2 - 2ax + a^2$
$(x + a)(x - a) = x^2 - ax + ax + a^2 = x^2 - a^2$

This last one is called the **difference of two squares**.

Example 34

Find each product.
a $(x + 4)^2$ **b** $(3x - 2)^2$
c $(2x + 3)(2x - 3)$

Answers

a $(x + 4)^2 = x^2 + 8x + 16$

Square the first term: $(x)^2 = x^2$
Double the product of the two terms: $2(4x) = 8x$
Square the last term: $(4x)^2 = 16$

b $(3x - 2)^2 = 9x^2 - 12x + 4$

Square the first term: $(3x)^2 = 9x^2$
Double the product of the two terms: $2(-6x) = -12x$
Square the last term: $(-2)^2 = 4$

c $(2x + 3)(2x - 3) = 4x^2 - 9$

Square the first term: $(2x)^2 = 4x^2$
Square the last term: $(-3x) = 9$
Write the difference of the squares: $4x^2 - 9$

Exercise 2C

Find each product and simplify your answer.

1 $(x + 5)^2$ **2** $(x - 4)^2$ **3** $(x + 2)(x - 2)$

4 $(3x - 4)^2$ **5** $(2x + 5)^2$ **6** $(2x + 7)(2x - 7)$

Factorizing quadratic expressions

The reverse is also possible – to express a quadratic expression as the product of two linear expressions.

$(x + 2)(x + 5) = x^2 + 7x + 10$

$(x + 6)(x - 4) = x^2 + 2x - 24$

To factorize quadratics of the form $x^2 + bx + c$, where the coefficient of x^2 is 1, look for pairs of factors of c whose sum is b.

> 10 is the product of 2 and 5 **and** 7 is the sum of 2 and 5

> −24 is the product of 6 and −4, 2 is the sum of 6 and −4

Example 35

Factorize
a $x^2 - 15x + 14$
b $x^2 + 5x + 6$
c $x^2 - 5x - 24$

Answers

a $x^2 - 15x + 14 = (x - 1)(x - 14)$

Factors of 14	Sum of factors
1 and 14	15
−1 and −14	−15 ←
2 and 7	9
−2 and −7	−9

b $x^2 + 5x + 6 = (x + 2)(x + 3)$

Factors of 6	Sum of factors
1 and 6	7
−1 and −6	−7
2 and 3	5 ←
−2 and −3	−5

c $x^2 - 5x - 24 = (x + 3)(x - 8)$

Factors of −24	Sum of factors
1 and −24	−23
−1 and 24	23
2 and −12	−10
−2 and 12	10
3 and −8	−5 ←
−3 and 8	5
4 and −6	−2
−4 and 6	2

Factorizing quadratics of the form $ax^2 + bx + c$, where $a \neq 0$

Use trial and error to find the correct pair of factors. Try factors that give the correct product for the first and last terms, until you find the one that gives the correct product for the middle term.

Example 36

Factorize
a $2x^2 + 5x + 3$
b $6x^2 + x - 15$

Answers

a $2x^2 + 5x + 3 = (2x + 3)(x + 1)$

Factors of $2x^2$: $2x, x$
Factors of 3: $1, 3; -1, -3$

Possible factors	Linear term
$(2x + 1)(x + 3)$	$6x + 1x = 7x$
$(2x - 1)(x - 3)$	$-6x - 1x = -7x$
$(2x + 3)(x + 1)$	$2x + 3x = 5x$ ←
$(2x - 3)(x - 1)$	$-2x - 3x = -5x$

b $6x^2 + x - 15 = (2x - 3)(3x + 5)$

Factors of $6x^2$: $6x, x; 2x, 3x$
Factors of -15: $1, -15; -1, 15; 3, -5; -3, 5$

Possible factors	Linear term
$(6x + 1)(x - 15)$	$-90x + 1x = -89x$
$(6x - 1)(x + 15)$	$90x - 1x = 89x$
$(6x + 3)(x - 5)$	$-30x + 3x = -27x$
$(6x - 3)(x + 5)$	$30x - 3x = 27x$
$(2x + 1)(3x - 15)$	$-30x + 3x = -27x$
$(2x - 1)(3x + 15)$	$30x - 3x = 27x$
$(2x + 3)(3x - 5)$	$-10x + 9x = -x$
$(2x - 3)(3x + 5)$	$10x - 9x = x$ ←

Factorizing the difference of two squares

Remember that $a^2 - b^2 = (a + b)(a - b)$.

Example 37

Factorize
a $x^2 - 16$
b $9x^2 - 25y^2$

Answers

a $x^2 - 16 = (x + 4)(x - 4)$

$a^2 = x^2$, so $a = x$
$b^2 = 16$, so $b = 4$
Substitute values into $(a + b)(a - b)$.

b $9x^2 - 25y^2 = (3x + 5y)(3x - 5y)$

$a^2 = 9x^2$, so $a = 3x$
$b^2 = 25y^2$, so $b = 5y$
Substitute values into $(a + b)(a - b)$.

Exercise 2D

1 Factorize these quadratic expressions.
- **a** $x^2 + 11x + 28$
- **b** $x^2 - 14x + 13$
- **c** $x^2 - x - 20$
- **d** $x^2 + 2x - 8$
- **e** $x^2 + 13x + 36$
- **f** $x^2 - 7x - 18$

2 Factorize these quadratic expressions.
- **a** $2x^2 - 9x + 9$
- **b** $3x^2 + 7x + 2$
- **c** $5x^2 - 17x + 6$
- **d** $4x^2 - x - 3$
- **e** $3x^2 - 7x - 6$
- **f** $14x^2 - 17x + 5$

3 Factorize these quadratic expressions.
- **a** $x^2 - 9$
- **b** $x^2 - 100$
- **c** $4x^2 - 81$
- **d** $25x^2 - 1$
- **e** $m^2 - n^2$
- **f** $16x^2 - 49y^2$

2.2 Formulae

Rearranging formulae

Example 38

The formula for the area of a circle is $A = \pi r^2$, where A is the area and r is the radius.
The **subject** of the formula is A.
Rearrange the formula to make r the subject.

The subject of a formula is the letter on its own on one side of the = sign.

Answer

$A = \pi r^2$

$r^2 = \dfrac{A}{\pi}$

$r = \sqrt{\dfrac{A}{\pi}}$

Use the same techniques as for solving equations. Whatever you do to one side of the formula, you must do to the other.
Divide both sides by π.
Take the square root of both sides.

You can use this formula to work out the radius of a circle when you know its area.

Example 39

a Einstein's theory of relativity gives the formula $E = mc^2$, where m is the mass, c is the speed of light, and E is the energy equivalent of the mass. Rearrange the formula to make m the subject.

b The formula for gross profit margin is:

$$\text{Gross profit margin} = \dfrac{\text{Gross profit}}{\text{Sales revenue}} \times 100.$$

Rearrange the formula so that *Sales revenue* is the subject.

Answers

a $E = mc^2$

$m = \dfrac{E}{c^2}$

▶ Continued on next page

b $\text{Gross profit margin} = \dfrac{\text{Gross profit}}{\text{Sales revenue}} \times 100$

$\dfrac{\text{Gross profit margin}}{100} = \dfrac{\text{Gross profit}}{\text{Sales revenue}}$

$\text{Sales revenue} \times \text{Gross profit margin} = \text{Gross profit} \times 100$

$\text{Sales revenue} = \dfrac{\text{Gross profit}}{\text{Gross profit margin}} \times 100$

Exercise 2E

Rearrange the following formulae to make the quantity shown in brackets the subject.

1 $v = u - gt$ (t) **2** $a = \sqrt{b^2 + c^2}$ (c) **3** $c = 2\pi r$ (r)

4 $\dfrac{\sin A}{a} = \dfrac{\sin B}{b}$ (b) **5** $a^2 = b^2 + c^2 - 2bc \cos A$ $(\cos A)$

6 To change temperature from degrees Fahrenheit, F, to degrees Celsius, C, you can use the formula $C = \dfrac{5}{9}(F - 32)$.

> Rearrange the formula to make *F* the subject.

7 The acid test ratio measures the ability of a company to use its current assets to retire its current liabilities immediately. The formula is given by:

$\text{Acid ratio test} = \dfrac{\text{Current assets} - \text{Stock}}{\text{Current liabilities}}$.

> Rearrange the formula to make *Stock* the subject.

Substituting into formulae

You can always use your GDC in Mathematical Studies.

When using formulae, let the calculator do the calculation for you. You should still show your working.

> 1. Find the formula you are going to use (from the formula booklet, from the question or from memory) and write it down.
> 2. Identify the values that you are going to substitute into the formula.
> 3. Write out the formula with the values substituted for the letters.
> 4. Enter the formula into your calculator. Use templates to make the formula look the same on your GDC as it is on paper.
> 5. If you think it is necessary, use brackets. It is better to have too many brackets than too few!
> 6. Write down, with units if necessary, the result from your calculator (to the required accuracy).

Example 40

x and y are linked by the formula $y = \dfrac{x^2+1}{2\sqrt{x+1}}$.
Find y when x is 3.1.

Answer

$y = \dfrac{3.1^2+1}{2\sqrt{3.1+1}}$ Write the formula with 3.1 instead of x.

$y = 2.62$

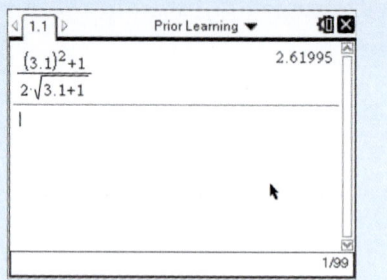

GDC help on CD: *Alternative demonstrations for the TI-84 Plus and Casio FX-9860GII GDCs are on the CD.*

Exercise 2F

1 If $a = 2.3$, $b = 4.1$ and $c = 1.7$, find d where
$$d = \dfrac{3a^2 + 2\sqrt{b}}{ac+b}$$

2 If $b = 8.2$, $c = 7.5$ and $A = 27°$, find a where
$$a = \sqrt{b^2 + c^2 - 2bc \cos A}$$

3 If $u_1 = 10.2$, $r = 0.75$ and $n = 14$, find the value of S, where
$$S = u_1 \dfrac{1-r^n}{1-r}$$

2.3 Solving linear equations

'Solve an equation' means 'find the value of the unknown variable' (the letter).

Rearrange the equation so that the unknown variable x becomes the subject of the equation. To keep the equation 'balanced' always do the same to both sides.

Example 41

Solve the equation $3x + 5 = 17$

Answer

$3x + 5 = 17$
$3x + 5 - 5 = 17 - 5$ subtract 5
$3x = 12$
$\dfrac{3x}{3} = \dfrac{12}{3}$ divide by 3
$x = 4$

Add, subtract, multiply or divide both sides of the equation until the x is by itself on one side. (This can be either the left- or the right-hand side.)

Example 42

Solve the equation $4(x - 5) = 8$

Answer

$4(x - 5) = 8$

$\dfrac{4(x-5)}{4} = \dfrac{8}{4}$ divide by 4

$x - 5 = 2$

$x - 5 + 5 = 2 + 5$ add 5

$x = 7$

> Always take care with − signs.

Example 43

Solve the equation $7 - 3x = 1$

Answer

$7 - 3x = 1$

$7 - 3x - 7 = 1 - 7$ subtract 7

$-3x = -6$

$\dfrac{-3x}{-3} = \dfrac{-6}{-3}$ divide by -3

$x = 2$

> An alternative method for this equation would be to start by *adding* 3x. Then the x would be positive, but on the right-hand side.

Example 44

Solve the equation $3(2 + 3x) = 5(4 - x)$

Answer

$3(2 + 3x) = 5(4 - x)$

$6 + 9x = 20 - 5x$

$6 + 9x + 5x = 20 - 5x + 5x$ add 5x

$6 + 14x = 20$

$6 + 14x - 6 = 20 - 6$ subtract 6

$14x = 14$

$\dfrac{14x}{14} = \dfrac{14}{14}$ divide by 14

$x = 1$

> Compare this method to the one in Example 24. Sometimes it can be quicker to **divide** first rather than expanding the brackets.

Exercise 2G

Solve these equations.

1. $3x - 10 = 2$
2. $\dfrac{x}{2} + 5 = 7$
3. $5x + 4 = -11$
4. $3(x + 3) = 18$
5. $4(2x - 5) = 20$
6. $\dfrac{2}{5}(3x - 7) = 8$
7. $21 - 6x = 9$
8. $12 = 2 - 5x$
9. $2(11 - 3x) = 4$
10. $4(3 + x) = 3(9 - 2x)$
11. $2(10 - 2x) = 4(3x + 1)$
12. $\dfrac{5x + 2}{3} = \dfrac{3x + 10}{4}$

2.4 Simultaneous linear equations

Simultaneous equations involve two variables.
There are two methods which you can use, called substitution and elimination.

Example 45

Solve the equations $3x + 4y = 17$ and $2x + 5y = 16$.

Answer

Geometrically you could consider these two linear equations as the equations of two straight lines. Finding the solution to the equation is equivalent to finding the point of intersection of the lines. The coordinates of the point will give you the values for x and y.

Substitution method

$3x + 4y = 17$
$2x + 5y = 16$
$5y = 16 - 2x$
$y = \dfrac{16}{5} - \dfrac{2}{5}x$

Rearrange one of the equations to make y the subject.

$3x + 4\left(\dfrac{16}{5} - \dfrac{2}{5}x\right) = 17$

Substitute for y in the other equation.

$3x + \dfrac{64}{5} - \dfrac{8}{5}x = 17$
$15x + 64 - 8x = 85$
$15x - 8x = 85 - 64$
$7x = 21$
$x = 3$

Solve the equation for x.

$3(3) + 4y = 17$
$9 + 4y = 17$
$4y = 8$
$y = 2$

Substitute for x in one of the original equations and solve for y.

The solution is $x = 3, y = 2$.

Elimination method

$3x + 4y = 17 \longrightarrow (1)$
$2x + 5y = 16 \longrightarrow (2)$
Multiply equation (1) by 2 and equation (2) by 3.

This is to make the coefficients of x equal.

$6x + 8y = 34 \longrightarrow (3)$
$6x + 15y = 48 \longrightarrow (4)$
Subtract the equations. [(4)−(3)]

Subtracting now eliminates x from the equations.

$7y = 14$
$y = 2$

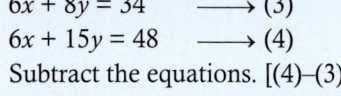
▶ Continued on next page

$3x + 4(2) = 17$
$3x + 8 = 17$
$3x = 17 - 8$
$3x = 9$
$x = 3$
The solution is $x = 3$, $y = 2$.

Substitute for y in one of the original equations and solve for x.

Exercise 2H

1 Solve these simultaneous equations using substitution.
  a $y = 3x - 2$ and $2x + 3y = 5$ **b** $4x - 3y = 10$ and $2y + 5 = x$
  c $2x + 5y = 14$ and $3x + 4y = 7$

2 Solve these simultaneous equations using elimination.
  a $2x - 3y = 15$ and $2x + 5y = 7$ **b** $3x + y = 5$ and $4x - y = 9$
  c $x + 4y = 6$ and $3x + 2y = -2$ **d** $3x + 2y = 8$ and $2x + 3y = 7$
  e $4x - 5y = 17$ and $3x + 2y = 7$

2.5 Exponential expressions

Repeated multiplication can be written as an **exponential** expression.
For example, squaring a number $3 \times 3 = 3^2$ or $5.42 \times 5.42 = 5.42^2$.

If we multiply a number by itself three times then the exponential expression is a cube. For example
$$4.6 \times 4.6 \times 4.6 = 4.6^3.$$

You can also use exponential expressions for larger integer values. So, for example, $3^7 = 3 \times 3 \times 3 \times 3 \times 3 \times 3 \times 3$.

Where the exponent is not a positive integer, these rules apply:
$$a^0 = 1, a \neq 0 \text{ and } a^{-n} = \frac{1}{a^n}$$

Index and **power** are other names for **exponent**.

You use squares in Pythagoras' Theorem $a^2 = b^2 + c^2$ or in the formula for the area of a circle $A = \pi r^2$.
You use a cube in the volume of a sphere $V = \frac{4}{3}\pi r^3$.

Example 46

Write down the values of 10^2, 10^3, 10^1, 10^0, 10^{-2}, 10^{-3}.

Answer
$10^2 = 10 \times 10 = 100$
$10^3 = 10 \times 10 \times 10 = 1000$
$10^1 = 10$
$10^0 = 1$
$10^{-2} = \frac{1}{10^2} = \frac{1}{100} = 0.01$
$10^{-3} = \frac{1}{10^3} = \frac{1}{1000} = 0.001$

To evaluate an exponential function with the GDC use either the ^ key or the template key and the exponent template.

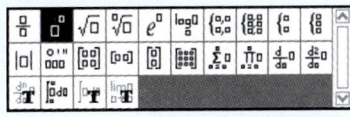

Exercise 2I

Evaluate these expressions.

1. **a** $2^3 + 3^2$ **b** $4^2 \times 3^2$ **c** 2^6

2. **a** 5^0 **b** 3^{-2} **c** 2^{-4}

3. **a** 3.5^5 **b** 0.495^{-2} **c** $2\dfrac{(1-0.02)^{10}}{1-0.02}$

2.6 Solving inequalities

Inequalities behave much like equations and can be solved in the same way.

Example 47

Solve the inequalities   **a** $2x + 5 < 7$ **b** $3(x - 2) \geq 4$

Answers

a $2x + 5 < 7$
$2x < 2$
$x < 1$

b $3(x - 2) \geq 4$
$x - 2 \geq 1\dfrac{1}{3}$
$x \geq 3\dfrac{1}{3}$

Take great care with + and – signs.

Example 48

Solve the inequalities $7 - 2x \leq 5$

Answer

$7 - 2x \leq 5$
$\quad -2x \leq -2$ *Divide by –2*
$\quad\;\; x \geq 1$ *Change $\leq$ to $\geq$*

> If you either multiply or divide an inequality by a negative value, the signs on both sides of the inequality will change. The inequality will also be reversed.

Example 49

Solve the inequalities $19 - 2x > 3 + 6x$

Answer

$19 - 2x > 3 + 6x$
$\quad 19 > 3 + 8x$
$\quad 16 > 8x$
$\quad\;\; 2 > x$
$\quad\;\; x < 2$ *Reverse the inequalities*

> Sometimes the *x* ends up on the right-hand side of the inequality. In this case reverse the inequality as in the example.

Exercise 2J

1. Solve the inequality for x and represent it on the number line.
   a. $3x + 4 \leq 13$ b. $5(x - 5) > 15$ c. $2x + 3 < x + 5$

2. Solve for x.
   a. $2(x - 2) \geq 3(x - 3)$ b. $4 < 2x + 7$ c. $7 - 4x \leq 11$

Properties of inequalities

→ When you add or subtract a real number from both sides of an inequality the direction of the inequality is unchanged.

For example:

- $4 > 6 \Rightarrow 4 + 2 > 6 + 2$
- $15 \leq 20 \Rightarrow 15 - 6 \leq 20 - 6$
- $x - 7 \geq 8 \Rightarrow x - 7 + 7 \geq 8 + 7$
- $x + 5 < 12 \Rightarrow x + 5 - 5 < 12 - 5$

→ When you multiply or divide both sides of an inequality by a positive real number the direction of the inequality is unchanged.
When you multiply or divide both sides of an inequality by a negative real number the direction of the inequality is reversed.

For example:

- $4 > 5 \Rightarrow 2(4) > 2(5)$
- $6 \leq 10 \Rightarrow -2(6) \geq -2(10)$
- $10 \leq 30 \Rightarrow \dfrac{10}{5} \leq \dfrac{30}{5}$
- $18 > 24 \Rightarrow \dfrac{18}{-3} < \dfrac{24}{-3}$
- $-12 > -20 \Rightarrow \dfrac{-12}{4} > \dfrac{-20}{4}$

2.7 Absolute value

The absolute value (or modulus) of a number $|x|$ is the numerical part of the number without its sign. It can be written as

$$|x| = \begin{cases} -x, \text{ if } x < 0 \\ x, \text{ if } x \geq 0 \end{cases}$$

Example 50

Write down $|a|$ where $a = -4.5$ and $a = 2.6$

Answer
If $a = -4.5$ then $|a| = 4.5$
If $a = 2.6$ then $|a| = 2.6$

Example 51

Write the value of $|p - q|$ where $p = 3$ and $q = 6$.

Answer
$|p - q| = |3 - 6|$
$= |-3| = 3$

Exercise 2K

1 Write the value of $|a|$ when a is
   a 3.25   **b** −6.18   **c** 0

2 Write the value of $|5 - x|$ when $x = 3$ and when $x = 8$.

3 If $x = 6$ and $y = 4$, write the values of
   a $|x - y|$ **b** $|x - 2y|$ **c** $|y - x|$

2.8 Adding and subtracting algebraic fractions

To add or subtract fractions, first write them over a common denominator.

Example 52

Combine these fractions, simplifying your answer.

a $\dfrac{x}{2x+1} + \dfrac{5x+3}{2x+1}$ **b** $\dfrac{2x-3}{4x-5} - \dfrac{6x-2}{4x-5}$

c $\dfrac{3x}{3x-1} + \dfrac{3x+1}{2x+5}$ **d** $\dfrac{5x}{x+3} - \dfrac{2x+1}{2x-1}$

Answers

a $\dfrac{x}{2x+1} + \dfrac{5x+3}{2x+1} = \dfrac{x + (5x+3)}{2x+1}$ *Keep the common denominator and add the numerators.*

$= \dfrac{6x+3}{2x+1}$ *Combine like terms.*

$= \dfrac{3(2x+1)}{2x+1}$ *Factorize and simplify whenever possible.*

$= 3$

▶ Continued on next page

b $\dfrac{2x-3}{4x-5} - \dfrac{6x-2}{4x-5} = \dfrac{(2x-3)-(6x-2)}{4x-5}$ Keep the common denominator and subtract the numerators.
Be sure to distribute the negative.

$= \dfrac{2x-3-6x+2}{4x-5}$

$= \dfrac{-4x-1}{4x-5}$ Combine like terms.

c $\dfrac{3x}{3x-1} + \dfrac{3x+1}{2x+5} = \dfrac{3x}{3x-1} \cdot \dfrac{2x+5}{2x+5} + \dfrac{3x+1}{2x+5} \cdot \dfrac{3x-1}{3x-1}$ Multiply each fraction by "one" to get a common denominator.

$= \dfrac{3x(2x+5)}{(3x-1)(2x+5)} + \dfrac{(3x+1)(3x-1)}{(2x+5)(3x-1)}$

$= \dfrac{(6x^2+15x)}{(3x-1)(2x+5)} + \dfrac{(9x^2-1)}{(3x-1)(2x+5)}$ Expand the brackets

$= \dfrac{15x^2+15x-1}{(3x-1)(2x+5)}$ Combine like terms.

d $\dfrac{5x}{x+3} - \dfrac{2x+1}{2x-1} = \dfrac{5x}{x+3} \cdot \dfrac{2x-1}{2x-1} - \dfrac{2x+1}{2x-1} \cdot \dfrac{x+3}{x+3}$ Multiply each fraction by "one" to get a common denominator.

$= \dfrac{5x(2x-1)}{(x+3)(2x-1)} - \dfrac{(2x+1)(x+3)}{(2x-1)(x+3)}$

$= \dfrac{(10x^2-5x)-(2x^2+7x+3)}{(x+3)(2x-1)}$ Expand the brackets

$= \dfrac{10x^2-5x-2x^2-7x-3}{(x+3)(2x-1)}$ Watch out for negative signs.

$= \dfrac{8x^2-12x-3}{(x+3)(2x-1)}$ Combine like terms.

Exercise 2L

Combine these fractions, simplifying your answer.

1 $\dfrac{2}{x+7} + \dfrac{3x-1}{x+7}$ **2** $\dfrac{4x}{2x+2} - \dfrac{3x-1}{2x+2}$ **3** $\dfrac{3x+9}{3x+4} + \dfrac{3x-1}{3x+4}$ **4** $\dfrac{2x}{x+5} + \dfrac{x+1}{2x-1}$

5 $\dfrac{4}{x} + \dfrac{2x+1}{x+2}$ **6** $\dfrac{2x-1}{x-2} - \dfrac{3x}{4x+3}$ **7** $\dfrac{x+1}{5x+1} + \dfrac{2x}{2x-5}$ **8** $\dfrac{x+5}{x-4} - \dfrac{x-2}{x+2}$

Solving equations with rational coefficients

To solve equations with rational coefficients, multiply both sides of the equation by the least common multiple of all the denominators.

Example 53

Solve these equations.

a $\dfrac{x}{6} = \dfrac{5}{4} - \dfrac{x}{2}$

b $\dfrac{1}{15} + \dfrac{1}{x} = \dfrac{1}{6}$

Answers

a $\dfrac{x}{6} = \dfrac{5}{4} - \dfrac{x}{2}$

$12\left(\dfrac{x}{6}\right) = 12\left(\dfrac{5}{4} - \dfrac{x}{2}\right)$ LCM of 6, 4 and 2 is 12

$2x = 14 - 6x$

$8x = 14$

$x = \dfrac{14}{8}$ or $\dfrac{7}{4}$

b $\dfrac{1}{15} + \dfrac{1}{x} = \dfrac{1}{6}$

$30x\left(\dfrac{1}{15} + \dfrac{1}{x}\right) = 30x\left(\dfrac{1}{6}\right)$ LCM of 15 and 6 is 30

$2x + 30 = 5x$

$-3x = -30$

$x = 10$

Exercise 2M

Solve the equations.

1 $\dfrac{x}{3} + \dfrac{1}{6} = \dfrac{x}{4} + \dfrac{1}{4}$

2 $\dfrac{1}{k} + \dfrac{1}{4} = \dfrac{9}{4k}$

3 $\dfrac{1}{6} = \dfrac{5}{6} - \dfrac{1}{x}$

4 $\dfrac{3}{5} - \dfrac{2x}{4} = \dfrac{x-1}{2}$

5 $\dfrac{3x}{4} + \dfrac{x+2}{3} = \dfrac{x-1}{8}$

3 Geometry

3.1 Pythagoras' theorem

→ In a right-angled triangle *ABC* with sides *a*, *b* and *c*, where *a* is the *hypotenuse*:

$$a^2 = b^2 + c^2$$

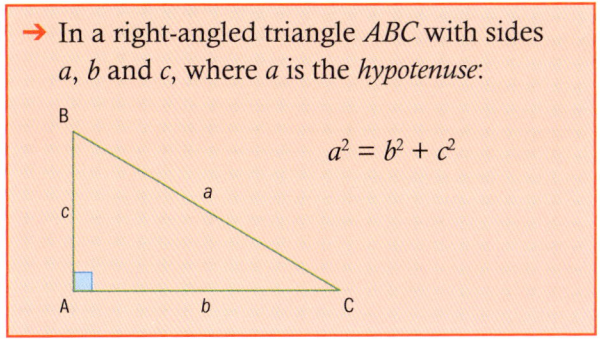

Although the theorem is named after the Greek mathematician Pythagoras, it was known several hundred years earlier to the Indians in their Sulba Sutras and thousands of years before to the Chinese as the Gougu Theorem.

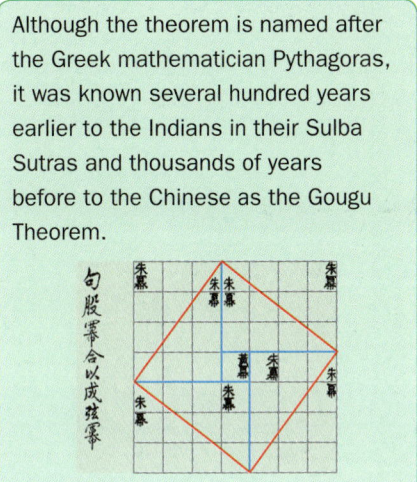

Example 54

Find the length marked *a*.

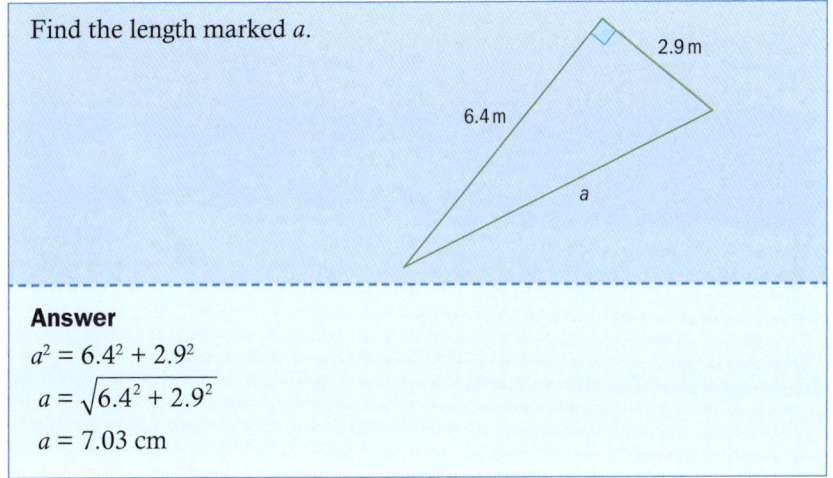

You can use Pythagoras' Theorem to calculate the length of one side of a right-angled triangle when you know the other two.

Answer
$a^2 = 6.4^2 + 2.9^2$
$a = \sqrt{6.4^2 + 2.9^2}$
$a = 7.03$ cm

Sometimes you have to find a shorter side.

Example 55

Find the length marked *b*.

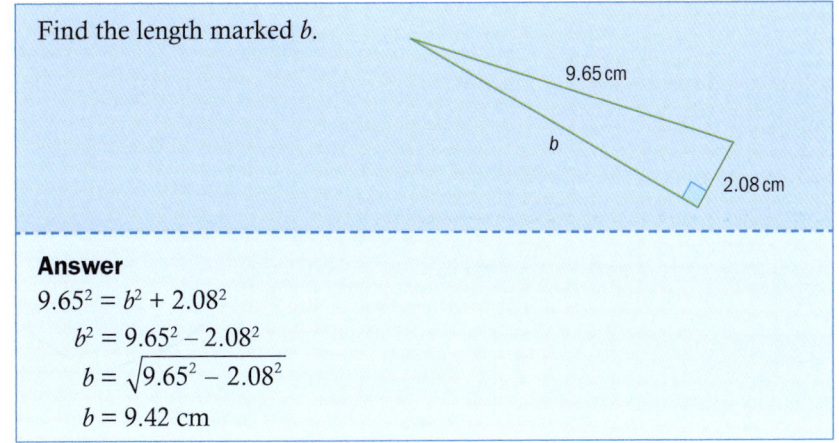

Answer
$9.65^2 = b^2 + 2.08^2$
$b^2 = 9.65^2 - 2.08^2$
$b = \sqrt{9.65^2 - 2.08^2}$
$b = 9.42$ cm

Check your answer by making sure that the hypotenuse is the longest side of the triangle.

Exercise 3A

In each diagram, find the length of the side marked *x*. Give your answer to 3 significant figures.

1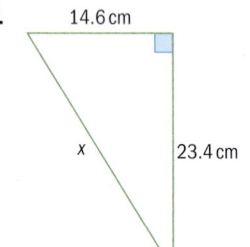

2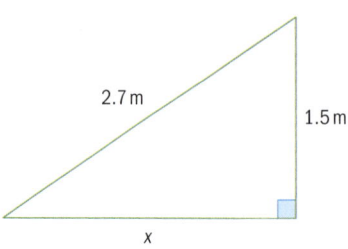

3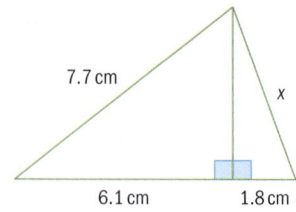

3.2 Geometric transformations

A transformation can change the position as well as the size of an object. A transformation maps an object to its image.

There are four main types of transformation:

- Reflection
- Rotation
- Translation
- Enlargement

Reflection

When an object is **reflected** in a mirror line, the object and its image are symmetrical about the mirror line. Every point on the image is the same distance from the mirror line as the corresponding point on the object.

To describe a reflection, state the equation of the mirror line.

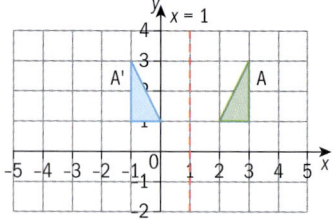

▲ Reflection in line $x = 1$

Rotation

A **rotation** moves an object around a fixed point called the center of rotation, in a given direction through a particular angle.

To describe a rotation give the coordinates of the center of rotation, and the direction and the angle of turn.

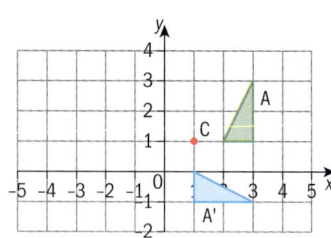

▲ Rotation of 90° clockwise about the point (1,1)

Translation

A **translation** moves every point a fixed distance in the same direction.

To describe a translation write the column vector $\begin{pmatrix} x \\ y \end{pmatrix}$, where *x* is the movement in the *x* direction and *y* is the movement in the *y* direction.

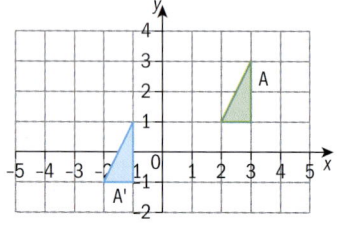

▲ Translation of $\begin{pmatrix} -4 \\ -2 \end{pmatrix}$

Enlargement

Enlargement increases or decreases the size of an object by a given scale factor.

To describe an enlargement give the coordinates of the center of enlargement and the scale factor.

The image after an enlargement is mathematically **similar** to the original object.

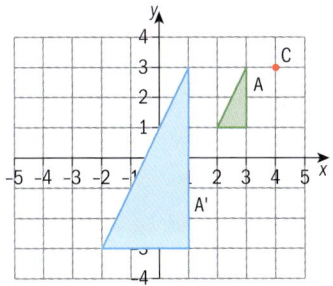

▲ Enlargement scale factor 3 center (4, 3)

Example 56

The grid contains five shapes A to E. Describe the single transformation that takes:

a A to B
b A to C
c A to D
d A to E
e C to D

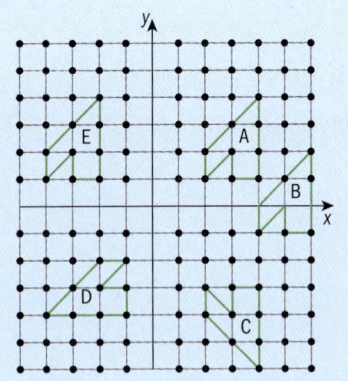

For more on similarity, see page 678

Answers

a A ⟶ B: Translation; vector $\begin{pmatrix} 2 \\ -2 \end{pmatrix}$

b A ⟶ C: Reflection; line $y = -1$

c A ⟶ D: Reflection; line $y = -x$

d A ⟶ E: Translation; vector $\begin{pmatrix} -6 \\ 0 \end{pmatrix}$

e C ⟶ D: Rotation; Center $(1, -1)$, 90° clockwise

Exercise 3B

1 The grid contains four shapes A to D.

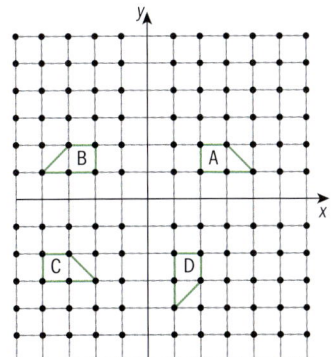

Describe the single transformation that takes:

a A to B b A to C c A to D e B to D

Chapter 18

2 Copy this diagram on to graph paper.

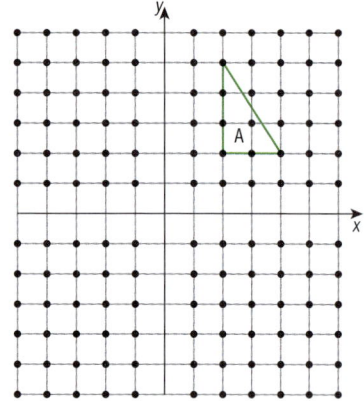

   a Reflect shape A in the line $y = -x$. Label the image B.
   b Reflect shape B in the line x-axis. Label the image C.
   c Describe fully a single transformation that would take A to C.

3 Draw a set of axes from −10 to 10 on both x and y axes.
   a Draw the triangle with vertices at (2, 1) (4, 1) (4, 4). Label it A.
   b Reflect A in the x axis. Label the image B.
   c Enlarge B by scale factor 2 center (0, 0). Label the image C.
   d Rotate C by 180° center (0, 0). Label the image D.
   e Reflect D in the x axis. Label the image E.
   f Rotate E by 180° center (0, 0). Label the image F.

   Describe the single transformation that maps

   g C ⟶ F
   h A ⟶ F
   i E ⟶ A
   j C ⟶ E

3.3 Congruence

> → Two figures that are exactly the same shape and size are **congruent**.
> In congruent shapes
> • Corresponding lengths are equal
> • Corresponding angles are equal

Objects and images after rotations, reflections or translations are congruent to each other.

To prove that two triangles are congruent, you need to show that they satisfy one of four sets of conditions.

▼ Three sides are the same (SSS)

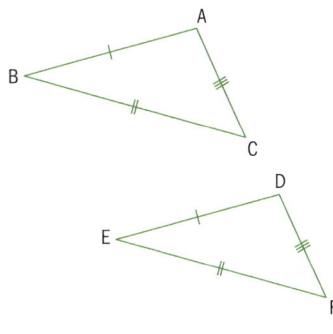

▼ Two sides and the included angle are the same (ASS)

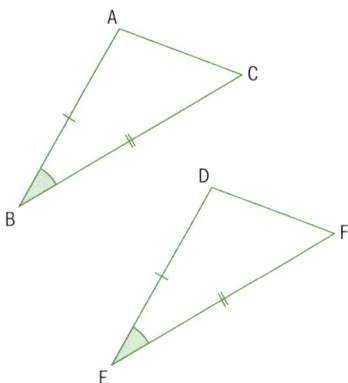

▼ Two angles and the included angle are the same (SAA)

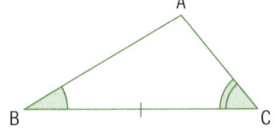

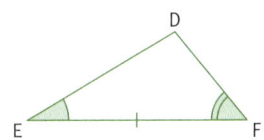

▼ Right-angled triangles with hypotenuse and one other side the same (RHS)

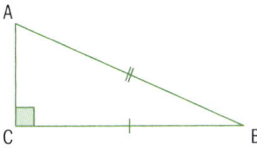

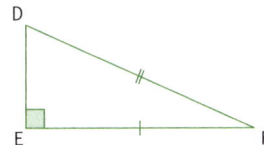

Example 57

State whether the shapes in each pair are congruent.
List the vertices in corresponding order and give reasons for congruence.

a

b

c

▶ Answers on next page

> **Answers**
> a Yes. ABC = XZY   so SAS
> b No. Only angles are equal; Corresponding sides may not be the same length.
> c No. parallelogram ABCD is not congruent to QRSP.
>   It is not clear whether AD = PQ or BC = RS

Exercise 3C

1 Show that △DEF is congruent to △ABC. Find the length of each of the sides.

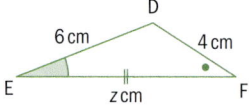 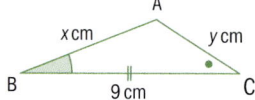

2 Give a brief reason why △DEF and △ABC are congruent. Find the value of each of the angles.

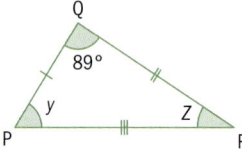 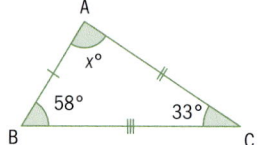

3 Prove that △DEF is congruent to △ABC. Find the values of x and y.

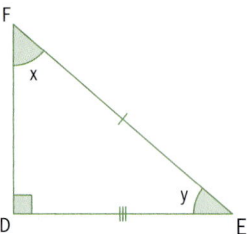 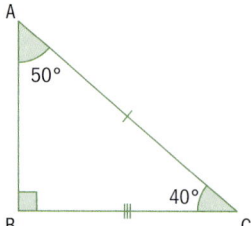

3.4 Similarity

> → Two figures are **similar** if they are the same shape.
>
> They are not necessarily the same size, so generally one is an enlargement of the other.

After an enlargement the image is always similar to the object. Enlarging a shape leaves the angles the same but changes all the lengths by the same scale factor.

> → The scale factor of an enlargement is the ratio of
>
> $$\frac{\text{length of a side on one shape}}{\text{length of corresponding side on other shape}}$$

Similar triangles

In similar triangles, corresponding angles are equal and corresponding sides are in the same ratio.

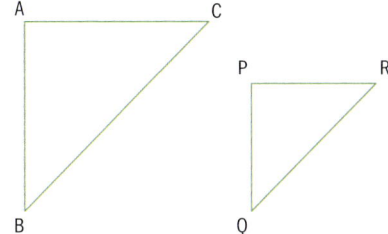

Triangles ABC and PQR are similar because

$\hat{A} = \hat{P}, \hat{B} = \hat{Q}, \hat{C} = \hat{P}$

$\dfrac{AB}{PQ} = \dfrac{BC}{QR} = \dfrac{AC}{PR} =$ scale factor

To prove that two triangles are similar, show that **one** of these three statements is true:

1 ▼ The three angles of one triangle are equal to the three angles of the other triangle

2 ▼ The corresponding sides of each triangle are in the same ratio

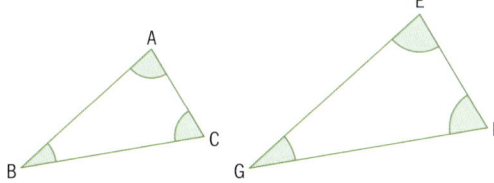

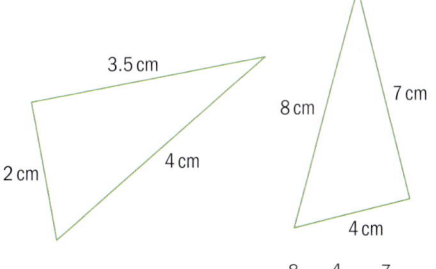

$\dfrac{8}{4} = \dfrac{4}{2} = \dfrac{7}{3.5} = 2$

3 ▼ There is one pair of equal angles and the sides containing these angles are in the same ratio.

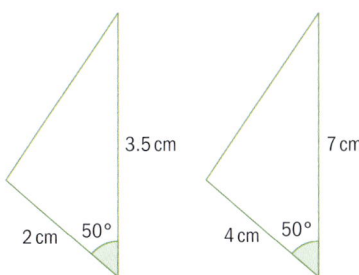

Chapter 18 679

Example 58

Find the length of the side marked *x*.

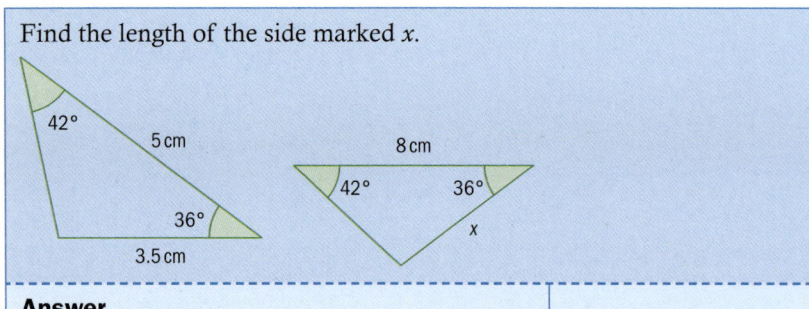

Answer

Two pairs of angles are equal, so the third pair must be equal.
Hence the triangles are similar.
The scale factor of the enlargement is $\frac{8}{5} = 1.6$
So $x = 3.5 \times 1.6 = 5.6$ cm

Prove similarity

Exercise 3D

1. Which pairs of rectangles are similar?

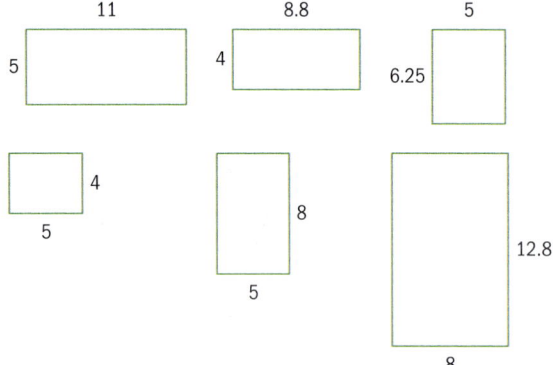

> Note the shapes in this exercise are not drawn to scale.

2. These shapes are similar. Calculate the lengths marked by letters.

   a

   b

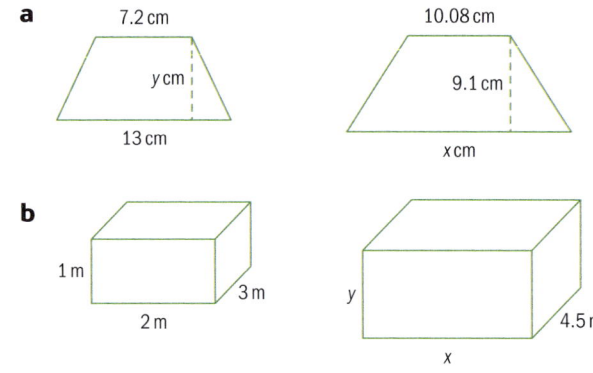

3 Which triangles are similar?

a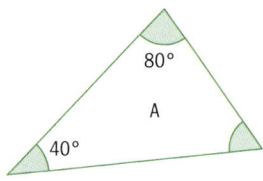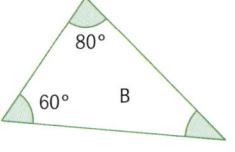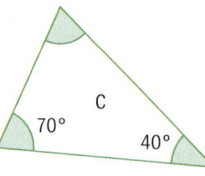

b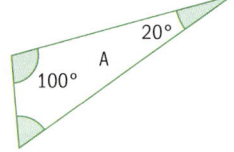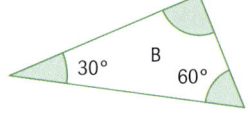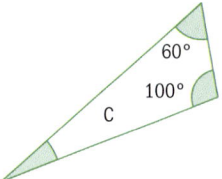

c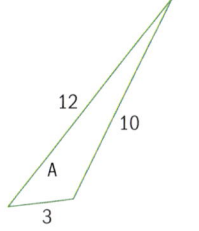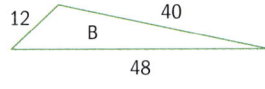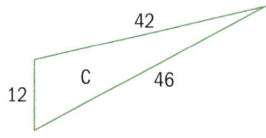

d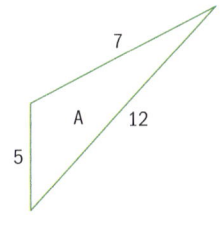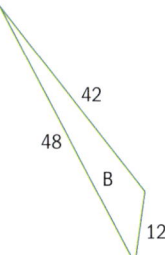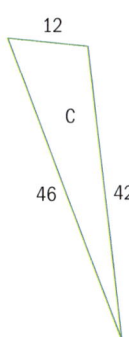

e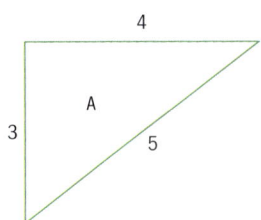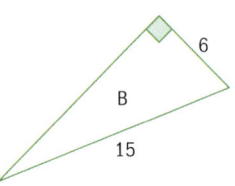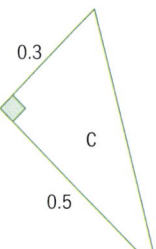

4 Show that triangles ABC and APQ are similar.
Calculate the length of AC and BP.

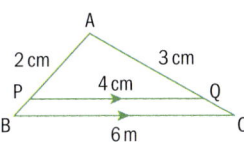

5 In the diagram AB and CD are parallel. AD and BC meet at X.

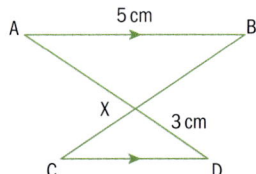

   a Prove that triangles ABX and DCX are similar.
   b Which side in triangle DCX corresponds to AX in triangle ABX?
   c Calculate the length of AX.

3.5 Points, lines, planes and angles

The most basic ideas of geometry are points, lines and planes. A **straight line** is the shortest distance between two points. Planes can be **finite** like the surface of a desk or a wall or can be **infinite**, continuing in every direction.

We say that a point has zero dimensions, a line has one dimension and a plane has two dimensions.

Angles are measured in degrees.

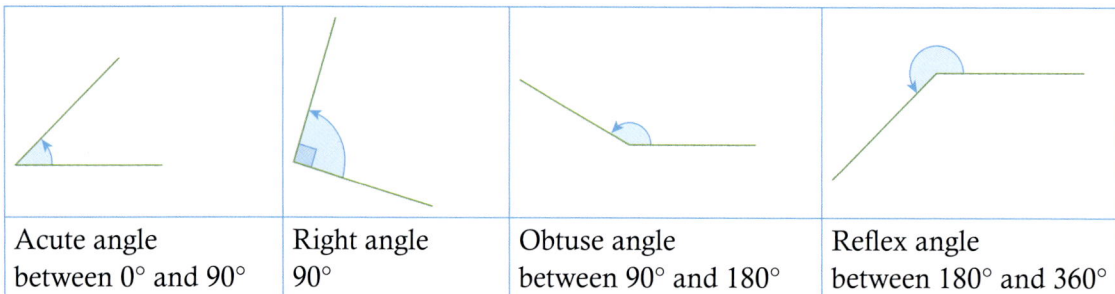

| Acute angle between 0° and 90° | Right angle 90° | Obtuse angle between 90° and 180° | Reflex angle between 180° and 360° |

Exercise 3E

1 Draw a sketch of:
   a a reflex angle   b an acute angle
   c a right angle   d an obtuse angle.

2 State whether the following angles are acute, obtuse or reflex.

   a   b   c

> The small lines on these diagrams show equal lines and the arrows show parallel lines.

3 State whether the following angles are acute, obtuse or reflex.
   a 173°   b 44°   c 272°
   d 82°   e 308°   f 196°

3.6 Two-dimensional shapes

Triangles

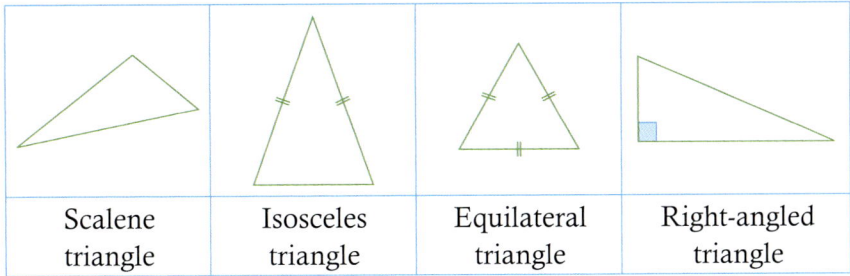

| Scalene triangle | Isosceles triangle | Equilateral triangle | Right-angled triangle |

Quadrilaterals

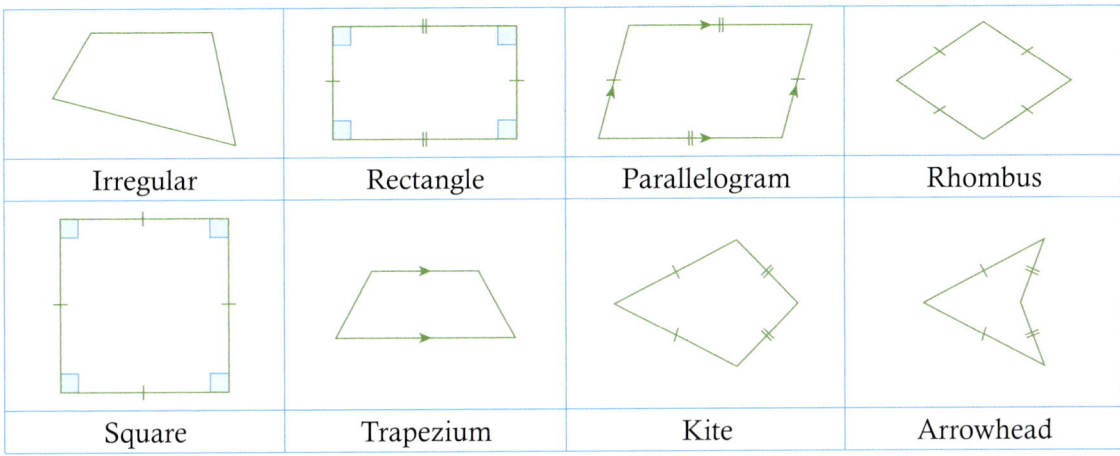

| Irregular | Rectangle | Parallelogram | Rhombus |
| Square | Trapezium | Kite | Arrowhead |

Polygons

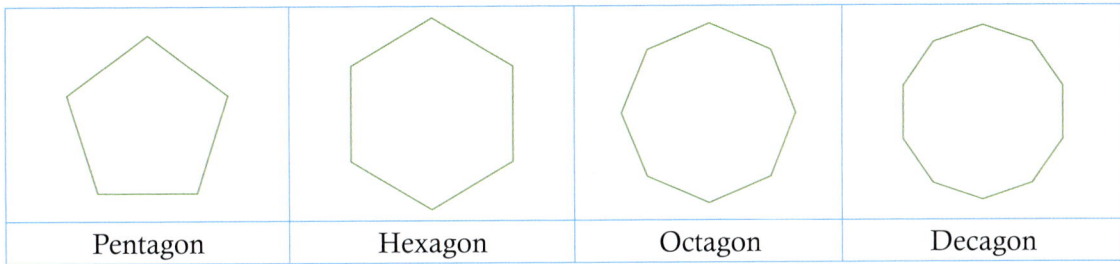

| Pentagon | Hexagon | Octagon | Decagon |

Exercise 3F

1. Sketch the quadrilaterals in the table above with their diagonals.
   Copy and complete the following table.

Diagonals	Irregular	Rectangle	Parallelogram	Rhombus	Square	Trapezium	Kite
Perpendicular					✓		
Equal					✓		
Bisect					✓		
Bisect angles					✓		

For example, the diagonals of a square are perpendicular to each other, equal in length, bisect each other and bisect the angles of the square.

2 List the names of all the shapes that are contained in the following figures.

a    b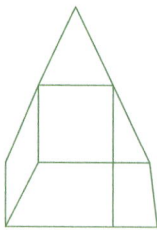

3.7 Circle definitions and properties

You should be familiar with these definitions related to circles.

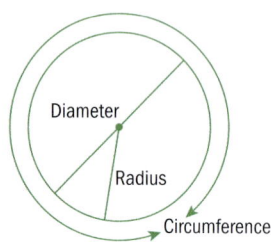

The distance from the center of the circle to any point of the circle is called the **radius**, usually denoted by r.

A **diameter** goes through the center and is twice as long as the radius. The diameter is usually denoted by D.

$D = 2r$

The distance around the circle is called the **circumference**. The circumference of a circle, C, is found using the formulae $C = 2\pi r$ or $C = \pi d$

Here are some other properties and definitions that you should know.

- The **area**, A, of a circle can be calculated using the formula $A = \pi r^2$
- A **chord** of a circle is a line drawn between two points on the circumference of the circle.
  A chord divides a circle into two segments – a **minor segment** and a **major segment**.

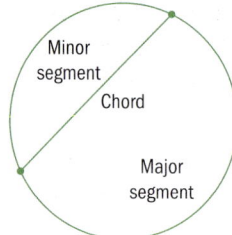

- Any part of the circumference of a circle is called an **arc**.

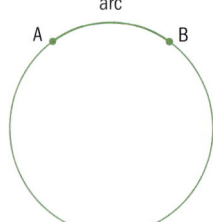

684 Prior learning

- A **semicircle** has an arc that is half the length of the circumference.

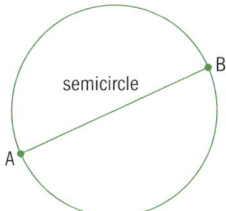

- The area lying between two radii is a **sector.**

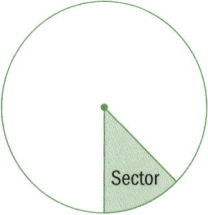

- A **tangent** to a circle is a straight line that touches the circumference of the circle at a single point called the **point of tangency.** The angle between a tangent and the radius at that point is 90°.

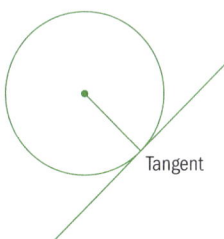

3.8 Perimeter

The **perimeter** of a figure is defined as the length of its boundary. The perimeter of a polygon is found by adding together the sum of the lengths of its sides.

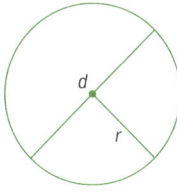

The perimeter of a circle is called its **circumference**.

In the circle on the left, r is the radius and d is the diameter. If C is the circumference.
$C = 2\pi r$
or
$C = \pi d$

$\pi = 3.141592653589793238462...$ Many maths enthusiasts around the world celebrate Pi day on March 14 (3/14). The use of the symbol π was popularized by the Swiss mathematician Leonhard Euler (1707–83).

Example 59

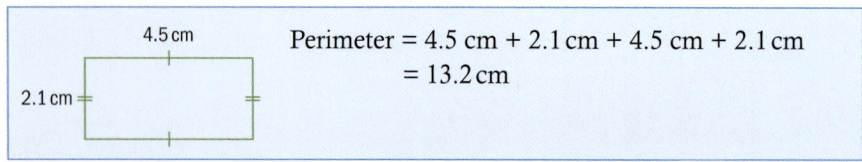

Perimeter = 4.5 cm + 2.1 cm + 4.5 cm + 2.1 cm
= 13.2 cm

Example 60

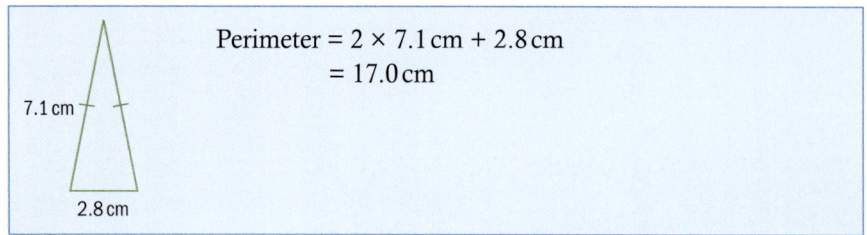

Perimeter = 2 × 7.1 cm + 2.8 cm
= 17.0 cm

Exercise 3G

Find the perimeters of these shapes.

a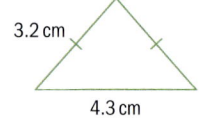

b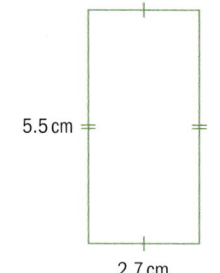

c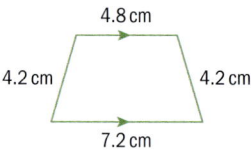

d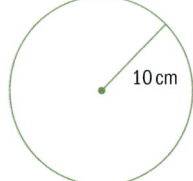

e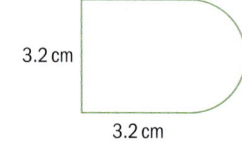

f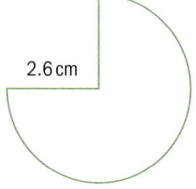

3.9 Area

These are the formulae for the areas of a number of plane shapes.

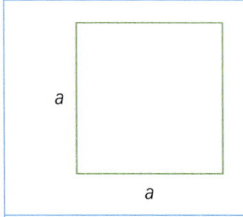
$A = a^2$

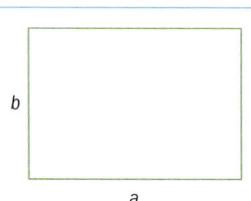
$A = ab$

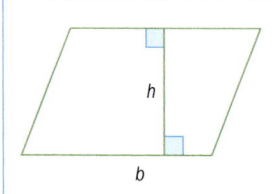
$A = bh$

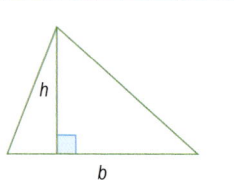
$A = \frac{1}{2}bh$

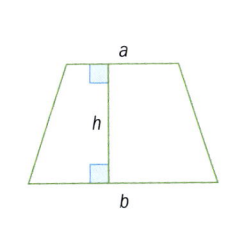	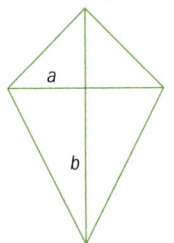	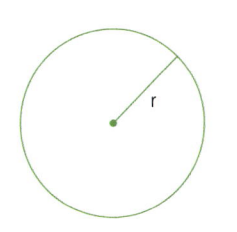
$A = \frac{1}{2}(a+b)h$	$A = \frac{1}{2}ab$	$A = \pi r^2$

Example 61

Find the area of this shape.

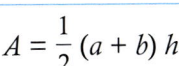

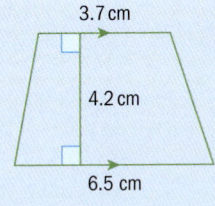

Answer

Area $= \frac{1}{2}(3.7 + 6.5)(4.2) = 21.42$ cm²

Example 62

Find the area of this shape giving your answer to 3 signifigant figures.

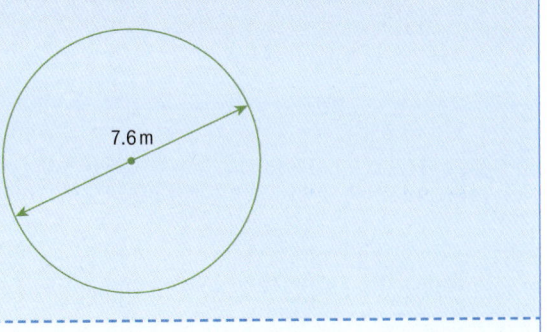

Use the π button on your calculator to enter π.

Answer

Area $= \pi(7.6)^2 = 181$ cm² (3 sf)

Exercise 3H

Find the areas of these shapes. Give your answer to 3 signifigant figures.

1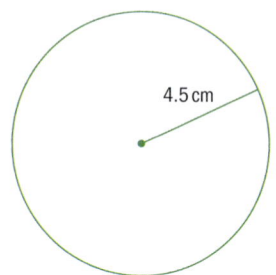

2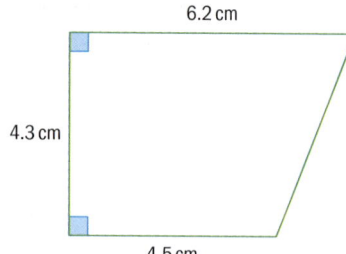

3

4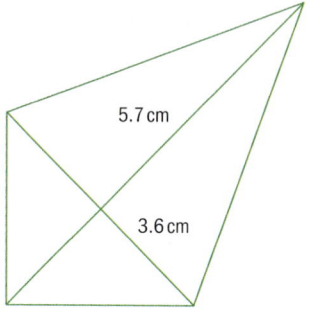

5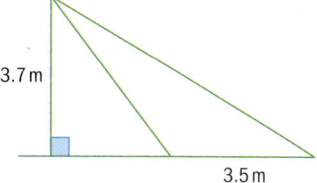

6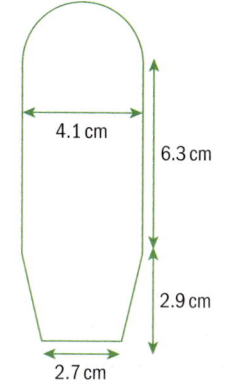

3.10 Volumes and surface areas of 3-dimensional shapes

Prism

→ A **prism** is a solid shape that has the same shape or cross-section all along its length.

A prism takes its name from the shape of its cross-section

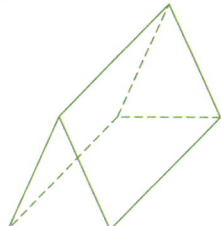

Triangular prism

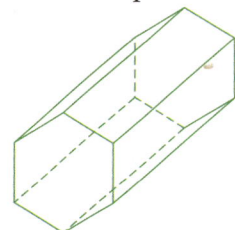

Hexagonal prism

→ To find the volume of a prism, use the formula

  $V = $ Area of cross-section $\times$ height

To find the surface area of a prism, calculate the area of each face and add them together.

Cylinder

A **cylinder** is a special case of a prism, with cross-section a circle.

→ The volume of a cylinder where the radius of the circular cross-section is r and the height is h is

  $V = \pi r^2 \times h$

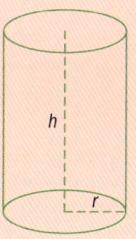

To calculate the surface area of a cylinder, open out the curved surface into a rectangle:

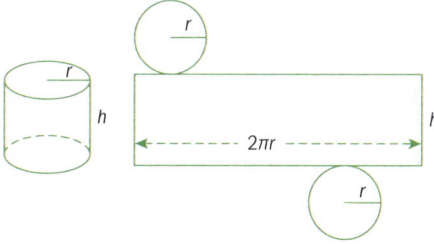

To find the curved surface area use the formula $CSA = 2\pi rh$

→ To find the surface area of a whole cylinder, find the curved surface area and add on the areas of the two circular ends:

  Total surface area $= 2\pi rh + 2\pi r^2$

Sphere

→ The formula for the volume of a **sphere** with radius r is

  $V = \dfrac{4}{3}\pi r^3$

  The formula for the surface area of a sphere is

  $SA = 4\pi r^2$

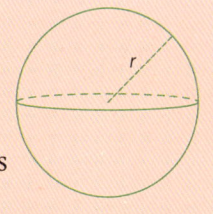

Pyramid

Any solid that has a flat base and which comes up to a point (the **vertex**) is a **pyramid**.

A pyramid takes its name from the shape of its base.

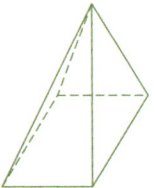
Square based prymid

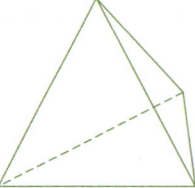
Triangle-based pyramid

→ To find the surface area of a pyramid, add together the areas of all the faces.

The volume of a pyramid with height h is
$$V = \frac{1}{3} \times \text{base area} \times h$$

Cone

A **cone** is a special type of pyramid with a circular base.

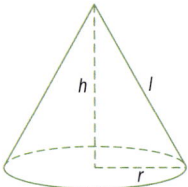

→ The volume of a cone with a circular base of radius r and **perpendicular height** h is given by the formula
$$V = \frac{1}{3} \times \pi r^2 \times h$$

The curved surface area of a cone uses the length of the **slanted height** l
$$CSA = \pi r \times l$$

To find the whole surface area of the cone, add the area of the circular base
$$SA = \pi r \times l + \pi r^2$$

Example 63

ABCDEF is a wedge
Angle ABC = 90°
AB = 5 cm, BC = 8 cm and CD = 12 cm
Calculate the volume of ABCDEF.

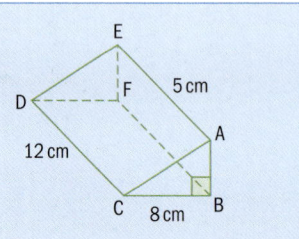

A wedge is a prism with triangular cross-section

▶ Answer on next page

Answer

Area of triangular cross section
= $\frac{1}{2}$ × 5 × 8 = 20

Volume of wedge
= 20 × 12 = 240 cm².

Calculate the area of the cross-section

Volume of prism = area of cross section × length

Exercise 3I

1. Find the surface area of each shape.

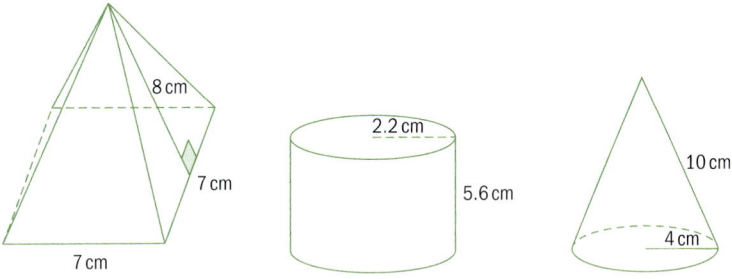

2. Calculate the volume of each shape.

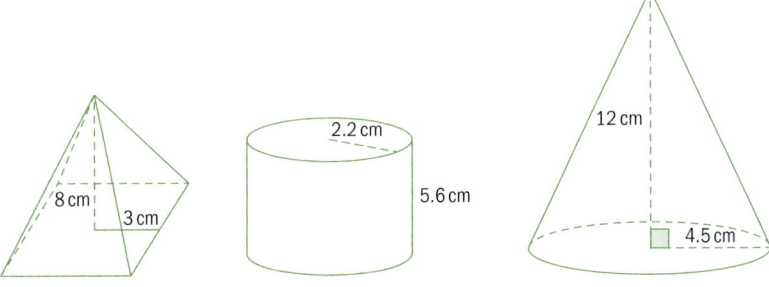

3. Find the height of a cone that has a radius of 2 cm and a volume of 23 cm³.

4. A cylinder has a volume of 2120.6 cm³ and a base radius of 5 cm. What is the volume of a cone with the same height but a base radius of 2.5 cm?

5. Determine the surface area and volume of each sphere.

   a 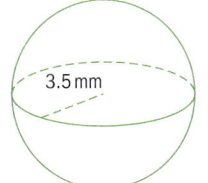 b

6 A hemisphere sits on top of a cylinder. Find the surface area and volume.

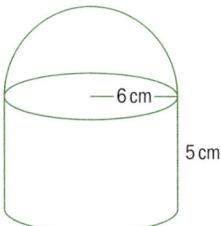

7 Eight basketballs are put into a holding container. The radius of each basketball is 10 cm. The container is shaped like a square-based pyramid with each side of the base measuring 40 cm and with a height of 70 cm. How much space is left in the container?

8 A cylindrical can has a diameter of 9 cm and is 14 cm high. Calculate the volume and surface area of the can to the nearest tenth of a centimetre.

9 Calculate the height of a cylinder that has a volume of 250 cm³ and a radius of 5.5 cm.

10 A cylindrical cardboard tube is 60 cm long and open at both ends. Its surface area is 950 cm². Calculate its radius to the nearest tenth of a centimetre.

3.11 Coordinate geometry

Coordinates

Coordinates describe the position of points in the plane. Horizontal positions are shown on the x-axis and vertical positions on the y-axis.

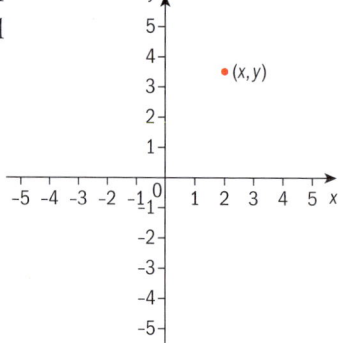

> René Descartes introduced the use of coordinates in a treatise in 1637. You may see axes and coordinates described as Cartesian axes and Cartesian coordinates.

Example 64

Draw axes for and $-10 \leq x \leq 10$ and $-10 \leq y \leq 10$.
Plot the points with coordinates: (4, 7), (3, −6), (−5, −2) and (−8, 4).

Answer

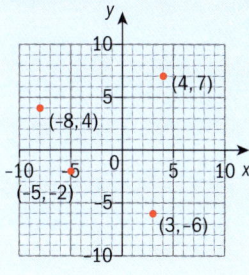

Exercise 3J

1. Draw axes for $-8 \leq x \leq 8$ and $-5 \leq y \leq 10$.
   Plot the points with coordinates:
   (5, 0), (2, −2), (−7, −4) and (−1, 9).

2. Write down the coordinates of the points shown in this diagram.

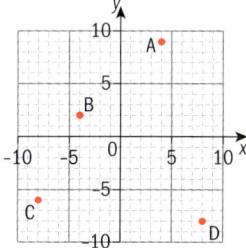

Midpoints

The midpoint of the line joining the points with coordinates (x_1, y_1) and (x_2, y_2) is given by $\left(\dfrac{x_1 + x_2}{2}, \dfrac{y_1 + y_2}{2}\right)$.

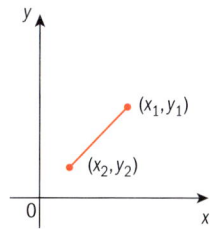

Chapter 18 693

Example 65

Find the midpoint of the line joining the points with coordinates (1, 7) and (−3, 3).

Answer

The midpoint is $= \left(\dfrac{1+(-3)}{2}, \dfrac{7+3}{2}\right) = (-1, 5)$

Exercise 3K

Calculate the midpoints of the lines joining the following pairs of points.

1 (2, 7) and (8, 3) **2** (−6, 5) and (4, −7) **3** (−2, −1) and (5, 6).

Distance between two points

The distance between points with coordinates (x_1, y_1) and (x_2, y_2) is given by $\sqrt{(x_2 - x_1)^2 + (y_2 - y_1)^2}$.

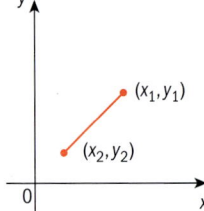

Example 66

Find the distance between the points with coordinates (2, −3) and (−5, 4).

Answer

Distance $= \sqrt{(-5-2)^2 + (4-(-3))^2} = \sqrt{(-7)^2 + 7^2} = 9.90$

Exercise 3L

Calculate the distance between the following pairs of points. Give your answer to 3 signifigant figures where appropriate.

1 (1, 2) and (4, 6)

2 (−2, 5) and (3, −3)

3 (−6, −6) and (1, 7)

The gradient of a straight line

The **gradient** of a straight line is a measure of how steep it is. It is also called the slope.

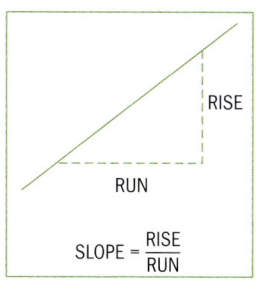

Another way of saying this is

$$\text{gradient} = \frac{\text{Change in } y \text{ values}}{\text{Change in } x \text{ values}}$$

> To find the gradient, measure the vertical increase (rise) between two points and divide by the horizontal increase (run).

Positive gradient

Negative gradient

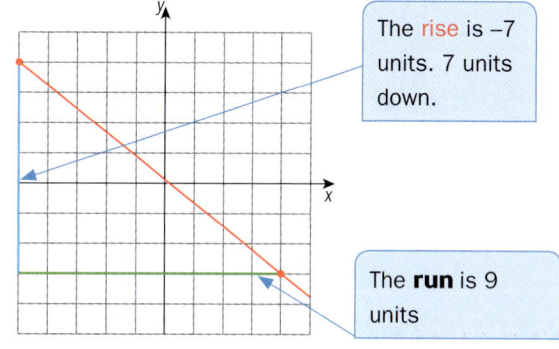

The rise is 4 units.

The run is 6 units.

The rise is −7 units. 7 units down.

The run is 9 units

▲ Gradient = $\dfrac{\text{Rise}}{\text{Run}} = \dfrac{4}{6} = \dfrac{2}{3}$

▲ Gradient = $\dfrac{\text{Rise}}{\text{Run}} = -\dfrac{7}{9}$

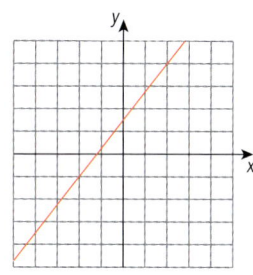

Positive Slope   Negative slope

▼ Horizontal lines have a gradient of zero because the rise is zero.

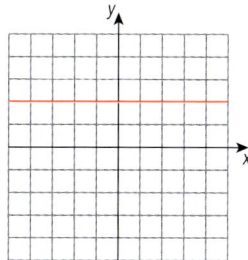

Zero Slope

▼ Vertical lines have an undefined gradient, as the run is zero

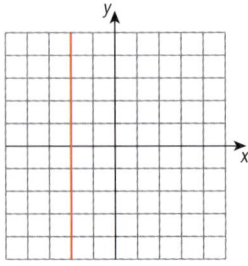

Undefined Slope

Exercise 3M

Find the gradient of each line.

1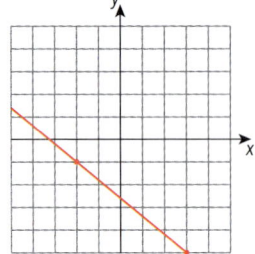

2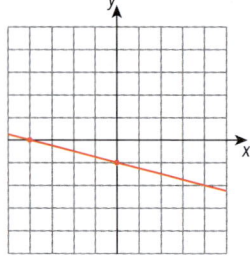

3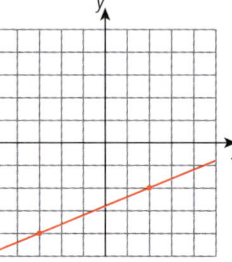

4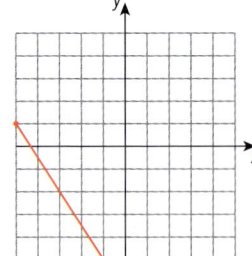

5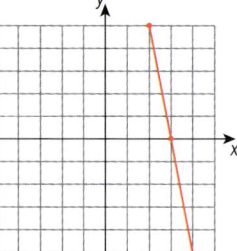

6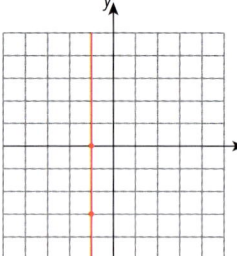

7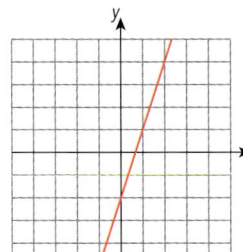

8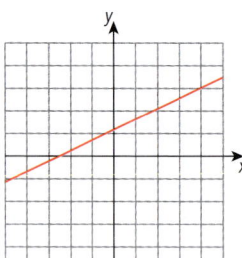

9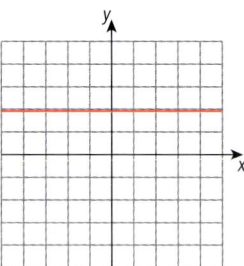

Finding the gradient of a line given two points

→ The gradient of a line is $\dfrac{Rise}{Run}$ which is $\dfrac{\text{The change in } y}{\text{The change in } x}$.

Given two points (x_1, y_1) and (x_2, y_2), $\dfrac{\text{The change in } y}{\text{The change in } x} = \dfrac{y_2 - y_1}{x_2 - x_1}$

Example 67

Find the gradient of the line joining (–3, –2) and (4, 1)

Answer

Gradient = $\dfrac{y_2 - y_1}{x_2 - x_1} = \dfrac{1-(-2)}{4-(-3)} = \dfrac{3}{7}$

Exercise 3N

Find the gradient of the line through each pair of points.

1. (19, –16) and (–7, –15)
2. (1, –19) and (–2, –7)
3. (–4, 7) and (–6, –4)
4. (20, 8) and (9, 16)
5. (17, –13) and (17, 7)
6. (14, 3) and (1, 3)
7. (3, 0) and (–11, –15)
8. (19, –2) and (–11, 10)
9. (6, –10) and (–15, 15)
10. (12, –18) and (18, –18)

Parallel and perpendicular lines

Parallel lines have the same gradient

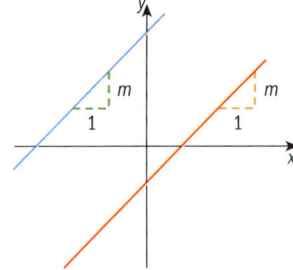

Both of these lines have slope m

Perpendicular lines have slope m and $-\dfrac{1}{m}$

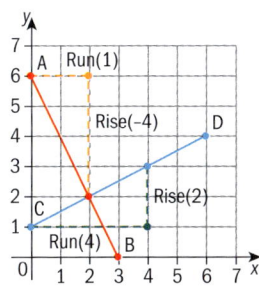

Line CD has slope $\dfrac{1}{2}$

Line AB has slope –2

Notice that the product of perpendicular gradients is –1.

$-2 \times \dfrac{1}{2} = -1$

Exercise 30

1. a Which of these gradients are parallel?
   b Which are perpendicular?

   $3, -3, \dfrac{1}{3}, 4.5, \dfrac{2}{3}, \dfrac{2}{9}, \dfrac{9}{2}, -\dfrac{2}{9}, -1.5, \dfrac{6}{2}$

2. State if the lines in each pair are parallel, perpendicular or neither.
   a Line A through (2, 5) and (0, 1) and line B though (4, 10) and (5, 12).
   b Line C through (3, 14) and (−2, −6) and line D though (12, −3) and (20, −5)
   c Line E through (1, 10) and (5, 15) and line F through (2, 2) and (4, 2).
   d Line G through (5, 7) and (2, 4) and line H through (8, −5) and (4, −1).
   e Line I through (4, 11) and (10, 20) and line J through (2, 1) and (6, 7)

Equations of lines

A straight line is defined by a linear equation of the form

$$y = \mathbf{m}x + \mathbf{c}$$

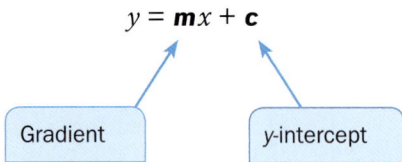

> This is called the gradient-intercept form. Some people use
> $y = ax + b$

Example 68

Find the equation of the line with gradient 3 passing through (0, 4)	
Answer The line is $y = 3x + 4$ 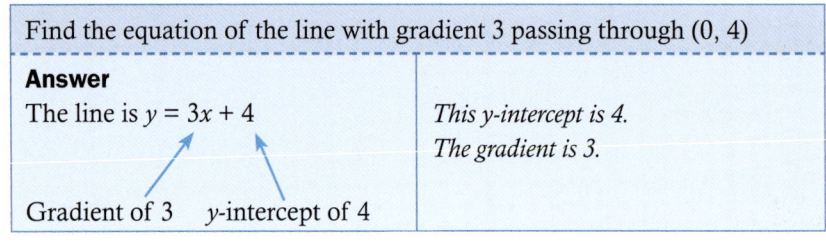 Gradient of 3 y-intercept of 4	*This y-intercept is 4. The gradient is 3.*

Using the gradient formula to find the equation of a line

Consider a line with a fixed point (x_1, y_1) and a general point (x, y).

Then $m = \dfrac{y - y_1}{x - x_1}$

or $y - y_1 = m(x - x_1)$.

Example 69

Find the equation of the line with gradient $m = 3$ passing through $(x_1, y_1) = (6, 12)$

$y - y_1 = m(x - x_1)$.
$y - 12 = 3(x - 6)$
$y - 12 = 3x - 18$
$\quad y = 3x - 6$

Exercise 3P

Find the equation of each line in gradient-intercept form

1   Gradient 3, passing through (1, 5)

2   Gradient 4, passing through (5, 11)

3   Gradient 2.5, passing through (4, 12)

4   Gradient $\frac{1}{2}$, passing through (12, 20)

5   Gradient 5, passing through (−2, −13)

6   Gradient −3, passing through (1, 1)

7   Gradient −2, passing through (−3, −1)

8   Gradient $-\frac{1}{2}$, passing though (−4, −3)

9   Find the equation of the line passing through (2, 7) and (5, 19).

10 Find the equation of the line passing through (−1, −3) and (−5, −11).

4 Statistics

4.1 Statistical graphs

In a statistical investigation we collect information, known as **data**. To represent this data in a clear way we can use graphs. Three types of statistical graph are bar charts, pie charts and pictograms.

Bar charts

A **bar chart** is a graph made from rectangles, or bars, of equal width whose length is proportional to the quantity they represent, or frequency. Sometimes we leave a small gap between the bars.

Example 70

Juliene collected some data about the ways in which her class travel to school.

Type of transport	Bus	Car	Taxi	Bike	Walk
Frequency	7	6	4	1	2

Represent this information in a bar chart.

Answer

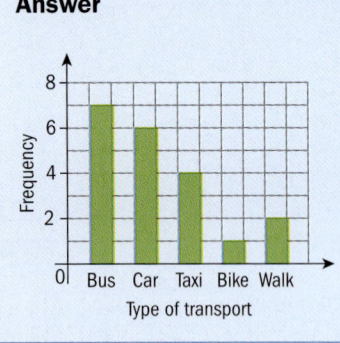

Example 71

Lakshmi collected data from the same class about the number of children in each of their families.

No. of children	1	2	3	4	6
Frequency	3	9	5	2	1

Represent this information in a bar chart.

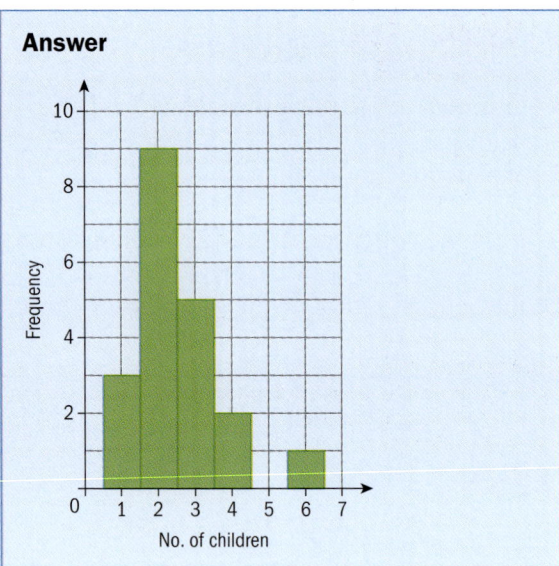

Pie charts

A **pie chart** is a circle divided into sectors, like slices from a pie.
The sector angles are proportional to the quantities they represent.

Example 72

Use Juliene's data from Example 70 to construct a pie chart.

Answer

Type of transport	Frequency		Sector angle
Bus	7	$\frac{7}{20} \times 360°$	126°
Car	6	$\frac{6}{20} \times 360°$	108°
Taxi	4	$\frac{4}{20} \times 360°$	72°
Bike	1	$\frac{1}{20} \times 360°$	18°
Walk	2	$\frac{2}{20} \times 360°$	36°

The total of the frequencies is 20. The total angle for the whole circle is 360°.

Start by drawing a radius and then measure, with your protractor, each angle in turn. The total of the sector angles should be 360°.

Pictograms

Pictograms are similar to bar charts, except that pictures are used. The number of pictures is proportional to the quantity they represent. The pictures can be relevant to the items they show or just a simple character such as an asterisk.

Example 73

Use Juliene's data from Example 70 to construct a pictogram.

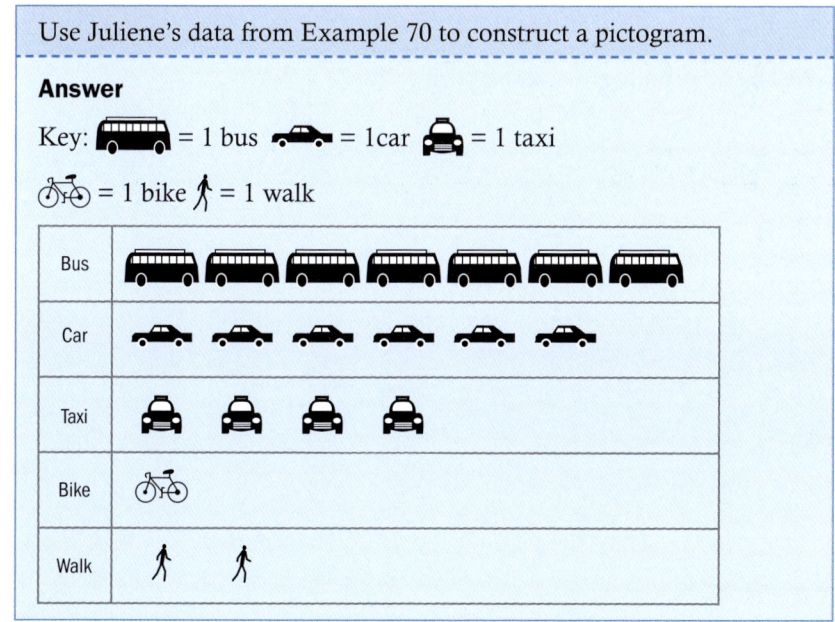

In this pictogram, different symbols are used for each category but the symbols describe the category as well.

Chapter 18   701

Example 74

Use this data on the number of children in a sample of families to construct a pictogram.

Number of children	1	2	3	4	6
Frequency	4	9	6	2	1

Answer

```
       No. of children
1      | △△ △△
2      | △△△△△△△△△
3      | △△△△△△
4      | △△
6      | △
   Key: △ = 1 child
```

Exercise 4A

1 Adam carried out a survey of the cars passing by his window on the road outside. He noted the colors of the cars that passed by for 10 minutes and collected the following data.

Color	Black	Red	Blue	Green	Silver	White
Frequency	12	6	10	7	14	11

Draw a bar chart, a pie chart and a pictogram to represent the data.

2 Ida asked the members of her class how many times they had visited the cinema in the past month. She collected the following data.

Number of times visited	1	2	3	4	8	12
Number of students	4	7	4	3	1	1

Draw a bar chart, a pie chart and a pictogram to represent the data.

Stem and leaf diagrams

Stem and leaf diagrams provide a simple means of organizing raw data without losing any of the detail.

Here is some data on the weights of 20 people (in kg).

50, 47, 53, 88, 75, 62, 49, 83, 57, 69, 71, 73, 73, 66, 51, 44, 78, 66, 54 and 80

You can draw a stem and leaf diagram for this data.

> They are also called stem plots.

The 'stem' is the tens, and the 'leaves' are the units.

You must give a key for a stem and leaf diagram.

```
4 | 4 7 9            Key
5 | 0 1 3 4 7        6|2 means 62 kg
6 | 2 6 6 9
7 | 1 3 3 5 8
```

The key explains what the stem and leaf data means

The leaves are the units digits written in order

The stem is the 10s digit

Exercise 4B

1. The test scores out of 50 for a math class are:
   21, 23, 25, 26, 28, 30, 30, 30, 33, 36, 37, 39, 39, 40, 41, 42, 42, 42, 42, 46, 49, 50, 54.
   Show this on a stem and leaf diagram.

2. The number of advertisements in different issues of a magazine are:
   164, 176, 121, 185, 148, 149, 177, 151, 157, 152, 163, 145, 123, 176
   Show this on a stem and leaf diagram.

   > Use this key:
   > 16|4 means 164 advertisements

3. The waiting time, in minutes, at the dentist's surgery was recorder for 24 patients as:
   55, 26, 27, 53, 19, 28, 30, 29, 22, 44, 48, 48, 37, 46, 62, 57, 49, 42, 25, 34, 58, 43, 52, 36.
   Show this on a stem and leaf diagram.

4. The number of tomatoes produced on different plants in a garden is given below:
   11, 34, 14, 23, 56, 36, 28, 19, 26, 35, 24, 30, 51, 18, 14, 16, 27, 29, 38, 26.
   Show this on a stem and leaf diagram.

5. The times, in seconds, for scouts to tie a knot were:
   4.6, 2.2, 3.1, 4.2, 5.2, 4.3, 6.0, 7.3, 7.4, 3.2, 3.3, 6.3, 3.2, 2.3, 2.5, 6.4, 5.2, 2.5, 2.9, 5.2, 5.4, 4.3, 4.8, 4.7
   Show this on a stem and leaf diagram.

   > Use the whole number part as the stem, and the tenths as the leaves.

4.2 Data analysis

> → **Discrete data** can only take specific values. Discrete data is often counted.

For example:
- the number of children in your family – the values can only be whole numbers.
- UK shoe sizes – 2, $2\frac{1}{2}$, 3, $3\frac{1}{2}$, 4, $4\frac{1}{2}$, 5, $5\frac{1}{2}$, 6, $6\frac{1}{2}$, ...

> → **Continuous data** can take any value within a certain range. Continuous data is measured, and its accuracy depends on the measuring instrument used.

For example:

- the time taken to run 100 m may be 14.4 seconds or 14.43 seconds or 14.428 seconds etc. depending on the measuring instrument.

Exercise 4C

State whether each set of data is discrete or continuous.

1 The number of cars in a school car park.

2 The number of books in a library.

3 The length of your pencil.

4 The time that it takes you to rum 400 m.

5 The speed of a car.

6 The number of friends that you have.

7 The number of shoes that you own.

8 The mass of a table.

9 The distance from the Earth to the Sun.

Measures of central tendency

A measure of central tendency, or **average**, describes a typical value for a set of data.

There are three common types of average:

- The **mode** – this is the data value that occurs most often.
- The **median** – this is the middle item when the data is arranged in order of size.
- The **mean** – this is what most people mean when they use the word "average". It is found by adding up all of the data and dividing by the number of pieces of data.

Example 75

Find **a** the mode **b** the median and **c** the mean of this data set:
2, 5, 4, 9, 1, 3, 2, 6, 9, 2, 5, 13, 4

Answers

a The mode is 2   *2 occurs the most often*

b 1, 2, 2, 2, 3, 4, 4, 5, 5, 6, 9, 9, 13   *write them in order and find the middle one*

   The median is 4

c Mean $= \dfrac{1+2+2+2+3+4+4+5+5+6+9+9+13}{13}$ *Add them all together. There are 13 pieces of data, so divide by 13.*

   $= \dfrac{65}{13} = 5$

Exercise 4D

1 Find **a** the mode **b** the median and **c** the mean of
   a 1, 4, 1, 5, 6, 7, 3, 1, 8
   b 4, 7, 5, 12, 5, −3, −2
   c 2, 3, 8, 2, 1, 7, 9, 8, 5
   d 25, 28, 29, 21, 25, 20, 27
   e 7.4, 10.2, 12.5, 6.8, 10.2

2 Fifteen students were asked how many brothers and sisters they had. The results were:
   2, 2, 1, 0, 3, 5, 2, 1, 1, 0, 1, 4, 1, 0, 2.

   Find **a** the mode, **b** the median and **c** the mean number of brothers and sisters.

3 My last nine homework scores, marked out of 10, were:
   8, 7, 9, 10, 8, 9, 6, 8, 7

   Find **a** the mode **b** the median and **c** the mean homework score.

4 A sprinter's times in seconds for the 40 m dash were:
   5.13, 4.82, 5.25, 4.94, 5.06, 4.82, 5.12

   Find **a** the mode, **b** the median and **c** the mean of the times.

5 Seven farmers own different numbers of chickens.
   These numbers are:
   253, 78, 497, 166, 710, 497 and 599

   Find **a** the mode, **b** the median and **c** the mean number of chickens.

Measures of dispersion

A measure of dispersion is a value that describes the spread of a set of data.

The **range** and **interquartile range** are two measures of dispersion.

The range shows how spread out the data is.

> → Range = highest value − lowest value

The **quartiles** divide a set of data into four equal amounts.

> → The **lower quartile** Q_1 is 25% of the way through the data and its position is found using the formula:
>
> $Q_1 = \left(\dfrac{n+1}{4}\right)^{th}$ where n is the number of items in the data set.
>
> The **upper quartile** Q_3 is 75% of the way through the data and its position is found using the formula
>
> $Q_3 = 3\left(\dfrac{n+1}{4}\right)^{th}$
>
> The **interquartile range** shows how spread out the middle 50% of the data is.
>
> Interquartile range = $Q_3 - Q_1$

Example 76

Here are the shoe sizes of fifteen boys:
42, 42, 38, 40, 42, 40, 34, 46, 44, 36, 38, 40, 42, 36, 42

Find **a** the range and **b** the interquartile range.

Answers

a 34, 36, 36, 38, 38, 40, 40, 40, 42, 42, 42, 42, 42, 44, 46

range = 46 − 34 = 12

To find the interquartile range, first arrange the data in order of size

b Lower quartile = $\dfrac{16}{4}$ th value = 4th value
= 38

$n = 15$

Upper quartile = 3 × 4th value = 12th value
= 42

So interquartile range
= 42 − 38 = 4

Exercise 4E

1. Here are the shoe sizes of fifteen girls:
   26, 28, 28, 36, 34, 32, 30, 34, 32, 28, 36, 38, 34, 32, 30

   Find a the range and   b the interquartile range of the shoe sizes.

2. 23 students were asked how many pets they had at home. Here are the replies:
   1, 4, 3, 5, 3, 2, 8, 0, 2, 1, 3, 2, 4, 2, 1, 0, 1, 2, 6, 7, 2, 8, 2

   Find a the range and   b the interquartile range for the number of pets.

3. The average daily temperatures in °C in Chillton during January were
   −6, −4, −4, −2, −1, 0, 4, 5, 7, 4, 2, 1, 0, −3, −4, −6,
   −7, −5, −3, −1, 1, 3, 4, 7, 7, 8, 3, −2, 0, −2, −5

   Find a the range and   b the interquartile range for the daily temperatures.

4. The grocer sells potatoes by the kilogram.
   I bought 1 kg of potatoes every day of the week and counted the number of potatoes each time. Here are the results:

Day	Monday	Tuesday	Wednesday	Thursday	Friday	Saturday	Sunday
Potatoes	18	15	20	17	14	12	15

   Find a the range and   b the interquartile range for the number of potatoes in 1 kilogram.

5. The time (in seconds) taken for eleven players in a soccer team to prepare for a free kick is given.
   12.4, 2.45, 3.75, 10, 3.5, 8.4, 9.6, 23.5, 2.48, 15.6, 5.2

   Find a the range and   b the interquartile range for the time taken.

19 Practice paper 1

Time allowed: 1 hour 30 minutes
- Answer all the questions
- Unless otherwise stated in the question, all numerical answers must be given exactly or correct to three significant figures.

Full marks are not necessarily awarded for a correct answer with no working. Answers must be supported by working and/or explanations. Where an answer is incorrect, some marks may be given for a correct method, provided this is shown by written working. You are therefore advised to show all working.

Practice exam papers on CD: *IB examination papers include spaces for you to write your answers. There is a version of this practice paper with space for you to write your answers on the CD. You can also find an additional set of papers for further practice.*

Worked solutions on CD: *Detailed worked solutions for this practice paper are given as a PowerPoint presentation on the CD.*

SECTION A

1 Let $f(x) = 2(x - p)(x - q)$. Part of the graph of f is shown below.

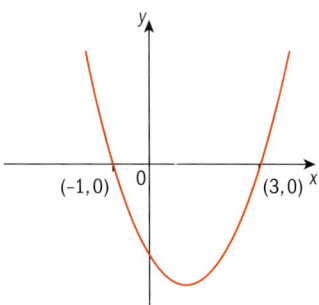

   a Write down the values of p and q. *[2 marks]*
   b **i** Write down the equation of the axis of symmetry. *[1 mark]*
       ii Find the coordinates of the vertex. *[3 marks]*

2 Given that $f(x) = e^{-2x}$, answer the following.
   a Find the first four derivatives of $f(x)$. *[4 marks]*
   b Write an expression for $f^{(n)}(x)$ in terms of x and n. *[3 marks]*

3 Consider the expansion of the expression $(x^4 - 2x)^5$.
   a Write down the number of terms in this expansion. *[1 mark]*
   b Find the term in x^{11}. *[5 marks]*

4 A straight line containing the points (0, 0) and (2, 3) makes an acute angle θ with the x-axis.
  a Find the value of
  i sin 2θ;
  ii cos 2θ. *[3 marks]*
  b Hence, write down the value of tan 2θ. *[4 marks]*

5 The Venn diagram below shows information about 40 students in a theater class. Of these, 11 take a dance class (D), 14 take a voice class (V) and 5 take both a dance and a voice class.

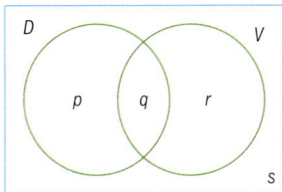

  a Write down the value of
  i p;
  ii q;
  iii r;
  iv s. *[4 marks]*
  b Find the value of $P(V \mid D')$. *[2 marks]*
  c Show that the V and D are **not** mutually exclusive. *[1 mark]*

6 The shaded region in the graph shown below is bounded by $f(x) = \dfrac{\sqrt{\sin\left(x^{\frac{1}{2}}\right)}}{x^{\frac{1}{4}}}$, x = 4, the x-axis and the y-axis.

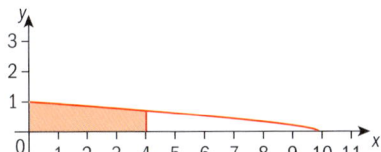

The shaded region is rotated 360° about the x-axis to form a solid.

  a Write down a definite integral that gives the volume of the solid. *[1 mark]*
  b Given that the volume of the solid is $p\pi(\cos(q) - 1)$, find the values of p and q. *[5 mark]*

7 Let $f(x) = 4^{-x}$.
  a Write down $\lim_{x \to \infty} f(x)$. *[1 mark]*
  b Show that $f^{-1}(x) = \log_4 \dfrac{1}{x}$. *[2 marks]*
  Let $g(x) = 2^x$.
  c Find the value of $(f^{-1} \circ g)(4)$, giving your answer as an integer. *[4 marks]*

SECTION B

8 Let $f(x) = 2x^3 - 1.5x^2 - 3x + 4.5$. Part of the graph of f is shown below.

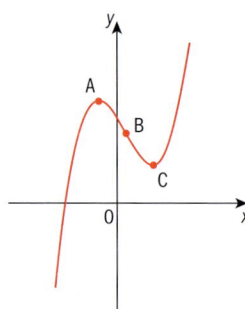

There is a relative maximum point at A, an inflection point at B and a relative minimum point at C(1, 2).

  a Find the x-coordinate of
    i point A;
    ii point B. *[10 marks]*
  b A certain transformation is defined by a reflection in the x-axis followed by a translation by the vector $\begin{pmatrix} 0 \\ -3 \end{pmatrix}$.
    i Write down the coordinates of the image of C after this transformation.
    ii The graph of a function g is obtained when function f undergoes this transformation.
    Write down the equation that defines the function g. *[4 marks]*

9 Let $f(x) = xe^{-x}$.
  a Use the product rule to show that $f'(x) = e^{-x}(1 - x)$. *[4 marks]*
  b Find $f''(x)$. *[3 marks]*
  c **i** Find the value of $f'(1)$ and the value of $f''(1)$.
    ii Hence explain why there is a relative minimum point, a relative maximum point or neither on the graph of f where $x = 1$. *[5 marks]*

10 The line L_1 is represented by the vector equation $r = \begin{pmatrix} 0 \\ 4 \\ 1 \end{pmatrix} + p \begin{pmatrix} 8 \\ -2 \\ 12 \end{pmatrix}$.

A second line L_2 is perpendicular to L_1 and represented by the vector equation $r = \begin{pmatrix} 4 \\ -2 \\ 15 \end{pmatrix} + s \begin{pmatrix} 2 \\ 2 \\ l \end{pmatrix}$.

a Show that $l = -1$. *[5 marks]*

The lines L_1 and L_2 intersect at the point A.

b Find $\overrightarrow{OA}$. *[6 marks]*

Let $\overrightarrow{OB} = \begin{pmatrix} 9 \\ 6 \\ 10 \end{pmatrix}$ and $\overrightarrow{BC} = \begin{pmatrix} 1 \\ -5 \\ 2 \end{pmatrix}$.

c i Find $\overrightarrow{BA}$.
  ii Hence find $A\hat{B}C$. *[7 marks]*

> Use the mark scheme in the Answer section at the back of this book to mark your answers to this practice paper.

Practice paper 2

Time allowed: 1 hour 30 minutes

- Answer all the questions
- Unless otherwise stated in the question, all numerical answers must be given exactly or correct to three significant figures.

Full marks are not necessarily awarded for a correct answer with no working. Answers must be supported by working and/or explanations. In particular, solutions found from a graphic display calculator should be supported by suitable working, e.g. if graphs are used to find a solution, you should sketch these as part of your answer. Where an answer is incorrect, some marks may be given for a correct method, provided this is shown by written working. You are therefore advised to show all working.

Practice exam papers on CD: *IB examination papers include spaces for you to write your answers. There is a version of this practice paper with space for you to write your answers on the CD. You can also find an additional set of papers for further practice.*

Worked solutions on CD: *Detailed worked solutions for this practice paper are given as a PowerPoint presentation on the CD.*

SECTION A

1 It is thought that the weight of a mango is related to its length. The length (x) in cm and the weight (y) in grams are shown in the table below.

Length x (cm)	14	21	10	22	15	17	12	25	22	18
Weight y (g)	70	95	58	112	77	92	63	130	121	100

    a Write down the correlation coefficient, r. *[1 mark]*
    b Comment on your value for r. *[2 marks]*
    c Write down the equation of the regression line of y on x. *[1 mark]*
    d Use your regression line to calculate the weight of a mango of length 20 cm. *[2 marks]*

2 Consider the arithmetic sequence 5, 9, 13, …, 329
    a Write down the common difference. *[1 mark]*
    b Find the number of terms in the sequence. *[3 marks]*
    c Find the sum of the sequence. *[2 marks]*

3 Let $f(x) = x \sin x$, for $0 \leq x \leq 6$.
  a Find $f'(x)$ [3 marks]
  b Sketch the graph of $y = f'(x)$ [4 marks]

4 The following table shows the number of computers that a class has owned. The mean was 4 computers.

Computers	1	2	3	4	5	6
Frequency	2	1	4	9	x	3

  a Show that the value of x is 6. [2 marks]
  b Write down the standard deviation. [1 mark]

A different school had a mean of 3.6 computers and a standard deviation of 1.2 computers. An old teacher gives every student a new computer.
  c What will be the new mean? [1 mark]
  d What effect will this have on the standard deviation? [1 mark]

5 The diagram below shows quadrilateral ABCD.

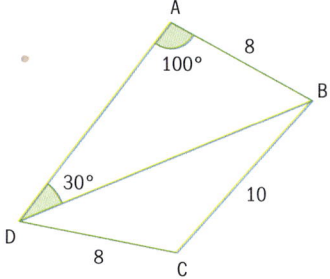

  a Find BD. [2 marks]
  b Find angle BCD. [3 marks]
  c Find the area of triangle BCD. [2 marks]

6 The acceleration, a ms^{-2}, of a particle at time t seconds is given by

$$a = \frac{1}{t} + 3\sin 2t, \text{ for } t \geq 1.$$

The particle is at rest when $t = 1$.
Find the velocity of the particle when $t = 5$. [6 marks]

7 The probability of winning in a game of chance is 0.25
  a If Wally plays 10 games, find the probability that he wins exactly 4. [3 marks]
  b What is the least number of games that Wally must play to ensure that the probability of winning at least twice is more than 0.9? [4 marks]

SECTION B

8 The following graph shows the depth of water in Fishfleet Harbor over a twelve-hour period from midnight to midday.

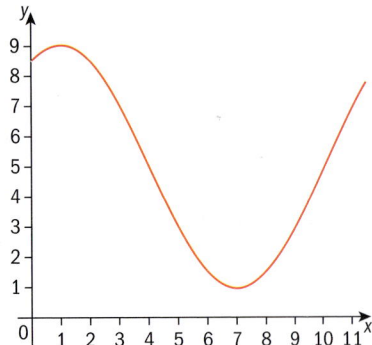

a Use the graph to estimate the time when
  i the water is at a maximum.
  ii the depth is increasing most rapidly. *[2 marks]*
b The depth of water can be modeled by the function
  $y = A\cos(B(x - C)) + D$
  i Show that $A = 4$
  ii Write down the value of C.
  iii Write down the value of D.
  iv Find the value of B.
  v Write down the function that models the depth of water. *[8 marks]*
c The Seahawk fishing trawler can only enter the harbor when the depth of the water is 4.5 m or more. Use your model to find earliest time after 7 am that the Seahawk can enter the harbor. *[3 marks]*

9 Let $f(x) = 4 - (1 - x)^2$, for $-2 \leq x \leq 4$ and $g(x) = \ln(x + 3) - 2$, for $-3 \leq x \leq 5$
  a Sketch the graphs of both functions on the same axis. *[4 marks]*
  b **i** Write down the equation of the vertical asymptote.
     ii Write down the x-intercept of g.
     iii Write down the y-intercept of g. *[4 marks]*
  c Find the values for which $f(x) = g(x)$. *[2 marks]*

  Let R be the region between the two curves where $x \geq 0$.
  d **i** Shade the region R on your graph.
     ii Write down an integral expression that represents the area of R.
     iii Evaluate the area of R. *[5 marks]*

10 In a large school, the heights of all fourteen-year-old students are measured.
The heights of the girls are normally distributed with mean 155 cm and standard deviation 10 cm.
The heights of the boys are normally distributed with mean 160 cm and standard deviation 12 cm.

  a Find the probability that a girl is taller than 170 cm. *[3 marks]*
  b Given that 10% of the girls are shorter than x cm, find x. *[3 marks]*
  c Given that 90% of the boys have heights between q cm and r cm where q and r are symmetrical about 160 cm, and $q < r$, find the value of q and of r. *[4 marks]*

In the group of fourteen-year-old students, 60% are girls and 40% are boys.
The probability that a girl is taller than 170 cm was found in part (a).
The probability that a boy is taller than 170 cm is 0.202.
A fourteen-year-old student is selected at random.

  d Calculate the probability that the student is taller than 170 cm. *[4 marks]*
  e Given that the student is taller than 170 cm, what is the probability the student is a girl? *[3 marks]*

Answers

Chapter 1

Skills check

1 a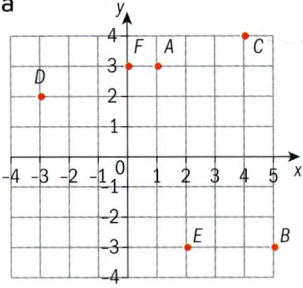

   b $A(0, 2)$, $B(1, 0)$, $C(-1, 0)$, $D(0, 0)$, $E(2, 1)$, $F(-2, -2)$, $G(3, -1)$, $H(-1, 1)$

2 a 34   b 82
   c 16   d $-\dfrac{13}{60}$

3 a 4   b -2 c 10

4 a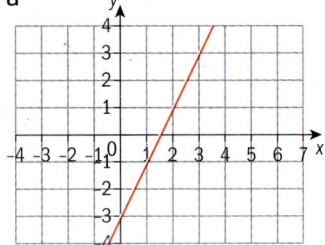

   b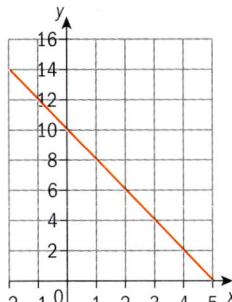

   c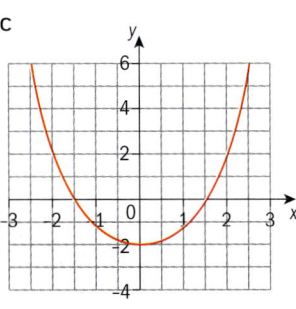

5 a $x^2 + 9x = 20$
   b $x^2 - 4x + 3$
   c $x^2 + x - 20$

Investigation – handshakes

a 6

b
Number of people	Number of handshakes
2	1
3	3
4	6
5	10
6	15
7	21
8	28
9	36
10	45

c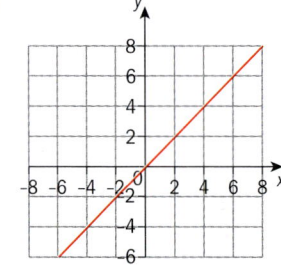

d $H = \dfrac{1}{2}n(n-1)$

Exercise 1A

1 Functions: **a b f**

2 a Function: domain {0, 1, 2, 3, 4} range {0, 1, 2}
   b Relation: domain {−1, 0, 1, 2, 3} range {−1, 0, 1, 2}

3 Yes, function.

Exercise 1B

1 a b d f h i

2 a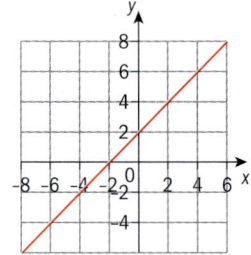

   b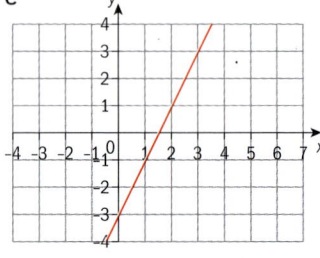

   c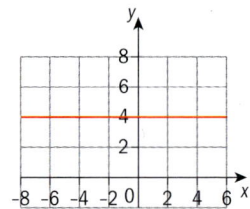

   d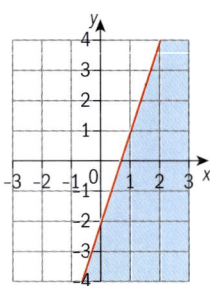

   e Yes, vertical line will only cross once.
   f No, vertical lines such as $x = 3$ are not functions.

3 Not a function as a vertical line crosses the region in many places

4 $y^2 = 4 - x^2$, $y = \pm\sqrt{4 - x^2}$

   When $x = 1$, $y = \pm\sqrt{3}$. two possible values so not a function.

Exercise 1C

1. Horizontal asymptote: $y = 0$
2. Horizontal asymptote: $y = 0$, Vertical asymptote: $x = 0$
3. Horizontal asymptote: $y = 0$, Vertical asymptote: $x = -1$
4. Horizontal asymptote: $y = 2$, Vertical asymptote: $x = -2$
5. Horizontal asymptote: $y = 2$, Vertical asymptote: $x = 1$
6. Horizontal asymptote: $y = 0$, Vertical asymptote: $x = -3$

Exercise 1D

1. Function, domain {2, 3, 4, 5, 6, 7, 8, 9, 10}, range {1, 3, 6, 10, 15, 21, 28, 36, 45}.
2.
   a. domain $\{x: -4 < x \le 4\}$, range $\{y: 0 \le y \le 4\}$
   b. domain $\{x: -1 \le x \le 5\}$, range $\{y: 0 \le y \le 4\}$
   c. domain $\{x: -\infty < x < \infty\}$, range $\{y: 0 \le y < \infty\}$
   d. domain $\{x: -\infty < x \le -2 < x < \infty\}$, range $\{y: -\infty < y \le 3\ 4 \le y < 8\}$
   e. domain $\{x: -5 \le x \le 5\}$, range $\{y: -3 \le y \le 4\}$
   f. domain $\{x: -\infty < x \le \infty\}$, range $\{y: -1 \le y \le 1\}$
   g. domain $\{x: -2 \le x \le 2\}$, range $\{y: -2 \le y \le 2\}$
   h. domain $\{x: -\infty < x \le \infty\}$, range $\{y: -\infty < x \le \infty\}$
   i. domain $x \in \mathbb{R}, x \ne 1$, range $y \in \mathbb{R}, y \ne 0$

3. a. $x \in \mathbb{R}, y \in \mathbb{R}$

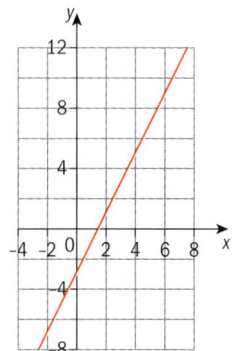

b. $x \in \mathbb{R}, y \ge 0$

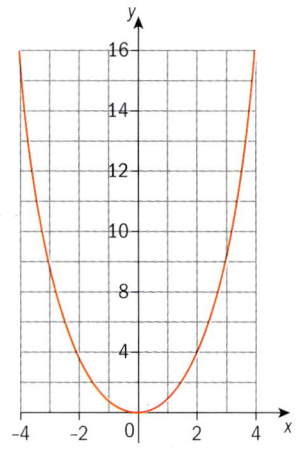

c. $x \in \mathbb{R}, y \ge -0.25$

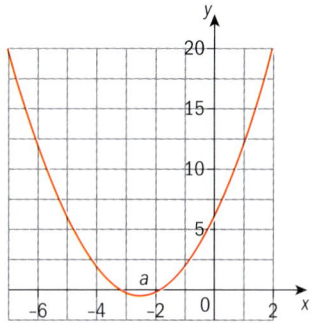

d. $x \in \mathbb{R}, y \in \mathbb{R}$

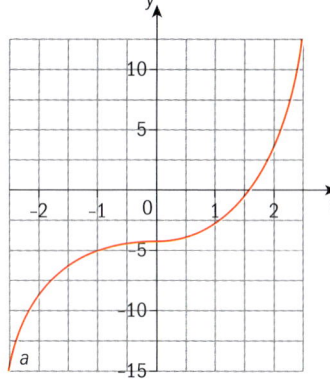

e. $x \ge 0, y \ge 0$

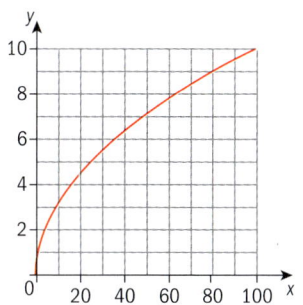

f. $x \le 4, y \ge 0$

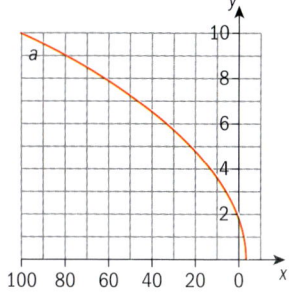

g. $x \in \mathbb{R}\ x \ne 0, y \in \mathbb{R}\ y \ne 0$,

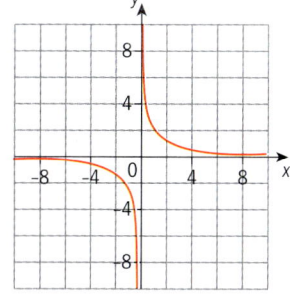

h. $x \in \mathbb{R}, y > 0$

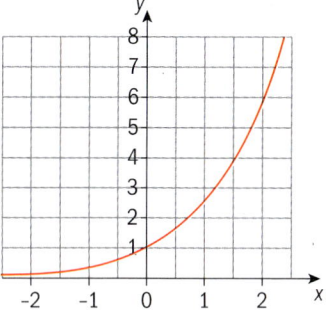

i. $x \in \mathbb{R}\ x \ne -2, y \in \mathbb{R}\ y \ne 0$,

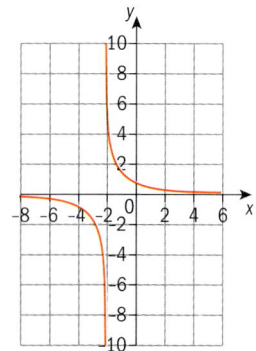

Answers 717

j $x \in \mathbb{R}\ x \neq 2,\ y \in \mathbb{R}\ y \neq 1$

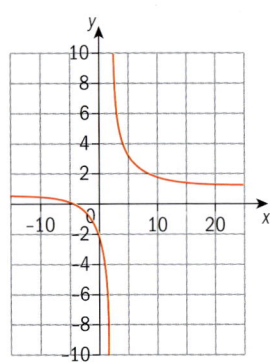

k $x \in \mathbb{R}\ x \neq -3,\ y \in \mathbb{R}\ y \neq -6$

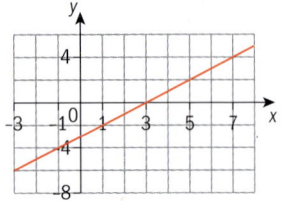

l $x \in \mathbb{R},\ 0 < y \leq 2$

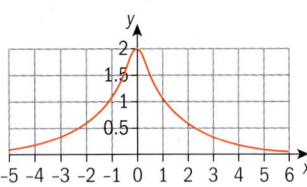

Exercise 1E

1 a i 5   ii −5   iii $-1\tfrac{1}{2}$
      iv −2, v $a - 2$
   b i 21   ii −9   iii $1\tfrac{1}{2}$
      iv 0   v $3a$
   c i $\tfrac{7}{4}$ ii $\tfrac{-3}{4}$ iii $\tfrac{1}{8}$
      iv 0   v $\tfrac{1}{4}a$
   d i 19   ii −1   iii 6
      iv 5   v $2a + 5$
   e i 51   ii 11   iii $2\tfrac{1}{4}$
      iv 2   v $a^2 + 2$

2 a $a^2 - 4$ b $a^2 + 10a + 21$
   c $a^2 - 2a - 3$ d $a^4 - 4a^2$
   e $21 - 10a + a^2$

3 a 2   b 11   c 2

4 a $-\tfrac{1}{9}$
   b $x = 6$, denominator $= 0$ and $h(x)$ undefined.

5 a 125
   b The volume of a cube of side 5.

6 a i $-\tfrac{1}{9}$ ii $\tfrac{5}{4}$
      iii $-\tfrac{1}{2}$ iv 0.
   b i −4   ii −11   iii −67
      iv −697   v −6997
      vi −69997
   c The value of $g(x)$ is getting increasingly smaller as x approaches 2.
   d 2
   e asymptote at $x = 2$.

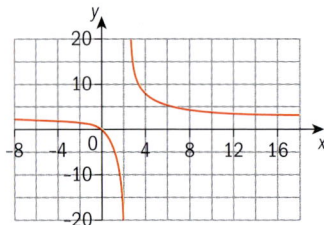

7 a $-9\ \text{m s}^{-1}$ b $7\ \text{m s}^{-1}$
   c $91\ \text{m s}^{-1}$ d 3 s

8 a $\dfrac{f(2 + 2h) - f(2 + h)}{h}$
   b $\dfrac{f(3 + 2h) - f(3 + h)}{h}$

Exercise 1F

1 a 12   b 3   c −15
   d $3x + 3$ e 13   f 16
   g −17   h $3x + 1$ i 18
   j 38   k $3x^2 + 6$
   l $9x^2 + 2$ m 12   n 18
   o $x^2 + 3$ p $x^2 + 2x + 3$

2 a 3   b 0   c −12
   d −1   e −5   f 48
   g $3 - 4x + x^2$ h $-2x + x^2$

3 a $x^2 + 4x + 4$ b 25

4 a $5x^2 + 5$ b $25x^2 + 1$

5 a $x^2 - 8x + 19$ b $x^2 - 1$
   c 2.5

6 $(r \circ s)(x) = x^2 - 4,\ x \in \mathbb{R},\ y \geq -4$

Exercise 1G

1 b, c

2 a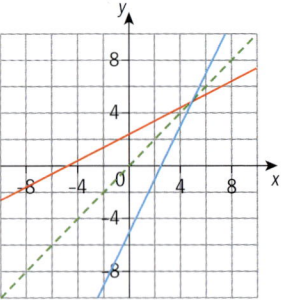

   b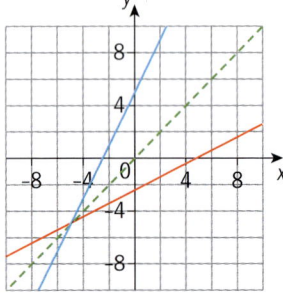

   c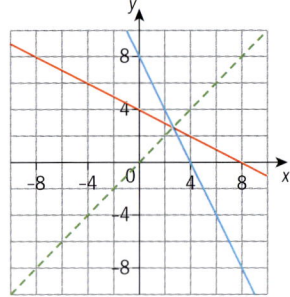

   d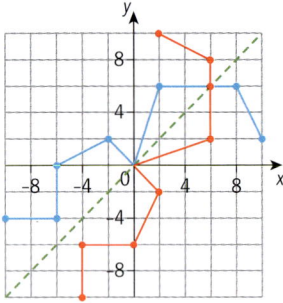

   e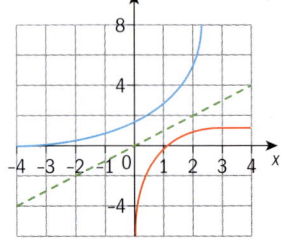

f

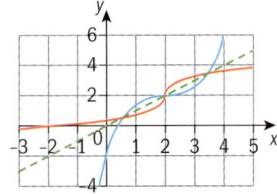

Exercise 1H

1. a i -2 and 1

   ii $\frac{1}{2}$ and -3

   iii x iv x

   b They are inverses of each other.

2. a $\frac{x+1}{3}$ b $\sqrt[3]{x+2}$

   c $4(x-5)$ d $(x+3)^3$

   e $\frac{1}{x+2}$ f $\sqrt[3]{\frac{x-3}{2}}$

   g $\frac{3x}{1-x}$ h $\frac{5x}{x+2}$

3. a $1-x$ b x c $\frac{1}{x}$

4. a 1 b -5 c $\frac{17}{20}$

5. $\frac{1+2x}{x-1}$

6. a–c

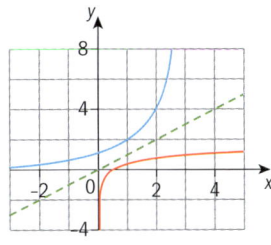

d $f(x): x \in \mathbb{R}, y > 0$
  $f^{-1}(x): x > 0, y \in \mathbb{R}$

7. $g^{-1}(x) = x^2$. The range of $g(x)$ is $x \geq 0$ so the domain of $g^{-1}(x)$ is $x \geq 0$. The domain of $f(x)$ is $x \in \mathbb{R}$ so $g^{-1}(x) \neq f(x)$

8. If $f(x) = mx + c$ then
   $f^{-1}(x) = \frac{1}{m}x - \frac{c}{m}$
   $m \times \frac{1}{m} = 1$ not -1 so not perpendicular.

Investigation – functions

1. Changing the constant term translates $y = x$ along the y-axis.

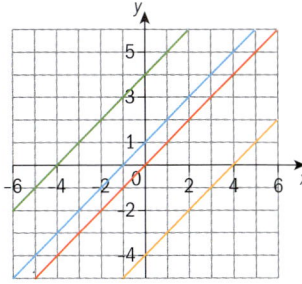

2. Changing the x-coefficient alters the gradient of the line.

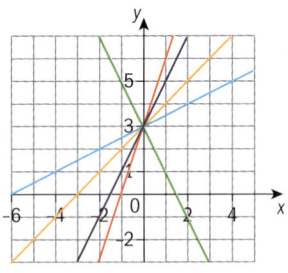

3. $y = |x + h|$ is a translation of $-h$ along the x-axis

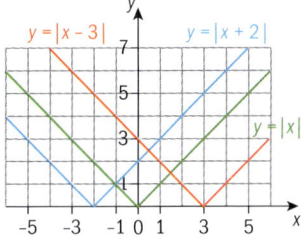

4. The negative sign reflects the graph in the x-axis. Increasing the value of a means the graph increases more steeply.

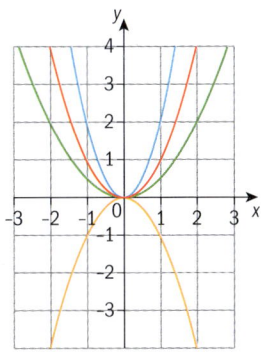

Exercise 1I

1. a

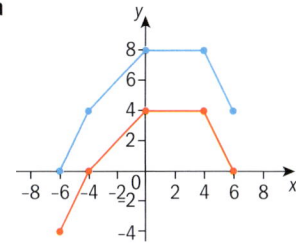

b

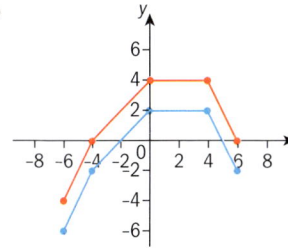

c

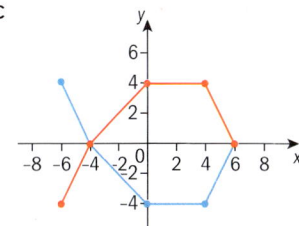

d

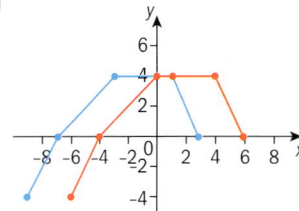

e

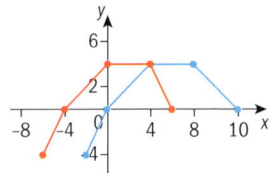

f

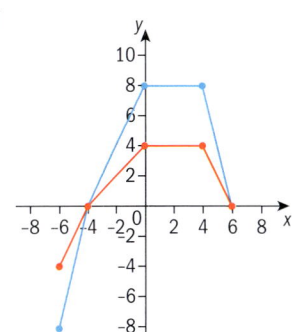

g

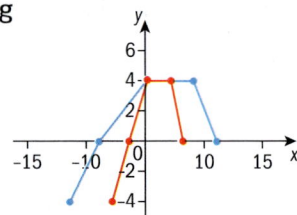

2 $g(x) = f(x) + 2$
   $h(x) = f(x) - 4$
   $q(x) = \frac{1}{2}f(x)$

3 $q(x) = f(x+4) - 2$
   $s(x) = f(x+4)$
   $t(x) = f(x-2)$

4 a Domain $-1 \le x \le 7$, range $-4 \le y \le 6$

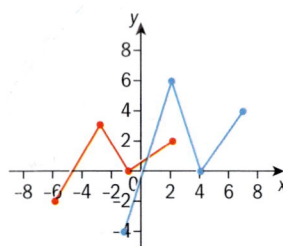

b Domain $-3 \le x \le 1$, range $0 \le y \le 5$

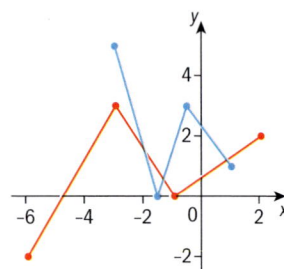

5 a

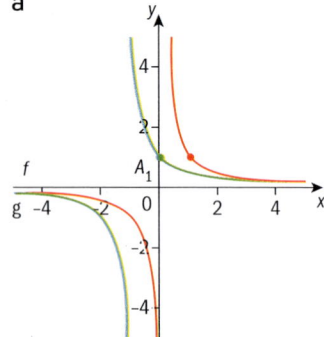

b

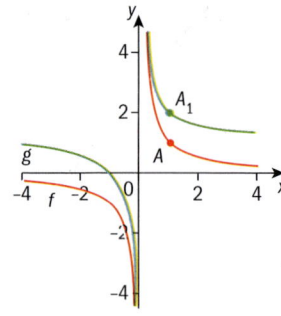

c

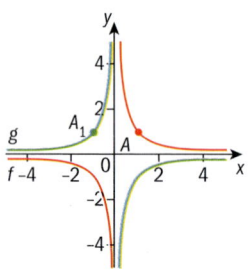

d

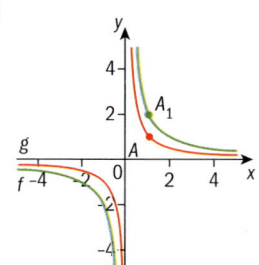

e

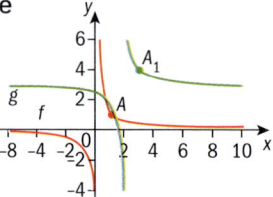

6 a Reflection in x-axis.
   b Horizontal translation 3 units.
   c Vertical stretch SF2, reflection x-axis, vertical translation of 5 units.

7 a, b

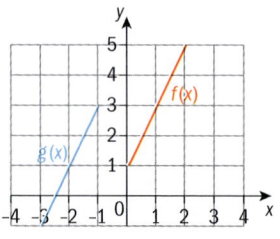

Review exercise non-GDC

1 a $4a - 13$ b $\frac{2-x}{x}$

2 a $2x^2 - 15x + 28$
   b $-2x^2 + 9$

3 a $\frac{2x-17}{3}$ b $\sqrt[3]{\frac{x-3}{2}}$

4 $f^{-1}(x) = -5x - 5$

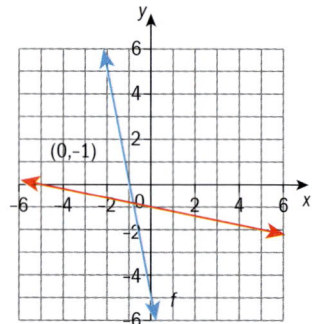

5 a $\frac{x-5}{3}$ b $x^3 - 2$

6 a

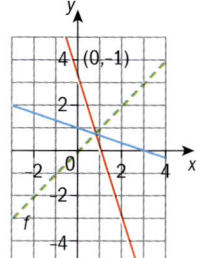

b

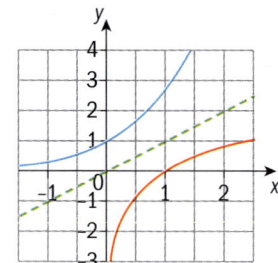

7 a Domain $x \in \mathbb{R}$, $y \ge 0$
   b Domain $x \in \mathbb{R}$, $x \ne 3$, Range $y \in \mathbb{R}$, $y \ne 0$

8 a $f(x) = 2\sqrt{-3x-9} + 2$
   b $f(x) = -\frac{1}{4}\left(3^{\frac{x-5}{3}}\right) - 1$

9 a Inverse function graph is the reflection in $y = x$.

b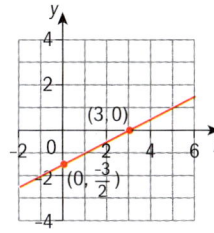

10 a −2 **b** −13

c $f^{-1}(x) = \sqrt[3]{\dfrac{x-3}{2}}$

11 a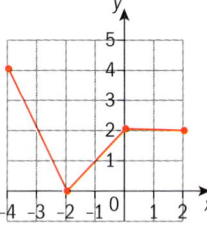

b P is (4, 1)

12 a $(f \circ g)(x) = 3x + 6$

b $f^{-1}(x) = \dfrac{x}{3}$ $g^{-1}(x) = x - 2$

$f^{-1}(12) = \dfrac{12}{3} = 4$

$g^{-1}(12) = 12 - 2 = 10$

$f^{-1}(12) + g^{-1}(12) = 4 + 10$

$f^{-1}(12) + g^{-1}(12) = 14$

13 a $(h \circ g)(x) = \dfrac{3(2x-1)}{(2x-1)-2}$

$= \dfrac{6x-3}{2x-3}$

b $x = \dfrac{1}{2}$

Review exercise GDC

1 Domain: $x \geq -2$, range: $y \geq 0$

2 Domain: $x \in \mathbb{R}$, range: $y \geq -4$

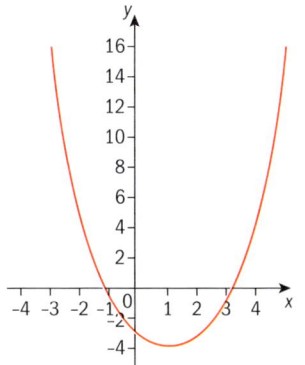

3 Domain $x \in \mathbb{R}, x \neq -2$, range $y \in \mathbb{R}, x \neq 0$

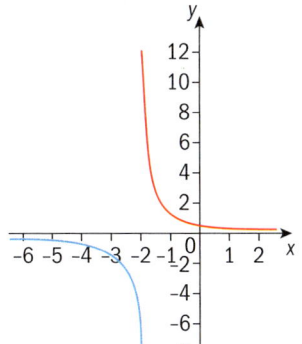

4 a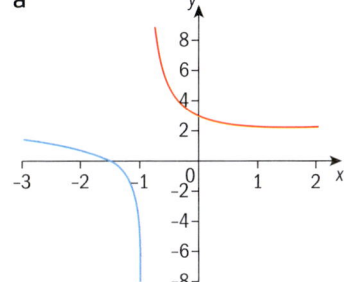

b x-intercept −1.5, y-intercept 3.

5 a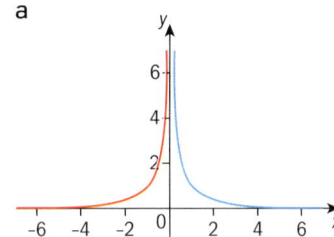

b 0

c Domain $x \in \mathbb{R}, x \neq \mathbb{R}$, range $y > 0$

6 a $x = -2, y = 2$

b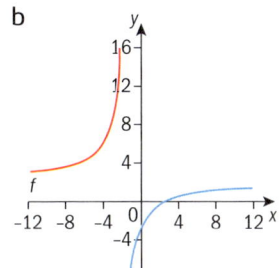

c (2.5, 0), (0, −2.5)

7 a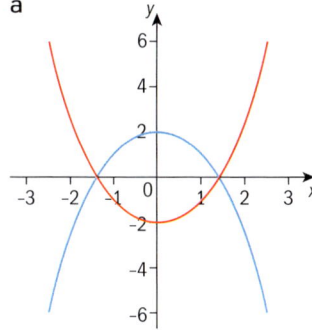

b $x = \pm\sqrt{2}$

8 a $\sqrt[3]{x+3}$

b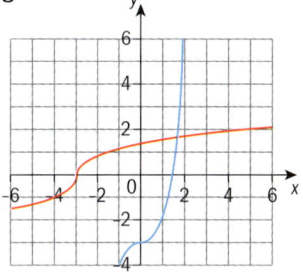

c 1.67

9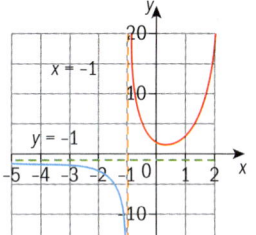

10 a $f^{-1}(x) = \dfrac{x+2}{3}$

b $(g^{-1} \circ f)(x) = (3x - 2) + 3$
$= 3x + 1$

c $(f^{-1} \circ g) = \dfrac{(x-3)+2}{3} = \dfrac{x-1}{3}$

$\dfrac{x-1}{3} = 3x + 1$
$x - 1 = 3(3x + 1)$
$x - 1 = 9x + 3$
$8x = -4$
$x = -\dfrac{1}{2}$

d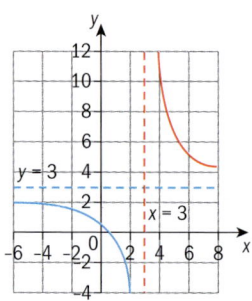

e $x = 3, y = 3$

Chapter 2
Skills check
1 a $a = 6$
   b $x = \pm\sqrt{5}$
   c $n = -11$
2 a $2k(k - 5)$
   b $7a(2a^2 + 3a - 7)$
   c $(2x + 3)(x + 2y)$
   d $(5a - b)(a - 2)$
   e $(n + 1)(n + 3)$
   f $(2x - 3)(x + 1)$
   g $(m + 6)(m - 6)$
   h $(5x + 9y)(5x - 9y)$

Exercise 2A
1 a 1, 2
   b $-8, 7$
   c 5, 6
   d $-5, 5$
   e $-8, 6$
   f -3
2 a $-\dfrac{4}{3}, \dfrac{1}{2}$
   b $-2, \dfrac{4}{5}$
   c $-1, \dfrac{5}{2}$
   d $-\dfrac{1}{2}, \dfrac{9}{2}$
   e $-4, -\dfrac{2}{3}$
   f $-\dfrac{3}{2}, \dfrac{4}{3}$

Exercise 2B
1 a $-5, 4$
   b $-2, -\dfrac{2}{3}$
   c $-\dfrac{3}{2}$
   d $-2, \dfrac{25}{2}$
   e $-9, 4$
   f $-\dfrac{1}{4}, 1$
2 -3 or 4
3 $\dfrac{2}{5}$ or 3

Investigation - perfect square trinomials
1 -5
2 -3
3 -7
4 4
5 9
6 10

Exercise 2C
1 $-4 \pm \sqrt{19}$
2 $\dfrac{5 \pm \sqrt{37}}{2}$
3 $3 \pm 2\sqrt{2}$
4 $\dfrac{-7 \pm \sqrt{65}}{2}$
5 $1 \pm \sqrt{7}$
6 $\dfrac{-1 \pm \sqrt{13}}{2}$

Exercise 2D
1 $-3 \pm 2\sqrt{3}$
2 $1 \pm \sqrt{2}$
3 $1 \pm \sqrt{\dfrac{3}{5}}$
4 $\dfrac{-3 \pm \sqrt{29}}{4}$
5 $-\dfrac{3}{2}, 2$
6 $\dfrac{-2 \pm 3\sqrt{6}}{10}$

Exercise 2E
1 $\dfrac{-9 \pm \sqrt{193}}{8}$
2 $-2, \dfrac{4}{3}$
3 $-1, -\dfrac{1}{5}$
4 $3 \pm \sqrt{5}$
5 no solution
6 $\dfrac{-5 \pm 2\sqrt{10}}{3}$
7 $\dfrac{3 \pm \sqrt{17}}{4}$
8 $\dfrac{9 \pm \sqrt{113}}{4}$
9 $x = \dfrac{-9 \pm \sqrt{129}}{4}$.
10 $\dfrac{3 \pm \sqrt{21}}{4}$

Exercise 2F
1 18, 32
2 24 m, 11 m
3 10
4 18 cm, 21 cm
5 2.99 seconds

Investigation - roots of quadratic equations
1 a 4   b $\dfrac{3}{2}$
   c $-\dfrac{1}{5}$
2 a $-7, 2$ b $\dfrac{4 \pm \sqrt{10}}{3}$
   c $\dfrac{3 \pm \sqrt{89}}{10}$
3 a No solution
   b No solution
   c No solution

Exercise 2G
1 a 37; two different real roots
   b 8; two different real roots
   c -79; no real roots
   d 0; two equal real roots
   e -23; no real roots
   f -800; no real roots

2 a $p < 4$ b $p < 3.125$
   c $|p| > 4\sqrt{2}$ d $|p| > \dfrac{2}{3}$
3 a $k = 25$ b $k = 1.125$
   c $k = \pm\sqrt{15}$ d $k = 0, -0.75$
4 a $m > 9$ b $-2 < m < 2$
   c $m > \dfrac{16}{3}$ d $m > 12$
5 $0 < q < 1$

Investigation – graphs of quadratic functions

a Discriminant, $\triangle = 29$

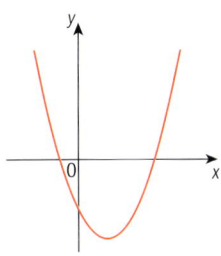

b $\triangle = -12$

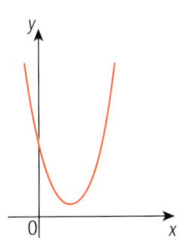

c $\triangle = -24$

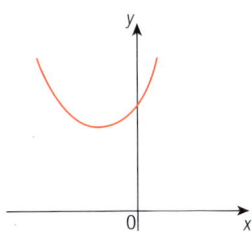

d $\triangle = -71$

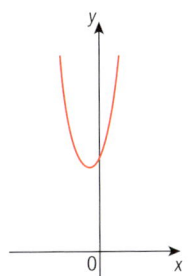

e $\triangle = 0$

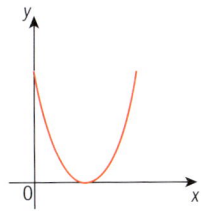

f $\triangle = 0$

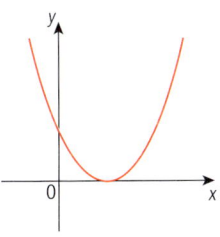

g $\triangle = 33$

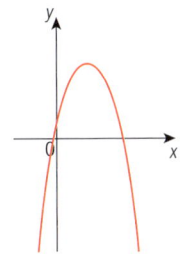

h $\triangle = 37$

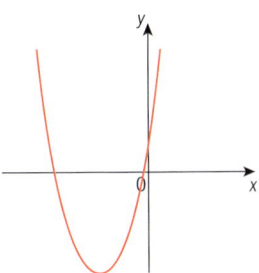

If $b^2 - 4ac > 0$, graph cuts x-axis twice; if $b^2 - 4ac = 0$, graph is tangential to x-axis; if $b^2 - 4ac < 0$, graph does not intersect x-axis.

Exercise 2H

1 a $x = -4; (0, 5)$
   b $x = 3; (0, -3)$
   c $x = -1; (0, 6)$
   d $x = \dfrac{5}{3}; (0, 9)$

2 a $(7, -2); (0, 47)$
   b $(-5, 1); (0, 26)$
   c $(1, 6); (0, 10)$
   d $(-2, -7); (0, 5)$

3 a $f(x) = (x + 5)^2 - 31$

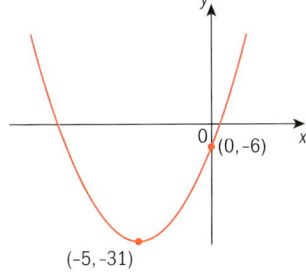

b $f(x) = (x - 2.5)^2 - 4.25$

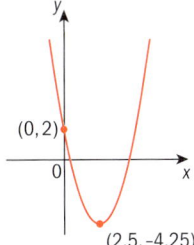

c $f(x) = 3(x - 1)^2 + 4$

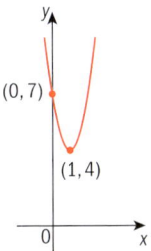

d $f(x) = -2(x - 2)^2 + 5$

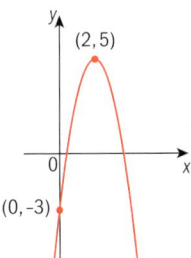

Exercise 2I

1. a (−3, 0); (7, 0); (0, −21)
   b (4, 0); (5, 0); (0, 40)
   c (−2, 0); (−1, 0); (0, −6)
   d (−6, 0); (2, 0); (0, −60)

2. a $y = (x − 8)(x + 1)$

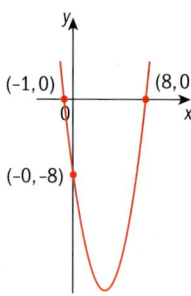

   b $y = (x − 3)(x − 5)$

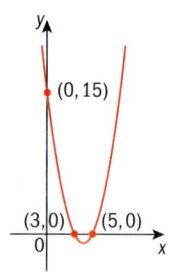

   c $y = −2(x + 1)(x − 2.5)$

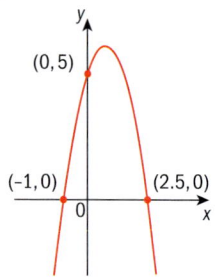

   d $y = 5(x + 2)\left(x − \dfrac{4}{5}\right)$

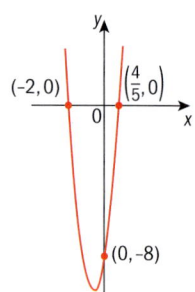

3. a $y = (x + 3)^2 − 25$;
      $y = (x + 8)(x − 2)$

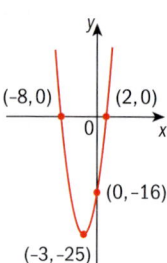

   b $y = −(x + 2)^2 + 25$;
      $y = −(x + 7)(x − 3)$

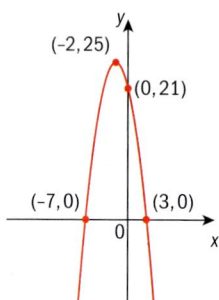

   c $y = −0.5(x − 3.5)^2 + 3.125$;
      $y = −0.5(x − 1)(x − 6)$

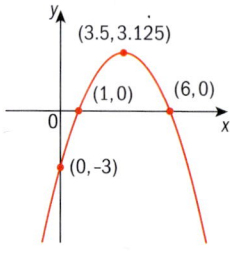

   d $y = 4(x − 2.25)^2 − 12.25$;
      $y = 4(x − 0.5)(x − 4)$

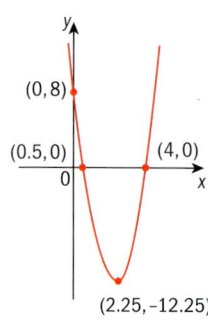

4. a i 0   ii 6
   b $x = 3$
   c (3, −18)

5. a $(f \circ g)(x) = (x − 2)^2 + 3$
   b (2, 3)
   c $h(x) = x^2 − 14x + 50$
   d 50

Exercise 2J

1. $y = x^2 − 4x + 5$
2. $y = x^2 − 4x − 12$
3. $y = −3x^2 − 6x + 5$
4. $y = \dfrac{1}{2}x^2 − \dfrac{5}{2}x − 3$
5. $y = 2x^2 + 7x + 4$
6. $y = −0.4x^2 + 8x$
7. $y = −x^2 + 4x + 21$
8. $y = 12x^2 − 12x + 3$

Exercise 2K

1. a 14.5 metres
   b 1.42 seconds
2. 14 cm, 18 cm
3. a $10 − x$
   c 50 cm²
4. 12.1 cm
5. 17 m, 46 m
6. 7, 9, 11
7. $\dfrac{1 + \sqrt{5}}{2}$
8. 28.125 m²
9. 60 km, 70 h⁻²
10. 6 hours

Review exercise non-GDC

1. a −6, 2
   b 8
   c $−\dfrac{7}{3}, 1$
   d 3, 4
   e $−1 \pm \sqrt{13}$
   f $\dfrac{7 \pm \sqrt{13}}{6}$

2. a −4
   b −4, 1
   c $x = −1.5$
   d −1.5

3 a −5, 1
   b −2
4 a (−3, −6)
   b $\frac{1}{2}$
   c 12
5 ±$\sqrt{3}$
6 a $f(x) = 2(x+3)^2 - 13$
   b (1, −5)
7 $y = \frac{1}{2}x^2 - x - 12$

Review exercise GDC

1 a −0.907, 2.57
   b −4.35, 0.345
   c −2.58, 0.581
   d −1.82, 0.220
2 a 20 m
   b 31.5 m
   c 3.06 s
   d 4.07 s
3 21, 68
4 $a = 0.4, b = 3, c = 2$
5 60 km h^{-1}

Chapter 3

Skills check

1 a $\frac{4}{7}$ b $1\frac{4}{35}$
   c $\frac{2}{15}$ d $\frac{22}{27}$
   e $\frac{19}{27}$ f $\frac{3}{7}$
2 a 0.625   b 0.7
   c 0.42   d 0.16
   e 15   f 4.84
   g 0.0096

Exercise 3A

1 a $\frac{1}{2}$ b $\frac{1}{4}$
   c $\frac{1}{4}$ d $\frac{3}{4}$
   e $\frac{3}{8}$

2 $\frac{1}{5}$

3 a i 0.21   ii 0.33
   b 252
4 a 0.27
   b No – the frequencies are very different
   c 450
5 a $\frac{2}{11}$ b 0
   c $\frac{5}{11}$
6 0.2
7 a $\frac{1}{2}$ b $\frac{13}{40}$

Exercise 3B

1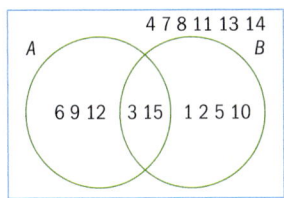

   $\frac{4}{7}$

2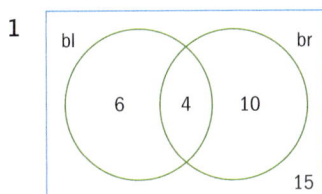

   $\frac{8}{25}$

3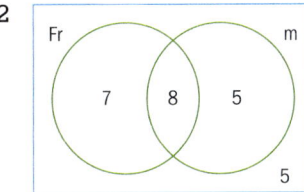

   Six have both activities.
   a $\frac{6}{25}$ b $\frac{11}{25}$

4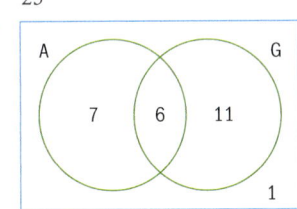

   Five play neither.

   a $\frac{11}{32}$ b $\frac{9}{32}$

5 a $A = \{3, 6, 9, 12, 15\}$
      $B = \{1, 2, 3, 5, 6, 10, 15\}$

   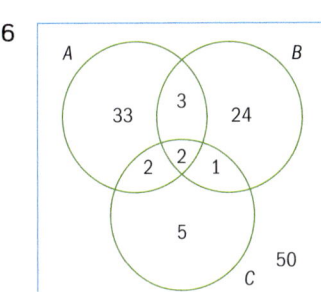

   c i $\frac{1}{5}$ ii $\frac{2}{5}$

6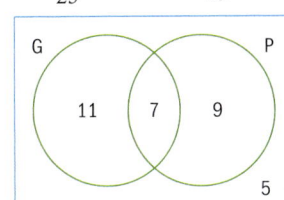

   a 0.33   b 0.24
   c 0.3

Exercise 3C

1 a $\frac{51}{250}$
   b $\frac{53}{100}$
   c $\frac{299}{500}$
2 a $\frac{2}{5}$
   b $\frac{3}{5}$
   c $\frac{1}{2}$
3 $\frac{17}{20}$
4 a $\frac{4}{13}$ b $\frac{9}{26}$
   c $\frac{2}{13}$ d $\frac{1}{2}$
5 a 0.5   b 0.5

6 $\dfrac{11}{60}$

7 a $\dfrac{1}{4}$ b $\dfrac{3}{4}$

8 a 0.6   b 0.4
   c 0.9

Exercise 3D

1 a N   b Y
   c N   d Y
   e N   f N
   g N

2 yes

3 $\dfrac{57}{89}$

4 a $\dfrac{7}{12}$ b $\dfrac{47}{60}$
   c $\dfrac{13}{60}$

Exercise 3E

1 HHH, HHT, HTH, HTT, THH, THT, TTH, TTT
   a $\dfrac{1}{2}$ b $\dfrac{3}{8}$
   c $\dfrac{1}{4}$

2

	BLUE			
	1	2	3	4
RED 1	(1, 1)	(1, 2)	(1, 3)	(1, 4)
RED 2	(2, 1)	(2, 2)	(2, 3)	(2, 4)
RED 3	(3, 1)	(3, 2)	(3, 3)	(3, 4)
RED 4	(4, 1)	(4, 2)	(4, 3)	(4, 4)

   a $\dfrac{3}{8}$ b $\dfrac{3}{8}$
   c $\dfrac{1}{4}$ d $\dfrac{9}{16}$

3

	Box 1		
	1	2	3
Box 2 2	(2, 1)	(2, 2)	(2, 3)
Box 2 3	(3, 1)	(3, 2)	(3, 3)
Box 2 4	(4, 1)	(4, 2)	(4, 3)
Box 2 5	(5, 1)	(5, 2)	(5, 3)

   a $\dfrac{1}{6}$ b $\dfrac{1}{3}$
   c $\dfrac{3}{4}$ d $\dfrac{5}{12}$
   e $\dfrac{2}{3}$

4

	First draw					
	0	1	2	3	4	5
Second draw 0	(0, 0)	(0, 1)	(0, 2)	(0, 3)	(0, 4)	(0, 5)
Second draw 1	(1, 0)	(1, 1)	(1, 2)	(1, 3)	(1, 4)	(1, 5)
Second draw 2	(2, 0)	(2, 1)	(2, 2)	(2, 3)	(2, 4)	(2, 5)
Second draw 3	(3, 0)	(3, 1)	(3, 2)	(3, 3)	(3, 4)	(3, 5)
Second draw 4	(4, 0)	(4, 1)	(4, 2)	(4, 3)	(4, 4)	(4, 5)
Second draw 5	(5, 0)	(5, 1)	(5, 2)	(5, 3)	(5, 4)	(5, 5)

   a $\dfrac{1}{6}$ b $\dfrac{23}{36}$
   c $\dfrac{13}{18}$ d $\dfrac{13}{36}$
   e $\dfrac{5}{9}$

5 a $\dfrac{1}{6}$ b $\dfrac{1}{9}$
   c $\dfrac{2}{9}$

Exercise 3F

1 $\dfrac{1}{25}$

2 $\dfrac{2}{169}$

3 $\dfrac{64}{125}$

4 0.6375

5 a P(B) = 0.2; P(B ∩ C) = 0.16
   b Not independent

6 $\dfrac{5}{12}$

7 $\dfrac{1}{59049}$

8 $\dfrac{1}{256}$

9 a 0.4
   b P(E) × P(F) = P(E ∩ F)
   c P(E ∩ F) ≠ 0
   d 0.64

10 $\dfrac{2}{27}$

11 $\dfrac{1}{27}$

12 a 0.27   b 0.63
    c 0.07

13 0.18, 0.28

14 a $\dfrac{1}{1296}$ b $\dfrac{1}{216}$

15 Rolling a 'six' on four throws of one dice

16 a 0.729   b 0.271

Exercise 3G

1 12 take both subjects
   a $\dfrac{8}{27}$ b $\dfrac{23}{27}$
   c $\dfrac{4}{5}$

2 a 0.2   b $\dfrac{1}{3}$
   c $\dfrac{7}{15}$

3 $\dfrac{39}{48}$

4 a $\dfrac{1}{3}$ b $\dfrac{2}{5}$
   c $\dfrac{3}{5}$ d $\dfrac{1}{2}$

5 $\dfrac{61}{95}$

6 $\dfrac{1}{6}$

7 a 0   b 0
   c 0.63

8 67.3%

9 $\dfrac{34}{47}$

10 a $\dfrac{1}{10}$ b $\dfrac{43}{50}$
    c $\dfrac{11}{13}$

11 0.3

12 $\dfrac{1}{3}$

Exercise 3H

1 a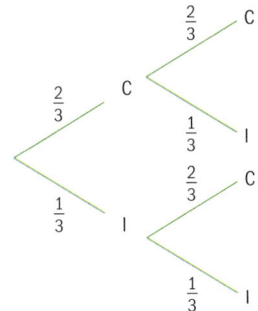

  b $\frac{4}{9}$

  c $\frac{8}{9}$

2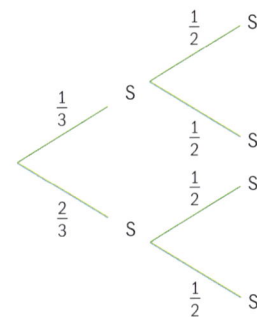

  $\frac{1}{3}$

3 a $\frac{1}{5}$

  b $\frac{49}{120}$

4 $\frac{4}{9}$

5 a 0.48

  b 0.64

6 a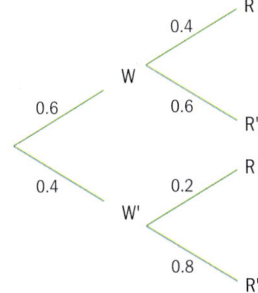

  b 0.32   **c** 0.4624

Exercise 3I

1 a $\frac{11}{1105}$ **b** $\frac{132}{1105}$

2 a $\frac{5}{33}$ **b** $\frac{15}{22}$

  c $\frac{1}{2}$

3 a $\frac{1}{12}$ **b** $\frac{5}{18}$

  c $\frac{5}{18}$ **d** $\frac{5}{12}$

4 $\frac{120}{1001}$

5 a $\frac{2}{5}$ **b** $\frac{8}{15}$

6 a $\frac{55}{63}$ **b** $\frac{9}{11}$

  c $\frac{7}{11}$ **d** $\frac{5}{11}$

Review exercise non-GDC

1 a $\frac{1}{5}$ **b** $\frac{1}{3}$

  c $\frac{49}{90}$ **d** $\frac{1}{15}$

2 $\frac{11}{30}$

3 a 0.55

  b $P(C \cap D) = 0.15$ $P(c) \times P(D)$
   $= 0.14$

4 a 0.02   **b** 0.78

  c 0.76   **d** $\frac{1}{30}$

5 a $6x$

  b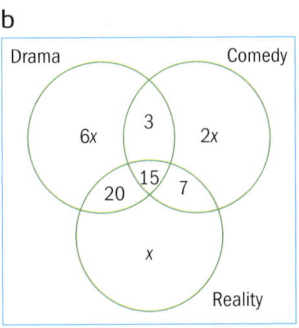

  c $x = 5$

Review exercise GDC

1 a 0.3

  b No, $P(C \text{ and } D) \neq 0$

  c No, $P(C) \times P(D) \neq P(C \text{ and } D)$

  d 0.6

  e 0.75

2 a 0.43

  b 0.316

3 a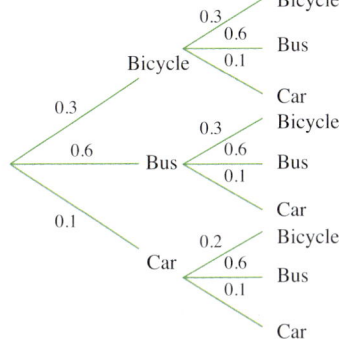

  b i 0.09

    ii 0.18

    iii 0.46

  c 0.343

  d 0.045

4 a $\frac{3}{8}$ **b** $\frac{2}{3}$

  c $\frac{2}{21}$

5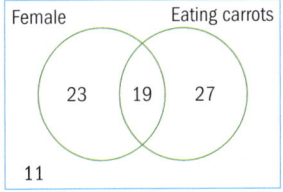

  Both female and eating carrots = 19.

  a $\frac{11}{70}$ **b** $\frac{19}{36}$

  c No, $P(F) \times P(C) \neq P(F \text{ and } C)$

Chapter 4

Skills check

1. a $\dfrac{1}{128}$ b $\dfrac{81}{256}$ c 1×10^{-9}

2. a 5 b 3 c 4

3.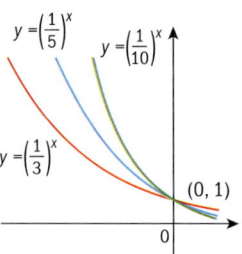

Investigation – folding paper

Number of folds	Number of layers	Thickness (km)	As thick as a
0	1	1×10^{-7}	Piece of paper
1	2	2×10^{-7}	
2	4	4×10^{-7}	Credit card
3	8	8×10^{-7}	
4	16	1.6×10^{-6}	
5	32	3.2×10^{-6}	
6	64	6.4×10^{-6}	
7	128	1.28×10^{-5}	Textbook
8	256	2.56×10^{-5}	
9	512	5.12×10^{-5}	

3. a 13 folds b 15 folds

4. $113\,000\,000$ km

Exercise 4A

1. a x^5
   b $6p^6q^2$
   c $\dfrac{1}{3}x^3y^3$
   d x^4y^6

2. a x^3
   b a^4
   c $\dfrac{a^4}{4}$
   d $2x^2y^3$

3. a x^{12} b $27t^6$
   c $3x^6y^4$ d $-y^6$

Exercise 4B

1. a 3 b 5 c 16
   d 4 e $\dfrac{4}{9}$

2. a $\dfrac{1}{8}$ b $\dfrac{1}{4}$ c $\dfrac{1}{3}$
   d $\dfrac{1}{16}$ e $\dfrac{25}{16} = 1\dfrac{9}{16}$

Exercise 4C

1. a $8a^3$ b $\dfrac{2}{x^2}$ c q^3
   d $\dfrac{d}{3c}$ e $\dfrac{P^{-\frac{4}{3}}}{4}$

2. a $\dfrac{a^{\frac{5}{2}}}{b}$ b $\dfrac{y}{5x^3}$ c $\dfrac{3x^3}{y^2}$

Exercise 4D

1. a $x = 5$ b $x = -2$
   c $x = 3, -1$ d $x = \dfrac{3}{2}$
   e $x = 3$

2. a $x = \dfrac{5}{2}$ b $x = -4$
   c $x = -\dfrac{3}{5}$ d $x = \dfrac{4}{5}$

3. $x = -6$

Exercise 4E

1. a $x = 3$ b $x = 2$ c $x = \dfrac{1}{4}$
   d $x = \dfrac{1}{2}$ e $x = 3^{-\frac{1}{3}}$ f $x = \dfrac{3}{4}$

2. a $x = 8$ b $x = 625$
   c $x = \dfrac{1}{256}$ d $x = 64$
   e $x = 32$ f $x = \dfrac{1}{16}$

3. a $x = \dfrac{1}{3125}$ b $x = \dfrac{1}{216}$
   c $x = 512$ d $x = \dfrac{27}{64}$

Investigation – graphs of exponential functions 1

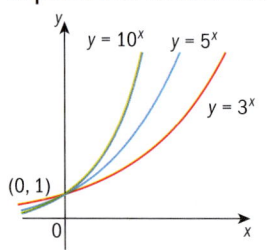

Investigation – graphs of exponential functions 2

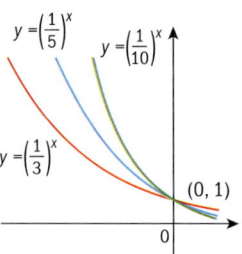

Investigation – compound interest

Half-yearly	$\left(1+\dfrac{1}{2}\right)^2$	2.25
Quarterly	$\left(1+\dfrac{1}{4}\right)^4$	2.441 406 25
Monthly	$\left(1+\dfrac{1}{12}\right)^{12}$	2.613 035 290 22…
Weekly	$\left(1+\dfrac{1}{52}\right)^{52}$	2.692 596 954 44…
Daily	$\left(1+\dfrac{1}{365}\right)^{365}$	2.714 567 482 02…
Hourly	$\left(1+\dfrac{1}{8760}\right)^{8760}$	2.718 126 690 63…
Every minute	$\left(1+\dfrac{1}{525600}\right)^{525600}$	2.718 279 215 4…
Every second	$\left(1+\dfrac{1}{31536000}\right)^{31536000}$	2.718 282 472 54…

Exercise 4F

1. Curves of
   a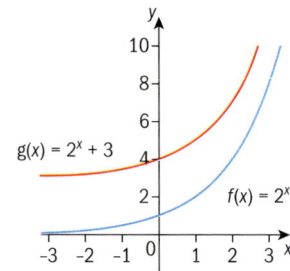
   b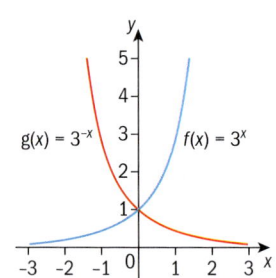

c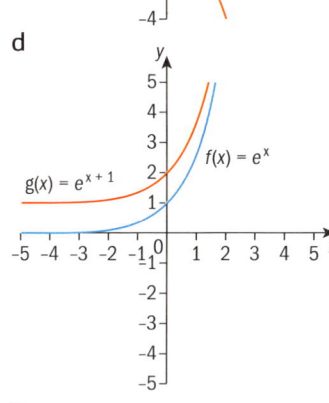

d

e

f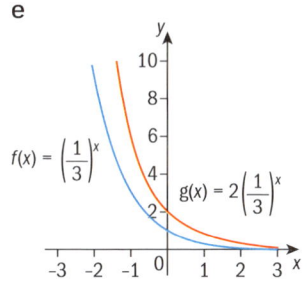

2
	Domain	Range
a	$x \in \mathbb{R}$	$f(x) > 3$
b	$x \in \mathbb{R}$	$f(x) > 0$
c	$x \in \mathbb{R}$	$f(x) < 0$
d	$x \in \mathbb{R}$	$f(x) > 0$
e	$x \in \mathbb{R}$	$f(x) > 0$
f	$x \in \mathbb{R}$	$f(x) > 0$

Exercise 4G

1 a 2   b $\frac{1}{2}$ c 6   d 0

2 a -4 b $\frac{3}{2}$ c $\frac{3}{5}$ d 4

Exercise 4H

1 a 1   b 1   c 1
   d 0   e 0   f 0

Exercise 4I

1 a $\log_2 x = 9$ b $\log_3 x = 5$
   c $\log x = 4$ d $\log_a x = b$
2 a $2^x = 8$ b $3^x = 27$
   c $10^x = 1000$ d $a^x = b$
3 a 64   b 81   c 8
   d 36   e $\frac{1}{32}$

Investigation – inverse functions

a The function $y = 2^x$

x	−3	−2	−1	0	1	2	3
y	$\frac{1}{8}$	$\frac{1}{4}$	$\frac{1}{2}$	1	2	4	8

b The inverse function of $y = 2^x$

x	$\frac{1}{8}$	$\frac{1}{4}$	$\frac{1}{2}$	1	2	4	8
y	−3	−2	−1	0	1	2	3

c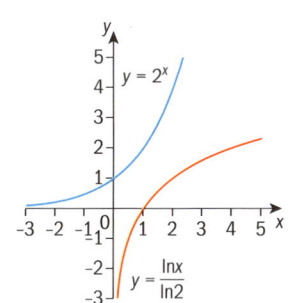

d The graphs are reflections of each other in the line $y = x$

Exercise 4J

1 a Curve is shifted down two units
   b Curve is translated right 2 units
   c Curve is stretched by factor 2 parallel with y-axis

2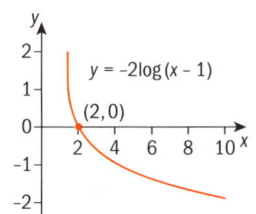

3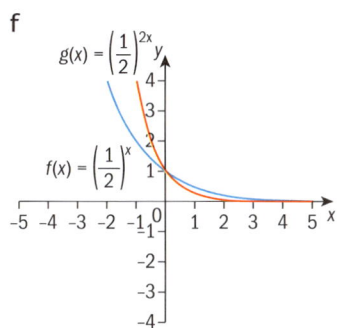

4 3

5 $f^{-1}(2) = 9$

Exercise 4K

1 a 0.477   b 1.20
   c 0.805   d 0.861
   e 0.861   f −0.0969
   g 0.228   h 0.954

Exercise 4L

1 a $x = 0.425$ b $x = -5.81$
   c $x = 0$ d $x = -0.693$
   e $x = -3.51$
2 a 0.367   b −0.222
   c 0   d −0.301
3 a 100   b $\frac{1}{10}$
   c 1   d 0.00000794
4 a 12   b 4
   c $\sqrt{3}$ d 4
5 a 5   b 2
   c 0   d 1
   e −3
6 $f^{-1}(x) = \dfrac{1 + \ln x}{2}, x > 0$
7 Domain $[e^{-0.5}, e]$; Range $[-2, 4]$
8 $f^{-1}(x) = \dfrac{1}{3} e^x$
9 $(g \circ f)(x) = 2x - 2$

Exercise 4M

1 a $\log 30$ b $\log 12$ c $\log 4$
   d $\log 7$ e $\log \dfrac{x^3}{y^2}$
   f $\log \dfrac{x}{yz}$ g $\log \dfrac{1}{x^2 y}$
2 a $\log_2\left(\dfrac{27}{2}\right)$ b $b = \dfrac{\pi}{2}$
   c $\log_a 6$ d $\ln\left(\dfrac{1}{2}\right)$ or $-\ln 2$
   e $\ln\left(\dfrac{8}{e^2}\right)$ f $\log_2\left(\dfrac{x^4 y^{\frac{1}{3}}}{z^5}\right)$

3 a 2   b 3   c 2
   d 3   e 1

Exercise 4N

1 a $p+q$ b $3p$ c $q-p$
   d $\dfrac{q}{2}$ e $2q-\dfrac{p}{2}$

2 $6x - 3y - 6z$

3 a $1 + \log x$ b $2 - 2\log x$
   c $\dfrac{1}{2} + \dfrac{1}{2}\log x$ d $-1 - \dfrac{1}{2}\log x$

4 $y = 3a - 4$ 5 $-3 - 2\log_3 x$

Exercise 4O

1 a 2.81 b -1.21 c -0.325
   d 0.514 e 12.4

2 $\dfrac{y}{2}$

3 a $\dfrac{y}{x}$ b $\dfrac{x}{y}$ c $\dfrac{2y}{x}$
   d $2x + y$ e $\dfrac{x+y}{y}$ f $\dfrac{y-x}{x}$

4 a

   <graph showing $y = \dfrac{\log x}{\log 4}$>

   b

   <graph showing $\dfrac{\log x^2}{\log 5}$>

5 a $2b$ b $\dfrac{b}{2}$
   c $-2b$ d $-\dfrac{b}{4}$

Exercise 4P

1 a 2.32 b 3.56 c -1.76
   d 0.425 e 0.229 f -3.64
   g 1.79 h -11.0

2 a 6.78   b 2.36
   c -3.88 d 0.263
   e 0.526   f 2.04
   g -999

Exercise 4Q

1 a 1.16 b 1.41 c -0.314
   d -0.0570 e 11.1

2 a $\dfrac{\ln 500}{\ln\left(\dfrac{5}{2}\right)}$ b $\dfrac{\ln\left(\dfrac{8}{5}\right)}{\ln\left(\dfrac{3}{7}\right)}$
   c $\dfrac{\ln\left(\dfrac{144}{5}\right)}{\ln 108}$ d $\dfrac{\ln 64}{\ln 3}$

3 a $x = 0$ b $x = \dfrac{\ln 3}{\ln 2}$

Exercise 4R

1 a $x = \dfrac{1}{5}$ b $x = 1$
   c $x = \dfrac{3}{7}$ d $x = \sqrt{2}$
   e $x = 1.62$

Exercise 4S

1 a $x = 83$ b $x = 14$
   c $x = \dfrac{95}{32}$

2 a $x = 9$ b $x = 6$
   c no solutions

3 $A = x(2x + 7) = 2x^2 + 7x$
   $x = 0.5$

4 $x = 4$

5 $x = 16$

Exercise 4T

1 a 450×1.032^n
   b 10 years

2 a i 121   ii 195
   b 9.6 days (10 days)

3 49.4 hours

4 a

   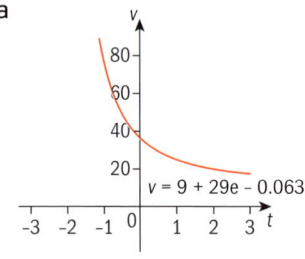

   b 38 m s^{-1} c 9 m s^{-1}
   d 10.7 m s^{-1} e 17

5 $a = 4, b = 3$

Review exercise GDC

1 3.52

2 a 0.548   b -0.954
   c -1.18

3 a 5   b 2
   c 3.60   d 1, 4
   e 100, $\dfrac{1}{100}$

4 a $f(x) > 0$, range of $g(x)$ is all real numbers
   b They are 1–1 functions;
      $f^{-1}(x) = \dfrac{1}{2}\ln x;\ g^{-1}(x) = e^{\frac{2}{3}x}$
   c $(f \circ g)(x) = x^3;$
      $(g \circ f)(x) = 3x$
   d $x = \sqrt{3}$

5 a 218 393 insects
   b 8.66 days

Review exercise non-GDC

1 0

2 $\dfrac{\log\left(\dfrac{3}{5}\right)}{\log\left(\dfrac{35}{9}\right)}$

3 4.5

4 $\log_3 \dfrac{x^4 \sqrt[3]{y}}{z^5}$

5 a $x = 7$ b $x = 2$
   c $x = 1, 4$ d $x = \dfrac{6}{7}$

6 a $\dfrac{n}{m}$ b $m - n$
   c $2m$ d $\dfrac{m+n}{n}$

7 Shift one unit to the right, stretch factor $\dfrac{1}{3}$ parallel to x-axis, shift 2 units up.

8 a $f^{-1}(x) = \dfrac{1}{2}\ln\left(\dfrac{x}{3}\right)$
   b $f^{-1}(x) = \dfrac{1}{3}\log x$
   c $f^{-1}(x) = \dfrac{2^x}{4} = 2^{x-2}$

9 $a = 2, b = 4$

Chapter 5

Skills check

1. a $-8x + 20$ b $12x - 18$
   c $-x^3 - 7x$
   d $x^4 + 6x^3 + 9x^2$
   e $x^3 + 5x^2 - 24x$

2.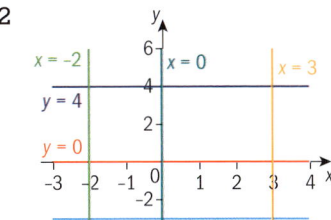

3. A is a horizontal shift of 4 units to the right. Function A is $y = (x - 4)^3$
   B is a vertical shift of 2 units down. Function B is $y = x^3 - 2$

Investigation – graphing product pairs

x	24	12	8	3	6	4	2	1
y	1	2	3	8	4	6	12	24

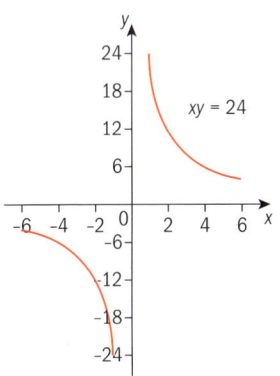

As y gets bigger, x gets smaller and vice versa.
The graph gets closer and closer to the axes as x- and y-values increase.

Exercise 5A

1. a $\dfrac{1}{2}$ b $\dfrac{1}{3}$ c $-\dfrac{1}{3}$
   d -1 e $\dfrac{3}{2}$ f $\dfrac{11}{7}$
   g $-\dfrac{2}{3}$ h $\dfrac{2}{7}$

2. a $\dfrac{2}{13}$ b $\dfrac{1}{x}$ c $\dfrac{1}{y}$
   d $\dfrac{1}{3x}$ e $\dfrac{1}{4y}$ f $\dfrac{9}{2x}$

   g $\dfrac{5}{3a}$ h $\dfrac{3d}{2}$ i $\dfrac{t}{d}$

   j $\dfrac{x-1}{x+1}$

3. a $6 \times \dfrac{1}{6} = 1$ b $\dfrac{3}{4} \times \dfrac{4}{3} = 1$
   c $\dfrac{2c}{3d} \times \dfrac{3d}{2c} = 1$

4. a 4   b x

5. a i 0.5   ii 0.05
      iii 0.005   iv 0.0005
   b y gets smaller, nearer to zero.
   c $y = \dfrac{24}{x}$ so it can never be zero.
   d i 0.5   ii 0.05
      iii 0.005   iv 0.0005
   e x gets smaller, nearer to zero.
   f $x = \dfrac{24}{y}$ so it can never be zero.

Investigation – graphs of reciprocal functions

1. a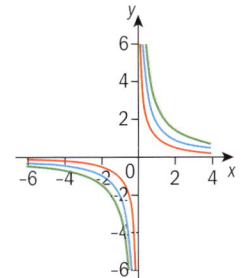
   The numerator indicates the scale factor of the stretch parallel to the y-axis.

2.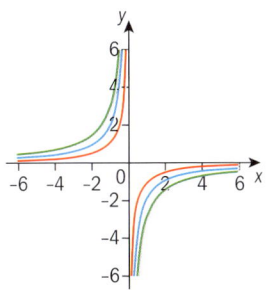
   Changing the sign of the numerator reflects the graphs of the original functions in the x-axis.

3. a
x	0.25	0.4	0.5	1	2	4	8	10	16
f(x)	16	10	8	4	2	1	0.5	0.4	0.25

   b The values of x and $f(x)$ are the same numbers but in reverse order.

   c d e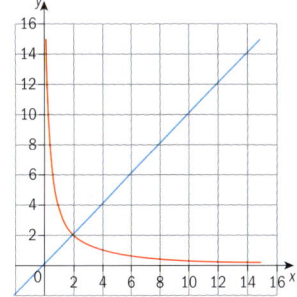

   f The function reflects onto itself.
   g The function is its own inverse.

Exercise 5B

1.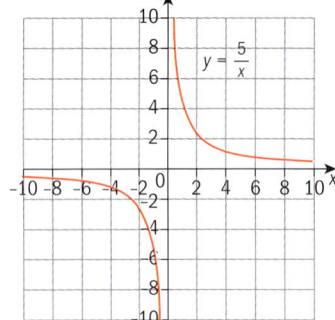

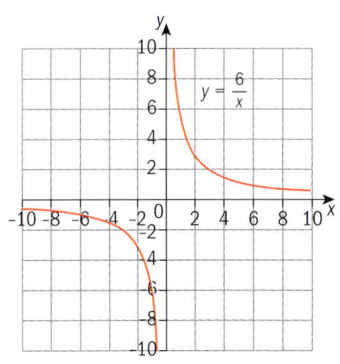

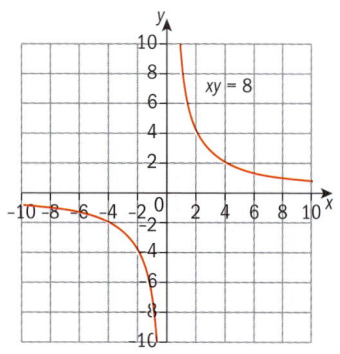

2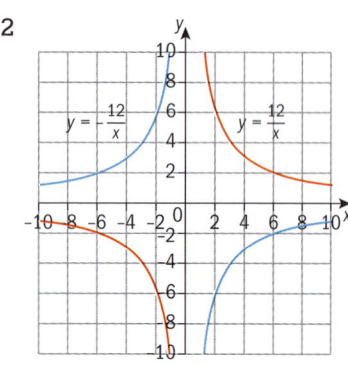

3 a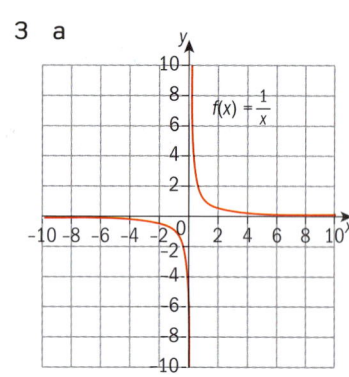

$x = 0$ and $y = 0$

b

$x = 0$ and $y = 2$

4 a $y = 0, x = 0$; Domain $x \in \mathbb{R}$, $x \neq 0$; Range $y \in \mathbb{R}, y \neq 0$

b $y = 2, x = 0$; Domain $x \in \mathbb{R}$, $x \neq 0$; Range $y \in \mathbb{R}, y \neq 2$

c $y = -2, x = 0$
Domain $x \in \mathbb{R}, x \neq 0$
Range $y \in \mathbb{R}, y \neq -2$

5 a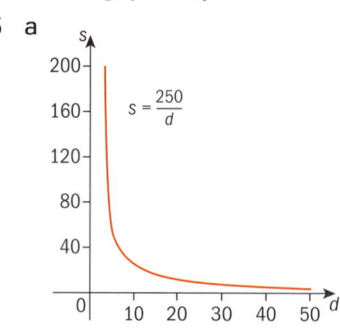

b 25 m   c 2.5 m s^{-1}

6 a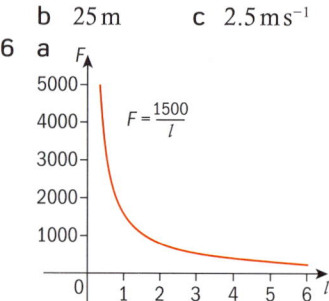

b 750 N
c i 1.5 m   ii 075 m
   iii 0.5 m

Investigation – graphing rational functions 1

a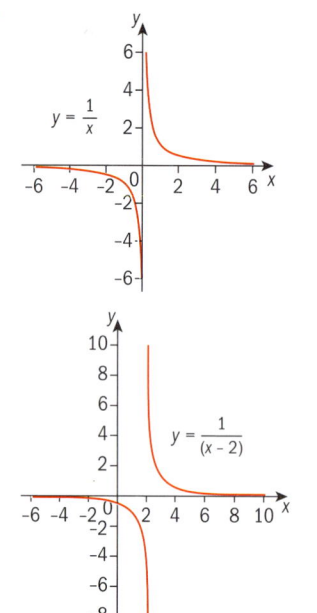

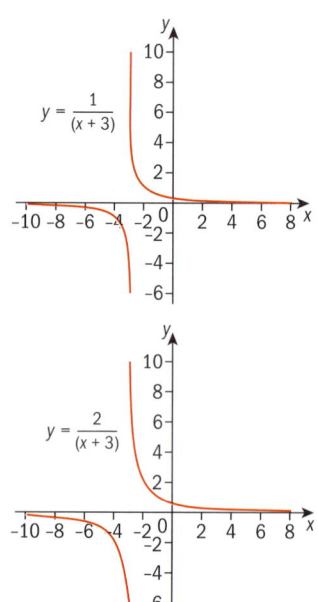

b

Rational function	Vertical asymptote	Horizontal asymptote	Domain	Range
$y = \frac{1}{x}$	$x = 0$	$y = 0$	$x \in \mathbb{R}, x \neq 0$	$y \in \mathbb{R}, y \neq 0$
$y = \frac{1}{x-2}$	$x = 2$	$y = 0$	$x \in \mathbb{R}, x \neq 2$	$y \in \mathbb{R}, y \neq 0$
$y = \frac{1}{x+3}$	$x = -3$	$y = 0$	$x \in \mathbb{R}, x \neq -3$	$y \in \mathbb{R}, y \neq 0$
$y = \frac{2}{x+3}$	$x = -3$	$y = 0$	$x \in \mathbb{R}, x \neq -3$	$y \in \mathbb{R}, y \neq 0$

c The vertical asymptote is the x-value that makes the denominator equal to zero.
d They are all $y = 0$
e $x \in \mathbb{R}, x \neq x$-value of asymptote
f $y \in \mathbb{R}, y \neq 0$, the y-value of the horizontal asymptote.

Exercise 5C

1 a $y = 0, x = -1$
Domain $x \in \mathbb{R}, x \neq -1$
Range $y \in \mathbb{R}, y \neq 0$

b $y = 0, x = 4$
Domain $x \in \mathbb{R}, x \neq 4$
Range $y \in \mathbb{R}, y \neq 0$

c $y = 0, x = 5$
Domain $x \in \mathbb{R}, x \neq 5$
Range $y \in \mathbb{R}, y \neq 0$

d $y = 0, x = -1$
Domain $x \in \mathbb{R}, x \neq -1$
Range $y \in \mathbb{R}, y \neq 0$

e $y = 2, x = -1$
Domain $x \in \mathbb{R}, x \neq -1$
Range $y \in \mathbb{R}, y \neq 2$

f $y = -2, x = -1$
Domain $x \in \mathbb{R}, x \neq -1$
Range $y \in \mathbb{R}, y \neq -2$

g $y = 2, x = 3$
Domain $x \in \mathbb{R}, x \neq 3$
Range $y \in \mathbb{R}, y \neq 2$

h $y = -2, x = -3$
Domain $x \in \mathbb{R}, x \neq -3$
Range $y \in \mathbb{R}, y \neq -2$

2 a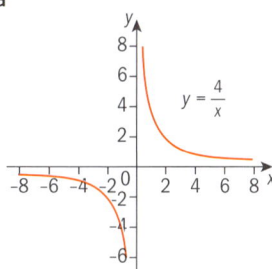

Domain $x \in \mathbb{R}, x \neq 0$
Range $y \in \mathbb{R}, y \neq 0$

b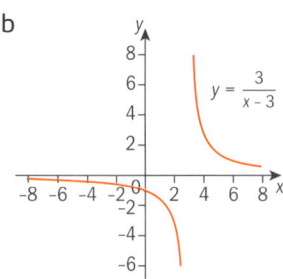

Domain $x \in \mathbb{R}, x \neq 3$
Range $y \in \mathbb{R}, y \neq 0$

c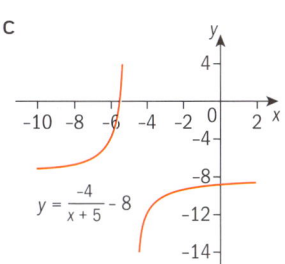

Domain $x \in \mathbb{R}, x \neq -5$
Range $y \in \mathbb{R}, y \neq -8$

d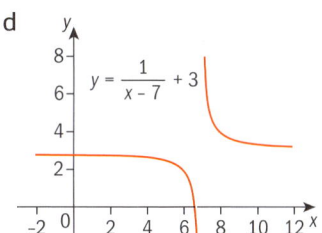

Domain $x \in \mathbb{R}, x \neq 7$
Range $y \in \mathbb{R}, y \neq 3$

e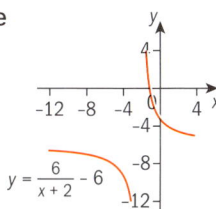

Domain $x \in \mathbb{R}, x \neq -2$
Range $y \in \mathbb{R}, y \neq -6$

f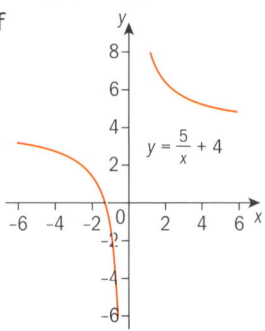

Domain $x \in \mathbb{R}, x \neq 0$
Range $y \in \mathbb{R}, y \neq 4$

g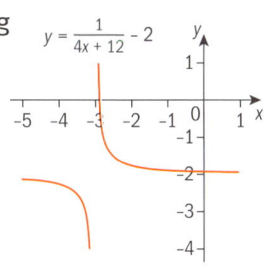

Domain $x \in \mathbb{R}, x \neq -3$
Range $y \in \mathbb{R}, y \neq -2$

h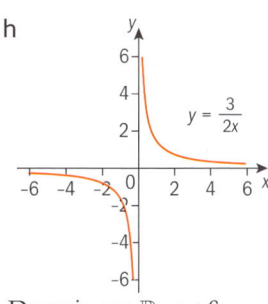

Domain $x \in \mathbb{R}, x \neq 0$
Range $y \in \mathbb{R}, y \neq 0$

i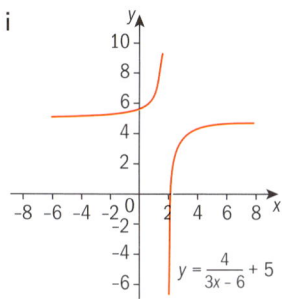

Domain $x \in \mathbb{R}, x \neq 2$
Range $y \in \mathbb{R}, y \neq 5$

3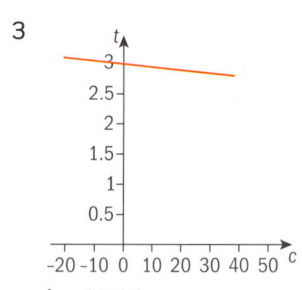

b 3.9 °C

4 a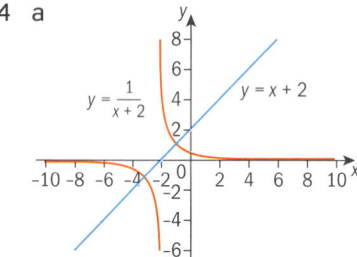

The linear function is a line of symmetry for the rational function. The linear function crosses the x-axis at the same place as the vertical asymptote.

b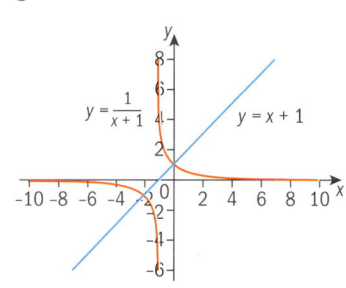

The linear function is a line of symmetry for the rational function. The linear function crosses the x-axis at the same place as the vertical asymptote of the rational function.

Answers 733

Investigation - graphing rational functions 2

a

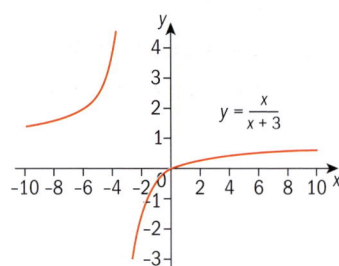

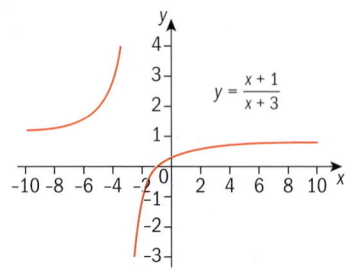

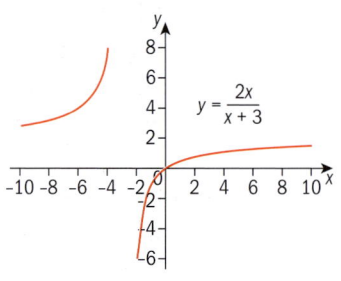

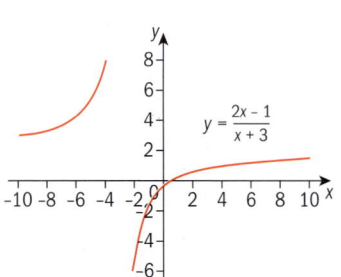

b

Rational function	Vertical asymptote	Horizontal asymptote	Domain	Range
$y = \frac{x}{x+3}$	$x = -3$	$y = 1$	$x \in \mathbb{R}, x \neq -3$	$y \in \mathbb{R}, y \neq 1$
$y = \frac{x+1}{x+3}$	$x = -3$	$y = 1$	$x \in \mathbb{R}, x \neq -3$	$y \in \mathbb{R}, y \neq 1$
$y = \frac{2x}{x+3}$	$x = -3$	$y = 2$	$x \in \mathbb{R}, x \neq -3$	$y \in \mathbb{R}, y \neq 2$
$y = \frac{2x-1}{x+3}$	$x = -3$	$y = 2$	$x \in \mathbb{R}, x \neq -3$	$y \in \mathbb{R}, y \neq 2$

c The horizontal asymptote is the quotient of the x-coefficients.

d The domain excludes the x-value of the vertical asymptote.

Exercise 5D

1 a $y = 1, x = 3$
Domain $x \in \mathbb{R}, x \neq 3$
Range $y \in \mathbb{R}, y \neq 1$

b $y = \frac{2}{3}, x = \frac{1}{3}$
Domain $x \in \mathbb{R}, x \neq \frac{1}{3}$
Range $y \in \mathbb{R}, y \neq \frac{2}{3}$

c $y = \frac{3}{4}, x = -\frac{5}{4}$
Domain $x \in \mathbb{R}, x \neq -\frac{5}{4}$
Range $y \in \mathbb{R}, y \neq \frac{3}{4}$

d $y = \frac{17}{8}, x = -\frac{1}{4}$
Domain $x \in \mathbb{R}, x \neq -\frac{1}{4}$
Range $y \in \mathbb{R}, y \neq \frac{17}{8}$

2 a iii, **b** i, **c** iv, **d** ii

3 a

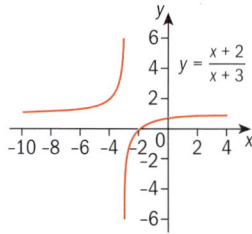

Domain $x \in \mathbb{R}, x \neq -3$
Range $y \in \mathbb{R}, y \neq 1$

b

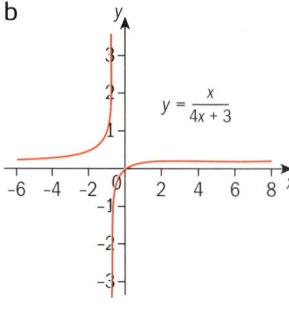

Domain $x \in \mathbb{R}, x \neq -\frac{3}{4}$
Range $y \in \mathbb{R}, y \neq \frac{1}{4}$

c

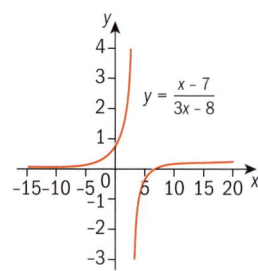

Domain $x \in \mathbb{R}, x \neq \frac{8}{3}$
Range $y \in \mathbb{R}, y \neq \frac{1}{3}$

d

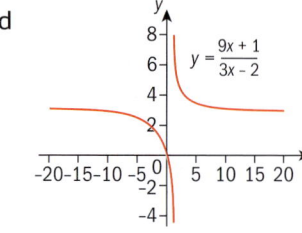

Domain $x \in \mathbb{R}, x \neq \frac{2}{3}$
Range $y \in \mathbb{R}, y \neq 3$

e

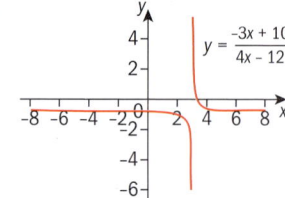

Domain $x \in \mathbb{R}, x \neq 3$
Range $y \in \mathbb{R}, y \neq -\frac{3}{4}$

f

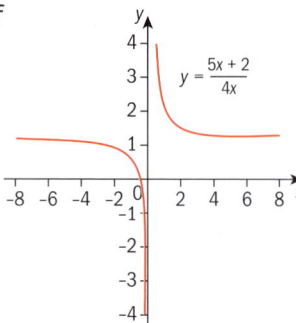

Domain $x \in \mathbb{R}, x \neq 0$
Range $y \in \mathbb{R}, y \neq \frac{5}{4}$

g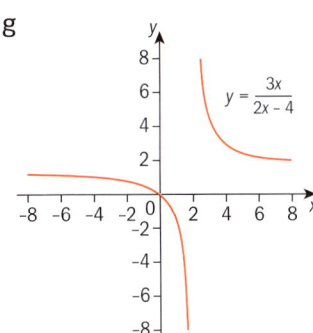

Domain $x \in \mathbb{R}, x \neq 2$
Range $y \in \mathbb{R}, y \neq \dfrac{3}{2}$

h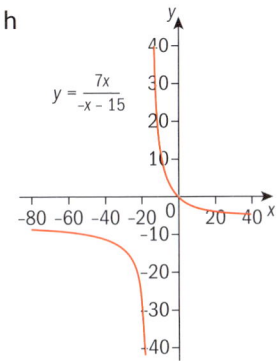

Domain $x \in \mathbb{R}, x \neq -15$
Range $y \in \mathbb{R}, y \neq -7$

i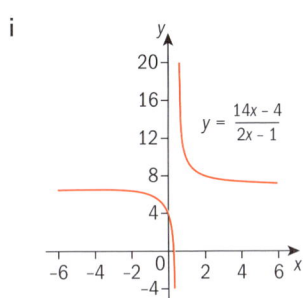

Domain $x \in \mathbb{R}, x \neq \dfrac{1}{2}$
Range $y \in \mathbb{R}, y \neq 7$

4 E.g. $y = \dfrac{1}{x+4} + 3$

5 a $C(x) = 450 + 5.5x$
   b $A(x) = \dfrac{450 + 5.5x}{x}$
   c Domain is $x > 0$. Since x represents the number of T-shirts produced, only positive values make sense. Exclude $x = 0$ since $A(x)$ is undefined for $x = 0$ and at $x = 0$ no T-shirts are made.
   d $x = 0$
   e The horizontal asymptote is $y = 5.5$. As the number of shirts produced increases, the set up costs become negligible.

6 a
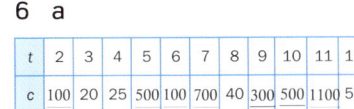

   b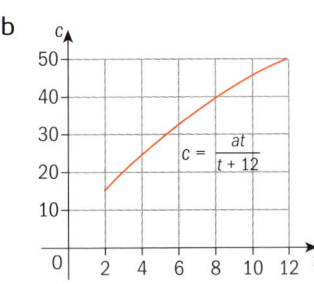

   c Approximately 38.5 mg
   d $c = 100$
   e The children's dose will not exceed 100 mg.

7 a $128.67
   b $C(n) = \dfrac{550 + 92n}{n}$ where n = number of years and C represents the annual cost.
   c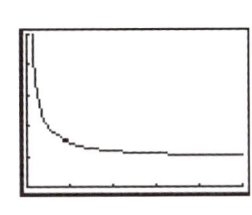

   d The vertical asymptote is $n = 0$ and the horizontal asymptote is $C = 92$.
   e The cost will never go below $92
   $C_2(n) = \dfrac{1200 + 92 \times 20}{20} = \$152$

   f No, as more expensive fridge is still more expensive in the long-run.

Review exercise non-GDC

1 i a, ii d, iii c, iv e, v b, vi f.

2 a i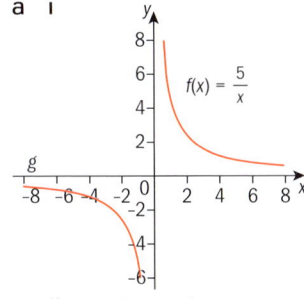

      ii $x = 0, y = 0$
      iii Domain $x \in \mathbb{R}, x \neq 0$
          Range $y \in \mathbb{R}, y \neq 0$

   b i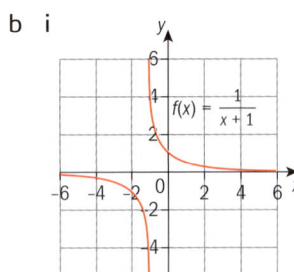

      ii $x = -1, y = 0$
      iii Domain $x \in \mathbb{R}, x \neq -1$
          Range $y \in \mathbb{R}, y \neq 0$

   c i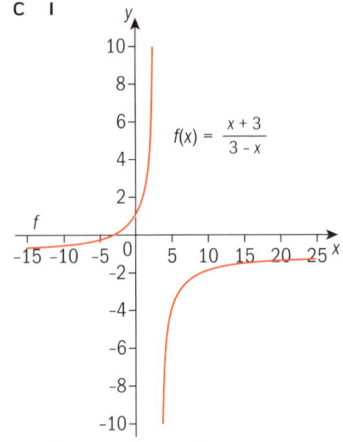

      ii $x = 3, y = -1$
      iii Domain $x \in \mathbb{R}, x \neq 3$
          Range $y \in \mathbb{R}, y \neq -1$

3 a $x = -4, y = 0$
       Domain $x \in \mathbb{R}, x \neq -4$
       Range $y \in \mathbb{R}, y \neq 0$

Answers 735

b $x = 0, y = -3$
Domain $x \in \mathbb{R}, x \neq 0$
Range $y \in \mathbb{R}, y \neq -3$

c $x = -6, y = -2$
Domain $x \in \mathbb{R}, x \neq -6$
Range $y \in \mathbb{R}, y \neq -2$

d $x = 1, y = 5$
Domain $x \in \mathbb{R}, x \neq 1$
Range $y \in \mathbb{R}, y \neq 5$

4 a $c = \dfrac{300}{s}$

b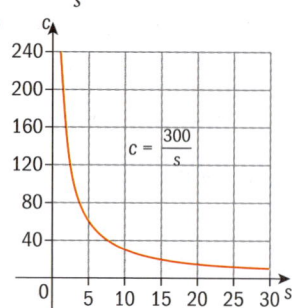

c The domain and range are limited to $\mathbb{R}^+$ and the domain to $\mathbb{Z}^+$

5 a i $y = \dfrac{2}{1} = 2$
   ii $x = -2$
   iii $(-2, 2)$

b $\left(0, -\dfrac{1}{2}\right) \left(\dfrac{1}{2}, 0\right)$

c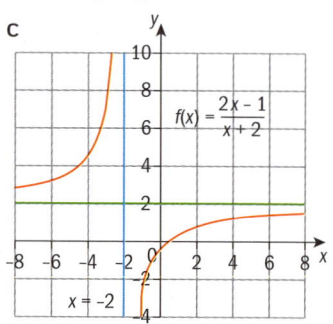

Review exercise GDC

1 a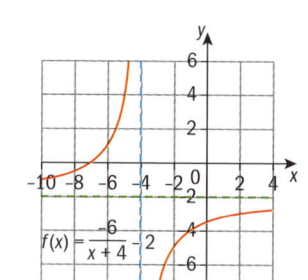

Domain $x \in \mathbb{R}, x \neq 0$
Range $y \in \mathbb{R}, y \neq 5$

b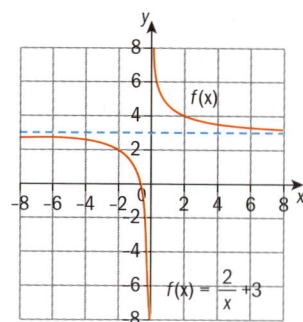

Domain $x \in \mathbb{R}, x \neq 0$
Range $y \in \mathbb{R}, y \neq 3$

c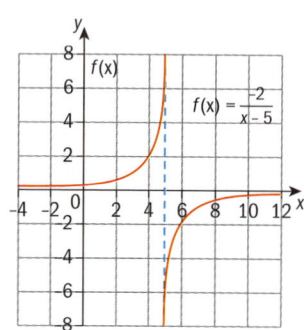

Domain $x \in \mathbb{R}, x \neq 5$
Range $y \in \mathbb{R}, y \neq 0$

d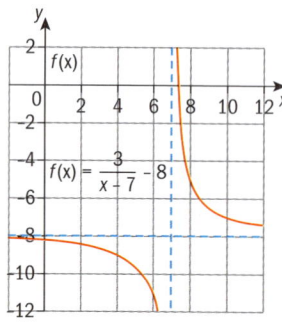

Domain $x \in \mathbb{R}, x \neq 7$
Range $y \in \mathbb{R}, y \neq -8$

e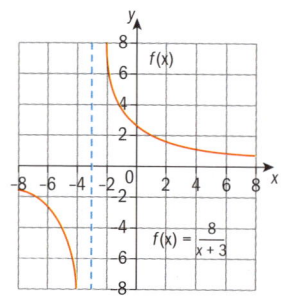

Domain $x \in \mathbb{R}, x \neq -3$
Range $y \in \mathbb{R}, y \neq 0$

f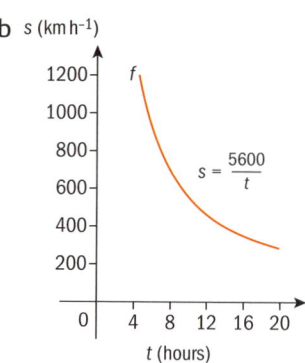

Domain $x \in \mathbb{R}, x \neq -4$
Range $y \in \mathbb{R}, y \neq -2$

2 a Using the equation

Speed $= \dfrac{\text{distance}}{\text{time}}$

distance $= 5600$, $s = \dfrac{5600}{t}$

b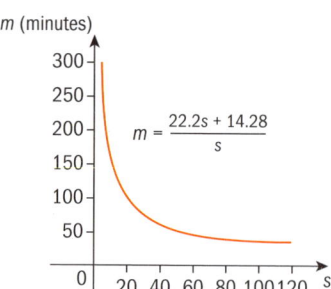

c $560 \, \text{km h}^{-1}$

3 a

m (minutes)

$m = \dfrac{22.2s + 14.28}{s}$

b i 165 min
   ii 57.9 min
   iii 36.5 min

c $m = 22.2$

d The number of minutes that can be spent in direct sunlight without skin damage on a day when $s = 1$.

4 a

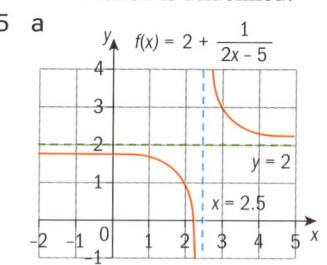

b i 187 500 Thai baht
   ii 750 000 Thai baht
   iii 6 750 000 Thai baht.

c No. When $m = 100$, the function is undefined.

5 a

b i $x = \dfrac{5}{2}, y = 2$
   ii 2.25
   iii 1.8

Chapter 6

Skills check

1 a -6 **b** $-3, 5$ **c** 5

2 a $k = \dfrac{15 - 3m}{4}$ **b** $k = \dfrac{4}{p}$

3 a 108 **b** -12.22

4 a 5 **b** 16 **c** $-\dfrac{3}{32}$

Investigation - saving money

a

Week number	Weekly savings	Total savings
1	20	20
2	25	45
3	30	75
4	35	110
5	40	150
6	45	195
7	50	345
8	55	300

b Savings in 10th week: $65;
Savings in 17th week: $100

c Total saved in 1st year (52 weeks): $7670

d $1000 saved after 17 weeks.

e $M = 20 + 5(n-1)$ or $M = 15 + 5n$

f $T = \dfrac{n(35 + 5n)}{2}$ or $T = \dfrac{5n(7+n)}{2}$

Exercise 6A

1 a 19, 23, 27 **b** 16, 32, 64
   c 18, 24, 31 **d** 80, -160, 320
   e $\dfrac{9}{14}, \dfrac{11}{17}, \dfrac{13}{20}$
   f 6.012 34, 6.012 345, 6.012 345 6

2 a 10, 30, 90, 270
   b 3, 7, 15, 31
   c $\dfrac{3}{4}, \dfrac{1}{2}, \dfrac{1}{3}, \dfrac{2}{9}$
   d x, x^2, x^4, x^8

3 a $u_1 = 2$ and $u_{n+1} = u_n + 2$
   b $u_1 = 1$ and $u_{n+1} = 3u_n$
   c $u_1 = 64$ and $u_{n+1} = \dfrac{u_n}{2}$
   d $u_1 = 7$ and $u_{n+1} = u_n + 5$

4 a 3, 9, 27, 81
   b $-3, -9, -15, -21$
   c 1, 2, 4, 8
   d 1, 4, 27, 256

5 a $u_n = 2n$ **b** $u_n = 3^{n-1}$
   c $u_n = 2^{7-n}$ **d** $u_n = 5n + 2$
   e $u_n = \dfrac{n}{n+1}$ **f** $u_n = nx$

6 a 610
   b $u_1 = 1, u_2 = 1$, and $u_{n+1} = u_n + u_{n-1}$

Exercise 6B

1 a i $u_{15} = 45$ ii $u_n = 3n$
   b i $u_{15} = 235$
       ii $u_n = 15n + 10$
   c i $u_{15} = 106$
       ii $u_n = 5n + 31$
   d i $u_{15} = -82$
       ii $u_n = 113 - 13n$
   e i $u_{15} = 14$
       ii $u_n = 0.6n + 5$
   f i $u_{15} = x + 14a$
       ii $u_n = x + an - a$

2 a 51 **b** 169 **c** 37
   d 15 **e** 27 **f** 10

Exercise 6C

1 $d = 0.9$
2 $d = -3, u_1 = 64$
3 5.5
4 8

Exercise 6D

1 a $r = \dfrac{1}{2}, u_7 = \dfrac{1}{4}$
   b $r = -3, u_7 = -2916$
   c $r = 10, u_7 = 1\,000\,000$
   d $r = 0.4, u_7 = 0.1024$
   e $r = 3x, u_7 = 1458x^6$
   f $r = \dfrac{b}{a}, u_7 = ab^7$

Exercise 6E

1 $r = 0.4, u_1 = 125$
2 $r = -2, u_1 = -4.5$
3 a $n = 12$ **b** $n = 9$
   c $n = 7$ **d** $n = 33$
4 $r = \pm 4, u_2 = \pm 36$
5 $p = \pm 27$
6 $x = 8$

Exercise 6F

1 a $\sum_{n=1}^{8} n$ **b** $\sum_{n=3}^{7} n^2$
   c $\sum_{n=1}^{6}(29 - 2n)$ **d** $\sum_{n=1}^{6} 240(0.5^{n-1})$
   e $\sum_{a=5}^{10} an$ **f** $\sum_{n=1}^{18}(3n+1)$
   g $\sum_{n=1}^{11} 3^{n-1}$ **h** $\sum_{n=1}^{5} na^n$

2 a $4 + 7 + 10 + 13 + 16 + 19 + 22 + 25$
   b $4 + 16 + 64 + 256 + 1024$
   c $40 + 80 + 160 + 320 + 640$
   d $x^5 + x^6 + x^7 + x^8 + x^9 + x^{10} + x^{11}$

3 a 315 **b** 363
   c 140 **d** 315

Exercise 6G

1 234
2 108
3 594
4 $40x + 152$

5 a $n = 24$ b 1776
6 2292

Exercise 6H

1 3
2 a $3n^2 - 2n$ b 17
3 a $1.75n^2 - 31.75n$ b 21
4 a 1600   b 12 600
5 a $n = 24$,   b $S_{24} = 1776$
6 $d = 2.5$, $S_{20} = 575$

Exercise 6I

1 a 132 860   b 1228.5
   c 42.656 25
   d $4095x + 4095$
2 a 435 848 050
   b ≈ 11819.58
   c $-1 048 575$
   d $\log(a^{1048575})$
3 a i 9   ii 76 684
   b i 6   ii 3685.5
   c i 8   ii 1.626 5375
   d i 11   ii 885.73

Exercise 6J

1 a 6   b 5
   c 19 d 6
2 $r = 3$, $S_{10} = \dfrac{59048}{15}$
3 $r = 3$
4 a 1.5   b 21
5 2059
6 3

Investigation – converging series

1 i a $r = \dfrac{1}{2}$
      b $r = \dfrac{2}{5}$
      c $r = \dfrac{-1}{4}$
   ii Inspect values on GDC
2 a The values are approaching 4 as $n \to \infty$
   b The values are approaching 125 as $n \to \infty$
   c The values are approaching 192 as $n \to \infty$
3 Results like $1 - \left(\dfrac{1}{2}\right)^{50}$ are beyond the limit of the display.

Exercise 6K

1 $|r| < 1$
2 a $S_4 = 213.\overline{3}$, $S_7 \approx 215.9$, and $S_\infty = 216$
   b $S_4 = 1476$, $S_7 = 1975.712$, and $S_\infty = 2500$
   c $S_4 = 88.88$, $S_7 = 88.888\,88$, and $S_\infty = 88.\overline{8}$
   d $S_4 = 10.8\overline{3}$, $S_7 \approx 12.71$, and $S_\infty = 13.5$
3 $13.\overline{4}$
4 192
5 16 or 48
6 150
7 4118

Exercise 6L

1 -20
2 a 26.25 cm   b 119
3 a $3984.62   b $4025.81
   c $4035.36
4 42
5 18
6 232
7 ≈ 19.6 years
8 a 1, 8, 21   b 1, 7, 13
   c $6n - 5$
9 a 4, 12, 28   b 4, 8, 16
   c $4(2^{n-1})$
10 ≈ 86 months
11 About $16.30

Exercise 6M

1 10
2 28
3 35
4 84
5 15
6 120

Investigation – patterns in polynomials

1 $a + b$
2 $a^2 + 2ab + b^2$
3 $a^3 + 3a^2b + 3ab^2 + b^3$
4 $a^4 + 4a^3b + 6a^2b^2 + 4ab^3 + b^4$
5 $a^5 + 5a^4b + 10a^3b^2 + 10a^2b^3 + 5ab^4 + b^5$

6 $a^6 + 6a^5b + 15a^4b^2 + 20a^3b^3 + 15a^2b^4 + 6ab^5 + b^6$

The coefficients are from Pascal's triangle.

$(a + b)^7 = a^7 + 7a^6b + 21a^5b^2 + 35a^4b^3 + 35a^3b^4 + 21a^2b^5 + 7ab^6 + b^7$

Exercise 6N

1 $y^5 + 15y^4 + 90y^3 + 270y^2 + 405y + 243$
2 $16b^4 - 32b^3 + 24b^2 - 8b + 1$
3 $729a^6 + 2916a^5 + 4860a^4 + 4320a^3 + 2160a^2 + 576a + 64$
4 $x^6 + 6x^3 + 12 + \dfrac{8}{x^3}$
5 $x^8 + 8x^7y + 28x^6y^2 + 56x^5y^3 + 70x^4y^4 + 56x^3y^5 + 28x^2y^6 + 8xy^7 + y^8$
6 $81a^4 - 216a^3b + 216a^2b^2 - 96ab^3 + 16b^4$
7 $243c^5 + \dfrac{810c^4}{d} + \dfrac{1080c^3}{d^2} + \dfrac{720c^2}{d^3} + \dfrac{240c}{d^4} + \dfrac{32}{d^5}$
8 $64x^6 + \dfrac{24x^4}{y} + \dfrac{3x^2}{y^2} + \dfrac{1}{8y^3}$

Exercise 6O

1 $336x^5$
2 $-1280y^4$
3 $4860a^2b^4$
4 -512
5 2
6 ± 4
7 17 920
8 4860
9 8
10 7

Review exercise non-GDC

1 a 4   b 283   c 25
2 a $\dfrac{1}{4}$ b 1   c $\dfrac{256}{3}$
3 a 4   b 5
4 a 30   b 262
5 120
6 a $\dfrac{1}{4}$ b $\dfrac{3200}{3}$
7 ± 4

8 $720x^3$
9 a 17   b 323

Review exercise GDC

1 a 3   b 52
2 a 96   b 32
3 a $u_1 = 7, d = 2$ b 720
4 a 2   b 11
5 18
6 $u_1 = 5, r = -3$
7 $\dfrac{-945x^4}{16}$
8 $\dfrac{1}{4}$
9 a ≈ 5.47 million   b 2056

Chapter 7
Skills check

1 a $3x(3x^3 - 5x^2 + 1)$
   b $(2x - 3)(2x + 3)$
   c $(x - 3)(x - 2)$
   d $(2x + 1)(x - 5)$
2 a $x^3 + 6x^2 + 12x + 8$
   b $81x^4 - 108x^3 + 54x^2 - 12x + 1$
   c $8x^3 + 36x^2y + 54xy^2 + 27y^3$
3 a x^{-6} b $4x^{-3}$
   c $5x^{\frac{1}{2}}$ d $x^{\frac{5}{7}}$
   e $7x^{-\frac{3}{2}}$

Investigation - creating a sequence

Round number	Portion of the paper you have at the end of the round Fraction	Decimal (3 sf)
1	$\dfrac{1}{3}$	0.333
2	$\dfrac{4}{9}$	0.444
3	$\dfrac{13}{27}$	0.481
4	$\dfrac{40}{81}$	0.494
5	$\dfrac{121}{243}$	0.498
6	$\dfrac{364}{729}$	0.499

1 The portion gets closer to $\dfrac{1}{2}$.
2 The portion gets closer and closer to $\dfrac{1}{2}$, yet never reaches $\dfrac{1}{2}$.

Exercise 7A

1 Divergent
2 Convergent; 3.5
3 Convergent; 0
4 Convergent; 0.75
5 Divergent

Exercise 7B

1 10
2 1
3 1
4 Does not exist
5 4
6 Does not exist

Investigation - secant and tangent lines

1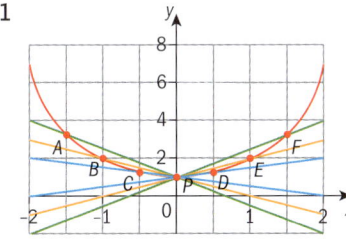

2

Point	Coordinates	Line	Gradient or slope
P	(0, 1)	–	–
A	(−1.5, 3.25)	AP	−1.5
B	(−1, 2)	BP	−1
C	(−0.5, 1.25)	CP	−0.5
D	(0.5, 1.25)	DP	0.5
E	(1, 2)	EP	1
F	(1.5, 3.25)	FP	1.5

3 0
4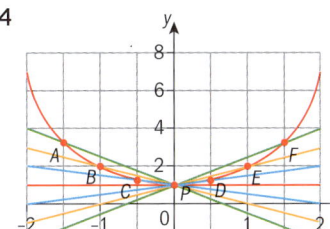

Exercise 7C

1 $\dfrac{[3(x+h)+4]-(3x+4)}{h} = 3$

2 $\dfrac{[2(x+h)^2-1]-(2x^2-1)}{h} = 4x+2h$

3 $\dfrac{[(x+h)^2+2(x+h)+3]-(x^2+2x+3)}{h}$
   $= 2x+h+2$

Exercise 7D

1 2; $m = 2$
2 $6x + 2$; $m = -16$
3 $2x - 1$; $m = 1$

Investigation - the derivative of $f(x) = x^n$

1 $f(x) = x^2$
   $f'(x) = \lim_{h \to 0} \dfrac{(x+h)^2 - x^2}{h}$
   $= \lim_{h \to 0}(2x+h)$
   $= 2x$

   $f(x) = x^3$
   $f'(x) = \lim_{h \to 0} \dfrac{(x+h)^3 - x^3}{h}$
   $= \lim_{h \to 0}(3x^2 + 3xh + h^2)$
   $= 3x^2$

   $f(x) = x^4$
   $f'(x) = \lim_{h \to 0} \dfrac{(x+h)^4 - x^4}{h}$
   $= \lim_{h \to 0}(4x^3 + 6x^2h + 4xh^2 + h^3)$
   $= 4x^3$

2 To find the derivative of $f(x) = x^n$, multiply x by the exponent n and subtract one from the exponent to get the new exponent. If $f(x) = x^n$, then $f'(x) = nx^{n-1}$

3 Prediction: $f'(x) = 5x^4$
   $f(x) = x^5$
   $f'(x) = \lim_{h \to 0} \dfrac{(x+h)^5 - x^5}{h}$
   $= \lim_{h \to 0}(5x^4 + 10x^3h + 10x^2h^2 + 5xh^3 + h^4)$
   $= 5x^4$

Exercise 7E

1. $5x^4$
2. $8x^7$
3. $-\dfrac{4}{x^5}$
4. $\dfrac{1}{3x^{\frac{2}{3}}}$ or $\dfrac{1}{3\sqrt[3]{x^2}}$
5. $-\dfrac{1}{2x^{\frac{3}{2}}}$ or $-\dfrac{1}{2\sqrt{x^3}}$
6. $\dfrac{3}{5x^{\frac{2}{5}}}$ or $\dfrac{3}{5\sqrt[5]{x^2}}$

Exercise 7F

1. $-\dfrac{16}{x^9}$
2. 0
3. $3x^2 + \dfrac{6}{x^3}$
4. $5\pi x^4$
5. $2x - 8$
6. $\dfrac{1}{2x^{\frac{1}{2}}} - \dfrac{4}{3x^{\frac{2}{3}}}$
7. $-\dfrac{3}{2x^3}$
8. $-\dfrac{3}{8x^3}$
9. $-4x^3$
10. $\dfrac{5}{6x^{\frac{1}{6}}} + \dfrac{3}{4x^{\frac{1}{4}}}$
11. $12x^3 - 4x$
12. $4x + 3$
13. $\dfrac{2}{3x^{\frac{1}{3}}} + \dfrac{2}{3x^{\frac{2}{3}}}$
14. $6x^2 - 12x$
15. $3x^2 + 4x - 3$

Exercise 7G

1. $y + 3 = 2(x - 3)$;
   $y + 3 = -\dfrac{1}{2}(x - 3)$

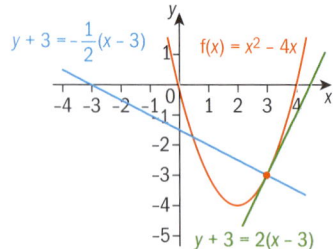

2. a $y - 4 = -4(x + 3)$
   b $y - 6 = 1(x - 1)$
   c $y - 5 = \dfrac{1}{3}(x - 3)$
   d $y - 9 = -\dfrac{15}{4}(x - 1)$

3. a $y - 3 = -\dfrac{1}{7}(x - 2)$
   b $y + 5 = \dfrac{1}{6}(x + 1)$
   c $y - 25 = -\dfrac{1}{20}(x - 2)$
   d $y + 2 = -\dfrac{3}{26}(x - 1)$

4. $x = 1; x = -1$
5. 5

Investigation - the derivatives of e^x and $\ln x$

1. Conjecture: $f'(x) = e^x$
2. Conjecture: $f'(x) = \dfrac{1}{x}$

Exercise 7H

1. $\dfrac{4}{x}$
2. $e^x + \dfrac{1}{2x^{\frac{1}{2}}}$
3. $12x^3 + \dfrac{1}{x}$
4. $8x + 3$
5. $2e^x + \dfrac{1}{x}$
6. $5e^x + 4$
7. $y - 5 = 12(x - \ln 3)$
8. $y - 9 = \dfrac{1}{6}(x + 3)$
9. $y - 1 = \dfrac{1}{e}(x - e)$
10. $y - 7 = -\dfrac{1}{9}(x - 2)$
11. $2e^3$; 40.2
12. $\dfrac{5}{24}$; 0.208

Investigation - the derivative of the product of two functions

1. 11
2. $f'(x) = 11x^{10}$
3. $u'(x) = 4x^3$; $v'(x) = 7x^6$
4. $u'(x) \cdot v'(x) = 28x^9$

5. No
6. $f'(x) = x^4 \cdot 7x^6 + x^7 \cdot 4x^3 = 11x^{10}$
7. $f'(x) = u(x) \cdot v'(x) + v(x) \cdot u'(x)$
8. $f(x) = (3x + 1)(x^2 - 1)$
   $= 3x^3 + x^2 - 3x - 1$
   $f'(x) = 9x^2 + 2x - 3$
   $f(x) = (3x + 1)(x^2 - 1)$
   $f'(x) = (3x + 1)(2x) + (x^2 - 1)(3)$
   $= 6x^2 + 2x + 3x^2 - 3$
   $= 9x^2 + 2x - 3$
   This supports the conjecture.

Exercise 7I

1. $\dfrac{x^2 - 8x}{(x - 4)^2}$
2. $10x^4 + 4x^3 + 9x^2 + 2x + 1$
3. $\dfrac{1 - \ln x}{x^2}$
4. $\dfrac{e^x}{x} + e^x \ln x$
5. $\dfrac{6}{(x + 4)^2}$
6. $\dfrac{e^x}{(e^x + 1)^2}$
7. $e^x(5x^3 + 15x^2 + 4x + 4)$
8. $\dfrac{x^4 - 6x^2 - 2x}{(x^3 + 1)^2}$
9. -1
10. $y - 2 = -\dfrac{1}{2}(x - 3)$; $y = -\dfrac{1}{2}(x + 1)$

Exercise 7J

1. $2x^2 - \dfrac{5}{3}$
2. $4x^3$
3. $4xe^x + 2x^2 e^x$
4. $\dfrac{2xe^x - 4e^x}{x^3}$
5. $3x^2 - \dfrac{16}{5x^{\frac{9}{5}}}$
6. $\dfrac{2x - x^2}{e^x}$
7. $\dfrac{2x}{(x^2 + 1)^2}$
8. $3 + 3\ln x$
9. $1 - \dfrac{1}{x^2}$

10 $\dfrac{5x^{\frac{3}{2}}}{2}+\dfrac{1}{2x^{\frac{1}{2}}}$

11 $-\dfrac{x+1}{(x-1)^3}$

12 $10x^4 + 12x^3 - 3x^2 - 18x - 15$

13 $y=-\dfrac{1}{e}(x-1)$

14 $y = x - 1$

15 $-9n + 3.5$

16 $4\pi r^2$

17 7

18 4

Investigation - finding the derivative of a composite function

1 a $f(x) = (2 - x)^3$
       $= 8 - 12x + 6x^2 - x^3$
       $f'(x) = -12 + 12x - 3x^2$
    b $f'(x) = 3(2 - x)^2 \cdot (-1)$

2 a $f(x) = (2x + 1)^2$
       $= 4x^2 + 4x + 1$
       $f'(x) = 8x + 4$
    b $f'(x) = 2(2x1)\cdot 2$

3 a $f(x) = (3x^2 + 1)^2$
       $= 9x^4 + 6x^2 + 1$
       $f'(x) = 36x^3 + 12x$
    b $f'(x) = 2(3x^2 + 1) \cdot (6x)$

4 The derivative of a composite function is the derivative of the outside function with respect to the inside function multiplied by the derivative of the inside function.

5 $f(x) = (x^4 + x^2)^3$
   $= x^{12} + 3x^{10} + 3x^8 + x^6$
   $f'(x) = 12x^{11} + 30x^9 + 24x^7 + 6x^5$
   $f'(x) = 3(x^4 + x^2)^2 \cdot (4x^3 + 2x)$
   $= 3(x^8 + 2x^6 + x^4)(4x^3 + 2x)$
   $= 3(4x^{11} + 10x^9 + 8x^7 + 2x^5)$
   $= 12x^{11} + 30x^9 + 24x^7 + 6x^5$

Exercise 7K

1 x^5; $3x^4 + 2x$;
   $5(3x^4 + 2x)^4 (12x^3 + 2)$

2 $4x^3$; $2x^2 + 3x + 1$;
   $12(2x^2 + 3x + 1)^2 (4x + 3)$

3 $\ln x$; $3x^5$; $\dfrac{5}{x}$

4 $\sqrt[3]{x}$; $2x+3$; $\dfrac{2}{3(2x+3)^{\frac{2}{3}}}$

5 e^x; $4x$; $4e^{4x}$

6 x^3; $\ln x$; $\dfrac{3(\ln x)^2}{x}$

7 $x^{\frac{2}{3}}$; $9x+2$; $\dfrac{6}{(9x+2)^{\frac{1}{3}}}$

8 $\sqrt[4]{x}$; $2x^2+3$; $\dfrac{x}{(2x^2+3)^{\frac{3}{4}}}$

9 $5x^4$; $x^3 + 3x$;
   $20(x^3 + 3x)^3 (3x^2 + 3)$

10 e^x; $4x^3$; $12x^2 e^{4x^3}$

Exercise 7L

1 $8x^2(2x - 3)^3 + 2x(2x - 3)^4$
   or $6x(2x - 1)(2x - 3)^3$

2 $\dfrac{-x^2 + 2x}{e^x}$

3 $\dfrac{-8x}{(x^2+3)^2}$

4 $\dfrac{-x}{(2x+1)^{\frac{3}{2}}} + \dfrac{1}{(2x+1)^{\frac{1}{2}}}$ or $\dfrac{x+1}{(2x+1)^{\frac{3}{2}}}$

5 $\dfrac{e^{2x} - e^{-2x}}{\left(e^{2x} + e^{-2x}\right)^{\frac{1}{2}}}$

6 $\dfrac{6x^2}{2x^3 - 1}$

7 $\dfrac{1}{x \ln x}$

8 $\dfrac{-2(e^x - e^{-x})}{(e^x + e^{-x})^2}$ or $\dfrac{-2e^x(e^{2x} - 1)}{(e^{2x} + 1)^2}$

9 $\dfrac{-2x+3}{(x^2 - 3x - 2)^2}$

10 $x^5(x^2+3)^{-\frac{1}{2}} + 4x^3(x^2+3)^{\frac{1}{2}}$
    or $\dfrac{5x^5 + 12x^3}{(x^2+3)^{\frac{1}{2}}}$

11 a $(2x-2)e^{x^2-2x}$
    b 2
    c $y - 1 = 2(x - 2)$

12 $e^{-\frac{1}{3}}$

13 $h'(x) = \dfrac{6}{(1-2x)^4}$. Since $6 > 0$ and $(1 - 2x)^4 > 0$ for all x where h is defined, the gradient of h is always positive.

14 a 6
    b 8

Exercise 7M

1 $\dfrac{3}{\sqrt{x}}$

2 $180x^2 + 24x$

3 $3e^{-3n}(6n + 5)$

4 $\dfrac{8}{x^3}$

5 $\dfrac{-3}{x^2}$

6 1

7 equals 0

8 $\dfrac{dy}{dx} = e^x - e^{-x}$
   $\dfrac{d^2 y}{dx^2} = e^x + e^{-x}$
   $\dfrac{d^3 y}{dx^3} = e^x - e^{-x}$
   $\dfrac{d^4 y}{dx^4} = e^x + e^{-x}$

   When n is odd
   $\dfrac{d^n y}{dx^n} = e^x - e^{-x}$ and when n is even
   $\dfrac{d^n y}{dx^n} = e^x + e^{-x}$.

9 $\dfrac{dy}{dx} = \dfrac{-1}{x^2}$
   $\dfrac{d^2 y}{dx^2} = \dfrac{2}{x^3}$
   $\dfrac{d^3 y}{dx^3} = \dfrac{-6}{x^4}$
   $\dfrac{d^4 y}{dx^4} = \dfrac{24}{x^5}$
   $\dfrac{d^n y}{dx^n} = \dfrac{(-1)^n n!}{x^{n+1}}$

10 $\dfrac{-18}{25 x^{\frac{8}{5}}}$

Exercise 7N

1. a 1.4 m; 21 m
   b $9.8\,\text{m s}^{-1}$
   c $9.8\,\text{m s}^{-1}$; $0\,\text{m s}^{-1}$; $-9.8\,\text{m s}^{-1}$; The ball is moving upward at 1 s, at rest at 2 s and downward at 3 s.

2. a 4000 litres; 1778 litres
   b −111 litres/min; During the time interval 0 to 20 minutes, water is being pumped out of the tank at an average rate of 111 litres per minute.
   c −89 litres/min; at 20 minutes, water is being pumped out of the tank at an average rate of 89 litres per minute.
   d $V'(t)$ is negative for $0 \le t < 40$ minutes, which means water is flowing out of the tank during this time interval. Therefore the amount of water in the tank is never increasing from $t = 0$ minutes to $t = 40$ minutes.

3. a 112 bacteria/day
   b $P'(t) = 25e^{0.25t}$
   c 305 bacteria/day; on day 10 the number of bacteria are increasing at a rate of 305 bacteria/day.

4. a 20.25 dollars/unit; 20.05 dollars/unit
   b $C'(n) = 0.1n + 10$
   c 20 dollars/unit; It costs 20 dollars per unit to produce units after the 100th unit.

Exercise 7O

1. a 0 cm; $9\,\text{cm s}^{-1}$
   b 1 s and 3 s
   c

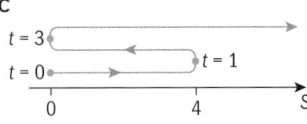

2. a 4 ft
   b $s(2) = -16(2)^2 + 40(2) + 4 = -64 + 80 + 4 = 20\,\text{ft}$
   c i $-16t^2 + 40t + 4 = 20$
      ii $t = \frac{1}{2}, 2\,\text{s}$
   d i $\frac{ds}{dt} = -32t + 40$
      ii $40\,\text{ft s}^{-1}$
      iii $\frac{5}{4}\,\text{s}$
      iv 29 ft

3. a $v(t) = s'(t)$
      $= \frac{e^t(1) - t(e^t)}{(e^t)^2}$
      $= \frac{e^t(1-t)}{e^{2t}}$
      $v(t) = \frac{1-t}{e^t}$
   b 1 second

Investigation – velocity, acceleration and speed

1. a Let acceleration be $2\,\text{m s}^{-2}$.

Time (s)	Velocity (m s⁻¹)	Speed (m s⁻¹)
0	10	10
1	12	12
2	14	14
3	16	16
4	18	18

   b Let acceleration be $-2\,\text{m s}^{-2}$.

Time (s)	Velocity (m s⁻¹)	Speed (m s⁻¹)
0	10	10
1	8	8
2	6	6
3	4	4
4	2	2

   c Let acceleration be $-2\,\text{m s}^{-2}$.

Time (s)	Velocity (m s⁻¹)	Speed (m s⁻¹)
0	−10	10
1	−12	12
2	−14	14
3	−16	16
4	−18	18

   d Let acceleration be $2\,\text{m s}^{-2}$.

Time (s)	Velocity (m s⁻¹)	Speed (m s⁻¹)
0	−10	10
1	−8	8
2	−6	6
3	−4	4
4	−2	2

2. a Speeding up
   b Slowing down
   c Speeding up
   d Slowing down

3. a Speeding up
   b Slowing down

Exercise 7P

1. a $v(t) = 8t^3 - 12t,\ t \ge 0$
      $a(t) = 24t^2 - 12,\ t \ge 0$
   b $84\,\text{cm s}^{-2}$; Velocity is increasing $84\,\text{cm s}^{-1}$ at time 2 seconds.
   c $v(t) = 0$ when $t = 0$ and 1.22 s; $a(t) = 0$ when $t = 0.707\,\text{s}$; speeding up for $0 < t < 0.707$ s and $t > 1.22$; slowing down for $0.707 < t < 1.22$

2. a $v(t) = -3t^2 + 24t - 36,\ 0 \le t \le 8$
      $a(t) = -6t + 24,\ 0 \le t \le 8$
   b $s(0) = 20\,\text{m}$;
      $v(0) = -36\,\text{m s}^{-1}$;
      $a(0) = 24\,\text{m s}^{-1}$;
   c $t = 2, 6\,\text{s}$; moving left on $0 \le t \le 2$ and $6 \le t \le 8$, moving right $2 \le t \le 6$
   d $t = 4$ s; speeding up on $2 \le t \le 4$ and $6 \le t \le 8$, slowing down on $0 \le t \le 2$ and $4 \le t \le 6$

3. a $v(t) = -9.8t + 4.9$
      $a(t) = -9.8$
   b 2.01 s
   c 0.5 s; 11.2 m
   d $v(0.3) = 1.96 > 0$ and $a(0.3) = -9.8 < 0$. Since the signs of $v(0.3)$ and $a(0.3)$ are different the particle is slowing down at 0.3 seconds.

4 a i $v(t) = \dfrac{1}{2}t - \dfrac{1}{t+1}$
   ii 1 second
   b i $a(t) = \dfrac{1}{2} + \dfrac{1}{(t+1)^2}$
   ii Since $\dfrac{1}{2} > 0$ and
       $\dfrac{1}{(t+1)^2} > 0$,
       $a(t) = \dfrac{1}{2} + \dfrac{1}{(t+1)^2} > 0$
       for $t \geq 0$ and so velocity is never decreasing.

Exercise 7Q

1 Decreasing $(-\infty, \infty)$
2 Increasing $(-\infty, 2)$; decreasing $(2, \infty)$
3 Increasing $(-1, 1)$; decreasing $(-\infty, -1)$ and $(1, \infty)$
4 Decreasing $(-\infty, 0)$; increasing $(0, \infty)$
5 Increasing $(-1, 0)$ and $(1, \infty)$; decreasing $(-\infty, -1)$ and $(0, 1)$
6 Decreasing $(-\infty, 3)$ and $(3, \infty)$
7 Decreasing $(0, \infty)$
8 Increasing $(-3, \infty)$; decreasing $(-\infty, -3)$
9 Increasing $(-\infty, -\sqrt{3})$ and $(\sqrt{3}, \infty)$; decreasing $(-\sqrt{3}, -1)$, $(-1, 1)$ and $(1, \sqrt{3})$
10 Increasing $(-\infty, -2)$ and $(4, \infty)$; decreasing $(-2, 4)$

Exercise 7R

1 relative minimum $(1, -5)$
2 relative minimum $(2, -21)$; relative maximum $(-2, 11)$
3 no relative extrema
4 relative minimum $(-1, -1)$ and $(1, -1)$; relative maximum $(0, 0)$
5 relative minimum $\left(-\dfrac{3}{4}, -\dfrac{2187}{256}\right)$
6 relative minimum $(0, 0)$; relative maximum $\left(2, \dfrac{4}{e^2}\right)$
7 no relative extrema
8 relative minimum $(1, 0)$; relative maximum $(-3, -8)$

Exercise 7S

1 concave up $(-\infty, \infty)$
2 concave up $(0, 2)$; concave down $(-\infty, 0)$ and $(2, \infty)$; inflexion points $(0, 0)$ and $(2, 16)$
3 concave up $(2, \infty)$; concave down $(-\infty, 2)$; inflexion point $(2, 8)$
4 concave up $(-\infty, \infty)$
5 concave up $(-2, \infty)$; concave down $(-\infty, -2)$; inflexion point $\left(-2, -\dfrac{4}{e^2}\right)$
6 concave up $\left(-\infty, -\dfrac{\sqrt{3}}{3}\right)$ and $\left(\dfrac{\sqrt{3}}{3}, \infty\right)$; concave down $\left(-\dfrac{\sqrt{3}}{3}, \dfrac{\sqrt{3}}{3}\right)$; inflexion points $\left(-\dfrac{\sqrt{3}}{3}, \dfrac{3}{4}\right)$ and $\left(\dfrac{\sqrt{3}}{3}, \dfrac{3}{4}\right)$

7 a $f'(x) = \dfrac{-48x}{(x^2+12)^2}$

   $f''(x)$
   $= \dfrac{(x^2+12)^2(-48) - (-48x)[2(x^2+12)(2x)]}{(x^2+12)^4}$
   $= \dfrac{(x^2+12)^2(-48) + 192x^2(x^2+12)}{(x^2+12)^4}$
   $= \dfrac{48(x^2+12)[-(x^2+12) + 4x^2]}{(x^2+12)^4}$
   $= \dfrac{48(x^2+12)(3x^2-12)}{(x^2+12)^4}$
   $= \dfrac{144(x^2+12)(x^2-4)}{(x^2+12)^4}$
   $= \dfrac{144(x^2-4)}{(x^2+12)^3}$

   b i relative maximum $(0, 2)$
      ii inflection points
         $\left(-2, \dfrac{3}{2}\right)$ and $\left(2, \dfrac{3}{2}\right)$

8 concave up $(-\infty, -2)$ and $(4, \infty)$; concave down $(-2, 4)$; inflection points at $x = -2, 4$

Exercise 7T

1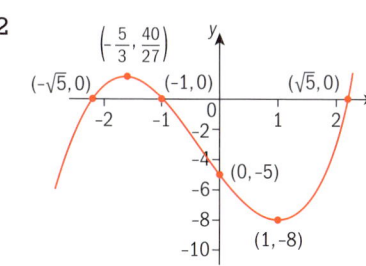

2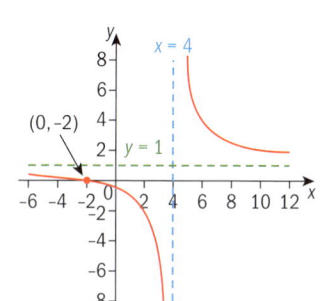

3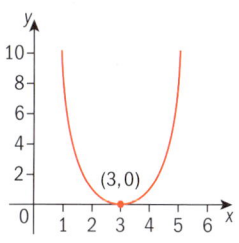

4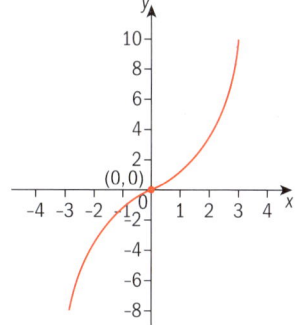

5

6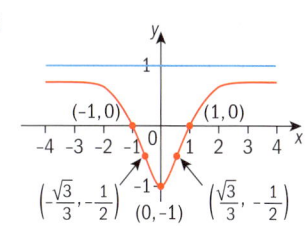

Exercise 7U

1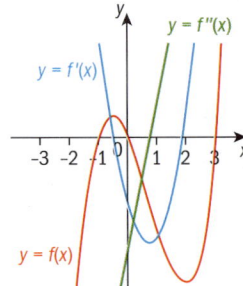

2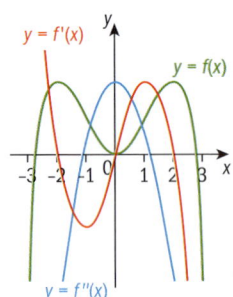

3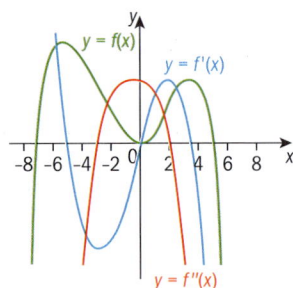

Exercise 7V

1 relative minimum $(3, -75)$
2 relative minimum $(1, 0)$ and $(-1, 0)$; relative maximum $(0, 1)$
3 relative minimum $(3, -27)$
4 relative minimum $\left(-1, -\dfrac{1}{e}\right)$
5 relative minimum $(1, 0)$
6 relative maximum $(0, 1)$

Exercise 7W

1 A – neither; B – relative and absolute minimum; C – absolute maximum
2 A – neither; B – relative minimum; C – relative and absolute maximum; D – absolute minimum
3 absolute maximum 8; absolute minimum -8
4 absolute maximum 16; absolute minimum -9
5 absolute maximum 2; absolute minimum $-\dfrac{5}{2}$

Exercise 7X

1 $\dfrac{79}{4}$ and $\dfrac{1}{4}$
2 100 and 50
3 $x = 50\,\text{ft}; y = \dfrac{200}{3}\,\text{ft}$

Exercise 7Y

1 40 cm by 40 cm by 20 cm
2 3 items
3 22
4 a $r = \dfrac{30 - 3h}{5}$
   b $V(h) = \pi\left(\dfrac{30 - 3h}{5}\right)^2 (h)$ or
     $V(h) = \dfrac{9\pi}{25}(100h - 20h^2 + h^3)$
   c $\dfrac{dV}{dh} = \dfrac{9\pi}{25}(100 - 40h + 3h^2);$
     $\dfrac{d^2V}{dh^2} = \dfrac{9\pi}{25}(-40 + 6h)$
   d $r = 4\,\text{cm}; h = \dfrac{10}{3}\,\text{cm}$
5 a $p(x) = 4\sqrt{x} - 2x^2$
   b $\dfrac{dp}{dx} = \dfrac{2}{x^{\frac{1}{2}}} - 4x; \dfrac{d^2p}{dx^2} = -\dfrac{1}{x^{\frac{3}{2}}} - 4$
   c 0.630 thousand units or 630 units

Review exercise non-GDC

1 a $12x^2 + 6x - 2$
  b $\dfrac{4}{3}x^{\frac{1}{3}}$
  c $-\dfrac{12}{x^5}$
  d $10x^4 - 4x^3 - 3x^2 + 2x - 1$
  e $\dfrac{11}{(x+7)^2}$
  f $4e^{4x}$
  g $12x^2(x^3 + 1)^3$
  h $\dfrac{2}{2x+3}$
  i $\dfrac{1 - 2\ln x}{x^3}$
  j $\dfrac{4}{3}x - \dfrac{1}{3}$
  k $e^x(3x^2 + 6x + 1)$
  l $-\dfrac{6e^x}{(e^x - 3)^2}$
  m $\dfrac{3}{\sqrt{2x - 5}}$
  n $2xe^{2x}(x + 1)$
  o $-\dfrac{1}{x}$

2 a $x^3 + 3x^2h + 3xh^2 + h^3$
  b
  $f'(x)$
  $= \lim_{h \to 0} \dfrac{[2(x+h)^3 - 6(x+h)] - (2x^3 - 6x)}{h}$
  $= \lim_{h \to 0} \dfrac{2x^3 + 6x^2h + 6xh^2 + 2h^3 - 6x - 6h - 2x^3 + 6x}{h}$
  $= \lim_{h \to 0} \dfrac{6x^2h + 6xh^2 + 2h^3 - 6h)}{h}$
  $= \lim_{h \to 0} \dfrac{h(6x^2 + 6xh - 2h^2 - 6)}{h}$
  $= \lim_{h \to 0} (6x^2 + 6xh - 2h^2 - 6)$
  $= 6x^2 - 6$
  c $p = -1; q = 1$
  d $f''(x) = 12x$
  e $(0, \infty)$

3 $y - 4 = -\dfrac{1}{12}(x - 1)$

4 $\left(\dfrac{2\sqrt{3}}{3}, \dfrac{9 - 2\sqrt{3}}{9}\right), \left(\dfrac{-2\sqrt{3}}{3}, \dfrac{9 + 2\sqrt{3}}{9}\right)$

5 a $f''(2) > f(2) > f'(2);$
  b $f''(2) > 0$; since the graph of f is concave up, $f(2) = 0$ and $f'(2) < 0$ since the graph of f is decreasing

6 a i $4x^3 - 12x^2$
     ii $12x^2 - 24x$
  b i $(0, 0), (4, 0)$
    ii $(3, -27)$
    iii $(0, 0), (2, -16)$
  c

7 a $v(t) = 20 - \dfrac{100}{t}$

b $t < 5$

c $v'(t) = a(t) = \dfrac{100}{t^2}$ and since $100 > 0$ and $t^2 > 0$, $v'(t) > 0$. Therefore velocity is always increasing.

Review exercise GDC

1 a Does not exist
b 1
c 8
d Does not exist

2 a i $y = \sqrt{x^2 + 100}$
  ii $z = \sqrt{(30-x)^2 + 625}$ or $\sqrt{x^2 - 60x + 1525}$
  iii $L(x) = \sqrt{x^2 + 100} + \sqrt{x^2 - 60x + 1525}$

b i $\dfrac{dL}{dx} = \dfrac{x}{\sqrt{x^2+100}} + \dfrac{x-30}{\sqrt{x^2-60x+1525}}$

  ii 8.57 ft

Chapter 8
Skills check

1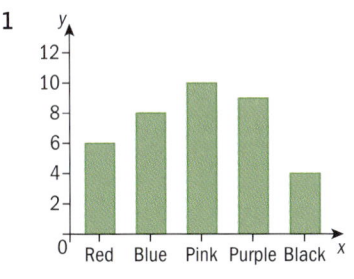

2 a 6.4 **b** 8
  c i 6 **ii** 10 **iii** 11

Exercise 8A
1 a Discrete **b** Continuous
  c Continuous **d** Discrete
2 Discrete

Exercise 8B
1 a Continuous

b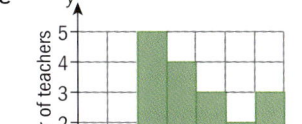

2 a Continuous
  b 17
  c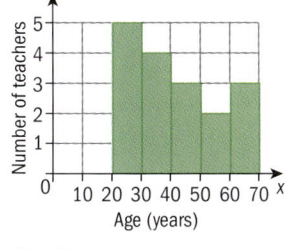

3 a Continuous
  b 96
  c

Mass (kg)	$1 \le w < 2$	$2 \le w < 3$	$3 \le w < 4$	$4 \le w < 5$
Number of chickens	8	24	50	14

4 a Continuous
  b

Time	$5 \le t < 10$	$10 \le t < 15$	$15 \le t < 20$	$20 \le t < 25$	$25 \le t < 30$	$30 \le t < 35$	$35 \le t < 40$	$40 \le t < 45$
f	1	2	4	4	2	2	1	1

  c 5 min

Exercise 8C
1 a 18 **b** 9
  c 18 and 24 **d** 0
  e $\dfrac{1}{2}$ and 2

2 a 1 **b** $170 \le h < 180$

Exercise 8D
1 62.5 km h^{-1}
2 \$1.86
3 a Discrete
  b $5.7\overline{6}$ calls per day
4 a Continuous
  b $90 \le m < 120$
  c 83.4 min per day
5 79
6 91.1 kg
7 255 km
8 568
9 103 points
10 \$315.20

Exercise 8E
1 a 4 **b** 5 **c** 3.5
  d 4 **e** 6
2 11
3 Mode 7, mean 5.25, median 5.5

Investigation: Measures of central tendency

	Data	Mean	Mode	Median
Data set	6, 7, 8, 10, 12, 14, 14, 15, 16, 20	12.2	14	13
Add 4 to each data set	10, 11, 12, 14, 16, 18, 18, 19, 20, 24	16.2	18	17
Multiply the original data set by 2	12, 14, 16, 20, 24, 28, 28, 30, 32, 40	24.4	28	26

a If you add 4 to each data value, you will add 4 to the mean, mode and median.

b If you multiply each data value by 2, you will multiply the mean, mode and median by 2.

Exercise 8F

1 a 95 cm   b 67.5   c 57.5
   d 92.5   e 35

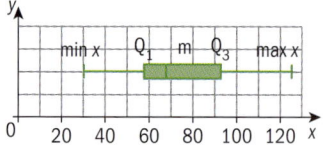

2 a 14   b 79   c 75
   d 82   e 7

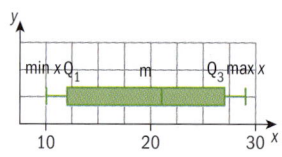

3 a 19   b 21   c 12
   d 27   e 15

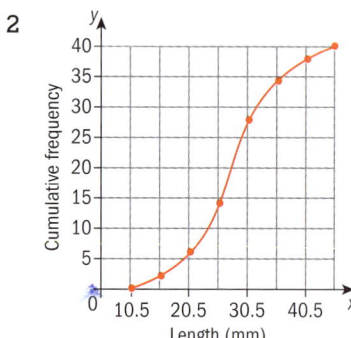

4 a 5   b 8   c 7
   d 10   e 3
5 a iii   b i   c ii

Exercise 8G

1 a 75 cm
   b (77.5 − 72) cm = 5.5 cm
   c The middle 50% of data has a spread of 5.5 cm.

2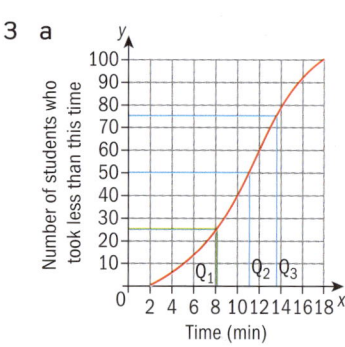

3 a

   i 11 min
   ii (13.6 − 8.2) min = 5.4 min
   b $p = 32$, $q = 8$

4 a

Marks	f	cf
$20 \leq m < 30$	2	2
$30 \leq m < 40$	3	5
$40 \leq m < 50$	5	10
$50 \leq m < 60$	7	17
$60 \leq m < 70$	6	23
$70 \leq m < 80$	4	27
$80 \leq m < 90$	2	29
$90 \leq m < 100$	1	30

b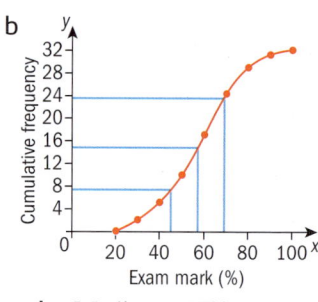

c i Median ≈ 57%
   ii Lower quartile ≈ 45%
       Upper quartile ≈ 69%
   iii Inter-quartile range ≈ 24%

5 a

Distance (d)	f	cf
$0 \leq d < 20$	4	4
$20 \leq d < 40$	9	13
$40 \leq d < 60$	15	28
$60 \leq d < 80$	10	38
$80 \leq d < 100$	2	40

b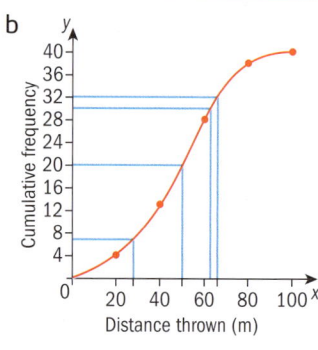

c Qualifying distance ≈ 66 m
d Inter-quartile range ≈ 28 m
e Median ≈ 50 m

6 a i 23 min   ii 16 min
      iii 37 min
   b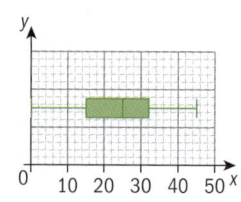

7 a 170 cm
   b 50 flowers between 135 cm and 163 cm
   c 22 flowers, 180 cm   d 110
   e

Exercise 8H

1 a Mean = 18, variance = 129.6, standard deviation = 11.4
   b Mean = 40, variance = 200, standard deviation = 14.1
2 a Variance = 78.5, standard deviation = 8.86
   b Variance = 80.18, standard deviation = 8.95
   c Variance = 449, standard deviation = 21.2
3 1.32
4 Mean = 2.5, standard deviation = 1.24
5 Standard deviation = 14.9
6 a Discrete   b 2.73
   c 1.34   d 23
7 Mean = 42.4, standard deviation = 21.6
8 a 51   b 69.5
   c i 21.8   ii None

Investigation: The effect of adding or multiplying the data set on a standard deviation

a 2.47
b The mean increases by 100 to 103.9.

c 2.47
d The standard deviation remains the same. This is because the standard deviation only measures the spread of the numbers, and that remains constant if the same number is added to each item in the list.
e The mean is doubled.
f 4.94
g The variance will be multiplied by 4 because the variance is the standard deviation squared.

Review exercise non-GDC

1 a 3   b 5   c 5
   d 9
2 a 4.2   b 4   c 4
3 Mean = 27.5 yrs, standard deviation = 0.4 yrs.
   Type A
4 a 52   b 14   c 8
   Type B
   a 52   b 8   c 3
5 a 426   b 72   c 62
6 a

   (cumulative frequency graph)

   b Median ≈ 163
   c IQR ≈ 6
7 a $k = 100 - 96 = 4$
   b i median = 3
      ii IQR = $5 - 1 = 4$
8 Median = 65 °F, IQR = 45 °F

Review exercise GDC

1 Median = 20, IQR = 14
2 a 6.48   b 1.31
3 a 6   b 6   c 5.92

4 a Mean = 2.57, median = 2, mode = 1, standard deviation = 1.68, variance = 2.82
   b Range = 6, lower quartile = 1, IQR = 3
5 a $160 \leq$ Height < 170
   b

Height	f
$140 \leq$ Height < 150	15
$150 \leq$ Height < 160	55
$160 \leq$ Height < 170	90
$170 \leq$ Height < 180	45
$180 \leq$ Height < 190	5

   Mean = 164 cm
6 a i $p = 65$ ii $q = 34$
   b Median = 18
   c Mean = 17.7

Chapter 9

Skills check

1 a $2 + 8 + 18 + 32 + 50$
   b $4 + 7 + 10 + 13 + 16$
   c $g(x_1) + 4g(x_2) + 9g(x_3) + 16g(x_4) + 25g(x_5)$
   d $f(x_1)(\Delta x_1) + f(x_2)(\Delta x_2) + f(x_3)(\Delta x_3)$
2 a 18 mm^2 b 8π cm^2
3 a 160π cm^3 b 42π ft^3

Investigation – antiderivatives of x^n

1

$f(x)$	Antiderivatives of f
x	$\frac{1}{2}x^2 + C$
x^2	$\frac{1}{3}x^3 + C$
x^3	$\frac{1}{4}x^4 + C$
x^4	$\frac{1}{5}x^5 + C$

2 $\frac{1}{n+1}x^{n+1}$

3 $\frac{1}{-3+1}x^{-3+1} = -\frac{1}{2}x^{-2}$ or $-\frac{1}{2x^2}$;

$\frac{d}{dx}\left[-\frac{1}{2}x^{-2}\right] = x^{-3}$

$\frac{1}{\frac{1}{2}+1}x^{\frac{1}{2}+1} = \frac{2}{3}x^{\frac{3}{2}}$; $\frac{d}{dx}\left[\frac{2}{3}x^{\frac{3}{2}}\right] = x^{\frac{1}{2}}$

4 $n = -1$

Exercise 9A

1 $\frac{1}{8}x^8 + C$
2 $\frac{1}{5}x^5 + C$
3 $-\frac{1}{x} + C$
4 $2x^{\frac{1}{2}} + C$
5 $\frac{3}{4}x^{\frac{4}{3}} + C$
6 $\frac{5}{7}x^{\frac{7}{5}} + C$
7 $-\frac{1}{3x^3} + C$
8 $-\frac{1}{11x^{11}} + C$
9 $\frac{3}{4}x^{\frac{4}{3}} + C$
10 $\frac{7}{10}x^{\frac{10}{7}} + C$
11 $\frac{5}{4}x^{\frac{4}{5}} + C$
12 $3x^{\frac{1}{3}} + C$

Exercise 9B

1 $\frac{1}{4}x^4 + C$
2 $-\frac{1}{t} + C$
3 $\frac{5}{9}x^{\frac{9}{5}} + C$
4 $2u + C$
5 $x^3 + x^2 + x + C$
6 $-\frac{2}{x^2} + C$
7 $\frac{1}{3}t^3 + \frac{4}{5}t^{\frac{5}{4}} + C$
8 $\frac{3}{5}x^{\frac{5}{3}} + x + C$
9 $x^5 + 3x^4 + 3x^2 - 2x + C$
10 $t + C$
11 a $3x^2 - \frac{8}{x^3}$
   b $\frac{1}{4}x^4 - \frac{4}{x} + C$
12 a $\frac{6}{x^{\frac{4}{5}}}$ b $25x^{\frac{6}{5}} + C$

Exercise 9C

1. $f(x) = \frac{2}{3}x^6 + 4x^2 + 8$
2. $y = \frac{1}{5}x^5 + \frac{4}{5}x^{\frac{5}{4}} + 9$
3. $s(t) = t^3 - t^2 - 6$
4. 115π cm^3
5. a -5 ms^{-2}
   b $s(t) = 5 + 20t - \frac{5}{2}t^2$

Exercise 9D

1. $2\ln x + C$, $x > 0$
2. $3e^x + C$
3. $\frac{1}{4}\ln t + C$, $t > 0$
4. $\frac{1}{2}x^2 + C$
5. $\frac{4}{3}x^3 + 6x^2 + 9x + C$
6. $\frac{2}{3}x^3 + 3x^2 + 5\ln x + C$, $x > 0$
7. $\frac{1}{3}u^3 + C$
8. $\frac{1}{4}x^4 - x^3 + \frac{3}{2}x^2 - x + C$
9. $\frac{1}{2}(e^x + x) + C$
10. $\frac{2}{5}x^{\frac{5}{2}} + \frac{2}{3}x^{\frac{3}{2}} + 2x^{\frac{1}{2}} + C$

Exercise 9E

1. $\frac{1}{6}(2x + 5)^3 + C$
2. $-\frac{1}{12}(-3x + 5)^4 + C$
3. $2e^{\frac{1}{2}x - 3} + C$
4. $\frac{1}{5}\ln(5x + 4) + C$, $x > -\frac{4}{5}$
5. $-\frac{3}{2}\ln(7 - 2x) + C$, $x > \frac{7}{2}$
6. $2e^{2x+1} + C$
7. $\frac{3}{16}(4x - 3)^8 + C$
8. $\frac{2}{21}(7x + 2)^{\frac{3}{2}} + C$
9. $\frac{1}{4}e^{4x} + \frac{4}{3}\ln(3x - 5) + C$, $x > \frac{5}{3}$
10. $-\frac{1}{12(4x - 5)^2} + C$
11. a $12(4x + 5)^2$
    b $\frac{1}{16}(4x + 5)^4 + C$
12. $s = -\frac{1}{3}e^{-3t} + 3t^2 + \frac{13}{3}$

Exercise 9F

1. $\frac{1}{3}(2x^2 + 5)^3 + C$
2. $\ln(x^3 + 2x) + C$, $x^3 + 2x > 0$
3. $\frac{2}{3}(3x^2 + 5x)^{\frac{3}{2}} + C$
4. $e^{x^4} + C$
5. $-\frac{1}{x^2 + 3x + 1} + C$
6. $e^{\sqrt{x}} + C$
7. $\frac{1}{30}(2x^3 + 5)^5 + C$
8. $\frac{4}{3}(x^2 + x)^{\frac{3}{4}} + C$
9. $\frac{1}{2}(x^4 - x^2)^4 + C$
10. $-\ln(x^3 - 4x) + C$, $x^3 - 4x > 0$
11. $f(x) = \ln(4x^2 + 1) + 4$
12. $f(x) = e^{x^3} + 4e$

Investigation - area and the definite integral

1. a i 0.5 ii 1; 1.25; 2; 3.25
      iii 3.75
   b i 0.5 ii 1.25; 2; 3.25; 5
      iii 5.75
   c 4.67; 3.75 < 4.67 < 5.75; the area of the shaded region
2. $\frac{1}{2}(3)(6) = 9$; $\int_{-1}^{2}(2x + 2)dx = 9$; they are equal
3. $\int_{b}^{a} f(x)dx$
4. a $\frac{1}{2}(2.5 + 1)(3) = 5.25$;
      $\int_{1}^{5}\left(-\frac{1}{2}x + 3\right)dx = 5.25$
   b $\frac{1}{2}\pi(4^2) \approx 25.1$;
      $\int_{-4}^{4}\sqrt{16 - x^2}\,dx \approx 25.1$

Exercise 9G

1. $\int_{-2}^{6}\left(\frac{1}{2}x + 1\right)dx = 16$;
   $\frac{1}{2}(8)(4) = 16$
2. $\int_{-2}^{0}(x^3 - 4x)dx = 4$; no area formula
3. $\int_{-1}^{3} 3\,dx = 12$; (4)(3) = 12
4. $\int_{0}^{3}\sqrt{9 - x^2}\,dx \approx 7.07$;
   $\frac{1}{4}\pi(3^2) \approx 7.07$
5. $\int_{1}^{3}\frac{1}{x}dx \approx 1.10$; no area formula
6. $\int_{0}^{6}\left(\frac{1}{3}x + 2\right)dx = 18$;
   $\frac{1}{2}(2 + 4)(6) = 18$

Exercise 9H

1. 12
2. 14
3. -4
4. -8
5. 12
6. 0
7. 11
8. -3
9. 20
10. 12
11. a 4 b 12
12. a 4
    b i $a = 3$; $b = 7$ ii $k = 3$

9.4 Exercise 9I

1. 1
2. $-\frac{10}{3}$
3. $\frac{1}{2}$
4. $-\frac{36}{5}$
5. $4(e^3 - 1)$
6. 1
7. $\frac{16}{3}$
8. 16
9. a 24 b $\frac{32}{3}$
10. 12

Exercise 9J

1. $\ln 3$
2. $\frac{1}{e^2} - \frac{1}{e^3}$
3. 0
4. $2\left(e - \frac{1}{e}\right)$

5 $\dfrac{56}{9}$

6 320

7 $2\ln\dfrac{18}{7}$

8 $2(e^4 - e^3)$

9 a $\displaystyle\int_0^2 -2x^2(x-2)\,dx$, b $\dfrac{8}{3}$

10 9

Investigation: Area between two curves

1

Interval	Width	Height	Area
$-1.5 \leq x \leq -0.5$	1	$f(-1) - g(-1) = -2 - (-3) = 1$	$1(1) = 1$
$-0.5 \leq x \leq 0.5$	1	$f(0) - g(0) = 0 - (-2) = 2$	$1(2) = 2$
$0.5 \leq x \leq 1.5$	1	$f(1) - g(1) = 4 - (-1) = 5$	$1(5) = 5$
$1.5 \leq x \leq 2.5$	1	$f(2) - g(2) = 10 - 0 = 10$	$1(10) = 10$
$2.5 \leq x \leq 3.5$	1	$f(3) - g(3) = 18 - 1 = 17$	$1(17) = 17$

2 Area ≈ 35; the area enclosed by part of the rectangles that extend beyond the region bounded by the two curves is greater than the gaps left by the rectangles in the region.

3 $\displaystyle\int_{-1.5}^{3.5} [(x^2 + 3x) - (x - 2)]\,dx$
≈ 35.4; The values are very close.

Exercise 9K

1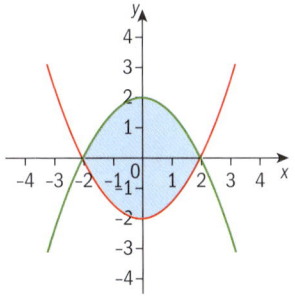

$\displaystyle\int_{-2}^{2}\left(\left(-\dfrac{1}{2}x^2 + 2\right) - \left(\dfrac{1}{2}x^2 - 2\right)\right)dx$
$= \dfrac{32}{3}$

2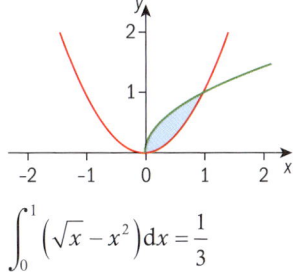

$\displaystyle\int_0^1 \left(\sqrt{x} - x^2\right)dx = \dfrac{1}{3}$

3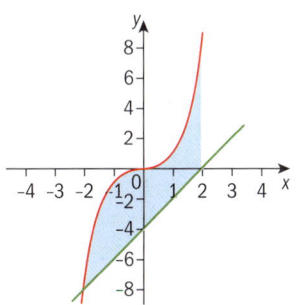

$\displaystyle\int_{-2}^{2} (x^3 - (2x - 4))\,dx = 16$

4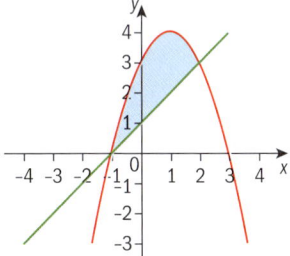

$\displaystyle\int_{-1}^{2} ((3 + 2x - x^2) - (x + 1))\,dx$
$= \dfrac{9}{2}$

5 a $(0,0), (-1, 0), (1,0)$

b i $f'(x) = 4x^3 - 2x$

ii Relative minimum points:
$\left(-\sqrt{\dfrac{1}{2}}, -\dfrac{1}{4}\right), \left(\sqrt{\dfrac{1}{2}}, -\dfrac{1}{4}\right)$

Relative maximum point: $(0, 0)$

c i and ii

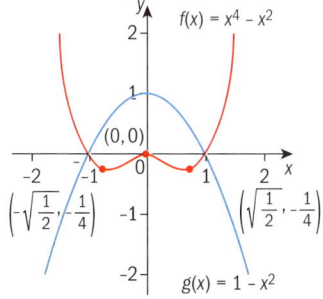

d $\displaystyle\int_{-1}^{1} ((1 - x^2) - (x^4 - x^2))\,dx = \dfrac{8}{5}$

6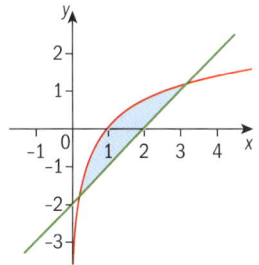

$\displaystyle\int_{0.1586}^{3.146} (\ln(x) - (x - 2))\,dx$
≈ 1.95

7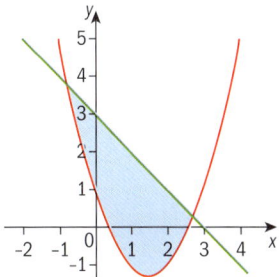

$\displaystyle\int_{-0.7321}^{2.732} ((-x + 3) - (x^2 - 3x + 1))\,dx$
≈ 6.93

8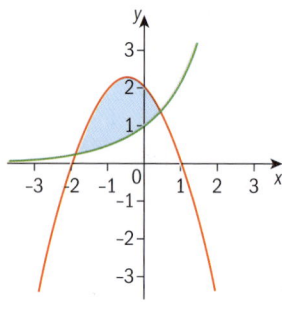

$\int_{-1.952}^{0.3841} ((2-x-x^2) - e^x) dx$

≈ 2.68

9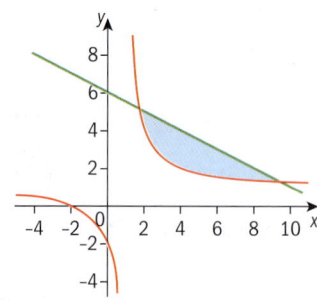

$\int_{1.725}^{9.275} \left(\left(-\frac{1}{2}x+6\right) - \frac{x+2}{x-1}\right) dx$

≈ 9.68

10 a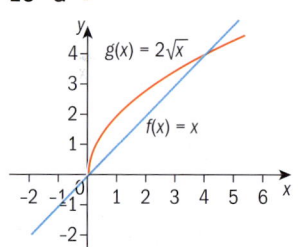

b i $\int_0^4 (2\sqrt{x} - x) dx$

   ii 2.67 or $\frac{8}{3}$

c i $\int_0^k (2\sqrt{x} - x) dx$ or

   $\frac{4}{3}k^{\frac{3}{2}} - \frac{1}{2}k^2$

   ii $k \approx 1.51$

Exercise 9L

1 $\int_0^1 ((x^3 - 2x^2) - (2x^2 - 3x)) dx +$

$\int_1^3 ((2x^2 - 3x) - (x^3 - 2x^2)) dx$

≈ 3.08

2 $\int_0^1 ((x-1)^3 - (x-1)) dx +$

$\int_1^2 ((x-1) - (x-1)^3) dx = 0.5$

3 $\int_{-1.131}^0 ((x^3 - x) - (xe^{-x^2})) dx +$

$\int_0^{1.131} ((xe^{-x^2}) - (x^3 - x)) dx$

≈ 1.18

4 $\int_{-3}^{-0.7071} ((-x^4 + 10x^2 - 9) -$

$(x^4 - 9x^2)) dx + \int_{-0.7071}^{0.7071} ((x^4 - 9x^2)$

$- (-x^4 + 10x^2 - 9)) dx +$

$\int_{0.7071}^3 ((-x^4 + 10x^2 - 9) -$

$(x^4 - 9x^2)) dx \approx 110$

5 a i (4, 4)

   ii $f'(x) = \frac{1}{2}x$

   $m = f'(4) = 2$

   $y - 4 = 2(x - 4)$

   $y = 2x - 4$

b i (1.236, −1.528)

   ii $\int_0^{1.236} \left(\frac{1}{4}x^2 - (-x^2)\right) dx +$

   $\int_{1.236}^4 \left(\frac{1}{4}x^2 - (2x-4)\right) dx$

   ≈ 2.55

Investigation: Volume of revolution

1

Interval	Radius	Height	Volume
$0 \le x \le 1$	$f(1) = 0.5$	$1 - 0 = 1$	$\pi(0.5)^2 (1) \approx 0.7854$
$1 \le x \le 2$	$f(2) = 1$	$2 - 1 = 1$	$\pi(1)^2 (1) \approx 3.142$
$2 \le x \le 3$	$f(3) = 1.5$	$3 - 2 = 1$	$\pi(1.5)^2 (1) \approx 7.069$
$3 \le x \le 4$	$f(4) = 2$	$4 - 3 = 1$	$\pi(2)^2 (1) \approx 12.57$
$4 \le x \le 5$	$f(5) = 2.5$	$5 - 4 = 1$	$\pi(2.5)^2 (1) \approx 19.63$
$5 \le x \le 6$	$f(6) = 3$	$6 - 5 = 1$	$\pi(3)^2 (1) \approx 28.27$

2 71.5; greater

3 $\int_0^6 \pi(0.5x)^2 dx \approx 56.5$

4 Volume $= \frac{1}{3}\pi(3)^2(6) \approx 56.5$

Exercise 9M

1 $\int_0^5 \pi(4^2) dx \approx 251$;

$V = \pi(4^2)(5) \approx 251$

2 $\int_0^3 \pi(6 - 2x)^2 dx \approx 113$;

$V = \frac{1}{3}\pi(6^2)(3) \approx 113$

3 $\int_{-2}^2 \pi(\sqrt{4-x^2})^2 dx \approx 33.5$;

$V = \frac{4}{3}\pi(2^3) \approx 33.5$

4 $\int_0^4 \pi(\sqrt{16-x^2})^2 dx \approx 134$;

$V = \frac{1}{2}\left(\frac{4}{3}\pi(4^3)\right) \approx 134$

5 $\int_2^4 \pi(x^2) dx \approx 58.6$; $V = \frac{1}{3}\pi(4^2)(4)$

$- \frac{1}{3}\pi(2^2)(2) \approx 58.6$

Exercise 9N

1 $\int_1^2 \pi(x^3)^2 dx = \frac{127\pi}{7}$

2 $\int_0^1 \pi(x^2 + 1)^2 dx = \frac{28\pi}{15}$

3 $\int_0^3 \pi(3x - x^2)^2 dx = \frac{81\pi}{10}$

4 $\int_1^4 \pi\left(\frac{1}{x}\right)^2 dx = \frac{3\pi}{4}$

5 a $\int_0^{\ln 4} \pi\left(e^{\left(\frac{1}{4}x\right)}\right)^2 dx$

   b 2

6 a $\int_1^a \pi\left(\frac{1}{\sqrt{x}}\right)^2 dx$

   b e^3

Exercise 9O

1 a $v(t) = 2t - 6$
   b
   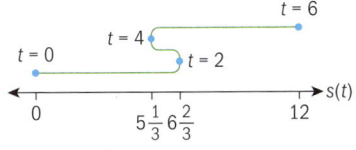

   c $\int_0^4 (2t-6)dt = -8$ m;

   $\int_0^4 |2t-6|dt = 10$ m

2 a $v(t) = t^2 - 6t + 8$
   b

   c $\int_0^6 (t^2 - 6t + 8)dt \approx 12$ m;

   $\int_0^6 |t^2 - 6t + 8|dt \approx 14.7$ m

3 a $v(t) = 3(t-2)^2$
   b

   c $\int_0^4 3(t-2)^2 dt = 16$ m;

   $\int_0^4 |3(t-2)^2|dt = 16$ m

4 a $\int_2^{12} v(t)dt = \frac{1}{2}(6)(6)$

   $-\frac{1}{2}(4+2)(2) = 12$ m

   $\int_2^{12} |v(t)|dt = \frac{1}{2}(6)(6)$

   $+\frac{1}{2}(4+2)(2) = 24$ m

   b $\int_0^5 v(t)dt = \frac{1}{2}(2)(2)$

   $+\frac{1}{2}(3)(6) = 11$ m

   $\int_0^5 |v(t)|dt = \frac{1}{2}(2)(2)$

   $+\frac{1}{2}(3)(6) = 11$ m

   c $\int_0^{12} v(t)dt = \frac{1}{2}(2)(2) + \frac{1}{2}(6)(6)$

   $-\frac{1}{2}(4+2)(2) = 14$ m

$\int_0^{12} |v(t)|dt = \frac{1}{2}(2)(2) + \frac{1}{2}(6)(6)$

$+\frac{1}{2}(4+2)(2) = 26$ m

5 a 2 m s^{-2}
   b $s(t) = \frac{1}{3}t^3 - 9t + 12$
   c $\int_2^8 |t^2 - 9|dt \approx 119$ m

6 a 2 m s^{-2}
   b $2 < t < 10$
   c 28

Exercise 9P

1 $\int_0^{10} 18.4e^{\frac{t}{20}}dt \approx 239$ billions of barrels

2 $\int_0^{1.5} (1375t^2 - t^3)dt \approx 1550$ spectators

3 $36.5 + \int_0^8 5te^{(-0.01t^4+0.13t^3-0.38t^2-0.3t+0.9)}dt \approx 240$ cm^3

4 $4000 + \int_0^{20} -133\left(1 - \frac{t}{60}\right)dt \approx$ 1780 gallons

Review exercise non-GDC

1 a $x^4 - 4x^2 + 6x + C$
   b $\frac{3}{7}x^{\frac{7}{3}} + C$
   c $-\frac{1}{x^3} + C$
   d $\frac{5}{18}x^3 - \frac{1}{2}\ln x + C, x > 0$
   e $\frac{1}{4}e^{4x} + C$
   f $\frac{1}{15}(x^3+1)^5 + C$
   g $\frac{1}{2}\ln(2x+3) + C, x > -\frac{3}{2}$
   h $\frac{1}{2}(\ln x)^2 + C, x > 0$
   i $\frac{1}{2}(3x^2+1)^2 + C$
   j $2\ln(e^x + 3) + C$
   k $(2x-5)^{\frac{3}{2}} + C$
   l $\frac{1}{2}e^{(2x^2)} + C$

2 a -4
   b 16
   c 8
   d $e^6 - e^3$
   e -20
   f $\frac{\ln 5}{2}$

3 a $\int_1^2 (x^2 - 1)dx$
   b $\frac{4}{3}$
   c $\int_1^2 (x^2-1)dx - \int_{-1}^1 (x^2-1)dx$
   d $\pi \int_1^2 (x^2-1)^2 dx$

4 $f(x) = \frac{3}{2}x^2 - 2x + 4$

5 a 5
   b 28

6 $s(t) = 2e^{2t} + 2t + 6$

7 13

Review exercise GDC

1 107

2 a $a(t) = 4t - 11$
   b $a = 1.5, b = 4$
   c 7.83 m

3 a $y = 3x$
   b $(2, 6)$
   c
   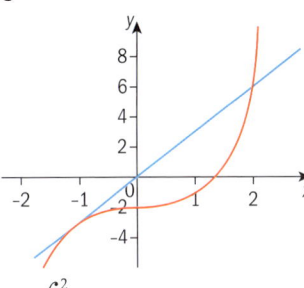
   d $\int_{-1}^2 (3x - (x^3 - 2))dx = 6.75$

Chapter 10

Skills check

1 a 32 b 27 c 343
   d $\frac{1}{128}$ e $\frac{81}{256}$
   f 0.000000001 or 1×10^{-9}

2 a $n = 4$ b $n = 5$ c $n = 3$
   d $n = 4$ e $n = 3$ f $n = 3$

Exercise 10A

1. a Positive, strong
   b Negative, weak
   c Negative, strong
   d Positive, weak
   e No correlation
2. a i Positive ii Linear
      iii Strong
   b i Negative ii Linear
      iii Strong
   c i Positive ii Linear
      iii Moderate
   d i No association
      ii Non-linear
      iii Zero
   e i Positive ii Linear
      iii Weak
   f i Negative
      ii Non-linear
      iii Strong
3. a Increases b Decreases
4. a

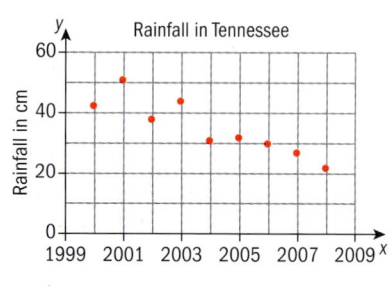

   b Strong, negative
   c As the year increases the rainfall decreases.
5. a

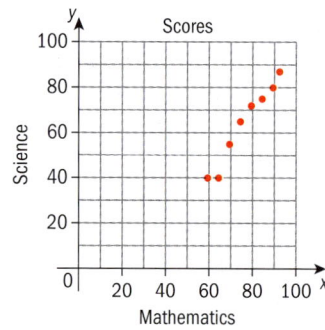

   b Strong, positive, linear

Investigation – leaning tower of Pisa (continued)

a

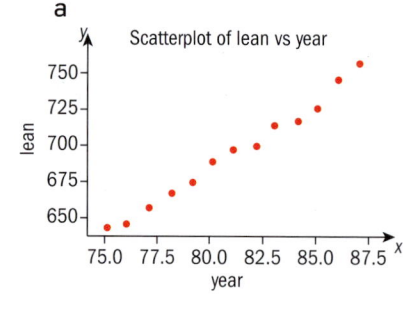

b Strong, positive

c The lean is increasing. The danger with extrapolation is that it assumes that the current trend will continue and this is not always the case.

Exercise 10B

1. a (96.7, 44.1)
   b

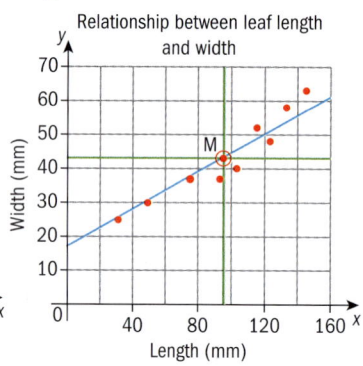

2. a i 175 cm
      ii 66 kg
   b

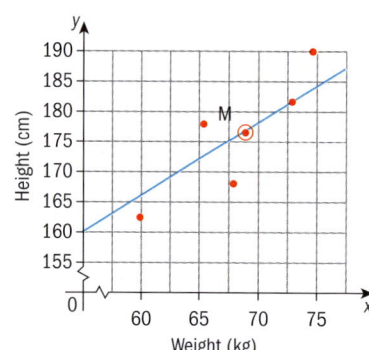

3. a (4, 6.67)
   b

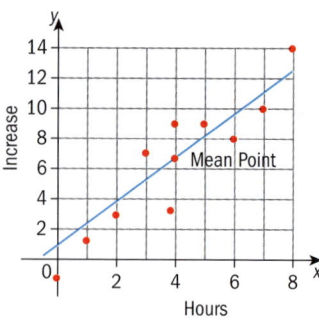

   c Strong, positive
   d An increase in the number of hours spent studying mathematics produces an increase in the grade.

Exercise 10C

1. a $(\bar{x}, \bar{y}) = (75, 7.03)$

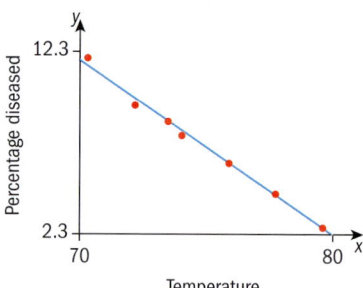

   b $y = -0.96x + 79$
   c 7
2. a £220000
   b 75.4
   c and d Note the values of m and b in the equation $y = mx + b$ are approximate.

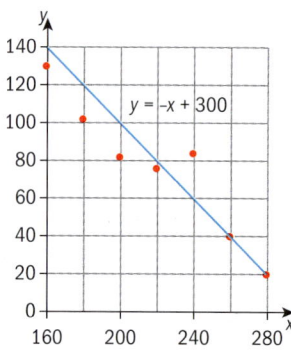

   e Approximately 70 houses

Exercise 10D

1 The slope is −0.3. As a student plays one more day of sport they do 18 minutes less homework. The y-intercept is 40, which means that the average student who does no hours of sport does 40 hours of homework.

2 The slope is 6. For every time a person has been convicted of a crime they know 6 more criminals.

The y-intercept is 0.5, which means that people who have not been convicted of a crime know 0.5 criminals on average.

3 The slope is 2.4. For every pack of cigarettes smoked per week there are 2.4 more sick days per year.

The y-intercept is 7, which means that the average person that does not smoke has 7 sick days per year.

4 The slope is 100. 100 more customers come to his shop every year.

The y-intercept is −5, which means that −5 people visited his shop in year zero; the y-intercept is not suitable for interpretation.

5 The slope is 0.8. Every 1 mark increase in mathematics results in a 0.8 increase in science.

The y-intercept is −10 which is not suitable for interpretation as a zero in mathematics would mean a −10 in science.

Exercise 10E

1 a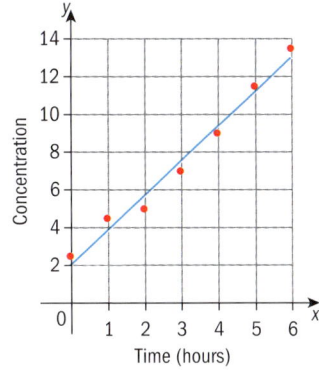

b $y = 1.84x + 1.99$

c 8.43 (3 sf)

2 a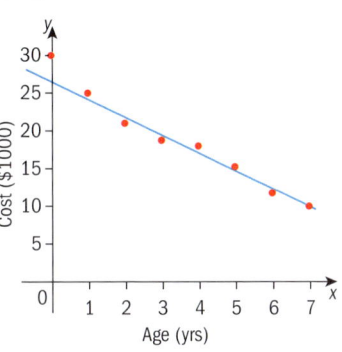

b $y = -2.67x + 28.1$

c $16085

d The relationship may not be linear. Antique cars are often more expensive after 50 years than when new.

3 a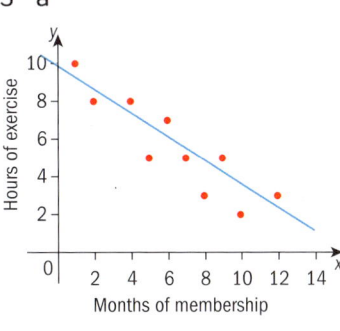

b $y = -0.665x + 9.86$

c 7.865 hours

d No. The equation gives −6.1 hours of exercise!

4 Fifty years = 600 months, and the line would predict Sarah's height at 50 years to be about 302 cm = 30.2 metres.

Clearly there is a major difficulty with extrapolation. In fact, most females reach their maximum height in their mid to late teens, and from then on their height is fairly constant. Therefore extrapolating with a linear function is unsuitable.

5 a (1981, 694)

b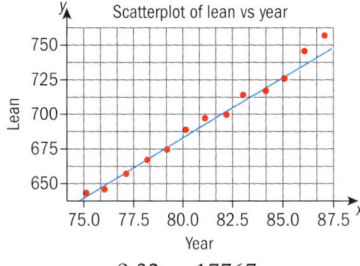

c $y = 9.32x - 17767$

d 780 m

Exercise 10F

1 $r = 0.863$. There is a strong, positive correlation.

2 a 0.789

b Strong, positive correlation

c The income increases as the number of years of education increases.

3 a 0.910

b The stopping distance increases as the car gets older.

c Strong correlation

4 a −0.887

b Strong, negative correlation

c Yes, Kelly's grade would increase if the chat time decreased.

5 a 0.0262

b Positive, weak correlation

c No, Mo's grade would not increase if the game time decreased.

6 0.994. Strong, positive correlation.

Review exercise non-GDC

1 a ii   b v
   c iii   d i

2 a and b

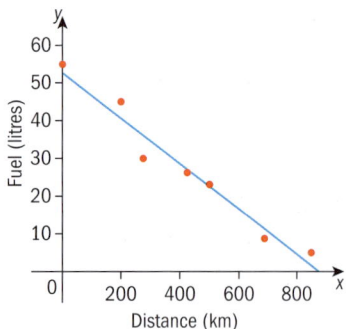

   c 32 litres

3 a and c

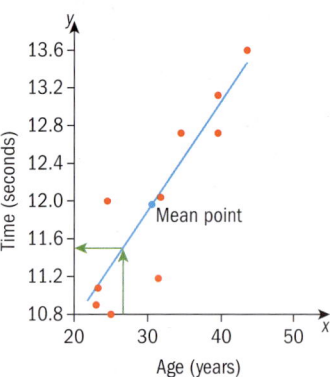

   b Mean age = 34 years, mean time = 12 seconds
   c Approximately 11.6 s

Review exercise GDC

1 a

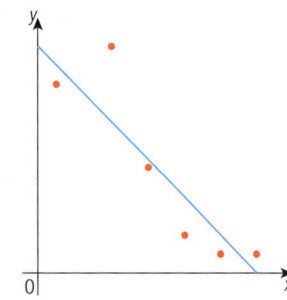

   b As the time increases, the number of push-ups decreases.

   c $y = -1.29x + 9$
   d $r = -0.929$. There is a strong, negative correlation.

2 a $w = -22.4 + 55.5h$
   b 66.4 kg

3 a $r = 0.785$
   b $y = 30.7 + 0.688x$
   c 99.5
   This should be reasonably accurate since the product–moment correlation coefficient shows fairly strong correlation.

4 a

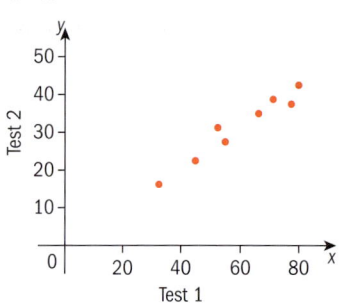

   b Positive, strong
   c high
   d $y = 0.50x + 0.48$
   e 20.48

5 a, c and f

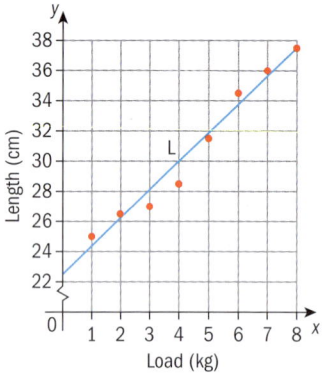

   b (4, 30)   d i $r = 0.986$
       ii (very) strong positive correlation
   e $y = 1.83x + 22.7$
   g 30.9 cm

   h Not possible to find an answer as the value lies too far outside the given data set.

6 a

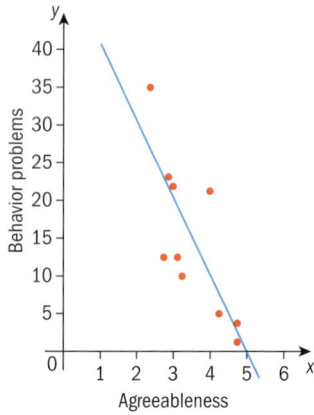

   b Behavior problems decrease.
   c −0.797
   d Strong, negative correlation
   e Fewer
   f $y = -10.2x + 51.0$
   g 5.1

7 a $y = 10.7x + 121$
   b i Every coat on average costs $10.65 to produce.
      ii When the factory does not produce any clothes it has to pay costs of $121.
   c $870
   d 14

Chapter 11

Skills check

1 a $x = 90$
   b $x = 50$
   c $x = 68$
   d $x = \dfrac{70}{3}$
   e $x = 6.09$ (3 sf)
   f $x = 14.7$ (3 sf)

Exercise 11A
1. $b = 16$, $\hat{A} = 36.9°$, $\hat{B} = 53.1°$
2. $\hat{B} = 50°$, $a = 31.0$, $c = 48.3$
3. $\hat{A} = 35°$, $a = 2.58$, $b = 3.69$
4. $a = 36$, $\hat{A} = 36.9°$, $\hat{B} = 53.1°$
5. $\hat{B} = 55°$, $b = 15.7$, $c = 19.2$
6. $c = 12.9$, $\hat{A} = 41.2°$, $\hat{B} = 48.8°$
7. $x = 5$, $\hat{A} = 22.6°$, $\hat{B} = 67.4°$

Exercise 11B
1. a $b = 12\sqrt{3}$, $\hat{A} = 30°$, $\hat{B} = 60°$
   b $\hat{B} = 45°$, $a = 9$, $c = 9\sqrt{2}$
   c $\hat{A} = 30°$, $a = 2.25$, $b = \dfrac{9\sqrt{3}}{4}$
   d $a = 2\sqrt{3}$, $\hat{A} = 30°$, $\hat{B} = 60°$
   e $b = 5\sqrt{2}$, $\hat{A} = 45°$, $\hat{B} = 45°$
2. $x = 8\sqrt{2}$, $y = 8\sqrt{3} - 8$, $z = 16$
3. $x = \dfrac{2\sqrt{3}+2}{\sqrt{3}}$, $AC = \dfrac{4\sqrt{3}+2}{3}$
4. $x = 1$, $AB = 3\sqrt{2}$ or $x = 3$, $AB = 11\sqrt{2}$
5. $w = 9.8$ cm, $x = 13.9$ cm, $y = 6.5$ cm, $z = 15.4$ cm

Exercise 11C
1. a $10\sqrt{2}$ cm
   b $B\hat{A}C = 70.5°$
      $A\hat{B}C = 38.9°$
2. a $AE = 29.1$, $BE = 34.4$
   b $A\hat{E}D = 74.1°$,
      $E\hat{B}A = 54.5°$,
      $A\hat{E}B = 51.5°$
3. 758 m
4. 71.5° and 108.5°
5. 4.78 km, N21.1°W
6. 70.7 m
7. 44.8 km, 243.5°
8. 135.7 m, 202.2 cm
9. 91.2 m
10. 40.7 m
11. 4.01 s
12. a 20.6°   b 26.6°
    c 35.1°   d 50.0°

Exercise 11D
1. a (0.940, 0.342)
   b (0.956, 0.292)
   c (0.5, 0.866)
   d (0.276, 0.961)
   e (0, 1)
2. a 66°   b 81°
   c 45°   d 14°
3. a 0.470   b 0.308
   c 0.203   d 0.25

Investigation – Obtuse angles
1.
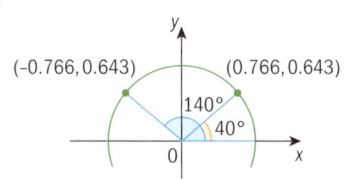

2.
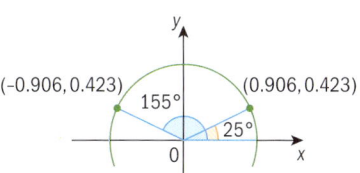

3.
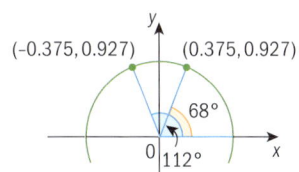

Exercise 11E
1. a $B(0.866, 0.5)$,
      $C(-0.866, 0.5)$
   b $B(0.545, 0.839)$,
      $C(-0.545, 0.839)$
   c $B(0.707, 0.707)$,
      $C(-0.707, 0.707)$
   d $B(0.974, 0.225)$,
      $C(-0.974, 0.225)$
   e $B(0.087, 0.996)$,
      $C(-0.087, 0.996)$
2. a 70.6°
   b 17.3°
   c 25.4°
   d 39.7°
3. a 0.2588, 165°
   b 0.5878, 144°
   c 0.9877, 99°
   d 0.8988, 116°
4. a 60.6°, 119.4°
   b 25.8°, 154.2°
   c 30.3°, 149.7°
   d 30°, 150°

Exercise 11F
1. a 1.50   b −1.92
   c −0.910   d 1
2. a $y = 1.09x$, $\theta = 48°$
   b $y = 1.87x$, $\theta = 62°$
   c $y = -2.80x$, $\theta = 110°$
   d $y = -1.21x$, $\theta = 129°$
   e $y = -0.75x$, $\theta = 143°$
   f $y = 2.36x$, $\theta = 113°$

Exercise 11G
1. a $\hat{C} = 50°$, $a = 17.7$ cm, $c = 18.5$
   b $\hat{B} = 68°$, $a = 1.69$ cm, $b = 2.44$ cm
   c $\hat{B} = 40.9°$, $\hat{C} = 84.1°$, $c = 5.46$ cm
   d $\hat{A} = 40°$, $a = 149$, $c = 190$
   e $\hat{C} = 110°$, $a = 2.80$, $b = 4.21$
2. 26.9 cm
3. 3.37 km, 2.24 km
4. 15.8 m

Investigation – Ambiguous triangles
1. $\hat{C}_1 = 62°$, $\hat{C}_2 = 118°$. The angles are supplementary.
2. $\hat{B}_1 = 86°$, $\hat{B}_2 = 30°$, $b_1 = 5.65$ cm, $b_2 = 2.83$ cm

Exercise 11H
1. a $\hat{C}_1 = 61.0°$, $\hat{B}_1 = 89.0°$, $b_1 = 8.0$ cm
      $\hat{C}_2 = 119.0°$, $\hat{B}_2 = 31.0°$, $b_2 = 4.1$ cm

b $\hat{C}_1 = 71.1°$, $\hat{A}_1 = 58.9°$,
   $a_1 = 19.0$ cm
   $\hat{C}_2 = 108.9°$, $\hat{A}_2 = 21.1°$,
   $a_2 = 8.0$ cm
c $\hat{B}_1 = 68.5°$, $\hat{A}_1 = 91.5°$,
   $a_1 = 7.3$ cm
   $\hat{B}_2 = 111.5°$, $\hat{A}_2 = 48.5°$,
   $a_2 = 5.5$ cm
d $\hat{C} = 30.5°$, $\hat{B} = 107.5°$,
   $b = 47.0$ cm
e Triangle does not exist
f $\hat{B}_1 = 77.8°$, $\hat{C}_1 = 32.2°$,
   $c_1 = 14.2$ cm
   $\hat{B}_2 = 102.2°$, $\hat{C}_2 = 7.8°$,
   $c_2 = 3.6$ cm
g $\hat{B} = 26.7°$, $\hat{C} = 108.3°$,
   $c = 29.5$ cm
h $\hat{C}_1 = 67.1°$, $\hat{A}_1 = 56.9°$,
   $a_1 = 45.5$ cm
   $\hat{C}_1 = 112.9°$, $\hat{A}_2 = 11.1°$,
   $a_2 = 10.4$ cm
2 a $BE = 8$ m, $CE = 6$ m,
     $DE = 15$ m
  b $EAB = 53.1°$,
     $B\hat{C}E = 53.1°$,
     $B\hat{C}D = 126.9°$,
     $A\hat{B}D = 98.8°$,
     $C\hat{B}D = 25.1°$
  c Given side $BD = 17$ m
     in $\triangle ABD$ and angle
     $\hat{D} = 28.1°$, and side
     $AB = 10$, then there are 2
     possible triangles, fitting
     this data, namely DBA and
     DBC.
3 b 5.80 km   c 24.9 km
  d 143.5°

Exercise 11I

1 a $a = 65.7$ m, $\hat{B} = 36.0°$,
     $\hat{C} = 80.0°$
  b $\hat{A} = 28.9°$, $\hat{B} = 52.8°$,
     $\hat{C} = 98.4°$

c $\hat{A} = 44.4°$, $\hat{B} = 107.8°$,
   $\hat{C} = 27.8°$
d $b = 7.48$ m, $\hat{A} = 43.5°$,
   $\hat{C} = 105.5°$
e $c = 92.8$ m, $\hat{A} = 49.4°$,
   $\hat{B} = 60.6°$
f $\hat{A} = 48.6°$, $\hat{B} = 56.4°$,
   $C = 75.0°$
2 12.1 km
3 4.07 cm, 6.48 cm
4 18.8 km
5 043.5° or 136.5°
6 a 45°
  b 71.8°
  c 63.8°

Exercise 11J

1 a 26.7 cm²
  b 40.8 cm²
  c 152 cm²
  d 34.1 cm²
  e 901 cm²
  f 435 cm²
2 47.8°
3 22.7 cm
4 a 76.7°
  b 81.4 cm²
5 $x = 2.5$ cm
6 5.31 mm, 18.5 mm

Exercise 11K

1 9.52 cm
2 39 cm
3 5 radians
4 3000 cm², 220 cm
5 22.95 cm², 21.3 cm
6 $\theta = 1.7$, $r = 16$
7 7.96 cm²

Exercise 11L

1 a $\dfrac{5\pi}{12}$
  b $\dfrac{4\pi}{3}$

  c $\dfrac{4\pi}{9}$
  d $\dfrac{11\pi}{6}$
2 a 0.977 rad
  b 1.87 rad
  c 5.65 rad
  d 4.01 rad
3 a 150°
  b 300°
  c 270°
  d 225°
4 a 85.9°
  b 20.6°
  c 136°
  d 206°

Exercise 11M

1 a $\dfrac{\sqrt{2}}{2}$
  b $-\dfrac{1}{2}$
  c $\dfrac{\sqrt{3}}{3}$
  d $\dfrac{\sqrt{3}}{2}$
2 a 0.892
  b 0.949
  c -1.12
  d 0.667
3 a 9.76 cm²
  b 5.45 cm
  c 50.5 cm²
4 10.9 m²
5 a 17.1 cm²   b 12.1 cm²
  c 2.63 rad   d 15.8 cm

Review exercise non-GDC

1 $7\sqrt{2}$ cm
2 a 30°   b $8\sqrt{3}$ cm
3 $\dfrac{2}{5}$
4 10 cm²
5 a 25 cm   b 125 cm²

Review exercise GDC

1. 72.7 m
2. a (0.848, 0.530)
   b 72.9°
   c (−0.600, 0.800)
3. a 54.7° b 10.9 cm
4. a 18.0 m b 34.3°
5. a 121° b 8.60 cm
6. 54.1 km
7. a 31.9 b 13.9 cm
   c 119 d 27.4 cm²
8. a 21.6 cm b 14.5 cm
   c 11.16 cm d 47.3 cm

Chapter 12

Skills check

1. a (3, 0, 0)
   b (3, 4, 0)
   c (3, 0, 2)
   d (3, 4, 2)
   e (1.5, 4, 2)
2. 6.71
3. a 20 cm
   b 101°

Exercise 12A

1. a $x = -2i + 3j$
   b $y = 7j$
   c $z = i + j - k$
2. a $\vec{AB} = \begin{pmatrix} 2 \\ 3 \end{pmatrix}$
   b $\vec{CD} = \begin{pmatrix} -1 \\ 6 \\ -1 \end{pmatrix}$
   c $\vec{EF} = \begin{pmatrix} 0 \\ 0 \\ 1 \end{pmatrix}$
3. $a = \begin{pmatrix} -3 \\ -5 \end{pmatrix} = -3i - 5j$

   $b = \begin{pmatrix} -2 \\ 4 \end{pmatrix} = -2i + 4j$

   $c = \begin{pmatrix} 3 \\ 8 \end{pmatrix} = 3i + 8j$

   $d = \begin{pmatrix} 0 \\ 6 \end{pmatrix} = 6j$

   $e = \begin{pmatrix} -3 \\ -6 \end{pmatrix} = -3i - 6j$

4. a 5
   b $\sqrt{10} = 3.16$
   c $\sqrt{29} = 5.39$
   d 5.3
   e $\sqrt{29} = 5.39$
5. a $\sqrt{38} = 6.16$
   b $\sqrt{26} = 5.10$
   c 3
   d 7
   e $\sqrt{2} = 1.41$

Exercise 12B

1. a $c = 3b$
   $d = \frac{1}{2}a$
   $e = -5b$
   $f = -2a$
   b They are perpendicular.
2. a, b, e
3. a $\frac{-24}{7}$
   b $\frac{28}{5}$
4. $t = -25, s = -\frac{8}{5}$
5. a $\vec{OG} = j + k$
   b $\vec{BD} = -i - j + k$
   c $\vec{AD} = -i + k$
   d $\vec{OM} = \frac{1}{2}i + j + k$
6. a $\vec{OG} = 4j + 3k$
   b $\vec{BD} = -5i - 4j + 3k$
   c $\vec{AD} = -5i + 3k$
   d $\vec{OM} = \frac{5}{2}i + 4j + 3k$

Exercise 12C

1. $\vec{PQ} = \begin{pmatrix} -5 \\ -1 \end{pmatrix}, \vec{QP} = \begin{pmatrix} 5 \\ 1 \end{pmatrix}$
2. a $\vec{AB} = \begin{pmatrix} -4 \\ -4 \end{pmatrix}$
   b $\vec{BA} = \begin{pmatrix} 4 \\ 4 \end{pmatrix}$
   c $\vec{AC} = \begin{pmatrix} -7 \\ 3 \end{pmatrix}$
   d $\vec{CB} = \begin{pmatrix} 3 \\ -7 \end{pmatrix}$.
3. a $2i - 3j + 5k$
   b $-i + 5j - 6k$
   c $-i + 5j - 6k$
   d $i - 5j + 6k$
4. $\vec{LM} = \begin{pmatrix} 5 \\ -4 \\ -3 \end{pmatrix}$
5. $\vec{US} = 2i + 8j - 3k$
6. $x = 0, y = 7, z = 9$

Exercise 12D

1. $\vec{AB} = \begin{pmatrix} -3 \\ 5 \\ -4 \end{pmatrix}, \vec{AC} = \begin{pmatrix} 3 \\ -5 \\ 4 \end{pmatrix},$

   $\vec{BC} = \begin{pmatrix} 6 \\ -10 \\ 8 \end{pmatrix}$. Any two of

   these are scalar multiples of each other

2. a $\vec{AB} = \begin{pmatrix} 3 \\ -2 \\ 8 \end{pmatrix}$
   b $\vec{AC} = \begin{pmatrix} 6 \\ -4 \\ 16 \end{pmatrix}$ so $\vec{AB} = \frac{1}{2}\vec{AC}$

   or $\vec{BC} = \begin{pmatrix} -3 \\ 2 \\ -8 \end{pmatrix}$ so $\vec{AB} = -\vec{BC}$

3. $\vec{P_1P_2} = \begin{pmatrix} -3 \\ -1 \\ 0 \end{pmatrix}, \vec{P_1P_3} = \begin{pmatrix} -6 \\ -2 \\ 0 \end{pmatrix},$

   $\vec{P_2P_3} = \begin{pmatrix} -3 \\ -1 \\ 0 \end{pmatrix}; P_4\left(2, \frac{7}{3}, 4\right)$

4. $x = \frac{5}{3}; AB : BC = 1 : 2$

Exercise 12E

1. $\vec{AB} = \begin{pmatrix} 5 \\ 0 \\ -2 \end{pmatrix}$; $\sqrt{29} = 5.39$

2. $|\vec{AB}| = \sqrt{129}$, $|\vec{AC}| = \sqrt{42}$
   $|\vec{BC}| = \sqrt{129}$. Two sides equal length therefore isosceles. Angle $CAB = 46.8°$

3. $t = \pm 6$
4. $x = \pm\sqrt{5}$
5. $a = \pm 2$
6. a 15
   b 10
   c 13

Exercise 12F

1. $\left(\dfrac{3}{5}\right)^2 + \left(\dfrac{4}{5}\right)^2 = 1$

2. $\left(\dfrac{1}{3}\right)^2 + \left(\dfrac{2}{3}\right)^2 + \left(\dfrac{2}{3}\right)^2 = 1$

3. $\dfrac{1}{5}(4\mathbf{i} - 3\mathbf{j})$

4. $\dfrac{1}{\sqrt{42}} \begin{pmatrix} -1 \\ -5 \\ 4 \end{pmatrix}$

5. $\dfrac{1}{3}(2\mathbf{i} + 2\mathbf{j} - \mathbf{k})$

6. $\dfrac{1}{\sqrt{5}}$

7. $\dfrac{5}{\sqrt{5}}(2\mathbf{i} - \mathbf{j})$

8. $\dfrac{7}{\sqrt{14}} \begin{pmatrix} -1 \\ -3 \\ 2 \end{pmatrix}$

9. a $\begin{pmatrix} \cos\theta \\ \sin\theta \end{pmatrix}$
   b $\begin{pmatrix} \cos\alpha \\ \sin\alpha \end{pmatrix}$

Exercise 12G

1. a $5\mathbf{i} + \mathbf{j}$
   b $2\mathbf{i} + 3\mathbf{j}$
   c $2\mathbf{i} + 4\mathbf{j}$
   d $8\mathbf{i} + 4\mathbf{j}$
   e $-\mathbf{i} - 3\mathbf{j}$
   f $2\mathbf{i}$

2. a $\begin{pmatrix} -2 \\ 2 \end{pmatrix}$
   b $\begin{pmatrix} 1 \\ 8 \end{pmatrix}$
   c $\begin{pmatrix} -1.5 \\ -3 \end{pmatrix}$
   d $\begin{pmatrix} -5 \\ 15 \end{pmatrix}$
   e $\begin{pmatrix} 3 \\ -34 \end{pmatrix}$

3. a $8\mathbf{i} - \mathbf{j} - 3\mathbf{k}$
   b $-\mathbf{i} + 2\mathbf{j} + 3\mathbf{k}$
   c $\mathbf{i} - 2\mathbf{j} - 3\mathbf{k}$
   d $8\mathbf{i} - 6\mathbf{j} - 10\mathbf{k}$

4. $\mathbf{x} = \begin{pmatrix} 4 \\ -5.5 \end{pmatrix}$, $\mathbf{y} = \begin{pmatrix} 19 \\ 3 \\ -16 \end{pmatrix}$,
   $\mathbf{z} = \begin{pmatrix} -6 \\ 10 \end{pmatrix}$

5. $x = -4.5$, $y = 10.5$
6. $s = 4.5$, $t = 9$, $u = 9$

Exercise 12H

4. a i $\mathbf{b} - \mathbf{a}$
      ii $\mathbf{b} - \mathbf{a}$
      iii $2\mathbf{b} - 2\mathbf{a}$
      iv $\mathbf{b} - 2\mathbf{a}$
      v $2\mathbf{b} - 3\mathbf{a}$
   b AB is parallel to and half the length of FC
   c FD and AC are parallel

5. d $\vec{MX} = 3\vec{MP}$

Investigation – cosine rule

Exercise 12I

1. a -18 b 5
   c 20   d -13
   e -13

2. a -9 b 20
   c 20   d -58
   e 13

3. a Perpendicular
   b Neither
   c Parallel
   d Neither
   e Perpendicular
   f Parallel
   g Parallel

4. -15

5. $\mathbf{d} = \begin{pmatrix} 2 \\ 1 \\ 3 \end{pmatrix}$

6. $45°$

7. a $94.8°$
   b $161.6°$
   c $136.4°$

8. a $\vec{AB} = \begin{pmatrix} -1 \\ 5 \end{pmatrix}$, $\vec{AC} = \begin{pmatrix} 1 \\ -2 \end{pmatrix}$
   b -11
   c $\dfrac{-11}{\sqrt{26}\sqrt{5}}$

9. a $79.0°$
   b $90°$
   c $118.1°$

10. a $AB = \sqrt{17}$; $AC = \sqrt{26}$
    b $\cos BAC = \dfrac{1}{\sqrt{17}\sqrt{26}}$
    c 10.5

11. $54.7°$

12. a $\vec{OA} \cdot \vec{OB} = 0$ therefore perpendicular
    b $\sqrt{62}$

13. $\lambda = 2.5$
14. $\lambda = \pm 9$
15. $p = \pm 3$

Exercise 12J

1. a $\mathbf{r} = \begin{pmatrix} -1 \\ 2 \end{pmatrix} + t\begin{pmatrix} 3 \\ 2 \end{pmatrix}$

   b $\mathbf{r} = \begin{pmatrix} -1 \\ 0 \end{pmatrix} + t\begin{pmatrix} 5 \\ -2 \end{pmatrix}$

   c $\mathbf{r} = \begin{pmatrix} 3 \\ 1 \\ -2 \end{pmatrix} + t\begin{pmatrix} 3 \\ -2 \\ 8 \end{pmatrix}$

   d $\mathbf{r} = 2\mathbf{j} - \mathbf{k} + t(3\mathbf{i} - \mathbf{j} + \mathbf{k})$

2. a E.g. $\mathbf{r} = \begin{pmatrix} 4 \\ 5 \end{pmatrix} + t\begin{pmatrix} -1 \\ -7 \end{pmatrix}$

   b E.g. $\mathbf{r} = \begin{pmatrix} 4 \\ -2 \end{pmatrix} + t\begin{pmatrix} 1 \\ 0 \end{pmatrix}$

c E.g. $\mathbf{r} = \begin{pmatrix} 3 \\ 5 \\ 2 \end{pmatrix} + t \begin{pmatrix} -1 \\ -9 \\ 3 \end{pmatrix}$

d E.g. $\mathbf{r} = \begin{pmatrix} 0 \\ 0 \\ 1 \end{pmatrix} + t \begin{pmatrix} 1 \\ -1 \\ -1 \end{pmatrix}$

3 a E.g. $\mathbf{r} = \begin{pmatrix} -1 \\ 6 \end{pmatrix} + t \begin{pmatrix} 2 \\ -3 \end{pmatrix}$

   b E.g. $\mathbf{r} = \begin{pmatrix} -1 \\ 0 \end{pmatrix} + t \begin{pmatrix} 2 \\ 5 \end{pmatrix}$

   c E.g. $\mathbf{r} = \begin{pmatrix} 4 \\ 2 \\ 1 \end{pmatrix} + t \begin{pmatrix} 1 \\ 0 \\ 3 \end{pmatrix}$

   d E.g. $r = 5\mathbf{k} + t(4\mathbf{i} - \mathbf{k})$

4 a Yes   b No
   c Yes   d No

5 $\mathbf{r} = \begin{pmatrix} 2 \\ 4 \\ 5 \end{pmatrix} + t \begin{pmatrix} -2 \\ 3 \\ 8 \end{pmatrix}$

   $p = -2, q = 21$

6 E.g. $\mathbf{r} = \begin{pmatrix} -6 \\ 5 \end{pmatrix} + t \begin{pmatrix} 0 \\ 1 \end{pmatrix}$

7 a Coincident
   b Perpendicular
   c Parallel
   d None
   e None

8 a 53.6°   b 115.2°

10 a i $2\mathbf{i} + 5\mathbf{j} + 3\mathbf{k}$
      ii $-2\mathbf{i} + 5\mathbf{j} + 3\mathbf{k}$
   b i $|\overrightarrow{OF}| = \sqrt{38}$
      ii $|\overrightarrow{AG}| = \sqrt{38}$
      iii $\overrightarrow{OF} \cdot \overrightarrow{AG} = 30$
   c 37.9°

11 a $\overrightarrow{AB} = 7\mathbf{i} - 8\mathbf{j} + 8\mathbf{k}$
   b $\cos O\hat{A}B = \dfrac{-49}{\sqrt{30}\sqrt{117}}$
   d $\mu = 3$
   e $(22, -19, 22)$

Exercise 12K

1 $(4, 2)$

2 $\begin{pmatrix} \frac{48}{5} \\ \frac{-3}{5} \end{pmatrix}$

3 $\left(\dfrac{23}{3}, \dfrac{1}{3}, \dfrac{2}{3}\right)$

4 $\left(-1, \dfrac{5}{3}\right)$

6 a $\begin{pmatrix} 5 \\ -8 \\ 15 \end{pmatrix}$

   b Dot product = 0

7 a $a = 5; b = 8$
   b $(4, 5, 7)$
   c $3\sqrt{10}$

8 a E.g. $\mathbf{r} = \begin{pmatrix} 2 \\ -1 \\ 2 \end{pmatrix} + t \begin{pmatrix} 1 \\ -1 \\ -3 \end{pmatrix}$

   b $(3, -2, -1)$
   c $\sqrt{11}$
   d 120.2°

Exercise 12L

1 a $\begin{pmatrix} 15 \\ 10 \end{pmatrix}$ or 10 km north and 15 km east
   b $5\sqrt{13}$ km

2 a $\sqrt{29}$ m s^{-1}
   b $\begin{pmatrix} 50 \\ -20 \end{pmatrix}$
   c 13 m s^{-1}
   d $8\sqrt{29}$ m
   e They will collide.

3 a 4 p.m.   b $7\mathbf{i} + 6\mathbf{j}$

4 a $3\sqrt{2}$ m s^{-1} and $\sqrt{86}$ m s^{-1}
   c 51.2 m

Review exercise non-GDC

1 a $\overrightarrow{AB} = \begin{pmatrix} -3 \\ 1 \\ 2 \end{pmatrix}, \overrightarrow{BC} = \begin{pmatrix} 9 \\ -3 \\ -6 \end{pmatrix},$

      $\overrightarrow{AC} = \begin{pmatrix} 6 \\ -2 \\ -4 \end{pmatrix}$

3 $(\mathbf{a} + \mathbf{b}) \cdot (\mathbf{a} - \mathbf{b}) = 0$

4 $(7, 9, 0)$

5 a $\overrightarrow{AB} = \begin{pmatrix} 3 \\ 3 \end{pmatrix}, \overrightarrow{AC} = \begin{pmatrix} -1 \\ -2 \end{pmatrix}$

   b $\overrightarrow{AB} \cdot \overrightarrow{AC} = -9$

6 a $\begin{pmatrix} -2 \\ 10 \\ 1 \end{pmatrix}$

   b $t = 2$

7 a $r = \begin{pmatrix} 2 \\ 2 \\ 4 \end{pmatrix} + s \begin{pmatrix} 1 \\ 3 \\ 2 \end{pmatrix}$

   b $(4, 8, 8)$

   d $\begin{pmatrix} 2 \\ 6 \\ 4 \end{pmatrix}$

   e $2\sqrt{14}$

8 a 12.30 p.m.; $\begin{pmatrix} -2 \\ 11.5 \end{pmatrix}$
   b 3 km

Review exercise GDC

1 122°

2 a $\overrightarrow{QR} = \begin{pmatrix} -1 \\ 0 \\ 5 \end{pmatrix}, \overrightarrow{QP} = \begin{pmatrix} 0 \\ -1 \\ 1 \end{pmatrix}$

   b 46.1°
   c 2.60

3 a i $4\mathbf{j}$ ii $\mathbf{i} + \sqrt{3}\mathbf{k}$
      iii $2\mathbf{i} + 4\mathbf{j}$
   b $\overrightarrow{BC} = -\mathbf{i} + 4\mathbf{j} - \sqrt{3}k$
      $\overrightarrow{BD} = \mathbf{i} + 4\mathbf{j} - \sqrt{3}k$
   c i $\sqrt{20}$ ii $\sqrt{20}$
      iii 18
   d 25.8°

4 a $0, 4, -2$
   b 82.9°

5 a $\overrightarrow{OP} \cdot \overrightarrow{PQ} = 0, \overrightarrow{PQ} = \begin{pmatrix} 0 \\ 6 \\ 2 \end{pmatrix}$

   b E.g. $\mathbf{r} = \begin{pmatrix} 1 \\ -1 \\ 3 \end{pmatrix} + \lambda \begin{pmatrix} 0 \\ 6 \\ 2 \end{pmatrix}$

   c $\begin{pmatrix} 1 \\ 2 \\ 4 \end{pmatrix}$

   d 158°

6 a $\overrightarrow{AB} = \begin{pmatrix} 6 \\ -2 \\ 0 \end{pmatrix}$

   c $(36, 18, 0)$ d 5.10 m s^{-1}
   e 6 seconds   f $(18, -6, 6)$

Chapter 13
Skills check

1. a $\dfrac{\sqrt{2}}{2}$ b $\sqrt{3}$
   c $-\dfrac{\sqrt{3}}{2}$ d $-\dfrac{\sqrt{2}}{2}$
2. a $\dfrac{\sqrt{3}}{2}$ b -1
   c -1 d -0.5
3. a -1.48 b ± 2
4. a $-0.182, 2.40$ b ± 1.14

Investigation – Sine, cosine and tangent on the unit circle

1. $\sin 90° = 1$, $\cos 90° = 0$, $\tan 90°$ does not exist
2. $\sin 180° = 0$, $\cos 180° = 1$, $\tan 180° = 0$
3. $\sin 270° = -1$, $\cos 270° = 0$, $\tan 270°$ does not exist
4. $\sin 360° = 0$, $\cos 360° = 1$, $\tan 360° = 0$
5. $\sin(-90°) = -1$, $\cos(-90°) = 0$, $\tan(-90°)$ does not exist
6. $\sin(-180°) = 0$, $\cos(-180°) = -1$, $\tan(-180°) = 0$
7. $\sin 0 = 0$, $\cos 0 = 1$, $\tan 0 = 0$
8. $\sin\dfrac{\pi}{2} = 1$, $\cos\dfrac{\pi}{2} = 0$, $\tan\dfrac{\pi}{2}$ does not exist
9. $\sin\pi = 0$, $\cos\pi = -1$, $\tan\pi = 0$
10. $\sin\dfrac{3\pi}{2} = -1$, $\cos\dfrac{3\pi}{2} = 0$, $\tan\dfrac{3\pi}{2}$ does not exist
11. $\sin-\dfrac{3\pi}{2} = 1$, $\cos-\dfrac{3\pi}{2} = 0$, $\tan-\dfrac{3\pi}{2}$ does not exist
12. $\sin 4\pi = 0$, $\cos 4\pi = 1$, $\tan 4\pi = 0$

Exercise 13A

1. a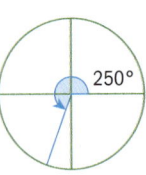
   b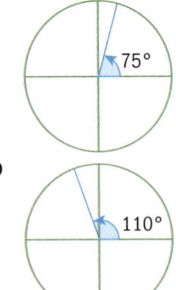
   c
   d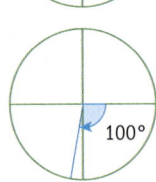
   e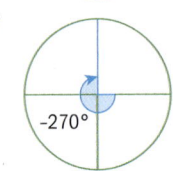
   f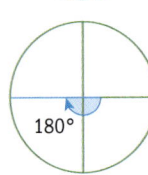
   g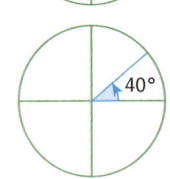
   h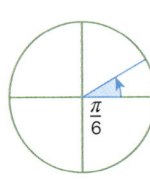

2. a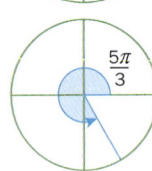
   b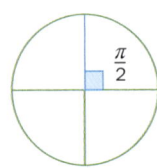
   c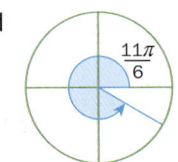
   d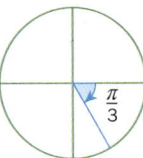
   e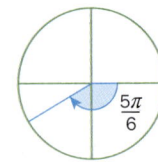
   f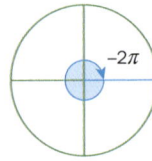
   g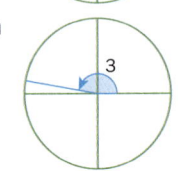
   h

For questions 3 to 8, there are many other possible correct answers.

3. a $120°, -240°, -300°$
   b $340°, -20°, -160°$
   c $255°, 285°, -105°$
   d $65°, -245°, -295°$
4. a $-35°, \pm 325°$
   b $-130°, \pm 230°$
   c $-295°, \pm 65°$
   d $240°, \pm 120°$
5. a $230°, -130°, -310°$
   b $280°, -80°, -260°$
   c $40°, -140°, -320°$
   d $155°, 335°, -205°$
6. a $\dfrac{2\pi}{3}, -\dfrac{4\pi}{3}, -\dfrac{5\pi}{3}$
   b $\dfrac{7\pi}{4}, -\dfrac{\pi}{4}, -\dfrac{3\pi}{4}$
   c $3\pi - 4.1, 4.1 - 2\pi, \pi - 4.1$
   d $\pi + 3, 2\pi - 3, 3 - \pi$
7. a $-\dfrac{\pi}{6}, \pm\dfrac{11\pi}{6}$
   b $-1, \pm(1 - 2\pi)$
   c $-2.5, \pm(2.5 - 2\pi)$
   d $\dfrac{3\pi}{5}, \pm\dfrac{7\pi}{5}$

8 a $\frac{5\pi}{4}, -\frac{3\pi}{4}, -\frac{7\pi}{4}$
   b $1.3 + \pi, 1.3 - \pi, 1.3 - 2\pi$
   c $\frac{12\pi}{7}, -\frac{2\pi}{7}, -\frac{9\pi}{7}$
   d $2\pi - 5, \pi - 5, -5 - \pi$

Exercise 13B

1 a 0.940   b 0.342
   c −0.342 d −0.940
2 a $-\frac{1}{2}$ b $-\frac{\sqrt{3}}{2}$
   c $-\frac{1}{2}$ d $\frac{\sqrt{3}}{2}$
3 a 0.8   b 0.6   c 0.6
   d −0.8 e $\frac{4}{3}$ f $-\frac{4}{3}$
   g −0.8 h $\frac{4}{3}$
4 a $\frac{a}{b}$ b a c $-b$
   d $\frac{a}{b}$ e $-a$ f b
   g $-a$ h $-b$

Exercise 13C

1 a −300°, −240°, 60°, 120°
   b ±120°, ±240°
   c −315°, −135°, 45°, 225°
   d −360°, −180°, 0°, 180°, 360°
   e ±45°, ±135°, ±225°, ±315°
   f ±30°, ±150°, ±210°, ±330°
2 a $-\frac{11\pi}{6}, -\frac{7\pi}{6}, \frac{\pi}{6}, \frac{5\pi}{6}$
   b $0, \pm\pi, \pm 2\pi$
   c $\pm\frac{\pi}{6}, \pm\frac{11\pi}{6}$
   d $-\frac{\pi}{2}, \frac{3\pi}{2}$
   e $\pm\frac{\pi}{3}, \pm\frac{2\pi}{3}, \pm\frac{4\pi}{3}, \pm\frac{5\pi}{3}$
   f $-\frac{7\pi}{4}, -\frac{3\pi}{4}, \frac{\pi}{4}, \frac{5\pi}{4}$
3 a 0°, 360°, 720°
   b −135°, −45°, 225°, 315°, 585°, 675°
   c −225°, −45°, 135°, 315°, 495°, 675°
   d ±60°, ±120°, 240°, 300°, 420°, 480°, 600°, 660°
4 a $\frac{\pi}{2}$ b $-\frac{5\pi}{6}, -\frac{\pi}{6}$
   c $\pm\frac{\pi}{4}, \pm\frac{3\pi}{4}$ d $\pm\frac{\pi}{6}, \pm\frac{5\pi}{6}$

Exercise 13D

1 a ±15°, ±165°
   b −165°, −105°, 15°, 75°
   c 90°
   d ±180°
2 a $-\frac{5\pi}{12}, -\frac{\pi}{12}, \frac{7\pi}{12}, \frac{11\pi}{12}$
   b $-\frac{11\pi}{12}, -\frac{7\pi}{12}, \frac{\pi}{4}, \frac{\pi}{12}, \frac{5\pi}{12}, \frac{3\pi}{4}$
   c $\pm\frac{\pi}{2}$ d $\pm\frac{3\pi}{4}$
3 a $\frac{2\pi}{3}, \frac{4\pi}{3}$ b $\frac{7\pi}{6}, \frac{3\pi}{2}, \frac{11\pi}{6}$
   c $\frac{3\pi}{4}, \frac{7\pi}{4}$ d $\frac{\pi}{2}$

Exercise 13E

1 a $\frac{5\sqrt{11}}{18}$ b $-\frac{7}{18}$ c $-\frac{5\sqrt{11}}{7}$
2 a $-\frac{4\sqrt{5}}{9}$ b $-\frac{1}{9}$ c $4\sqrt{5}$
3 a $\frac{\sqrt{11}}{5}$ b $\frac{5\sqrt{11}}{18}$
   c $\frac{7}{18}$ d $\frac{5\sqrt{11}}{7}$
4 a $\frac{\sqrt{63}}{32}$ b $\frac{31}{32}$
   c $\frac{\sqrt{63}}{31}$ d $\frac{31\sqrt{63}}{512}$
5 a $\frac{3}{5}$ b $\frac{4}{5}$
   c $\frac{24}{25}$ d $\frac{7}{25}$
6 a $-\frac{7}{25}$ b $-\frac{24}{7}$
   c $-\frac{336}{625}$ d $-\frac{527}{625}$
7 a $\frac{a}{\sqrt{a^2+b^2}}$ b $\frac{b}{\sqrt{a^2+b^2}}$
   c $\frac{2ab}{a^2+b^2}$ d $\frac{b^2-a^2}{a^2+b^2}$

Exercise 13F

1 a 30°, 90°, 150°
   b 22.5°, 112.5°
   c 135°
   d 45°, 135°
2 a −150°, −120°, 30°, 60°
   b 90°
   c ±150°, ±30°
   d −90°, 30°, 150°
3 a $0, \pi$
   b $\frac{\pi}{8}, \frac{7\pi}{8}$
   c $0, \frac{2\pi}{3}$
   d $0, \frac{\pi}{6}, \frac{\pi}{2}, \frac{5\pi}{6}, \pi$
4 a $\frac{\pi}{8}, \frac{5\pi}{8}$ b $\frac{\pi}{2}$
   c $0, \pi$ d $\frac{\pi}{4}, \frac{3\pi}{4}$
6 $k = 6$
7 $b = 8$

Exercise 13G

1 −346°, −194, 14°, 166°
2 ±27°, 333°
3 244°, 296°
4 55°, 235°, 415°
5 −5.33, −4.10, 0.955, 2.19
6 ±1.71, 4.58
7 −0.739
8 −0.637, 1.41

Investigation: graphing tan x

1

Angle measure (x) (degrees)	Tangent value ($\tan x$)
0	0
−30, +30	$-\frac{1}{\sqrt{3}}, \frac{1}{\sqrt{3}}$
−45, +45	−1, 1
−60, +60	$-\sqrt{3}, \sqrt{3}$
120	$-\sqrt{3}$
135	−1
150	$-\frac{1}{\sqrt{3}}$
180	0
210	$\frac{1}{\sqrt{3}}$
225	1
240	$\sqrt{3}$
300	$-\sqrt{3}$
315	−1
330	$-\frac{1}{\sqrt{3}}$
360	0

3 tan ±90° and tan ±270° are undefined. The limit of the tangent as the angle approaches ±90° or ±270° is infinite. Asymptotes are often shown on graphs for values that do not exist.

Exercise 13H

1 −297°, −117°, 63°, 243°
2 −107°, 73°, 253°
3 124°, 304°
4 38°, 142°, 398°, 502°
5 −5.88, −2.74, 0.405, 3.55
6 −1.88, 1.26
7 4.55
8 −4.66, 1.20, 2.28, 4.77

Investigation: transformations of sin x and cos x

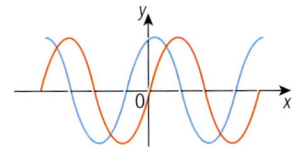

1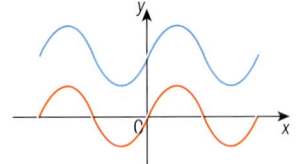

2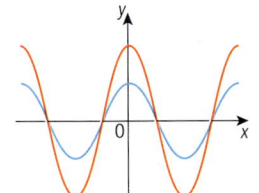

3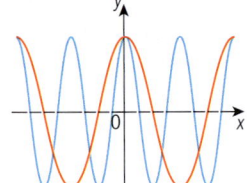

4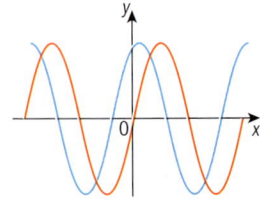

5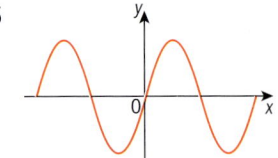

Exercise 13I

1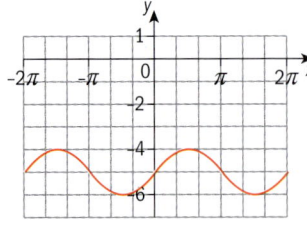

2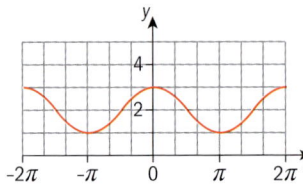

3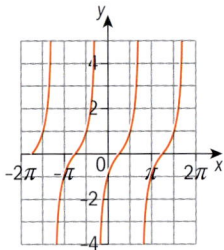

4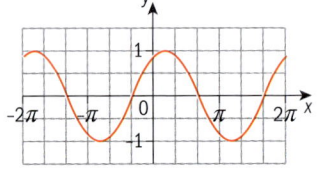

5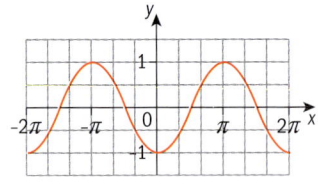

6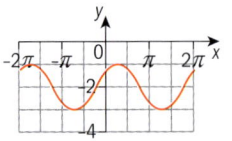

7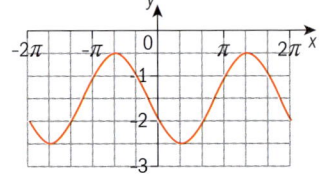

8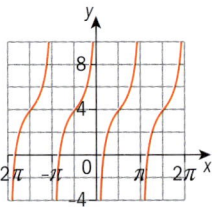

9 $y = \cos\left(x - \dfrac{2\pi}{3}\right)$ or $y = \sin\left(x - \dfrac{\pi}{6}\right)$

10 $y = \sin x + 1$

11 $y = \tan\left(x - \dfrac{\pi}{4}\right)$

12 $y = \cos\left(x - \dfrac{\pi}{4}\right) - 1.5$

Exercise 13J

1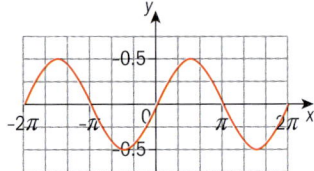

2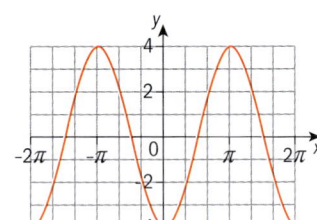

3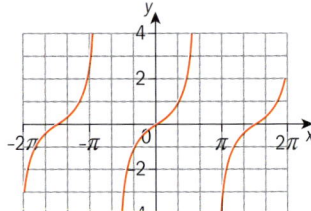

4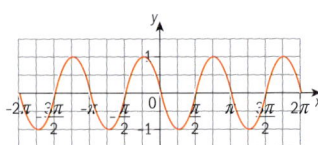

5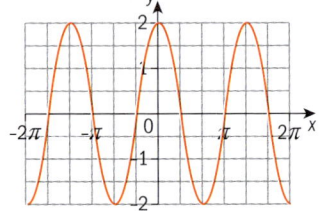

6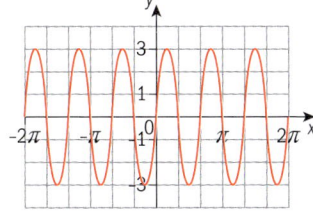

7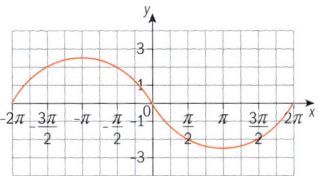

8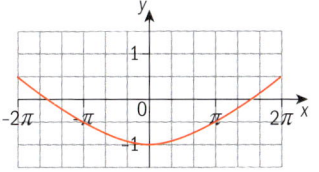

9 $y = 7.5\sin x$

10 $y = \cos(0.25x)$

11 $y = \tan(0.25x)$

12 $y = -3\cos(0.5x)$ or
$y = \sin\left(x - \dfrac{\pi}{4}\right) - 1.5$

Exercise 13K

For questions 1 to 4, answers may vary.

1 $y = 3.5\sin\left(x - \dfrac{2\pi}{3}\right) - 1.5$,
$y = 3.5\cos\left(x + \dfrac{5\pi}{6}\right) - 1.5$

2 $y = \sin\left(\dfrac{1}{2}\left(x + \dfrac{4\pi}{3}\right)\right) - 2$,
$y = \cos\left(\dfrac{1}{2}\left(x + \dfrac{\pi}{3}\right)\right) - 2$

3 $y = 2\sin(2x) + 1$,
$y = 2\cos\left(2\left(x - \dfrac{\pi}{4}\right)\right) + 1$

4 $y = 5\sin\left(\dfrac{2}{3}\left(x - \dfrac{\pi}{2}\right)\right)$,
$y = -5\cos\left(\dfrac{2}{3}\left(x + \dfrac{\pi}{4}\right)\right)$

5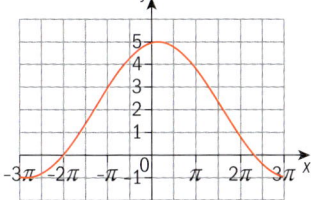

6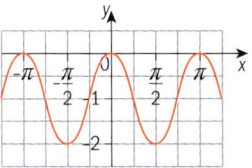

7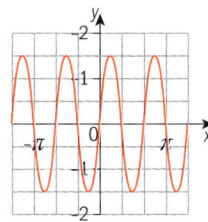

8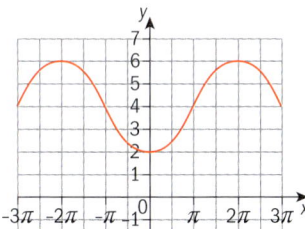

Exercise 13L

1 a, b
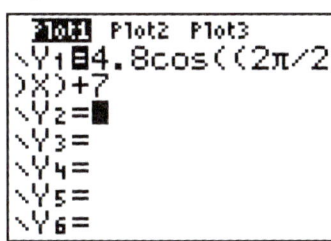

c
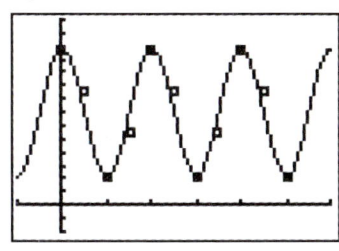

d
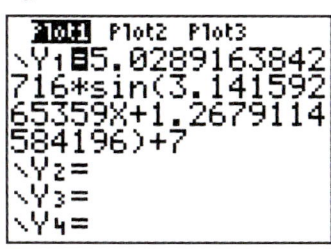

2 a, b
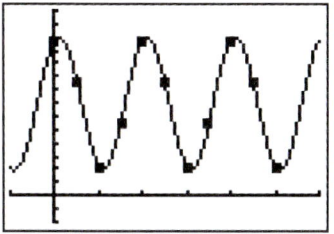

c
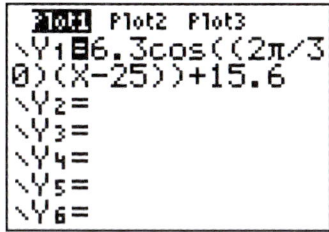
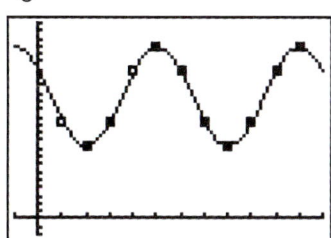

d
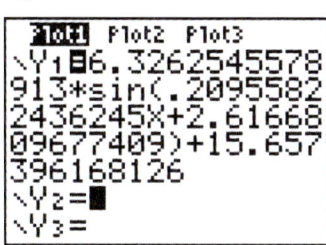
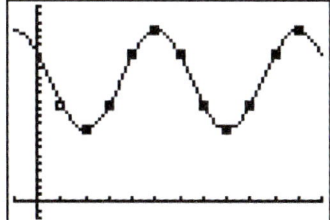

Answers 763

3 a, b

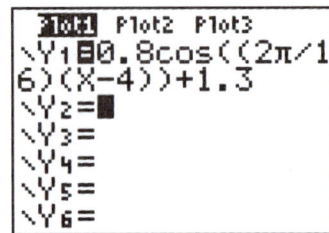

c

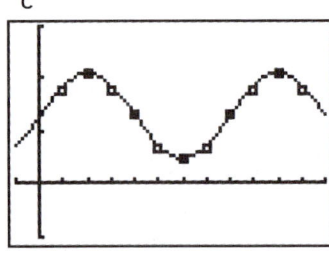

d

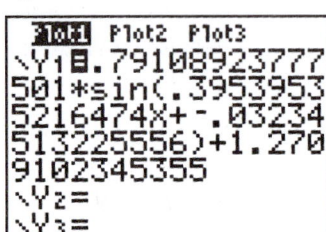

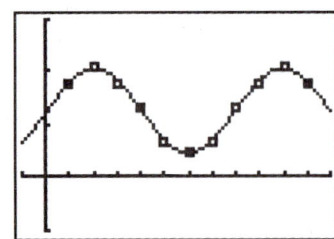

Exercise 13M

1 a   Approximately 12 hours
   b   9.49 m
   c   13.5 m
   d   05:30
2 a $-3.06\,°C$
   b $30\,°C$, day 187 (about 6 July)
   c   about 90 days: days 1–49 inclusive and days 325–365 inclusive
3 a   46 m
   b $h(t) = 22.5\sin\left(\dfrac{2\pi}{20}(x-5)\right) + 23.5$

c   10.3 m
d   4.75 minutes

4 a $g(x) = -16\cos\left(\dfrac{2\pi}{12}(x-1)\right) + 21$
   b   21 gallons
   c   Early May and late August

Review exercise non - GDC

1 a -0.342
   b -0.342
   c 0.342
2 a 0.643
   b -0.643
   c -0.643
3 a $\pm 120°, \pm 240$
   b $-330°, -150°, 30°, 210°$
   c $-270°, -150°, -30°, 90°, 210°, 330°$
4 $0, \dfrac{2\pi}{3}, \pi$
5 a i $a = -5, c = 0, d = 6$
      ii $b = \dfrac{2\pi}{\text{period}}$, and the period is 8. $b = \dfrac{2\pi}{8} = \dfrac{\pi}{4}$
   b $4 < x < 8$
6 a $\dfrac{\sqrt{21}}{5}$ b $\dfrac{\sqrt{21}}{2}$ c $\dfrac{4\sqrt{21}}{25}$
7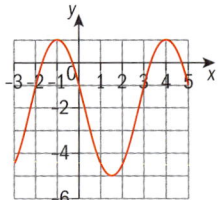

Review exercise GDC

1 a   48.6, 131.4
   b $\pm 129, 231$
   c $-70.3, 109.7, 289.7$
2 a $-3.36, 0.515, 2.85, 6.06$
   b 0.607
   c $\pm 1.89, 0$
3 a $a = -4, b = \dfrac{\pi}{2}, c = 3$
   b   0.667, 3.33, 4.67

4 a $P = 4, Q = 7$
   b

   c $t = 2$, at 2:00
   d   8 hours
5 a $A = 2.825, B = 12.175$
   b   9.91

Chapter 14

Skills check

1 a $\dfrac{\sqrt{2}}{2}$
   b -1
   c $-\dfrac{1}{\sqrt{3}}$ or $-\dfrac{\sqrt{3}}{3}$
   d $-\dfrac{\sqrt{3}}{2}$
2 a $x = 0, \pi, 2\pi$
   b $x = \dfrac{\pi}{6}, \dfrac{\pi}{2}, \dfrac{5\pi}{6}, \dfrac{3\pi}{2}$
   c $x = \dfrac{\pi}{2}, \pi, \dfrac{3\pi}{2}$
3 a $2x^3 e^x + 6x^2 e^x$
   b $2 + \ln(x^2)$
   c $\dfrac{-x^2 + 10x + 4}{(x^2+4)^2}$
   d $\dfrac{1 - \ln x}{x^2}$

Exercise 14A

1   $3\cos x + 2\sin x$
2   $\dfrac{3}{\cos^2 3x}$
3   $-\dfrac{2\cos x}{\sin^2 x}$
4   $-2\cos t \sin t$ or $-\sin(2t)$
5   $\dfrac{\cos\sqrt{x}}{2\sqrt{x}}$
6   $\dfrac{2\tan x}{\cos^2 x}$
7   $-\dfrac{1}{2}\sin\dfrac{x}{2} + 4\cos(4x)$

8. $\dfrac{2\sin(2x)}{\cos^2(2x)}$

9. $\dfrac{-8\pi\cos(\pi x)}{\sin^3(\pi x)}$

10. $[\cos(\sin x)]\cos x$

11. a $\dfrac{3x^2}{\cos^2(x^3)}$

    b $-4\cos^3 x \sin x$

12. a $3\cos(3x-4)$

    b $-9\sin(3x-4)$

Exercise 14B

1. $y-1=1\left(x-\dfrac{\pi}{2}\right);\ y-1=-1\left(x-\dfrac{\pi}{2}\right)$

2. $y-2=4\left(x-\dfrac{\pi}{4}\right);\ y-2=-\dfrac{1}{4}\left(x-\dfrac{\pi}{4}\right)$

3. -2

4. a $-\dfrac{1}{2}$ b $-2\sin(2x)$

   c $y+\dfrac{1}{2}=-\sqrt{3}\left(x-\dfrac{\pi}{3}\right)$

5. $\dfrac{\pi}{3},\dfrac{5\pi}{3}$

Exercise 14C

1. $-12\sin\left(2x-\dfrac{\pi}{3}\right)+3$

2. $\dfrac{1}{(1+\cos x)}$

3. xe^x

4. $e^{\sin 2t}\cos 2t$

5. $2e^x \sin x$

6. $\dfrac{t}{\cos^2 t}+\tan t$

7. $3e^{3x}\cos 4x - 4e^{3x}\sin 4x$

8. $\dfrac{1}{\cos^2 2x\sqrt{\tan 2x}}$

9. $\dfrac{\cos x}{x}-\ln x \sin x$

10. $-\dfrac{\sin x}{\cos x}$ or $-\tan x$

11. a $\dfrac{2}{x}$ b $\dfrac{1}{2}\cos\dfrac{x}{2}$

    c $\dfrac{1}{2}\ln 3x^2 \cos\dfrac{x}{2}+\dfrac{2}{x}\sin\dfrac{x}{2}$

12. $a=1,\ b=2$

Exercise 14D

1. Relative minimum: $\left(\dfrac{4\pi}{3},-2\right)$;

   relative maximum: $\left(\dfrac{\pi}{3},2\right)$

2. relative minimums: $\left(\dfrac{\pi}{2},1\right),\left(\dfrac{3\pi}{2},-3\right)$; relative

   maximums: $\left(\dfrac{\pi}{6},\dfrac{3}{2}\right),\left(\dfrac{5\pi}{6},\dfrac{3}{2}\right)$

3. decreasing: $\dfrac{\pi}{2}<x<\pi$;

   increasing $0<x<\dfrac{\pi}{2}$; concave

   down: $0<x<\pi$; relative

   maximum: $\left(\dfrac{\pi}{2},1\right)$

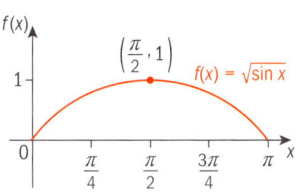

4. decreasing:

   $0<x<\dfrac{\pi}{4},\dfrac{\pi}{2}<x<\dfrac{3\pi}{4}$;

   increasing

   $\dfrac{\pi}{4}<x<\dfrac{\pi}{2},\dfrac{3\pi}{4}<x<\pi$;

   concave up:

   $\dfrac{\pi}{8}<x<\dfrac{3\pi}{8},\dfrac{5\pi}{8}<x<\dfrac{7\pi}{8}$;

   concave down:

   $0<x<\dfrac{\pi}{8},\dfrac{3\pi}{8}<x<\dfrac{5\pi}{8},\dfrac{7\pi}{8}<x<\pi$;

   relative maximum: $\left(\dfrac{\pi}{2},1\right)$;

   relative minimums:

   $\left(\dfrac{\pi}{4},0\right),\left(\dfrac{3\pi}{4},0\right)$;

   x points:

   $\left(\dfrac{\pi}{8},\dfrac{1}{2}\right),\left(\dfrac{3\pi}{8},\dfrac{1}{2}\right),\left(\dfrac{5\pi}{8},\dfrac{1}{2}\right),\left(\dfrac{7\pi}{8},\dfrac{1}{2}\right)$

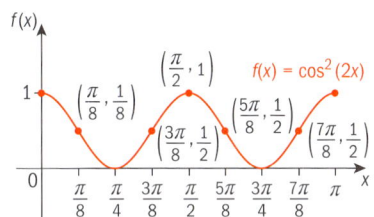

5. a $f'(x) = -2\sin 2x + 2\cos x(-\sin x)$

   $= -2\sin 2x - 2\sin x \cos x$

   $= -2\sin 2x - \sin 2x$

   $= -3\sin 2x$

   b $\left(\dfrac{\pi}{2},-1\right)$

   c $f''(x) = -6\cos 2x$

   d $\left(\dfrac{\pi}{4},\dfrac{1}{2}\right),\left(\dfrac{3\pi}{4},\dfrac{1}{2}\right)$

6. a i $f'(x) = x\cos x + \sin x$

      ii $a=-1,\ b=2$

   b i $x \approx 2.03, 4.91$

      ii $f''(2.03) \approx -2.71 < 0$
          $\Rightarrow$ relative maximum
          at $x=2.03$

          $f''(4.91) \approx 5.21 > 0 \Rightarrow$
          relative minimum at
          $x=4.91$

7. a $f'(x) = -x^2 \sin x + 2x\cos x$

   b minimum: -11.6;
      maximum: 7.09

8. a $d'(\theta) = -2\sin\theta - \dfrac{4\sin\theta\cos\theta}{\sqrt{25-4\sin^2\theta}}$

   or $-2\sin\theta - \dfrac{2\sin 2\theta}{\sqrt{25-4\sin^2\theta}}$

   b

   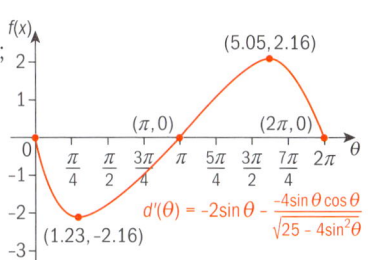

   c i The blade is closest
         the center of the
         wheel when $d(\theta)$ has
         a relative minimum or
         at an endpoint. There
         is a relative minimum
         when $d'(\theta)$ changes from
         negative to positive at
         $\theta = \pi$. Testing the
         endpoints and critical

numbers we find
d(0) = 7, d(2π) = 7
and d(π) = 3. So the
closest distance is 3
metres and it occurs
when the angle of
rotation is π.

ii The distance is
changing fastest
when d'(θ) has a
relative minimum or
maximum. This occurs
when θ is 1.23 radians
or 5.05 radians.

Exercise 14E

1. $2\sin x - 3\cos x + C$
2. $\frac{1}{3}x^3 + 3\sin\left(\frac{1}{3}x\right) + C$
3. $-\cos(\pi x) + C$
4. $-\frac{1}{2}\cos(2x+3) + C$
5. $\sin(5x^4) + C$
6. $\frac{1}{4}\sin(4x^2 - 4x) + C$
7. $\frac{1}{3}e^{\tan 3x} + C$
8. $\sin(\ln x) + C$
9. $\frac{1}{3}\sin^3 x + C$
10. $-\ln(\cos x) + C$, $\cos x > 0$
11. a $-e^{\sin x}\sin x + e^{\sin x}\cos^2 x$
    b $e^{\sin x} + C$
12. a $f'(x) = \frac{1}{\cos x}(-\sin x)$
    $= -\frac{\sin x}{\cos x}$
    $= -\tan x$
    b $-\frac{1}{2}[\ln(\cos x)]^2 + C$

Exercise 14F

1. $\sqrt{3}$; 1.73
2. 4; 4
3. $\frac{3\sqrt{3}}{4}$; 1.30

4. $\frac{\sqrt{3} - \sqrt{2}}{2}$; 0.159

Exercise 14G

1. 12.1
2. 6.31
3. $\frac{\pi}{6}$
4. a 3.97
   b 38.3
5. a $a = 2, b = \frac{1}{2}$
   b $\int_0^{2\pi} 2\sin\left(\frac{1}{2}x\right)dx = 8$
6. a i $c = 1, d = 2$
      ii $\frac{\pi}{2}$ and $\frac{7\pi}{6}$
   b i 2
      ii 4.25
   c 9.12

Exercise 14H

1. a $v = e^t\cos t + e^t\sin t$
   b $a = 2e^t\cos t$
2. a -2 m s^{-1}
   b $\frac{\pi}{2}$ s
   c -1 m
3. a i $\frac{\pi}{2}, \frac{3\pi}{2}$
      ii $\frac{\pi}{2} < t < \frac{3\pi}{2}$
   b $a(t) = -e^{\sin t}\sin t + e^{\sin t}\cos^2 t$
   c $s(t) = e^{\sin t} + 3$
4. a $\int_0^4 (4\sin t + 3\cos t)dt$
   b 4.34 m
5. a i -2.52 m s^{-2}
      ii speeding up
   b 2.51 s and 3.54 s
   c 7.37 m
6. a 5.82 m s^{-2}

b i

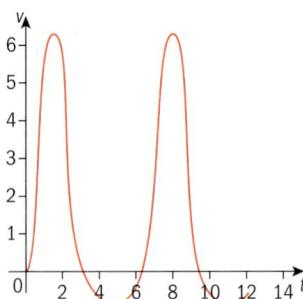

ii 1.11 s, 2.03 s, 7.39 s, 8.31 s

iii No, the particle does not return to the origin. Looking at the area between the curve and the t-axis, there is more area above the axis than below, indicating that the particle moves to the right a greater distance than to the left, so it never returns to the origin.

c 24.1 m

Review exercise non-GDC

1. a $2\sin(1 - 2x)$
   b $3\sin^2 x \cos x$
   c $\frac{e^{\tan t}}{\cos^2 t}$
   d $\frac{x\cos x^2}{\sqrt{\sin x^2}}$
   e $-x^2\sin x + 2x\cos x$
   f $\frac{1}{\tan x \cos^2 x}$ or $\frac{1}{\sin x \cos x}$
   g $(\ln x)(\cos x) + \frac{\sin x}{x}$
   h $-2\sin^2 x + 2\cos^2 x$ or $2\cos 2x$

2. a $x^4 + \cos x + C$
   b $\frac{1}{3}\sin(3x) + C$
   c $-\frac{1}{4}\cos(4x+1) + C$
   d $\frac{1}{4}\sin(2x^2) + C$

e $\dfrac{1}{2\cos(2t+1)} + C$

f $-\cos(\ln x) + C$

g $\dfrac{1}{2}e^{\sin x^2} + C$

h $-\dfrac{6}{2+\sin x} + C$

3 a 0   b $2+\pi$
   c 2   d 2

4 $x = 2$

5 $\left(\dfrac{2\pi}{3}, \dfrac{\sqrt{3}}{2}\right)$

6 $y = \dfrac{1}{2}x^2 + \cos x + 1$

7 a $p = 2, q = 2$
   b $3\pi + 2$

Review exercise GDC

1 a 4.53
   b 1.36

2 a 4.93
   b 45.0

3 1.23

4 a i $s'(t) = -10\sin(5t)\,e^{\cos(5t)}$

    ii $s''(t) = -10\sin(5t)$
       $\times [e^{\cos(5t)})(-\sin(5t))(5)]$
       $+ e^{\cos(5t)}[-10(\cos(5t))(5)]$
       $= 50\sin^2(5t)[e^{\cos(5t)}$
       $- 50\cos(5t)(e^{\cos(5t)})]$
       $= 50\,e^{\cos(5t)}(\sin^2(5t)$
       $- \cos(5t))$

    iii
$s'\left(\dfrac{\pi}{5}\right) = 0$ and $s''\left(\dfrac{\pi}{5}\right) \approx 18.4 > 0$

Therefore by the second derivative test s has a relative minimum at $t = \dfrac{\pi}{5}$.

   b 14.2 m

Chapter 15

Skills check

1 a 5.5
   b $\dfrac{8516}{39} = 14.6$ (3sf)

2 a 15
   b 56
   c 0.267

3 a 1.71875
   b 2.98
   c 8.68

Exercise 15A

1 a Discrete
   b Continuous
   c Discrete
   d Continuous

2 a

s	P(S = s)
2	$\dfrac{1}{36}$
3	$\dfrac{2}{36}$
4	$\dfrac{3}{36}$
5	$\dfrac{4}{36}$
6	$\dfrac{5}{36}$
7	$\dfrac{6}{36}$
8	$\dfrac{5}{36}$
9	$\dfrac{4}{36}$
10	$\dfrac{3}{36}$
11	$\dfrac{2}{36}$
12	$\dfrac{1}{36}$

b

n	P(N = n)
0	$\dfrac{25}{36}$
1	$\dfrac{10}{36}$
2	$\dfrac{1}{36}$

c

n	P(N = n)
1	$\dfrac{11}{36}$
2	$\dfrac{9}{36}$
3	$\dfrac{7}{36}$
4	$\dfrac{5}{36}$
5	$\dfrac{3}{36}$
6	$\dfrac{1}{36}$

d

p	P(p)
1	$\dfrac{1}{36}$
2	$\dfrac{2}{36}$
3	$\dfrac{2}{36}$
4	$\dfrac{3}{36}$
5	$\dfrac{2}{36}$
6	$\dfrac{4}{36}$
8	$\dfrac{2}{36}$
9	$\dfrac{1}{36}$
10	$\dfrac{2}{36}$
12	$\dfrac{4}{36}$
15	$\dfrac{2}{36}$
16	$\dfrac{1}{36}$
18	$\dfrac{2}{36}$
20	$\dfrac{2}{36}$
24	$\dfrac{2}{36}$
25	$\dfrac{1}{36}$
30	$\dfrac{2}{36}$
36	$\dfrac{1}{36}$

3 a

T	2	3	4	5	6
P(T = t)	$\dfrac{1}{36}$	$\dfrac{4}{36}$	$\dfrac{10}{36}$	$\dfrac{12}{36}$	$\dfrac{9}{36}$

   b $P(T > 4) = \dfrac{21}{36} = \dfrac{7}{12}$

4 a

s	1	2	3	6	10
P(S = s)	$\frac{1}{6}$	$\frac{1}{3}$	$\frac{1}{6}$	$\frac{1}{6}$	$\frac{1}{6}$

b $\frac{1}{2}$

5 a $\frac{1}{6}$ **b** $\frac{1}{2}$

6 $\frac{1}{36}$

7 0.2

8 $\frac{27}{40}$

9 a $a = \frac{1}{8}, b = \frac{5}{24}$

b $\frac{25}{96}$

10 b

c

c	2	3	4	5	6
P(C = c)	$\frac{1}{18}$	$\frac{5}{18}$	$\frac{6}{18}$	$\frac{5}{18}$	$\frac{1}{18}$

Investigation – dice scores

1

d	0	1	2	3	4	5
P(D = d)	$\frac{6}{36}$	$\frac{10}{36}$	$\frac{8}{36}$	$\frac{6}{36}$	$\frac{4}{36}$	$\frac{2}{36}$

2

d	0	1	2	3	4	5
Expected frequency	6	10	8	6	4	2

3 Mean = $\frac{35}{18}$

4

d	Expected frequency
0	$\frac{150}{9}$
1	$\frac{250}{9}$
2	$\frac{200}{9}$
3	$\frac{150}{9}$
4	$\frac{100}{9}$
5	$\frac{50}{9}$

mean = $\frac{35}{18}$

5 Same mean

6 $\frac{35}{18}$

Exercise 15B

1 $\frac{91}{6} = 15.2 \,(3\,\text{sf})$

2 $x = \frac{3}{8}, y = \frac{1}{8}$

3 $5\frac{1}{3}$

4 $5\frac{2}{3}$

5 a $k = \frac{1}{25}$
b $E(X) = 5$

6 a

X	1	2	3
P(X = x)	0.2	1 – k	k – 0.2

b $0.2 \leq k \leq 1$,
c $k + 1.6$

7 0.2

8 a

r	P(R = r)
1	$\frac{18}{90}$
2	$\frac{16}{90}$
3	$\frac{14}{90}$
4	$\frac{12}{90}$
5	$\frac{10}{90}$
6	$\frac{8}{90}$
7	$\frac{6}{90}$
8	$\frac{4}{90}$
9	$\frac{2}{90}$

b $3\frac{2}{3}$
c 1

9 b $\frac{16}{125}$, **c** $\left(\frac{4}{5}\right)^{n-1}\left(\frac{1}{5}\right)$ **d** 1

10 a P(Z = 0) = 0.7489
b E(Z) = 70, The expected amount to be won on a ticket
c Lose $0.30

Investigation: The binomial quiz

1 T **2** T **3** F **4** F **5** F

You would expect to get 2.5 questions right

Probability that you get exactly 3 right out of 5 = 0.3125

Exercise 15C

1 a $\frac{1}{4}$ **b** $\frac{1}{16}$

c $\frac{5}{16}$ **d** $\frac{15}{16}$

2 a 0.329
b 0.351 P(X < 2)
c 0.680
d 0.649

3 a 0.0389
b 0.952
c 0.00870
d 0.932

Exercise 15D

1 1; 0.421

2 a 0.257
b 0.260

3 a 0.851
b 0.000491
c 0.0109

4 a 0.0584
b 0.9996

5 0.913

6 a 0.224
b 0.399

7 a i 0.0307
   ii 0.463
   iii 0.171
b i 0.215
   ii 0.0292
   iii 0.158

Exercise 15E
1. $n = 4$
2. 68
3. $n = 7$
4. 9 attempts
5. 7 times

Exercise 15F
1. a 20
   b $6\frac{2}{3}$
   c 10
2. $n = 25$
3. a $X \sim B(15, 0.25)$
   b 3.75
   c 0.000795
4. a 0.51
   b 38.2

Exercise 15G
1. Mean = 0
   Variance = 0
2. Mean = 7.2
   Standard deviation = 1.70 (3 sf)
3. Mean = 20
   Standard deviation = 3.16 (3 sf)
3. a $E(X) = \frac{5}{3}$
   b $Var(X) = \frac{25}{18}$
   c $P(X < \mu) = 0.485$ (3 sf)
5. a $E(X) = \frac{22}{5}$
   b $Var(X) = \frac{88}{25}$
   c $P(X < 4) = 0.332$ (3 sf)
6. $P(X \geq 3) = 0.873$ (3 sf)
7. a $n = 26$
   b $Var(X) = 5.46$
8. $n = 12, p = 0.8$, $P(X = 6) = 0.0155$

Exercise 15H
1. a $P(-1 < Z < 1) = 0.683$
   b $P(-2 < Z < 2) = 0.954$
   c $P(-3 < Z < 3) = 0.997$
2. a 0.272 b 0.483

3. a 0.159
   b 0.00820
4. a 0.159
   b 0.0401
5. a 0.742
   b 0.236
   c 0.0359
   d 0.977
   e 0.390
6. a 0.306
   b 0.595
   c 0.285
7. a 0.311
   b 0.215

Exercise 15I
1. a 0.655 b 0.841
   c 0.186 d 0.5
2. a 0.672 b 0.748
   c 0.345
3. a 0.994 b 0.977
   c 0.494

Exercise 15J
1. a 0.933 b 0.691
   c 0.736
2. 477
3. a 0.0668 b 15.9%
4. 53.5%
5. a 0.106 b 0.00118

Exercise 15K
1. a 1.42 b 0.407
   c 2.58
2. a 1.77 b −1.00
   c −0.841
3. a 0.385 b 1.60
4. a 1.64 b 0.842

Exercise 15L
1. 5.64
2. a 413.4
   b 432.8
3. a 0.106 b 0.864
   c 498.9 and 505.1
4. a 0.673 b 582 g
5. a 79.7 marks
   b 35.8 marks

Exercise 15M
1. 8.33
2. 15.4
3. $\mu = 49.9$ and $\sigma = 4.23$
4. $\mu = 71.4$ and $\sigma = 13.8$.
5. 7.66
6. 546.5 g
7. a 0.389   b 34.9%
8. 54.3
9. 0.260
10. a 126; 33.7
    b Yes (60.5%)
11. $\mu = 507.2$ and $\sigma = 7.41$.

Review exercise non-GDC
1. a 6   b $-\frac{7}{15}$
2. a $\frac{1}{35}$ b 3
3. $x = \frac{3}{8}, \frac{13}{64}$
4. a 2, 4, 6, 8, 12, 16
   b $\frac{1}{8}, \frac{2}{8}, \frac{1}{8}, \frac{2}{8}, \frac{1}{8}, \frac{1}{8}$
   c 7.5   d £62.50
5. $\frac{40}{243}$
6. 0.2
7. a 85   b 0.023

Review exercise GDC
1. a $\frac{19}{27}$
   b

x	−5	1
$P(X = x)$	$\frac{8}{27}$	$\frac{19}{27}$

   c i $-\$\frac{7}{9}$
      ii −$7
2. a 0.254 b 0.448
3. 0.0243
4. a i 0.0881
      ii 0.00637
   b 2   c 14
5. 1.44
6. a 8.68
   b 0.755
7. 38.9; 8.63
8. a 33.3 b 0.328
   c 0.263

Chapter 18

Exercise 1A
1. a 11 b 10 c 8
   d 4 e 5 f 3
   g 16 f 3
2. a 5 b $1\frac{1}{2}$
   c $\frac{5}{4}$ d 24
3. a 12 b 540
   c 16 d 5
4. a 5 b 8 c 8 d 2
5. a 2 b 4 c 34

Exercise 1B
1. a $\frac{\sqrt{2}}{2}$ b $2\sqrt{3}$ c $\sqrt{5}$
   d $2\sqrt{10}$ e $\frac{\sqrt{10}}{5}$
2. a $2\sqrt{3}$ b $5\sqrt{3}$ c $6\sqrt{2}$
   d $6\sqrt{2}$ e $15\sqrt{3}$
3. a 6 b 9 c $16\sqrt{3}$
   d $6\sqrt{6}$ e $75\sqrt{15}$
4. a $5\sqrt{5}$ b $2\sqrt{2}$ c $4\sqrt{3}$
   d $-\sqrt{2}$ e 0
5. a $11+6\sqrt{2}$ b $5+2\sqrt{6}$
   c $1-2\sqrt{2}$
   d $4+\sqrt{3}-4\sqrt{2}-\sqrt{6}$
   e 2
6. a $\frac{(\sqrt{21}+\sqrt{7})}{7}$
   b $\frac{(1+2\sqrt{3})}{11}$
   c $\frac{(5-\sqrt{5})}{4}$
   d $16+11\sqrt{2}$
7. a $\frac{11\sqrt{3}}{3}$ b $\frac{13\sqrt{3}}{6}$
   c $\frac{12\sqrt{5}}{5}$

Exercise 1C
1. a 1, 2, 3, 6, 9, 18
   b 1, 3, 9, 27
   c 1, 2, 3, 5, 6, 10, 15, 30
   d 1, 2, 4, 7, 14, 28
   e 1, 2, 3, 6, 13, 26, 39, 78

2. a $2^2 \times 3^2$ b $2^2 \times 3 \times 5$
   c 2×3^3 d 2^5 e $2^4 \times 7$
3. a 4 b 2
4. a 336 b 540

Exercise 1D
1. a $\frac{11}{12}$ b $\frac{16}{15}$ c 1
   d $\frac{211}{81}$ or $2\frac{49}{81}$
2. a $\frac{4}{9}$ b $\frac{7}{20}$
   c $\frac{2}{3}$ d $\frac{5}{8}$
3. a $\frac{18}{5}$ b $\frac{22}{7}$
   c $\frac{93}{4}$ d $\frac{167}{72}$
4. a $4\frac{4}{7}$ b $33\frac{1}{3}$
   c $4\frac{1}{4}$ d $14\frac{8}{11}$
5. a 0.32 b 0.714
   c 3.8 d 2.647

Exercise 1E
1. a 52% b 70%
2. a 2.24 CHF b 0.54 GBP
   c 187.57 EUR d 10400 JPY

Exercise 1F
1. 576 GBP
2. 14875 JPY
3. 7%
4. 26.5%
5. 26542100
6. 32 USD
7. 3.40 to 4.00 GBP so 0.60 GBP
8. No, 10% of 50.00 AUD is 5.00 AUD making a total of 55.00 AUD 10% of the new price is 5.50 AUD so would be 49.50 AUD.

Exercise 1G
1. 5:4
2. 105:100
3. 21:160
4. 15.6×72=1123.3 cm or 11.232 m
5. 3 km = 3000 m = 300 000 cm so the scale is 1:300 000 ÷1.5 = 1:200 000 800 ÷ 200 000 = 0.004 m or 0.4 cm
6. 72 USD = 5 + 3 = 8 parts, so 1 part = 9 USD Hence 45 USD: 27 USD is donated
7. 5:3:2 so 5 + 3 + 2 = 10 parts 1 part = 15 items So 75:45:30 items, i.e. 75 brownies, 45 chocolate chip cookies, 30 flapjacks

Exercise 1H
1. 5000:7000:4000 simplifies to 5:7:4 5 + 7 + 4 = 16 parts = 24000 so 1 part is 1500 USD so they receive Josh = 1500 × 5 = 7500 USD Jarrod = 1500 × 7 = 10500 USD Se Jung = 1500 × 4 = 6000 USD
2. 12 + 18 + 20 = 50 marks = 75 minutes so 1 mark = 1.5 minute So 12 × 1.5 = 18 minutes 18 × 1.5 = 27 minutes 20 × 1.5 = 30 minutes

Exercise 1I
1. a rational b rational
   c irrational d rational
   e rational f irrational
   g rational h irrational
   I rational j irrational
2. a a and g b a
3. a $83 = \frac{83}{1}$ b $\frac{4}{9}$
   c - d $\frac{-24}{25}$
   e $-0.45 = \frac{-5}{11}$ f -
   g $-4 \times 9 = -36$ h -
   I $\frac{1123}{900} = 1\frac{223}{900}$
   j -

Exercise 1J
1. a 2180 b 400 c 4000
   d 21 e 13
2. a 0.69 b 28.8 c 1.00
   d 77.985 e 0.06
3. a 2200 b 440 c 3500
   d 21 e 13
4. a 0.69 b 28.8 c 1.00
   d 78.0 e 0.06

5 a 0.67   b 0.07   c 0.39
6 a 50÷10=5   b $\frac{(3\times 4)}{2} = 6$
  c $\frac{(8-1)}{10^2} = 0.07$
7 a 5.46   b 5.77   c 0.084

Exercise 1K
1 a 1.475×10^3 b 2.31×10^5
  c 2.8×10^9 d 3.5×10^4
  e 7.35×10^6
2 a 62500   b 420 000 000
  c 355.4
3 a 1.232×10^{-4}
  b 4.515×10^{-5}
  c 6.17×10^{-1}
  d 7.5×10^{-6}
  e 3.49×10^{-4}
4 a 0.00000035
  b 0.000000089
  c 0.01253
5 1 sec = 3×10^5 m
  So $\frac{1}{3}$ sec = 10^5 m
  So 1 m = $0.33\times10^{-5} = 3.3\times10^{-4}$

Exercise 1L
1 a A = {1, 2, 3, 4, 6, 8, 12, 18, 24, 36, 72}
  b B = {2, 3}
  c C = {2}
  d D = {14, 28, 42, 56, 70,... }
  e E = {−3, −2, −1, 1, 2, 3}
  f F = {20, 21, 22, 23, 24,...}
  g G = {}
2 a 11
  b 2
  c 1
  d infinite
  e 6
  f infinite
  g 0

Exercise 1M
1 a Yes, all the elements of B are contained in A
  b No, they have elements in common
  c {4, 5}
  d {1, 2, 3, 4, 5, 6}
2 a A = {1, 2, 3, 4, 6, 9, 12, 18, 36} and
      B = {1, 3, 5, 15}
  b No, they have different elements
  c No, they have some elements in common
  d {1, 3}
  e {1, 2, 3, 4, 5, 6, 9, 12, 15, 18, 36}
3 a A = {17, 18, 19, 20, 21, 22...} and
      B = {20, 40, 60, 80...}
  b Yes
  c No, they have some elements in common
  d {20, 40, 60, 80...} = B
  e {17, 18, 19, 20, 21, 22...} = A
4 {x | x is all positive integers which are not multiples of 3}
5 {40, 50, 60, 70,... }
6 (different answers possible)
  a A = {1, 3, 5, 7,...} and B = {2, 4, 6, 8,...}
  b A = {1, 2, 3, 4, 5, 6, 7, 8, 9, 10} and B = {4, 7, 10, 13, 16,...}
  c A = {1, 2, 3} and B = {4, 5}
  d A = {1, 2, 3, 4, 5} and B = {2, 4, 6, 8}
  e A = {1, 3, 5, 7} and B = {2, 4, 6, 8}
  f A = {1, 2, 3, 4} and B = {1, 2, 4, 6, 7, 8}
  g A = {1, 2, 3, 4, 5, 6} and B = {2, 4, 6}

Exercise 1N
1 a $x < 2$
  b $-1 \leq x < 5$
  c $x > 2$
  d $-4 \leq x \leq 3$

2 a
  b
  c
  d

Exercise 1O
1 a Input / Output (1→2, 2→4, 3→6, 4→8, 5→10)
  b
x	1	2	3	4	5
y	2	4	6	8	10
  c (1, 2), (2, 4), (3, 6), (4, 8), (5, 10)
  d

2 a Input / Output
  b
x	−3	−2	−1	0	1	2	3
y	3	2	1	0	1	2	3
  c (−3, 3), (−2, 2), (−1, 1), (0, 0), (1, 1), (2, 2), (3, 3)
  d

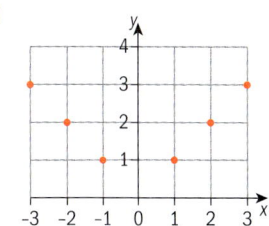

Exercise 2A
1. a $3x^2 - 2$ b $\dfrac{X^3 - XY + X^2}{Y}$
   c $ab - 2ac + 2ab + b^2$
2. a $3pq(1 - 2pq^2r)$
   b $3c(4ac + 5b - 1c)$
   c $abc(2a + 3b - 5c)$

Exercise 2B
1. $x^2 + 3x - 28$
2. $x^2 - 5x + 6$
3. $3x^2 + 2x - 8$
4. $6x^2 - 11x - 10$
5. $9x^2 + 3x + 2$

Exercise 2C
1. $x^2 + 10x + 25$
2. $x^2 - 8x - 16$
3. $x^2 - 4$
4. $9x^2 - 24x - 16$
5. $4x^2 + 20x + 25$
6. $4x^2 - 49$

Exercise 2D
1. a $(x + 4)(x + 7)$
   b $(x - 1)(x - 13)$
   c $(x + 4)(x - 5)$
   d $(x + 4)(x - 2)$
   e $(x + 4)(x + 9)$
   f $(x + 2)(x - 9)$
2. a $(2x - 3)(x - 3)$
   b $(3x + 1)(x + 2)$
   c $(5x - 2)(x - 3)$
   d $(4x + 3)(x - 1)$
   e $(3x + 2)(x - 3)$
   f $(7x - 5)(2x - 1)$
3. a $(x - 3)(x + 3)$
   b $(x - 10)(x + 10)$
   c $(2x - 9)(2x + 9)$
   d $(5x + 1)(5x - 1)$
   e $(m + n)(m - n)$
   f $(4x - 7)(4x + 7)$

Exercise 2E
1. $t = \dfrac{(u - v)}{g}$
2. $c = \sqrt{(b^2 - a^2)}$
3. $r = \dfrac{c}{2\pi}$
4. $b = \dfrac{a \sin B}{\sin A}$
5. $\cos A = \dfrac{(b^2 + c^2 - a^2)}{2bc}$
6. $F = \dfrac{32 + 9c}{5}$
7. Stock=Acid ratio test - (Current liabilities)(Current assets)

Exercise 2F
1. 2.487
2. 3.728
3. 40.073

Exercise 2G
1. $x = 4$
2. $x = 4$
3. $x = -3$
4. $x = 3$
5. $x = 5$
6. $x = 9$
7. $x = -2.5$
8. $x = -2$
9. $x = 3$
10. $x = 1.5$
11. $x = 1$
12. $x = 2$

Exercise 2H
1. a $y = 3x - 2$ (1)
   $3y = 5 - 2x$ (2)
   Multiply Eqn (1) by 3 to give
   $3y = 9x - 6$ (3)
   Combining Eqns (2) and (3) gives
   $5 - 2x = 9x - 6$
   $11 = 11x$, hence $x = 1$
   Substituting back into Eqn (1) gives
   $y = 3 - 2 = 1$
   Thus $x = 1$ and $y = 1$
   b $4x - 3y = 10$ (1)
   $2y + 5 = x$ (2)
   Substituting Eqn (2) in Eqn (1) gives
   $4(2y + 5) - 3y = 10$
   $8y + 20 - 3y = 10$
   $5y = -10$, hence $y = -2$
   Substituting back into Eqn (1) gives
   $4x - 3(-2) = 10$
   $4x = 4, x = 1$
   Thus $x = 1$ and $y = -2$
   c $2x + 5y = 14$ (1)
   $3x + 4y = 7$ (2)
   Multiplying Eqn (1) by 3 and Eqn (2) by 2 gives
   $6x + 15y = 42$ (3)
   $6x + 8y = 14$ (4)
   Thus
   $42 - 15y = 14 - 8y$
   $28 = 7y$ $y = 4$
   Substituting back into Eqn (1) gives
   $2x + 20 = 14$
   $2x = -6, x = -3$
   Thus $x = -3$ and $y = 4$
2. a $2x - 3y = 15$ (1)
   $2x + 5y = 7$ (2)
   Substituting Eqn (2) from Eqn (1) gives
   $-8y = 8$
   $y = -1$
   Substituting back into Eqn (1) gives
   $2x - 3(-1) = 15$
   $2x = 12, x = 6$
   $x = 6, y = -1$
   b $3x + y = 5$ (1)
   $4x - y = 9$ (2)
   Adding Eqn (2) to Eqn (1) gives
   $7x = 14$
   $x = 2$
   Substituting back into Eqn (1) gives
   $3(2) + y = 5$
   $y = -1$
   $x = 2, y = -1$

c $x + 4y = 6$ (1)
$3x + 2y = -2$ (2)
Multiplying Eqn (1) by 3 gives
$3x + 12y = 18$ (3)
Subtracting Eqn (2) from Eqn (3) gives
$10y = 20$
$y = 2$
Substituting back into Eqn (1) gives
$x + 8 = 6$
$x = -2$
$x = -2, y = 2$

d $3x + 2y = 8$ (1)
$2x + 3y = 7$ (2)
Multiplying Eqn (1) by 2 gives
$6x + 4y = 16$ (3)
Multiplying Eqn (2) by 3 gives
$6x + 9y = 21$ (4)
Subtracting Eqn (3) from Eqn (4) gives
$5y = 5$
$y = 1$
Substituting back into Eqn (1) gives
$3x + 2 = 8$
$x = 2$
$x = 2, y = 1$

e $4x - 5y = 17$ (1)
$3x + 2y = 7$ (2)
Multiplying Eqn (1) by 3 gives
$12x - 15y = 51$ (3)
Multiplying Eqn (2) by 4 gives
$12x + 8y = 28$ (4)
Subtracting Eqn (3) from Eqn (4) gives
$23y = -23$
$y = -1$
Substituting back into Eqn (1) gives

$4x + 5 = 17$
$x = 3$
$x = 3, y = -1$

Exercise 2I

1 a 17 b 144 c 64
2 a 1 b $\frac{1}{9}$ c $\frac{1}{16}$
3 a 525.219 b 4.081
c 2.488

Exercise 2J

1 a $3x + 4 \leq 13$
$3x \leq 9$
$x \leq 3$
  -4 -3 -2 -1 0 1 2 3 4 5 6 x

b $5(x - 5) > 15$
$x - 5 > 3$
$x > 8$
  -2 -1 0 1 2 3 4 5 6 7 8 9 x

c $2x + 3 < x + 5$
$x + 3 < 5$
$x < -2$
  -5 -4 -3 -2 -1 0 1 2 3 4 x

2 a $2(x - 2) \geq 3(x - 3)$
$2x - 4 \geq 3x - 9$
$-4 \geq x - 9$
$5 \geq x$ or $x \leq 5$

b $4 < 2x + 7$
$-3 < 2x$
$x > -\frac{3}{2}$

c $7 - 4x \leq 11$
$-4x \leq 4$
$x \geq -1$

Exercise 2K

1 a 3.25 b 6.18 c 0
2 when $x = 3$ $|5 - x|$ is 2, when $x = 8$ $|5 - x|$ is 3
3 a 2 b 4 c 2

Exercise 2L

1 $\frac{3x+1}{x+7}$
2 $\frac{x+1}{2x+2}$
3 $\frac{6x+8}{3x+4}$

4 $\frac{2x(2x-1)+(x+1)(x+5)}{(x+5)(2x-1)}$
$= \frac{2x^2 - 2x + x^2 + 6x + 5}{2x^2 + 9x - 5}$
$= \frac{3x^2 + 4x + 5}{2x^2 + 9x - 5}$

5 $\frac{4(x+2)+x(2x+1)}{x(x+2)}$
$= \frac{4x + 8 + 2x^2 + x}{x^2 + 2x}$
$= \frac{2x^2 + 5x + 8}{x^2 + 2x}$

6 $\frac{(2x-1)(4x+3) - 3x(x-2)}{(x-2)(4x+3)}$
$= \frac{8x^2 + 2x - 3 - 3x^2 + 6x}{4x^2 - 5x - 6}$
$= \frac{5x^2 + 8x - 3}{4x^2 - 5x - 6}$

7 $\frac{(x+1)(2x-5) + 2x(5x+1)}{(5x+1)(2x-5)}$
$= \frac{2x^2 + 2x - 5x - 5 + 10x^2 + 2x}{10x^2 + 2x - 25x - 5}$
$= \frac{12x^2 - x - 5}{10x^2 - 23x - 5}$

8 $\frac{(x+5)(x+2) - (x-4)(x-2)}{(x-4)(x+2)}$
$= \frac{x^2 + 7x + 10 - x^2 + 6x - 8}{x^2 + 2x - 8}$
$= \frac{13x + 2}{x^2 + 2x - 8}$

Exercise 2L

Please see copy attached for handwritten answers

Exercise 3A

1 27.6 cm
2 2.24 cm
3 5.032 cm

Exercise 3B

1 a Reflection in $x = 0$ (y-axis)
b Translation of $[-6, -4]$
c Rotation in $(0,0)$ by 90 degrees clockwise
d Reflection in $y = x$
2 TODO

Answers 773

Exercise 3C

1. Angle DFE = Angle ACB
   Angle DEF = Angle ABC
   EF = BC
   Two angles and included side so SAA and congruent
   Hence $x = 6$ cm and $y = 4$ cm

2. QP = AB
   PR = BC
   QR = AC
   Three sides are the same (SSS) so congruent
   $y = 58°$, $z = 33°$

3. Angle FDE = Angle ABC = 90°
   DE = BC
   FE = AC = Hypotenuse
   One side and the hypotenuse are the same in a right angled triangle (RHS) so congruent
   $x = 50°$, $y = 40°$

Exercise 3D

1. Rectangles with sides 5, 11 and 4, 8.8
   Rectangles with sides 5, 6.25 and 4, 5
   Rectangles with sides 5, 8 and 8, 12.8

2. a Scale factor is $10.08 \div 7.2$ so 7/5
      $y = 9.1 \times \dfrac{5}{7} = 6.5$ cm
      $x = 13 \times \dfrac{7}{5} = 18.2$ cm
   b Scale factor is $4.5 \div 3$ so 1.5
      $y = 1 \times 1.5 = 1.5$ cm
      $x = 2 \times 1.5 = 3$ cm

3. a A and B
   b A and C
   c A and B
   d None
   e None

4. Angle PAQ = Angle BAC
   Angle ABC = APQ (parallel lines and corresponding angles)
   Angle ACB = AQP (parallel lines and corresponding angles)
   Hence similar triangles
   Scale factor is 6/4 or 1.5
   So AB = $2 \times 1.5 = 3$ cm and
   BP = AB − AP = 3 − 2 = 1 cm
   So AC = $4 \times 1.5 = 4.5$ cm

5. a Angle AXB = Angle CXD (vertically opposite angles)
      Angle BAX = XDC (parallel lines and alternate angles)
      Angle ABX = XCD (parallel lines and alternate angles)
      Hence similar triangles
   b XD
   c Query: Can't work this out, need to know length of CD to give scale factor.

Exercise 3E

1. a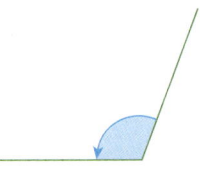

   b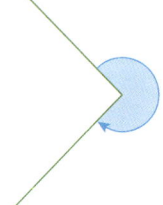

   c

   d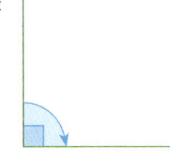

2. a Reflex   b Obtuse
   c Acute

3. a Obtuse   b Acute
   c Reflex   d Acute
   e Reflex   f Reflex

Exercise 3F

1.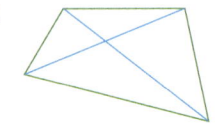

   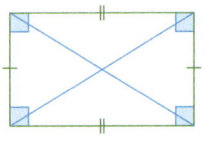

   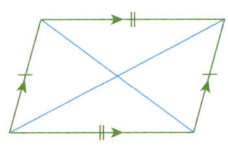

   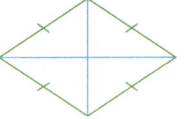

   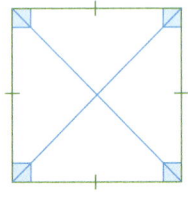

   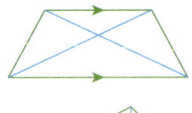

   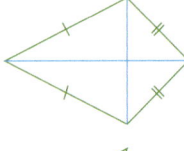

   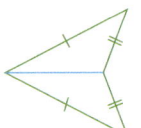

Diagonals	Irregular	Rectangle	Parallelogram	Rhombus	Square	Trapezium	Kite
Perpendicular	×	×	×	✓	✓	×	✓
Equal	×	✓	×	×	✓	×	×
Bisect	×	✓	✓	✓	✓	×	✓
Bisect angles	×	×	✓	✓	✓	×	✓

2 **a** Isosceles triangle, parallelogram, right-angled triangle, scalene triangle, rhombus, quadrilateral, kite

b Equilateral triangle, square, parallelogram, right-angled triangle, trapezium

Exercise 3G

1 **a** 3.2 + 3.2 + 4.3 = 10.9 cm

b 5.5 + 2.7 + 5.5 + 2.7 = 16.4 cm

c 7.2 + 4.2 + 4.8 + 4.2 = 20.4 cm

d 20π = 62.8 cm

e $3.2 + 3.2 + 3.2 + 1.6\pi$ = 14.6 cm

f $3(5.2\pi)/4 + 2.6 + 2.6$ = 17.5 cm

Exercise 3H

1 $4.5^2\pi$ = 63.6 cm²

2 $\dfrac{(6.2+4.5)}{2\times 4.3}$ = 23.005 cm²

3 6.5×5.8 = 37.7 cm²

4 5.7×3.6 = 20.52 cm²

5

6 $\dfrac{2.9(2.7+4.1)}{2} + (6.3\times 4.1) + (2.05^2\pi)$
  = 42.3 cm²

Exercise 3I

1 Pyramid: $\dfrac{(7\times 7)+4(7\times 8)}{2}$ = 161 cm²

Cylinder: $2(2.2^2\pi) + (4.4\pi \times 5.6)$ = 107.8 cm²

Cone: $(\pi \times 4 \times 10) + (4^2\pi)$ = 175.9 cm²

2 Pyramid: $\dfrac{(3^2\times 8)}{3}$ = 24 cm³

Cylinder: $2.2^2\pi \times 5.6$ = 85.15 cm³

Cone: $\dfrac{(4.5^2\pi \times 12)}{3}$ = 254.47 cm³

3 Volume = $\pi r^2 \dfrac{h}{3}$

$23 = \dfrac{4\pi h}{3}$

$69 = 4\pi h$

h = 5.49 cm

4 Volume = $\pi r^2 h$

$H = \dfrac{2120.6}{25\pi}$ = 27.000 cm

New volume = $\pi 2.5^2 \times 27$ = 530.144 cm³

5 **a** surface area = $4\pi r^2$
  = $4\pi 3.5^2$
  = 153.938 cm²

Volume = $4\dfrac{\pi r^3}{3} = 4\dfrac{\pi 3.5^3}{3}$
  = 179.594 cm³

b surface area = $4\pi r^2 = 4\pi 15^2$
  = 2827.433 cm²

Volume = $4\dfrac{\pi r^3}{3} = 4\dfrac{\pi 15^3}{3}$
  = 14137.167 cm³

6 surface area = $2\pi r^2 + \pi r^2 + 2\pi rh$
  = $2\pi 6^2 + \pi 6^2 + 2\pi 6 \times 5$
  = 527.788 cm²

Volume = $2\dfrac{\pi r^3}{3} + \pi r^2 h$
  = $2\dfrac{\pi 6^3}{3} + \pi 6^2 \times 5$
  = 1017.876 cm³

7 Volume of container
  = $\dfrac{(40^2\times 70)}{3}$ = 37333.33 cm³

Volume of one ball
  = $\dfrac{4\pi r^3}{3} = \dfrac{4\pi 10^3}{3}$ = 4188.79 cm³

Volume of eight balls = 33510.32 cm³

Space left in the container is 37333.33 − 33510.32 = 3823.012 cm³

8 Surface area = $2\pi r^2 + 2\pi rh$
  = $2\pi 4.5^2 + 2\pi 4.5 \times 14$
  = 523.1 cm²

Volume = $\pi r^2 h = \pi 4.5^2 \times 14$
  = 890.6 cm³

9 Volume = $\pi 5.5^2 h = 250$

$\dfrac{250}{\pi 5.5^2} = h$

h = 2.63 cm

10 Surface area = $2\pi rh = 950$

$\dfrac{950}{2\pi \times 60} = r$

r = 2.6 cm

Exercise 3J

1

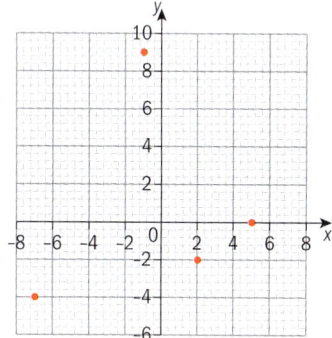

A(4, 9), B(−4, 2), C(−8, −6), D(8, −8)

Exercise 3K

1 (5, 5)

2 (−1, 1)

3 (1.5, 2.5)

Exercise 3L

1 5

2 9.43

3 14.8

Exercise 3M

1 $\dfrac{-4}{5} = -0.8$

2 $\dfrac{-1}{4} = -0.2$

Answers 775

3 $\frac{2}{5} = 0.4$

4 $\frac{-6}{4} = -1.5$

5 $\frac{-5}{1} = -5$

6 undefined

7 $\frac{3}{1} = 3$

8 $\frac{2}{4} = \frac{1}{2}$

9 0

Exercise 3N

1 $\frac{(-15--16)}{(-7-19)} = \frac{1}{(-26)} = -0.039$

2 $\frac{(-7--19)}{(-2-1)} = \frac{12}{-3} = -4$

3 $\frac{(-4-7)}{(-6--4)} = \frac{-11}{-2} = 4.5$

4 $\frac{(16-8)}{(9-20)} = \frac{8}{-11} = -0.73$

5 $\frac{(7--13)}{(17-17)} = \frac{20}{0}$ = undefined

6 $\frac{(3-3)}{(1-14)} = \frac{0}{-13} = 0$

7 $\frac{(-15-0)}{(-11-3)} = \frac{-15}{-14} = 1.071$

8 $\frac{(10--2)}{(-11-19)} = \frac{12}{-30} = -0.4$

9 $\frac{(15--10)}{(-15-6)} = \frac{(25)}{(-21)} = -1.19$

10 $\frac{(-18--18)}{(18-12)} = \frac{0}{6} = 0$

Exercise 3O

1 a 3 and $\frac{6}{2}$, 4.5 and $\frac{9}{2}$

   b -3 and $\frac{1}{3}$, 4.5 and $\frac{-2}{9}$, $\frac{2}{3}$ and -1.5

2 a Parallel (both have gradient of 2)

   b Neither (one has a gradient of $\frac{1}{4}$ and the other)

   c Neither (one has a gradient of $\frac{5}{4}$ and the other $\frac{-1}{4}$ of 0)

   d Perpendicular (gradients of 1 and -1)

   e Parallel (gradients of 1.5)

Exercise 3P

1 $y - 5 = 3(x - 1)$
   $y - 5 = 3x - 3$
   $y = 3x + 2$

2 $y - 11 = 4(x - 5)$
   $y - 11 = 4x - 20$
   $y = 4x + 9$

3 $y - 12 = 2.5(x - 4)$
   $2y - 24 = 5(x - 4)$
   $2y - 24 = 5x - 20$
   $2y = 5x + 4$

4 $y - 20 = 0.5(x - 12)$
   $y - 20 = 0.5x - 6$
   $y = 0.5x + 14$

5 $y - -13 = 5(x - -2)$
   $y + 13 = 5(x + 2)$
   $y + 13 = 5x + 10$
   $y = 5x - 3$

6 $y - 1 = -3(x - 1)$
   $y - 1 = -3x + 3$
   $y = -3x + 4$

7 $y - -1 = -2(x - 3)$
   $y + 1 = -2x + 6$
   $y = 2x - 5$

8 $y - -3 = \frac{-1}{2(x - -4)}$
   $y + 3 = \frac{-1}{2x - 2}$
   $y = \frac{-1}{2x - 5}$

9 Gradient is $\frac{19-7}{5-3} = 6$
   $y - 7 = 6(x - 2)$
   $y - 7 = 6x - 12$
   $y = 6x - 5$

10 Gradient is $\frac{-11--3}{-5--1} = \frac{-8}{-4} = 2$
   $y - -3 = 2(x - 1)$
   $y + 3 = 2(x + 1)$
   $y + 3 = 2x + 2$
   $y = 2x - 1$

Exercise 4A

1 Bar graph to show colors of cars

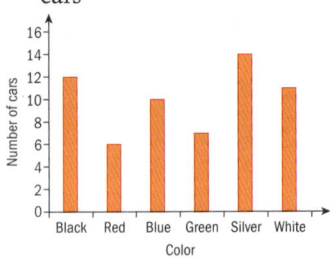

Pie Chart to show colours of cars

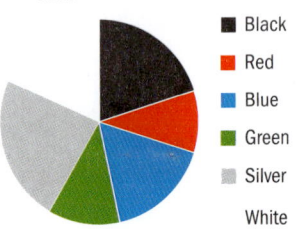

Pictogram to show colours of cars

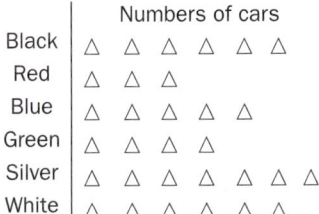

Key △ is 2 students

2 Bar graph to show number of times Ida's classmates had visited the cinema per month

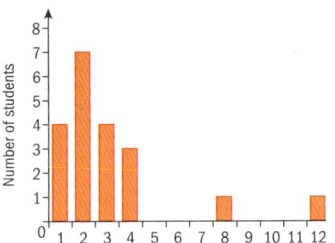

Pictogram to show number of visits to the cinema by Ida's classmates

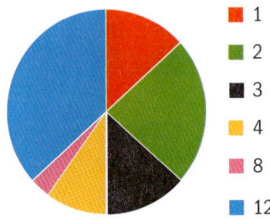

Times visited per month	Numbers of students
1	△ △
2	△ △ △ △
3	△ △
4	△ △
8	△
12	△

Key △ is 2 students

Exercise 4B

1

```
2 | 1 3 5 6 8
3 | 0 0 0 3 6 7 9 9
4 | 0 1 2 2 2 2 6 9
5 | 0 4
```
Key 2|1 means 21

2

```
12 | 1 3
14 | 5 8 9
15 | 1 2 7
16 | 3 4
17 | 6 6 7
18 | 5
```
Key 16|4 means 164

3

```
1 | 9
2 | 2 5 6 7 8 9
3 | 0 4 6 7
4 | 2 3 4 6 8 8 9
5 | 2 3 5 7 8
6 | 2
```
Key 4|2 means 42

4

```
1 | 1 4 4 6 8 9
2 | 3 4 6 6 7 8 9
3 | 0 4 5 6 8
4 |
5 | 1 6
```
Key 2|4 means 24

5

```
2 | 2 3 5 5 9
3 | 1 2 2 3
4 | 2 3 3 6 7 8
5 | 2 2 2 4
6 | 0 3 4
7 | 3 4
```
Key 6|3 means 6.3

Exercise 4C

1 Discrete
2 Discrete
3 Continuous
4 Continuous
5 Continuous
6 Discrete
7 Discrete
8 Continuous
9 Continuous

Exercise 4D

1 a mode=1
      median=4
      mean=4
   b mode=5
      median=5
      mean=4
   c mode=2 and 8
      median=5
      mean=5
   d mode=25
      median=25
      mean=25
   e mode=10.2
      median=10.2
      mean=9.42

2 a 1
   b 1
   c 1.67

3 a 8
   b 8
   c 9

4 a 4.82
   b 5.06
   c 5.02

5 a 497
   b 497
   c 400

Exercise 4E

1 a 38-26=12
   b 34-28=6
2 a 8-0=8
   b 4-1=3
3 a 8--7=15
   b 4--4=8
4 a 20-12=8
   b 18-14=4
5 a 23.5-2.45=21.05
   b 12.4-3.5=8.9

Mark scheme

Practice Paper 1

SECTION A

1. **a** $p = -1$, $q = 3$ (or vice versa) — A1 A1 N2
   b i $x = 1$ (must be an equation) — A1 N1
   ii correctly substituting the values for x, p and q
   $f(1) = 2(1 + 1)(1 - 3)$ — M1
   vertex $(1, -8)$ — A1 A1 N2

2. **a** $f'(x) = -2e^{-2x}$ — A1
   $f''(x) = 4e^{-2x}$ — A1
   $f'''(x) = -8e^{-2x}$ — A1
   $f^{(4)}(x) = 16e^{-2x}$ — A1 N4
   b generalization of alternating signs — (A1)
   $f^{(n)}(x) = (-1)^n 2^n e^{-2x}$ or $f^{(n)}(x) = (-2)^n e^{-2x}$ — A1 A1 N3

3. **a** 6 — A1 N1
   b evidence of using binomial expansion — (M1)
   evidence of calculating the factors $\binom{5}{2}(x^4)^2(-2x)^3$ — A1 A1 A1
   $-80x^{11}$ — A1 N2

4. **a i** $\sin\theta = \dfrac{3}{\sqrt{13}}$, $\cos\theta = \dfrac{2}{\sqrt{13}}$ — (A1) (A1)
   correct substitution — A1
   e.g. $\sin 2\theta = 2\left(\dfrac{3}{\sqrt{13}}\right)\left(\dfrac{2}{\sqrt{13}}\right)$
   $\sin 2\theta = \dfrac{12}{13}$ — A1 N3
   ii correct substitution — A1
   e.g. $\cos 2\theta = 2\left(\dfrac{2}{\sqrt{13}}\right)^2 - 1$, $\left(\dfrac{2}{\sqrt{13}}\right)^2 - \left(\dfrac{3}{\sqrt{13}}\right)^2$
   $\cos 2\theta = -\dfrac{5}{13}$ — A1 N1
   b $\tan 2\theta = -\dfrac{12}{5}$ — A1 N1

5. **a i** $p = 6$ — A1 N1
   ii $q = 5$ — A1 N1
   iii $r = 9$ — A1 N1
   iv $s = 20$ — A1 N1
   b $P(V \mid D') = \dfrac{\frac{9}{40}}{\frac{29}{40}}$ — (M1)
   $P(V \mid D') = \dfrac{9}{29}$ — A1 N2
   c valid reason — R1
   e.g. $P(A \cap B) \neq 0$ or $P(A \cup B) \neq P(A) + P(B)$ or correct numerical equivalent thus, V and D are not mutually exclusive — AG N0

6 a correct expression A1 N1

$$\pi \int_0^4 \left(\frac{\sqrt{\sin\left(x^{1/2}\right)}}{x^{1/4}}\right)^2 dx, \quad \pi \int_0^4 \frac{\sin\left(x^{1/2}\right)}{x^{1/2}} dx$$

b uses a correct substitution

e.g. $\pi \int_0^4 \frac{\sin\left(x^{1/2}\right)}{x^{1/2}} dx = 2\pi \int_0^2 \sin u \, du$ (M1)

correct antiderivative

$2\pi \int_0^2 \sin u \, du = -2\pi \left[\cos u\right]_0^2$ or $-2\pi\left[\cos\left(x^{1/2}\right)\right]_0^4$ A1

correct evaluation
$2\pi(\cos 2 - \cos 0) = -2\pi(\cos 2 - 1)$ A1

$p = -2, q = 2$ A1 A1 N0

7 a 0 A1 N1

b interchanging x and y (may be seen at any time) (M1)

e.g. $x = 4^{-y}$

evidence of correct manipulation A1

e.g. $-y = \log_4 x, y = \log_4 x^{-1}$

$f^{-1}(x) = \log_4 \frac{1}{x}$ AG N0

c finding $g(4)$ (seen anywhere) A1
attempt to substitute M1

$\left(f^{-1} \circ g\right)(4) = \log_4 \frac{1}{2^4}$

$\left(f^{-1} \circ g\right)(4) = \log_4 \frac{1}{16}$ (A1)

$\left(f^{-1} \circ g\right)(4) = -2$ A1 N1

SECTION B

8 a i finds the derivative of f A1 A1 A1
$f'(x) = 6x^2 - 3x - 3$ (M1)
sets the derivative equal to 0
e.g. $6x^2 - 3x - 3 = 0, f'(x) = 0$ A1
solves the equation

e.g. $3(2x^2 - x - 1) = 0 \Rightarrow 2(2x+1)(x-1) = 0 \Rightarrow x = -\frac{1}{2}, 1$

chooses the negative value

$x = -\frac{1}{2}$ A1 N0

ii find the second derivative of f A1 A1
$f''(x) = 12x - 3$ (M1)
sets the second derivative equal to 0
e.g. $12x - 3 = 0, f''(x) = 0$
solves the equation A1 N0

$x = \frac{1}{4}$

b i reflection gives (1, −2)	(A1)	
translation gives (1, −5)	A1	N2
ii reflection gives $y = -2x^3 + 1.5x^2 + 3x - 4.5$	(A1)	
translation gives $g(x) = -2x^3 + 1.5x^2 + 3x - 7.5$	A1	N2

9 a Shows evidence of using product rule M1
$f'(x) = (x)(-e^{-x}) + (e^{-x})(1)$ A1 A1
$= e^{-x}(-x + 1)$ A1
$= e^{-x}(1 - x)$ AG N0

b $f''(x) = (e^{-x})(-1) + (1 - x)(-e^{-x})$ A1 A1
$= -2e^{-x} + xe^{-x} (= e^{-x}(x - 2))$ A1 N3

c i $f'(1) = 0$ A1
$f''(1) = -\dfrac{1}{e}$ A1 N2

ii applies the second derivative test A1
There is a relative maximum at $x = 1$ R2 N0
since $f'(1) = 0$ and $f''(1) = -\dfrac{1}{e} < 0$

10 a recognizing scalar product must be zero (seen anywhere) R1
e.g. $a \cdot b = 0$

evidence of choosing direction vectors $\begin{pmatrix} 8 \\ -2 \\ 12 \end{pmatrix}, \begin{pmatrix} 2 \\ 2 \\ l \end{pmatrix}$ (A1) (A1)
(A1)

correct calculation of scalar product
e.g. $8(2) + (-2)(2) + 12l$
simplification that clearly leads to a solution A1
e.g. $12 + 12l = 0$ AG N0
$l = -1$

b i evidence of equating vectors M1
e.g. $\begin{pmatrix} 0 \\ 4 \\ 1 \end{pmatrix} + p\begin{pmatrix} 8 \\ -2 \\ 12 \end{pmatrix} = \begin{pmatrix} 8 \\ -2 \\ 15 \end{pmatrix} + s\begin{pmatrix} 2 \\ 2 \\ -1 \end{pmatrix}$

any **two** correct equation A1 A1
e.g. $8p = 4 + 2s,\ 4 - 2p = -2 + 2s,\ 1 + 12p = 15 - s$
attempting to solve equations (M1)
finding **one** correct parameter ($p = 1,\ s = 2$) A1

$\overrightarrow{OA} = \begin{pmatrix} 8 \\ 2 \\ 13 \end{pmatrix}$ A1 N3

c i evidence of approach A1

e.g. $BA = OA - OB,\ BA = \begin{pmatrix} 8 \\ 2 \\ 13 \end{pmatrix} - \begin{pmatrix} 9 \\ 6 \\ 10 \end{pmatrix}$

$BA = \begin{pmatrix} -1 \\ -4 \\ 3 \end{pmatrix}$ A1 N2

ii choosing correct vectors, $\overrightarrow{BA}$ and $\overrightarrow{BC}$ (A1)
calculating $\overrightarrow{BA} \cdot \overrightarrow{BC}$ A1

$\begin{pmatrix} -1 \\ -4 \\ 3 \end{pmatrix} \cdot \begin{pmatrix} 1 \\ -5 \\ 2 \end{pmatrix} = 25$

calculating $|\vec{BA}|$ and $|\vec{BC}|$ A1 A1
$|\vec{BA}| = \sqrt{26}$ and $|\vec{BC}| = \sqrt{30}$
evidence of using the formula to find cosine
$\cos\theta = \dfrac{25}{\sqrt{26}\sqrt{30}} = 26.5°$ A1 N4

Mark scheme

Practice Paper 2

SECTION A

1. a 0.966 (A1)
    b Strong, positive correlation. (A1)(A1)
    c $y = 4.96x + 4.84$ (A1)
    d $y = 4.96(20) + 4.84 = 104$ gms (M1)(A1)

2. a 4 A1
    b Evidence of appropriate approach e.g. $u_n = 329$ (M1)
       Correct working e.g $329 = 5 + (n-1)4$ A1
       $n = 82$ A1
    c Evidence of correct substitution
       e.g. $S_{82} = \dfrac{82}{2}(2(5) + (82-1)4)$ (M1)
       $S_{82} = 13694$.

3. a Evidence of choosing the product rule (A1)
       e.g. $(x \cos x) + (1 \sin x)$
       $f'(x) = \sin x + x \cos x$ (M1)
    b (A1)(A1)

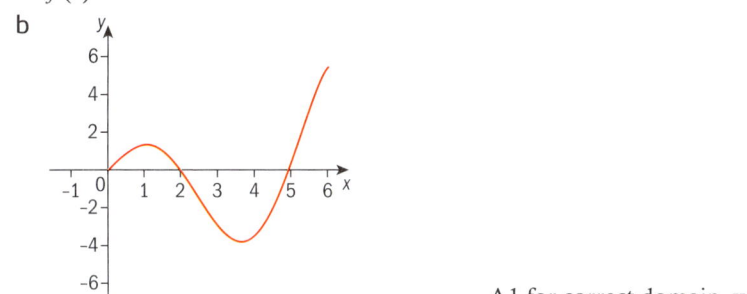

       A1 for correct domain, with endpoints in the correct place.
       A1 for approximately correct shape.
       A1 for local minimum in the correct place.
       A1 for local maximum in the correct place.

4. a Evidence of substituting into the mean $= \dfrac{\sum fx}{\sum f}$ (M1)
       Correct substitution e.g. $\dfrac{70 + 5x}{19 + x} = 4$ (A1)
    b 1.33 (A1)
    c 4.6 (A1)
    d No change. (A1)

5. a Using the Sine rule e.g. $\dfrac{BD}{\sin 100°} = \dfrac{8}{\sin 30°}$ (M1)
       BD = 15.8 (A1)
    b Using the Cosine rule e.g. $\cos BCD = \dfrac{8^2 + 10^2 - BD^2}{2 \times 8 \times 10}$ (M1)
       Angle BCD = 122°. (A1)

c Using the area formula *e.g.* $A = \frac{1}{2} \times 8 \times 10 \times \sin BCD$ (M1)

Area BCD = 34.0 (A1)

6 a Evidence of integrating the acceleration function. (M1)

e.g. $\int \left(\frac{1}{t} + 3\sin 2t \right) dt$

correct expression *e.g.* $\ln t - \frac{3}{2} \cos 2t + c$

evidence of substituting (1,0) *e.g.* $0 = \ln 1 - \frac{3}{2} \cos 2 + c$ (A1)(A1)

$c = -0.624$ (M1)

$v = \ln 1 - \frac{3}{2} \cos 2 - 0.624$ (A1)

$v(5) = 2.24$ (A1)
(A1)

7 a Evidence of using binomial probability. (M1)

Correct substitution *e.g.* $\binom{10}{4}(0.25)^4 (0.75)^6$ (A1)

0.146 (A1)

b $P(X \geq 2) > 0.9 = P(X < 2) < 0.1$

$P(X < 2) = P(X = 0) + P(X = 1)$

$= \binom{n}{0}(0.25)^0 (0.75)^n + \binom{n}{1}(0.25)^1 (0.75)^{n-1}$ (M1)
(A1)

$= (0.75)^n + 0.25n(0.75)^{n-1} < 0.1$

Use of graphical or table function. (M1)

The game must be played at least 15 times. (A1)

SECTION B

8 a i   1am (A1)
   ii 10am (A1)

b The depth of water can be modeled by the function
$y = A\cos(B(x - C)) + D$

i   Amplitude $= \frac{9-1}{2} = 4$ (M1) (A1)

ii 1 (A1)

iii 5 (A1)

iv $B = \frac{2\pi}{12} = \frac{\pi}{6}$ (M1) (A1)

v $y = 4\cos\left(\frac{\pi}{6}(x-1)\right) + 5$ (A1) (A1)

c Correct use of the model $4.5 = 4\cos\frac{\pi}{6}(x-1) + 5$ (M1)

Evidence of using a graphical method (M1)

The Seahawk can enter after 10:15am. (A1)

9 a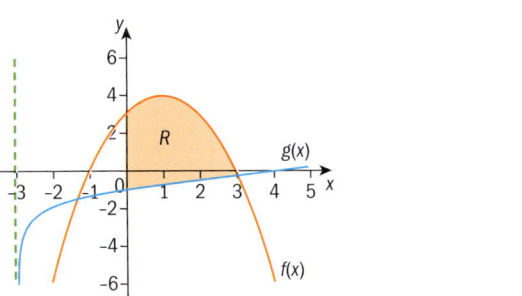

(A1)(A1)

(A1) for showing the basic shape of $f(x)$.
(A1) for showing both the vertical asymptote
(A1) the basic shape of $g(x)$.
(A1) for the correct x-intercepts.
(A1) for the correct y-intercepts

b i $x = -3$ is the vertical asymptote. (A1)(A1)
ii x-intercept: $x = 4.39\ (= e^2 - 3)$ (G1)
iii y-intercept: $y = -0.901\ (= \ln 3 - 2)$ (G1)

c $f(x) = g(x)$
$x = -1.34$ or $x = 3.05$ (G1)(G1)

d i See graph
ii Area of $R = \int_0^{3.05} (4 - (1-x)^2) - (\ln(x+3) - 2)\, dx$ (M1)(A1)(A1)
iii Area of $R = 10.6$ (G1)(G1)

10 a $P(G > 170) = 1 - P(G < 170)$ (A1)
$P(G > 170) = P\left(Z < \dfrac{170 - 155}{10}\right)$ (A1)
$P(G > 170) = 1 - \Phi(1.5) = 1 - 0.9332$ A1 N3
$= 0.0668$

b $z = -1.2816$ (A1)
Correct calculation (A1)
(e.g. $x = 155 + -1.282 \times 10$)
$x = 142$ A1 N3

c Calculating one variable (A1)
e.g. $P(B < r) = 0.95,\ z = 1.6449$
$r = 160 + 1.645(12) = 179.74$
$= 180$ A1 N2
Any valid calculation for the second variable, including (A1)
use of symmetry
e.g. $P(B < q) = 0.05,\ z = -1.6449$
$q = 160 - 1.645(12) = 140.26$
$= 140$ A1 N2

d $P(M \cap (B > 170)) = 0.4 \times 0.2020,\ P(F \cap (G > 170))$ (A1)(A1)
$= 0.6 \times 0.0668$ A1
$P(H > 170) = 0.0808 + 0.04008$ A1 N2
$= 0.12088 = 0.121$ (3 sf)

> The following symbols are used in this mark scheme:
> Girls' height $G \sim N(155, 10^2)$,
> boys' height $B \sim N(160, 12^2)$.
> Height H, Female F, Male M.

Subject index

absolute extrema, 242
absolute values, 669–70
Abū al-Wafā Būzjānī (c.940–c.998), 17
Abū Kāmil Shujā (c.850–c.930), 38
academic honesty, explorations and, 562–3
acceleration, 226, 227–9, 251, 510
  average, 226
  instantaneous, 226
accuracy, graphs, 31
addition
  algebraic fractions, 670–2
  vectors, 420–5, 443
addition rule, 72–4
adjacent sides, 364
agendas, hidden, 555
Agnesi, Maria Gaetana (1718–99), 217
Ahmes Papyrus, 165
aims, explorations, 558
al-Khwārizmī, Muḥammad ibn Mūsā (c.780–c.850), 657
algebra, 657–72
  and geometry, 444
algebraic fractions
  addition, 670–2
  subtraction, 670–2
algebraic functions, 500
algebraic proofs, 445
altitude, 380
ambiguous case, 384–5
ambiguous triangles, 384
amplitude, 464, 470, 475, 490, 491
analysis
  bivariate, 332–61
  data, 703–7
  univariate, 256–7, 284, 333
angles, 682
  between vectors, 427
    on GDCs, 610–11
  of depression, 369
  of elevation, 369
  obtuse, 375–6
  subtended, 391
Anscombe, Francis (1918–2001), 361
Anscombe's Quartet, 361
antiderivatives, 291–7, 328
  of x^n, 292
Apollonius of Perga (c.262–c.190 BCE), 46, 60
applied mathematics, vs. pure mathematics, 492–3
Archimedes of Samos (287–212 BCE), 146
arcs, 391–7, 401, 684
area, 686–8
  between two curves, 313–17, 329
  circles, 684
  and definite integrals, 302–9, 329
  surface, 688–92
  triangles, 389–91, 401
  under curves, 303
    on GDCs, 607–8
Argand, Jean-Robert (1768–1822), 423
arguments, 129
Aristotle (384 BCE–322 BCE), 423
arithmetic patterns, applications, 181–3
arithmetic sequences, 164–7, 190
arithmetic series, 172–5, 191
arrowheads, 683
Aryabhata (476–550), 365
associative law, 657
associative property, 648
asymptotes, 8, 9–10, 28, 144–6, 157
  horizontal, on GDCs, 584–5
authentic work, explorations, 562
average acceleration, 226
average velocity, 221–2
averages, 262, 704
  see also mean; median; mode
axes
  coordinate, 373–80, 400
  of revolution, 318
  of symmetry, 44
Babington Smith, Bernard (1905–93), 520
bacteria, 161
bar charts, 257–8, 700
  see also histograms
base formulae, changes of, 125–6, 137
base vectors, 409, 442
base-10 system, 402
base-60 system, 402
bases, 103
bean machine, 535
bearings, three-figure, 370
BEDMAS rule, 633
behavior, human, statistics of, 554–5
BEMDAS rule, 633
Bernoulli, Jacob (1654–1705), 112
best fit, lines of, 339–44, 357–8
BIDMAS rule, 633
bimodal data, 261

binomial distributions, 527–38, 553
  definition, 527–34
  expectation of, 535–6
  variance, 536–8
binomial expansion, 184–9, 191
binomial experiments, 528
binomial probabilities, calculating, on GDCs, 621–4
binomial quiz, 527–30
binomials, 658
birthday problem, 99
bivariate analysis, 332–61
BODMAS rule, 633
BOMDAS rule, 633
Boole, George (1815–64), 493
Boolean logic, 493
boundary conditions, 295
box and whisker plots, 269, 286
  drawing, on GDCs, 614–15, 616–17
Boyle, Robert (1627–91), 139
Boyle's law, 139
brackets
  curly, 10
  expanding, 657–62
  square, 10

calculations, 633–4
calculus, 195
  fundamental theorem of, 309–13, 329, 507
  GDCs and, 598–606
  trigonometric functions, 494–517
  see also differentiation; integration
Cardan, Jerome (1501–75), 64
Cartesian planes, 6, 230
  domain and range of relations on, 8–12, 28
causation, 336–9
  vs. correlation, 360–1
central limit theorem, 538
central tendency, measures of, 260–7, 285, 704–5
chain rule, 216–17, 496
  and higher order derivatives, 215–21, 251
change
  constant, 287
  rates of, 221–9, 251
charts
  bar, 257–8, 700
  drawing, on GDCs, 613–17

pie, 700–1
see also box and whisker plots; diagrams; graphs; histograms
chocolate factories, 495
chords, 684
circles, 60
   area, 684
   definitions, 684–5
   properties, 684–5
   see also unit circle
circular functions, 446–93
   graphs, 462–9, 490–1
circumference, 684, 685
coincident vectors, 428, 443
column vector form, 408
combinations, 184
combined transformations, 478–82, 491
common difference, 165, 190
common errors, 265
common fractions, 638
common ratio, 167, 191
communication, in explorations, 557–8
commutative law, 657
commutative property, 648
compass points, 370
complementary events, 68–9
complements, 654
completing the square, 36–8
complicated functions, on GDCs, 591–2
components
   horizontal, 408
   vertical, 408
composite functions, 14–16, 29
   derivatives, 216–17
compound interest, 111
computers, early, 493, 520
concave down, 234, 251
concave up, 234, 251
conclusions, explorations, 558
conditional probability, 85–8, 91–3, 97
conditions
   boundary, 295
   initial, 295
cones
   perpendicular height, 690
   slanted height, 690
   volume, 690–2
confounding factors, 336
congruence, 676–8
conic sections, 46, 60–1
conjectures, 516–17
constant of integration, 293
constant multiple rule, 204, 250, 293, 328, 496, 505
constant rule, 204, 250, 293, 328, 496, 505

continuous data, 256, 284, 704
continuous random variables, 520
controllability, 555
convergence, limits and, 196–200
convergent sequences, 196
convergent series, 178–81
coordinate axes, in trigonometry, 373–80, 400
coordinate geometry, 692–9
coordinates, 692–3
correlation, 334, 357
   measuring, 349–53, 359
   negative, 335, 357
   vs. causation, 360–1
   see also positive correlation
correlation coefficients, on GDCs, 627–31
cosine
   derivative, 497
   double-angle identities for, 457
   integrals, 505–10, 515
cosine functions
   graphs, 462–7
   modeling with, 483–8, 491
   and sine functions, combined transformations, 478–82, 491
   transformations, 469–70
   translations, 470–4
cosine ratio, 364, 365–7, 400
cosine rule, 386–9, 401, 426–7
cosine values, unit circle, 449–51
critical numbers, 231
cumulative frequencies, 271–6, 286
curly brackets, 10
currencies, international, 641
curves
   area between two, 313–17, 329
   area under, 303
   on GDCs, 607–8
   family of, 539
   tangents to, on GDCs, 599–600
   see also hyperbolas; parabolas
cylinders, volume, 689

data, 256, 284
   bimodal, 261
   continuous, 256, 284, 704
   discrete, 256, 284, 703
   dynamic, 554
   entering, on GDCs, 612
   presenting, 257–60, 284
   qualitative, 256, 284
   quantitative, 256, 284
Data & Statistics pages, scatter diagrams from, 627–9
data analysis, 703–7
de Moivre, Abraham (1667–1754), 538
decagons, 683

decay, exponential, 102, 131, 133–4
decay functions, exponential, 110
decimal places, 648
decimals
   and fractions, 638–40
   recurring, 639
   terminating, 639
deductive reasoning, 253
definite integrals
   and area, 302–9, 329
   with linear motion, 321–6, 329
   properties, 307, 329
degree mode, 366, 381, 465
degrees, on GDCs, 589
dependent variables, 334, 357
depression, angles of, 369
derivatives, 194–253
   composite functions, 216–17
   cosine, 497
   exponential functions, 209–10, 250
   first, 220
   on GDCs, 602–6
   and gradients of tangent lines, 202
   and graphs, 230–40, 251
   higher order, 220–1
   and chain rule, 215–21, 251
   natural logarithms, 209–10, 250
   practice with, 500–4
   product of two functions, 210–11
   rules for, 203–5, 208–15, 250
   sine, 496–500
   tangent functions, 497
   trigonometric functions, 496–500, 515
   of x^n, 200–7, 250
   see also antiderivatives; numerical derivatives; second derivatives
Descartes, René (1596–1650), 6, 230, 444, 692
descriptive statistics, 254–89
diagrams
   sample space, 77–84, 97
   stem and leaf, 702–3
   Venn, 68–77, 96
   see also charts; graphs; scatter diagrams; tree diagrams
diameters, 684
dice
   rolling, 64
   scores, 524
difference, common, 165, 190
difference of two squares, 659
   factorization, 661–2
differential calculus, 195
   GDCs and, 598–606

differentiation, 204, 292
  see also derivatives
direction, vectors, 407, 442
direction vectors, 431, 443
discontinuities, 199
discrete data, 256, 284, 703
discrete random variables, 520
discriminants, 41–2
disjoint sets, 653
dispersion, measures of, 267–71, 286, 706–7
displacement, 407
displacement functions, 224, 510
distance, 407
  between two points, 418–19, 694
  total, 322
distance traveled, 510
distributions see binomial distributions; normal distributions; probability distributions
distributive law, 657
distributive property, 648
divergent sequences, 196
division, exponents, 104
domains, 5, 28, 110, 136
  of Cartesian planes, 8–12, 28
double-angle identities
  for cosine, 457
  for sine, 458–62
dynamic data, 554

Egyptian fractions, 158
Einstein, Albert (1879–1955), 492
  equation, 139
elementary functions, 500
elements, 651
elevation, angles of, 369
ellipses, 60
empty sets, 651
end behavior, 142
energy equations, 139
enlargements, 675
equal roots, 34
equal vectors, 411–14
equation solving, unit circle, 454–6
equations
  energy, 139
  of lines, 698–9
  of normal lines, 205–7
  with rational coefficients, solving, 672
  of regression lines, 345–8
  simple, 139
  simultaneous, 574–6
  of tangent lines, 205–7
  vector, 430–6, 443
  see also exponential equations; linear equations; logarithmic equations; quadratic equations

equilateral triangles, 683
equivalent fractions, 639
errors, common, 265
estimation, 648–50
Euclid (c.325–c.265 BCE), 142
events, 64, 96
  complementary, 68–9
  independent, product rule, 80–2, 97
  intersections of, 69
  mutually exclusive, 76
  repeated, 89–90
  unions of, 70–1
exhaustion, method of, 330–1
expansion, binomial, 184–9, 191
expectation, 523–7
  of binomial distributions, 535–6
expected values, 523, 553
experimental probability, 65–6
experiments, 64, 96
  binomial, 528
  human behavior, 554
  random, 64, 96
explorations, 556–69
  and academic honesty, 562–3
  aims, 558
  authentic work, 562
  communication in, 557–8
  conclusions, 558
  getting started, 568–9
  internal assessment criteria, 557–61
  introduction, 558
  and malpractice, 563
  marking, 562
  mathematical presentation, 558–9
  mathematics, use of, 561
  overview, 556–7
  personal engagement, 559–60
  rationales, 558
  record keeping, 563–4
  reflections, 560
  sources, acknowledging, 563
  topic choice, 564–7
exponential decay, 102
exponential decay functions, 110
exponential equations, 127–31
  solving, 107–9, 127–9
exponential expressions, 667–8
exponential functions, 100–39
  applications, 131–4
  definition, 136
  derivatives, 209–10, 250
  on GDCs, 583–5
  graphs, 109–10
  integrals, 505
  modeling, using sliders, 596–8
  natural, 111–12

  solving, on GDCs, 591–2
  transformations, 112–14
exponential graphs, drawing, on GDCs, 583–4
exponential growth, 101, 131–2
exponential growth functions, 103, 109–10
exponents, 103–7, 667
  division, 104
  fractional, 105
  laws of, 103–7, 136
  multiplication, 103
  negative, 106
  roots, 105
  see also powers
expressions
  exponential, 667–8
  involving roots, simplifying, 634–6
  see also quadratic expressions
extrapolation, 339, 347
extrema, 240–8, 251
  absolute, 242
  global, 242
  see also maxima; minima

Facebook, 101
factorials, 184
factorization, 34–6, 657–62
  difference of two squares, 661–2
  quadratic expressions, 660
  quadratics, 660–1
factors, 637–8
  confounding, 336
  scale, 23
family of curves, 539
Fibonacci, Leonardo of Pisa (c.1170–c.1250), 164, 193
Fibonacci sequences, 193
finite planes, 682
first derivative test, 233, 251
first derivatives, 220
first quartile, 268, 286
Fisher, Sir Ronald Aylmer (1890–1962), 264
folding paper, 102–3
formulae, 662–4
  base, 125–6, 137
  quadratic, 38–41, 58
  rearranging, 662–3
  recursive, 163
  subjects of, 662
  substituting into, 663–4
Fourier, Jean Baptiste Joseph (1768–1830), 498
fractional exponents, 105
fractions
  algebraic, 670–2
  common, 638
  and decimals, 638–40

Egyptian, 158
equivalent, 639
improper, 638
proper, 638
unit, 158, 638
frequencies
  cumulative, 271–6, 286
  relative, 66
frequency histograms, drawing, on GDCs, 613–14
frequency tables, 257
  calculating statistics from, on GDCs, 618–19
  data entering, on GDCs, 612
  drawing frequency histograms from, on GDCs, 614
  grouped, 258, 284
function notation, 13–14, 29
functions, 2–31
  algebraic, 500
  complicated, on GDCs, 591–2
  decreasing, 230
  definition, 5, 28
  displacement, 224, 510
  elementary, 500
  GDCs and, 572–98
  increasing, 230
  integrable, 304
  introducing, 4–8, 28
  inverse, 118–19, 137
  limits of, 197–200
  linear, 572
  periodic, 464, 468, 490
  probability, 522
  product of two, derivatives, 210–11
  reciprocal, 143–6, 157
  reflections, 23
  and relations, 4–6
  self-inverse, 144, 157
  stretches, 23–4
  transcendental, 500
  transformations, 21–5, 29
  translations, 22
  velocity, 224, 251
  see also circular functions; composite functions; exponential functions; inverse functions; logarithmic functions; quadratic functions; rational functions; reciprocal functions; trigonometric functions
fundamental theorem of calculus, 309–13, 329, 507

Gabriel's horn, 331
Galton, Francis (1822–1911), 288, 535

Galton board, 535
Gapminder, 554
Gauss, Carl Friedrich (1777–1855), 172, 346, 538
Gaussian curve *see* normal distributions
GDCs *see* graphic display calculators (GDCs)
general solution, 295
genetic fingerprinting, 80
geometric patterns, applications, 181–3
geometric proofs, 423–5, 445
geometric sequences, 167–70, 191
geometric series, 175–8, 179, 191
geometric transformations, 674–6
geometry, 673–99
  and algebra, 444
  coordinate, 692–9
Gladwell, Malcolm (b.1963), 102
global extrema, 242
golden ratio, 56
golden section, 193
Gougu Theorem, 673
gradians, 403
gradients
  finding, on GDCs, 573–4, 598–9
  of lines, 696–7
  negative, 695
  positive, 695
  of straight lines, 695–6
  of tangent lines, 202
graphic display calculators (GDCs)
  degree mode, 366, 381, 465
  differential calculus, 598–606
  functions, 572–98
  integral calculus, 606–8
  probability, 612–31
  radian mode, 396, 466
  statistics, 612–31
  using, 570–631
  vectors, 608–11
graphs, 30–1
  accuracy, 31
  circular functions, 462–9, 490–1
  cosine functions, 462–7
  and derivatives, 230–40, 251
  exponential functions, 109–10
  finding quadratic equations from, 49–52
  of inverse functions, 18–19
  logarithmic, on GDCs, 588
  of quadratic functions, 43–52, 59
  sine functions, 462–7
  statistical, 699–703
  tangent functions, 467–9
  trigonometric, on GDCs, 590
  see also charts; diagrams
Graphs pages, scatter diagrams using, 629–31

gravitation, laws of, 139
grids, 31
grouped frequency tables, 258, 284
growth
  exponential, 101, 131–2
  population, 182–3
growth functions, exponential, 103, 109–10
guitar strings, 195

handshakes, 4
harmonic motion, 498
Hawthorne effect, 554
HCF (highest common factor), 638
Hero of Alexandria (c.10–70), 390
hexagonal prisms, 688
hexagons, 683
hidden agendas, 555
higher order derivatives, 220–1
  and chain rule, 215–21, 251
highest common factor (HCF), 638
histograms, 258–9
  frequency, 613–14
Hogben, Lancelot (1895–1975), 517
honesty, academic, 562–3
horizontal asymptotes, on GDCs, 584–5
horizontal components, 408
horizontal line test, 16–17, 29
horizontal stretches, 23, 476–8, 491
horizontal translations, 470–2, 491
human behavior
  experiments, 554
  statistics of, 554–5
Hypatia (c.350/370–415), 60
hyperbolas, 60, 144, 157
hypotenuse, 364

Ibn al-Haytham (965–1040), 320
icosahedrons, 65
identities
  trigonometric, 456–62, 490
  see also double-angle identities
improper fractions, 638
Incas, 158
indefinite integrals, 291–302, 328
  on GDCs, 606–7
indefinite integration, 293
independent events, product rule, 80–2, 97
independent variables, 334, 357
indices *see* exponents
inductive reasoning, 252
inequalities, 10
  properties, 669
  and sets, 655–6
  solving, 668–9
infinite planes, 682

infinity, sums to, 178–81, 191
inflexion points, 234, 251
initial conditions, 295
initial position, 224
initial side, 373
initial velocity, 224
instantaneous acceleration, 226
instantaneous velocity, 221–2
integers, 646
integrable functions, 304
integral calculus, 195
  GDCs and, 606–8
integrals
  cosine, 505–10, 515
  exponential functions, 505
  with linear composition, 505
  reciprocal functions, 505
  sine, 505–10, 515
  see also definite integrals;
    indefinite integrals
integrands, 293
integration, 290–331
  constant of, 293
  indefinite, 293
  lower limit of, 304
  upper limit of, 304
  variables of, 293
interest, compound, 111
internal assessment criteria,
  explorations, 557–61
international currencies, 641
International Space Station (ISS), 3
internationalism, of symbols, 10
interpolation, 342
interquartile range (IQR), 269, 286, 706
  calculating, on GDCs, 619–20
intersection points, vectors, 434–6
intersections, 652–5
  of events, 69
interval notation, 10, 28
intuition, probability and, 99
inverse functions, 16–21, 29, 118–19, 137
  finding algebraically, 19
  on GDCs, 585–7
  graphs of, 18–19, 29
inverse normal distributions, 544–51
IQR see interquartile range (IQR)
irrational numbers, 112, 634, 639, 646
irregular quadrilaterals, 683
isosceles triangles, 683
ISS (International Space Station), 3

Jeffreys, Alec (b.1950), 80
jokes, mathematics, 253
Jones, William (1675–1749), 455

Kendall, Sir Maurice George (1907–83), 520
Khayyám, Omar (c.1048–1131), 60, 192
kinematics, 224
kites, 683
Koch snowflakes, 176

Lagrange, Joseph Louis (1736–1813), 444
Lancaster, Henry Oliver (1913–2001), 333
Laplace, Pierre-Simon (1749–1827), 538
LCM (lowest common multiple), 637
leading questions, 555
learning, prior, 632–707
least squares regression, 345–8, 358
Legendre, Adrien-Marie (1752–1833), 346
Leibniz, Gottfried Wilhelm (1646–1716), 13, 214, 217, 330, 493
Leibniz notation, 214
lies, 555
  and statistics, 289
limits, 194–253
  and convergence, 196–200
  of functions, 197–200
  of sequences, 196–7
linear dependence, variables, 349, 359
linear equations
  solving, 664–5
  see also simultaneous linear equations
linear functions, graphing, on GDCs, 572
linear motion, 510–13
  definite integrals with, 321–6, 329
linear regression, on GDCs, 627–31
linear relationships, 337
lines, 682
  of best fit, 339–44, 357–8
  equations of, 698–9
  gradients of, 696–7
  normal, 205–7
  parallel, 697–8
  perpendicular, 697–8
  vector equations of, 430–6, 443
  see also number lines; regression lines; secant lines; straight lines; tangent lines
lists
  calculating statistics from, on GDCs, 617–18

drawing frequency histograms from, on GDCs, 613
  entering in GDCs, 612
logarithmic equations, 127–31
  solving, 129–31
logarithmic functions, 100–39
  applications, 131–4
  definition, 137
  on GDCs, 585–8
  overview, 118–22
  transformations, 119
logarithmic graphs, drawing, on GDCs, 588
logarithms
  evaluating, on GDCs, 585
  laws of, 122–6, 137
  properties, 115–18, 137
  to base 10, 120
  see also natural logarithms
London Eye, 447
lower limit of integration, 304
lower quartiles, 706
lowest common multiple (LCM), 637

magnitude, vectors, 410–11
major segments, 684
malpractice, explorations and, 563
mappings, 656–7
mathematical presentation, explorations, 558–9
mathematical relations, 656
mathematical representation, 30–1
mathematical symbols, 517
mathematics
  applied vs. pure, 492–3
  beauty of, 138–9
  jokes, 253
  topics, 444–5
  truth in, 252–3
  see also pure mathematics
maxima
  finding, on GDCs, 579–83, 600–1
  relative, 233
maximum function, on GDCs, 582–3
maximum points see maxima
mean, 260, 262–5, 266, 285, 523, 704
mean points, 339, 341–2, 358
measurement, units of, 402–3
measures
  of central tendency, 260–7, 285, 704–5
  of dispersion, 267–71, 286, 706–7
median, 260, 265–6, 285, 704
midpoints, 693–4

mind maps, 566–7
minima
   finding, on GDCs, 579–83, 600–1
   relative, 233
minimum function, 580–1
minimum points *see* minima
minor segments, 684
mixed numbers, 638
mode, 260–1, 266, 285, 704
modeling
   with cosine functions, 483–8, 491
   on GDCs, 592–8
   with sine functions, 483–8, 491
Monte Carlo methods, 65
Monty Hall dilemma, 84, 88
motion
   harmonic, 498
   laws of, 139, 428
   in a line, 221–9, 251
   see also linear motion
MP3 players, 141
multiples, 637–8
multiplication, exponents, 103
mutually exclusive events, 76

natural exponential functions, 111–12
natural logarithms, 120–2
   derivatives, 209–10, 250
natural numbers, 646
nature, patterns in, 193
nCr, using, 621–2
negative correlation, 335, 357
negative exponents, 106
negative gradients, 695
negative vectors, 411–14
Newton, Isaac (1642–1727), 139, 217, 230, 330, 428
Nightingale, Florence (1820–1910), 288
non-linear relationships, variables, 336
normal distributions, 538–51, 553
   curves, area beneath, 539
   inverse, 544–51
   probabilities, 542–4
   standard, 540–1
normal lines, equations of, 205–7
normal probabilities
   calculating
     from X-values, 624–5
     on GDCs, 624–6
   calculating X-values from, on GDCs, 625–6
notation
   function, 13–14, 29
   interval, 10, 28
   Leibniz, 214
   prime, 214

set builder, 10–11, 652
sigma, and series, 170–1, 191
nth terms, of sequences, general formula, 163–4
number lines
   real, 655
   and sets, 655–6
number sequences, 162, 190
number systems, 158–9, 646–8
numbers, 633–57
   critical, 231
   irrational, 112, 634, 639, 646
   mixed, 638
   natural, 646
   prime, 637
   rational, 646
   real, 648
numerical derivatives
   on GDCs, 602
   graphing, on GDCs, 603–4

obtuse angles, 375–6
octagons, 683
octopi, 519
ogives, 271
opposite sides, 364
optimization problems, 240–8, 251
Oresme, Nicole (1323–82), 3
oscillations, 492, 498
outliers, 269

paper, folding, 102–3
parabolas, 33, 44, 60
   origin of term, 46
paradoxes, 178, 331
parallel lines, 697–8
parallel vectors, 411–14, 428, 443
parallelograms, 683
parameters, 539
particular solutions, 295
Pascal, Blaise (1623–62), 184, 192
Pascal's triangles, 184–9, 191, 192, 193
patterns, 160–93
   arithmetic, applications, 181–3
   geometric, applications, 181–3
   in nature, 193
   in polynomials, 185
   and sequences, 162–4, 190
Pearson, Karl (1857–1936), 349
Pearson product-moment correlation coefficient (r), 349, 359
PEMDAS rule, 633
pendulums, 498
pentagons, 683
percentage decrease, 641–2
percentage increase, 641–2
percentages, 640–3

perfect square trinomial, 36
perimeters, 685–6
periodic functions, 464, 468, 490
periods, 464, 470, 476
perpendicular lines, 697–8
perpendicular vectors, 428, 443
personal engagement, explorations, 559–60
pictograms, 701–2
pie charts, 700–1
Pisa, leaning tower of, 334, 339
planes, 682
   finite, 682
   infinite, 682
   see also Cartesian planes
Platonist School, 60
playing cards, 73
Plimpton tablet, 402
points, 682
   compass, 370
   distance between two, 418–19, 694
   inflexion, 234, 251
   intersection, 434–6
   mean, 339, 341–2, 358
   midpoints, 693–4
   stationary, 231
   see also maxima; minima
polyhedrons, 65
polynomials, patterns in, 185
population growth, 182–3
population standard deviation, 287
population variance, 287
populations, 284, 333
   and samples compared, 257
position vectors, 414, 442
positive correlation, 335, 357
   strong, 337
positive gradients, 695
power rule, 203, 250, 293, 328, 505
powers, 667
   raising to, 104
   zero, 104–5
   see also exponents
practice papers, 708–15
prejudice, 555
prime notation, 214
prime numbers, 637
primes, 637–8
prior learning, 632–707
prisms
   hexagonal, 688
   triangular, 688
   volume, 688–9
probability, 62–99
   conditional, 85–8, 91–3, 97
   definitions, 64–8, 96
   experimental, 65–6

GDCs and, 612–31
and intuition, 99
normal distributions, 542–4
subjective, 66–8
theoretical, 64–5
uses and abuses, 98–9
see also normal probabilities
probability distributions, 518–55
of random variables, 520–3
probability functions, 522
probability tree diagrams, 89–93
problems
birthday, 99
optimization, 240–8, 251
product pairs, graphing, 142
product rule, 77–84, 97, 211, 250, 496
independent events, 81–2, 97
products
and quadratic expressions, 658–9
see also scalar products
proofs, 516–17
algebraic, 445
geometric, 423–5, 445
vector, 445
proper fractions, 638
proper subsets, 653
prophecies, self-fulfilling, 555
proportion, 643–5
Ptolemy (*c*.90–168), 383
pure mathematics
in applications, 493
applications of, 493
vs. applied mathematics, 492–3
pyramids, volume, 689–90
Pythagoras (569–500 bc), 634, 673
Pythagoras' theorem, 388, 673–4
proving, 444–5

quadrants, 374
quadratic equations, 32–61
finding, from graphs, 49–52
roots, 41–3, 58
solving, 34–8, 58
by completing the square, 36–8
by factorization, 34–6
on GDCs, 578, 591–2
quadratic expressions
factorization, 660
products and, 658–9
quadratic formula, 38–41, 58
quadratic functions, 32–61
finding equations of, from graphs, 49–52
on GDCs, 577–83
graphs of, 43–52, 59
modeling, via transformations, 594–6
quadratic graphs, on GDCs, 577–8
quadratics

applications, 53–6
factorization, 660–1
quadrilaterals, 683
irregular, 683
qualitative data, 256, 284
quantitative data, 256, 284
quartiles, 267–71
first, 268, 286
lower, 706
second, 268, 286
third, 268, 286
upper, 706
see also interquartile range (IQR)
questions
leading, 555
sensitive, 98–9
quincunx, 535
quotient rule, 211, 250, 496

r (Pearson product-moment correlation coefficient), 349, 359
radian mode, 396, 466
radians, 391–7, 401, 403
on GDCs, 589
radii, 684
random experiments, 64, 96
random samples, 257
random variables, 520–7, 553
continuous, 520
discrete, 520
probability distributions of, 520–3
randomized response method, 98–9
range, 5, 28, 110, 267, 286, 706
of Cartesian planes, 8–12, 28
see also interquartile range (IQR)
rates of change, 221–9, 251
rational coefficients, solving equations with, 672
rational functions, 140–59
graphing, 148, 150–1
rational numbers, 646
rationales, in explorations, 558
ratios, 643–5
common, 167, 191
golden, 56
trigonometric, 364–7, 400
unitary, 643
real number lines, 655
real numbers, properties, 648
reasoning
deductive, 253
inductive, 252
reciprocal functions, 143–6, 157
graphs of, 143
integrals, 505
reciprocals, 142–3, 157
use of term, 143

record keeping, explorations, 563–4
rectangles, 683
recurring decimals, 639
recursive formulae, 163
reflections, 674
functions, 23
regression
least squares, 345–8, 358
linear, on GDCs, 627–31
sinusoidal, on GDCs, 592–4
regression lines, 340, 341–2, 343–4, 358
equations of, 345–8
relations, 28
on Cartesian planes, domain and range of, 8–12, 28
and functions, 4–6
mathematical, 656
relationships
linear, 337
non-linear, 336
relative frequency, 66
relative maxima, 233
relative minima, 233
repeatability, 554
repeated events, 89–90
representation, mathematical, 30–1
residuals, 345, 358
resultant vectors, 414–17, 442
revolution
axes of, 318
solids of, 318
volume of, 318–21, 329
Rhind Mathematical Papyrus, 158
rhombuses, 683
Richter, Charles Francis (1900–85), 134
Riemann, Georg (1826–66), 313
right-angled triangle trigonometry, 363–9, 400
applications, 369–73, 400
right-angled triangles, 683
special, 367–9
rolling dice, 64
roots
equal, 34
exponents, 105
expressions involving, simplifying, 634–6
of quadratic equations, 41–3, 58
Rosling, Hans (b.1948), 554
rotations, 674
rounding, 648–50
Russell, Bertrand (1872–1970), 493

sample space, 65
sample space diagrams, 77–84, 97
samples, 284, 333

and populations compared, 257
random, 257
scalar products, 426–30, 443
   calculating, on GDCs, 608–10
   properties, 428–30
scalars, 406, 442
scale factors, 23
scalene triangles, 683
scatter diagrams, 334–9, 357
   on GDCs, 627–31
   using Data & Statistics pages, 627–9
   using Graphs pages, 629–31
Schrödinger, Erwin (1887–1961), 139
secant lines, 200
   gradients, 201
second derivative test, 240
second derivatives, 220
   on GDCs, 605–6
second quartile, 268, 286
sectors, 391–7, 401, 685
segments
   major, 684
   minor, 684
self-fulfilling prophecies, 555
self-inverse functions, 144, 157
semicircles, 685
sensitive questions, asking, 98–9
sequences, 103, 160–93
   arithmetic, 164–7, 190
   convergent, 196
   creating, 196
   divergent, 196
   Fibonacci, 193
   geometric, 167–70, 191
   limits of, 196–7
   nth terms of, general formula, 163–4
   number, 162, 190
   and patterns, 162–4, 190
series, 160–93
   arithmetic, 172–5, 191
   convergent, 178–81
   geometric, 175–8, 179, 191
   and sigma notation, 170–1, 191
set builder notation, 10–11, 652
sets, 651–7
   disjoint, 653
   empty, 651
   and inequalities, 655–6
   and number lines, 655–6
   universal, 651
   see also subsets
sexagesimal number system, 402
shapes
   in real world, 60–1
   two-dimensional, 683–4

sides
   adjacent, 364
   initial, 373
   opposite, 364
   terminal, 373
sigma notation, and series, 170–1, 191
significant figures, 649
similar triangles, 364, 679–82
similarity, 678–82
simultaneous equations, solving, on GDCs, 574–6
simultaneous linear equations, 666–7
   solving, on GDCs, 576–7
sine
   derivative, 496–500
   double-angle identities for, 458–62
   integrals, 505–10, 515
sine functions
   and cosine functions, combined transformations, 478–82, 491
   graphs, 462–7
   modeling with, 483–8, 491
   transformations, 469–70
   translations, 470–4
sine ratio, 364, 365–7, 400
sine rule, 380–5, 401
sine values, unit circle, 449–51
sinusoidal regression, on GDCs, 592–4
size, vectors, 407, 442
snowflakes, Koch, 176
Soccer World Cup, 519
SOHCAHTOA, 365
solids, of revolution, 318
solutions
   particular, 295
   simple, 138–9
sources, acknowledging, 563
speed, 227–9, 407
   see also velocity
spheres, volume, 689
springs, 492
square brackets, 10
squares, 683
   difference of two, 659
      factorization, 661–2
standard deviation, 276–81, 287
   data sets, adding or multiplying, 281
   population, 287
   properties, 278–80
standard form, 650–1
standard normal distribution, 540–1
standard position, 373
stationary points, 231

   see also extrema; maxima; minima
statistical graphs, 699–703
statistical summaries, 268
statistics, 699–707
   calculating, on GDCs, 617–20
   descriptive, 254–89
   facts and misconceptions, 288–9
   GDCs and, 612–31
   lies and, 289
   using, on GDCs, 620
stem and leaf diagrams, 702–3
Stevin, Simon (1548–1620), 423
straight lines, 682
   gradients of, 695–6
stretches
   functions, 23–4
   horizontal, 23, 476–8, 491
   trigonometric functions, 469–78
   vertical, 23, 475–6, 491
strong positive correlation, 337
subjective probability, 66–8
subjects, of formulae, 662
subsets, 652–5
   proper, 653
substitution, into formulae, 663–4
substitution method, 300–2
subtraction
   algebraic fractions, 670–2
   vectors, 420–5, 443
Sulba Sutras, 673
sum or difference rule, 204, 250, 293, 328, 496, 505
summation, 346
sums, to infinity, 178–81, 191
supply and demand curves, 24
surds, 634
surface area, three-dimensional shapes, 688–92
surveying, 363
symbols
   internationalism of, 10
   mathematical, 517
symmetry, axes of, 44

tables
   on GDCs, 579, 581–2
   see also frequency tables
tangency, points of, 685
tangent functions
   derivatives, 497
   graphs, 467–9
tangent lines, 200–7, 250
   equations of, 205–7
   gradients of, 202
tangent ratio, 364, 365–7, 400
tangent values, unit circle, 449–51
tangents, 685

to curves, on GDCs, 599–600
terminal side, 373
terminating decimals, 639
terms, 163, 190
  nth, of sequences, 163–4
test scores, 256
theoretical probability, 64–5
third quartile, 268, 286
Thomson, James (1822–92), 403
three-dimensional shapes
  surface area, 688–92
  volume, 688–92
three-figure bearings, 370
Tippett, Leonard Henry Caleb (1902–85), 520
topics, choosing, 564–7
total distance, 322
transcendental functions, 500
transformations
  combined, 478–82, 491
  cosine functions, 469–70
  exponential functions, 112–14
  of functions, 21–5, 29
  geometric, 674–6
  logarithmic functions, 119
  quadratic function modeling, 594–6
  sine functions, 469–70
translations, 674
  cosine functions, 470–4
  functions, 22
  horizontal, 470–2, 491
  sine functions, 470–4
  trigonometric functions, 469–78
  vertical, 470–2, 490–1
trapeziums, 683
tree diagrams
  probability, 89–93
  with replacement and repeated events, 89–90
  without replacement and conditional probability, 91–3
triangles, 683
  ambiguous, 384
  area, 389–91, 401
  equilateral, 683
  isosceles, 683
  Pascal's, 184–9, 191, 192, 193
  scalene, 683
  similar, 364, 679–82
  see also right-angled triangles
triangular prisms, 688
trigonometric functions
  calculus, 494–517
  derivatives, 496–500, 515
  on GDCs, 589–90
  stretches, 469–78
  translations, 469–78
  see also cosine functions; sine functions; tangent functions
trigonometric graphs, drawing, on GDCs, 590
trigonometric identities, 456–62, 490
trigonometric ratios, 364–7, 400
trigonometry, 362–403
  coordinate axes in, 373–80, 400
  see also right-angled triangle trigonometry
trinomial, perfect square, 36
truth, in mathematics, 252–3
two-dimensional shapes, 683–4

unions, 652–5
  of events, 70–1
unit circle, 374
  cosine values, 449–51
  equation solving, 454–6
  sine values, 449–51
  tangent values, 449–51
  using, 448–53, 490
unit fractions, 158, 638
unit vector form, 409
unit vectors, 419–20
unitary method, 645–6
units, of measurement, 402–3
univariate analysis, 256–7, 284, 333
universal sets, 651
upper limit of integration, 304
upper quartiles, 706

values
  absolute, 669–70
  expected, 523, 553
  see also X-values
variables
  dependent, 334, 357
  independent, 334, 357
  of integration, 293
  linear dependence, 349, 359
  non-linear relationships, 336
  see also random variables
variance, 276–81, 286
  binomial distributions, 536–8
  population, 287
vector equations, of lines, 430–6, 443
vector proofs, 445
vectors, 404–45
  addition, 420–5, 443
  angles between, 427
    on GDCs, 610–11
  applications, 437–8
  base, 409, 442
  coincident, 428, 443
  concepts of, 407–20, 442
  direction, 407, 431, 442, 443
  equal, 411–14

GDCs and, 608–11
  intersection points, 434–6
  magnitude, 410–11
  negative, 411–14
  parallel, 411–14, 428, 443
  perpendicular, 428, 443
  position, 414, 442
  representation, 408–9
  resultant, 414–17, 442
  size, 407, 442
  subtraction, 420–5, 443
  unit, 419–20
  zero, 422–3
velocity, 227–9, 407, 510
  average, 221–2
  initial, 224
  instantaneous, 221–2
  see also speed
velocity function, 224, 251
Venn, John (1834–1923), 68
Venn diagrams, 68–77, 96
vertical components, 408
vertical line test, 6–8, 28
vertical stretches, 23, 475–6, 491
vertical translations, 470–2, 490–1
vertices, 44, 689
volume
  cones, 690–2
  cylinders, 689
  prisms, 688–9
  pyramids, 689–90
  of revolution, 318–21, 329
  spheres, 689
  three-dimensional shapes, 688–92

Wallis, John (1616–1703), 517
Wells, Herbert George (1866–1946), 288
Wessel, Caspar (1745–1818), 423
World War II Memorial (Washington, DC), 33

X-values
  calculating from normal probabilities, 625–6
  calculating normal probabilities from, 624–5
x^n
  antiderivatives of, 292
  derivatives of, 200–7, 250

zero, 142, 335, 357
  concept of, 159
  finding, on GDCs, 572–3
zero power, 104–5
zero product property, 34
zero vector, 422–3